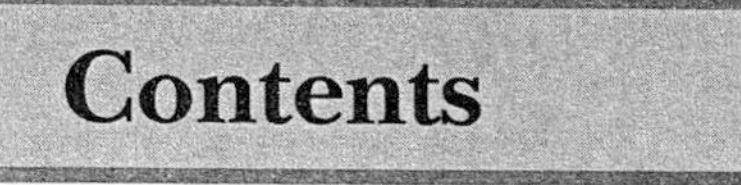

Contents

CHAPTER 12 **543**

HEAT, WORK, AND ENERGY

CHAPTER 20 903

METALLURGY

CHAPTER 21 933

ORGANIC CHEMISTRY

chapter twelve

12

HEAT, WORK, AND ENERGY

12.1 CHEMISTRY AND ENERGY

In the last two chapters we explored chemical equilibrium and the essential theme that all processes are approaching equilibrium although sometimes imperceptibly slowly. We also noted that equilibrium must be related to temperature since the position of equilibrium for a process depends on the temperature, one aspect of Le Chatelier's principle. If the temperature of a system at equilibrium is changed, the equilibrium position will change, and the direction of this change will depend on whether the process absorbs heat (endothermic) or gives heat off (exothermic). In this chapter we look more carefully at the heat absorbed or given off in chemical reactions: (1) experimentally by measuring the quantity of heat in chemical reactions; and (2) without experiment by calculating the heat of reaction. Then, in the next chapter we will tie together ideas about heat and equilibrium. By the end of Chapter 13 we will be able to predict whether a reaction moves in a given direction without having to test the reaction in the laboratory.

Two ideas reappear in this chapter as key concepts:

1. The temperature of an object will rise if heat is absorbed and will fall if heat is given off unless (a) there is a phase change taking place, in which case the temperature stays constant as long as both phases are present, or (b) the object is a gas and all the heat is causing it to expand.
2. The energy of an object rises with its temperature.

TABLE 12.1 Humanpower, Horsepower, and Machine Power: Rate of Energy Delivery*

Energy Source	Horsepower	Watts
Human	1/30	25
Horse	1	750
Waterwheel	5	3750
Windmill	8	6000
Early steam engine	100	75,000
Modern electric power station	1300	1000 megawatt (1 megawatt = 1×10^6 watts)

*The power of a machine is a statement of how fast it can work.

For the first time, we will be able to rigorously define and discuss work and its relationship to heat and energy.

One of the great lessons learned by humans is that energy can be usefully converted into work, but that you need a machine to do it and fuel to run the machine. Some machines and some fuels, however, are better than others. A strong back, bulging biceps, and a meal are fine for doing a day's work, but a good horse and a bag of oats work better, a steam engine and a cord of wood, better still, and a coal-fired power plant, better yet (Table 12.1). There is no question that energy is basic to life, an understanding that develops very early in everybody's experience. Today, the world's energy resources are largely chemical, and almost all of these are based on combustion—the burning of coal, oil, and wood and, to some extent, organic waste. Wind, water, geothermal, solar, and, especially, nuclear power, are real energy resources, but together they account for no more than about a 9% piece of the annual energy pie (Fig. 12.1).

Predicting how long our energy resources, especially petroleum reserves, will last continues to prove difficult. What does seem clear is that within the

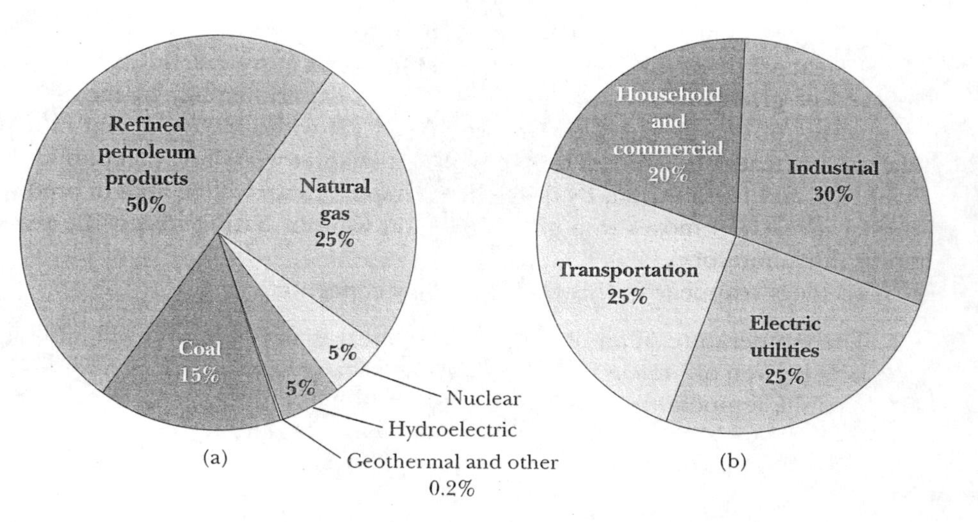

FIGURE 12.1 The Energy Pie. (a) U.S. consumption of fuels and energy from available resources. (b) Sector breakdown of energy consumption.

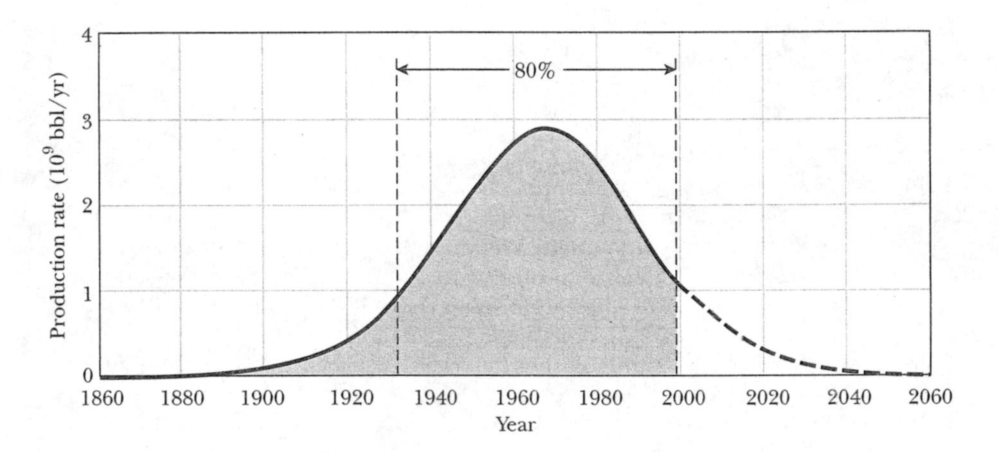

FIGURE 12.2 Crude oil production in the United States. Complete cycle of crude oil production in billions of barrels per year. The harsh reality is that a few generations will have consumed most of the oil that ever was.

short span of less than 200 years from the first discovery of petroleum in the United States in 1859, our estimated total recoverable resources may well be exhausted. In Figure 12.2, note that roughly 80% of our total resources has been exhausted in only 30% of the 200-year time span, suggesting the importance of the petroleum to our industrial and economic base.

Fuels are the vital connection between energy and work in their practical applications. It is important for each of us to know something about how energy is stored, released, transferred, and used; also, we each need to understand the nature and consequences of the transfer of energy. These are the subjects of thermodynamics. **Thermodynamics** is the study of the transfer of heat, the doing of work, and the change of energy. There is an essential connection between temperature, heat, and energy, and the motion or movement of matter at the molecular level. The laws of thermodynamics are based on more than 200 years of experience drawn from basic science. The great triumph of thermodynamics lies in the explanation of the macroscopic or bulk properties of matter —especially the *thermal* properties of matter—in terms that are consistent with our microscopic view of a material world made up of atoms and molecules.

We begin with some basic terms and concepts in need of careful definition and a few fundamental ideas that will help us understand two important principles that are the backbone of this first chapter on thermodynamics. Here are the two principles: (1) Energy is conserved, or, to put it another way, the amount of energy in the universe is constant. (2) Energy can be converted from one form to another, for example, from the energy content of gasoline to motive power.

12.2 TERMS AND CONCEPTS

Heat and Temperature as Thermodynamic Concepts

Temperature indicates the relative *hotness* of an object. The familiar Fahrenheit, Celsius, and Kelvin scales of temperature assign number values to *hotness* so we can say more than something is *warmer* or *cooler* than something else.

Among our most basic experience is what happens when two objects at different temperatures are brought together. The qualitative observation is that the

PROFILES IN CHEMISTRY

It All Began with a $500 Consulting Fee In the summer of 1854, a Dartmouth alumnus by the last name of Bissell visiting his alma mater happened upon a bottle of oil from a spring in western Pennsylvania. Thinking there might be wealth in the spring that had yielded the bottle's contents, this young, entrepreneurial gentleman formed the Pennsylvania Rock Oil Company and bought the spring without quite knowing what to do with its product. The fledgling company sought out Benjamin Silliman, Jr., a Yale chemistry professor, to analyze a sample of their rock oil and see what it might be good for. Silliman was to be paid $500. After performing tests, he reported that the oil had wonderful lubricating properties and was "chemically identical with illuminating gas in liquid form . . . that the lamp burning this fluid gave as much light as any they had seen . . . the oil spent more economically . . . and the uniformity of the light was greater than in camphene, burning for twelve hours without a sensible diminution, and without smoke." Unfortunately, the Pennsylvania spring yielded very little oil. What there was had oozed up into the spring from below ground level. Was there more below? If so, why not drill for it? And drill they did, using technology then available for drilling into salt beds. In 1859, the first well was brought in by "Colonel" Edwin Drake. Other wells followed in Ohio, Indiana, and Illinois.

Neither Drake nor those who backed him got much out of their discovery. Leaving what quickly became the Pennsylvania Oil Fields, Edwin Drake went to Wall Street where he became a broker in oil stocks and lost everything he had in speculation. And although a grateful Pennsylvania legislature voted him a pension of $1500 a year, Drake died destitute and in obscurity (in 1880) in spite of the fact that the oil mania he started was second only to the California Gold Rush.

Oil was discovered in Texas in 1901 at just about the time the automobile was becoming more than just a toy gadget. The site, called "Spindletop" (for the area was covered with spindly pine trees), was beside a pond on the surface of which boys were fond of throwing matches "to see the lake catch fire." Flammable gas, it seemed, just bubbled up to the pond's surface from the limestone roof of a salt dome only 10 to 12 feet below the Earth's surface. Drilling through the roof of the dome began late in 1900. By January 10, 1901, the drill was down 800 feet . . . and then things began to happen. As one of the drilling crew described that day . . .

> We put the new bit on and had 700 feet of the drill pipe back in the hole when the rotary mud began flowing up through the rotary table. It came so fast and with such force that [the crew] had a hard time getting out of danger. [Then] the four-inch pipe started up through the derrick, knocking off the crown block. It shot through the top of the derrick, breaking off in lengths of several joints at a time as it shot skyward. It all happened in much less time than it can be told.
>
> After the water, mud and pipe were blown out, gas followed, but only for a short time. Then all became quiet . . . and . . . we boys ventured back, after the wild scramble for safety, to find things in a terrible mess—at least six inches of mud on the derrick floor and damage to our equipment. We were disgusted. As we started shoveling the mud away . . . without any warning a lot of heavy mud shot out of the well with the report of a cannon! Gas followed for a short time. Then oil showed up. In [seconds] oil was going up through the top of the derrick. Rocks shot hundreds of feet in the air. In a very few minutes the oil was shooting at a steady flow at more than twice the height of the derrick.

For nine days the oil rained down before it could be capped, gushing 40,000 barrels a day. Spindletop proved to be the Klondike of oil.

Literally overnight, the Texas oil rush had begun and with it the twentieth century oil ride. It hasn't ended yet. It will, someday in the foreseeable future, closing out a couple of centuries of thermodynamic and economic dominance of the world's energy resources. As for Professor Silliman and his fee of $508.30, including expenses? There is no record of his ever being paid!

Questions

Name some of the most important changes that have been made to our society as a result of Bissell's discovery. What did Professor Silliman mean when he said the oil was chemically identical to illuminating gas?

When the oil finally runs out, what likely alternatives do you foresee?

cooler object will warm up and the warmer one will cool down. Suppose, for example, that we have two objects, such as the blocks labeled 1 and 2 in Figure 12.3. Now suppose the blocks are at different temperatures and are separated from each other by a perfectly insulating material, but that both blocks are sitting on the same *third* block. The only connection between blocks 1 and 2 is block 3. Without removing the insulating wall separating block 1 from block 2, their respective temperatures will change. This changing of temperature will continue until both blocks 1 and 2 are at the same temperature. In this case we say that heat has been transferred through block 3 from block 1 to block 2. **Heat** is simply the transfer of energy between two objects at different temperatures, and the temperature of an object changes because heat has flowed. Note that the concept of heat has no meaning except when it is being transferred.

Although there is no meaning to "the amount of heat" in an object, we can determine the quantity of heat that is transferred when an object undergoes a change in temperature. The quantity of heat q depends on the size of the object and the specific heat, or molar heat capacity, of the object or of the substance of which the object is composed. **Specific heat** is the quantity of heat required to raise the temperature of 1 gram of a substance by 1K (or 1°C). The **molar heat capacity** of a substance is the amount of heat necessary to raise the temperature of one mole of a substance by 1K (or 1°C) and is given the symbol C. Molar heat capacity is an intensive property of matter since it does not depend on the size of the sample. It is a measure of the quantity of energy that can be stored up in a substance that has been heated. Its value is approximately constant for a particular substance if the range of the temperature change is reasonably small. For the change of temperature of a pure substance, then

$$q = nC\Delta T$$

Quantity of heat: The quantity of heat q transferred is the number of moles of the substance n multiplied by its molar heat capacity C and the change in temperature ΔT (°C or K). The molar heat capacity has the SI units of J/mol·K, q is in joules, and the change in temperature ΔT is the final temperature T_f minus the initial temperature T_i.

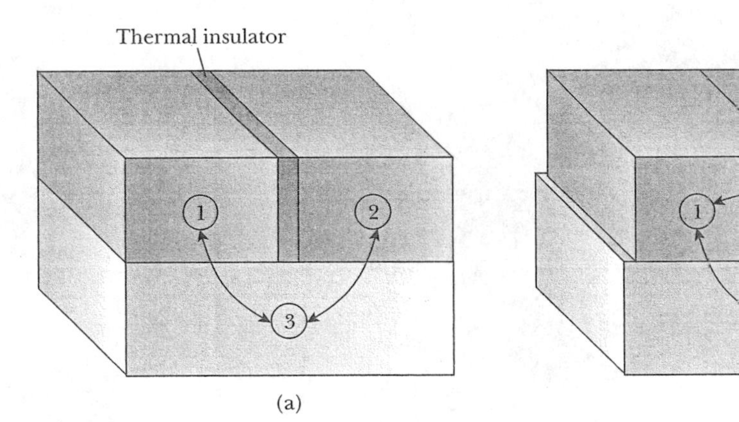

FIGURE 12.3 Thermal equilibrium. (a) After blocks 1 and 2 come to thermal equilibrium with block 3, they are in thermal equilibrium with each other. (b) The same end would be realized if the insulating wall were to be removed and the two blocks pushed together.

When the temperature of a substance increases, ΔT and q have positive values because heat has been transferred into the substance. When the temperature of a substance decreases, ΔT and q have negative values because heat has been transferred out of the substance. It is important to remember that q is positive when heat is transferred into the substance and negative when heat flows out. The molar heat capacities of some solids and liquids are given in Table 12.2. These values will allow you to calculate the heat transfer in raising or lowering the temperature of these substances.

TABLE 12.2 Heat Capacities at Constant Pressure of Selected Solids, Liquids, and Gases

Substance	Molar Heat Capacity, (J/mol·K)
air	20.8
Ag(s)	25.48
Al(s)	24.3
Ar(g)	20.8
Au(s)	25.2
$Br_2(\ell)$	75.6
C(s) (graphite)	8.65
$C_2H_5OH(\ell)$ (ethanol)	113.
$(C_2H_5)_2O(\ell)$ (diethyl ether)	172.
$C_6H_6(\ell)$ (benzene)	136.
Ca(s)	26.2
$CCl_4(\ell)$ (carbon tetrachloride)	132.
$Cl_2(g)$	33.8
CO(g)	29.0
$CO_2(g)$	37.5
Cu(s)	24.7
Fe(s)	25.1
$H_2(g)$	28.6
$H_2O(\ell)$	75.4
$H_2O(s)$	37.7
HBr(g)	29.1
HCl(g)	29.6
He(g)	20.9
$Hg(\ell)$	27.7
Mg(s)	24.3
$N_2(g)$	29.0
Ne(g)	20.7
$NH_3(g)$	35.1
$O_2(g)$	29.1
$O_3(g)$	39.2

EXAMPLE 12.1

Calculate the heat that is transferred into 100.0 g of iron when it is heated from 25.0°C to 75.0°C.

COMMENT Look up the molar heat capacity of iron, then determine n and ΔT. Then apply the equation $q = nC\Delta T$. Note that ΔT has the same value in Celsius degrees and in kelvin units because the size of the degree is the same for both scales.

SOLUTION

$$q = (100.0 \text{ g})\left(\frac{1 \text{ mol Fe}}{55.85 \text{ g Fe}}\right)(25.1 \text{ J/mol·K})(75.0 - 25.0°\text{C}) = 2250 \text{ J}$$

The quantity of heat is positive, and, therefore, heat is transferred into the iron.

EXERCISE 12.1

How much heat is transferred when 50.0 g of water is cooled from 55.0°C to 53.0°C?

ANSWER −420 J

If calculations are made using specific heat, then the equation for quantity of heat is

$$q = m(\text{specific heat})\Delta T$$

where m is the mass in grams.

Defining Thermodynamic Systems and States

Descriptions of molar heat capacity and other thermal properties focus on a single substance or one object. Think of a nail or a mole of table salt. There are instances, however, in which it is necessary to encompass more than one object or substance. A liquid in its container is a good example. For this reason, we introduce the term *system* to define objects or substances contained within some arbitrarily but precisely defined volume. The *boundary* of the system separates it from its *surroundings* and can be the walls of a room, the sides of a picnic cooler, or simply an imaginary surface trapping the objects or substances under consideration. The **system**, then, is the limited part of the universe being studied; the rest of the universe is the **surroundings**, and the **boundary** of the system separates it from the surroundings. The boundary can be flexible and movable. The description of a system can be as simple as the gas within a balloon and the boundary could be the rubber wall of the balloon. Thus, the surroundings would be all of the universe except for the balloon. In other cases, the description of the system and its surroundings can be much more complex, as illustrated by some of the more important thermodynmic systems encountered by chemists and engineers (Fig. 12.4).

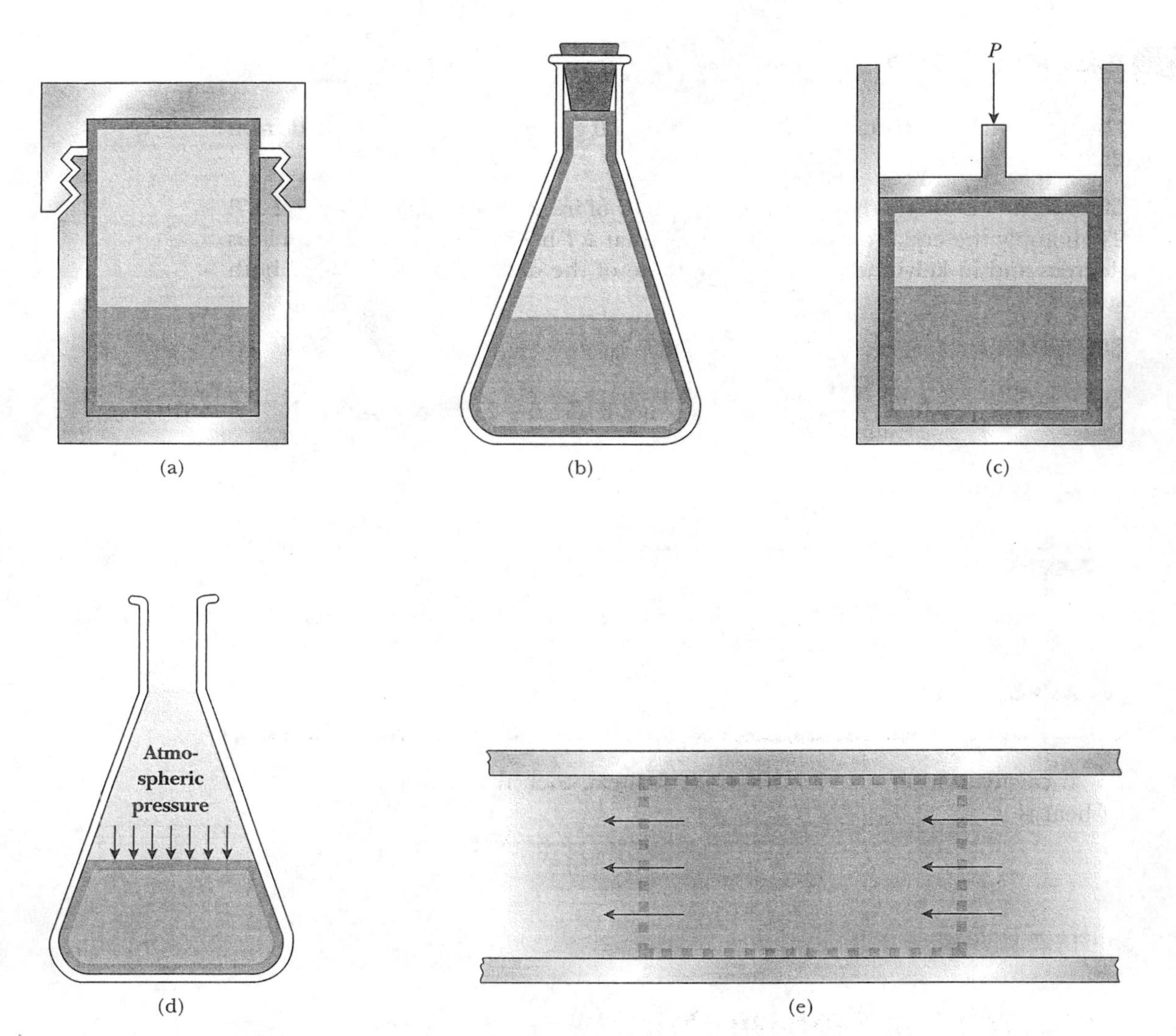

FIGURE 12.4 Thermodynamic systems encountered by chemists and engineers. (a) A closed (sealed) bomb experiment in which the volume of the system remains constant as a chemical reaction takes place within. (b) The stoppered Erlenmeyer flask is the chemist's equivalent of a closed bomb and can be safely used so long as the pressure does not increase too much due to chemical reactions taking place (in the flask). (c) A piston is capable of moving within a cylinder in response to a chemical change. (d) The Erlenmeyer flask is open to the atmosphere, which acts as the piston in (c) by exerting a constant pressure on the system. (e) A pipe through which a fluid can flow is an open system with matter crossing the boundaries of the system, which is in marked contrast to (a)–(d), where there was no movement of matter across the boundaries of the systems.

Matter and heat can flow across the boundaries of some systems, but not all systems. The nature of the boundary determines which, if either, can flow across it. Boundaries that are well-insulated prevent the flow of heat, and boundaries that are tightly closed prevent the transfer of matter. We will consider systems in which heat can flow across the boundaries, but generally will avoid systems in

APPLICATIONS OF CHEMISTRY

Measuring Temperature The actual value of the temperature of an object can be determined in a number of ways. For example, in our treatment of gases in Chapter 4 we discussed Charles's law and stated that the volume of an ideal gas held at constant pressure is directly proportional to the absolute (Kelvin) temperature. Therefore, measuring the volume of a gas is one way of determining the temperature, and one such device for doing that is the gas thermometer. The functioning principle of ordinary household and laboratory thermometers is the property of thermal expansion expanding when heated and contracting when cooled. Mercury thermometers are the most common.

A typical mercury thermometer consists of a capillary tube—a tube with a very narrow channel—that is sealed at the top and has an enlarged cylindrical or spherical bulb at the bottom. The bulb is filled with mercury that, when heated, expands and rises up the channel. Because the channel is very narrow, even a small change in the volume of mercury causes it to rise significantly. The thermometer is calibrated between two fixed reference points, typically the freezing and boiling point of water at atmospheric pressure. The difference in level in the mercury column between these fixed points is divided into 100 equal parts for the Celsius scale or Kelvin scale, each division being 1°C. On the Fahrenheit scale, the difference between the two fixed points is divided into 180 parts, each division being 1°F. The boiling points and freezing points in the Celsius and Fahrenheit scales are the familiar values of 0°C, 100°C, 32°F, and 212°F.

A thermocouple can be made by soldering together the ends of two wires of dissimilar metals or metal alloys, for example copper and iron, or copper and constantan. If one soldered junction is kept at a constant temperature while the other is heated, an electric potential difference (voltage) develops between the two junctions. This potential difference increases as the difference in temperature between the junctions increases. The difference can be read on a voltmeter calibrated to give temperature readings.

Bimetallic thermometers are constructed of strips of dissimilar metals soldered together. Remember the bimetallic disks described in Chapter 1? The two metals have different coefficients of expansion, which means they undergo different increases in length on heating. To illustrate, consider the operating principle of the temperature-measuring device embodying a bimetallic spiral whose curvature varies with the temperature and causes a pointer to be deflected. The scale is calibrated by establishing the positions of the pointer at certain known temperatures and then marking the scale so that each division corresponds to a degree.

But keep in mind that for literally thousands of years before there was any kind of *thermal transducer* for converting the condition of temperature into some other physical effect, craftspeople needed to *know* the temperature. A Chinese porcelain kiln master had to judge 1300 ± 30°C by eye, and commercial success depended on the ability to get it right or be left with ceramic pots reduced to stone masses. As a visit to the ceramics collections at the world's great museums reveals, some got it right. More recently, the town blacksmith and the furnace master in a steel mill could not be color-blind since both were taught to get the temperature right by the color of the object being *tempered* in the furnace. An experienced furnace master could distinguish 430°C (very pale yellow) from 460°C (straw yellow).

Questions

In the Arctic and Antarctic, thermometers are normally filled with colored alcohol instead of mercury. Why is this done?

Suppose you decide the Celsius scale is not satisfactory and instead assign 100 and 200 to the ice and steam points. You call your new scale the °N (for New) scale. Convert °N to °C.

which mass can be transferred. We will also consider systems that are not tightly closed and yet allow no transfer of matter across the system boundary. To illustrate what we mean, examine the thermos bottles in Figure 12.5. We see an open system, a system that can exchange matter and energy with its surroundings; an isolated system, a system not capable of exchanging matter or energy; and a closed system, a system that can exchange energy but not matter. If our system is a beaker of water boiling on a stove, we can define the boundary so that the

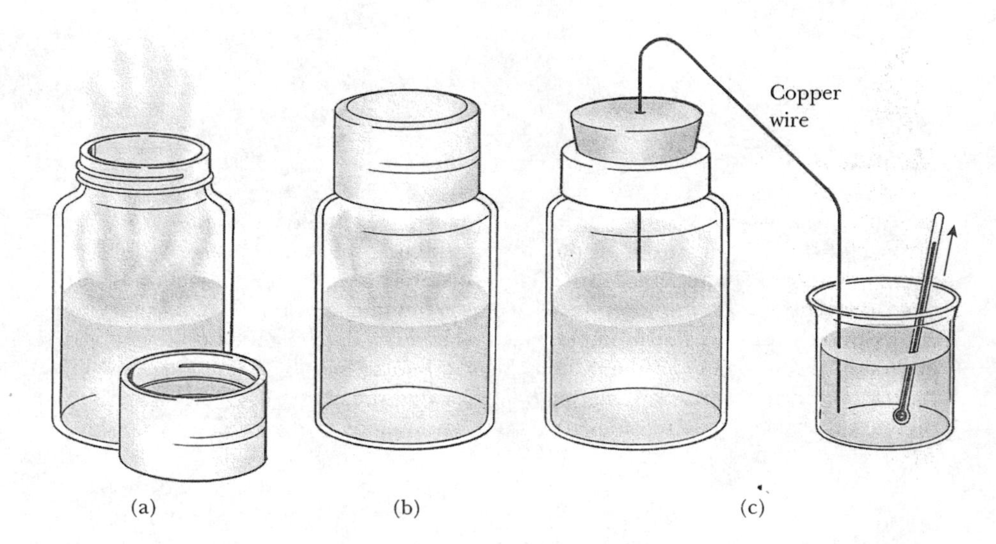

FIGURE 12.5 Matter and heat can flow across the boundaries of some systems. Here three thermos bottles illustrate what we mean by open, isolated, and closed systems: (a) An open system can exchange matter and energy with its surroundings. (b) An isolated system can exchange neither matter nor energy. (c) A closed system can exchange energy, but not matter.

system includes the beaker, the water, and the total volume of the steam produced. Thus, the boundaries of the system can move so that no transfer of matter occurs.

A system is described by its state. The **state** of a system is defined by the macroscopic variables we can measure: composition, volume, temperature, and pressure. Thus, one possible state of a system composed of air in a cylinder might be stated as 20% O_2 and 80% N_2 (composition), 100.0 L (volume), 295K (temperature), and 50.0 atm (pressure). An **equilibrium state** is a state in which none of the variables change over time. We will consider equilibrium states exclusively.

In thermodynamics we study processes. A **process** consists of a change from an initial state to a final state. Later we will look at chemical reactions as processes and will use thermodynamics to predict equilibrium compositions of reaction products. For the moment consider the simpler process of only changing the pressure and volume of a gas of fixed composition and temperature. The change in each of the variables is the difference between the initial state and the final state. Here is a description of the initial and final states for a possible process and the change each variable undergoes:

Initial State	Final State	Change
$V_i = 3.00$ L	$V_f = 4.00$ L	$\Delta V = V_f - V_i = 1.00$ L
$P_i = 1.00$ atm	$P_f = 0.750$ atm	$\Delta P = P_f - P_i = -0.25$ atm
$T_i = 298$K	$T_f = 298$K	$\Delta T = T_f - T_i = 0$
$n_i = 0.123$ mol O_2	$n_f = 0.123$ mol O_2	$\Delta n = n_f - n_i = 0$

There are many ways or *paths* by which a process like the one just discussed can actually occur. For example, the pressure and the volume could be changed in one step, or in a number of very small steps. The **path** of a process is the series of steps from the initial state to the final state. The variables V, P, T, and the composition in number of moles n are called **state functions** because they define the state of the system. These are the variables in the equation of state for a gas. The changes in a state function, ΔV, ΔP, ΔT, and Δn depend on only the initial and final states and do not depend on the actual path of the process.

Thermodynamic Work and Energy

In our study of gases (Chapter 4) we saw that there was a relationship between temperature and **kinetic energy**, the energy of particles due to their motion. In fact, we observe motion causing increased temperature when rubbing our hands together or rapidly bending a piece of metal back and forth. When the temperature of a gas rises, there is an increase in the translational energy of its atoms or molecules. **Translational energy** is the energy involved in moving a single particle in a straight line and can be generally expressed as follows:

$KE = \frac{3}{2}kT$ **average translational kinetic energy of a single particle in an ideal gas,** where k is the Boltzmann constant, $k = R/N_A$.

For a mole of particles we multiply by Avogadro's number N_A,

$KE = \frac{3}{2}RT$ **translational kinetic energy of a mole of particles in an ideal gas**

For real gases, liquids, solids, and mixtures the general result is also the same: Kinetic energy increases with rising temperature as the atoms or molecules in the substance move more rapidly.

The translational kinetic energy is an important component of the total internal energy of a sample of a substance. The **internal energy (*E*)** is all of the energy contained in a substance, including both potential energy and kinetic energy:

$E = KE + PE$ **internal energy (*E*)** is the total of kinetic energy (KE) and potential energy (PE)

Kinetic energy is the energy of motion, which includes translational motion and other motions such as rotational and vibrational. **Potential energy** is energy that is available because of position and existing attractions or repulsions. A weight held above the Earth has potential energy because of its position and the gravitational attraction. Potential energy and kinetic energy can be interconverted as when the weight held above the Earth's surface is dropped and goes into motion.

When the temperature of a substance changes, the internal energy changes, and the magnitude of the change is the difference between the final and initial values, that is,

$$\Delta E = E_f - E_i$$

If heat flows into a system with a constant volume and there is not a phase change, the temperature increases and ΔE is positive. When heat flows out without a phase change, the temperature decreases and ΔE is negative. In fact, heat and internal energy have the same units. However, it is very important to remember that heat and energy are not the same thing. Heat is the *transfer* of energy from one object to another, whereas internal energy is a property of the system being studied. Heat flows during a process, whereas the internal energy is a property of a system.

Heat (q) and change in internal energy (ΔE) are involved in almost every kind of chemical reaction, including the familiar burning of fuels such as coal, gasoline, and heating oil. Heat is given off in the burning of these fuels, and these processes are exothermic. In an **exothermic reaction**, heat is given off to the surroundings, and by convention we assign a negative value to q. For example, burning a hydrocarbon fuel such as octane is exothermic:

$$2C_8H_{18}(\ell) + 25O_2(g) \longrightarrow 16CO_2(g) + 18H_2O(g) + \text{heat}$$

Conversion of a metal such as lithium to the fluoride salt is exothermic:

$$2Li(s) + F_2(g) \longrightarrow 2LiF(s) + \text{heat}$$

Likewise, the displacement of certain metals by others that are more reactive is exothermic:

$$2K(s) + MgCl_2(s) \longrightarrow 2KCl(s) + Mg(s) + \text{heat}$$

On the other hand, if the system absorbs heat from the surroundings, the process is endothermic. In an **endothermic reaction** heat is absorbed from the surroundings, and by convention we assign a positive value to q. The decomposition of limestone is an endothermic reaction and heat must be supplied to drive it forward:

$$CaCO_3(s) + \text{heat} \longrightarrow CaO(s) + CO_2(g)$$

Likewise, the decomposition of red mercury(II) oxide is an endothermic reaction:

$$2HgO(s) + \text{heat} \longrightarrow 2Hg(\ell) + O_2(g)$$

Figure 12.6 gives a schematic representation of an exothermic reaction and an endothermic reaction.

Heat exchange between a system and its surroundings, then, is one way of changing the energy of a system. Besides heat, there is also work, which can achieve the same effects. Both heat and work involve energy, and both have significance only while energy is being transferred. The concept of **work** (w) comes from classical mechanics and is defined operationally as the product of the force (F) acting on an object, causing it to move, times the distance (d) that the object moves in the direction of the force:

$$w = Fd \qquad \text{mechanical work}$$

Work done *by* a system on the surroundings, as when the system expands and pushes back the atmosphere, is conventionally given a negative sign. If work is done *on* the system by the surroundings, as when the system is made to contract, then the sign of w is positive.

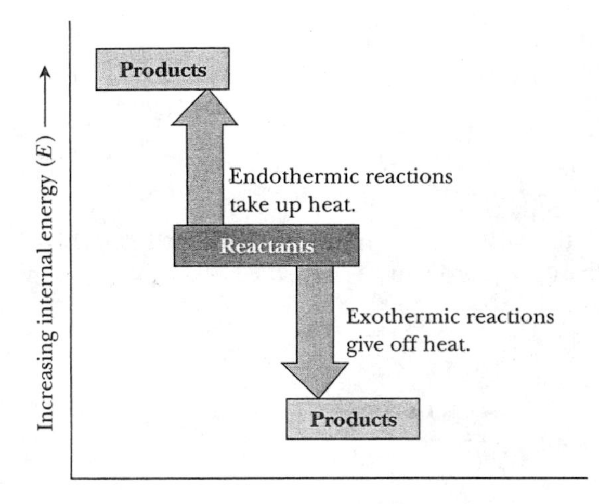

FIGURE 12.6 Chemical reactions exchange energy with their surroundings. An exothermic reaction is accompanied by a release of energy to the surroundings; an endothermic reaction is accompanied by an uptake of heat from the surroundings.

The kind of work most frequently encountered in chemical systems is the change in volume of a gas *working* against some constant external pressure. This is referred to as pressure–volume (PV) work. We can define a system as one mole of an ideal gas trapped in the cylinder shown in Figure 12.7 and held at a constant temperature. In order to characterize the system completely, the pressure and the volume must be fixed. Now if the piston is pushed and displaced by a distance d, the pressure on the gas within the cylinder is increased as a direct result of the work done. By convention, w has a positive sign since work is being done on the system. As the piston is pushed, the gas is compressed, and the volume decreases. A force has acted through a distance.

Here is another way of looking at this same compression. Before anything happens, consider the piston to be locked in place so it cannot move. Call the pressure outside the cylinder P_{ex} (external pressure), which will be equal to the pressure of the trapped gas after compression is complete. Before the compression, the pressure of the trapped gas P is less than P_{ex}. The lock is released, and the piston is pushed to the point where $P = P_{ex}$. The piston ceases to move at this point since the pressures are the same on both sides. Now, the gas in the cylinder has been compressed from an initial volume V_i to a final volume V_f in response to the external compressing force F. The external force is related to the external pressure P_{ex} and the cross-sectional area A of the piston,

$$P_{ex} = F/A$$

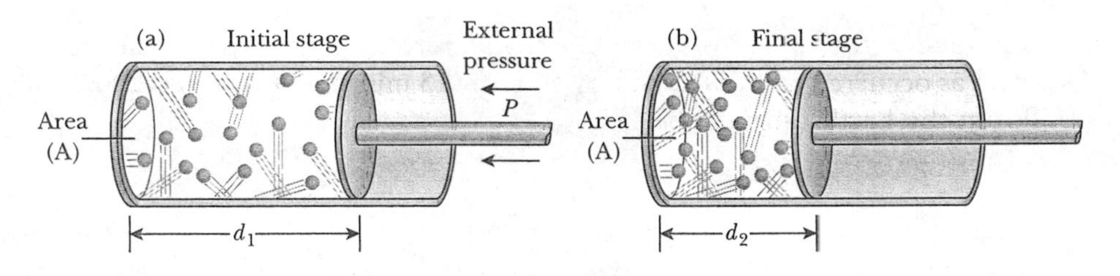

FIGURE 12.7 Pressure–volume work. The gas-filled cylinder (a) with a movable piston of area A acting against an external force changes from V_i to V_f. In the process leading to (b), the gas within has been compressed by the force of the external (applied) pressure, as the distance d_1 changes to d_2.

This rearranges to

$$F = P_{ex}A$$

Multiplying force times distance to get work gives

$$w = P_{ex}Ad$$

The product Ad corresponds to the volume change that takes place as the piston of surface area A sweeps through the distance $d = d_2 - d_1$ in moving from its initial volume, V_i , to its final volume, V_f;

$$Ad = V_f - V_i$$

With this result the expression for work becomes

$$w = -P_{ex}(V_f - V_i)$$

The negative sign is necessary to give the conventional sign for w:

$$w = -P_{ex}\Delta V$$

Work accompanying a volume change at constant pressure: Work (w) equals the product of the external pressure (P_{ex}) and the change in volume ($\Delta V = V_f - V_i$). Note the negative sign.

If the surroundings do work on the system, compressing it, then V_f is less than V_i, ΔV is negative, and w will be positive. If the external pressure is less than the pressure within the system, the gas within the cylinder expands so that work is done by the system on the surroundings. In that case V_f is larger than V_i, ΔV is positive, and the sign of w will be negative.

If it is not clear why the external pressure—not the pressure of the gas within the cylinder—determines the work done on or by the system, consider this example. A gas-filled cylinder has a hypothetical friction-free piston and is placed in deep space where, for all practical purposes, there is no external pressure at all. That is, $P_{ex} = 0$. No useful work can be done, no matter how great the internal pressure, because the expanding gas is pushing against nothing, and $-P_{ex}\Delta V$ is zero. In this case, $w = 0$, and this is called a free expansion. A **free expansion** is any gas expansion that does no work. The work done by the system during a constant volume process must also equal zero, since $\Delta V = 0$. The experimental arrangement in Figure 12.8a results in a free expansion when the stopcock separating the two globes at different pressures is opened. Another example of a free expansion would be the venting of a gas from a space station or vehicle as occurred in 1970 during the Apollo 13 mission while it was 200,000 miles from the Earth (Fig. 12.8b).

EXAMPLE 12.2

(a) An ideal gas is compressed by a constant external pressure of 1.50 atm from an initial volume of 40.0 L to a final volume of 30.0 L. What is the work done?

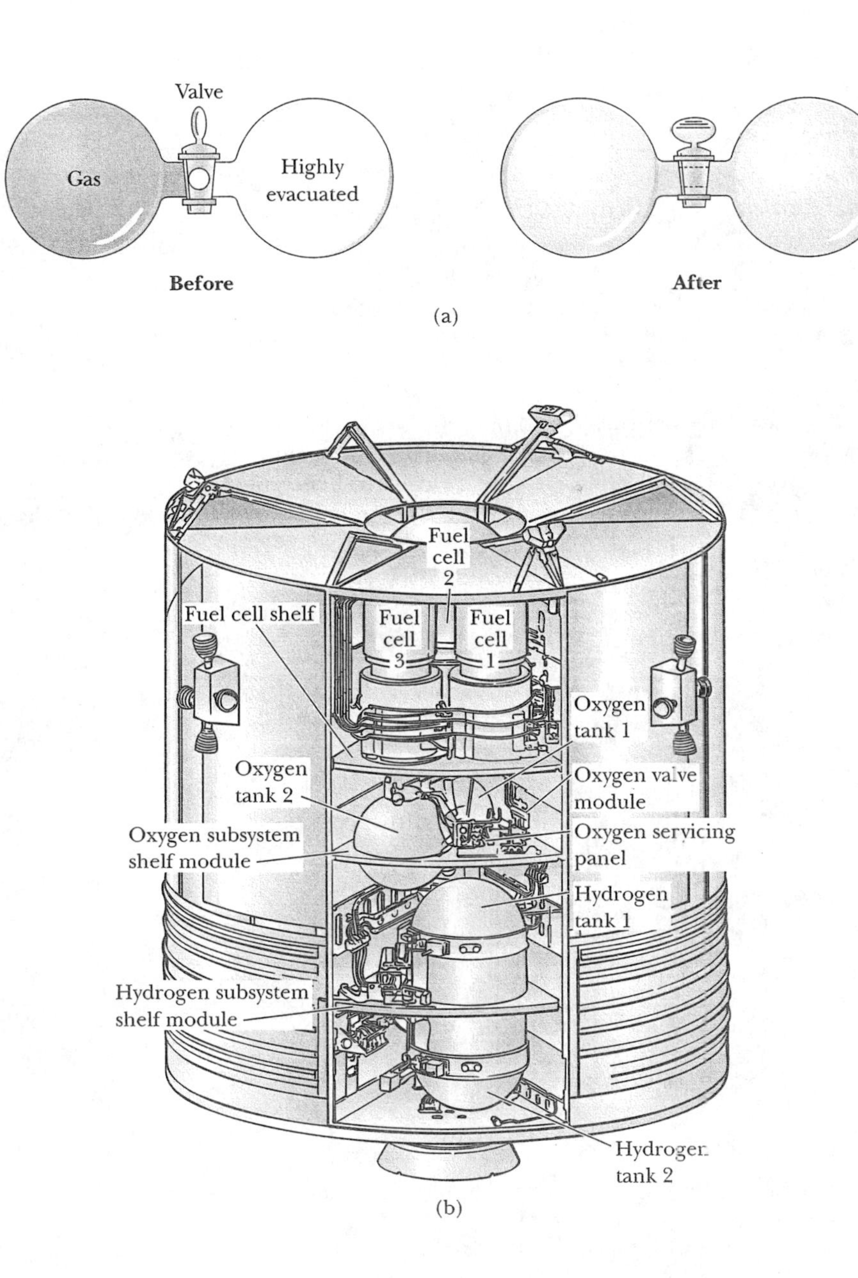

FIGURE 12.8 Free expansion of a gas does no work. (a) *Before:* The left globe contains a compressed gas, whereas the right globe is highly evacuated. *After:* When the stopcock is opened, the gas expands into the evacuated globe. (b) Sketch of the near-catastrophic Apollo 13 event in space 200,000 miles beyond the Earth's atmospheric envelope of gases. One of the oxygen tanks (mounted externally) had ruptured, damaging the adjacent back-up tank and removing the oxygen supply to the main fuel cells that produced power to the command module. These events led to a week in which the world watched as the astronauts attempted (and succeeded) to escape from danger. Check out the Apollo 13 website: <http://nssdc.gsfc.nasa.gov/planetary/lunar/ap13acc.html>

(b) What is the work done when a sample of an ideal gas expands from 1.00 L to 1.33 L against an external pressure of 2.00 atm?

COMMENT Since the external pressure is constant in each case, the equation $w = -P_{ex}\Delta V$ applies for both (a) and (b). In using this equation, be certain that the result has the correct sign of w. It should be positive if work is being done on the system and negative if the system is doing work on the surroundings. Note that with P in atmospheres and V in liters, work is calculated in liter·atmospheres. The conversion factor from L·atm to joules, the usual energy unit, is 1 L·atm = 101.3 J = 0.1013 kJ. A conversion factor composed of different values of the universal gas constant can also be used.

PROFILES IN CHEMISTRY

The Mechanical Theory of Heat Count Rumford (Benjamin Thompson, British, born in America, 1753–1814), while superintending the boring of cannon in the workshops of the military arsenal at Munich (1798), observed and documented the intense heat a brass gun acquired in the short time of being bored. Over 30 minutes of observation, he obtained a quantitative measure of the heating effect caused by rotating a hardened steel boring tool against a solid casting for "a brass six-pounder" from which he could only conclude that the heat produced had to do with the boring motion. The main action in cannon-boring was sliding friction, which was responsible for the heating effect and the temperature rise.

Others were stimulated by Rumford's conclusions, particularly two physicists. Julius Robert Mayer (German, 1814–1878) noted in 1842 that heat and motion (work) were quantitatively equivalent. At about the same time, James Prescott Joule (British, 1818–1889) was busy studying the heating effect of an electrical current generated by a rotating coil in an electromagnetic field. By modifying his apparatus so that the coil was driven by falling weights, he was able to show that . . .

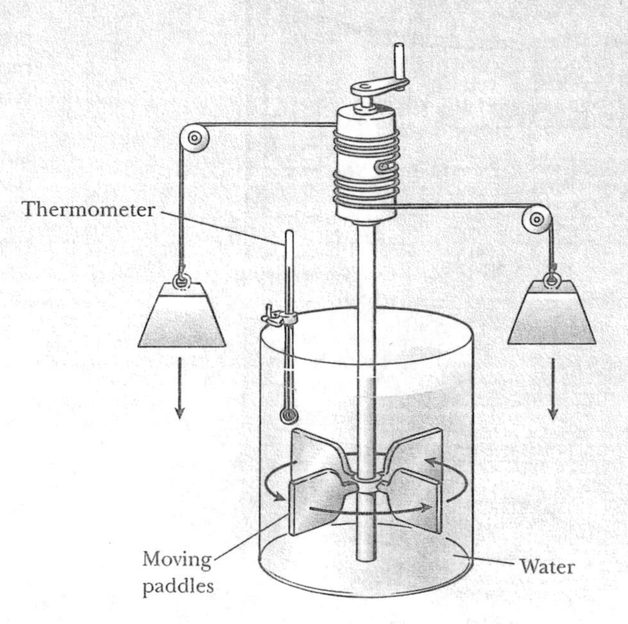

Joule's experiment for determining the mechanical equivalent of heat. Joule succeeded in raising the temperature of a well-insulated quantity of water by the motion of a paddle wheel driven by a falling weight. He found a proportional relationship between the amount of work done on the liquid by the paddle wheel and the increase in temperature of the liquid.

SOLUTION

(a) $V_i = 40.0\ \text{L},\ V_f = 30.0\ \text{L},\ P_{ex} = 1.50\ \text{atm}$

$$w = -P_{ex}\Delta V = -P_{ex}(V_f - V_i)$$

$$= -1.50\ \text{atm}\ [(30.0 - 40.0)\ \text{L}] = +15.0\ \text{L·atm}$$

The result is converted to joules:

$$15.0\ \text{L·atm} \times \frac{101.3\ \text{J}}{\text{L·atm}} = 1520\ \text{J}$$

the quantity of heat capable of increasing the temperature of a pound of water by one degree of Fahrenheit's scale is equal to, and may be converted into, a mechanical force capable of raising 830 lb to the perpendicular height of one foot.*

The experiments of Rumford, Mayer, and Joule involved measurements of the heating effects of work—the product of a force and the distance through which it moves. Joule succeeded in raising the temperature of a well-insulated quantity of water by the motion of a paddle wheel driven by a falling weight. The amount of work performed on the water could be calculated from $w = Fd$, where d is the distance through which the weight fell, and the force F is equal to the mass of the falling weight times the acceleration of gravity. He had found a proportional relationship between the amount of work done on the liquid by the paddle wheel and the increase in temperature of the water. Thus, he was able to determine the molar heat capacity of the water in terms of the work performed on it.

Joule's experiments convincingly showed that heat and work are manifestations of the same thing, and that if work is done and heat is not permitted to flow, the internal energy of the system must change. In fact, heat and work provide the only means for changing internal energy. Heat is motion or motion is a form of heat. Finally, since the elevation of a weight to some height and its acceleration to a steady velocity are mechanical effects, this concept has come to be known as the mechanical theory of heat. It is one of nineteenth-century science's great conclusions. Though Mayer's work preceded Joule's, his published results aroused little interest, and in the end it was Joule with his imposing experiments who received the lion's share of the credit for working out the mechanical equivalent of heat.

Questions

Would cannon-boring be a good experiment for determining the mathematical relationship between calories and joules? Why or why not?

Rumford was after the source of the heating that resulted from cannon-boring. What was the basic conclusion of his careful experiments?

If work is done and heat is not permitted to flow, then does the internal energy of the system change or stay the same? Why?

*Sanborn C. Brown, Editor: *The Collected Works of Count Rumford, Volume 1: The Nature of Heat,* pp. 3–26. Cambridge, MA: The Belknap Press of Harvard University Press, 1968.

The same result is obtained using the two values of the gas constant for the conversion:

$$15.0\ \text{L·atm} \times \frac{8.314\ \text{J/mol·K}}{0.08206\ \text{L·atm/mol·K}} = 1520\ \text{J}$$

The value of w is positive, as it should be since work is being done on the system.

(b) To determine w, proceed as above:

$$w = -P_{ex}\Delta V$$
$$= -P_{ex}(V_f - V_i)$$
$$= -(2.00 \text{ atm})(1.33 - 1.00) \text{ L} = -0.66 \text{ L·atm}$$
$$w = (-0.66 \text{ L·atm}) \times \frac{101.3 \text{ J}}{\text{L·atm}} = -67 \text{ J}$$

In this case, the system is doing work by pushing back the surrounding atmosphere, and the sign of w is negative.

EXERCISE 12.2

Calculate w in liter atmospheres and joules for a gas that expands from 1.0 L to 3.0 L against a constant external pressure of 2.0 atm.

ANSWERS −4.0 L·atm; −400 J

12.3 THE FIRST LAW OF THERMODYNAMICS

Energy Conservation and the First Law of Thermodynamics

Recognizing that heat and work were different manifestations of the same thing was the last step in extending the idea of conservation of energy to yield the mid-nineteenth century principle known today as the first law of thermodynamics. The **first law of thermodynamics** is based on three simple ideas that arise from observations:

1. Energy is conserved.
2. Heat and work can produce equivalent effects (as Joule demonstrated —see *Profiles in Chemistry: The Mechanical Theory of Heat*).
3. The only way that energy can be transferred is through heat and work.

To say that energy is conserved means that it can be accounted for just like the mass balance in stoichiometric calculations that determine masses of products from reactants in chemical reactions. Therefore, the change in the internal energy of a system during a process must be accounted for by the heat that flows and the work that is done during this process. This leads to the mathematical statement of the first law of thermodynamics:

$$\Delta E = q + w$$

First law of thermodynamics: The change in internal energy (ΔE) of a system is equal to the sum of the heat (q) that is exchanged with the surroundings and the work (w) that is done by or on the system. The sign of ΔE depends on q and w. According to the conventions already described, q and w are positive when heat flows into the system and when the surroundings do work on the system. No confirmed contradictions to these first law experiences have ever been observed.

Like pressure, temperature, and volume, the internal energy (E) is a state function. Therefore, the change in the internal energy (ΔE) depends only on the initial and final states of the system and is independent of the path connecting the states. By contrast, functions like heat (q) and work (w) have no meaning in an equilibrium state and are only observed while the process is occurring. Therefore, they are not state functions but are called path-dependent functions or path functions. The values of **path functions** vary, depending on the path between the initial and final states. Uppercase letters are used to designate state functions; lowercase letters are reserved for path functions.

Keep in mind that the first law of thermodynamics does not reveal how the change in internal energy between two states, E_i and E_f, is divided up between q and w any more than two pairs of coordinates of longitude and latitude on a map tell how you might get from the first point to the second. As you see from Figure 12.9, the coordinates on the map are analogous to state functions, and the result is the same no matter which of the many possible pathways is followed in navigating from the initial state to the final state.

We can demonstrate that E is a state function and that q and w are path functions. Suppose, for example, we convert the internal energy in a gallon of fuel oil to heat and work, using a hot water heater or the engine of a truck. Most of the energy is given off as heat in the hot water heater and the work done is negligible. In the truck, the energy stored in the fuel is used for work in the form of locomotion, but some clearly ends up as heat, dissipated by the engine's cooling system. In fact, approximately 80% of the stored energy that has been released radiates away uselessly as the truck moves along. In either case, the first law holds:

$$\Delta E = q + w$$

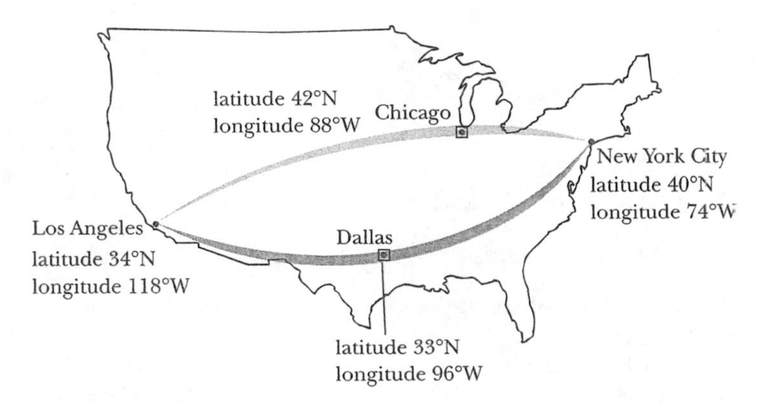

FIGURE 12.9 State functions are analogous to coordinates on a map. It does not matter, with respect to the change of the coordinates of latitude and longitude between New York (40°N by 74°W) and Los Angeles (34°N by 118°W), whether one is navigating from one to the other via Chicago (42°N by 88°W) or Dallas (33°N by 96°W). Therefore, latitude and longitude are state functions.

The more heat that is given off, the less work is done. But ΔE is the same, as long as the initial and final states are the same. The change in the internal energy ΔE is the difference between E of the fuel oil reactants and E of the combustion products. But q and w are different in each case (Fig. 12.10).

To summarize the terms and ideas embodied in the first law of thermodynamics, consider an experiment in which we add 40.67 kJ of heat to 1.00 mol of water, causing it to vaporize at constant pressure at its normal boiling point. We wish to know how much work is done and what is the change in the internal energy of the water. In our experiment (Fig. 12.11a), the water is trapped in a cylinder fitted with an ideal friction-free piston of negligible mass, and the base of the cylinder is in contact with a furnace capable of supplying heat. A weight is placed on top of the piston to provide an external pressure P_{ex}, which, we will say, equals 1.00 atm. Set the temperature of the furnace just slightly higher than the boiling point. The liquid is at the same temperature as the furnace, and the 1.00 atm pressure produced by the mass on the piston is a little less than the pressure produced by the vapor of the liquid in the cylinder at the boiling point. Heat flows from the furnace to the liquid until vaporization is complete (Fig. 12.11b) and the vapor forces the piston upward. Our model amounts to a single stroke of a steam engine. Since the process is carried out under the constant pressure of the weight, we may write,

$$w = -P_{ex}(V_f - V_i) = -P_{ex}\,\Delta V$$

Here, V_f equals the volume of 1.00 mol of water vapor at 100°C and V_i is the volume of a 1.00 mol of liquid water at that same temperature, at which the density of water is 0.96 g/mL:

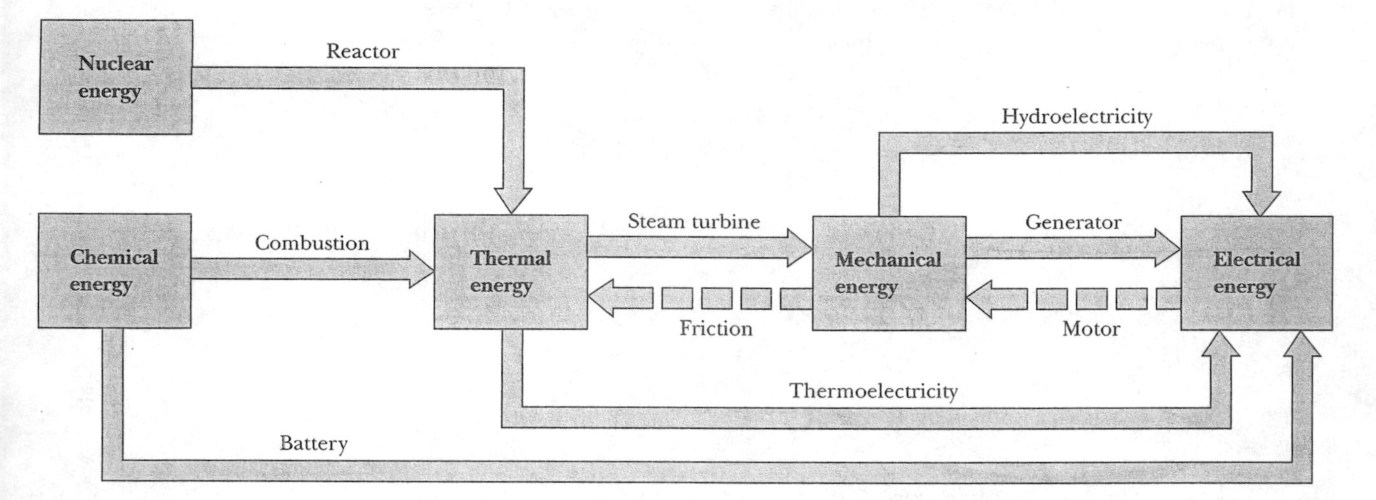

FIGURE 12.10 Summarizing energy conversion processes. Energy conversion takes place by nuclear, chemical, thermal, mechanical, and electrical processes generally driven by local availability of inexpensive resources, such as nuclear power, to produce heat that can be used to produce turbine electric power. Regardless of the process or the available energy resource, ΔE must be the same as long as the initial and final states are the same while heat and work are coupled to each other—more of one, less of the other. Thus, ΔE in a gallon of motor fuel produces heat (removed by radiator fluid) and work (transferred to the wheels).

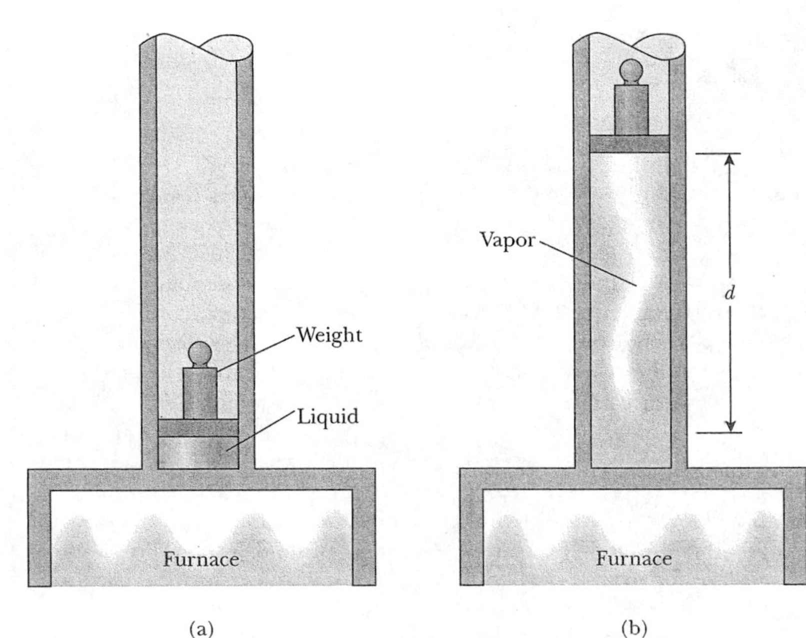

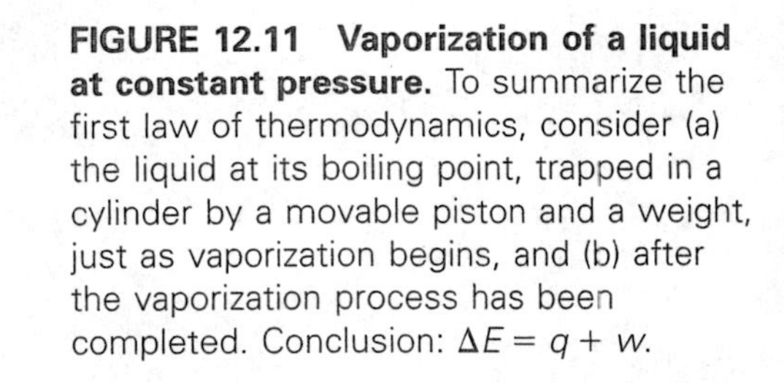

FIGURE 12.11 Vaporization of a liquid at constant pressure. To summarize the first law of thermodynamics, consider (a) the liquid at its boiling point, trapped in a cylinder by a movable piston and a weight, just as vaporization begins, and (b) after the vaporization process has been completed. Conclusion: $\Delta E = q + w$.

$$V_f = \frac{nRT}{P} = \frac{(1.00\ \text{mol})(0.0821\ \text{L·atm/mol·K})(373\text{K})}{1.00\ \text{atm}} = 30.6\ \text{L}$$

$$V_i = \frac{\text{mass}}{\text{density}} = (18.0\ \text{g})\left(\frac{1/\text{ml H}_2\text{O}}{0.96\ \text{g H}_2\text{O}}\right) = 19\ \text{mL} = 0.019\ \text{L}$$

$$V_f - V_i = \Delta V = (30.6 - 0.019)\ \text{L} = 30.6\ \text{L}$$

$$w = -P_{ex}\Delta V = -(1.00\ \text{atm})(30.6\ \text{L}) \times 0.1013\frac{\text{kJ}}{\text{L·atm}} = -3.10\ \text{kJ}$$

Note that the volume of the liquid is negligible compared to the volume of the vapor. In general, the difference in volume between a liquid and its vapor (or a solid and its vapor) can be calculated simply as the volume of the vapor. To calculate the change in internal energy (ΔE) requires using the known factor that it takes 40.7 kJ to vaporize 1.00 mol of water. This is the value of q, the heat that flows into the system. Then,

$$\Delta E = q + w = (40.7 - 3.10)\ \text{kJ}$$

$$\Delta E = 37.6\ \text{kJ}$$

Note the consequences of the result:

- Less than 10% of the energy (3.10 kJ) added to the system from the furnace was used to do useful work, the mechanical work of expansion that can lift or move things.
- More than 90% of the energy (37.6 kJ) input to the system was used for changing liquid into vapor.

APPLICATIONS OF CHEMISTRY

The "Snow Tube" Carbon Dioxide Fire Extinguisher . . . And Making Artificial Snow Here is an illustration of the first law of thermodynamics in action. The BC-type carbon dioxide fire extinguishers commonly in use in chemical laboratories typically are charged with several liters of liquid carbon dioxide under very high pressure. If you shake an extinguisher that is fully charged, you can literally hear and feel the liquid sloshing around inside. When the safety pin is pulled and the triggering device is squeezed, the carbon dioxide is released and rapidly expanded to atmospheric pressure in the megaphone-shaped "snow tube" where the greater part rapidly vaporizes. Because of the huge pressure difference within the system (the contents of the extinguisher) and the surroundings (the atmosphere into which the carbon dioxide is ejected) the response is instantaneous, and we can reasonably assume that there is no heat exchanged between system and surroundings. There simply is not time for heat to flow. Therefore, q is zero, ΔE is equal to w, and the sign of w is negative because the system is doing work on the surroundings. The net result is therefore a decrease in the energy of the system. We also assume that E is proportional to the absolute temperature T, and so ΔE is proportional to ΔT. Because ΔE is negative, ΔT must also be negative. It is just this very cooling effect due to the negative ΔT that results in some of the carbon dioxide (about 30%) being cooled to $-78°C$, at which temperature it turns solid at atmospheric pressure. This "snow" of solid carbon dioxide is directed at the base of the fire, where it serves to displace the oxygen needed to support combustion, suffocating the fire, and, in many cases, lowering the temperature of the burning material to below its ignition temperature.

Interestingly, this same principle serves to explain how the snow-making equipment that has revolutionized the ski resort and winter vacation industry works, freeing the people who depend on snow-covered slopes for their livelihood from the vagaries and uncertainties of the weather. As long as the atmospheric temperature is around the freezing point of water, a mixture of compressed air and water sprayed into the atmosphere results in the production of something akin to snow.

Questions

Explain why the gas coming out of the snow tube is so cold.

When the contents of a CO_2 extinguisher are expended, no heat is exchanged between system and surroundings. Why?

To put out a fire, direct the CO_2 exiting the funnel at the base of the flames. How are the flames extinguished?

Modern snow-making equipment sends a mixture of compressed air and water into the air, producing snow as long as the atmospheric temperature is (well below, well above, at about) freezing. Why?

What would happen if a CO_2 fire extinguisher were used in outer space?

EXAMPLE 12.3

Calculate q, w, and ΔE for the vaporization of 2.00 mol of liquid ammonia at its normal boiling point of $-33.4°C$ and a pressure of 1.00 atm. At this temperature and pressure, one mole of liquid ammonia has a volume of 0.04 L and to vaporize it requires 23.3 kJ of heat.

COMMENT Whenever a list of different quantities (q, w, and ΔE) must be calculated from a set of data, a useful rule is always to start with the quantity from the list that is easiest to calculate. In this case, first calculate q from the heat per mole required to vaporize ammonia. To find w, ΔV is needed, so next calculate the volume of vapor formed. The volume of the liquid is only 0.08 L in this case, a negligible quantity. So ΔV is effectively equal to the volume of the vapor and can be calculated from the ideal gas equation of state ($\Delta V = V_f = nRT/P$). The work (w) can then be calculated and ΔE determined from the first law relationship.

SOLUTION

$$q = (2.00 \text{ mol})\frac{(23.3 \text{ kJ})}{1 \text{ mol}} = 46.6 \text{ kJ}$$

$$\Delta V = \frac{(2.00 \text{ mol})(0.0821 \text{ L·atm/mol·K})(240.\text{K})}{(1.00 \text{ atm})} = 39.4 \text{ L}$$

$$w = -P_{ex}\,\Delta V = -(1.00 \text{ atm})(39.4 \text{ L}) \times \frac{0.1013 \text{ kJ}}{1 \text{ L·atm}} = -4.00 \text{ kJ}$$

$$\Delta E = q + w = (46.6 - 4.00) \text{ kJ} = 42.6 \text{ kJ}$$

EXERCISE 12.3

A gas expands against a constant pressure of 5 atm, from 10. to 20. L, while absorbing 2 kJ of heat. Calculate the work done and the change in the internal energy of the gas.

ANSWERS $w = -5$ kJ; $\Delta E = -3$ kJ

12.4 THE HEAT OF REACTION

Enthalpy and ΔH

Chemical systems are usually more interesting for the properties of their substances than for the useful work that can be extracted from them. The principal exceptions to that statement are the pressure–volume work of expansion and compression that drives cars and trucks, and the electrical work of batteries. Pressure–volume work can be calculated using

$$w = -P_{ex}\,\Delta V$$

where P_{ex} is the constant external pressure against which the system is working. Consequently, if there is no volume change, then no useful work can be done. When that is the case, as for a reaction run in a closed container, $\Delta V = 0$, $w = 0$, and, from the first law ($\Delta E = q + w$), q is the change in internal energy:

$$q_v = \Delta E \qquad (q_v \text{ is } q \text{ at constant volume})$$

Heat for a constant volume process: The heat exchanged between a system and its surroundings at constant volume (q_v) is equal to the change in internal energy of the system (ΔE).

The important meaning of this equation is that for a chemical system in which change takes place under constant volume conditions, any increase or decrease in the internal energy (ΔE) must equal (q_v) the heat absorbed or released.

Most chemical experiments, however, are carried out at constant pressure, not at constant volume. For these conditions we define another state function, the **enthalpy** H:

$$H = E + PV$$

Definition of enthalpy: Enthalpy (H) is equal to the internal energy of a system plus the product of the pressure (P) and volume (V) of that system.

To see why we define this new function, consider a process that changes the state of the system at constant pressure:

$$\Delta H = \Delta E + \Delta(PV)$$

If pressure is held constant, then $\Delta(PV)$ equals

$$PV_f - PV_i = P\Delta V$$

and

$$\Delta H = \Delta E + P\Delta V$$

Change in enthalpy at constant pressure: The change in enthalpy equals the change in internal energy of the system plus the pressure times the change in volume of the system.

Like E, P, and V, the enthalpy H is a state function, so the value of ΔH is independent of the path.

Using the statement of the first law of thermodynamics and substituting for ΔE gives the following:

$$\Delta E = q + w$$

$$\Delta H = q_p + w + P\Delta V$$

where q_p is q at constant pressure. Since $P\Delta V$ equals $-w$ (Section 12.2), the term $P\Delta V$ is equal to the negative of the work done by or to the system, and

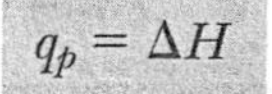

$$q_p = \Delta H$$

Heat for a constant pressure process: The heat exchanged between a system and its surroundings at constant pressure (q_p) is equal to the change in enthalpy (ΔH).

To illustrate the importance of the enthalpy function ΔH, consider the following chemical reaction. It has been written as a thermochemical equation. A **thermochemical equation** is a chemical equation that includes thermodynamic data such as ΔH or ΔE for the reaction:

$$CS_2(g) + 3O_2(g) \longrightarrow CO_2(g) + 2SO_2(g) \qquad \Delta H = -1110 \text{ kJ}$$

The value of ΔH accompanying a thermochemical equation applies to the chemical process as written, and since under constant pressure ΔH equals q_p it is often referred to as the *heat of reaction.* In this case, the equation shows that when one mole of CS_2 gas reacts with three moles of oxygen gas at constant pressure, the enthalpy of the system decreases by 1110 kJ as this much heat is liberated from the system to the surroundings.

For processes involving only solids and liquids, ΔH and ΔE are essentially the same because PV is negligible and no work is done. That is not the case, however, for gas phase reactions or for reactions in which one or more reactants or products are gases. Here ΔH and ΔE can be significantly different because of the PV term. As we showed above, at constant pressure,

$$\Delta H = \Delta E + P\Delta V$$

$$P\Delta V = PV_f - PV_i$$

Substituting nRT for PV from the ideal gas equation of state gives

$$P\Delta V = n_f RT_f - n_i RT_i$$

If the temperature as well as the pressure is held constant,

$$P\Delta V = (n_f - n_i)RT = \Delta nRT$$

where Δn is equal to the change in the number of moles of *gases* that results from the reaction. Now substitute for $P\Delta V$ in our original expression:

$$\Delta H = \Delta E + P\Delta V$$

This gives an equation for the enthalpy of reaction at constant pressure and temperature in terms of moles of gases:

$$\Delta H = \Delta E + \Delta nRT$$

Relationship between ΔH and ΔE for a reaction involving gases: The enthalpy change (ΔH) is equal to the change in internal energy (ΔE) plus a term equal to the negative of the pressure–volume work (ΔnRT) resulting from a change in the number of moles of gas ($\Delta n = n_f - n_i$). When using this equation, it is important to remember that Δn refers only to those reactants and products that are in the gaseous state.

Consider the combustion of one mole of benzene (C_6H_6) at constant volume and at 298K. Under these conditions, benzene is a liquid, the heat of reaction is -3264 kJ/mol, and the reaction is exothermic:

$$C_6H_6(\ell) + \frac{15}{2}O_2(g) \longrightarrow 6CO_2(g) + 3H_2O(\ell) + \text{heat}$$

Since this is a constant volume process, the heat of reaction is ΔE, the change in the internal energy, not ΔH, the change in enthalpy:

$$q = q_v = \Delta E = -3264 \text{ kJ}$$

To calculate ΔH for this reaction, keep in mind that since a gas has a volume so much larger than that of an equal mass of liquid it is necessary only to consider the volumes of the gases. In this reaction the reactants include 15/2 moles of gas whereas the products include only 6 moles of gas:

$$\Delta H = \Delta E + \Delta nRT = -3264 \text{ kJ} + (n_f - n_i)RT$$
$$= -3264 \text{ kJ} + \left(6 - \frac{15}{2}\right)RT$$
$$\Delta H = -3264 \text{ kJ} + (-1.5 \text{ mol}) \times \frac{8.314 \text{ J}}{\text{mol·K}} \times \frac{1 \text{ kJ}}{1000 \text{ J}} \times 298\text{K}$$
$$\Delta H = -3264 \text{ kJ} + (-4 \text{ kJ}) = -3268 \text{ kJ} = q_p$$

EXAMPLE 12.4

A mole of methane is oxidized to carbon dioxide and water as either a gas or a liquid at 25°C, according to the following equation:

$$CH_4(g) + 2O_2(g) \longrightarrow CO_2(g) + 2H_2O$$

For this reaction,

$$q_v = -887 \text{ kJ} \qquad q_p = -891 \text{ kJ}$$

Calculate Δn for the moles of gas in this reaction and determine whether the H_2O produced is a gas or liquid.

COMMENT Remember that $\Delta E = q_v$ and $\Delta H = q_p$, and the difference between ΔE and ΔH is negligible unless different numbers of moles of gaseous reactants and products are involved. If H_2O is produced as a gas in this case, then Δn is zero and ΔE equals ΔH. But that is not true. Use the equation $\Delta H = \Delta E + \Delta nRT$ to calculate Δn.

SOLUTION

$$\Delta H = \Delta E + \Delta nRT$$
$$\Delta nRT = \Delta H - \Delta E$$
$$\Delta n = \frac{\Delta H - \Delta E}{RT} = \frac{(-891 \text{ kJ/mol} + 887 \text{ kJ/mol})(1000 \text{ J/kJ})}{(8.314 \text{ J/mol·K})(298\text{K})} = -2$$

Note that the final answer had to be rounded to one significant figure since the difference between ΔH and ΔE yields only one significant figure. However, Δn of -2 is consistent with the reaction yielding liquid H_2O and only one mole of gaseous product.

EXERCISE 12.4

The constant volume combustion of toluene takes place according to the following reaction at 298K:

$$C_6H_5CH_3(\ell) + 9O_2(g) \longrightarrow 7CO_2(g) + 4H_2O(\ell)$$

The value of q for the reaction is -3912 kJ. Determine the values for q_v, ΔE, and ΔH.

ANSWERS $q_v = -3912$ kJ; $\Delta E = -3912$ kJ; $\Delta H = -3917$ kJ

We will now consider measuring the enthalpy changes for chemical reactions. The enthalpy change of a reaction is often called the *heat of reaction* since ΔH is identical to q under conditions of constant pressure.

Calorimetry and Combustion

Combustion reactions are especially good for making measurements of q since they go quickly to completion with the generation of easily characterized products and give up lots of heat. The thermochemical data recorded for combustion reactions are called **heats of combustion (ΔH_c)** and turn out to be very useful for characterizing fuels for cars and trucks (gasoline-range petroleum fractions such as pentanes and octanes); fuels for burners (methane, propane, and acetylene); and fuels for people (glucose and starch derived from foods). Some values are given in Table 12.3. Note especially that the *ordinary* calorie that is equal to 4.184 joules is not the same as the *food* Calorie commonly listed on the labels of products purchased at the super market. The Calorie cited for foods and beverages is equivalent to 1000 ordinary calories, or 1 kilocalorie.

TABLE 12.3 Standard Enthalpies (Heats) of Combustion (ΔH_c°) at 298K

Substance/State		ΔH_c° at 298K, kJ/mol
acetic acid	$CH_3COOH(\ell)$	−875
acetylene	$C_2H_2(g)$	−1300
benzene	$C_6H_6(g)$	−3293
benzene	$C_6H_6(\ell)$	−3268
benzoic acid	$C_6H_5COOH(s)$	−3227
n-butane	$C_4H_{10}(g)$	−2879
carbon (diamond)	$C(s)$	−395.388
carbon (graphite)	$C(s)$	−393.505
ethane	$C_2H_6(g)$	−1560
ethanol	$C_2H_5OH(\ell)$	−1367
ethylene	$C_2H_4(g)$	−1411
glucose	$C_6H_{12}O_6(s)$	−2803
hydrogen	$H_2(g)$	−286
methane	$CH_4(g)$	−890
methanol	$CH_3OH(\ell)$	−727
naphthalene	$C_{10}H_8(s)$	−5153.9
n-octane	$C_8H_{18}(\ell)$	−5450
n-pentane	$C_5H_{12}(g)$	−3507
n-pentane	$C_5H_{12}(\ell)$	−3487
propane	$C_3H_8(g)$	−2220
sucrose	$C_{12}H_{22}O_{11}(s)$	−5641
urea	$CO(NH_2)_2(s)$	−632

Heats of reactions can be determined with a calorimeter. A **calorimeter** is a vessel in which heats of combustion or heats of other reactions are measured. One of the most useful devices for measuring the heat released during a chemical reaction is a well-insulated vessel known as a bomb calorimeter, in which combustion reactions (like the burning of toluene in an oxygen atmosphere in Exercise 12.4) can be carried out. Work plays no role in calorimeters designed to be used under constant volume conditions (Fig. 12.12), and so the measured heat of reaction q_v is the energy change ΔE. For a calorimeter designed to be used under constant pressure conditions, both heat and work can be exchanged with the surroundings, and the heat of reaction q_p is ΔH. In either case—constant volume or constant pressure—it is desirable to conduct calorimetric measurements under nearly isothermal (constant temperature) conditions since that simplifies many thermochemical calculations and allows us to make comparisons of heats of combustion of many reactions. Thus, the calorimeter must be able to absorb the heat released by the reaction with an increase in temperature of no more than a few degrees. Since the recorded temperature change of the calorimeter and its contents is small, the temperature measurements must be accurately made, and especially accurate thermometers have been developed for use in calorimetry.

The heat flow q for a reaction in a calorimeter can be calculated from the temperature change and the heat capacity C or calorimeter constant of the calorimeter, the amount of heat it takes to raise the temperature of the calorimeter and its contents by one kelvin. Since no heat flows from the calorimeter into the surroundings,

$$q_{\text{reaction}} + q_{\text{calorimeter}} = 0$$

The value of $q_{\text{calorimeter}}$ is calculated from the heat capacity of the calorimeter and the change in temperature,

$$q_{\text{calorimeter}} = C_{\text{calorimeter}}\Delta T$$

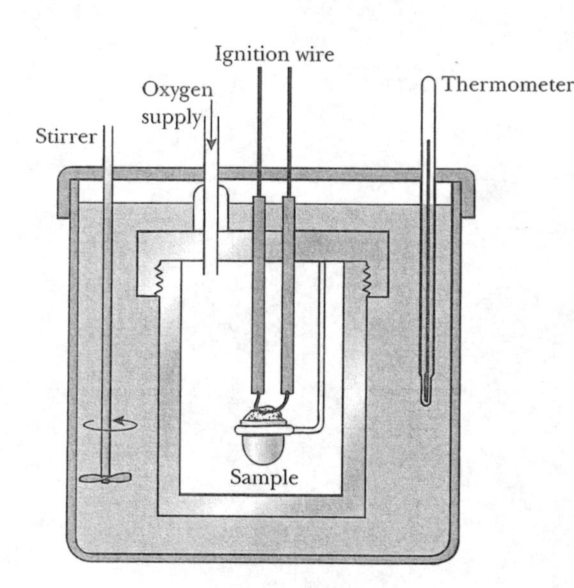

FIGURE 12.12 Constant-volume calorimeter experiments. The sample in the cup is ignited by a hot wire in an oxygen atmosphere. Heat generated by the combustion process that follows is measured by the rise in temperature of the calorimeter, its contents and its immediate surroundings, in this case, the water bath. For combustion experiments carried out at constant volume, $\Delta E = q_v$.

Substituting this into the previous equation and rearranging gives the fundamental calorimeter equation:

$$q_{reaction} = -C_{calorimeter}\Delta T$$

Calorimetry equation and calorimeter constant: The heat capacity or *calorimeter constant* of the calorimeter and its contents is $C_{calorimeter}$, and the measured temperature change is ΔT. The heat that flows is proportional to the temperature change, and the heat capacity of the calorimeter is the proportionality constant.

The negative sign in the equation gives the correct sign for q of the reaction. For example, ΔT has a positive value in an exothermic process in which q is negative. For a constant volume calorimeter,

$$q = q_v = \Delta E$$

and for a constant pressure calorimeter,

$$q = q_p = \Delta H$$

EXAMPLE 12.5

The heat of combustion of benzoic acid ($C_7H_6O_2$) to $H_2O(\ell)$ and $CO_2(g)$ is known:

$$C_6H_5COOH(s) + \frac{15}{2}O_2(g) \longrightarrow 7CO_2(g) + 3H_2O(\ell) \qquad \Delta H_c = -3227 \text{ kJ/mol}$$

A constant pressure calorimeter is calibrated by burning 1.000 g of benzoic acid, and the temperature is observed to rise from 23.00°C to 25.12°C. Next, a 3.100-g sample of solid citric acid ($C_6H_8O_7$) is completely burned in the same calorimeter, and a temperature change from 23.00°C to 25.57°C is observed. Calculate ΔH and ΔE in kJ/mol for the combustion of citric acid.

COMMENT The first experiment is necessary to calculate the heat capacity of the calorimeter from the heat evolved, the known heat of combustion of benzoic acid, and the value of ΔT. Then the value of q for the citric acid experiment can be calculated. Since the calorimeter operates at constant pressure $q = q_p = \Delta H$, then ΔE can be calculated.

SOLUTION The heat released by the benzoic acid calibration reaction is

$$(-3227 \text{ kJ/mol}) \times 1.000 \text{ g} \times \frac{1 \text{ mol}}{122.1 \text{ g}} = -26.43 \text{ kJ}$$

Therefore, the heat capacity of the calorimeter is

$$C_{calorimeter} = -\frac{q}{\Delta T} = -\frac{-26.43 \text{ kJ}}{2.12 \text{ K}} = 12.5\frac{\text{kJ}}{\text{K}}$$

Since the citric acid is burned in the same calorimeter, the value of C, the calorimeter constant, is the same. The heat released by combustion of the citric acid sample is

$$q = -C_{\text{calorimeter}}\Delta T = -(12.5 \text{ kJ/K})(2.57\text{K}) = -32.1 \text{ kJ}$$

The heat of combustion per mole of citric acid is

$$q_p = \Delta H = \frac{32.1 \text{ kJ}}{3.100 \text{ g}} \times \frac{192.1 \text{ g}}{1 \text{ mol}} = -1.99 \times 10^3 \text{ kJ/mol}$$

To determine ΔE we need to find Δn, which we get from the balanced equation for burning one mole of citric acid:

$$C_6H_8O_7(s) + \frac{9}{2}O_2(g) \longrightarrow 6CO_2(g) + 4H_2O(\ell)$$

$$\Delta n = \left(6 - \frac{9}{2}\right) = \frac{3}{2} = 1.5$$

$$\Delta E = \Delta H - \Delta nRT = (-1990 \text{ kJ/mol}) - (1.5)(8.314 \text{ J/mol·K})\left(\frac{1 \text{ kJ}}{1000 \text{ J}}\right)(296\text{K})$$

$$= (-1990 \text{ kJ/mol}) - (3.69 \text{ kJ/mol})$$

$$= -1990 \text{ kJ/mol}$$

The ΔnRT term is so small compared to the ΔH term in this example that ΔE equals ΔH to three significant figures even though the heats of combustion were measured under constant pressure conditions

EXERCISE 12.5

In order to calibrate a constant pressure calorimeter, a 2.000-g sample of naphthalene ($C_{10}H_8$) is burned in it, and the temperature is observed to change from 26.503°C to 27.683°C. We will assume that complete combustion of compounds composed only of hydrogen and carbon leads only to $CO_2(g)$ and $H_2O(\ell)$. The heat of combustion of naphthalene at constant pressure is -5153.9 kJ/mol. A 3.210-g sample of liquid pentene (C_5H_{10}) is then burned to liquid water and carbon dioxide in the same calorimeter, and the temperature rises from 25.506°C to 26.696°C. Determine ΔE and ΔH for the combustion of pentene.

ANSWERS $\Delta H = -1772$ kJ/mol; $\Delta E = -1766$ kJ/mol. Note that since this is a constant pressure calorimeter, $q = q_p = \Delta H$.

12.5 THERMODYNAMICS AND THERMOCHEMICAL METHODS

Hess's Laws

Thermochemistry is the study of the heat transfer that takes place between a system and its surroundings when a chemical reaction or phase change takes place. The quantities of heat transferred are related to the energy and enthalpy changes occurring in the system. Because H and E are state functions,

- The sum of the changes in ΔH or in ΔE must be zero over a complete cycle, one that returns to a state identical to the starting state (Fig. 12.13).
- The difference between two states is independent of the path taken. As a result, ΔH and ΔE can be determined from the most conveniently calculated path for many processes that are otherwise difficult, sometimes impossible, to actually carry out experimentally.

Although Lavoisier had measured heats of combustion half a century earlier, it remained for a Swiss-born Russian chemist named Germain Hess (1830) to make the first thermodynamic measurements of sufficient accuracy to establish relationships now recognized as **Hess's laws**, all of which are the result of H being a state function. From Hess's laws we are able to conclude the following:

1. The enthalpy change associated with a chemical reaction is directly proportional to the amounts of substances present, that is, ΔH is an extensive reaction. For example, one-half mole of H_2 reacts with one-half mole of Cl_2, giving one mole of HCl, according to the following equation:

$$\frac{1}{2}\mathrm{H_2(g)} + \frac{1}{2}\mathrm{Cl_2(g)} \longrightarrow \mathrm{HCl(g)} \qquad \Delta H = -92.5 \text{ kJ}$$

The change in enthalpy is -92.5 kJ, and the sign of ΔH is negative because the reaction is exothermic. If two moles of HCl are formed, twice the quantity of heat is released by the system:

$$\mathrm{H_2(g)} + \mathrm{Cl_2(g)} \longrightarrow \mathrm{2HCl(g)} \qquad \Delta H = -185 \text{ kJ}$$

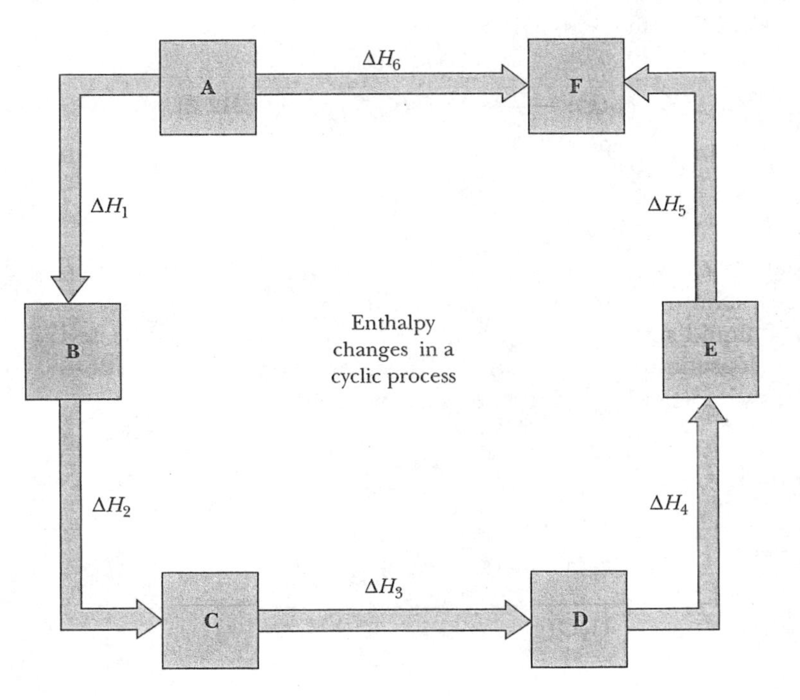

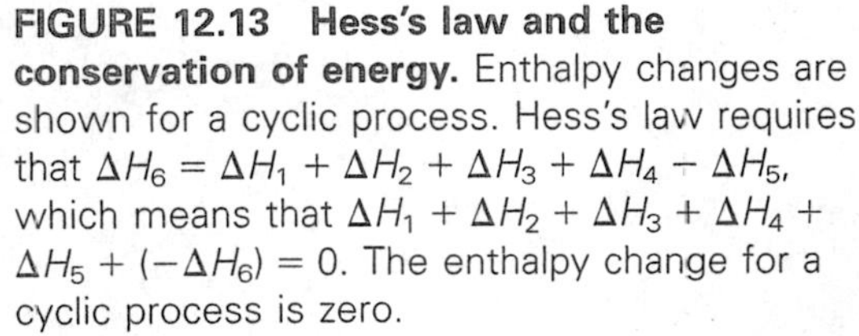
FIGURE 12.13 Hess's law and the conservation of energy. Enthalpy changes are shown for a cyclic process. Hess's law requires that $\Delta H_6 = \Delta H_1 + \Delta H_2 + \Delta H_3 + \Delta H_4 - \Delta H_5$, which means that $\Delta H_1 + \Delta H_2 + \Delta H_3 + \Delta H_4 + \Delta H_5 + (-\Delta H_6) = 0$. The enthalpy change for a cyclic process is zero.

2. When reactions are run in reverse, ΔH has the same numerical value but the opposite sign. That is,

$$\Delta H_{\text{forward}} = -\Delta H_{\text{reverse}}$$

$$H_2(g) + \frac{1}{2}O_2(g) \longrightarrow H_2O(\ell) \qquad \Delta H = -286 \text{ kJ}$$

$$H_2O(\ell) \longrightarrow H_2(g) + \frac{1}{2}O_2(g) \qquad \Delta H = +286 \text{ kJ}$$

3. When two or more equations are added to give the equation for an overall or net reaction, the value of ΔH is equal to the sum of the values of ΔH for the equations that have been combined. For example, a process that cannot be performed in the laboratory uses chlorine to produce oxygen from water as stated by reaction (3) below:

$$H_2O(\ell) \longrightarrow H_2(g) + \frac{1}{2}O_2(g) \qquad \Delta H_1 = +286 \text{ kJ} \quad (1)$$

$$H_2(g) + Cl_2(g) \longrightarrow 2HCl(g) \qquad \Delta H_2 = -185 \text{ kJ} \quad (2)$$

$$H_2O(\ell) + Cl_2(g) \longrightarrow \frac{1}{2}O_2(g) + 2HCl(g) \qquad \Delta H_3 = +101 \text{ kJ} \quad (3)$$

Fortunately, it is not necessary to perform reaction (3) in order to calculate its change in enthalpy ΔH_3, since it is the sum of equations (1) and (2). Hess's law has rescued us from an impossible demand.

4. The heat associated with a chemical reaction is influenced by the physical states of the reactants and products. Therefore, when writing thermochemical equations, it is necessary to identify the physical states of each reactant and product at the pressure and temperature of the reaction. Consider the combustion of hydrogen:

$$H_2(g) + \frac{1}{2}O_2(g) \longrightarrow H_2O(g) \qquad \Delta H = -242 \text{ kJ}$$

$$H_2(g) + \frac{1}{2}O_2(g) \longrightarrow H_2O(\ell) \qquad \Delta H = -286 \text{ kJ}$$

The heat of reaction associated with formation of one mole of H_2O in the gaseous state differs from that for the formation of one mole of H_2O in the liquid state. The 44 kJ/mol difference is the heat released during condensation of one mole of water or, in the reverse reaction, the heat required to vaporize one mole of water:

$$H_2O(\ell) \longrightarrow H_2(g) + \frac{1}{2}O_2(g) \qquad \Delta H = 286 \text{ kJ}$$

$$H_2(g) + \frac{1}{2}O_2(g) \longrightarrow H_2O(g) \qquad \Delta H = -242 \text{ kJ}$$

$$H_2O(\ell) \longrightarrow H_2O(g) \qquad \Delta H = 44 \text{ kJ}$$

Standard Enthalpies of Formation

Chemical processes performed in calorimeters result in ΔH values—calorimetric data—which we place at the end of the chemical equations for these reactions. But since there are many, many reactions that interest us, compiling thermochemical data for all of them is a formidable task, which would produce an enormous table of data. Instead, we compile thermochemical data for the individual substances of which these chemical reactions are composed. Their number is large, but manageable.

In discussing enthalpy in chemical processes, it is important to recognize that we cannot measure the actual enthalpy H, only the *change* in enthalpy ΔH associated with a chemical reaction, that is,

$$\Delta H = \Sigma H_{\text{products}} - \Sigma H_{\text{reactants}}$$

When we state

$$2O_2(g) + CH_4(g) \longrightarrow 2H_2O(g) + CO_2(g) \qquad \Delta H = -890 \text{ kJ}$$

what we are really saying is

$$\Delta H = [2H_{H_2O(g)} + H_{CO_2(g)}] - [2H_{O_2(g)} + H_{CH_4(g)}] = -890 \text{ kJ}$$

Since we cannot determine values of H for these substances, we need a different quantity to list in tables of data.

The change in the enthalpy (ΔH) that takes place when a compound forms from its elements is a measure of the enthalpy content of the compound compared to its elements. We call this enthalpy (heat) of formation ΔH_f° for a compound. The **enthalpy of formation** is the ΔH value accompanying the formation of one mole of a compound from its stable elements. For HCl the pertinent reaction is

$$\frac{1}{2} H_2(g) + \frac{1}{2} Cl_2(g) \longrightarrow HCl(g) \qquad \Delta H = -92.5 \text{ kJ}$$

and the ΔH for this reaction is the standard enthalpy of formation of HCl. It is important that the state of a compound be unambiguously defined by writing (g) for gas, (ℓ) for liquid, and (s) for solid along with the chemical formula. Ambiguities about the state of the elements and the compounds that are formed are avoided by specifying their most stable states under standard conditions: one atmosphere pressure and at a definite temperature, usually 298K. This is called the **standard state**. For example, oxygen can exist in a number of forms, including O atoms, O_2 molecules, and O_3 molecules. Oxygen can also be solid, liquid, or gas, depending on conditions. However, at 1 atm and 298K, its stable form is $O_2(g)$ molecules. On the other hand, Al(s) is the stable form for aluminum at the same temperature. Standard states for some other common elements are $H_2(g)$, $N_2(g)$, $F_2(g)$, $Cl_2(g)$, $Br_2(\ell)$, $I_2(s)$; all metals are solids in their standard states at 298K except mercury, which is a liquid. The standard state of carbon is considered to be graphite. Standard enthalpies or heats of formation are identified by the symbol ΔH_f°, in which the ΔH_f tells us that the values are for the formation of the compound from the elements, and the superscript $^\circ$ means that all substances are in their standard states.

Using this definition results in the very important conclusion that the standard enthalpies of formation of elements are zero. This follows from the fact that there is no reaction and therefore no enthalpy change in the formation of an element in its standard state from itself. For example,

$$O_2(g) \longrightarrow O_2(g) \qquad \Delta H = 0$$

The assignment of zero for the values of ΔH_f° of elements in their standard states is arbitrary but useful. The tabulated value of ΔH_f° for a compound is relative to this arbitrary assignment of a zero value.

The most important use of standard enthalpies of formation is calculating standard enthalpy changes of reactions using the general equation

$$\Delta H^\circ = \Sigma\Delta H^\circ_{f\,\text{products}} - \Sigma\Delta H^\circ_{f\,\text{reactants}}$$

Calculating ΔH° from ΔH_f° values: The heat of a reaction at constant pressure and a given temperature under standard state conditions can be found by subtracting the sum of the heats of formation of the reactants from the sum of the heats of formation of the products.

Standard enthalpies of formation are listed in Table 12.4. Because the tabulated ΔH_f° values are for formation of one mole of a compound, each ΔH_f° value must be multiplied by the coefficient for that substance in the balanced chemical equation.

Values for standard heats of formation can be used to find the change in enthalpy for any chemical reaction involving compounds listed in Table 12.4. For example, consider the reaction whereby urea (NH_2CONH_2) is produced from simple compounds such as ammonia, methane, and oxygen, as might have occurred under the prebiotic conditions existing during early Earth history. Remember that the standard enthalpy of formation of $O_2(g)$, an element in its standard state, is zero:

$$2NH_3(g) + CH_4(g) + 2O_2(g) \longrightarrow NH_2CONH_2(s) + 3H_2O(\ell)$$

$$\Delta H^\circ = [\Delta H_f^\circ(NH_2CONH_2(s)) + 3\Delta H_f^\circ(H_2O(\ell))] - [2\Delta H_f^\circ(NH_3(g)) + \Delta H_f^\circ(CH_4(g))]$$

$$\Delta H^\circ = [(-330) + 3(-286)] - [2(-46) + (-75)] \text{ kJ}$$

$$\Delta H^\circ = -1021 \text{ kJ}$$

EXAMPLE 12.6

Calculate the standard enthalpy change ΔH° for the following reaction:

$$4FeS_2(s) + 11O_2(g) \longrightarrow 8SO_2(g) + 2Fe_2O_3(s)$$

using the enthalpies of formation in Table 12.4. This reaction is an industrial process in which pyrite (fool's gold) is roasted to form an iron oxide.

TABLE 12.4(a) Standard Enthalpies (Heats) of Formation (ΔH_f°) at 298K—Inorganic Substances

Substance/State	ΔH_f°, kJ/mol
Br(g)	111.9
Br_2(g)	30.91
C(diamond)	1.90
Cl(g)	121
CO(g)	−110
CO_2(g)	−393.505
H_2O(g)	−242
H_2O(ℓ)	−286
H_2O_2(ℓ)	−187
H_2S(g)	−20.1
H_2SO_4(ℓ)	−811
HBr(g)	−36.4
HCl(g)	−92.5
HF(g)	−271.1
HI(g)	+25.9
HNO_3(ℓ)	−173
H_3PO_4(s)	1280
I(g)	75.5
I_2(g)	63
ICl(g)	17
N_2O(g)	82.05
NH_3(g)	−46.0
NO(g)	+90.4
NO_2(g)	+33.9
PCl_3(ℓ)	−636
PCl_5(g)	−374.9
SO_2(g)	−297
SO_3(g)	−395

TABLE 12.4(b) Standard Enthalpies (Heats) of Formation (ΔH_f°) at 298K—Ionic Compounds

Substance/State	ΔH_f°, kJ/mol
Ag_2O(s)	−30.5
AgCl(s)	−127
$AgNO_3$(s)	−123
Al_2Cl_6(s)	−1291
Al_2O_3(s)	−1676
$BaCO_3$(s)	−1216
BaO(s)	−556.9
$BaSO_4$(s)	−1444.6
$Ca(OH)_2$(s)	−986.09
$CaCl_2$(s)	−795
$CaCO_3$(s)	−1207
CaO(s)	−635.09
Cu_2O(s)	−168.06
CuO(s)	−155
CuS(s)	−53
Fe_2O_3(s)	−822
Fe_3O_4(s)	−1121
FeO(s)	−269
FeS_2(s)	−178
$MgCO_3$(s)	−1113
MgO(s)	−602
Na_2CO_3(s)	−1131
Na_2SO_4(s)	−1384
NaCl(s)	−411
PbO(s)	−217
WO_3(s)	−1675
ZnO(s)	−348.28

COMMENT Remember that the value of ΔH for any process depends on the amount of material. Therefore, the coefficients of substances in equations must be accounted for in the calculation.

SOLUTION

$$\Delta H^\circ = \Sigma\Delta H^\circ_{f\,\text{products}} - \Sigma\Delta H^\circ_{f\,\text{reactants}}$$

$$\Delta H^\circ = 8\Delta H_f^\circ(SO_2(g)) + 2\Delta H_f(Fe_2O_3(s)) - 4\Delta H_f^\circ(FeS_2(s))$$

$$\Delta H^\circ = 8(-297) + 2(-822) - 4(-178)\text{ kJ} = -3308\text{ kJ}$$

TABLE 12.4(c) Standard Enthalpies (Heats) of Formation (ΔH_f°) at 298K—Organic Compounds

Substance/State		ΔH_f°, kJ/mol	Substance/State		ΔH_f°, kJ/mol
acetic acid	$CH_3COOH(\ell)$	−484.5	lactic acid	$C_3H_6O_3(s)$	−694
acetylene	$C_2H_2(g)$	+227	methane	$CH_4(g)$	−74.9
benzene	$C_6H_6(\ell)$	+49.0	methanol	$CH_3OH(\ell)$	−238
n-butane	$C_4H_{10}(g)$	−125	*n*-octane	$C_8H_{18}(\ell)$	−208
ethane	$C_2H_6(g)$	−84.5	propane	$C_3H_8(g)$	−103.8
ethanol	$C_2H_5OH(\ell)$	−277.7	propane	$C_3H_8(\ell)$	−119
ethylene	$C_2H_4(g)$	52.26	propylene	$C_3H_6(g)$	20.0
glucose	$C_6H_{12}O_6(s)$	−1274	urea	$CO(NH_2)_2(s)$	−330

EXERCISE 12.6

Calculate ΔH° for this reaction. This is an industrial process in which iron oxide is being reduced to the metal by carbon monoxide:

$$Fe_3O_4(s) + 4CO(g) \longrightarrow 3Fe(s) + 4CO_2(g)$$

ANSWER −11 kJ

Checking the tabulated standard heat of formation ΔH_f° for CO_2 in Table 12.4 reveals the value to be −393 kJ/mol. But only a few standard heats of formation (such as that of CO_2) can be determined directly by calorimetry:

$$C(\text{graphite}) + O_2(g) \longrightarrow CO_2(g) \qquad \Delta H = -393 \text{ kJ} = \Delta H_f^\circ$$

For the most part, ΔH_f° values must be determined indirectly, as we will now demonstrate by finding the ΔH_f° value for glucose ($C_6H_{12}O_6$), a substance whose formation is not feasible from its elements in any possible calorimetric study:

$$6C(\text{graphite}) + 6H_2(g) + 3O_2(g) \longrightarrow C_6H_{12}O_6(s) \qquad \Delta H_f^\circ = ?$$

It is often the case that organic compounds can be burned to carbon dioxide and water, so let us begin with the standard heat of combustion for glucose from Table 12.3:

$$C_6H_{12}O_6(s) + 6O_2(g) \longrightarrow 6CO_2(g) + 6H_2O(\ell) \qquad \Delta H = -2803 \text{ kJ}$$

Reversing the reaction gives the *formation* rather than the combustion of glucose from carbon dioxide and water:

$$6CO_2(g) + 6H_2O(\ell) \longrightarrow C_6H_{12}O_6(s) + 6O_2(g) \qquad \Delta H = +2803 \text{ kJ}$$

What we want, however, is the formation of glucose from the elements, not from $CO_2(g)$ and $H_2O(\ell)$, a result that can be obtained by adding in the heats of combustion of graphite and hydrogen so as to cancel out the CO_2 and H_2O:

$$6CO_2(g) + 6H_2O(\ell) \longrightarrow C_6H_{12}O_6(s) + 6O_2(g) \qquad \Delta H = +2803 \text{ kJ}$$

$$6[C(\text{graphite}) + O_2(g) \longrightarrow CO_2(g)] \qquad \Delta H = 6(-393.5) \text{ kJ}$$

$$6[H_2(g) + \frac{1}{2}O_2(g) \longrightarrow H_2O(\ell)] \qquad \Delta H = 6(-286) \text{ kJ}$$

$$6C(\text{graphite}) + 6H_2(g) + 3O_2(g) \longrightarrow C_6H_{12}O_6(s) \qquad \Delta H = -1274 \text{ kJ/mol}$$

Using Hess's law, the resulting equation for the formation of glucose from its elements in their standard states is obtained by adding the chemical equations for the three steps algebraically.

EXAMPLE 12.7

What is the standard enthalpy of formation ΔH_f° of naphthalene, $C_{10}H_8(s)$?

COMMENT You cannot put carbon and hydrogen in a calorimeter and measure the heat of formation of naphthalene. But by combining thermochemical equations having known values for ΔH°, the desired ΔH_f° can be found. The standard enthalpy of combustion ΔH_c° for naphthalene is given as -5153.9 kJ/mol in Table 12.3. This means that ΔH° for the reverse direction will be $+5153.9$ kJ/mol. Adding the thermochemical equation for this reaction to the thermochemical equations for the formation of 10 moles of CO_2 and 4 moles of H_2O (found by multiplying the coefficients and the ΔH° values by 10 and 4, respectively) gives the equation for the formation of naphthalene. Notice how amounts of substances on opposite sides of the equations are canceled when the equations are added.

SOLUTION

$$10CO_2(g) + 4H_2O(\ell) \longrightarrow C_{10}H_8(s) + 12O_2(g) \qquad 5153.9 \text{ kJ}$$

$$10[C(s) + O_2(g) \longrightarrow CO_2(g)] \qquad 10(-393) \text{ kJ}$$

$$4[H_2(g) + \frac{1}{2}O_2(g) \longrightarrow H_2O(\ell)] \qquad 4(-286) \text{ kJ}$$

$$10C(s) + 4H_2(g) \longrightarrow C_{10}H_8(s) \qquad 80. \text{ kJ/mol}$$

EXERCISE 12.7

What is ΔH_f° for benzoic acid, $C_6H_5COOH(s)$?

ANSWER -382 kJ/mol

Allotropic Modifications

Enthalpy changes are associated with changes between the allotropes of an element. **Allotropes** are different forms of the same element. For example, diamond and graphite are allotropes of carbon. Conversions of one allotrope into another are called allotropic modifications, and $\Delta H°$ values can be calculated for these transformations.

EXAMPLE 12.8

Calculate $\Delta H°$ for the conversion of graphite to diamond.

COMMENT Apply Hess's laws, the principles enumerated at the beginning of this section governing thermochemical equations and their manipulation. In this instance, heat of combustion data establish the enthalpy differences between diamond and graphite.

SOLUTION

$$C(\text{diamond}) + O_2(g) \longrightarrow CO_2(g) \qquad \Delta H_c° = -395.388 \text{ kJ/mol (diamond)}$$

$$C(\text{graphite}) + O_2(g) \longrightarrow CO_2(g) \qquad \Delta H_c° = -393.505 \text{ kJ/mol (graphite)}$$

Graphite is the lower energy form at room temperature. The enthalpy change associated with the transition from one form to the other can be calculated from the sum of the two thermochemical equations:

$$C(\text{graphite}) + O_2(g) \longrightarrow CO_2(g) \qquad \Delta H_c° = -393.505 \text{ kJ/mol}$$

$$CO_2(g) \longrightarrow C(\text{diamond}) + O_2(g) \qquad \Delta H° = -(-395.388) \text{ kJ/mol}$$

$$C(\text{graphite}) \longrightarrow C(\text{diamond}) \qquad \Delta H° = +1.883 \text{ kJ/mol}$$

EXERCISE 12.8

Given the following thermochemical data, calculate the enthalpy change accompanying the conversion of one allotrope of sulfur to another, rhombic sulfur to monoclinic sulfur:

$$S(\text{rhombic}) + O_2(g) \longrightarrow SO_2(g) \qquad \Delta H_c° = -296.980 \text{ kJ/mol}$$

$$S(\text{monoclinic}) + O_2(g) \longrightarrow SO_2(g) \qquad \Delta H_c° = -297.148 \text{ kJ/mol}$$

Which is the lower energy form under standard conditions?

ANSWERS $\Delta H° = 0.168$ kJ/mol = 168 J/mol; rhombic sulfur

Phase Changes

We know that it takes heat to melt a solid or to vaporize a liquid and that heat is given off when a gas is condensed or a liquid is frozen. These processes are normally carried out at constant pressure, and so the concept of enthalpy applies to phase changes just as it does to chemical reactions and allotropic modifications. Transforming any solid to a liquid requires energy to disrupt the interactions that hold the atoms or molecules in the solid state. The heat required to overcome intermolecular attractions and transform one mole of solid to liquid at the melting point is called the **enthalpy of fusion** or ΔH_{fus}. Often called heat of fusion, the enthalpy of fusion is a positive quantity since fusion is an endothermic process requiring heat to be transferred to the system. The effectiveness of ice for cooling as the ice itself melts is due to its large enthalpy of fusion:

$$H_2O(s) \longrightarrow H_2O(\ell) \qquad \Delta H_{fus} = 6.00 \text{ kJ/mol}$$

For the reverse of the fusion of one mole of solid, one mole of liquid is frozen to solid at the melting point. The reverse process is exothermic, and the heat involved is the negative of the enthalpy of fusion:

$$H_2O(\ell) \longrightarrow H_2O(s) \qquad \Delta H = -\Delta H_{fus} = -6.00 \text{ kJ/mol}$$

Conversion of one mole of liquid to gas also requires the input of heat as **enthalpy of vaporization** or **ΔH_{vap}** to disrupt the intermolecular attractions that hold the molecules in the liquid phase. In fact, the change from liquid to gas in which the molecules are completely separated requires much more energy than the partial separation of the molecules that occurs in fusion. Thus, ΔH_{vap} is much greater than ΔH_{fus} for any substance (Table 12.5):

$$H_2O(\ell) \longrightarrow H_2O(g) \qquad \Delta H_{vap} = 44 \text{ kJ/mol}$$

The heat for the reverse and exothermic process or condensation is −44 kJ, the negative of the enthalpy of vaporization.

Likewise, the input of heat to sublime one mole of a substance is its **enthalpy of sublimation** or **ΔH_{sub}**, which has a positive value. The opposite process, sometimes called deposition, has an enthalpy with the same numerical value, but the opposite sign. For example, solid iodine can pass directly to vapor,

$$I_2(s) \longrightarrow I_2(g) \qquad \Delta H_{sub} = 63 \text{ kJ/mol}$$

For any substance, the value of ΔH_{sub} is always the highest for the three phase changes because the disruption of intramolecular forces is the greatest. The sublimation of $CO_2(s)$ (dry ice) to $CO_2(g)$ is very familiar.

TABLE 12.5 Selected Enthalpies (Heats) of Fusion and Vaporization at the Normal Phase Transition Temperatures

Substance	Melting Point (°C)	ΔH_{fus} (kJ/mol)	Boiling Point (°C)	ΔH_{vap} (kJ/mol)
aluminum	660	10.7	2450	284
bromine	−7.2	5.42	58.8	7.8
copper	1083	8.5	1187	322
ethanol	−114	4.79	78	39.3
gallium	29.8	5.57	1983	258
gold	1063	12.6	2660	1311
helium	−269.7	0.021	−268	36.9
lead	327	5.08	1750	178
mercury	−38.9	2.26	356	58.0
oxygen	−218.8	0.442	−183.0	6.82
potassium	63.2	2.33	760	78.0
silver	961	9.53	2193	254
sodium	97.5	2.64	880	98.0
sulfur (thombic)	119	1.22	445	10.5
water	0	6.00	100	44

PROCESSES IN CHEMISTRY

Geothermal Electric Power from Hot Dry Rock (HDR), Hydrothermal Energy Technologies

Geothermal electric power is produced from the kinetic energy stored in the Earth's crust, where it is close enough to be extracted. Use of geothermal energy is thermodynamically based on the temperature difference between rock formations and water beneath the Earth's surface and water at the surface. Temperatures in the Earth increase to about 1000°C at the base of the crust and to perhaps 4000°C at the center of the Earth. The heat sources that produce these temperatures come from the flow of heat from the deep crust and mantle, and energy largely generated in the upper crust by radioactive decay of uranium, thorium, and potassium isotopes.

Scientists at Los Alamos National Laboratory have recently demonstrated that heat in sufficient quantity can be extracted from hot dry rocks under the surface of the Earth to produce a limited but steady flow of electric power. By pumping water at rates up to 100 gallons per minute through cracks made in rocks at 235°C and 12,000 feet below the nearby Jemez Mountains, they were able to return water to the surface as steam at 175°C. That temperature differential, in turn, delivered a heat equivalent of 4 megawatts of power. While hot dry rock (HDR) technology offers the potential to vastly increase the supply of power derived directly from the Earth's natural heat sink, hydrothermal energy is already being extracted from hot waters and steam circulating naturally through underground rock formations by power plants throughout California, Nevada, Utah, and Hawaii—some 70 plants in all, with a generating capacity of about 3000 megawatts, or enough for about 1 million people—more commercial electricity than is at present being produced from the two more popular forms of renewable energy, solar and wind, combined.

The advantages geothermal technologies offer are impressive: no oil spills, no carbon dioxide added to the atmosphere or ash left over from burning coal and petroleum resources, no acid run-off from open-pit coal mining, and no radioactive wastes. But, as always, there is no such thing as a free lunch. Geothermal may be less polluting, but it may not be cheaper, largely because commercial developers will have to bore holes a mile or perhaps two miles deep through bedrock. Current estimates are triple the current costs of gas turbine electricity from existing fossil fuel–based technologies. Nevertheless, the possibilities of geothermal energy cannot be denied and must be included in any mix of energy technologies that will see us through the twenty-first century and beyond.

Questions

What problems do you anticipate when circulating hot water and steam from underground rock formations through the heating system of your house?

Compare the advantages of geothermal energy technologies to coal, petroleum, and nuclear energy.

What principal engineering problems will have to be dealt with before HDR technologies can be added to the mix of useful energies for human use?

SUMMARY

In this introduction to thermodynamics, we described the absorption or release of heat that accompanies chemical change or changes of state, and we introduced several relationships between these thermal effects and other important quantities such as work and energy. Internal energy E is a property of an object and is a function of its state. Heat (q) and work (w) represent ways of transferring energy.

We defined a thermodynamic system as one that is separated from its surroundings by carefully specified boundaries. The system is described by its state, the value of a number of variables such as temperature, volume, pressure, and composition. A state of equilibrium exists when none of these variables is changing. The change from one equilibrium state to another is a thermodynamic process. Changes in state functions such as ΔE, ΔV, ΔP, and ΔT depend only on the initial and final states. Path functions such as q and w depend on the actual steps of the path taken to get from the initial to the final state. The first law of thermodynamics is based on the observations that energy is conserved and that q and w are the only ways in which it can be transferred. Thus, $\Delta E = q + w$.

The change in enthalpy ΔH equals q_p, the heat of reaction at constant pressure. Since constant pressure conditions are much employed for chemical reactions, the enthalpy function is very useful in chemistry. The heat of reaction at constant volume is q_v, which is equal to ΔE. Values of ΔH and ΔE are essentially the same for chemical reactions in which there are no gaseous reactants or products, or where there is no difference between the number of moles of gaseous reactants and products. Calorimetry provides an experimental method for determining ΔH and ΔE values for chemical reactions.

Values of the enthalpy changes for chemical reactions at standard conditions can be obtained using tables of standard enthalpies of formation and combustion. These calculations utilize the properties of state functions, in particular that the sum of the $\Delta H°$ values for any series of steps leading from the initial state to the desired final state will give the correct value of $\Delta H°$ for the overall process.

The thermodynamic meaning of heat, work, and energy requires precise definition, which may be different from the meanings of the terms in everyday usage. Table 12.6 gives some of the special definitions used in this chapter.

TABLE 12.6 Selected Definitions

System: The part of the universe that we are studying at a particular time. The system is separated from the rest of the universe by boundaries that we define.

Surroundings: The surroundings are all of the universe except the system. The way a system can affect its surroundings is determined by the nature of its boundaries.

Internal Energy: The total of the kinetic energy and the potential energy for a system or part of a system.

Kinetic Energy: Energy related to movement of atoms and molecules. Kinetic energy increases with increase of temperature.

Potential Energy: Energy resulting from attractions and repulsions of an object as a result of its position.

State: A situation within a system that can be defined by the properties volume, pressure, temperature, and chemical composition. The important states in our study of thermodynamics are equilibrium states in which none of the properties is changing.

Process: A change in a system from one equilibrium state to another.

Path: The route taken during a process from one equilibrium state to another.

State Function: A function like T, V, P, E, or H that helps define the state of a system. The value of a change of a state function depends only on the initial and final states of the system and not on the pathway. The change in internal energy ΔE is an example.

TERMS

Thermodynamics (12.1)
Heat (12.2)
Specific heat (12.2)
Molar heat capacity (12.2)
System (12.2)
Surroundings (12.2)
Boundary (12.2)
State (12.2)
Equilibrium state (12.2)
Process (12.2)
Path (12.2)
State function (12.2)
Kinetic energy (12.2)
Translational energy (12.2)
Internal energy (12.2)
Potential energy (12.2)
Exothermic reaction (12.2)
Endothermic reaction (12.2)
Work (12.2)
Free expansion (12.2)
First law of thermodynamics (12.3)
Path function (12.3)
Enthalpy (12.4)
Thermochemical equation (12.4)
Heat of combustion (12.4)
Calorimeter (12.4)
Thermochemistry (12.5)
Hess's laws (12.5)
Enthalpy of formation (12.5)
Standard state (12.5)
Allotrope (12.5)
Enthalpy of fusion (12.5)
Enthalpy of vaporization (12.5)
Enthalpy of sublimation (12.5)

IMPORTANT EQUATIONS AND THE CONDITIONS UNDER WHICH THEY APPLY

Equation	Conditions
$q = nC\Delta T$	Heat transfer; moderate temperature interval
$KE = \frac{3}{2}RT$	Translational kinetic energy for 1 mol of ideal gas
$w = -P_{ex}\Delta V$	Work; constant external pressure
$\Delta E = q + w$	First law of thermodynamics; always applies
$\Delta H = \Delta E + P\Delta V$	Change in enthalpy; constant pressure
$\Delta E = q_v$	Internal energy change; constant volume
$\Delta H = q_p$	Enthalpy change; constant pressure
$\Delta H = \Delta E + \Delta nRT$	Constant pressure and temperature; ideal gas behavior; changes in solid and liquid volumes are assumed negligible
$q = -C_{calorimeter}\,\Delta T$	Constant pressure or constant volume
$\Delta H^\circ = \Sigma\Delta H_f^\circ(\text{products}) - \Sigma\Delta H_f^\circ(\text{reactants})$	Calculation of the standard ΔH for a reaction from standard enthalpies of formation, at constant pressure

QUESTIONS

Conceptual questions are denoted by a square screen. Extra-credit questions are denoted by a circular screen.

1. Using illustrations drawn from your common experience, explain what is meant by the following:
   (a) an exothermic process
   (b) an endothermic process
   (c) a cyclic process.
2. Distinguish between each of the following:
   (a) energy and enthalpy
   (b) heat and work
   (c) heat and temperature
   (d) work and energy
3. Fakirs (and fakers) have walked barefooted on burning coals without apparent injury. How is that possible?
4. Why is it desirable that a calorimeter have a large heat capacity (calorimeter constant)?
5. What is the essential requirement of a thermometer used for calorimetry? Why?
6. Which kind of functions obey Hess's laws, state functions or path functions? Why?
7. Why are Hess's laws so useful?
8. From your own experience, how would you judge temperature without a thermometer? Offer two or three methods.
9. A blacksmith seems to pay unusual attention to the sound or "ring" of the horse shoe as it is hammered. Why is that?
10. Ethyl alcohol has about half the heat capacity of water. If equal moles of alcohol and water in separate containers are supplied with the same amount of heat, how will their respective temperature changes compare?
11. Two objects in thermal equilibrium with each other are at the same temperature; two objects in thermal equilibrium with a third object must be in thermal equilibrium with each other. How would you explain these two statements by making use of a simple algebraic argument?
12. Noting that thermodynamics says nothing about the time required to establish equivalence of temperature between two systems in thermal contact, what examples of each of the following can you think of?
    (a) Equilibrium is established very quickly (in less than a second).
    (b) Equilibrium is established in a few hours.
    (c) Millions of years pass, and equilibrium still hasn't been established.
13. What are the sign conventions regarding the transfer of heat and work carried out between a thermodynamic system and its surroundings?
14. What other thermodynamic quantities besides temperature tend toward equivalence when two systems are in direct and intimate contact?
15. Why is it necessary to keep liquid nitrogen in a thermos bottle?

PROBLEMS

Problems marked with a bullet (•) are answered in Appendix A, in the back of the text.

First Law Calculations: w, q, ΔE, ΔH [1–18]

•1. What quantity of heat is necessary:
   (a) To raise the temperature of 1.50 kg of liquid water from 30°C to 60°C?
   (b) To heat 2.45 g of solid iron from 25°C to 28°C?

2. How much heat is required
   (a) To increase the temperature of 205 kg of benzene [$C_6H_6(\ell)$] from 40°C to 50°C?
   (b) To heat 0.00450 g of Al(s) from 48.0°C to 50.5°C?

•**3.** If the internal energy of a thermodynamic system is decreased by 500. J when 150. J of work was done on the system, how much heat was transferred? Was energy transferred to or from the system?

4. A system undergoes a process in which its internal energy drops by 3660 kJ. The system gives off 2950 kJ of heat during the process. Give the values of w, q, and ΔE for the process. Does the system do work on the surroundings, or do the surroundings do work on the system, as the process proceeds?

•**5.** As a gas expands against a constant pressure of 5.0 atm, its volume changes from 10.0 L to 20.0 L while it absorbs 2.00 kJ of heat. Calculate the change in internal energy experienced by the gas.

6. An ideal gas expanded against a constant external pressure of 700. torr and as it did so its volume changed from 50.0 L to 150. L. During the process, 6485 J of heat was absorbed. Calculate the change in the internal energy of the gas.

•**7.** One liter of an ideal gas at 0°C and 10.0 atm was allowed to expand against a constant external pressure of 1 atm. The temperature of the gas remained constant throughout. (*Hint:* ΔE is zero if ΔT is zero, since E is proportional to T for an ideal gas.)
(a) Calculate q, w, ΔE, and ΔH.
(b) If the expansion were instead a free expansion, how would these values be changed?

8. Calculate w, q, ΔE, and ΔH for the expansion of 2.00 L of an ideal gas to a final volume of 5.00 L against a constant external pressure of 2.50 atm. The process takes place in a well-insulated container, so you can assume that $q = 0$.

•**9.** A gas is allowed to expand from 10.00 L to 20.00 L at a constant pressure of 1.500 atm. During the course of the expansion, the gas absorbs 500. J. Calculate the work done on the gas and the change in the internal energy of the gas.

10. A gas is compressed from 10.00 L to 5.00 L at a constant pressure of 0.500 atm. During the course of the compression, the gas gives off 400. J of heat. Calculate the work done on the gas and the change in the internal energy of the gas.

•**11.** During the course of a certain constant volume process, 200. J of heat is transferred to 2.0 mol of Ne gas, initially at 25°C. Calculate each of the following:
(a) The work done by the gas
(b) The final temperature of the gas (under these conditions C is 12.4 J/mol·K)

12. Calculate the change in the temperature for a mole of helium that has absorbed 121 J of heat.

•**13.** Given the thermochemical equation for the following reaction at 298K and 1.00 atmosphere, determine q_p and q_v:

$$C(s) + \frac{1}{2}O_2(g) \longrightarrow CO(g) \qquad \Delta H = -110.\ kJ$$

14. Given the following equation for the indicated process at 298K and 1.00 atmosphere pressure, determine ΔH and ΔE if q_p is known to be -393 kJ/mol:

$$C(s) + O_2(g) \longrightarrow CO_2(g)$$

15. A mole of ice melts at 273K and 1.00 atm pressure, and 6.00 kJ of heat are absorbed by the system. Given the density of ice and water under these conditions to be 0.915 and 1.00 g/mL, respectively, calculate ΔH and ΔE.

16. A mole of water freezes at 273K and 1.00 atm, and 6.00 kJ of heat are given off to the surroundings. Using the data in the preceding problem, calculate ΔH and ΔE.

17. A 1.00-kg sample of $K(\ell)$ is allowed to vaporize at a constant temperature of 760.°C and an unknown but constant pressure. During the vaporization process, 216.2 kJ of work was done, and the internal energy increased by 1530. kJ. The volume of K(g) produced is 2134 L. Assume that the volume of $K(\ell)$ is negligible.
(a) Calculate the heat of vaporization for potassium.
(b) Find the constant pressure against which the system did work.

18. Calculate the work done in vaporizing 1.00 mol of water under the following conditions, assuming ideal behavior ($P_{ex} = 1.00$ atm):
(a) at its normal boiling point (100°C)
(b) by letting it evaporate at room temperature (25°C) and one atmosphere pressure.

Calorimetry [19–20]

•**19.** A sample of pure benzoic acid (C_6H_5COOH), a combustible organic solid, weighed 1.221 g. It was placed in a constant volume calorimeter, ignited in a pure oxygen atmosphere, and a temperature rise from 25.240°C to 31.668°C was noted. The heat capacity of the calorimeter was 5.020 kJ/K. The combustion products were carbon dioxide and water.
(a) Write the balanced chemical equation for the combustion reaction.
(b) Calculate ΔE, q, and w for the calorimeter process.
(c) Calculate ΔH for the reaction.

20. A constant volume calorimeter has a heat capacity of 32.606 kJ/K. Combustion of a 2.316-g sample of menthol, $C_{10}H_{20}O(s)$, causes the temperature to increase from 25.010°C to 27.897°C. Calculate ΔE and ΔH for the combustion of 1 mol of menthol.

Thermochemistry; Hess's Laws [21–34]

•**21.** Using the data in Table 12.4, calculate the enthalpy change for the hydration of ethylene, producing ethanol according to the following equation:

$$C_2H_4(g) + H_2O(\ell) \longrightarrow C_2H_5OH(\ell)$$

22. Using the appropriate thermochemical data from Table 12.4, calculate $\Delta H°$ and $\Delta E°$ for the following reaction at 298K:

$$C(\text{graphite}) + 2H_2O(g) \longrightarrow CO_2(g) + 2H_2(g)$$

•**23.** From the thermochemical data in Table 12.4, calculate the heats of reaction for each of the following transformations carried out at 298K:
(a) $CO(g) + \frac{1}{2}O_2(g) \longrightarrow CO_2(g)$
(b) $H_2(g) + \frac{1}{2}O_2(g) \longrightarrow H_2O(\ell)$
(c) $H_2O(\ell) \longrightarrow H_2O(g)$
(d) Then, using these data, calculate the heat of reaction at 298K for the following reaction:

$$CO(g) + H_2O(g) \longrightarrow CO_2(g) + H_2(g)$$

24. You are given the following thermochemical data:

$$P(s) + \frac{3}{2}Cl_2(g) \longrightarrow PCl_3(\ell) \qquad \Delta H° = -636 \text{ kJ}$$

$$PCl_3(\ell) + Cl_2(g) \longrightarrow PCl_5(s) \qquad \Delta H° = -138 \text{ kJ}$$

(a) Write the applicable equation for the formation of $PCl_5(s)$, then calculate its $\Delta H_f°$.
(b) Obtain any necessary data from Table 12.4, and calculate $\Delta H°$ for the following reaction:

$$PCl_5(s) + 4H_2O(\ell) \longrightarrow H_3PO_4(s) + 5HCl(g)$$

•**25.** The Goldschmidt process is used industrially to produce metallic iron.

$$2Al(s) + Fe_2O_3(s) \longrightarrow Al_2O_3(s) + 2Fe(s)$$

Calculate the heat liberated per gram of metal produced.

26. Methane can be used to convert iron(II) oxide (FeO) to iron metal. The other products of the reaction are carbon monoxide gas and water vapor:

$$CH_4(g) + 3FeO(s) \longrightarrow 3Fe(s) + CO(g) + 2H_2O(g)$$

How much heat is required to produce 1 kg of Fe by this process at constant pressure?

•**27.** Taking the necessary thermochemical data from Table 12.4, calculate the enthalpy change of the following reaction at 298K (note that the standard state of sulfur is the rhombic form):

$$SO_2(g) + 2H_2S(g) \longrightarrow 3S(\text{rhombic}) + 2H_2O(\ell) \qquad \Delta H° = ?\ \text{kJ}$$

28. Using the following reaction and thermochemical data from Table 12.4, calculate ΔH_f° for AgBr(s):

$$AgBr(s) + \frac{1}{2}Cl_2(g) \longrightarrow AgCl(s) + \frac{1}{2}Br_2(g) \qquad \Delta H^\circ = -27.5 \text{ kJ}$$

•29. Write the formation equation, then calculate ΔH_f° for $Ca(OH)_2(s)$ using data from Table 12.4 and

$$CaO(s) + H_2O(\ell) \longrightarrow Ca(OH)_2(s) \qquad \Delta H^\circ = -64 \text{ kJ}$$

30. Write the equation for the formation of $ICl(\ell)$ from its elements in their standard states, and use the data in Table 12.4 and the following thermochemical data to calculate ΔH_f° for $ICl(\ell)$:

$$ICl(\ell) \longrightarrow I(g) + Cl(g) \qquad \Delta H^\circ = +211 \text{ kJ}$$

•31. Calculate the ΔE_c° for the combustion of liquid *n*-pentane (C_5H_{12}), a gasoline fraction obtained from petroleum, on carrying out the reaction in a constant volume calorimeter at room temperature (25°C). First write out the balanced equation, assuming that $CO_2(g)$ and $H_2O(\ell)$ are the only products.

32. On a mass basis, determine which is the preferred rocket fuel, assuming they both cost the same per unit mass: dimethyl hydrazine or hydrogen.

$$(CH_3)_2NNH_2(\ell) + 4O_2(g) \longrightarrow N_2(g) + 4H_2O(\ell) + 2CO_2(g) \qquad \Delta H^\circ = -1694 \text{ kJ}$$

$$H_2(g) + \frac{1}{2}O_2(g) \longrightarrow H_2O(\ell) \qquad \Delta H^\circ = -286 \text{ kJ}$$

•33. Sufficient heat is liberated on ignition and burning of 1.00 g of rhombic sulfur (the standard form of sulfur) to $SO_2(g)$ to raise the temperature of 100. g of water by 22.2°C at constant pressure. What is ΔH_c° for rhombic sulfur?

34. On burning 1.00 kg of a high-carbon coal in a water-jacketed furnace at constant pressure, the 47 liters of water in the jacket increased in temperature from 25.0°C to 38.3°C. (Omit consideration of the heat capacity of the metal of the furnace for this calculation.) If we assume that the coal is 90.% carbon and 10.% inert ingredients, what is ΔH_c° for carbon?

Additional Problems [35–47]

35. A bomb calorimeter (constant volume) with a heat capacity of 4.05 kJ/K is used to burn 1370 mg of a white crystalline solid believed to be urea, a major source of fertilizer nitrogen. The temperature rises from 24.12°C to 27.70°C during the combustion. Assuming the reaction produces gaseous carbon dioxide, nitrogen, and liquid water, does the experiment confirm the belief?

36. For the bomb calorimeter experiment described in the previous example, determine each of the following:

(a) The volume change that would have been observed if the sample was urea and 1.00 mol was burned under a constant pressure of 1.00 atm.

(b) The value of w if the combustion had taken place in a constant pressure calorimeter at 1.00 atm.

(c) The quantity of heat absorbed by the calorimeter in such a constant pressure process.

(d) The enthalpy change (ΔH) for the process.

•37. In an experiment designed to determine the heat of vaporization of water at 1 atm and 100°C, it was found that 40.66 kJ are required to cause 18.02 grams to "boil" away. Calculate ΔE and ΔH for the vaporization.

38. Taking values for the standard heats of formation of glucose and lactic acid—two biologically important molecules—from Table 12.4, find the heat of reaction for the formation of lactic acid from glucose. The chemical reaction can be written as follows:

$$C_6H_{12}O_6(s) \longrightarrow 2C_3H_6O_3(s)$$

39. Diborane has been considered for use as a jet fuel. The combustion reaction can be written according to the following equation:

$$B_2H_6(g) + 3O_2(g) \longrightarrow B_2O_3(s) + 3H_2O(g) \qquad \Delta H = -1940 \text{ kJ}$$

The combustion of elemental boron proceeds according to the following equation:

$$2B(s) + \frac{3}{2}O_2(g) \longrightarrow B_2O_3(s) \qquad \Delta H = -2370 \text{ kJ}$$

What is the standard heat of formation of diborane?

40. Calculate the heat of formation of sucrose using data from Table 12.3 and Table 12.4.

•**41.** Calculate the heat of formation of $PbCl_2(s)$ from the following thermochemical reaction and data in Table 12.4.

$$2Ag(s) + PbCl_2(s) \longrightarrow 2AgCl(s) + Pb(s) \qquad \Delta H° = +105 \text{ kJ}$$

42. A gas expands against a constant external pressure of 0.500 atm from an initial volume of 1.500 L to a final volume of 6.25 L. The container in which this process is taking place is well insulated, and you may assume no heat enters or leaves the system.

(a) Determine the work w done by the system.

(b) Calculate the energy change ΔE for the system.

(c) What happens to the temperature of the gas: Does it rise, fall, or stay the same? Briefly explain.

•**43.** Both acetylene (C_2H_2) and ethylene (C_2H_4) yield ethane (C_2H_6) on hydrogenation, according to the following thermochemical reactions:

$$C_2H_2(g) + 2H_2(g) \longrightarrow C_2H_6(g) \qquad \Delta H = -311 \text{ kJ}$$

$$C_2H_4(g) + H_2(g) \longrightarrow C_2H_6(g) \qquad \Delta H = -136 \text{ kJ}$$

Calculate ΔH for the formation of ethylene from acetylene and hydrogen, and write the equation for this process.

44. (a) Calculate the work done when 1.00 mol of ethyl ether ($C_4H_{10}O$) is allowed to vaporize completely at its normal boiling point (34.5°C) and 1.00 atm. Assume ideal behavior on the part of the gas.

(b) Based on your answer to part (a) and the known enthalpy change for the vaporization for ethyl ether (20.0 kJ/mol), find ΔE for the process.

•**45.** Here is a laboratory experiment on calorimetry, designed to determine the heat capacity of the calorimeter and the heat of combustion of an unknown compound. A 605-mg sample of naphthalene ($C_{10}H_8$) was burned in a constant volume calorimeter. The recorded change in temperature was +2.255°C. A 1.67-g sample of the unknown compound, burned in the same way and in the same calorimeter, resulted in a temperature change of 2.030°C. Calculate the heat capacity of the calorimeter and the heat of combustion of the unknown compound if its formula mass is known (from a separate experiment) to be 176 g/mol.

46. The volume of a sample of methane was measured at 23.4°C and 0.987 atm and found to be 1.29 L. The sample was then burned in a constant volume bomb calorimeter with a heat capacity of 5432.10 J/K and the temperature was observed to rise 8.57K. Calculate the ΔH_c for methane in kJ/mol.

47. In a classic experiment, the heat of combustion of benzoic acid was determined by burning it in a constant-volume calorimeter. Initially, 1.870 g of benzoic acid was placed in the calorimeter, which was at 25.010°C. After combustion, the temperature of the calorimeter was observed to have risen to 29.315°C.

(a) If the heat capacity or calorimeter constant is 11.485 kJ/K, what is the standard heat of combustion at 298K in kJ/mol?

(b) Compare this result to that obtained from Table 12.3.

(c) Calculate the heat of combustion of benzoic acid if $H_2O(g)$ is a product.

Cumulative Problems [48–52]

48. The industrial process for the synthesis of acetylene is based on the partial combustion of liquefied petroleum gas (LPG), which is a mixture of 95% propane (C_3H_8) and 5% methane (CH_4) by mass. The methane cannot be converted to acetylene. The acetylene synthesis from propane is shown by the following equation:

$$C_3H_8(g) + 2O_2(g) \longrightarrow C_2H_2(g) + CO(g) + 3H_2O(\ell)$$

(a) Calculate the standard enthalpy of reaction at 298K for the acetylene synthesis.
(b) Starting with a kilogram of LPG, what is the maximum yield in grams of acetylene?
(c) Calculate the heat for the 5% methane side-reaction:

$$2CH_4(g) + 3O_2(g) \longrightarrow 2CO(g) + 4H_2O(\ell)$$

•**49.** Beginning with a kilogram of the LPG mixture described in Problem 48 and assuming that only the two indicated reactions occur, what fraction of the total heat from the reaction is due to the acetylene synthesis? What is the total overall heat of the reactions at 298K as measured in kJ/kg?

50. If a kilogram of water is produced in the synthesis of acetylene from LPG as described in Problem 48, how many kilograms of LPG were consumed?

•**51.** Consider the industrial synthesis of methanol from synthesis gas:

$$CO(g) + 2H_2(g) \longrightarrow CH_3OH(\ell)$$

A 10.0-L constant-volume reactor is initially charged with $CO(g)$ and $H_2(g)$ in a 1:2 mole ratio. The initial and final reaction temperatures are 298K, and the initial pressure is 22.0 atm. If the reaction is complete, what is the final pressure? How much heat was transferred into or out of the reactor? Be sure to state which it is—into or out of the reactor.

52. A sample of iron pyrite ore, known to be 87.3% FeS_2, with the remainder being inert materials (mainly dirt and gravel), was roasted in an excess of oxygen. The reaction is as follows:

$$4FeS_2(s) + 11O_2(g) \longrightarrow 2Fe_2O_3(s) + 8SO_2(g)$$

If 93.6% of the $FeS_2(s)$ reacts, what is the standard enthalpy of reaction at 298K per kilogram of crude ore?

Applied Problems [53–59]

•**53.** An important industrial application of Hess's Law is the determination of enthalpies of hydrogenation from heats of combustion, such as the enthalpy of hydrogenation of ethylene. Using the data in Tables 12.3 and 12.4, determine $\Delta H°$ at 298K and express your answer in kJ/mol.

$$C_2H_4(g) + H_2(g) \longrightarrow C_2H_6(g)$$

54. The enthalpy change that occurs during the following reaction is −1172 kJ.

$$Al_2Cl_6(s) + 6Na(s) \longrightarrow 2Al(s) + 6NaCl(s)$$

Calculate the heat of formation of $Al_2Cl_6(s)$ in kJ/mol. Check your answer against Table 12.4.

•**55.** Tungsten carbide (WC) is a refractory material of considerable industrial interest and has been studied extensively. On the basis of the following data and any other thermodynamic data you might need from Tables 12.3 and 12.4, calculate the standard enthalpy of formation ($\Delta H_f°$) of tungsten carbide at 298K:

$$2WC(s) + 5O_2(g) \longrightarrow 2WO_3(s) + 2CO_2(g) \quad \Delta H° = -2392 \text{ kJ}$$

56. In making the necessary calculations to answer Problem 55, you needed to obtain the standard enthalpy of formation of carbon dioxide:

$$C\ (\text{graphite}) + O_2(g) \longrightarrow CO_2(g) \quad \Delta H° = -393.505 \text{ kJ/mol}$$

Would your answer be the same or different if you had chosen diamond instead of graphite? Give reasons and calculations to support your answer.

•**57.** The standard enthalpy of combustion of animal fats per gram is an important value that is commonly referred to by nutritionists as the "calorific value." It can be determined by measuring the heat produced on burning a known sample in a constant volume calorimeter. If a 1.00-g sample of a C_{20} component of human fat known as arachidonic acid released 42.0 kJ at 98.6°F, what is its calorific value, that is, its $\Delta H_c°$ in kJ/g? Assume the unbalanced combustion reaction to be the following:

$$C_{20}H_{32}O_2(s) + O_2(g) \longrightarrow CO_2(g) + H_2O(\ell)$$

58. The heat required to sustain animals that hibernate—polar bears and hamsters alike—arises from the combustion of fatty acids such as arachidonic acid (see Problem 57). Calculate the mass of fatty acid needed to warm a 500.-kg bear from 5°C to 25°C, assuming the average specific heat of bear tissue is 4.18 J/g·K.

•**59.** Liquefied propane (C_3H_8) is widely used as a portable heating fuel. Use the heats of formation in Table 12.4 to carry out the following calculations:
(a) The heat of combustion of gaseous propane
(b) The heat of combustion per cubic foot of propane gas at 298K and 101.3 kPa. A cubic foot is 28.3 L.
(c) Based on the assumption that the hot water needs of an average home, for a family of four, are about 100 gallons per day, estimate the volume of propane gas needed per day. (1 gallon = 3.8 L)

ESTIMATES AND APPROXIMATIONS [60–68]

60. Estimate the heat of combustion of the kerosene-range diesel fuel hydrocarbon known as *n*-dodecane ($C_{12}H_{26}$), using only the data in Tables 12.3 and 12.4. What assumptions have you made? Estimate the heat of combustion that would be determined experimentally in a constant volume calorimeter.

•**61.** The CO content in an enclosed or poorly ventilated space can be determined by converting the CO to CO_2 by oxidizing a sample of the polluted air and measuring the rise in the temperature that results from the oxidation. Estimate how sensitive a thermometer would have to be in order to detect parts per million (ppm) by volume of CO in air samples taken at tunnel check points such as the Brooklyn Battery Tunnel, the Houston Baytown Tunnel, the Baltimore Harbor Tunnel, or the tunnel of your choice. What assumptions have you made?

62. Approximately what mass of coal is equivalent in energy to the flow of Niagara Falls for one hour?

63. Estimate the temperature that will be reached if one milliliter of gasoline, initially at room temperature, is very quickly combusted to products.

64. How much fuel oil will be burned in one year by an electric power station that serves an entire city the size of Cedar Rapids, Iowa?

65. An iron horseshoe is taken from a furnace at red heat, about 900°F, and plunged into a bathtub-sized trough of water at room temperature. What is the approximate final thermal equilibrium temperature after this "quenching" operation?

66. Early in the American Space Program, the Apollo 13 mission had to be aborted some 200,000 miles from Earth due to the rupture of an oxygen tank used for powering certain fuel cells. Considering the oxygen to be an ideal gas, freely escaping into the vacuum of deep space, estimate q, w, and ΔE for the process, assuming constant temperature.

67. It has been estimated that an adult human being expends about 2 kJ of energy per kg of body mass per km during a brisk walk. On this basis, compare the energy requirements for a walking human being and an automobile. Select as your representative automobile a 2400-lb auto and a fuel consumption of 22 miles per gallon. A gallon is 3.78 L and the density of gasoline is about 0.7 g/mL. A pound is 454 grams.

68. Ethanol is being used to *extend* gasolines in automotive fuel mixtures in many states and nations at the present time. Engines can also be designed to run on pure ethanol, thereby reducing still further our dependence on gasoline. Estimate the amount of heat released when a liter of ethanol is burned. Assume the only reaction products are carbon dioxide and water, and assume a density for ethanol of 0.8 g/mL. Compare ethanol or methanol as a possible gasoline substitute.

The values of ΔH_c° for ethanol and methanol are given in Table 12.3.

FOR COOPERATIVE STUDY [69–70]

69. How do you make an atom bomb? Take 100 mol of ^{235}U surrounded by 5000 mol of iron bomb casing. On fission, each ^{235}U atom releases 3.2×10^{-11} J. About 90% of this energy is used up vaporizing and ionizing the bomb casing, creating a gas of ions and electrons.

(a) Suppose that 2% of the ^{235}U nuclei suddenly undergo fission, that the process creates a monatomic ideal gas with 25,000 mol of particles in it, and that 10% of the fission energy becomes kinetic energy of the gas. Calculate the temperature of the gas after fission, before the gas begins to expand and cool.

(b) Calculate the speed of a typical ^{235}U atom in cm/sec, at that temperature.

(c) At the instant of fission, the gas doesn't take up much more room than the original bomb—say, 10,000 liters. Calculate the pressure of the gas in atmospheres.

70. Consider vaporizing the following liquid substances at their boiling points and using the change in volume resulting from the vaporization to do work against a constant external pressure of 1 atm. Thus, each of these substances is being used as the working fluid in a one-stroke equivalent of a steam engine. In each case use one mole of substance. Calculate the efficiency of each process by determining the amount of work realized for the total amount of heat absorbed. For each substance, the boiling point and the heat required to vaporize the liquid at its boiling point in kJ/mol are given. What conclusions can you draw regarding the choice of working fluid?

Substance	Boiling Point (°C)	Enthalpy of Vaporization (kJ/mol)
water	100.0	40.66
ammonia	−33.4	23.26
nitrogen	−195.6	5.58
mercury	356.6	59.27

WRITING ABOUT CHEMISTRY [71–73]

71. When you are presented with a problem in thermodynamics, the first task is to define the system and describe the location of its boundaries. Do the results of thermodynamic calculations depend on the boundaries chosen? That is, will two different people get different results if they define the system differently? Give a couple of examples. Does your answer to the preceding question make you more or less confident in thermodynamic calculations? Explain with examples.

72. Heat and energy have the same units, but they are not the same things, and temperature is often confused with heat and energy. Explain with examples how heat, temperature, and energy are different. Then give your own opinion why so many students and other people cannot keep heat, temperature, and energy straight.

73. Is thermodynamics different from many of the other topics studied in general chemistry class? Explain your opinion with examples. Then explain whether you are a student who finds thermodynamics to be one of the hardest topics in general chemistry or whether you find thermodynamics to be relatively easy to learn. What kind of a person would find thermodynamics to be hard, and what kind would find it to be relatively easy? Why?

chapter thirteen

SPONTANEOUS CHANGE

13.1 DRIVING CHEMICAL REACTIONS

How do chemical reactions happen? What drives a chemical reaction in one direction or another? Answers to these questions are essential to understanding chemistry, but answers to these questions are complex. Just look at the world for a moment. The leaves on the trees and the trees themselves flourish on the photosynthetic conversion of CO_2 and H_2O to the cellulosic materials of which they are composed. Yet right alongside them you are likely to find decaying plant matter returning CO_2 and H_2O to the atmosphere. Both processes happen spontaneously and occur at the same time in apparent opposition to each other. Meanwhile, the blast furnace at the steel mill frees iron from the oxide, while in the parking lot a steelworker's car rusts away, turning iron and oxygen back to iron oxide. If rusting is clearly spontaneous, at least it is mercifully slow—you do not have to replace your car each year!

In this chapter we will develop the means to predict which chemical reactions occur spontaneously. To do this we must know what drives chemical reactions. Perhaps it is the heat of reaction. Within our own experience many reactions that are observed to proceed spontaneously are exothermic—have negative values of ΔH and give off heat. This includes many familiar chemical reactions such as the burning of coal, gasoline, and wood and the rusting of iron. There are, however, spontaneous chemical reactions that are endothermic, or heat absorbing. For example, if barium hydroxide and ammonium thiocyanate are mixed in a flask and the flask and contents are then set on a wet surface, the flask will soon be frozen to the surface. Clearly spontaneous processes can have values of ΔH that are either negative (exothermic) or positive (endothermic). So if ΔH is not a dependable predictor of whether a process is spontaneous, then what is?

We continue our discussion of thermodynamic principles begun in Chapter 12. The first law of thermodynamics summarized a long history of human experience in converting energy into heat and work, the embodiment of which is the equation

$$\Delta E = q + w$$

This equation is an algebraic statement of energy conservation. It states that the energy content of a system that is totally separated from its surroundings is constant, and because the system may be as large as the universe, the first law of thermodynamics implies that the energy of the universe is constant. The universe is really a discontinuous collection of smaller systems, each with its own energy content, capable of exchanging energy with other systems with which it is in contact. When systems are in contact, they can exchange energy either by the flow of heat between them or by one performing work on the other. Unless two interacting systems are in equilibrium, transfer of energy between them is spontaneous—one will be the energy donor and the other will be the energy receiver. By the end of this chapter you will be able to answer such questions as the following:

- Is it possible to make diamonds from graphite?
- Should you expect aluminum to react with water the way sodium does?
- Can NH_3 be prepared from its elements, N_2 and H_2?
- Can charcoal be converted to methane?
- Why isn't H_2O_2 prepared from its elements, H_2 and O_2?
- How can limestone be converted to quicklime for making mortar?

To continue our discussion of spontaneous processes, we first need to understand what is actually meant by "spontaneous."

13.2 REVERSIBLE AND IRREVERSIBLE PROCESSES

Spontaneity, Reversibility, and Equilibrium

The second law of thermodynamics deals with predicting the spontaneous direction of chemical change. To understand this law we need to distinguish between systems that change reversibly and systems that change irreversibly. Natural processes are not reversible. You cannot unscramble eggs or reverse time!

Reversibility in the thermodynamic sense is not at all like watching a motion picture moving backward. Thermodynamic reversibility does not simply mean that a process can somehow be stopped, placed in reverse, and forced to go in the opposite direction. For a process to be thermodynamically **reversible**, the reversibility must be microscopic. It must be so delicately balanced between forward and reverse that an infinitely small change in some single feature of the process is enough to change the direction of the process. As on a perfectly balanced teeter-totter, adding the smallest weight on either side would put it out of balance and cause it to move one way or the other.

What kinds of processes are so delicately balanced? As an example of a thermodynamically reversible process, recall the phase transformation taking place between ice and water:

$$H_2O(s, 0°C, 1 \text{ atm}) \rightleftharpoons H_2O(\ell, 0°C, 1 \text{ atm})$$

In our example the system consists of ice floating in water at the equilibrium temperature of 0°C, which is the melting point of ice at 1 atm. Because this is a dynamic process, molecules in the solid state are constantly entering the liquid phase as molecules of liquid return to the solid phase. At equilibrium, the forward process (melting) and reverse process (freezing) occur at equal rates, and there is no net change in the amount of solid or liquid. However, an infinitesimal change in a variable such as heat can disrupt the equilibrium. For example, if even a tiny amount of heat is put into the mixture, some ice will melt, with the temperature remaining at 0°C. Another small input of heat will cause still more of the ice to melt. Conversely, if a similarly small amount of heat is removed to the surroundings, a corresponding amount of the water will freeze. Thus, the change in state taking place at the normal melting point is a reversible phase change. The same holds true for vaporization, the phase change between liquid and vapor at the boiling point, and for sublimation, the phase change between gas and solid.

Now consider the change in state from liquid to solid at some temperature other than the normal freezing point, a process that is not reversible. In freezing supercooled water at 1 atm and −5°C,

$$H_2O(\ell) \longrightarrow H_2O(s)$$

liquid converts to the solid spontaneously, and no infinitesimal change in any outside variable can reverse this. Liquid water cannot be formed from ice at −5°C and 1 atm! It would take a change of temperature of 5°C to reverse the process, a change that is far from infinitesimal. Such a process is thermodynamically **irreversible** and is spontaneous. This is in contrast to a reversible process in which there is no spontaneous change because the process is already at equilibrium.

Chemical reactions such as the nitrogen dioxide dimerization can also be carried out reversibly. Infinitesimal changes in the conditions that exist at equilibrium can cause the reaction to proceed in one direction or the other, favoring dinitrogen tetroxide formation or dissociation:

$$2NO_2(g) \rightleftharpoons N_2O_4(g)$$

For this particular reaction, these changes include temperature, total pressure, and the partial pressures of $NO_2(g)$ and $N_2O_4(g)$.

The cardinal principles demonstrated by these examples are as follows:

- Reversible processes are at equilibrium and do not proceed spontaneously in one direction or the other.

- A process is spontaneous if the system is not at equilibrium. A spontaneous process proceeds in a direction that takes it closer to the equilibrium condition. When equilibrium is finally reached, the process occurs reversibly.

Reversibility, Work, and Heat

A process is defined by its initial and final states. That's the law! But, there is nothing in the first law of thermodynamics that limits the number of ways systems can change from some initial state to some final state. Although ΔE and ΔH are independent of the path taken, q and w for different paths are different and are related by $\Delta E = q + w$, the first law. In the compression of an ideal gas, for example, the amounts of heat (q) and work (w) each depend on the path followed from the initial state to the final state. Consider the simple constant temperature process of compressing an ideal gas from 1.0 L at 1 atm to 0.5 L at 2 atm at 298K. Figure 13.1 describes three possibilities.

The most obvious path for the process would be to abruptly change the external pressure from 1 atm to 2 atm and allow the compression to occur in one quick step. This is clearly an irreversible process because once the pressure is changed to 2 atm, there is no infinitesimal change in the outside conditions that can reverse the direction of the process. A second possible path would be to perform the compression in two steps. In the first step the external pressure could be changed to 1.5 atm, and the gas would compress to 0.67 L, according to the ideal gas law. In the second step the external pressure could be changed to 2 atm, and the compression to 0.5 L would complete the process. Although this path is also irreversible, the two-step process results in a different value for w, a smaller value. The value for w is even smaller for the three-step process.

This compression from 1 L to 0.5 L could be performed in any number of steps, not just one or two, but three, four, five or many, many more. Each of these represents a different path for the process, and there are an infinite number of different paths the process can take. Calculations of the work w for compressing the gas by each of these processes show that w decreases as the number of steps is increased. The minimum w is the path with an infinite number of steps, and each step is infinitely small. When the steps are infinitely small, the process is reversible because an infinitesimal change in the external pressure could change the direction of the process. Therefore, the minimum w occurs for the reversible path. Figure 13.2 is a graph of pressure as a function of volume for the fully reversible process. Note how the smooth curve representing the instantaneous change in P as a function of V is developing in Figure 13.1 by connecting the four points. There is nothing in ΔE to indicate whether the process proceeds by the reversible or irreversible path, because ΔE depends only on the final and initial states. Because $\Delta E = q + w$ and w has the smallest positive value for the reversible path, the heat absorbed q is the maximum for the reversible path. This result is very important in our further study of spontaneous processes.

A macroscopic model of microscopic reversibility that comes to mind is that of a gas trapped by a piston weighted down by a pile of sand (Fig. 13.3). Removing one grain of sand produces a seemingly microscopic change, but even removing (or adding) one grain at a time is far from the infinitely small change

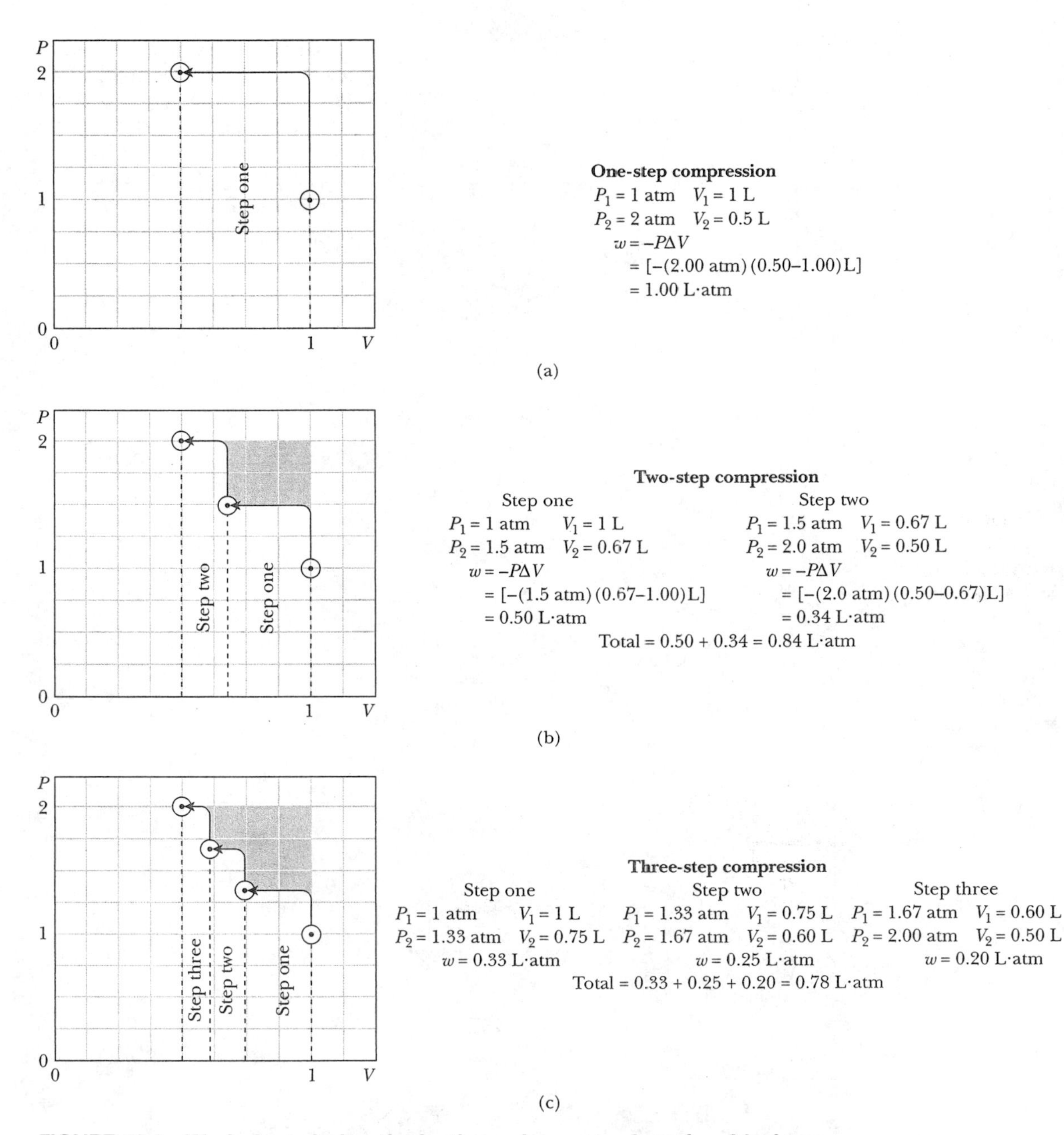

FIGURE 13.1 Work done during the isothermal compression of an ideal gas. (a) The one-step pathway is irreversible. (b) The two-step pathway is also irreversible, but w is less than in the one-step process, which means q is correspondingly greater. (c) Likewise, the three-step process is irreversible, but the net effect is to produce a still smaller value for w, and a still larger value for q.

we need to imagine. Even removing a particle of finely divided flour or dust particles is no better than an approximation of reversibility. Finally, keep in mind that it is just not possible to compress a gas in a truly reversible fashion in any finite time. Microscopic reversibility can only be approached as a limit.

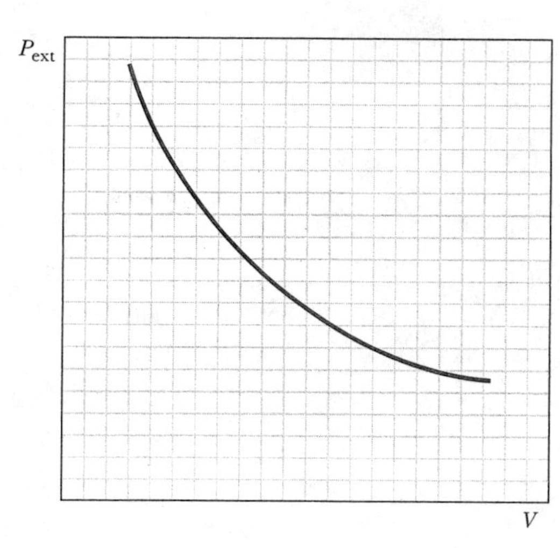

FIGURE 13.2 Graph of *P* versus *V* for the reversible compression of 1.0 L of an ideal gas at 1.0 atm to 0.50 L at 2.0 atm. At any point along the curve, the compression process can be reversed by an infinitely small change in the external pressure.

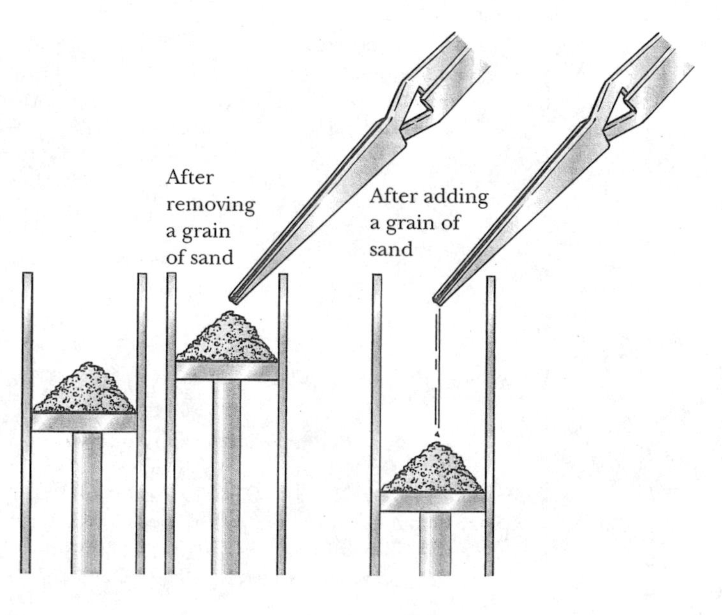

FIGURE 13.3 Adding or removing grains of sand, a macroscopic model of microscopic reversibility. Even removing one grain at a time is far too large a change to fairly model what we mean by an infinitely small change in pressure, which results in an infinitely small, and therefore reversible, change in the volume of the trapped gas.

13.3 DISORDER, ENTROPY, AND THE SECOND LAW OF THERMODYNAMICS

A process is spontaneous if q for the process is less than q_{rev}, the value of q if the process is performed reversibly. If the process is a chemical reaction that is not reversible, we know that it will *spontaneously* proceed in one direction or the other. But which way will it go? To answer the question, consider what directs any process. Many reactions tend to be exothermic, but that is not the whole story, because some spontaneous reactions are endothermic. Chemical reactions also tend to yield products that are more *disordered* than the reactants that produced them. So what we need now is a measure of disorder.

The thermodynamic term for disorder is entropy. The **entropy** of a system is associated with freedom of motion and ability to assume many different orientations. The symbol for entropy is S and the change in entropy is ΔS, where $\Delta S = S_f - S_i$. Entropy is a state function, and therefore the change in entropy ΔS for a particular process depends only on the final and initial states. We will need to be able to calculate ΔS for a process, but first let us make some qualitative comparisons of the relative disorder between different states of matter:

- **Liquids are more disordered than solids**. In liquids the molecules tumble around freely, whereas in solids the molecules are restricted to fixed locations. That is why liquids flow.
- **Gases are more disordered than liquids**. Though both gases and liquids are fluids, gases have more freedom of motion than liquids, flow more freely, and fill their containers completely.
- **A given substance is more disordered at a higher temperature**. The higher the temperature the faster and more random the movement of the particles.
- **A sample of gas in a large space is more disordered than it is in a small space**. In the larger space, the gas particles have more freedom to occupy different positions.

In each of these cases, adding heat to the system increases the disorder. Heat eventually converts a solid to a liquid and a liquid to a gas; heat raises the temperature of a substance; and heat increases the volume of a gas held at constant pressure. Thus, there is a fundamental relationship between the flow of heat q into (or out of) a system and ΔS. The more heat that flows in, the greater is ΔS. There is, however, no specific q for a process, because q depends on the path. But there is a specific q_{rev} for a process. It is the maximum q that the process can have. The **change in entropy ΔS** is defined as q_{rev}/T:

$$\Delta S = \frac{q_{rev}}{T}$$

Entropy change for a process: The ratio of q to T for the process carried out reversibly at constant temperature is the change in entropy, a thermodynamic function that measures the change of disorder for the process. As usual in thermodynamics, T is in kelvin units.

The entropy change is a state function and depends only on the initial and final states, no matter what path is taken. Therefore, this equation holds for any isothermal (constant temperature) change between two states, whether the actual process is performed reversibly or not.

Knowing the entropy change for a process makes it possible for us to formulate a rule for determining whether the process can take place spontaneously. This rule is the second law of thermodynamics. The best demonstration of this law from a macroscopic point of view is that experiences and processes forbidden by it are never observed. For example, observing the spontaneous flow of heat from a colder body to a warmer body or the spontaneous compression of an ideal gas without any change in the external conditions would violate this

SEEING THINGS

Time's Arrow and the Spontaneous Direction of Events, and a Sure Bet Suppose you have before you a block made of copper, a good conductor of heat, into which a machinist has drilled identical holes in which identical thermometers have been fitted snugly. Rapidly heat the bar at one end. If you could successfully isolate the apparatus from loss or gain of heat from any other source or to the surroundings, and you could take a snapshot, here is what you might see.

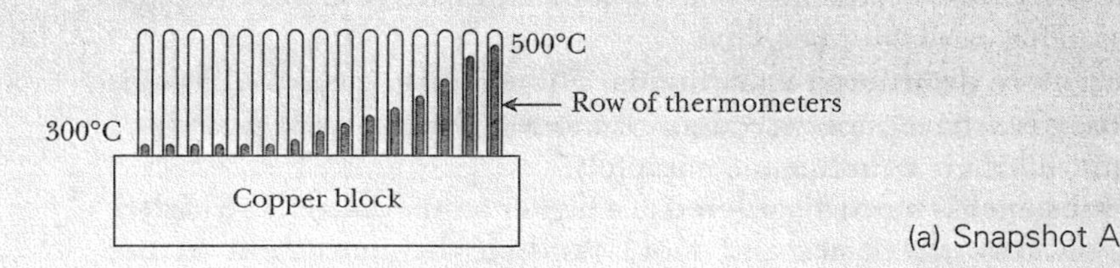

(a) Snapshot A

There is a "hot" region at one end, registering 500°C, and the other is only warmed above room temperature, in this case, 300°C. And after a while, here is what you would see if you took another snapshot:

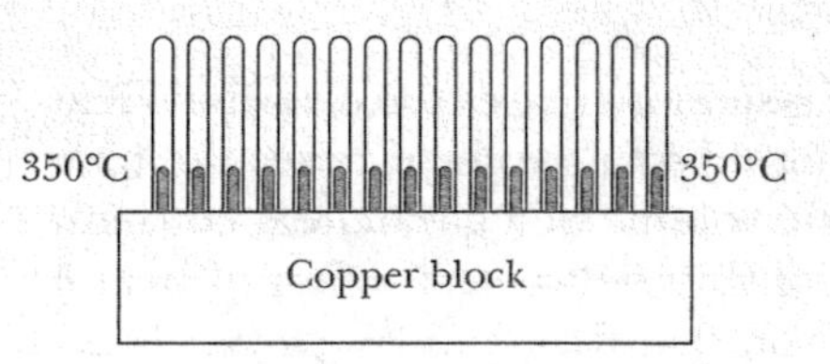

(b) Snapshot B

All the thermometers register the identical intermediate temperature. Now show your two snapshots to a friend who did not witness the earlier events and see if she can tell which photo depicts the first event. Any "experienced" human being should bet her life on the certainty that (a) preceded (b). The prediction of the direction of the spontaneous process is the second law of thermodynamics in action. Guaranteed. As Sir Arthur Eddington* spoke of it, "Times Arrow."

second law and demand a rethinking of not only the second law of thermodynamics but the very nature of the universe. You can bet on it!

The first law of thermodynamics states that energy is conserved, and so the energy of the universe is constant. The **second law of thermodynamics** states that the entropy of the universe tends toward a maximum. That is,

- Any spontaneous process must increase the entropy of the universe.
- Any process that would spontaneously decrease the entropy of the universe is impossible and therefore must be spontaneous in the opposite direction.
- Reversible or equilibrium situations result in no change of the entropy of the universe.

The principle of conservation of energy tells us that the energy content of the universe is constant. However, there is no conservation of entropy in the universe. In fact, the entropy of the universe is always increasing. The two key

Likewise, the direction of events depicted in consecutive frames of the filmstrip below should be immediately, almost intuitively, obvious. A layer of flour sits on top of a layer of soot in a mixing bowl. Once the mixture has been stirred in a clockwise direction until fully dispersed, it is impossible to restore the original arrangement by stirring in a counterclockwise manner. Although the filmstrip can be run in reverse, that sequence does not in fact represent physical reality.

Questions

Give a couple of other examples of processes for which the direction of the process can be inferred by looking at the states on either end of the process without being told which is initial and which is final.

Because there is a natural (spontaneous) tendency for heat to flow from a "hotter" to a "colder" body, will the universe eventually cool to absolute zero?

List several other "snapshot" combinations on which you would be willing to "bet the farm" as a thermodynamic time line.

*Sir Arthur Eddington (British, 1882–1944) was the Astronomer Royal of England in the early years of the twentieth century. He was leader of the scientific team that measured the bending of light in the vicinity of the Sun during a total eclipse in late November 1919 that confirmed the predictions of Einstein and his relativity theory. Eddington was a great popularizer of science in his writings.

conclusions of the second law are that the entropy of the universe increases for any spontaneous process and that the entropy of the universe is constant for any reversible process.

13.4 ENTROPY CALCULATIONS

The Entropy Change of the Universe

The second law of thermodynamics describes spontaneous and reversible processes in terms of the change in entropy for the entire universe resulting from a process. Calculating the entropy change of the universe for a process sounds like a very difficult job but actually is not when we take advantage of the precise nature of thermodynamics. First, remember that in thermodynamics we define our system as that part of the universe in which we are interested. Our system could be a flask in which a reaction or a phase transformation is taking place. All of the rest of the universe is the surroundings. As a result, the entropy change of the entire universe that results from a process is the sum of the entropy change of the system plus the entropy change of the surroundings,

$$\Delta S_{univ} = \Delta S_{sys} + \Delta S_{surr}$$

Entropy change for the universe: The sum of ΔS_{sys}, the entropy change of the system, and ΔS_{surr}, the entropy change of the surroundings.

The entropy change of the system depends on the process that occurs. We will look at several of these calculations in the next section.

Calculation of the entropy change of the surroundings is straightforward. Because the surroundings are so vast compared to the system, the surroundings can be viewed as an enormous heat sink that can reversibly absorb or give off any amount of heat at any temperature. Therefore, q for the transfer of heat into or out of the surroundings is always considered reversible. Using our definition for the entropy change ΔS we can write the following:

$$\Delta S = \frac{q_{surr}}{T}$$

In this equation T is the temperature of the surroundings. The value of q_{surr}, q for the surroundings, is simply the negative of the value of q for the process since heat flowing *out* of the system flows *into* the surroundings and vice versa. Thus,

$$\Delta S_{surr} = \frac{-q}{T}$$

Entropy change for the surroundings: The absolute value of the heat is the same for the system as for the surroundings but with the opposite direction of flow. That is, if heat is flowing out of the system it is flowing into the surroundings and vice versa.

We need to calculate ΔS_{sys} to complete the calculation of ΔS_{univ}. ΔS_{sys} is commonly called simply ΔS, and we will now look at calculations of ΔS.

Entropy Change and Phase Transformations

The entropy change for a phase transformation reflects the relative change in organization and disorder of a system that results. The quantity ΔS_{sys} is positive for transformations from solid to liquid to gas, and is negative for the reverse of those processes.

For phase changes at the normal melting, boiling, and sublimation points, ΔS_{univ} is zero because the transformation is at equilibrium, or reversible, in those cases. The enthalpy of the process is ΔH_{fus}, ΔH_{vap} or ΔH_{sub} and is equal to q_{rev} under the usual conditions of constant pressure. Therefore, ΔS is easily calculated using the following equation:

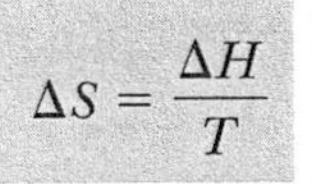

$$\Delta S = \frac{\Delta H}{T}$$

Entropy change for a phase transformation at the equilibrium temperature: The temperature of the transformation is T and ΔH is the enthalpy change.

EXAMPLE 13.1

Find the ΔS_{sys} and ΔS_{univ} for the transformation of 1.00 mol of liquid benzene to benzene vapor at 1.00 atm pressure and at its normal boiling point, which is 353K. The enthalpy of vaporization of benzene at its normal boiling point is 3.08×10^4 J/mol.

COMMENT Because this is a process carried out at constant pressure, q equals the enthalpy of vaporization, ΔH_{vap}. Furthermore, vaporizing a substance at its boiling point is a reversible process, and so q is q_{rev}. Now we can apply the equation $\Delta S = q_{rev}/T$ in a straightforward manner.

SOLUTION

$$q_{rev} = \Delta H_{vap} = (1.00 \text{ mol})\left(\frac{3.08 \times 10^4 \text{ J}}{1 \text{ mol}}\right) = 3.08 \times 10^4 \text{ J}$$

$$\Delta S_{sys} = \frac{q_{rev}}{T} = \frac{3.08 \times 10^4 \text{ J}}{353\text{K}} = 87.2 \text{ J/K}$$

$$\Delta S_{surr} = \frac{-q}{T} = -\frac{3.08 \times 10^4 \text{ J}}{353\text{K}} = -87.2 \text{ J/K}$$

$$\Delta S_{univ} = \Delta S_{sys} + \Delta S_{surr} = (87.2 - 87.2) \text{ J/K} = 0$$

Note that ΔS_{univ} is zero, as must be the case for a reversible phase change such as this phase transformation taking place at the normal boiling point. However, ΔS_{sys} is positive in this process because the resulting gas phase is more disordered than the initial liquid phase.

EXERCISE 13.1

Calculate ΔS for the vaporization of 1.00 mol of diethyl ether at its normal boiling point of 308K. The enthalpy of vaporization is 27.2 kJ/mol.

ANSWER $\Delta S = 88.3$ J/K

If a phase transformation is conducted at a temperature at which the phases are not at equilibrium, ΔS_{univ} will not be zero. For example, if ice changes to liquid water at 1 atm and 10°C, the process is spontaneous because the temperature is above the normal melting point. Therefore, ΔS_{univ} must be positive. The reverse process, liquid water changing to ice at 1 atm and 10°C, would have a negative value of ΔS_{univ} and is impossible.

PROFILES IN CHEMISTRY

Heat Engines and Thermal Pollution The dynamics of heat and the power of the steam engine were essential themes in nineteenth-century science and technology. It has been said of the century before nuclear power that the steam engine was its most impressive symbol. Here was heat being continuously used to create power on command. Could anyone doubt the majesty of the steam locomotive? But the problem then, as today, is a very practical one—namely, the *efficiency of a heat engine.* Devoting himself to the study of physics, engineering, and economics, the young Sadi Carnot (French, 1796–1832) anticipated much of the applied science of thermodynamics in his 118-page booklet, published in 1824 under the title *Reflections on the Motive Power of Heat and on the Machines Equipped to Develop this Power.* His developments on the thermal efficiency of heat engines are still very much in use.

The production of motion is due to the transfer, not the consumption, of heat. If you plan to get work out of heat, a temperature difference is necessary. A classic steam engine has two heat reservoirs, one in the boiler at a higher temperature T_2 and one where the waste steam is exhausted at a lower temperature T_1. In this kind of engine steam is the working fluid and is in contact with both heat reservoirs. The ratio of the net work done by the system and the heat put into the working fluid is the efficiency of the engine. The maximum possible efficiency is related to the temperature of the reservoirs by

$$\text{Percentage efficiency} = \frac{T_2 - T_1}{T_2} \times 100$$

The heat delivered to the low-temperature reservoir represents wasted heat that is not converted to work. Thus, the efficiency of a heat engine is increased by having as great a difference as possible between the high-temperature reservoir and the low-temperature reservoir. Because the low-temperature reservoir cannot usually be varied by much, operating temperatures are made as high as possible. However, the materials available limit the higher temperatures. Much research is being performed to produce materials for turbines that can withstand increasingly high temperatures. Efficiencies of about 50% are realized in commercial steam-turbine electrical plants, where the low-temperature reservoir can be maintained at a fairly low temperature. Where it is necessary to discharge the waste heat at a relatively high temperature, as in an automobile engine, the efficiency drops to about 15 to 20%. Consider that in terms of gasoline consumption—80 to 85% of your fuel dollar is waste heat, taken away through the car's radiator. Note that the general process described by the steam engine could be run backward, in which case we would have a refrigerator instead of a heat engine.

The operation and efficiency of a modern electric power plant, whether one of conventional design or the nuclear variety, are no more complicated than this discussion of heat engines suggests. The input heat at T_2 is supplied to the high-temperature reservoir by burning coal, oil, natural gas, or by nuclear fission in uranium fuels; useful work produced is used to turn electric generators; and finally, the unused heat at T_1 exits to the environment via the low-temperature reservoir. This waste heat being exhausted into the environment is what we refer to as thermal pollution.

Thermal pollution is particularly insidious. Unlike pollution of air and water by solid, liquid, and gaseous wastes, which, in principle, can be largely removed by chemical and mechanical means, heat is an inevitable product of power generation. That is, the heat must go somewhere. Thermal pollution of the atmosphere can contribute to changing weather patterns. Thermal pollution of the aquatic environment is even more serious because of its effects on existing life cycles. Some of the effects of increased temperatures in lakes and rivers include the following:

13.5 ABSOLUTE ENTROPY AND THE THIRD LAW

Having calculated ΔS for phase transformations, we now turn our attention to chemical reactions. The third law of thermodynamics gives us a method for determining ΔS for chemical processes. Its experimental basis arises from the study

- Diminished concentrations of dissolved oxygen needed to sustain aquatic life
- Temperature stratification due to density increase, affecting species distribution and migration patterns and life cycles such as spawning patterns
- Acceleration of chemical reactions, especially rates of eutrophication, the overproduction of plant matter in a lake

Nuclear power plants offer some environmental advantages because there are no combustion products. However, that does not alter the heat-pollution problem associated with the operation of the second law of thermodynamics. The exhausted heat must be spread through a sufficiently large mass of air or water to minimize the resulting temperature rise.

Questions

What determines the limiting temperature of the low-temperature reservoir of a heat engine?

Imagine a room that is perfectly insulated so heat cannot enter or leave. In the room is a refrigerator plugged into a wall socket. If the door of the refrigerator is left open, what happens to the temperature in the room? Why?

There is a tremendous amount of energy in the oceans. What do you think about the potential for an invention designed around sucking in seawater and dumping out blocks of ice, making use of the heat released to run the ship's engine?

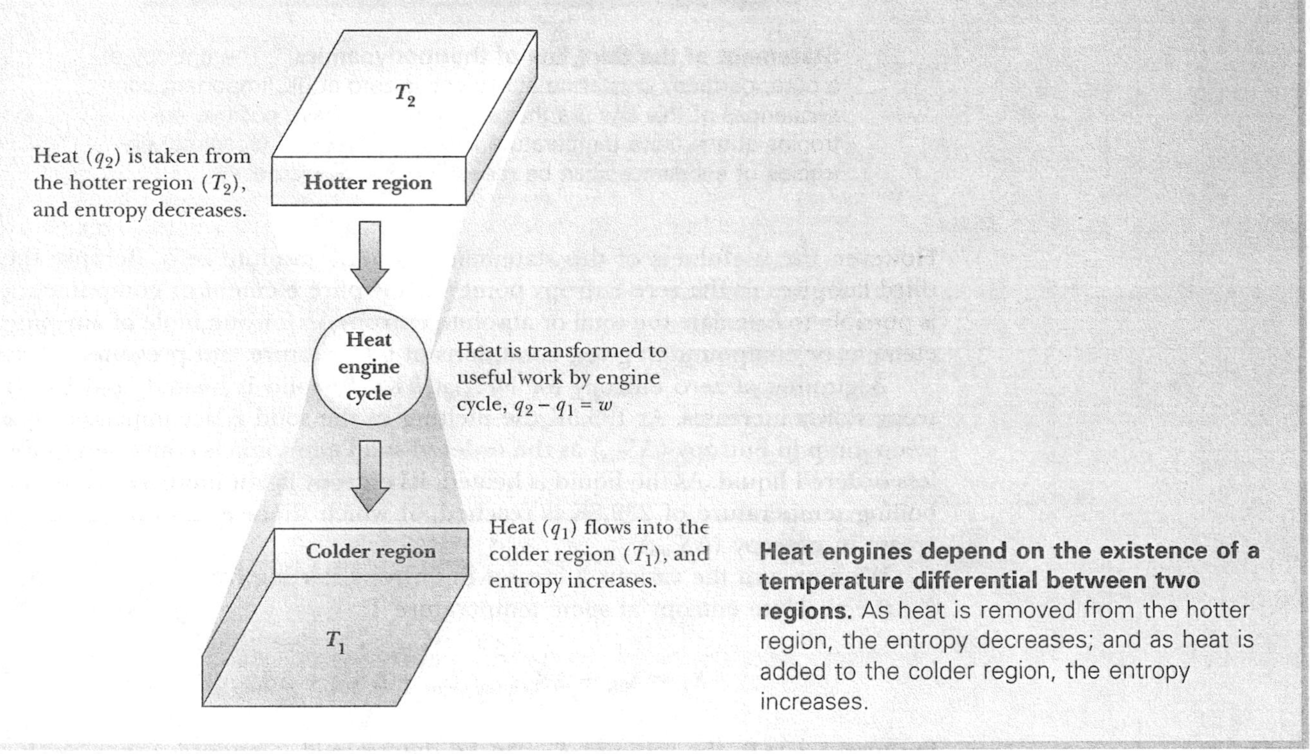
T_2
Heat (q_2) is taken from the hotter region (T_2), and entropy decreases.
Hotter region
Heat engine cycle
Heat is transformed to useful work by engine cycle, $q_2 - q_1 = w$
Colder region
Heat (q_1) flows into the colder region (T_1), and entropy increases.
T_1

Heat engines depend on the existence of a temperature differential between two regions. As heat is removed from the hotter region, the entropy decreases; and as heat is added to the colder region, the entropy increases.

of the properties of substances at low temperatures, especially heat capacities of substances.

It is not possible to establish the zero point for H, the enthalpy. With entropy, however, it *is* possible to set an absolute zero. A system with no disorder would have an entropy of zero. To establish a real chemical situation, think of a system composed of atoms. The gaseous state would have the most disorder and the solid state would be the most ordered. In the solid or crystalline state each atom has its own location in the structure. However, this is not a system with zero entropy because the atoms are in thermal motion, vibrating about some average position within the crystal. And every crystal has some defects. In addition, even the presence of impurities of the smallest imaginable quantities serves to introduce disorder, because the impurity atoms could be situated in different locations, which would allow for alternative arrangements. If the crystal had no impurities and if there were no defects and if each atom were at its regular site and at minimum energy, then there would be only one possible arrangement, and there would be no disorder. Under these circumstances we would have a zero-entropy situation. This would occur for a perfect crystal at the absolute zero of temperature.

The **third law of thermodynamics** defines the zero of entropy for a system. It states that perfect crystals of all pure substances at absolute zero of temperature have an entropy of zero:

$$S = 0 \text{ for a perfect crystal of a pure substance at 0K.}$$

Statement of the third law of thermodynamics: The entropy of a pure, perfectly crystalline substance is zero at 0K. Important consequences of this law are that all substances have positive entropies at any finite temperature, and that individual (absolute) entropies of substances can be measured and tabulated.

However, the usefulness of this statement is not at absolute zero. Because the third law gives us the zero-entropy point for any pure element or compound, it is possible to calculate the total or absolute entropy $S°$ for one mole of any pure element or compound at given conditions of temperature and pressure.

Beginning at zero entropy for NH_3 at 0 K, the solid is heated, and its entropy slowly increases. At 195.5K the melting of the solid is accompanied by a steep jump in entropy (ΔS_{fus}) as the ordered solid ammonia is converted to the less ordered liquid. As the liquid is heated, its entropy again increases, until the boiling temperature of 239.8K is reached, at which there is another steep increase in entropy (ΔS_{vap}).

We now sum the various increases in entropy starting from 0 K to arrive at S_T, the absolute entropy at some temperature T:

$$S_T = S_{0K} + \Delta S_{\text{heating steps}} + \Delta S_{fus} + \Delta S_{vap}$$

Because S_{0K} is 0, the value of S_T can be determined. Standard values are for 1 atm pressure and a specified temperature, usually 298K. For NH_3, $S°$ is 192 J/mol·K at 25°C.

Absolute entropies of pure compounds and elements are tabulated in Table 13.1. These values can be used to calculate $\Delta S°$, the entropy change for a reaction conducted at standard conditions. Because the change in entropy is a state function, it is equal to the entropy in the final state minus the entropy in the initial state:

$$\Delta S° = (\text{entropy of products}) - (\text{entropy of reactants})$$

TABLE 13.1 Absolute Entropy (J/mol·K) Values for Compounds and Elements at 298K

Solid Elements		Solid Compounds		Liquids	
Ag	42.7	BaO	70.3	Br_2	152
B	7.1	$BaCO_3$	112	H_2O	70.0
Ba	63.2	$BaSO_4$	132	Hg	76.0
C(graphite)	5.7	CaO	40	CH_3OH	127
C(diamond)	2.5	$Ca(OH)_2$	72.8	C_2H_5OH	161
Ca	41.6	$CaCO_3$	92.9	CH_3COOH	160
Cu	33.3	CuO	43.5	C_6H_6	203
Fe	27.2	Fe_2O_3	90.0		
S(rhombic)	31.9	ZnO	43.9		
Zn	41.6	ZnS	57.7		
Mg	33	$MgCO_3$	66		
		MgO	27		
Monatomic Gases		**Diatomic Gases**		**Polyatomic Gases**	
He	126.1	H_2	130.6	H_2O	189
Ne	146.2	D_2	145	CO_2	214
Ar	154.7	F_2	203	SO_2	248
Kr	164.0	Cl_2	223	H_2S	205
Xe	169.6	Br_2	245	NO_2	240.
H	114.6	CO	198	NH_3	192
F	158.6	NO	210	O_3	238
				N_2O	219.7
				H_2S	205.7
Cl	165.1	N_2	191	CH_4	186
Br	174.9	O_2	205	C_2H_6	229
I	180.7	HF	174	C_3H_8	270.
N	153.2	HCl	187	*n*-C_4H_{10}	310.
C	158.0	HBr	198	*i*-C_4H_{10}	294.6
O	161	HI	206	C_2H_2	167
				C_2H_4	219.4
				C_3H_6	266.9
				C_4H_8 (*cis*-2-butene)	301
				C_4H_8 (*i*-butene)	294
				C_4H_8 (1-butene)	307.4
				C_4H_8 (*trans*-2-butene)	297

The entropies of the products and reactants can be determined from the table of absolute entropies. Thus, for a general reaction of the form

$$aA + bB \longrightarrow cC + dD$$

$$\Delta S^\circ = (cS^\circ_C + dS^\circ_D) - (aS^\circ_A + bS^\circ_B)$$

$$\Delta S^\circ = \Sigma S^\circ_{products} - \Sigma S^\circ_{reactants}$$

ΔS° from absolute entropies: The difference of the sums of the absolute entropies (S°) for the products and reactants for a chemical reaction gives the entropy change ($\Delta S°$) for the chemical reaction. The absolute entropy of each substance must be multiplied by its stoichiometric coefficient in the balanced equation.

EXAMPLE 13.2

Calculate the standard entropy change $\Delta S°$ for the reaction

$$CaCO_3(s) \longrightarrow CaO(s) + CO_2(g)$$

COMMENT Use the values of absolute entropy, $S°$, from Table 13.1 and the previous equation.

SOLUTION

$$\Delta S° = (40. + 214 - 92.9)\ \text{J/mol·K} = 161\ \text{J/mol·K}$$

It is reasonable that this process should show an increase in entropy, because we started with a solid reactant and ended with a gas as one of the reaction products.

EXERCISE 13.2

Calculate $\Delta S°$ for the reaction

$$2H_2(g) + O_2(g) \longrightarrow 2H_2O(\ell)$$

ANSWER $\Delta S° = -326\ \text{J/mol·K}$

13.6 FREE ENERGY

The Free Energy Change

There is no question that the thermodynamic quantity ΔS_{univ} can be used to show the direction in which a process will proceed spontaneously. However, knowledge about both the system and its surroundings is required:

$$\Delta S_{univ} = \Delta S_{sys} + \Delta S_{surr}$$

A more convenient measure of spontaneity is available that depends only on knowledge of the system, ignoring the surroundings, a feature that makes it particularly attractive for chemistry. The measure is the Gibbs free energy or simply the **free energy *G***, which is defined as

$$G = H - TS$$

Josiah Willard Gibbs (American, 1839–1903) entered Yale College in 1854. He received the doctor of philosophy degree in 1863 in mathematics and physics, only the fifth to be granted in the United States. His work is essentially a unified theory of physical and chemical behavior. He proposed that the conditions for equilibrium derived from the first two laws of thermodynamics were universally applicable.

The free energy can be used to predict spontaneity under conditions of constant temperature and constant pressure, the most common conditions encountered for chemical reactions. When G is calculated for two different states, the initial calculation (G_i) will be

$$G_i = H_i - T_i S_i$$

and the final calculation (G_f) will be

$$G_f = H_f - T_f S_f$$

If the temperature remains constant, T_i and T_f will be the same temperature T, and we can subtract the initial equation from the final equation, resulting in an equation that can be used to calculate the change in free energy ΔG:

$$G_f - G_i = [H_f - TS_f] - [H_i - TS_i]$$

$$G_f - G_i = [H_f - H_i] - T[S_f - S_i]$$

$$\Delta G = \Delta H - T\Delta S$$

Free energy change ΔG: Measured under constant temperature conditions, ΔG for a process is the difference between the change in enthalpy ΔH and the entropy term $T\Delta S$. In other words, it is the difference between the heat exchanged with the surroundings at constant pressure and the entropy factor. For chemistry this is arguably the most important equation in all of thermodynamics.

For a system under standard-state conditions at temperature T,

$$\Delta G^\circ = \Delta H^\circ - T\Delta S^\circ$$

Standard free energy change: The change in free energy with reactants and products in their standard states and at a fixed temperature, most often 298K.

The standard free energy change refers to reactants and products at partial pressures of 1 atm for gases and concentrations of 1 M for dissolved solutes. Because the standard free energy changes can be determined at different temperatures, the appropriate absolute temperature often appears as a subscript, as in ΔG°_{298}.

We need to know how this new function ΔG is related to spontaneity. For a process at constant temperature and pressure, $\Delta H = q$ and

$$\Delta G = \Delta H - T\Delta S = q - T\Delta S.$$

In this equation, ΔS is the entropy change of the system. Dividing each side of the equation by T gives the following result:

$$\frac{\Delta G}{T} = \frac{q}{T} - \Delta S_{sys}$$

Because

$$\frac{q}{T} = -\frac{-q}{T} = -\Delta S_{surr}$$

$$\frac{\Delta G}{T} = -\Delta S_{surr} - \Delta S_{sys} = -\Delta S_{univ}$$

Finally, we can clear the fraction and write

$$\Delta G = -T\Delta S_{univ}$$

Thus, at constant temperature and pressure:

- For a reversible process, ΔS_{univ} is 0 and $\Delta G = 0$.
- For a spontaneous process, ΔS_{univ} is positive and ΔG is negative.
- A process with a positive value for ΔG is not spontaneous in the direction written, but is spontaneous in the opposite direction.

The value of ΔG is negative for spontaneous processes, and so the equation $\Delta G = \Delta H - T\Delta S$ tells us that chemical reactions can be driven by energy (if ΔH is negative) or by entropy (if ΔS is positive) or by both. The change in free energy is defined so that it is useful at conditions of constant pressure and constant temperature. If T and P are not both constant, ΔG is *not* a predictor of spontaneity. At constant pressure the change in enthalpy ΔH is easily determined because it equals q_p. The actual values of the enthalpy change ΔH and the change in internal energy ΔE are usually very similar for chemical reactions, because the pressure–volume work of a chemical reaction is normally much less than the heat involved in breaking and forming bonds. We will generally refer to the two driving forces in chemical reactions as "decreasing enthalpy" and "increasing entropy." The equation for free energy allows us to calculate the effects of these two driving forces for a particular chemical reaction carried out at constant temperature and pressure.

Calculations Using Free Energy Changes

We now have a single, convenient thermodynamic function that can be used to show the direction in which a process will spontaneously proceed. To calculate $\Delta G°$ values for chemical reactions we can use the relationship $\Delta G° = \Delta H° - T\Delta S°$, along with the data in tables of standard enthalpies of formation ($\Delta H°$, Table 12.4) and absolute entropies ($S°$, Table 13.1).

EXAMPLE 13.3

Calculate $\Delta G°$ for the following reaction at 298K and determine whether it is spontaneous:

$$2NO(g) + O_2(g) \longrightarrow 2NO_2(g)$$

COMMENT Use the standard enthalpies of formation in Table 12.4 and the equation

$$\Delta H° = \Sigma\Delta H_f°(\text{products}) - \Sigma\Delta H_f°(\text{reactants})$$

to calculate the standard enthalpy change of the reaction. Use the absolute entropies in Table 13.1 and the equation

$$\Delta S^\circ = \Sigma S^\circ\,(\text{products}) - \Sigma S^\circ\,(\text{reactants})$$

to calculate the standard entropy change. Then use $\Delta G^\circ = \Delta H^\circ - T\Delta S^\circ$ to calculate the standard free energy change.

SOLUTION

$$\begin{aligned}\Delta H^\circ &= \Sigma\Delta H_f^\circ(\text{products}) - \Sigma\Delta H_f^\circ(\text{reactants})\\ &= 2(33.9\text{ kJ}) - 2(90.4\text{ kJ})\\ &= -113.0\text{ kJ}\end{aligned}$$

$$\begin{aligned}\Delta S^\circ &= \Sigma S^\circ(\text{products}) - \Sigma S^\circ(\text{reactants})\\ &= 2(240.\text{ J/K}) - 205\text{ J/K} - 2(210\text{ J/K})\\ &= -145\text{ J/K}\end{aligned}$$

Each of these values is calculated for the reaction as written—that is, 2 mol of NO reacting with 1 mol of O_2 to produce 2 mol of NO_2:

$$\Delta G^\circ = -113.0\text{ kJ} - (298\text{K})(-145\text{ J/K})\left(\frac{1\text{ kJ}}{1000\text{ J}}\right) = -69.8\text{ kJ}$$

Because $\Delta G^\circ < 0$, the reaction is spontaneous under standard conditions. Note that J must be converted to kJ for ΔS° to be consistent with ΔH°.

EXERCISE 13.3

Calculate ΔG° for the ammonia synthesis at 298K, and determine the spontaneous direction of this reaction:

$$3H_2(g) + N_2(g) \rightleftharpoons 2NH_3(g)$$

ANSWER $\Delta G^\circ = -33.2$ kJ; spontaneous as written under standard conditions

---

In much the same way as standard enthalpies of formation ΔH_f° simplified calculations of ΔH° in Chapter 12, a considerable simplification is introduced into ΔG° calculations for chemical processes by making use of ΔG_f° data. The **standard free energy of formation** ΔG_f° is the free energy of reaction for the formation of the compound from its elements in their standard states—that is, at partial pressures of 1 atm for gases and concentrations of 1 M for solutes. Taking ΔG_f° for the elements in their standard states to be zero, we may write the following:

$$\Delta G^\circ = \Sigma\Delta G_f^\circ(\text{products}) - \Sigma\Delta G_f^\circ(\text{reactants})$$

Table 13.2 includes standard free energies of formation for some compounds and for the monatomic forms of elements that are normally diatomic. Using the previous equation and ΔG_f° values from Table 13.2, ΔG° for the oxidation of carbon monoxide can be calculated, as in this example.

TABLE 13.2 Standard Free Energy of Formation (ΔG_f°) Values, kJ/mol at 298K

Solid Compounds		Simple Gaseous Molecules		Monatomic Gases	
BaO	−528.4	HCl	−95.27	H	203.2
$BaCO_3$	−1139	H_2O	−228.6	F	59.4
$BaSO_4$	−1465	H_2O_2	−103.3	Cl	105.4
CaO	−604.2	CO_2	−394.4	Br	82.38
$Ca(OH)_2$	−896.6	CO	−137.3	C	672.95
$CaCO_3$	−1129	SO_2	−300.4	I	70.16
Fe_2O_3	−741	SO_3	−370.4	N	340.9
Al_2O_3	−1582	H_2S	−33.56	O	161
CuO	−127	NO_2	51.84		
Cu_2O	−146.0	N_2O	104.18		
ZnO	−318.2	NH_3	−16.6		
SiO_2	−805	O_3	163.4		
PbO_2	−219	NO	86.69		
PbS	−98.7				

Liquids		Hydrocarbon Gases	
CH_3OH	−166.2	CH_4 (methane)	−50.79
C_2H_5OH	−174.8	C_2H_6 (ethane)	−32.8
CH_3COOH	−392	C_3H_8 (propane)	−23.6
C_6H_6	+124.5	*n*-C_4H_{10} (*n*-butane)	−17.2
H_2O	−237.18	*i*-C_4H_{10} (*i*-butane)	−20.9
H_2O_2	−120.42	C_2H_2 (acetylene)	209.2
		C_2H_4 (ethylene)	68.24
		C_3H_6 (propylene)	62.84
		C_4H_8 (1-butene)	77.3
		C_4H_8 (*cis*-2-butene)	65.86
		C_4H_8 (*trans*-2-butene)	62.97
		C_4H_8 (*i*-butene)	58.07

Note that as with ΔH_f°, the value of ΔG_f° for an element in its standard state is zero.

$$2CO(g) + O_2(g) \longrightarrow 2CO_2(g)$$

$$\Delta G^\circ = (2 \text{ mol})(-394.4 \text{ kJ/mol}) - (2 \text{ mol})(-137.3 \text{ kJ/mol})$$

$$= -514.2 \text{ kJ (for the reaction as written using 2 mol of CO)}$$

If the reaction were written

$$CO(g) + \frac{1}{2}O_2(g) \longrightarrow CO_2(g)$$

ΔG would be half of this value and would equal

$$\frac{-514.2 \text{ kJ}}{2} = -257.1 \text{ kJ}$$

EXAMPLE 13.4

The following reaction is believed to affect the ozone layer in the upper atmosphere.

$$O_3(g) + NO(g) \longrightarrow O_2(g) + NO_2(g)$$

Calculate $\Delta G°$ for this reaction. Is it spontaneous at the conditions 1 atm and 25°C?

COMMENT Use the standard free energies of formation and the equation

$$\Delta G° = \Sigma \Delta G_f°(\text{products}) - \Sigma \Delta G_f°(\text{reactants})$$

to calculate $\Delta G°$ for the reaction. Note that $\Delta G_f°$ for $O_2(g)$ is zero because it is an element in its standard state.

SOLUTION

$$\Delta G° = \Delta G_f°[NO_2(g)] - \Delta G_f°[O_3(g)] - \Delta G_f°[NO(g)]$$

$$= (1 \text{ mol})(51.84 \text{ kJ/mol}) - (1 \text{ mol})(163.4 \text{ kJ/mol}) - (1 \text{ mol})(86.69 \text{ kJ/mol})$$

$$= -198.2 \text{ kJ}$$

$\Delta G°$ is negative, and the reaction is clearly spontaneous under standard conditions.

EXERCISE 13.4(A)

Use $\Delta G_f°$ values to calculate the standard free energy change for the following reaction:

$$2H_2S(g) + 3O_2(g) \longrightarrow 2H_2O(g) + 2SO_2(g)$$

ANSWER $\Delta G° = -990.9$ kJ

EXERCISE 13.4(B)

Suppose you are planning a chemical plant for the manufacture of hydrogen peroxide. There are two possible routes to the synthesis.

1. Direct reaction of oxygen with hydrogen:

$$H_2(g) + O_2(g) \longrightarrow H_2O_2(\ell)$$

2. Reaction of oxygen with water:

$$2H_2O(\ell) + O_2(g) \longrightarrow 2H_2O_2(\ell)$$

Using the appropriate data from the tables, predict which process might be possible.

ANSWER The second process has a positive value of $\Delta G°$. However, $\Delta G° < 0$ for the first process, which makes it possible. It is, in fact, the basis for a commercial process.

Free Energy Change and the Equilibrium Constant

As we mentioned earlier, the standard free energy change $\Delta G°$ refers to reactants and products at partial pressures of 1 atm for gases or for solutes in 1-M solutions. For example, from Table 13.2, $\Delta G°_{298K}$ is positive for the process

$$2N_2(g) + O_2(g) \longrightarrow 2N_2O(g) \qquad \Delta G° = 208.36 \text{ kJ}$$

Because of this, we know that if nitrogen gas, oxygen gas, and dinitrogen monoxide gas are mixed at 298K so that each is at a partial pressure of 1 atm, the reaction proceeds spontaneously from right to left. The partial pressures of nitrogen and oxygen increase at the expense of the partial pressure of dinitrogen monoxide. To state it another way, this is a reaction with an equilibrium constant less than one. What we would like to establish is how to relate $\Delta G°$ values directly to equilibrium constants.

The value of ΔG for the reaction will be zero in all equilibrium situations. However, the free energy change ΔG will generally not be $\Delta G°$, because conditions at equilibrium will not be standard. That is, at equilibrium it is unlikely that all partial pressures are 1 atm and all concentrations are 1 M. Thus, it is necessary to find out how to calculate ΔG from $\Delta G°$. For the general reaction,

$$\text{aA} + \text{bB} \longrightarrow \text{cC} + \text{dD}$$

it can be shown that the difference between ΔG and $\Delta G°$ is a term governed by the reaction quotient Q, the ratio of the product and reactant partial pressures or concentrations raised to powers equal to the coefficients in the balanced chemical equation (Chapter 10). If all reactants and products are gases, Q is

$$Q = \frac{p_C^c p_D^d}{p_A^a p_B^b}$$

The relationship between ΔG and $\Delta G°$ is

$$\Delta G = \Delta G° + RT \ln Q$$

If the system is at equilibrium, then ΔG equals zero and the reaction quotient Q equals the equilibrium constant K. This leads us to the equation

$$0 = \Delta G° + RT \ln K$$

$$\Delta G° = -RT \ln K$$

$\Delta G°$ and the equilibrium constant: The difference between ΔG and $\Delta G°$ is a term that depends on Q, the reaction quotient. At equilibrium, ΔG is zero and $\Delta G°$ determines the equilibrium constant K. This is the fundamental equation relating $\Delta G°$ and the equilibrium constant.

When the reactants and products are in their standard states, spontaneous reactions have negative $\Delta G°$ values and equilibrium constants greater than one. Reactions that move spontaneously in the reverse direction have positive $\Delta G°$ values and equilibrium constants less than one. For the special case in which $\Delta G°$ is zero at equilibrium, the equilibrium constant is one.

APPLICATIONS OF CHEMISTRY

Thermite Welding on the Railroad For as long as anyone can remember, railway workers have used aluminum to reduce iron oxide to the molten metal by a process known as thermite welding. It is an especially convenient application of thermodynamic principles to chemistry because the process can be performed by relatively unskilled laborers working in distant locations without heavy equipment. Here is how thermite welding works.

Aluminum powder and powdered iron oxide are stirred together into a uniformly dispersed mixture. But intimate contact is insufficient to bring about the desired reaction, even though the free energy change is highly favorable (Table 13.2):

$$2Al(s) + Fe_2O_3(s) \longrightarrow Al_2O_3(s) + 2Fe(s) \qquad \Delta G^\circ = -1582 - (-741) = -841 \text{ kJ}$$

It takes the ignition of a magnesium flare to produce the local rise in temperature to about 1000°C necessary to jump-start the chemical reaction. Once started, however, the iron-oxide-reduction reaction is self-sustaining, because the reaction is highly exothermic (Table 12.4); $\Delta H^\circ = -1676 - (-822) = -854$ kJ. At the high temperature produced by such an exothermic reaction, the overall reaction produces molten iron and enough heat to vaporize the molten metal. However, thermal conductivity comes to the rescue as the rails themselves limit the temperature rise by carrying away much of this heat.

Questions

What are the advantages of using thermite welding for laying railroad track compared with other forms of welding? What are the disadvantages?

Scrap aluminum can be used in a thermite type of reaction to reduce Cr_2O_3 to Cr metal. Write the chemical reaction and balance the equation. Explain why (in view of its being energetically favorable) does the reaction not take place directly.

EXAMPLE 13.5

Calculate the equilibrium constant at 298K for the reaction

$$2SO_2(g) + O_2(g) \longrightarrow 2SO_3(g)$$

COMMENT Calculate the standard free energy change using the values in Table 13.2. Then rearrange the equation $\Delta G^\circ = -RT \ln K$ and solve for K.

SOLUTION From the table of free energies of formation, ΔG° is -140.0 kJ. Rearrangement of the equation relating ΔG° and the equilibrium constant gives

$$-RT \ln K = \Delta G^\circ$$

$$\ln K = -\frac{\Delta G^\circ}{RT}$$

$$\ln K = -\frac{(140.0 \text{ kJ})(1000 \text{ J/kJ})}{(8.314 \text{ J/mol·K})(298\text{K})} = 56.5$$

$$K = e^{56.5} = 3.5 \times 10^{24}$$

This very large value for K has huge environmental implications, because SO_2 is a by-product of burning coal and petroleum-based industrial fuels with high sulfur contents. Where SO_2 is present, there is a significant driving force to produce

SEEING THINGS

The Thermite Reaction and Thermite Welding: A Classic Lecture and Laboratory Demonstration The 6-mm-diameter hole drilled through the bottom of an unglazed 250-mL clay crucible is covered with a disk of very thin sheet copper so that the crucible can be half filled with commercial thermite mixture. Invented by the German chemist Hans Goldschmidt (1861–1923) about 1900, thermite is a mixture of iron(III) oxide and powdered aluminum and was originally used as an incendiary bomb.

In the center of the thermite-filled clay crucible is placed a 10-cm fuse made of two strips of magnesium ribbon bound together by loosely wrapping a third strip of the ribbon around them. The base of this fuse is surrounded by a small heap (about 5-mL volume) of finely powdered magnesium to ensure ignition. The crucible is then supported by a small ring into which the crucible just fits. The bottom of the crucible will be exposed to the molten iron formed during the reaction. The iron will melt through the copper and fall from the hole in the bottom.

Under the crucible is placed a box about 40 × 45 × 15 cm filled with dry sand and two iron rods of about 12-mm diameter and 20-cm length. A shallow trough is made in the sand into which the two metal rods are laid so that they are end to end, with about 12 mm of space between their ends. The crucible is centered above the adjoining ends of the rod, about 10 cm above the rods.

When everything is ready, the magnesium ribbon fuse is ignited with a burner. A great deal of light and heat is produced and the molten iron that is produced melts its way through the copper plug in the bottom of the crucible, falling into the space between and around the ends of the rods. The reaction generates enough heat to raise the temperature above 2200°C, temperatures sufficiently high to soften steel. Note that the oxide–metal reaction provides its own oxygen supply and is thus very difficult to stop once started. When the metal has solidified, the rods may be lifted with a pair of tongs so all can view the welded joint.

Questions

What would be the likely values (positive, near zero, or negative) for ΔH°, ΔS°, and ΔG° for the thermite reaction?

Why is the thermite process so hard to stop once it gets going compared to some other exothermic reactions such as burning wood?

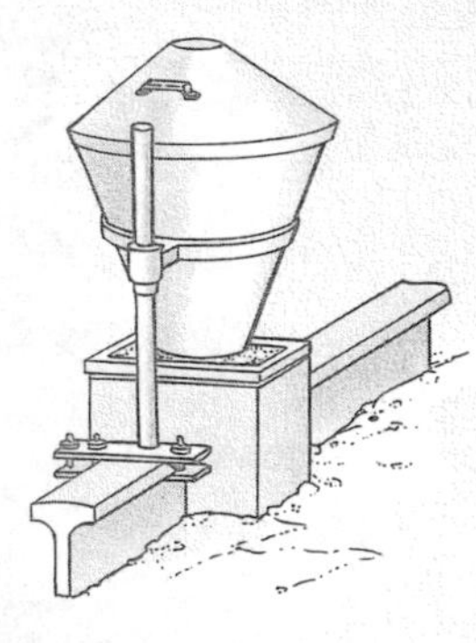

Sketch of a crucible and mold in position for welding a steel rail with thermite

Thermite welding of a failed section of railroad track. Schematic diagram shows the classic procedure used until very recently along isolated stretches of the line. Crucible and mold are shown in position for welding.

Warning: The thermite reaction is an impressive lecture demonstration. However, it is a potentially very dangerous reaction that, if performed, should be run only by an experienced demonstrator employing extreme safety precautions. *This reaction should not be run under other circumstances!*

SO_3, and, as an acidic oxide, SO_3 becomes a major contributor to the acid rain problem. We can take a little comfort, however, in that the conversion of SO_2 to SO_3 takes place slowly in the absence of a catalyst, a topic in Chapter 15.

EXERCISE 13.5

Determine the equilibrium constant for the following reaction at 298K:

$$H_2O(g) + O_3(g) \longrightarrow H_2O_2(g) + O_2(g)$$

ANSWER $K_{eq} = 4.8 \times 10^6$

13.7 EFFECT OF TEMPERATURE ON EQUILIBRIUM

Temperature and the Free Energy Change

We showed that the change in free energy $\Delta G°$ for a process under standard-state conditions with a temperature of 298K can be obtained using the relationship

$$\Delta G° = \Delta H° - T\Delta S°$$

Then from $\Delta G°$ values, equilibrium constants can be obtained. However, because not all reactions are carried out at 298K, we need to be able to determine $\Delta G°$ at other temperatures. If we can calculate the value of $\Delta G°$ at temperatures other than 298K, we can also calculate the value of the equilibrium constant of the reaction at other temperatures with the equation

$$\Delta G° = -RT \ln K$$

Using the equation $\Delta G° = \Delta H° - T\Delta S°$ to make accurate calculations of $\Delta G°$ at temperatures other than 298K would require that the values of $\Delta H°$ and $\Delta S°$ be constant within the temperature range. In fact, $\Delta H°$ and $\Delta S°$ do vary somewhat with temperature, but for many reactions they change so little that the value of $\Delta G°$ calculated from $\Delta G° = \Delta H° - T\Delta S°$ at a temperature other than 298K is sufficiently accurate, and such calculations are commonly performed.

EXAMPLE 13.6

Calculate $\Delta G°$ at 400.K for the reaction

$$2NO(g) + O_2(g) \longrightarrow 2NO_2(g)$$

COMMENT Use the standard enthalpies of formation at 298K in Table 12.4 to calculate the standard enthalpy change of the reaction, and use the absolute entropies at 298K in Table 13.1 to calculate the standard entropy change. Then use $\Delta G° = \Delta H° - T\Delta S°$ to calculate the standard free energy change at 400.K.

SOLUTION First obtain ΔH° and ΔS° values from the tables of ΔH_f° and S°:

$$\Delta H^\circ = -113.0 \text{ kJ}; \Delta S^\circ = -145 \text{ J/K}$$

Then calculate ΔG° from $\Delta G^\circ = \Delta H^\circ - T\Delta S^\circ$:

$$= -113.0 \text{ kJ} - (400.\text{K})(-145 \text{ J/K})\left(\frac{1 \text{ kJ}}{1000 \text{ J}}\right) = -55.0 \text{ kJ}$$

EXERCISE 13.6

Calculate ΔG° for the following reaction at 500K:

$$BaCO_3(s) \longrightarrow BaO(s) + CO_2(g)$$

ANSWER $\Delta G^\circ = 180$ kJ

The sign of ΔG° tells the spontaneous direction of a reaction starting with a mixture in which all of the reactants and products are in their standard states. From $\Delta G^\circ = \Delta H^\circ - T\Delta S^\circ$ it is clear that processes that are driven by both enthalpy (negative ΔH°) and entropy (positive ΔS°) will have negative values of ΔG° and thus will be spontaneous at all temperatures. The sign combinations for ΔH and ΔS and the resulting sign of ΔG are as follows:

- If ΔH° is negative and ΔS° is positive, ΔG° is negative for all values of T.
- If ΔH° and ΔS° are both negative, ΔG° can be negative at low T.
- If ΔH° and ΔS° are both positive, ΔG° can be negative at high T.
- If ΔH° is positive and ΔS° is negative, ΔG° is always positive, no matter what the value of T.

The spontaneous direction of reactions favored by one factor and not the other will depend on the temperature. Processes driven by enthalpy but not entropy are spontaneous if the temperature is low enough, and those driven by entropy but not enthalpy are spontaneous if the temperature is sufficiently high. Processes driven by neither enthalpy nor entropy have positive values of ΔG° and cannot be made spontaneous by changing the temperature.

The direction of a spontaneous reaction can be changed by changing the temperature in those cases in which ΔH° and ΔS° are both positive or they are both negative. For these reactions, the temperature at which ΔG° is zero can be considered the temperature at which the change between nonspontaneity and spontaneity is made. At the temperature at which ΔG° for a reaction is zero, the system is at equilibrium if all reactants and products are in their standard states—all gases at a partial pressure of 1 atm and all concentrations at 1 M.

The temperature at which ΔG° is zero can be found by setting ΔG° equal to zero in the equation $\Delta G^\circ = \Delta H^\circ - T\Delta S^\circ$ and solving for T.

$$0 = \Delta H^\circ - T\Delta S^\circ$$

$$T = \frac{\Delta H^\circ}{\Delta S^\circ}$$

EXAMPLE 13.7

Once again we turn our attention to an environmentally important reaction. Calculate the temperature at which ΔG° is zero for the reaction

$$2NO(g) + O_2(g) \longrightarrow 2NO_2(g)$$

COMMENT The values of ΔH° were calculated in Example 13.6 from the table of standard enthalpies of formation (Table 12.4) and ΔS° from the table of standard entropies (Table 13.1). Set ΔG° equal to zero and rearrange $\Delta G^\circ = \Delta H^\circ - T\Delta S^\circ$ to solve for T. Note that the use of this equation assumes that ΔH° and ΔS° are constant over the pertinent range.

SOLUTION $0 = \Delta H^\circ - T\Delta S^\circ$

$$T = \frac{\Delta H^\circ}{\Delta S^\circ}$$

$$T = \frac{(-113.0 \text{ kJ})(1000 \text{ J/kJ})}{-145 \text{ J/K}} = 779\text{K}$$

EXERCISE 13.7

Calculate the temperature at which the reaction in Exercise 13.3 would be at equilibrium with all gases at partial pressures of 1 atm.

ANSWER $T = 463\text{K}$

Note that these calculations will give a positive value for the temperature only for reactions in which ΔH° and ΔS° have the same sign. If ΔH° and ΔS° have opposite signs, the process is spontaneous at all temperatures or at no temperatures. In these cases the negative value found for the absolute temperature by setting ΔG° equal to zero has no physical significance.

Temperature and the Equilibrium Constant

The temperature dependence of the equilibrium constant can be obtained by combining two equations:

$$\Delta G^\circ = \Delta H^\circ - T\Delta S^\circ$$

and

$$\Delta G^\circ = -RT \ln K$$

Setting the two equalities equal to each other gives the result

$$-RT \ln K = \Delta H^\circ - T\Delta S^\circ$$

which can be rearranged by dividing both sides of the equation by $-RT$ to give

$$\ln K = -\frac{\Delta H^\circ}{RT} + \frac{\Delta S^\circ}{R}$$

Dependence of the equilibrium constant *K* on *T*: The utility of this equation depends on the extent to which ΔH° and ΔS° are independent of temperature.

From this equation it can be seen that if ΔH° is negative, then K will decrease with increasing temperature. If ΔH° is positive, then an increase in temperature will increase K. This is in agreement with Le Chatelier's principle, which predicts that heating an endothermic reaction drives it forward, whereas heating an exothermic reaction has the opposite effect.

Because this equation is a linear equation of the general form $y = mx + b$, it can be used to determine ΔH° and ΔS° by a graphing technique.

$$y = \ln K$$

$$m = \text{the slope of the line} = -\frac{\Delta H^\circ}{R}$$

$$x = \frac{1}{T}$$

$$b = \text{the intercept of the line on the } y\text{-axis} = \frac{\Delta S^\circ}{R}$$

A plot of $\ln K$ versus $1/T$ gives a straight line from which ΔH° can be calculated from the slope of the line $-\Delta H^\circ/R$, and ΔS° can be calculated from the intercept $\Delta S^\circ/R$.

If the equilibrium constant for a reaction is known at two temperatures T_1 and T_2, the equation can be used to determine the average ΔH° value for the reaction over the temperature range T_1 to T_2:

$$\text{at } T_1: \quad \ln K_1 = -\frac{\Delta H^\circ}{RT_1} + \frac{\Delta S^\circ}{R}$$

$$\text{at } T_2: \quad \ln K_2 = -\frac{\Delta H^\circ}{RT_2} + \frac{\Delta S^\circ}{R}$$

Subtracting the first equation from the second gives

$$\ln K_2 - \ln K_1 = -\frac{\Delta H^\circ}{RT_2} + \frac{\Delta H^\circ}{RT_1}$$

And because

$$\ln K_2 - \ln K_1 = \ln \frac{K_2}{K_1}$$

$$\ln \frac{K_2}{K_1} = -\frac{\Delta H^\circ}{R}\left(\frac{1}{T_2} - \frac{1}{T_1}\right)$$

***K* values at two temperatures:** Relates the equilibrium constant for a reaction at one temperature to the equilibrium constant for the same reaction at another.

The standard enthalpy and entropy changes of the reaction are assumed to remain constant between the two temperatures, a factor that limits the accuracy of the result.

EXAMPLE 13.8

Equilibrium constants for the combustion of sulfur dioxide are 3.98×10^{24} at 298K and 1.56×10^7 at 600.K:

$$2SO_2(g) + O_2(g) \longrightarrow 2SO_3(g)$$

Calculate the average $\Delta H°$ for this reaction.

COMMENT Substitute appropriate values for K_1, T_1, K_2, and T_2 into the equation for K values at two temperatures.

SOLUTION

$$\ln\frac{K_2}{K_1} = -\frac{\Delta H°}{R}\left(\frac{1}{T_2} - \frac{1}{T_1}\right) = -\frac{\Delta H°}{R}\left(\frac{T_1 - T_2}{T_2 T_1}\right)$$

$$\Delta H° = -R\ln\frac{K_2}{K_1}\left(\frac{T_2 T_1}{T_1 - T_2}\right)$$

$$= -\left(8.314\frac{\text{J}}{\text{mol·K}}\right)\left(\ln\frac{1.56 \times 10^7}{3.98 \times 10^{24}}\right)\left(\frac{[600\text{ K·}298\text{K}]}{[298 - 600]\text{K}}\right)$$

$$= -1.97 \times 10^5\text{ J} = -197\text{ kJ}$$

EXERCISE 13.8

For the reaction

$$H_2O(g) + O_3(g) \longrightarrow H_2O_2(g) + O_2(g)$$

the equilibrium constant is 4.89×10^6 at 298K and 1.28×10^4 at 500.K. What is the average $\Delta H°$ for this reaction over the given temperature range?

ANSWER -36.5 kJ

For the equilibrium between a liquid and its vapor, the equilibrium constant is simply the equilibrium vapor pressure, because pure liquids do not appear in equilibrium expressions:

$$\text{Liquid} \rightleftharpoons \text{Vapor} \qquad K = p_{vap}$$

If the two values of p_{vap} are substituted into the equation for K at two temperatures, the result is known as the **Clausius–Clapeyron equation**:

$$\ln\frac{p_2}{p_1} = -\frac{\Delta H_{vap}}{R}\left(\frac{1}{T_2} - \frac{1}{T_1}\right)$$

Clausius–Clapeyron equation: Gives the enthalpy of vaporization from data showing the variation of vapor pressures with temperature.

PROCESSES IN CHEMISTRY

Substitute Natural Gas The thermodynamics of chemical reactions provides essential guidelines for planning chemical processes for manufacturing many basic chemicals. To begin with, thermodynamics tells us whether a particular reaction can happen at all. If it can, thermodynamics is necessary to help us find the best conditions for carrying out the reaction. Therefore, the chemical reactions of many industrial processes have been subjected to careful thermodynamic scrutiny. As an example, consider the production of substitute natural gas (SNG) from coal.

Production of a natural gas substitute from coal is attractive because of uncertainties in the supplies of petroleum and natural gas and because of the huge existing coal reserves. In particular, let us examine the synthesis of methane (CH_4), the principal constituent of natural gas. Coal is a complicated material containing carbon, hydrogen, oxygen, nitrogen, sulfur, and other elements in a complex structure, as well as interspersed mineral matter. On heating in the absence of air, coal is converted to coke, which consists principally of carbon in the form of microscopic graphite crystals and mineral matter (ash). Treatment of coke with steam produces carbon monoxide and hydrogen, a mixture known as synthesis gas:

$$C(s) + H_2O(g) \longrightarrow CO(g) + H_2(g) \qquad \text{preparation of synthesis gas} \qquad (1)$$

Synthesis gas can be converted to methane. However, the synthesis gas is first enriched with hydrogen, because to produce CH_4, four hydrogen atoms are needed per carbon atom. The enrichment is done by taking some of the carbon monoxide produced by reaction 1 and letting it react with steam in what is called the shift reaction:

$$CO(g) + H_2O(g) \longrightarrow CO_2(g) + H_2(g) \qquad \text{shift reaction} \qquad (2)$$

The carbon dioxide produced in reaction 2 is absorbed by CaO (lime) and disposed of as calcium carbonate ($CaCO_3$). In the final step of the process, carbon monoxide and hydrogen react to form methane:

$$CO(g) + 3H_2(g) \longrightarrow CH_4(g) + H_2O(g) \qquad \text{methane formation} \qquad (3)$$

That's the chemistry! Now, the thermodynamics.

The values of $\Delta H°$, $\Delta S°$, and $\Delta G°$ can be calculated for all of these reactions from tables in this chapter and Chapter 12. Here are the thermodynamic data for reaction 1, the preparation of synthesis gas:

$$C(s) + H_2O(g) \longrightarrow CO(g) + H_2(g)$$
$$\Delta H_1^\circ = 132 \text{ kJ}$$
$$\Delta S_1^\circ = 134 \text{ J/K}$$
$$\Delta G_1^\circ = 91.3 \text{ kJ}$$

This is a strongly endothermic reaction. The value of $\Delta G°$ shows that the reaction is not spontaneous at 298K. In fact, the equilibrium constant at that temperature is 9.91×10^{-17}. However, $\Delta S°$ is positive because two moles of gas are produced starting with one mole of gas and one of solid. Because the reaction is endothermic, raising the temperature favors the forward reaction. Calculation of the temperature at which $\Delta G°$ is zero, where the reaction is at equilibrium with all partial pressures at 1 atm, gives a value of 980K. Therefore, at temperatures above 980K, conversions should start to become significant. In fact, this reaction is normally run at temperatures in excess of 1090K.

Using this equation it is possible to calculate ΔH_{vap} if the vapor pressures p_1 and p_2 of the liquid are known at the corresponding temperatures T_1 and T_2. With the same equation, given the heat of vaporization and the vapor pressure at a certain temperature, the vapor pressure at another temperature can be determined. Note that because the Clausius–Clapeyron equation is written as a ratio of pressures, measurements can be made in any convenient units.

Reaction 2 is a spontaneous reaction, and it is exothermic. The thermodynamic data for the reaction are as follows:

$$CO(g) + H_2O(g) \longrightarrow CO_2(g) + H_2(g)$$

$$\Delta H_2^\circ = -42 \text{ kJ}$$

$$\Delta S_2^\circ = -42 \text{ J/K}$$

$$\Delta G_2^\circ = -28.5 \text{ kJ}$$

Because the standard entropy change is negative, raising the temperature will not increase the equilibrium constant and the yield. In fact, raising the temperature drives the reaction in the reverse direction. To get the reaction to proceed at a reasonable rate, a catalyst of chromium and iron and a somewhat raised temperature of about 400°C are employed. It is often the case in chemical processes that temperatures are raised somewhat to increase the speed of an exothermic reaction, even though this makes the equilibrium constant less favorable.

Reaction 3 is an exothermic and spontaneous reaction for which we have the following thermodynamic data:

$$CO(g) + 3H_2(g) \longrightarrow CH_4(g) + H_2O(g)$$

$$\Delta H_3^\circ = -207 \text{ kJ}$$

$$\Delta S_3^\circ = -215. \text{ J/K}$$

$$\Delta G_3^\circ = -142.1 \text{ kJ}$$

To increase the rate, a nickel–iron catalyst is used, and the temperature is raised to about 575 to 600K.

The overall process for production of substitute natural gas is the sum of twice reaction 1, reaction 2, and reaction 3:

$$2C(s) + 2H_2O(g) \longrightarrow CO_2(g) + CH_4(g) \qquad \Delta H^\circ = 15 \text{ kJ}$$

The total ΔH° for the entire sequence is 15 kJ, only modestly endothermic. This means that the entire process would require the input of very little energy if the heat evolved in reactions 2 and 3 could be put into reaction 1. Unfortunately, reactions 2 and 3 are carried out at reasonably low temperatures to keep their equilibrium constants favorable. Therefore, waste heat from these reactions is not at nearly the high temperature needed to sustain reaction 1. As a consequence, a considerable amount of coal must also be burned to provide the energy requirements of the first reaction. Such examples clearly show the trade-offs that are typically made when engineers put chemical principles into commercial practice.

Questions

Why is the heat from reactions 2 and 3 of the substitute natural gas process of limited value for driving reaction 1? Show that the sum of twice reaction 1, reaction 2, and reaction 3, is the net reaction

$$2C(s) + 2H_2O(g) \longrightarrow CO_2(g) + CH_4(g)$$

Is the reaction favorable? In other words, find ΔG° for the process.

EXAMPLE 13.9

The vapor pressure of water at 373K is 760 torr and the average heat of vaporization over the temperature range 363K to 373K is 40.8 kJ/mol. Calculate the vapor pressure of water at 363K.

COMMENT Substitute the values of ΔH_{vap}, T_1, and T_2, and p_1 into the Clausius–Clapeyron equation and solve for p_2.

SOLUTION

$$\ln \frac{p_2}{760 \text{ torr}} = -\frac{(40.8 \text{ kJ/mol})(1000 \text{ J/kJ})}{8.314 \text{ J/mol}\cdot\text{K}}\left(\frac{1}{363\text{K}} - \frac{1}{373\text{K}}\right) = -0.362$$

$$\frac{p_2}{760 \text{ torr}} = e^{-0.362} = 0.696$$

$$p_2 = (0.696)(760 \text{ torr}) = 529 \text{ torr}$$

EXERCISE 13.9

The vapor pressure of ethanol is 100.0 torr at 34.9°C and 400.0 torr at 63.5°C. Calculate the average heat of vaporization in this range.

ANSWER 4.18×10^4 J/mol

SUMMARY

The ultimate purpose of this chapter is to develop an understanding of ΔG, the change in free energy. It is a state function that can be used to predict whether a chemical process will occur spontaneously. A spontaneous process is defined as any process that is not reversible, and we note that only processes at equilibrium are reversible. All other processes proceed spontaneously toward equilibrium. The path function q can be used to determine whether a process is conducted reversibly. The value of q for the reversible path is q_{rev} and is the maximum value of q for the process.

The change in entropy ΔS is a measure of the change in disorder and for a process is related to the maximum heat that can be absorbed. For a process at constant temperature the value of ΔS is given by q_{rev}/T. Both the change of enthalpy ΔH and the entropy change ΔS are factors that can drive chemical processes. The second law of thermodynamics states that the entropy change of the universe is related to the spontaneity of processes. All spontaneous processes increase the total entropy of the universe and the total entropy of the universe can never decrease.

The third law of thermodynamics provides a zero point of entropy for elements and compounds. This allows for a method of calculating the absolute entropies of substances, which in turn provides a method of calculating the entropy changes of chemical processes. The free energy change at constant temperature and pressure can be calculated using $\Delta G = \Delta H - T\Delta S$. Under standard conditions of pressure and concentration and at a fixed temperature (usually 298K), this equation becomes $\Delta G° = \Delta H° - T\Delta S°$. At constant temperature and pressure, the value of ΔG is zero for a reversible process and negative for a spontaneous process. The standard free energy change $\Delta G°$ can be calculated from ΔH° and ΔS° values or from standard free energies of formation. The standard free energy change is related to the equilibrium constant and can be used to calculate equilibrium constants and vapor pressures at different temperatures.

TERMS

Reversible process (13.2)
Irreversible (spontaneous) process (13.2)
Entropy (13.3)
Change in entropy (13.3)
Second law of thermodynamics (13.3)
Third law of thermodynamics (13.5)
Free energy (13.6)
Standard free energy of formation (13.6)
Clausius–Clapeyron equation (13.7)

IMPORTANT EQUATIONS

Equation	Description
$q_{rev} > q_{irrev}$	Relationship between reversible and nonreversible heat; all conditions
$\Delta S = q_{rev}/T$	Change of entropy at constant temperature
$\Delta S_{univ} = \Delta S_{sys} + \Delta S_{surr}$	Entropy change of the universe; all conditions
$\Delta S_{surr} = -q/T$	Entropy change of surroundings; constant temperature
$\Delta G = \Delta H - T\Delta S$	Free energy change; constant pressure and temperature
$\Delta G^\circ = \Delta H^\circ - T\Delta S^\circ$	Free energy change; standard conditions
$\Delta G = \Delta G^\circ + RT \ln Q$	Free energy at nonstandard conditions; constant pressure and temperature
$\Delta G^\circ = -RT \ln K$	Relationship between free energy and equilibrium constant; constant pressure and temperature
$\ln K = -\frac{\Delta H^\circ}{RT} + \frac{\Delta S^\circ}{R}$	Variation of equilibrium constant with temperature; constant pressure
$\ln \frac{p_2}{p_1} = -\frac{\Delta H_{vap}}{R}\left(\frac{1}{T_2} - \frac{1}{T_1}\right)$	Variation of vapor pressure with temperature; constant ΔH_{vap}

QUESTIONS

Conceptual questions are denoted by a square screen. Extra-credit questions are denoted by a circular screen.

1. What four or five processes can you think of that are spontaneous but proceed very slowly?

2. Why is the following phase transformation referred to as a "reversible process"?

$$H_2O(s, 0°C) \rightleftharpoons H_2O(\ell, 0°C)$$

3. Why is the following process not considered to be a reversible process?

$$H_2O(\ell, -5°C) \rightleftharpoons H_2O(s, -5°C)$$

4. Why is entropy defined as a state function?

5. With the passage of time, what happens to the total energy of the universe? What happens to the total entropy of the universe?

6. The politics of oil can produce what is popularly called an "energy crisis" if oil-producing countries cut supplies. Can there really be an energy crisis? What might be a better name for the crisis that they have in mind?

7. What assumptions do we make about the nature of "the surroundings" in our discussions of the second law?

8. What is the basis of our belief in the second law of thermodynamics? Is this the same as our basis for believing the ideal gas law? Briefly explain.

9. If you stretch a rubber band suddenly and press it quickly against your lips, you will notice an increase in the temperature of the band after the stretching. Offer a thermodynamic explanation of this effect.

10. Addition of thionyl chloride to hexaaquacobalt(II) chloride is an example of an endothermic but spontaneous chemical reaction. There is an immediate drop in temperature, accompanied by liberation of gaseous HCl and SO_2:

$$Co(H_2O)_6Cl_2(s) + 6SOCl_2(\ell) \longrightarrow CoCl_2(s) + 12HCl(g) + 6SO_2(g)$$

Briefly explain the spontaneous drive from reactants to products.

11. What does it mean if the value of $\Delta G°$ for a chemical reaction is zero?

12. Give a few examples of processes for which the value of $\Delta G°$ is zero.

13. What is the basis for the assumption that for many reactions $\Delta G°$ can be calculated at a variety of temperatures from $\Delta G° = \Delta H° - T\Delta S°$?

14. When solid ammonium chloride is dissolved in water, the temperature of the solution falls. Offer a brief explanation of this spontaneous change in temperature in terms of the second law of thermodynamics.

15. When sodium hydroxide pellets dissolve, the temperature of the solution rises spontaneously. Briefly explain in terms of the second law of thermodynamics.

16. Which of the following situations results in an increase in the entropy of the system and why?
   (a) $H_2O(\ell) \longrightarrow H_2O(g)$
   (b) $2NO_2(g) \longrightarrow N_2O_4(g)$
   (c) $CaCO_3(s) \longrightarrow CaO(s) + CO_2(g)$
   (d) $3H_2(g) + N_2(g) \longrightarrow 2NH_3(g)$
   (e) $HCl(g) + NH_3(g) \longrightarrow NH_4Cl(s)$
   (f) $C(s) + H_2O(g) \longrightarrow CO(g) + H_2(g)$

17. What is the sign of q, the heat for each of the following using the sign conventions and ideas established in this chapter?
   (a) The vaporization of liquid butane in a lighter:

$$C_4H_{10}(\ell, 298K) \longrightarrow C_4H_{10}(g, 298K)$$

   (b) The burning of liquid *iso*-octane in a high-compression internal combustion engine cylinder at normal operating temperatures:

$$C_8H_{18}(g) + \frac{25}{2}O_2(g) \longrightarrow 8CO_2(g) + 9H_2O(g)$$

18. What is the sign of q, the heat for each of the following using the sign conventions and ideas established in this chapter?
   (a) water freezes (b) water vaporizes
   (c) water sublimes

PROBLEMS

Problems marked with a bullet (•) are answered in Appendix A, in the back of the text.

Spontaneous and Reversible Processes [1–4]

•1. Calculate the work for the expansion of a gas from 2.0 L to 3.0 L against a constant external pressure of 1.5 atm. How would your calculated value of work for this process compare with the work if the process were reversible?

2. What is the work for the contraction of a gas from 7.5 L to 5.0 L resulting from an external pressure of 5.0 atm? What can you say about the work involved in the reversible contraction of the gas from 7.5 L to 5.0 L without making any further calculations?

•3. Calculate the total work w for the expansion of an ideal gas from 2.00 L to 3.00 L in one, two, and three irreversible steps. The initial gas pressure is 1.00 atm and the temperature of 298K is kept constant throughout the process. The expansion in one step is performed against a constant external pressure equal to the final pressure of the gas in the cylinder. The two- and three-step expansions are done by equal stepwise decreases in the external pressure, reaching the same final pressure. Calculate the value for the work for the three expansions and compare the results.

4. You are given 10.0 L of an ideal gas contained in a cylinder with a movable piston at 298K at an initial pressure of 2.00 atm. The gas is then compressed to 1.00 L in one, two, and three irreversible steps. Calculate the total value of the work for each process. See the previous problem.

Entropy Calculations [5–8]

•5. What is ΔS_{fus} for one mole of ice changing to water at 0°C, the normal melting point? What is ΔS_{univ} for this process? (The enthalpy of fusion of ice is 6025 J/mol.)

6. Calculate ΔS_{vap} for water at the boiling point, 100.°C. The enthalpy of vaporization of water is 40,660 J/mol. What is ΔS_{univ} for this process?

•7. Calculate the entropy change for the system and the universe that accompanies the melting of 1.00 kg of lead at its normal melting point of 327°C. (The heat of fusion of lead is 21.3 kJ/mol.)

8. Methane boils at −159°C and its heat of vaporization is 9240 J/mol. Calculate the entropy of vaporization of methane at its boiling point.

Free Energy Change and the Equilibrium Constant [9–20]

•9. Calculate $\Delta G°$ at 298K for a process for which $\Delta H°$ is −1220 kJ/mol and $\Delta S°$ is −130 J/mol·K. Is this reaction spontaneous under standard conditions?

10. What is the value of $\Delta G°$ at 298K for a process if $\Delta H°$ is +930 kJ/mol and $\Delta S°$ is +340 J/mol·K? What will be the spontaneous direction of reaction under standard conditions?

•11. Use Tables 12.4 and 13.1 and $\Delta G° = \Delta H° - T\Delta S°$ to calculate $\Delta S°$, $\Delta G°$, and $\Delta H°$ for each of the following reactions at 298K, assuming all substances to be in their standard states:
(a) $2Ca(s) + O_2(g) \longrightarrow 2CaO(s)$
(b) $3C(graphite) + 4H_2(g) \longrightarrow C_3H_8(g)$
(c) $HCl(g) \longrightarrow H(g) + Cl(g)$
(d) $CH_4(g) + \frac{1}{2}O_2(g) \longrightarrow CH_3OH(\ell)$
(e) $H_2(g) + CO_2(g) \longrightarrow H_2O(g) + CO(g)$
(f) $2H_2S(g) + SO_2(g) \longrightarrow 2H_2O(g) + 3S$ (orthorhombic)

12. Use Tables 12.4 and 13.1 to calculate $\Delta S°$, $\Delta G°$, and $\Delta H°$ for each of the following reactions at 298K; all substances in their standard states:
(a) $2F_2(g) + 2H_2O(\ell) \longrightarrow 4HF(g) + O_2(g)$
(b) $3Zn(s) + Fe_2O_3(s) \longrightarrow 3ZnO(s) + 2Fe(s)$

(c) $CuO(s) + H_2(g) \longrightarrow Cu(s) + H_2O(\ell)$
(d) $Ca(OH)_2(s) \longrightarrow CaO(s) + H_2O(g)$
(e) $2C_6H_6(\ell) + 15O_2(g) \longrightarrow 12CO_2(g) + 6H_2O(g)$
(f) $2HBr(g) + Cl_2(g) \longrightarrow 2HCl(g) + Br_2(\ell)$

•**13.** Determine which of the reactions in Problem 11 are spontaneous in the indicated direction.

14. Which of the reactions in Problem 12 are spontaneous in the indicated direction?

•**15.** Use Table 13.2 to calculate $\Delta G°$ for the following reactions at 298K:
(a) $2Cu_2O(s) + O_2(g) \longrightarrow 4CuO(s)$
(b) $C_2H_5OH(\ell) \longrightarrow C_2H_4(g) + H_2O(g)$
(c) $C_3H_6(g) + H_2(g) \longrightarrow C_3H_8(g)$

16. Use Table 13.2 to calculate $\Delta G°$ for the following reactions at 298K:
(a) $BaO(s) + SO_3(g) \longrightarrow BaSO_4(s)$
(b) $2Al(s) + Fe_2O_3(s) \longrightarrow Al_2O_3(s) + 2Fe(s)$
(c) $O_3(g) \longrightarrow O_2(g) + O(g)$

•**17.** Calculate the equilibrium constants for the three reactions in Problem 15.

18. Calculate the equilibrium constants for the three reactions in Problem 16.

•**19.** Calculate $\Delta G°$ and the equilibrium constant for the following reaction performed at 410.°C,

$$2N_2O(g) \longrightarrow 2N_2(g) + O_2(g)$$

20. Determine the standard free energy change and the equilibrium constant at −100.°C for the reaction using data from the appropriate table:

$$H_2S(g) + \frac{3}{2}O_2(g) \longrightarrow H_2O(g) + SO_2(g)$$

Effect of Temperature on the Free Energy Change [21–32]

•**21.** What is the value of $\Delta G°$ at 150.K for a process if $\Delta H°$ is +1170 kJ/mol and $\Delta S°$ is +445 J/mol·K? What will be the spontaneous direction of reaction at this temperature?

22. Calculate the value of $\Delta G°$ at 450.K for a process for which $\Delta H°$ is −2270 kJ/mol and $\Delta S°$ is −250. J/mol·K. Is this reaction spontaneous under standard conditions at this temperature?

•**23.** Determine the temperature at which each of the following reactions would be at equilibrium with all participating species in their standard states:
(a) $C_3H_6(g) + H_2(g) \longrightarrow C_3H_8(g)$
(b) $CH_3COOH(\ell) \longrightarrow CH_4(g) + CO_2(g)$

24. Determine the temperature at which $K = 1$ for the following reactions:
(a) $C_2H_5OH(\ell) \longrightarrow C_2H_4(g) + H_2O(g)$
(b) $Br_2(\ell) \longrightarrow 2Br(g)$

•**25.** The equilibrium constant for the reaction

$$I_2(s) + Br_2(\ell) \longrightarrow 2IBr(g)$$

was found to be 4.52×10^{-2} at 25°C and 3.05×10^3 at 177°C. What is $\Delta H°$ for this reaction?

26. The equilibrium constant for the reaction

$$2NO_2(g) \rightleftharpoons N_2O_4(g)$$

is 7.49 at 300K and 0.270 at 350K. What is $\Delta H°$ for this reaction?

•**27.** Lime for mortar is made from limestone by the following two reactions:

$$CaCO_3(s) \longrightarrow CaO(s) + CO_2(g)$$

$$CaO(s) + H_2O(\ell) \longrightarrow Ca(OH)_2(s)$$

Use thermodynamic data to suggest the best conditions for performing these reactions.

28. Consider the thermodynamics for the possible preparation of benzene (C_6H_6) from carbon monoxide and hydrogen gas. Would this process be possible and, if so, what do you think would be the best conditions?

•**29.** The Haber process for the preparation of ammonia from hydrogen and nitrogen gases was formulated after a careful thermodynamic study. Consider the thermodynamics of this process and tell what you think would be the best reaction conditions.

30. When $MgCO_3(s)$ is heated, it decomposes to $MgO(s)$ and $CO_2(g)$.
(a) Use thermodynamic data to determine the temperature at which the decomposition first becomes feasible.

(b) By considering the data further, indicate why the reaction stops at MgO(s) and does not proceed further to Mg(s) and $\frac{1}{2}$ O_2(g).

•**31.** The vapor pressure of hexane (C_6H_{14}) is 40.0 torr at −2.3°C and 100.0 torr at 15.8°C. Calculate the average enthalpy of vaporization for hexane within this range.

32. The vapor pressure of ethyl acetate ($C_4H_8O_2$) is 400. torr at 59.3°C. Calculate the normal boiling point of ethyl acetate. (The average enthalpy of vaporization in that range is 34,900 J/mol.)

ADDITIONAL PROBLEMS [33–37]

33. Determine whether these statements are true or false:

(a) $G = S - TH$
(b) $G = H - TS$
(c) $G = HS - T$
(d) $H = E - PV$
(e) $\Delta G^\circ_f = \Delta H^\circ_f - T\Delta S^\circ_f$
(f) $\Delta E = q + w$
(g) At constant T and P, $\Delta G > 0$ in an irreversible process.
(h) At constant T, $\Delta S = q_{rev}/T$.
(i) If $q = 0$, $\Delta S \geq 0$ in any process.
(j) If a reversible process and an irreversible process connect the same initial and final states, then $q_{irrev} < q_{rev}$.

34. Calculate the work for the expansion of 1.0 L of an ideal gas to a final volume of 5.0 L, against a constant external pressure of 1.5 atm. What can you say about q_{rev} for this process?

•**35.** Calculate q, w, ΔE, ΔH, ΔS, and ΔG for the reversible vaporization of one mole of water at 100.°C and 1.00 atm. ΔH_{vap} for water is 40,660 J/mol.

36. Calculate ΔS_{sys}, ΔS_{univ}, ΔH, and ΔG for the melting of 50.0 g of ice at 0°C. The heat of fusion of water is 6025 J/mol.

37. (a) *n*-pentane is C_5H_{12} and the carbon atoms in the molecule are arranged in a chain: $CH_3—CH_2—CH_2—CH_2—CH_3$. At 298K and $P = 1$ atm, ΔH°_f of gaseous *n*-pentane = −146 kJ/mol, $\Delta G^\circ_f = -8.4$ kJ/mol. What is ΔS°_f, the entropy change in forming *n*-pentane from its elements, solid carbon and gaseous hydrogen?

(b) The absolute entropy of *n*-pentane at 298K and 1 atm is $S^\circ = 349$ J/K mol, and you calculated the value of ΔS°_f in (a). Solid carbon has very little entropy; neglecting it entirely, estimate the absolute entropy of a mole of H_2 gas at 298K and 1 atm.

(c) C_5H_{12} comes in two other forms, 1-methylbutane, which is

$$(CH_3)_2CH—CH_2—CH_3$$

and 2,2-dimethylpropane, which is $C(CH_3)_4$. Here are some thermodynamic data at 298K and 1 atm. Find the missing data:

	S°, (J/mol·K)	ΔH°_f, (kJ/mol)	ΔG°_f, (kJ/mol)
n-pentane	349	−146	−8.4
2-methylbutane	344	____	−13.4
2,2-dimethylpropane	____	−168	−17.6

(d) The three isomers of C_5H_{12} can interconvert,

From the thermodynamic data in (c), calculate the equilibrium constants for the reactions:

$$\text{2,2-dimethylpropane} \rightleftharpoons n\text{-pentane}$$
$$\text{2-methylbutane} \rightleftharpoons n\text{-pentane}$$
$$\text{2,2-dimethylpropane} \rightleftharpoons \text{2-methylbutane}$$

Multiple Principles [38–43]

38. For $MnCO_3$, the standard enthalpy of solution, ΔH°_{sol}, is −12.1 kJ/mol, and the standard entropy of solution, ΔS°_{sol}, is 66.1 J/mol·K. Calculate the solubility product of $MnCO_3$ at 298K.

•**39.** The solubility product of $PbBr_2$ at 18°C is 5.0×10^{-5} and the standard enthalpy of solution at this temperature is −42.0 kJ/mol. Calculate the standard entropy of solution of $PbBr_2$ at 18°C.

40. Calculate the equilibrium constant at 400.°C for the reaction

$$CO(g) + H_2O(g) \longrightarrow CO_2(g) + H_2(g)$$

assuming that $\Delta H°$ and $\Delta S°$ do not change significantly with temperature. Calculate the equilibrium partial pressures of all components at this temperature, starting with $CO(g)$ and $H_2O(g)$ each at a partial pressure of 0.100 atm.

•**41.** Determine the equilibrium constant at 25°C for the reaction

$$2N_2O(g) + 3O_2(g) \longrightarrow 4NO_2(g)$$

In an equilibrium mixture of these gases at 25°C, the partial pressures of both N_2O and NO_2 were each 0.500 atm. What was the partial pressure of oxygen?

42. The vapor pressures in torr for water near room temperature are as follows:

T (K)	273	283	293	303	313	323	333
p_{vap} **(torr)**	4.58	9.21	17.54	31.82	55.32	92.51	149.4

Create a table of T (K), $1/T$, and ln p. Then graph the data as ln p versus $1/T$, and find $\Delta H°$ and $\Delta S°$ for the vaporization process, recording your answers in kJ/mol and J/K, respectively.

•**43.** Here are some data for the following gas-phase equilibrium at several different temperatures:

$$SO_2(g) + \frac{1}{2}O_2(g) \rightleftharpoons SO_3(g)$$

T(K)	800	850	900	950	1000	1100
K_p	31.3	13.8	6.55	3.24	1.85	0.628

(a) Write the equilibrium expression for K_p.
(b) Calculate $\Delta H°$ and $\Delta S°$ algebraically over the range of T values.
(c) Calculate $\Delta H°$ and $\Delta S°$ graphically over the range of T values. (Plot ln K_p vs $1/T$.)

Applied Principles [44–47]

44. Carbon monoxide and steam can be used as a source of hydrogen, according to the following reaction:

$$CO(g) + H_2O(g) \longrightarrow CO_2(g) + H_2(g)$$

Taking $\Delta H°$ and $\Delta S°$ to be constants over the temperature range under consideration, calculate the fraction of CO converted to CO_2 at 100°C intervals from 300 to 800°C. In each case the initial partial pressure of H_2O is 0.900 atm and that of CO is 0.100 atm.

•**45.** Calculate the values of the equilibrium constant at 0°C, 200.°C, and 400.°C for the ammonia synthesis, assuming that $\Delta H°$ and $\Delta S°$ do not change significantly with temperature:

$$N_2(g) + 3H_2(g) \longrightarrow 2NH_3(g)$$

Comment on the effect of increasing the temperature on the extent of conversion of reactants into products.

46. It has been estimated that in every cubic kilometer of seawater there are about 6 g of gold in ions. At current gold prices of about $270/ounce, that would suggest a sizable fortune for anyone clever enough to mine it successfully. Should you try to exploit this gold mine or not? Use thermodynamic arguments to make your case.

47. In Chapter 9, we described an important water purification system called reverse osmosis, in which saltwater is forced through tubes that are made of semipermeable materials that allow pure water to pass through the walls. Explain why this highly effective and commercially successful process is not in conflict with the second law of thermodynamics.

ESTIMATES AND APPROXIMATIONS [48–51]

48. What is the approximate absolute entropy of a baseball?

49. Explain why the random, ceaseless motion of the molecules of a gas, although an example of "perpetual motion" of a kind, does not a contradict the second law of thermodynamics.

50. If it is true that every spontaneous process is accompanied by an increase in the entropy of the universe, then how do you account for the spontaneous freezing of supercooled water, for which the calculated entropy change must be negative?

51. Two separated blocks at different temperatures are connected by a wire that has an infinitesimally small thermal conductivity, so that heat flows infinitely slowly from the warmer to the cooler until temperature equilibrium is established. Has the process been carried out reversibly or not? How so?

FOR COOPERATIVE STUDY [52–53]

52. The value of the equilibrium constant for the reaction

$$2NO_2(g) \longrightarrow N_2O_4(g)$$

at several temperatures is the following:

Temperature (K)	Equilibrium Constant
300	7.49
350	2.70×10^{-1}
400	2.23×10^{-2}
450	3.21×10^{-3}

Use a plot of ln K versus $1/T$ to determine $\Delta H°$ and $\Delta S°$ of the reaction.

53. Suppose that while walking along a lonely stretch of beach you encounter a half-obscured footprint in the sand. Subsequent patient observation gradually reveals a slow increase in entropy of that region of the beach as wind and tidal water erode the shape of the footprint. Clearly, the entropy of the local area of the beach increases in this fashion. But it should also be clear that prior to your encountering the footprint, the footprint was more sharply defined—that is, in a lower entropy state. Reconcile the possibilities that the footprint appeared from either a sudden indentation by a passer-by prior to your walk on the beach or from some chance arrangement of sand particles by wind and water.

WRITING ABOUT CHEMISTRY [54–56]

54. This chapter deals heavily with the two driving forces in chemical reactions, enthalpy (or energy) and entropy. Many times these two driving forces are pulling in opposite directions, and calculation of the change in free energy tells which driving force is dominant under the given conditions. Construct an approximate model using concepts parallel to energy and entropy to explain some aspects of human behavior. For example, consider how close people spread their towels to each other on the beach. You may want to substitute some other concept for temperature. In this model the particles will be individual humans rather than atoms and molecules. Explain how your model works and where it fails.

55. Explore the concept of entropy (disorder or chaos) in a macroscopic sense. Some of the many possible topics are building a house, driving a truck, digging a hole, and countless others. Select three examples. Tell how you define the system and then use second-law concepts to explain why the process is or is not spontaneous. Does the second law provide a useful predictor for everyday activities? Explain why or why not.

56. Here is yet another statement of the second law of thermodynamics: The total entropy of a closed system is a monotonic function of time. In physical terms, this means that if we encounter a closed system *not* at equilibrium, then some time later—never earlier—it will be closer to equilibrium, and the entropy will have increased with the passage of time. Using as an example a chance encounter with a marble precariously positioned near the rim of a hemispherical bowl at a particular instant in time, write about the sequence of events that follow in terms of entropy as a function of time.

chapter fourteen 14

ELECTROCHEMISTRY

14.1 INTRODUCTION TO ELECTROCHEMISTRY

This chapter is concerned with the 200-year relationship between electrical phenomena and chemical reactions called **electrochemistry**. Despite its age, electrochemistry is a modern science of great interest. In their many forms, dry cells and storage batteries give us portable sources of electric current for purposes that range from driving toys to driving trains. There are cells that produce some of our most important commodities, such as magnesium, chlorine, and pure copper for electrical transmission. At the same time, electrochemical cells can provide the means to the simplest yet most precise method for determining free energy changes (ΔG) and equilibrium constants (K) in chemical reactions. The electrochemical determination of pH is the basis for one of the most important scientific instruments developed in the twentieth century, the Beckman pH meter. On an industrial scale, electrochemical reactions produce the aluminum needed for molding soft drink and beer cans and the nylon used for spinning textile fibers. Compact disks are metal-coated by electrochemical deposition, and electrochemical rusting and corrosion result in structural damage that can severely limit the lifetime of bridges and the cars that cross them.

The twenty-first century will be electricity's third and most exciting with the promise of a fuel-cell-driven energy economy based on hydrogen produced from the electrochemical decomposition of water. In the early years of the nineteenth century, Michael Faraday, one of the giants on whose shoulders electrochemistry rests, was asked by a politician of what value his new discovery would be. He quipped, "Why Sir, you could tax it," as indeed we have.

Oxidation–Reduction Reactions

Electrochemical reactions involve the flow of electrons from one chemical species to another. Such electron transfer reactions are called **oxidation–reduction reactions**, or simply **redox reactions**. The concept of **oxidation states** is particularly useful in analyzing redox reactions. Recall from Chapter 8 that for monatomic ions, the charge assigned to the ion is its oxidation state, and that pure elements are assigned the oxidation state of zero. Thus, when sodium atoms and chlorine molecules produce monatomic Na^+ and Cl^- ions, the transfer of electrons can be traced by following the change in oxidation states:

$$2Na(s) + Cl_2(g) \longrightarrow 2Na^+ + 2Cl^- \qquad \text{redox reaction}$$

In this example, the oxidation state of Na is zero; it becomes Na^+ with an oxidation state of $+1$, signifying a loss of one electron. Similarly, the oxidation state of each Cl atom in the Cl_2 molecule is zero. Electrons from two Na atoms are transferred, forming two Cl^- ions, each with an oxidation state of -1, signifying a gain of one electron per Cl atom. The following are useful rules regarding the oxidation states of elements in compounds:

- Group IA metals are always $+1$.
- Group IIA metals are always $+2$.
- Hydrogen is always $+1$ except in hydrides of active metals such as NaH and CaH_2, where it is -1.
- Fluorine is always -1.
- Oxygen is always -2 except when bonded to itself (H_2O_2) or to fluorine (OF_2), where it is -1 or $+2$ respectively.
- The sum of the oxidation states in a neutral compound is always zero; for an ion, the sum is always the charge on the ion.

Oxidation involves a loss of electrons and corresponds to an increase in oxidation state. **Reduction** involves a gain of electrons and corresponds to a decrease in oxidation state. In oxidation–reduction reactions, one reactant is oxidized as another is reduced. The reactant that is oxidized is called the **reducing agent**, or *reductant,* because it causes the other reactant to be reduced, whereas the reactant that is reduced is called the **oxidizing agent**, or *oxidant.* Thus, when Na reacts with Cl_2, Na is oxidized and is called the reducing agent, and Cl_2 is reduced and is called the oxidizing agent.

In the following aqueous redox reaction, the oxidation state of each element is shown beneath the equation:

H_2O	$+$ SO_3^{2-}	$+$ $2Fe^{3+}$	$\longrightarrow$ SO_4^{2-}	$+$ $2Fe^{2+}$	$+$ $2H^+$	redox reaction
$+1 -2$	$+4 -2$	$+3$	$+6 -2$	$+2$	$+1$	oxidation states

Notice that the oxidation states of two elements have changed. Sulfur atoms were oxidized from $+4$ in SO_3^{2-} to $+6$ in SO_4^{2-}. At the same time, iron atoms were reduced from $+3$ in the Fe^{3+} ion to $+2$ in the Fe^{2+} ion. We say that SO_3^{2-} is oxidized to SO_4^{2-}, and therefore SO_3^{2-} is the reducing agent or reductant. Similarly, Fe^{3+} is said to be the oxidizing agent or oxidant.

EXAMPLE 14.1

Identify the oxidizing and reducing agents in the following reaction:

$$ClO^-(aq) + 2H^+(aq) + Cu(s) \longrightarrow Cl^-(aq) + H_2O(\ell) + Cu^{2+}(aq)$$

COMMENT Begin by identifying the oxidation state of each atom:

$$\underset{+1-2}{ClO^-(aq)} + \underset{+1}{2H^+(aq)} + \underset{0}{Cu(s)} \longrightarrow \underset{-1}{Cl^-(aq)} + \underset{+1-2}{H_2O(\ell)} + \underset{+2}{Cu^{2+}(aq)}$$

Note that the oxidation state for Cu increased from 0 to +2, whereas that of Cl decreased from +1 for ClO^- to −1 for Cl^-.

SOLUTION Cu is oxidized and acts as the reducing agent, reducing the Cl atom in ClO^- from the +1 to −1 state. Because ClO^- is reduced, it is the oxidizing agent.

EXERCISE 14.1

Identify the oxidizing and reducing agents in the following reaction:

$$7H_2O + 2Cr^{3+} + XeF_6 \longrightarrow Cr_2O_7^{2-} + 14H^+ + Xe + 6F^-$$

ANSWER Oxidizing agent, XeF_6; reducing agent, Cr^{3+}

Half-Reactions in Redox Processes

It is often useful in studying redox processes to break them into pairs of **half-reactions**, one representing oxidation and the other reduction. A half-reaction cannot exist on its own because it includes electrons as reactants or products, and electrons cannot exist free under ordinary circumstances. As an example, consider the oxidation of Cl^- by Au(III):

$$6Cl^-(aq) + 2Au^{3+}(aq) \longrightarrow 3Cl_2(g) + 2Au(s)$$

Two chloride ions are oxidized to atoms and become a chlorine molecule by giving up one electron each, a process that can be represented by the following half-reaction:

$$2Cl^-(aq) \longrightarrow Cl_2(g) + 2e^- \qquad \text{oxidation half-reaction}$$

Au(III) accepts three electrons in a process that can be represented by the following half-reaction:

$$Au^{3+}(aq) + 3e^- \longrightarrow Au(s) \qquad \text{reduction half-reaction}$$

The number of electrons produced in the oxidation half-reaction must be made equal to the number of electrons consumed in the reduction half-reaction. In this case multiply the oxidation half-reaction by 3 and the reduction half-reaction by 2. Then, combine the two balanced half-reactions to obtain the overall reaction:

$$\begin{array}{ll} 3[2Cl^-(aq) \longrightarrow Cl_2(g) + 2e^-] & \text{oxidation half-reaction} \\ 2[Au^{3+}(aq) + 3e^- \longrightarrow Au(s)] & \text{reduction half-reaction} \\ \hline 6Cl^-(aq) + 2Au^{3+}(aq) \longrightarrow 3Cl_2(g) + 2Au(s) & \text{overall reaction} \end{array}$$

The six electrons transferred in the overall reaction were generated in the oxidation half-reaction and consumed in the reduction half-reaction.

Balancing Oxidation–Reduction Reactions

Some redox reactions can be balanced easily by simple inspection or trial-and-error. However, many redox reactions are much more difficult to balance. The situation is further complicated by the fact that balancing many redox reactions in aqueous solutions involves incorporation of H_2O, H^+, or OH^- from water into the reaction. Generally, in oxidation–reduction reactions, $H^+(aq)$ is written instead of $H_3O^+(aq)$, simplifying the balancing process, but remember that an H^+ ion in water(aq) is equivalent to $H_3O^+(aq)$. Here is a typical example of how to balance one such redox reaction:

$$I_2(s) + PbO_2(s) \longrightarrow IO_3^-(aq) + Pb^{2+}(aq)$$

Elements, charges, and electrons all need to be balanced. That is, the number of electrons produced in the oxidation half-reaction must be the same as the number of electrons used up in the reduction half-reaction. Balancing the equation will also depend on whether the reaction is taking place in acidic or basic solution. Proceed as follows:

Step 1: Write out as much of the reaction as is known. Then, write the oxidation state of every atom on each side of the reaction. In this way it is possible to tell if the reaction is an oxidation–reduction and, if it is, what is oxidized and what is reduced:

$$\underset{0}{I_2(s)} + \underset{+4-2}{PbO_2(s)} \longrightarrow \underset{+5-2}{IO_3^-(aq)} + \underset{+2}{Pb^{2+}(aq)}$$

I_2 is oxidized, with the I atoms going from the 0 to the +5 state. PbO_2 is reduced, with Pb going from the +4 to the +2 state.

Step 2: Separate the original reaction into those species involved in the oxidation half-reaction and those involved in the reduction half-reaction:

$$I_2(s) \longrightarrow IO_3^-(aq) \qquad \text{oxidation}$$

$$PbO_2(s) \longrightarrow Pb^{2+}(aq) \qquad \text{reduction}$$

Step 3: Balance all elements except oxygen and hydrogen by adding suitable coefficients:

$$I_2(s) \longrightarrow 2IO_3^-(aq)$$

$$PbO_2(s) \longrightarrow Pb^{2+}(aq)$$

Step 4: Now we need to balance the oxygen atoms and then the hydrogen atoms. At this point we will balance the equation, assuming that the reaction is taking place in acidic solution. Add water molecules as necessary to one side of the equation or the other to balance the oxygen atoms:

$$6H_2O(\ell) + I_2(s) \longrightarrow 2IO_3^-(aq)$$

$$PbO_2(s) \longrightarrow Pb^{2+}(aq) + 2H_2O(\ell)$$

and then add H^+ ions to balance the hydrogen atoms. Here is the result:

$$6H_2O(\ell) + I_2(s) \longrightarrow 2IO_3^-(aq) + 12H^+(aq)$$

$$4H^+(aq) + PbO_2(s) \longrightarrow Pb^{2+}(aq) + 2H_2O(\ell)$$

Step 5: Add electrons to each half-reaction to balance the charges on either side. In the oxidation, 10 negative charges are needed on the right side, so add 10 electrons to that side:

$$6H_2O(\ell) + I_2(s) \longrightarrow 2IO_3^-(aq) + 12H^+(aq) + 10e^-$$

In the reduction, 2 negative charges are needed on the left side, so add 2 electrons to that side:

$$4H^+(aq) + PbO_2(s) + 2e^- \longrightarrow Pb^{2+}(aq) + 2H_2O(\ell)$$

Step 6: The half-reactions now can be combined, but the number of electrons produced in the oxidation half-reaction must equal the number consumed in the reduction half-reaction. This can be accomplished by multiplying the reduction half-reaction by 5 and then adding the half-reactions together:

$$[6H_2O(\ell) + I_2(s) \longrightarrow 2IO_3^-(aq) + 12H^+(aq) + 10e^-]$$

$$5[4H^+(aq) + PbO_2(s) + 2e^- \longrightarrow Pb^{2+}(aq) + 2H_2O(\ell)]$$

$$6H_2O(\ell) + I_2(s) + 20H^+(aq) + 5PbO_2(s) \longrightarrow 2IO_3^-(aq) + 12H^+(aq) + 5Pb^{2+}(aq) + 10H_2O(\ell)$$

Step 7: Note the presence of H_2O and H^+ on both sides of the equation and cancel the excess of each. There are $6H_2O$ on the left and $10H_2O$ on the right, leaving a net $4H_2O$ on the right. Canceling H^+ in a similar fashion leaves a net $8H^+$ on the left:

$$I_2(s) + 8H^+(aq) + 5PbO_2(s) \longrightarrow 2IO_3^-(aq) + 5Pb^{2+}(aq) + 4H_2O(\ell)$$

Step 8: Check to see if the equation is completely balanced in terms of atoms and charges.The net reaction has 8+ charges and the same number of each of the atoms on each side.

This equation has been balanced for acidic solution. If the solution were basic, H^+ ions would react with OH^- to form H_2O. Therefore, for each H^+ ion in the balanced, acidic solution equation, add one OH^- to each side. This will convert each H^+ to H_2O and leave the same number of OH^- ions on the other side of the equation. Rewriting the reaction balanced in acidic solution,

$$I_2(s) + 8H^+(aq) + 5PbO_2(s) \longrightarrow 2IO_3^-(aq) + 5Pb^{2+}(aq) + 4H_2O(\ell)$$

acidic solution

add $8OH^-$ ions to each side, converting the $8H^+$ ions to H_2O.

$$I_2(s) + 8H_2O(\ell) + 5PbO_2(s) \longrightarrow 2IO_3^-(aq) + 5Pb^{2+}(aq) + 4H_2O(\ell) + 8OH^-(aq)$$

Now cancel out any H_2O molecules appearing on both sides of the equation, as in Step 7,

$$I_2(s) + 4H_2O(\ell) + 5PbO_2(s) \longrightarrow 2IO_3^-(aq) + 5Pb^{2+}(aq) + 8OH^-(aq)$$

and this is the balanced equation for basic solution.

EXAMPLE 14.2

Use the half-reaction method to balance the following reaction, which occurs in basic solution:

$$ClO_3^-(aq) + Zn(s) \longrightarrow ZnO_2^{2-}(aq) + ClO_2^-(aq)$$

COMMENT Assignment of oxidation states gives us the start of two half-reactions:

$$ClO_3^-(aq) \longrightarrow ClO_2^-(aq) \qquad \text{reduction}$$

$$+5-2 \qquad\qquad +3-2$$

$$Zn(s) \longrightarrow ZnO_2^{2-}(aq) \qquad \text{oxidation}$$

$$0 \qquad\qquad +2\;-2$$

SOLUTION First, balance oxygen and hydrogen as though in acidic solution for each half-reaction:

$$2H^+(aq) + ClO_3^-(aq) \longrightarrow ClO_2^-(aq) + H_2O(\ell)$$

$$2H_2O(\ell) + Zn(s) \longrightarrow ZnO_2^{2-}(aq) + 4H^+(aq)$$

Add electrons:

$$2H^+(aq) + ClO_3^-(aq) + 2e^- \longrightarrow ClO_2^-(aq) + 2H_2O(\ell)$$

$$2H_2O(\ell) + Zn(s) \longrightarrow 2e^- + ZnO_2^{2-}(aq) + 4H^+(aq)$$

In this case, the half-reactions can be added directly because each involves two electrons. Addition and cancellation of redundant H_2O and H^+ gives the acidic solution result:

$$H_2O(\ell) + ClO_3^-(aq) + Zn(s) \longrightarrow ClO_2^-(aq) + ZnO_2^{2-}(aq) + 2H^+(aq)$$

To convert this to the basic solution reaction, add two OH^- ions to each side,

$$H_2O(\ell) + ClO_3^-(aq) + Zn(s) + 2OH^-(aq) \longrightarrow ClO_2^-(aq) + ZnO_2^{2-}(aq) + 2H_2O(\ell)$$

and cancel the redundant H_2O,

$$ClO_3^-(aq) + Zn(s) + 2OH^-(aq) \longrightarrow ClO_2^-(aq) + ZnO_2^{2-}(aq) + H_2O(\ell)$$

EXERCISE 14.2(A)

Complete and balance the following redox reaction in basic solution;

$$Al(s) + IO_3^-(aq) \longrightarrow H_2AlO_3^-(aq) + I^-(aq)$$

ANSWER $2Al(s) + 2OH^-(aq) + IO_3^-(aq) + H_2O(\ell) \longrightarrow 2H_2AlO_3^-(aq) + I^-(aq)$

EXERCISE 14.2(B)

Complete and balance the following reaction in acid solution:

$$Cl^-(aq) + MnO_4^-(aq) \longrightarrow Cl_2(g) + Mn^{2+}(aq)$$

ANSWER $10Cl^-(aq) + 2MnO_4^-(aq) + 16H^+(aq) \longrightarrow$
$5Cl_2(g) + 2Mn^{2+}(aq) + 8H_2O(\ell)$

In some redox reactions an ion or molecule will oxidize an identical ion or molecule, and itself be reduced. This type of redox process is called a disproportionation reaction. A **disproportionation** has the same reactant in both half-reactions and is illustrated by the following example.

EXAMPLE 14.3

Complete and balance the following redox equation for reaction of $HS_2O_3^-$ ion in acid solution;

$$HS_2O_3^-(aq) \longrightarrow S(s) + HSO_4^-(aq)$$

COMMENT The oxidation state of sulfur changes from +2 in $HS_2O_3^-$ to 0 in S and to +6 in HSO_4^-. Therefore, $HS_2O_3^-$ is both oxidized and reduced and is the reactant in each half-reaction:

$HS_2O_3^-(aq) \longrightarrow S(s)$ reduction half-reaction

$HS_2O_3^-(aq) \longrightarrow HSO_4^-(aq)$ oxidation half-reaction

Completing and balancing each half-reaction gives

$$4e^- + 5H^+(aq) + HS_2O_3^-(aq) \longrightarrow 2S(s) + 3H_2O(\ell)$$

$$5H_2O(\ell) + HS_2O_3^-(aq) \longrightarrow 2HSO_4^-(aq) + 9H^+(aq) + 8e^-$$

Multiply the reduction half-reaction by 2, add the two half-reactions, and then remove excess H_2O and H^+.

SOLUTION $3HS_2O_3^-(aq) + H^+(aq) \longrightarrow 4S(s) + H_2O(\ell) + 2HSO_4^-(aq)$

The products of this reaction are produced as a result of two $HS_2O_3^-$ ions oxidizing a third.

EXERCISE 14.3

Complete and balance the equation for the disproportionation of phosphorus in basic solution:

$$P_4(s) \longrightarrow PH_3(g) + H_2PO_2^-(aq)$$

ANSWER $3OH^-(aq) + P_4(s) + 3H_2O(\ell) \longrightarrow 3H_2PO_2^-(aq) + PH_3(g)$

14.2 ELECTROCHEMICAL CELLS

It is possible to physically separate the oxidation and reduction halves of a redox reaction by conducting the electrons passing between them through a wire in an apparatus called an **electrochemical cell.** Electrochemical cells separate the oxidation and reduction half-reactions and can be used in two ways, galvanic and electrolytic, depending on whether the electrons are flowing in the spontaneous direction.

- In a **galvanic cell**, the flow of electrons is spontaneous, and it can be used as a source of stored energy, available on demand. Here potential energy from a chemical reaction is converted to electrical energy. Such cells are everywhere, from the lead–acid batteries used in the automobile to the common dry cell batteries that power flashlights, cameras, and calculators.
- In an **electrolytic cell**, electrical energy from an outside source that has been placed in the cell circuit can be used to drive nonspontaneous redox reactions. Here, electrical energy is converted into potential energy of chemical reaction. Such cells are tremendously important for the industrial-scale production of aluminum, for example, and many other metals, as well as sodium hydroxide, and chlorine.

Discussion of electrochemistry requires a few definitions. A **negative charge** is the result of an excess of electrons and a **positive charge** is the result of a deficiency of electrons. Electrons will flow along a conductor from the negative charge to the positive charge creating an electric current. We are familiar with the fact that unlike charges attract each other and like charges repel. The SI unit of charge is the coulomb C. The charge on a single electron is 1.6022×10^{-19} C. The **faraday** (F), the total charge of one mole of electrons, is 9.6485×10^4 C/mol:

$$\frac{1.6022 \times 10^{-19}\ \text{C}}{e^-} \times \frac{6.022 \times 10^{23}\ e^-}{\text{mol}} = 9.648 \times 10^4\ \frac{\text{C}}{\text{mol}} = 1\ \text{F}$$

The flow of charge is the **current** and is given in amperes (A), where one ampere is a current of one coulomb per second. Finally, the energy per coulomb is called the **electrical potential** and is given in volts, where one volt is an electrical potential of one joule per coulomb:

Summary of Electrical Units

The coulomb (C) is the SI unit of charge.

The electron charge is 1.6022×10^{-19} C.

The faraday (F) is the total charge of one mole of electrons and is 96,485 C (typically rounded off to 96,500 for problem solving).

The ampere (A) measures the flow of current. Amperes (A) = coulombs (C)/sec (s)

The volt (V) measures electrical potential. Volts (V) = joules(J)/coulomb (C)

Note that V × C has units of energy (joules).

Galvanic Cells

If a bar of zinc is placed in a copper sulfate solution, a redox reaction takes place for which the following half-reactions and overall equation can be written:

$Zn(s) \longrightarrow Zn^{2+}(aq) + 2e^-$	oxidation half-reaction
$2e^- + Cu^{2+}(aq) \longrightarrow Cu(s)$	reduction half-reaction
$Zn(s) + Cu^{2+}(aq) \longrightarrow Zn^{2+}(aq) + Cu(s)$	net redox reaction

There is a transfer of electrons, but under the described conditions, the transfer occurs directly from zinc atoms to copper ions at the surface of the bar of zinc. There is clearly no flow of electrons along any external conductor, and although this is a redox reaction, it is not an electrochemical reaction. The energy is transferred as heat. It can become an electrochemical reaction if the oxidation and reduction halves of the reaction are separated so that the transferred electrons flow from one half to the other by conduction along a wire. Such an arrangement is the galvanic cell shown in Figure 14.1.

Let us examine the various parts of this two-compartment cell. One side contains a bar of Zn dipping into a 1 M solution of Zn^{2+} formed by dissolving a zinc salt such as $ZnSO_4$ in water. The other side contains a bar of Cu dipping into a 1 M solution of Cu^{2+} prepared by dissolving a soluble copper salt such as $CuSO_4$ in water. The bars are called **electrodes**. They are dipping into **electrolyte** solutions made conducting by the presence of ions.

The zinc electrode is made negative by the electrons produced from the oxidation half-reaction while in the other compartment electrons are consumed, making the copper electrode positive:

$$Zn(s) \longrightarrow Zn^{2+}(aq) + 2e^-$$

Oxidation produces electrons; Zn electrode is negative.

$$Cu^{2+}(aq) + 2e^- \longrightarrow Cu(s)$$

Reduction consumes electrons; Cu electrode is positive.

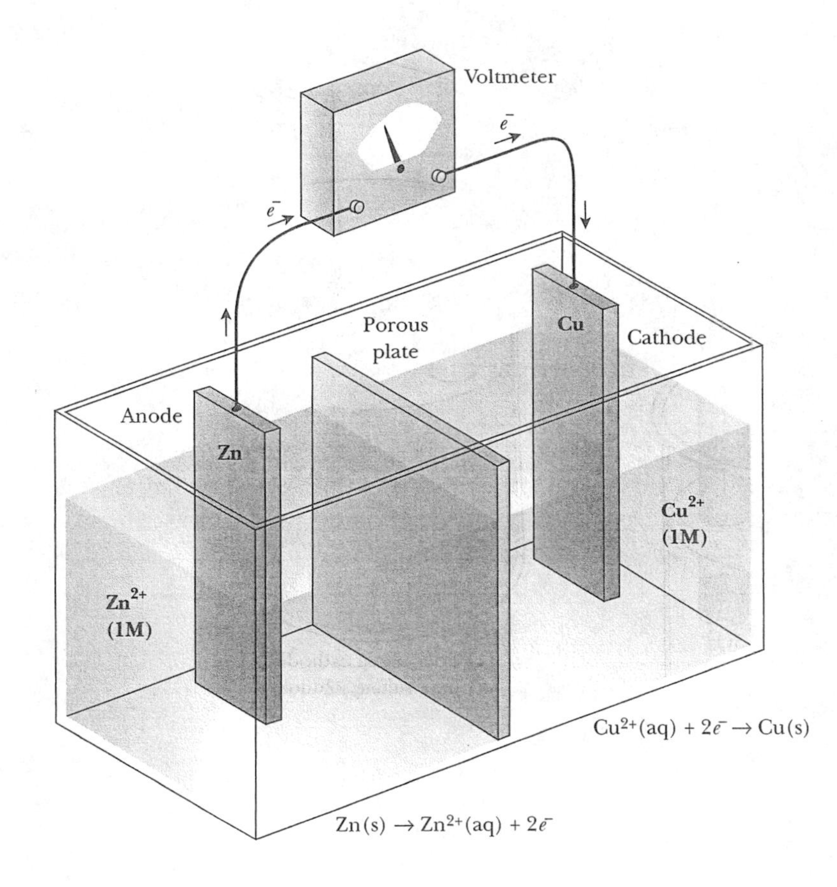

FIGURE 14.1 An electrochemical cell operating spontaneously is called galvanic. Here, the zinc electrode corrodes and copper plates out of solution onto the copper electrode. When the electrolyte solutions in each compartment are 1 M in Cu^{2+} and Zn^{2+} ions, the cell potential is 1.10 V. A clay plate or semipermeable barrier separates the two chambers, or half-cells, completing the circuit. Electrons flow along the wire in the external circuit, and ions move across the barrier, internally.

To balance this electrical inequality, electrons from the negative Zn electrode flow along the wire to the positive Cu electrode. During the process, the difference in the electrical potential between the two electrodes registers on a voltmeter placed in the circuit.

The compartments are separated by a porous plate that allows the passage of ions. If that were not the case, no current could flow because the electrical circuit would be incomplete. Because the plate separating the compartments is porous, migration of positive ions from the Zn compartment to the Cu compartment—and negative ions in the opposite direction—balances the charges and allows current to flow through the wire. There cannot be an excess of positively charged ions in the Zn compartment or a corresponding shortage of positively charged ions in the Cu compartment.

A salt bridge (Fig. 14.2) is another device for completing the electrical circuit. The **salt bridge** in this figure is a tube containing a solution of a salt such as potassium chloride that does not react under the cell conditions. The migration of ions between the salt bridge and the electrolyte solutions maintains electrical neutrality, and this bridge can be used in place of a porous plate. Glass wool or a cotton plug prevents the salt solution in the bridge from running freely into the electrolyte compartments.

Electrochemical Cell Notation

In any electrochemical cell, the **anode** is the electrode at which oxidation takes place; the **cathode** is the electrode at which reduction takes place. During the

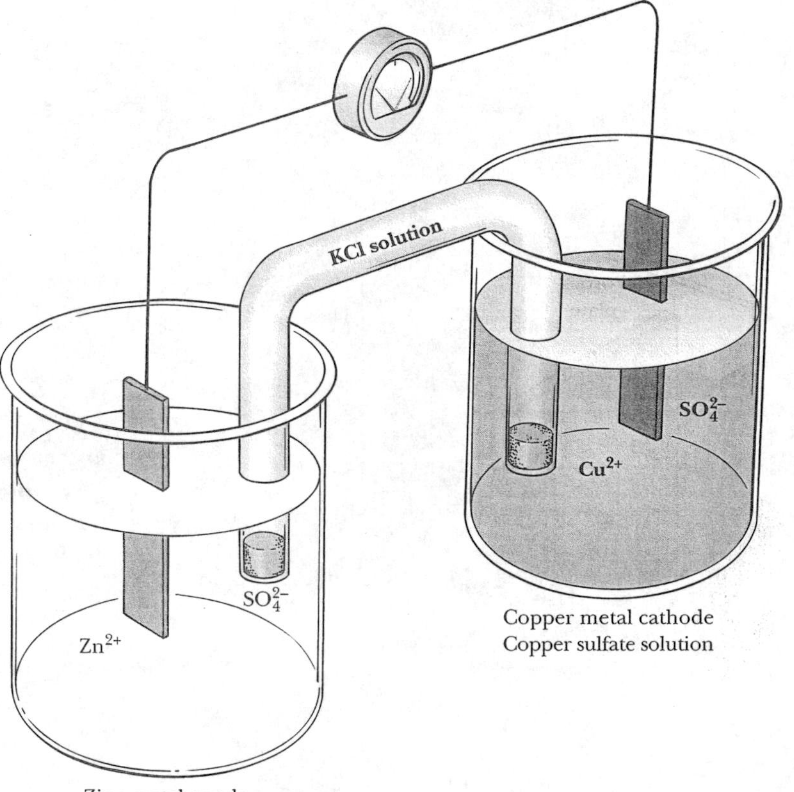

FIGURE 14.2 A salt bridge connecting two separated half-cells. In this version of the galvanic cell described in Figure 14.1, two half-reactions have been established in physically separated beakers. The salt bridge, filled with a potassium chloride solution and plugged at the ends with tightly packed glass wool or cotton, has been placed across the two beakers, allowing ions to pass freely between half-cells, completing the electrical circuit, and allowing current to flow.

spontaneous electrochemical reaction, anions migrate to the anode and cations migrate to the cathode. To describe a half-cell reaction we write half-cell reactants separated from half-cell products by a vertical line if the reactants and products are in different phases and by a comma if they are components of a solution:

half-cell reactants | half-cell products in different phases

half-cell reactants, half-cell products in solution

Thus, when we place a copper electrode in a $CuSO_4$ solution, we write the reduction half-cell reaction taking place in the cathode compartment as follows, using the vertical line because the Cu^{2+} is in solution and the Cu is a solid:

$$Cu^{2+}(aq) + 2e^- \longrightarrow Cu(s) \qquad \text{cathode half-cell reaction}$$

$$Cu^{2+}(aq) \mid Cu(s) \qquad \text{cathode half-cell notation}$$

To complete the galvanic cell, an oxidation half-cell reaction is needed. For example, place the reaction involving zinc metal and a $ZnSO_4$ solution in the anode compartment:

$$Zn(s) \longrightarrow Zn^{2+}(aq) + 2e^- \qquad \text{anode half-cell reaction}$$

$$Zn(s) \mid Zn^{2+}(aq) \qquad \text{anode half-cell notation}$$

The complete notation scheme for the cell is written with the anode on the left and a double vertical line denoting the presence of a porous plate or salt bridge. Finally, the concentrations of ions in solution, or pressures of gases, are given. Thus,

$$\underbrace{Zn(s) \mid Zn^{2+}\ (1\ M)}_{\text{anode (oxidation)}} \parallel \underbrace{Cu^{2+}\ (1\ M) \mid Cu(s)}_{\text{cathode (reduction)}} \qquad \text{cell notation}$$

If none of the components in a half-reaction is a metal, it is necessary to make use of an inert electrode material such as platinum (or much cheaper carbon) that serves only as a surface across which electrons are transferred. For example, in the half-cell reaction

$$Fe^{2+} \longrightarrow Fe^{3+} + e^-$$

the electrolyte would contain Fe^{3+} and Fe^{2+} ions. The electron produced would be transferred to an inert electrode. If the inert electrode is platinum, then the notation for this half-cell at standard conditions is written as follows.

$$Pt(s) \mid Fe^{2+}(1\ M),\ Fe^{3+}(1\ M)$$

In this case the Fe^{2+} and Fe^{3+} are separated by a comma because they are both components of a single solution.

Cell Potentials

Current flows spontaneously in galvanic cells, and a voltmeter wired into the circuit measures the cell potential E_{cell} in volts. By definition, a spontaneous cell reaction has a positive potential. The greater the potential, the more complete the spontaneous conversion of reactants to products. If the concentrations of the ions in the cell reaction are 1 M, and partial pressures of any gases are

1 atm, then the potential is the special case called the **standard cell potential** (E°_{cell}).

The experimental value of E°_{cell} = for the Zn/Cu cell is 1.100 V:

$$Zn(s) + Cu^{2+}(1\ M) \longrightarrow Cu(s) + Zn^{2+}(1\ M) \qquad E^\circ_{cell} = 1.100\ V$$

The potential of a cell is an *intensive* property. It does not depend on the size of the cell or the total number of electrons transferred because potential is in units of energy/coulomb of charge. Therefore, the standard cell potential is 1.100 V no matter how large the cell or how we balance the equation. If we change the coefficients, the potential does not change, for example,

$$2Zn(s) + 2Cu^{2+}(1\ M) \longrightarrow 2Cu(s) + 2Zn^{2+}(1\ M) \qquad E^\circ_{cell} = 1.100\ V$$

$$3Zn(s) + 3Cu^{2+}(1\ M) \longrightarrow 3Cu(s) + 3Zn^{2+}(1\ M) \qquad E^\circ_{cell} = 1.100\ V$$

$$Zn(s) + Cu^{2+}(1\ M) \longrightarrow Cu(s) + Zn^{2+}(1\ M) \qquad E^\circ_{cell} = 1.100\ V$$

The sign of the standard cell potential changes from plus to minus if we write the reaction in reverse, indicating that the reverse reaction is nonspontaneous.

$$Cu(s) + Zn^{2+}(1\ M) \longrightarrow Zn(s) + Cu^{2+}(1\ M) \qquad E^\circ_{cell} = -1.100\ V$$

nonspontaneous process

A reaction is spontaneous in one direction and nonspontaneous in the other.

E° Values for Half-Reactions

In our Zn/Cu electrochemical cell the half-cell processes in the two compartments can be considered independently. Each half-cell has its own E° value, with E° for the overall cell process being simply the sum of the values of E° of each half-cell:

$$E^\circ_{cell} = E^\circ_{oxidation} + E^\circ_{reduction}$$

$$E^\circ_{cell} = 1.100\ V = E^\circ_{Zn/Zn^{2+}} + E^\circ_{Cu^{2+}/Cu}$$

However, although it is easy to determine the sum of the E° values of two half-reactions, one cannot determine E° for either one alone. The problem is that the states of the two electrons in the half-reactions are not specified:

$$Zn(s) \longrightarrow Zn^{2+}(aq) + 2e^-$$

$$Cu^{2+}(aq) + 2e^- \longrightarrow Cu(s)$$

For the overall reaction, this is not a problem because the two electrons generated in the oxidation half-reaction are simply picked up by the reduction half-reaction, and the states of all reactants and products are known.

In order to determine useful values of E° for half-reactions, it is necessary to establish a standard half-reaction against which all other half-reactions can be measured. The universally agreed-upon standard half-reaction is the hydrogen half-cell, and its E° is set as exactly 0 V in either direction:

$$H_2(1\ atm) \longrightarrow 2H^+(1\ M) + 2e^- \qquad E^\circ = 0\ V \qquad \text{oxidation half-reaction}$$

$$2H^+(1\ M) + 2e^- \longrightarrow H_2(1\ atm) \qquad E^\circ = 0\ V \qquad \text{reduction half-reaction}$$

The hydrogen half-cell most often used is shown on the left-hand side of Figure 14.3. A piece of platinum acts as the conducting surface for electron transfer. Platinum is inert under these conditions because $H_2(g)$ is more readily oxidized. Therefore, the cell reaction is written as follows:

$$H_2(g) + Cu^{2+}(aq) \longrightarrow 2H^+(aq) + Cu(s) \qquad E° = 0.337 \text{ V}$$

The complete notation scheme representing such a cell is written as follows:

$$Pt(s) \mid H_2(1 \text{ atm}) \mid H^+(1 \text{ M}) \parallel Cu^{2+}(1 \text{ M}) \mid Cu(s)$$

The standard potential for the cell reaction is

$$E°_{cell} = E°_{H_2/H^+} + E°_{Cu^{2+}/Cu}$$

And because $E°_{H_2/H^+}$ is 0 V, it follows that $E°_{Cu^{2+}/Cu} = 0.337$ V.

Substituting into the equation for $E°$ for the Zn/Cu cell and solving for $E°_{Zn/Zn^{2+}}$ gives the following result (remember that $E°_{cell} = 1.100$ V):

$$E°_{cell} = E°_{Zn/Zn^{2+}} + E°_{Cu^{2+}/Cu}$$

$$E°_{Zn/Zn^{2+}} = E°_{cell} - E°_{Cu^{2+}/Cu}$$

$$E°_{Zn/Zn^{2+}} = 1.100 \text{ V} - 0.337 \text{ V} = 0.763 \text{ V}$$

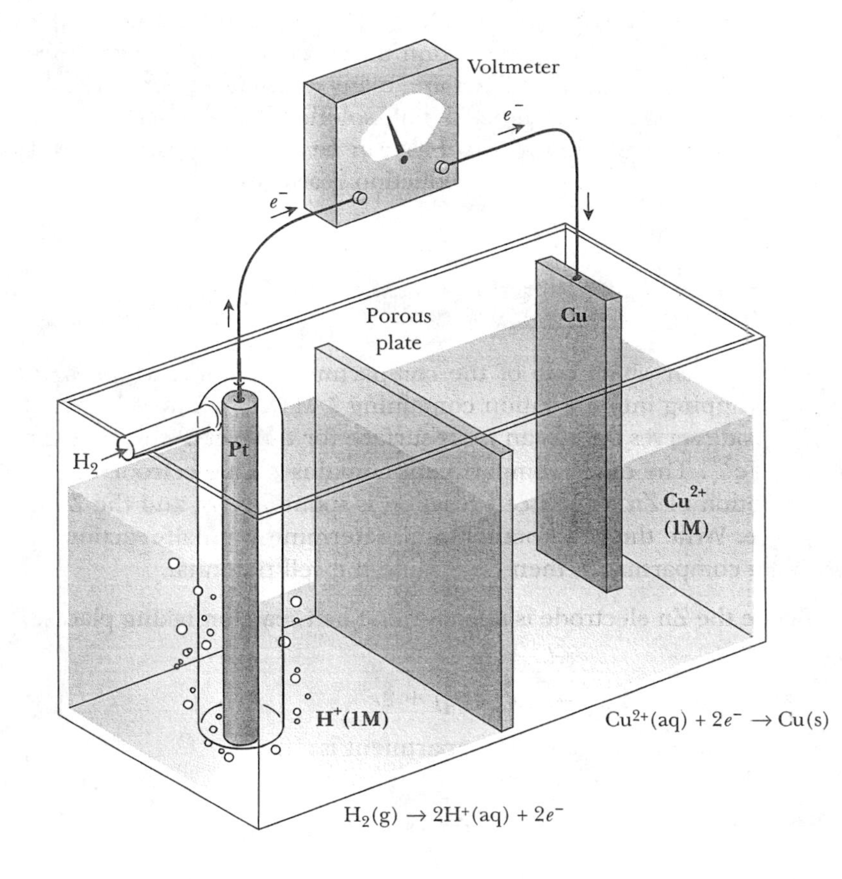

FIGURE 14.3 Standard hydrogen electrode, used for measuring half-cell potentials. Here is a Cu^{2+} (1 M) | Cu(s) half-cell coupled to a standard hydrogen electrode. Such an arrangement can be used to determine the half-cell potential and the spontaneous direction of the reaction, which, in this case, leads to the reduction of Cu^{2+} ions by hydrogen gas and the assignment of a standard half-cell potential of 0.337 V.

We now have $E°$ values for three half-reactions, which can be listed as follows:

Half-Reaction	$E°$	Process
$Zn(s) \longrightarrow Zn^{2+}(aq) + 2e^-$	0.763 V	Oxidation half-reaction
$Cu^{2+}(aq) + 2e^- \longrightarrow Cu(s)$	0.337 V	Reduction half-reaction
$H_2(g) \longrightarrow 2H^+(aq) + 2e^-$	0 V	Oxidation half-reaction

For purposes of comparison it is useful to have all reactions written in the same direction. Current practice is to tabulate half-reactions as reduction potentials. This is simply done by reversing the two oxidation half-reactions and changing the signs for their corresponding $E°$ values:

Half-Reaction	$E°$	Process
$Zn^{2+}(aq) + 2e^- \longrightarrow Zn(s)$	−0.763 V	Reduction half-reaction
$2H^+(aq) + 2e^- \longrightarrow H_2(g)$	0 V	Reduction half-reaction
$Cu^{2+}(aq) + 2e^- \longrightarrow Cu(s)$	0.337 V	Reduction half-reaction

The $E°$ value in each of these cases is the standard reduction potential for the half-reaction. The **standard reduction potential** is the electrical potential assigned to a reduction half-reaction under standard conditions (all partial pressures = 1 atm, all concentrations = 1 M). Note that the negative value of the $E°$ for the reduction of Zn^{2+} only means that the reduction of Zn^{2+} is 0.763 V less spontaneous than the reduction of H^+. There are many other important half-reactions, and a number of these in acidic and basic solutions are given in Table 14.1. The standard potentials given in Table 14.1 can be used to calculate cell potentials for a large number of oxidation–reduction reactions.

EXAMPLE 14.4

Figure 14.4 shows a cell in which one of the compartments contains an inert platinum electrode dipping into a solution containing 1 M Fe^{3+} and 1 M Fe^{2+}. The platinum electrode serves only as an inert surface for the electron transfer between Fe^{2+} and Fe^{3+}. The other compartment contains a Zn electrode dipping into a 1 M solution of Zn^{2+}. The cell reaction is spontaneous, and the Zn electrode is negative. Write the cell notation and determine the half-reactions taking place in each compartment, then determine the cell potential.

COMMENT Because the Zn electrode is negative, the half-reaction taking place in that compartment is

$$Zn(s) \longrightarrow Zn^{2+}(aq) + 2e^-$$

The reduction taking place in the other compartment is

$$Fe^{3+}(aq) + e^- \longrightarrow Fe^{2+}(aq)$$

TABLE 14.1 Standard Reduction Potentials

Acid Solutions

Half-Reaction	$E°$(V)	Half-Reaction	$E°$(V)
$Li^+ + e^- \rightleftharpoons Li$	−3.045	$Fe(CN)_6^{3-} + e^- \rightleftharpoons Fe(CN)_6^{4-}$	0.36
$Ca^{2+} + 2e^- \rightleftharpoons Ca$	−2.866	$Cu^+ + e^- \rightleftharpoons Cu$	0.521
$Na^+ + e^- \rightleftharpoons Na$	−2.714	$I_2 + 2e^- \rightleftharpoons 2I^-$	0.535
$La^{3+} + 3e^- \rightleftharpoons La$	−2.52	$I_3^- + 2e^- \rightleftharpoons 3I^-$	0.536
$Mg^{2+} + 2e^- \rightleftharpoons Mg$	−2.36	$MnO_4^- + e^- \rightleftharpoons MnO_4^{2-}$	0.558
$AlF_6^{3-} + 3e^- \rightleftharpoons Al + 6F^-$	−2.07	$H_3AsO_4 + 2H^+ + 2e^- \rightleftharpoons HAsO_2 + 2H_2O$	0.560
$Al^{3+} + 3e^- \rightleftharpoons Al$	−1.66	$O_2(g) + 2H^+ + 2e^- \rightleftharpoons H_2O_2$	0.695
$SiF_6^{2-} + 4e^- \rightleftharpoons Si + 6F^-$	−1.24	$PtCl_4^- + 4e^- \rightleftharpoons Pt + 4Cl^-$	0.73
$V^{2+} + 2e^- \rightleftharpoons V$	−1.19	$Fe^{3+} + e^- \rightleftharpoons Fe^{2+}$	0.77
$Mn^{2+} + 2e^- \rightleftharpoons Mn$	−1.18	$Hg_2^{2+} + 2e^- \rightleftharpoons 2Hg$	0.788
$Zn^{2+} + 2e^- \rightleftharpoons Zn$	−0.763	$Ag^+ + e^- \rightleftharpoons Ag$	0.799
$Cr^{3+} + 3e^- \rightleftharpoons Cr$	−0.744	$2Hg^{2+} + 2e^- \rightleftharpoons Hg_2^{2+}$	0.920
$Fe^{2+} + 2e^- \rightleftharpoons Fe$	−0.44	$Br_2 + 2e^- \rightleftharpoons 2Br^-$	1.087
$Cr^{3+} + e^- \rightleftharpoons Cr^{2+}$	−0.41	$2IO_3^- + 12H^+ + 10e^- \rightleftharpoons I_2 + 6H_2O$	1.19
$PbSO_4 + 2e^- \rightleftharpoons Pb + SO_4^{2-}$	−0.359	$O_2 + 4H^+ + 4e^- \rightleftharpoons 2H_2O$	1.23
$Co^{2+} + 2e^- \rightleftharpoons Co$	−0.277	$Cr_2O_7^{2-} + 14H^+ + 6e^- \rightleftharpoons 2Cr^{3+} + 7H_2O$	1.33
$Ni^{2+} + 2e^- \rightleftharpoons Ni$	−0.250	$Cl_2 + 2e^- \rightleftharpoons 2Cl^-$	1.36
$Sn^{2+} + 2e^- \rightleftharpoons Sn$	−0.138	$PbO_2 + 4H^+ + 2e^- \rightleftharpoons Pb^{2+} + 2H_2O$	1.45
$Pb^{2+} + 2e^- \rightleftharpoons Pb$	−0.126	$Au^{3+} + 3e^- \rightleftharpoons Au$	1.50
$Fe^{3+} + 3e^- \rightleftharpoons Fe$	−0.037	$MnO_4^- + 8H^+ + 5e^- \rightleftharpoons Mn^{2+} + 4H_2O$	1.51
$2D^+ + 2e^- \rightleftharpoons D_2$	−0.0034	$Ce^{4+} + e^- \rightleftharpoons Ce^{3+}$	1.61
$2H^+ + 2e^- \rightleftharpoons H_2$	0 (definition)	$MnO_4^- + 4H^+ + 3e^- \rightleftharpoons MnO_2 + 2H_2O$	1.679
$Cu^{2+} + e^- \rightleftharpoons Cu^+$	0.153	$H_2O_2 + 2H^+ + 2e^- \rightleftharpoons 2H_2O$	1.776
$AgCl + e^- \rightleftharpoons Ag + Cl^-$	0.222	$O_3 + 2H^+ + 2e^- \rightleftharpoons O_2 + H_2O$	2.07
$Hg_2Cl_2 + 2e^- \rightleftharpoons 2Hg + 2Cl^-$	0.2676	$F_2 + 2e^- \rightleftharpoons 2F^-$	2.87
$Cu^{2+} + 2e^- \rightleftharpoons Cu$	0.337	$H_4XeO_6 + 2H^+ + 2e^- \rightleftharpoons XeO_3 + 3H_2O$	3.0

Basic Solutions

Half-Reaction	$E°$(V)	Half-Reaction	$E°$(V)
$H_2AlO_3^- + 2H_2O + 3e^- \rightleftharpoons Al + 4OH^-$	−2.33	$IO_3^- + 3H_2O + 6e^- \rightleftharpoons I^- + 6OH^-$	0.26
$CrO_2^- + 2H_2O + 3e^- \rightleftharpoons Cr + 4OH^-$	−1.27	$ClO_3^- + H_2O + 2e^- \rightleftharpoons ClO_2^- + 2OH^-$	0.33
$ZnO_2^{2-} + 2H_2O + 2e^- \rightleftharpoons Zn + 4OH^-$	−1.21	$ClO_4^- + H_2O + 2e^- \rightleftharpoons ClO_3^- + 2OH$	0.36
$Sn(OH)_6^{2-} + 2e^- \rightleftharpoons HSnO_2^- + H_2O + 3OH^-$	−0.93	$O_2 + 2H_2O + 4e^- \rightleftharpoons 4OH^-$	0.40
$HSnO_2^- + H_2O + 2e^- \rightleftharpoons Sn + 3OH^-$	−0.91	$HO_2^- + H_2O + 2e^- \rightleftharpoons 3OH^-$	0.88
$2H_2O + 2e^- \rightleftharpoons H_2 + 2OH^-$	−0.8277	$ClO^- + H_2O + 2e^- \rightleftharpoons Cl^- + 2OH^-$	0.89
$HPbO_2^- + H_2O + 2e^- \rightleftharpoons Pb + 3OH^-$	−0.54	$HXeO_4^- + 3H_2O + 6e^- \rightleftharpoons Xe + 7OH^-$	0.9
$Co(OH)_3 + e^- \rightleftharpoons Co(OH)_2 + OH^-$	0.17		

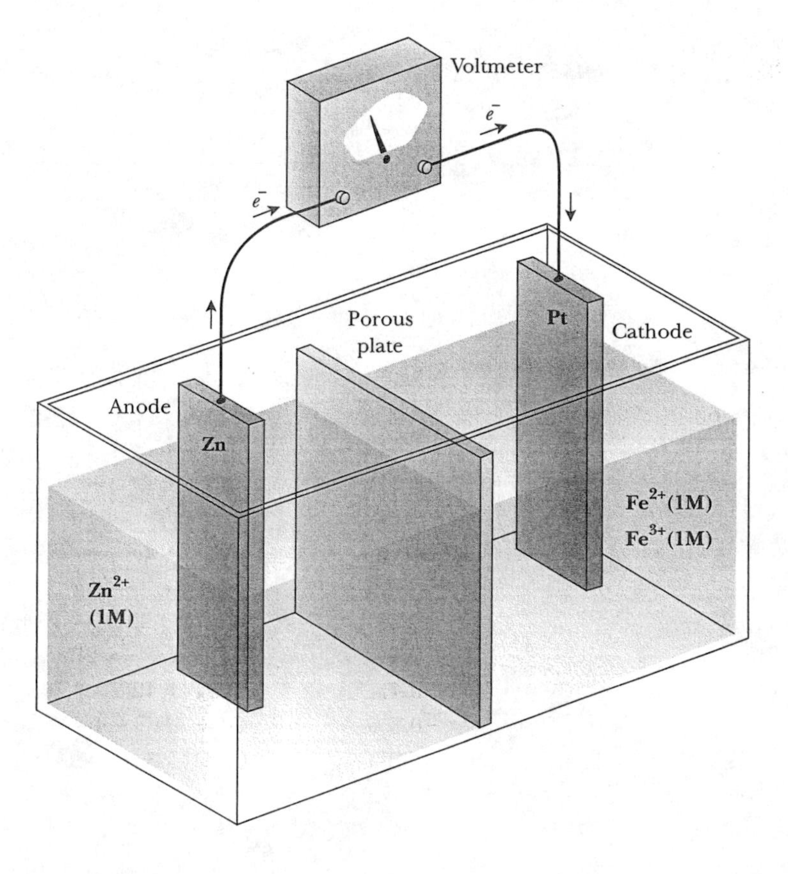

FIGURE 14.4 Galvanic cell operating with an inert electrode. Here, an inert electrode provides a surface across which electrons are transferred during the Fe^{3+}/Fe^{2+} half-cell reaction taking place in the right-hand chamber. Either platinum or carbon could typically be used as the inert electrode, in contrast to the zinc electrode in the other chamber, which corrodes as the cell reaction proceeds.

The Pt electrode is the cathode, the site of the reduction, and the overall cell reaction is

$$Zn(s) + 2Fe^{3+}(aq) \longrightarrow Zn^{2+}(aq) + 2Fe^{2+}(aq)$$

SOLUTION The cell notation is

$$Zn(s) \mid Zn^{2+}(1\ M) \parallel Fe^{3+}(1\ M),\ Fe^{2+}(1\ M) \mid Pt(s)$$

Adding up the standard potentials for the half-reactions, one written for oxidation and the other for reduction, gives the measured standard potential for the cell:

$$E^\circ_{cell} = E^\circ_{Zn/Zn^{2+}} + E^\circ_{Fe^{3+}/Fe^{2+}}$$

Because E° is an intensive property, not dependent upon the amount of matter undergoing reaction, $E^\circ_{Fe^{3+}/Fe^{2+}}$ is *not* multiplied by a coefficient, even though two moles of Fe^{3+} must be reduced for every one mole of Zn oxidized. Now, we can use the standard reduction potentials in Table 14.1 to solve for E°_{cell}. Remember to change the sign of the potential for the oxidation half-reaction:

$$E^\circ_{cell} = E^\circ_{Zn/Zn^{2+}} + E^\circ_{Fe^{3+}/Fe^{2+}} = 0.763\ V + 0.77\ V = 1.53\ V$$

EXERCISE 14.4

A cell contains one compartment with a Pt electrode dipping into a solution containing Hg^{2+} and Hg_2^{2+}, each at 1 M concentration. The other compartment contains a Cu electrode in a 1 M Cu^{2+} solution. The Pt electrode is positive in this spontaneous cell reaction. Give the half-reactions taking place at the anode and the cathode, and the cell notation. Determine the overall cell reaction and $E°$, the standard potential of the cell.

ANSWERS

$Cu(s) \longrightarrow Cu^{2+}(aq) + 2e^-$	anode (oxidation)
$2Hg^{2+}(aq) + 2e^- \longrightarrow Hg_2^{2+}(aq)$	cathode (reduction)
$Cu(s) + 2Hg^{2+}(aq) \longrightarrow Cu^{2+}(aq) + Hg_2^{2+}(aq)$	overall

$Cu(s) \mid Cu^{2+}(1\ M) \parallel Hg^{2+}(1\ M), Hg_2^{2+}(1\ M) \mid Pt(s)$

$E°_{cell} = 0.583\ V$

EXAMPLE 14.5

Consider a cell consisting of two compartments. One has a Pb electrode dipping into a 1 M solution of Pb^{2+}. The other has a Pt electrode dipping into a solution that is 1 M in both Cr^{3+} and Cr^{2+}. The two compartments are joined by a salt bridge, and the electrodes are connected through a voltmeter. First, write the balanced equations for the half-reactions as they take place in each cell and for the spontaneous cell reaction. Give the notation of the cell. Then, identify the anode and the cathode and their respective polarities (charges). Finally, indicate in which direction electrons are flowing, and determine $E°_{cell}$. The required half-reactions are found in Table 14.1:

$$Cr^{3+}(aq) + e^- \longrightarrow Cr^{2+}(aq) \qquad E° = -0.41\ V$$

$$Pb^{2+}(aq) + 2e^- \longrightarrow Pb(s) \qquad E° = -0.126\ V$$

COMMENT In order to produce the complete cell reaction, one of these half-reactions will have to be reversed, and the resulting overall $E°$ must be positive for the reaction to be spontaneous.

SOLUTION To get a positive value of $E°$, reverse the half-reaction that is less positive or more negative. In this case, the Cr^{3+}/Cr^{2+} half-reaction is the more negative, and reversal of this half-reaction gives $E°_{cell} = 0.41 - 0.126 = +0.28\ V$. Thus, an overall spontaneous reaction results from reversing the Cr^{3+}/Cr^{2+} reduction half-reaction to make it an oxidation half-reaction. The oxidation is taking place at the Pt electrode, making it the anode. The half-reaction taking place at the anode is

$$Cr^{2+}(aq) \longrightarrow Cr^{3+}(aq) + e^-$$

and the electrons produced there cause it to be negatively charged. Similarly, the Pb electrode is the cathode with the half-reaction

$$Pb^{2+}(aq) + 2e^- \longrightarrow Pb(s)$$

and a positive charge. Multiplying the Cr^{2+}/Cr^{3+} half-reaction by 2 to balance the electrons and then adding the half-reactions gives the overall reaction:

$$Pb^{2+}(aq) + 2Cr^{2+}(aq) \longrightarrow Pb(s) + 2Cr^{3+}(aq)$$

The cell notation is $Pt(s) \mid Cr^{2+}(1\ M), Cr^{3+}(1\ M) \parallel Pb^{2+}(1\ M) \mid Pb(s)$

The electrons in the external circuit flow from the Pt electrode through the voltmeter to the Pb electrode. The cell potential E°_{cell} is +0.28 V.

EXERCISE 14.5

A galvanic cell has two compartments connected by a salt bridge and an external circuit through a voltmeter. One compartment contains a Mn electrode dipping into a 1 M solution of Mn^{2+} and the other contains a Pt electrode in a solution that is 1 M in both Cr^{3+} and Cr^{2+}. State the half-reactions in each compartment, identify the spontaneous cell reaction, and calculate E°_{cell}. Give the cell notation. Tell which electrode is the anode, which the cathode, what are the electric charges on the electrodes, and in what direction the electrons flow in the external circuit.

ANSWERS $Mn(s) \longrightarrow Mn^{2+}(aq) + 2e^-$

$Cr^{3+}(aq) + e^- \longrightarrow Cr^{2+}(aq)$

$Mn(s) + 2Cr^{3+}(aq) \longrightarrow Mn^{2+}(aq) + 2Cr^{2+}(aq)$

$Mn(s) \mid Mn^{2+}\ (1\ M) \parallel Cr^{3+}\ (1\ M), Cr^{2+}\ (1\ M) \mid Pt(s)$

$E^\circ_{cell} = 0.77\ V$

Mn electrode, anode, negative; Pt electrode, cathode, positive.

Electrons flow in the external circuit from Mn to Pt.

The standard reduction potentials in Table 14.1 can be used to determine if a number of different kinds of reactions are spontaneous, provided that the reactions are conducted under standard conditions. The following examples and exercises all assume standard conditions, that is, all reactants and products are 1 M if dissolved in water solution, and all gases have partial pressures of 1 atm.

EXAMPLE 14.6

Consider the displacement of one metal by another. Will aluminum displace copper ions from an aqueous 1 M Cu^{2+} solution?

COMMENT From Table 14.1 we see that the Al^{3+}/Al half-reaction is the more negative, and so the Al/Al^{3+} oxidation half-reaction is the more positive. This means that aluminum atoms lose electrons more easily than do copper atoms, and so aluminum atoms will be oxidized to Al^{3+} ions and will displace the Cu^{2+} ions.

SOLUTION In terms of cell potentials, the solution is

$$Al(s) \longrightarrow Al^{3+}(aq) + 3e^- \qquad E^\circ = +1.66\ V$$

$$Cu^{2+}(aq) + 2e^- \longrightarrow Cu(s) \qquad E^\circ = +0.337\ V$$

$$2Al(s) + 3Cu^{2+}(aq) \longrightarrow 2Al^{3+}(aq) + 3Cu(s) \qquad E^\circ_{cell} = 2.00\ V$$

Aluminum displaces Cu^{2+} ions from solution under standard-state conditions.

EXERCISE 14.6

Will lead atoms displace 1 M zinc ions from solution?

ANSWER No. The reverse process is spontaneous, namely zinc atoms displace lead ions.

EXAMPLE 14.7

Now consider the reactivity of metals with acids. Will aluminum generate hydrogen gas from 1 M aqueous solutions of strong hydrogen acids?

COMMENT Again, from Table 14.1 we see that aluminum atoms lose electrons more easily than do hydrogen atoms, and so the H^+ ions will be reduced to $H_2(g)$:

SOLUTION

$$Al(s) \longrightarrow Al^{3+}(aq) + 3e^- \qquad E° = +1.66\ V$$
$$2H^+(aq) + 2e^- \longrightarrow H_2(g) \qquad E° = 0$$

Now add appropriate multiples of these equations to obtain the *net* reaction:

$$2Al(s) + 6H^+(aq) \longrightarrow 2Al^{3+}(aq) + 3H_2(g) \qquad E°_{cell} = 1.66\ V$$

Aluminum atoms will displace hydrogen ions generating hydrogen from acidic solutions such as hydrochloric and sulfuric acids.

EXERCISE 14.7

Will copper atoms displace hydrogen from 1 M hydrochloric acid solutions to form Cu^{2+} ions?

ANSWER No. The spontaneous process is the reduction of Cu^{2+} ions by hydrogen.

EXAMPLE 14.8

We will now consider the displacement of one halogen by another. Will chlorine displace bromide ions in a 1 M aqueous solution?

COMMENT Again, from Table 14.1 we see that electrons are more easily lost by bromide ions than chloride ions, and so Br^- will be oxidized to Br_2 and be replaced by Cl^- ions.

SOLUTION

$$2Br^-(aq) \longrightarrow Br_2(aq) + 2e^- \qquad E° = -1.087\ V$$
$$Cl_2(aq) + 2e^- \longrightarrow 2Cl^-(aq) \qquad E° = +1.36\ V$$

Adding the equations gives the net reaction:

$$Cl_2(aq) + 2Br^-(aq) \longrightarrow Br_2(aq) + 2Cl^-(aq) \qquad E°_{cell} = 0.27\ V$$

Chlorine will displace bromide ions in aqueous solution.

EXERCISE 14.8

What will happen if aqueous iodine is added to a 1 M aqueous solution of bromide ions?

ANSWER Nothing. The table of standard reduction potentials tells us that electrons are more readily lost by iodide ions, so the spontaneous process is bromine displacing iodide ions.

14.3 THERMODYNAMICS OF ELECTROCHEMICAL CELLS

Electrical Work, Free Energy, and Cell Potentials

In this section we will relate the standard potential E°_{cell} of an electrochemical cell to the thermodynamic quantities we learned about in Chapters 12 and 13, particularly ΔG°, the standard Gibbs free energy change, and K, the equilibrium constant. Recall the fundamental relationship between ΔG° and K, the equilibrium constant:

$$\Delta G^\circ = -RT \ln K$$

Once we know the relationship between E°_{cell} and ΔG°, we will be able to relate K to the potential of the cell. Remember that E°_{cell} and ΔG° are the values of E_{cell} and ΔG at standard conditions with all concentrations at 1 M and all partial pressures at 1 atm.

The standard cell potential E° for an electrochemical reaction is clearly related to spontaneity. A positive value means that the process is spontaneous under standard-state conditions. The standard free energy ΔG° is also related to spontaneity; a negative value means that the process is spontaneous, once again under standard-state conditions. Therefore, E°_{cell} and ΔG° must be related to each other. By looking at the thermodynamics of an electrochemical reaction we can see the relationship between the two. Electrochemical reactions are particularly susceptible to thermodynamic analysis because, by placing a large resistance in the circuit, the process can be slowed to near reversible conditions. Also, an electrochemical process is capable of doing work. This work can be of the ordinary pressure–volume type, but more important to this discussion, the potential generated in an electrochemical process can be used for electrical work, such as running a motor. Thus, for an electrochemical process the total work w is given by the following:

$$w = w_{elec} + w_{PV}$$

We considered work of the pressure–volume variety in Chapter 12. At constant pressure,

$$w_{PV} = -P\Delta V$$

Now we can consider electrical work. An operating galvanic cell proceeds spontaneously and has a positive potential, that is, $E_{cell} > 0$. Such a cell can do work on the surroundings, giving a negative w_{elec}. Electrical work equals the negative of charge times potential:

$$w_{elec} = -EQ$$

If E is in volts, or joules/coulomb, and Q is in coulombs, w_{elec} is calculated in joules. The negative sign in this equation is the result of the convention that work done by the system is negative. For an electrochemical reaction under standard conditions, E_{cell} becomes E°_{cell}. The charge is equal to the number of moles of electrons transferred per mole of reaction (n) times the faraday constant (F), which is 96,500 C/mol. Therefore,

$$Q = nF$$

and

$$w_{elec} = -nFE_{cell}$$

Electrical work: The work done by an electrochemical cell is the product of the total charge nF and the cell potential E_{cell}. The number of moles of electrons flowing through the circuit per mole of reaction is n.

Now we can examine electrical work in terms of the laws of thermodynamics. For an electrochemical cell, the first law of thermodynamics can be stated as follows:

$$\Delta E = q + w = q + w_{PV} + w_{elec}$$

At constant pressure,

$$\Delta H = \Delta E + P\Delta V = \Delta E - w_{PV}$$

Substituting for ΔE gives

$$\Delta H = q + w_{PV} + w_{elec} - w_{PV} = q + w_{elec}$$

The free energy change at constant temperature is

$$\Delta G = \Delta H - T\Delta S$$

Substituting for ΔH gives

$$\Delta G = q + w_{elec} - T\Delta S$$

For a reversible process, $q = q_{rev}$ and $q_{rev} = T\Delta S$. This reduces the equation for the free energy change to

$$\Delta G = +w_{elec}$$

Combining this result with $w_{elec} = -nFE_{cell}$ gives

$$\Delta G = -nFE_{cell}$$

which can be expressed as follows under standard conditions:

$$\Delta G^\circ = -nFE^\circ_{cell}$$

Free energy and cell potential: Relates the standard free energy change $\Delta G°$ to the standard cell potential $E°_{cell}$ and the total charge nF.

EXAMPLE 14.9

Calculate $\Delta G°$ for the reaction

$$Cu(s) + 2Ag^+(aq) \longrightarrow Cu^{2+}(aq) + 2Ag(s)$$

COMMENT Oxidation of one mole of copper in this reaction requires two moles of electrons. The standard cell potential for this process calculated from Table 14.1 is 0.462 V.

SOLUTION Use of $\Delta G° = -nFE°_{cell}$ gives

$$\Delta G° = -(2 \text{ mol } e^-/\text{mol Cu})(96{,}500 \text{ C/mol } e^-)(0.462 \text{ J/C})\left(\frac{1 \text{ kJ}}{1000 \text{ J}}\right)$$

$$= -89.2 \text{ kJ/mol Cu}$$

$$= -89.2 \text{ kJ for one mole of the reaction as written}$$

EXERCISE 14.9

Calculate $\Delta G°$ for the reaction

$$2Cr(s) + 3PbSO_4(s) \longrightarrow 2Cr^{3+}(aq) + 3Pb(s) + 3SO_4^{2-}(aq)$$

ANSWER $\Delta G° = -223$ kJ

To find the relationship between $E°$ and the equilibrium constant K we start with the relationship between $\Delta G°$ and K established in Chapter 13:

$$\Delta G° = -RT \ln K$$

Substituting the value for $\Delta G°$ from the relationship

$$\Delta G° = -nFE°$$

gives

$$-nFE°_{cell} = -RT \ln K$$

which rearranges to an equation in terms of the standard potential and equilibrium constant:

$$E°_{cell} = \frac{RT \ln K}{nF}$$

Solving for ln K gives

$$\ln K = \frac{nFE°}{RT}$$

and taking the inverse natural logarithm of each side of this equation leaves us finally with the desired result, an expression for the equilibrium constant in terms of the standard cell potential:

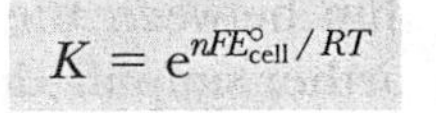

Equilibrium constant and cell potential: Relates the standard cell potential and the equilibrium constant for the cell reaction at a given temperature, usually taken to be 298K (25°C), where *n* is the number of moles of electrons transferred for the reaction as written.

For a spontaneous process, $E°$ is greater than 0 and K is greater than 1. For a nonspontaneous process, $E°$ is less than 0 and K is less than 1. A reaction for which $E°$ equals 0 has an equilibrium constant of 1 and would be a situation where equilibrium prevails at standard conditions: all concentrations are 1 M and all partial pressures are 1 atm. We can summarize as follows:

Spontaneous process:	$E° > 0$ and $K > 1$
Nonspontaneous process:	$E° < 0$ and $K < 1$
Equilibrium process:	$E° = 0$ and $K = 1$

EXAMPLE 14.10

Determine the equilibrium constant at 298K for the reduction of V^{2+} by Cr^{2+} according to the following reaction:

$$V^{2+}(aq) + 2Cr^{2+}(aq) \longrightarrow V(s) + 2Cr^{3+}(aq)$$

COMMENT $E°_{cell}$ for this reaction can be calculated from the data in Table 14.1. The value is −0.78 V; therefore, this is not a spontaneous reaction. Use the equation relating the standard cell potential and the equilibrium constant.

SOLUTION

$$\ln K = \frac{nFE°_{cell}}{RT}$$

$$\ln K = \frac{(2 \text{ mol e}^-/\text{mol V})(96{,}500 \text{ C/mol e}^-)(-0.78 \text{ J/C})}{(8.314 \text{ J/mol·K})(298\text{K})} = -60.8$$

$$K = e^{-60.8} = 4 \times 10^{-27}$$

Note that the modestly negative potential of −0.78 V leads to this very small value for the equilibrium constant. Electrochemistry provides a simple way to determine equilibrium constants, including those that are very small or very large.

EXERCISE 14.10

Calculate the equilibrium constant for the following reaction at 298K:

$$2Fe^{3+}(aq) + 3I^-(aq) \longrightarrow 2Fe^{2+}(aq) + I_3^-(aq)$$

ANSWER $K = 8.3 \times 10^7$

The fundamental relationship between the equilibrium constant and the standard cell potential can be further simplified by incorporating the values for the constants and the absolute temperature at the arbitrarily established value of 298K (25°C), and then converting from natural to common (base 10) logarithms by multiplying by 2.303. Here is the result:

$$E^\circ_{cell} = \frac{0.0591}{n} \log K$$

$$K = 10^{nE^\circ_{cell}/0.0591}$$

Simplified equation relating the equilibrium constant for a chemical reaction and the cell potential: If the electrochemical cell operates at 298K (25°C), *R* is 8.314 J/mol·K, the faraday constant *F* is taken to be 96,500 C/mol, and the logarithm of *K* is expressed to the base 10, then the combined terms give this simplified form of the equation relating the equilibrium constant and the standard cell potential.

EXAMPLE 14.11

Determine the equilibrium constant *K* for this reaction at 298K:

$$2Cr^{3+}(aq) + Fe(s) \longrightarrow 2Cr^{2+}(aq) + Fe^{2+}(aq)$$

COMMENT E°_{cell} is calculated from Table 14.1 to be 0.03 V, and 2 mol of electrons are transferred per mole of reaction.

SOLUTION $K = 10^{2\times0.03/0.059} = 1.0 \times 10^1 = 10.$

EXERCISE 14.11

Determine *K* for the reaction at 298K:

$$Hg_2Cl_2(s) + 2Cu^+(aq) \longrightarrow 2Hg(\ell) + 2Cl^-(aq) + 2Cu^{2+}(aq)$$

ANSWER 7.55×10^3

The Nernst Equation

What is the potential of an electrochemical cell operating at *other* than standard conditions? We have already developed an equation for ΔG under non-standard-state conditions in Chapter 13:

$$\Delta G = \Delta G^\circ + RT \ln Q$$

The reaction quotient *Q* takes the same form as the equilibrium constant, but applies whether or not the reaction is at equilibrium. Thus, the reaction quotient equals K at equilibrium. The relationship between free energy change and the half-cell or cell potential is

$$\Delta G = -nFE$$

Now substitute $-nFE$ for ΔG and $-nFE^\circ$ for ΔG° to get the expression for calculating half-cell and cell potentials under conditions other than standard state:

$$\Delta G = \Delta G^\circ + RT \ln Q$$

$$-nFE = -nFE^\circ + RT \ln Q$$

which rearranges to give the **Nernst equation**, an important relationship named for the turn-of-the-century thermodynamicist, Walther Nernst.

$$E = E^\circ - \frac{RT}{nF} \ln Q$$

Nernst equation for cell potentials under nonstandard conditions: This equation shows the dependence of the cell potential on the concentrations of reactants and products as expressed by the reaction quotient Q.

A simpler form of the equation is obtained by entering values of the constants, taking the temperature to be 298K, and converting to common logarithms by $\ln Q = 2.303 \log_{10} Q$:

$$E = E^\circ - \frac{0.0591}{n} \log Q$$

This form of the Nernst equation applies at 298K.

We can apply the Nernst equation to the reduction of Pb^{2+} by Cr^{2+} ions, according to the following reaction:

$$Pb^{2+}(aq) + 2Cr^{2+}(aq) \longrightarrow Pb(s) + 2Cr^{3+}(aq) \qquad E^\circ_{cell} = 0.28 \text{ V}$$

Because two moles of electrons are transferred per mole of reaction, $n = 2$, and the Nernst equation for this reaction is expressed as follows:

$$E_{cell} = E^\circ_{cell} - \frac{0.0591}{n} \log \frac{[Cr^{3+}]^2}{[Pb^{2+}][Cr^{2+}]^2}$$

Note that Pb(s) does not appear in the expression for Q. Once again pure solids and pure liquids do not appear in expressions for reaction quotients or equilibrium constants. Under standard-state conditions, all concentrations are 1 M, Q becomes equal to one, $\log Q$ reduces to zero, and E_{cell} equals E°_{cell}. If we increase the concentrations of the reactants, Q will be less than 1, $\log Q$ will be negative, and E_{cell} will be greater than E°_{cell}. The change in E_{cell} resulting from an increase in the driving force for the reaction in the spontaneous direction is in agreement with Le Chatelier's principle. If instead the concentration of the product is increased, Q will be greater than 1, $\log Q$ will be positive, and E_{cell} will be less than E°_{cell}.

EXAMPLE 14.12

Calculate E_{cell} for the reaction above at 298K if $[Cr^{3+}] = 1.25 \times 10^{-3}$, $[Pb^{2+}] = 1.00 \times 10^{-1}$, and $[Cr^{2+}] = 9.00 \times 10^{-1}$ M.

COMMENT Use the Nernst equation to calculate the cell potential E_{cell} under non-standard-state conditions.

SOLUTION

$$E_{cell} = 0.28 \text{ V} - \frac{0.0591}{2} \log \frac{(1.25 \times 10^{-3})^2}{(1.00 \times 10^{-1})(9.00 \times 10^{-1})^2}$$

$$= 0.28 \text{ V} - \frac{0.0591}{2} \log 1.93 \times 10^{-5}$$

$$= 0.42 \text{ V}$$

EXERCISE 14.12

One of the two compartments of a galvanic cell contains a strip of Co dipping into a 0.00235 M solution of Co^{2+}. The other compartment has a bar of Ag dipping into a 0.61 M solution of Ag^{+}. What is the value of E_{cell} for this cell at 298K?

ANSWER 1.141 V

Concentration Cells

Because the potential for a cell reaction is simply the sum of the potentials of its half-reactions, the Nernst equation applies just as well to half-cell reactions. Thus, two half-cells having the same components but different ion concentrations will have different half-cell potentials (see Fig. 14.5). Under standard-state conditions, the reduction potential for the Ag/Ag^{+} half-reaction is as follows:

$$Ag^{+}(aq) + e^{-} \longrightarrow Ag(s) \qquad E° = 0.799 \text{ V}$$

However, if the Ag^{+} concentration is only 1.00×10^{-3} M, then the half-cell potential is

$$E = E° - \frac{0.0591}{n} \log \frac{1}{[Ag^{+}]}$$

$$= 0.799 \text{ V} - \frac{0.0591}{1} \log \frac{1}{1.00 \times 10^{-3}}$$

$$= 0.622 \text{ V}$$

Furthermore, because these two identical half-cells have different potentials, they could be connected in such a way as to produce a spontaneous process and a positive cell potential. Such a cell is called a **concentration cell** and is shown in Figure 14.5. Here are the two reduction half-reactions:

$$Ag^{+}(1 \text{ M}) + e^{-} \longrightarrow Ag(s) \qquad E° = 0.799 \text{ V}$$

$$Ag^{+}(1.00 \times 10^{-3} \text{ M}) + e^{-} \longrightarrow Ag(s) \qquad E = 0.622 \text{ V}$$

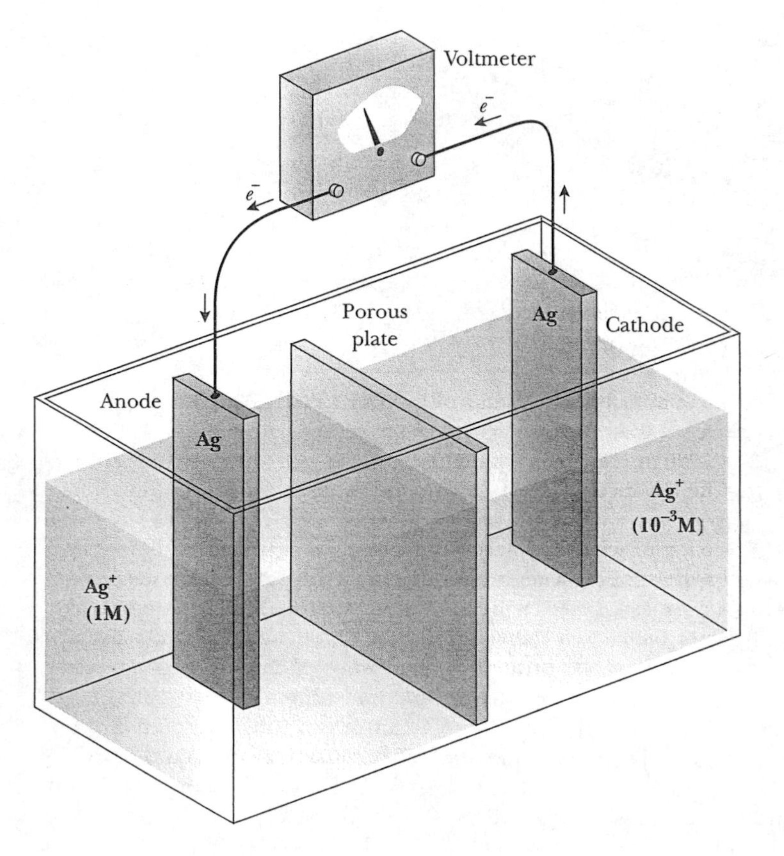

FIGURE 14.5 Concentration cell. Such a cell is capable of producing a cell voltage based only on a difference in concentration of the ions in two otherwise identical half-cells.

To establish a concentration cell, the second half-reaction must be reversed. Adding the two produces a net positive potential for the cell:

$$Ag^+(1\ M) \longrightarrow Ag^+(1.00 \times 10^{-3}\ M) \qquad E_{cell} = [0.799 - 0.622]\ V = 0.177\ V$$

An alternative approach to calculating the potential for this cell is to use the Nernst equation for the complete reaction. In a concentration cell like this, E°_{cell} will be zero because the reaction products and reactants are the same and

$$E_{cell} = 0 - \frac{0.0591}{1} \log \frac{1.00 \times 10^{-3}}{1} = 0.177\ V$$

This result is identical to the result obtained if the Nernst equation is used on the half-reactions, and then the half-reaction potentials are added.

As the reaction taking place in the concentration cell proceeds, the more dilute solution increases in concentration while the more concentrated decreases. The value of E_{cell} approaches zero as Q moves closer to one. Finally, when the concentrations are equal, the cell potential reaches zero, and the process stops. The overall spontaneous process that occurs in the concentration cell is the same as would result from mixing the two solutions directly and obtaining a solution of intermediate concentration. Predictions based on an understanding of entropy and the second law of thermodynamics are in full agreement because one would expect the spontaneous process to be dilution of the more concentrated solution, which is the equivalent of a tendency toward randomization and disorder.

PROFILES IN CHEMISTRY

Walther Nernst: The Man and His Equations A towering genius of physical chemistry in the late nineteenth and early twentieth centuries, Walther Nernst (1864–1942) made his mark at age 25 when he applied the principles of thermodynamics to the electrochemical cell and was able to give a reasonable explanation of the potential it produced. His simple equation related the cell potential to the concentrations of reactants and products and obtained for him a highly prized professorship at the 700-year-old University of Göttingen, which was at the time challenging Berlin and the Kaiser Wilhelm Institute as the central science faculty on the continent of Europe. The textbook of physical chemistry that Nernst wrote while he was there as a young professor was without equal for 40 years and served to introduce generations of students to the field.

Nernst's greatest contributions, however, were to come after he advanced to Berlin in 1905. There, he proposed what has become known as the third law of thermodynamics, that ΔS approaches zero as temperature approaches absolute zero, a concept that won him the Nobel prize in 1920, a year before Einstein and two years before Neils Bohr. Nernst also did important work in photochemistry, and he advanced a theory for the ready dissociation of ionic compounds in water.

His life story was a peculiar counterpoint to his huge log of scientific successes. Nernst was a practical fellow, very much given to invention and technology. He succeeded in developing a ceramic lamp about the time Edison was looking for a proper combination of filament and globe to serve as an electric lightbulb. He patented his invention and sold it to European industrialists for a million dollars (at a time when there were no income taxes). Nernst's invention turned out to be hardly worth the paper this million dollars was printed on, and word of this costly failure set Edison back significantly in his quest for universal support for his successful lightbulb invention.

During World War I, Nernst served Germany as a science and technology adviser, lost two sons who died in the trenches in France, and was decorated for all these sacrifices by his grateful nation. But the marriage of his two daughters to men of Jewish extraction changed all of that as the National Socialist (Nazi) party took control of Germany in the 1930s, leaving a broken and disappointed Walther Nernst to resign all his offices in protest and to live out the last decade of his life in disgust of the country for which he believed he had sacrificed so much.

Question

Which of Nernst's accomplishments do you consider the most significant. Why?

EXAMPLE 14.13

Determine the potential for a concentration cell containing Co electrodes dipping into solutions of 0.500 M Co^{2+} and 1.25×10^{-5} M Co^{2+} ions.

COMMENT Use the Nernst equation, remembering that $E°$ must be zero for two identical half-cell reactions.

SOLUTION

$$E_{cell} = 0 - \frac{0.0591}{2} \log \frac{0.0000125}{0.500} = 0.136 \text{ V}$$

EXERCISE 14.13

What is the potential of a concentration cell containing Cr^{3+} at 0.100 M and 2.00×10^{-3} M and Cr electrodes? Give the equation for the overall process.

ANSWER 0.0335 V

$$Cr^{3+}(0.100 \text{ M}) \longrightarrow Cr^{3+}(0.00200 \text{ M})$$

14.4 APPLICATIONS OF GALVANIC CELLS

Batteries

Galvanic cells, either used singly or coupled together in series, are very important portable sources of electric current. They are the batteries in all their many forms that power so much of modern technology at home, at work, and at play. The most familiar of these is the common dry cell (Fig. 14.6). Here, a Zn casing serves as the anode, and the electrolyte is a paste of MnO_2 and NH_4Cl in water with starch added as a thickener. The reduction half-reaction occurs at the surface of a graphite rod that serves as an inexpensive inert cathode. The cell produces a potential difference of about 1.5 V that drives the process in the direction indicated by the following reactions:

$$\mathrm{Zn(s)} \longrightarrow \mathrm{Zn^{2+}(aq)} + 2e^- \qquad \text{anode reaction}$$

$$2\mathrm{NH_4^+(aq)} + 2\mathrm{MnO_2(s)} + 2e^- \longrightarrow \mathrm{Mn_2O_3(s)} + 2\mathrm{NH_3(aq)} + \mathrm{H_2O}(\ell) \qquad \text{cathode reaction}$$

$$\mathrm{Zn(s)} + 2\mathrm{NH_4^+(aq)} + 2\mathrm{MnO_2(s)} \longrightarrow \mathrm{Zn^{2+}(aq)} + \mathrm{Mn_2O_3(s)} + 2\mathrm{NH_3(aq)} + \mathrm{H_2O}(\ell) \qquad \text{net reaction}$$

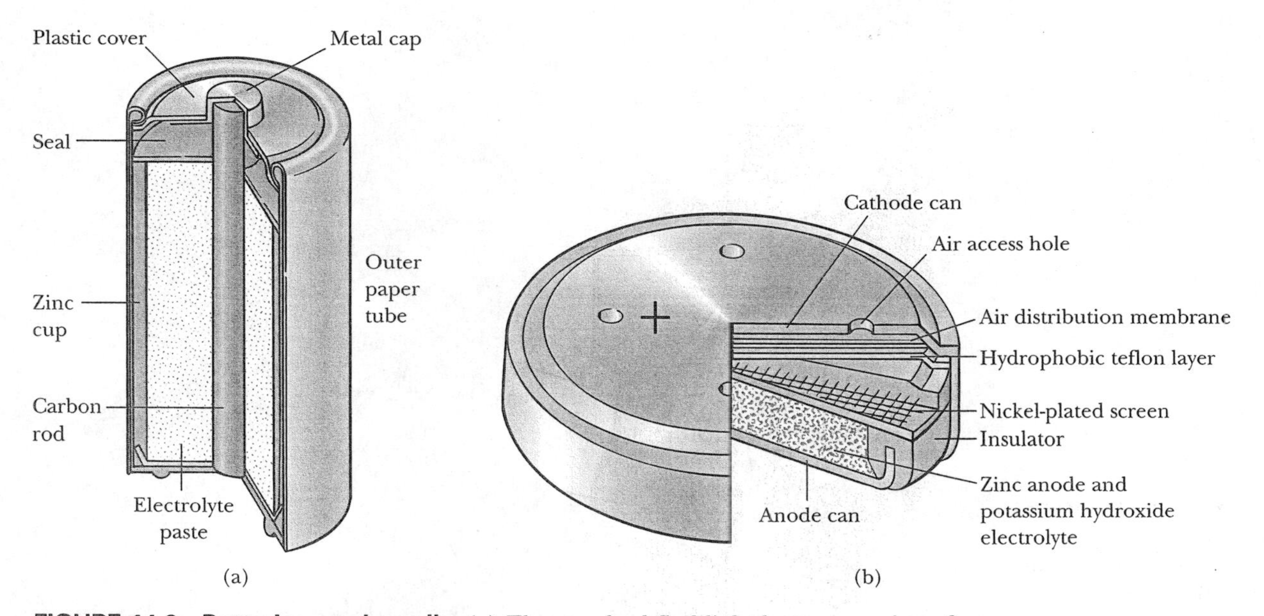

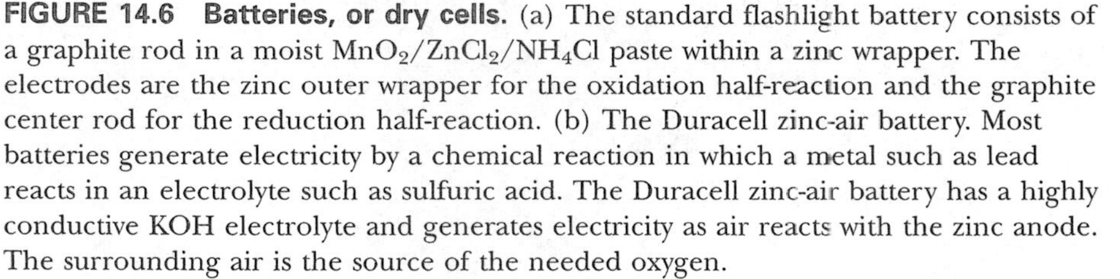

FIGURE 14.6 Batteries, or dry cells. (a) The standard flashlight battery consists of a graphite rod in a moist $MnO_2/ZnCl_2/NH_4Cl$ paste within a zinc wrapper. The electrodes are the zinc outer wrapper for the oxidation half-reaction and the graphite center rod for the reduction half-reaction. (b) The Duracell zinc-air battery. Most batteries generate electricity by a chemical reaction in which a metal such as lead reacts in an electrolyte such as sulfuric acid. The Duracell zinc-air battery has a highly conductive KOH electrolyte and generates electricity as air reacts with the zinc anode. The surrounding air is the source of the needed oxygen.

A modern version of this classic cell is the alkaline battery, with a paste of NaOH or KOH for the electrolyte rather than NH_4Cl:

$$Zn(s) + 2OH^-(aq) \longrightarrow Zn(OH)_2(s) + 2e^- \qquad \text{anode reaction}$$

$$2MnO_2(s) + H_2O(\ell) + 2e^- \longrightarrow Mn_2O_3(s) + 2OH^-(aq) \qquad \text{cathode reaction}$$

$$Zn(s) + 2MnO_2(s) + H_2O(\ell) \longrightarrow Zn(OH)_2(s) + Mn_2O_3(s) \qquad \text{net reaction}$$

This alkaline version of the battery lasts longer than the acidic NH_4Cl version because the Zn electrode corrodes more slowly under the alkaline conditions.

The galvanic cells described so far cannot be recharged once their components are used up; such cells are called **primary cells**. Rechargeable, or **secondary cells**, are particularly useful because the same cell can be recharged and used again after recharging. The best-known rechargeable cell is the familiar lead-acid storage battery. Discovered about the middle of the nineteenth century, the lead-acid storage battery has played a major role in transportation since about 1915, when self-starters began to appear in automobiles. The modern 12-V automobile battery, powering the starters of literally millions of internal combustion engines on the road today, produces a potential of about 2 V at the electrolyte concentration employed from each of six cells connected in series. One lead-acid storage battery cell is shown in Figure 14.7. The anode consists of Pb in a porous form so as to maximize its surface area. The cathode is composed

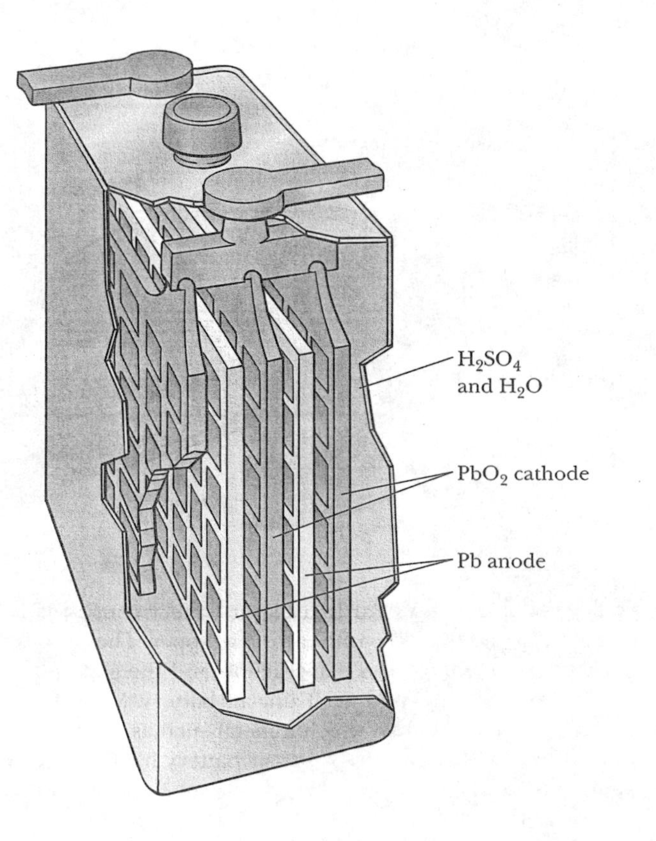

FIGURE 14.7 Lead storage battery. The lead storage battery consists of alternating plates of lead and lead(IV) oxide, immersed in a sulfuric acid solution. The positive plates consist of a lead grid filled with the lead(IV) oxide; the alternating negative plates are lead grids filled with "spongy" lead.

of $PbO_2(s)$, and the electrolyte is a solution of sulfuric acid. Here are the electrode reactions:

$$Pb(s) + SO_4^{2-}(aq) \longrightarrow PbSO_4(s) + 2e^- \qquad \text{anode reaction}$$

$$PbO_2(s) + 4H^+(aq) + SO_4^{2-}(aq) + 2e^- \longrightarrow PbSO_4(s) + 2H_2O \qquad \text{cathode reaction}$$

$$Pb(s) + PbO_2(s) + 4H^+(aq) + 2SO_4^{2-}(aq) \longrightarrow 2PbSO_4(s) + 2H_2O \qquad \text{net reaction}$$

Current is drawn from the cell as the spontaneous reaction proceeds. Note that this uses up H_2SO_4 (written in the above equation as $4H^+$ ions and $2SO_4^{2-}$ ions), lowering its concentration, the cell potential, and the density of the electrolyte. Consequently, measuring the electrolyte density is a direct way of determining the strength and reliability of a battery. The $PbSO_4(s)$ produced in each half-reaction precipitates and deposits onto the electrodes. This allows the cell to be operated as an electrolytic cell in the recharging process, where each of the half-reactions is made to run in the opposite direction. Lead-acid storage batteries can undergo many cycles of charging and discharging before significant flaking off of the deposited $PbSO_4(s)$ or internal short circuits cause irreversible failure.

Recently, there has been a great deal of interest in electric vehicles with batteries that can be recharged during periods when the vehicle is idle. Because of its relatively low cost and high dependability, the lead-acid storage battery is a good candidate for such applications. However, it is very heavy for the amount of energy it can generate. The ratio of energy to mass for batteries is generally called **specific energy**, given in units of watt-hours per kilogram, W·hr/kg. A watt is a unit of power that equals potential times current,

$$\text{Power (watts)} = \text{volts} \times \text{amperes} = (\text{J/C})(\text{C/s}) = \text{J/s} \qquad \text{electrical power}$$

A watt-hour is the power of one watt applied for one hour and is actually a unit of energy:

$$(1\text{ W})(1\text{ hr}) = (1\text{ J/s})\left(\frac{3600\text{ s}}{1\text{ hr}}\right) = 3600\text{ J}$$

We can calculate the specific energy for the lead-acid storage battery, considering only the reactants in the overall chemical process. The total mass of one mole of $Pb(s)$, one mole of $PbO_2(s)$, and two moles of H_2SO_4 is 642 g. This mass of starting materials will produce 2 mol of electrons:

$$2\text{ mol} \times \frac{96{,}500\text{ C}}{\text{mol}} = 193{,}000\text{ C}$$

The potential of the cell steadily decreases as sulfuric acid is used, and the average potential is about 1 V, that is, 1 J/C. Thus, the total energy is

$$193{,}000\text{ C} \times 1\,\frac{\text{J}}{\text{C}} = 200{,}000\text{ J}$$

Only one significant figure results from this calculation because of the uncertainty in the value of the potential. The specific energy is

$$\frac{200{,}000\text{ J}}{0.642\text{ kg}} = 3 \times 10^5\,\frac{\text{J}}{\text{kg}}$$

Conversion to watt-hours gives

$$3 \times 10^5 \frac{\text{J}}{\text{kg}} \times \frac{1 \text{ W·hr}}{3600 \text{ J}} = 80 \frac{\text{W·hr}}{\text{kg}}$$

This number, however, considers only the actual reacting material. Inclusion of the casing, electrolyte, and supporting material contained in the electrodes themselves increases the mass enough to lower the specific energy to 35 W·hr/kg.

Higher specific energies can be achieved by using lighter electrode materials. The nickel-cadmium battery has a higher specific energy and produces 1.3 V per cell. Here are the electrode reactions:

$$Cd(s) + 2OH^-(aq) \longrightarrow Cd(OH)_2(s) + 2e^- \quad \text{anode reaction}$$

$$NiO_2(s) + 2H_2O + 2e^- \longrightarrow Ni(OH)_2(s) + 2OH^-(aq) \quad \text{cathode reaction}$$

$$Cd(s) + NiO_2(s) + 2H_2O \longrightarrow Cd(OH)_2(s) + Ni(OH)_2(s) \quad \text{net reaction}$$

Note that OH^- is not used up as the cell is discharged. Therefore, there is no voltage drop through the useful life of the battery, which is in marked contrast to the lead storage battery, which consumes sulfuric acid as it discharges.

Very high specific energies require light-weight electrodes. The sodium–sulfur cell delivers about 2.0 V and achieves a value near 200 W·hr/kg. Here are the electrode reactions:

$$Na(\ell) \longrightarrow Na^+ + e^- \quad \text{anode reaction}$$

$$S(\ell) + 2e^- \longrightarrow S^{2-} \quad \text{cathode reaction}$$

$$2Na(\ell) + S(\ell) \longrightarrow 2Na^+ + S^{2-} \quad \text{net reaction}$$

However, an operating temperature above 300°C is required to obtain liquid sodium and sulfur, which limits the usefulness of the cell.

Fuel Cells

A **fuel cell** is a galvanic cell to which the chemical reactants are constantly added as the electrochemical reaction proceeds. Oxygen is often the oxidizing agent of choice, whereas traditional fuels derived from natural gas and petroleum, such as methane (CH_4), ethane (C_2H_6), propane (C_3H_8), and butane (C_4H_{10}) have all been used as reducing agents, serving as fuels for these cells. Carbon monoxide, methanol (CH_3OH), ammonia, and hydrazine (N_2H_4) have also been used as fuels. However, hydrogen is potentially the most important commercial fuel candidate. Most fuel cells produce low voltages, commonly less than one volt, so a number of them are connected together in series, producing "fuel batteries." The attractive aspects of fuel cells are their theoretical high efficiencies as well as their ability to convert fuel to electric energy without the use of moving parts. Conversion of fuel to mechanical work by means of a heat engine is subject to efficiency limits of 20% or so. Conversion of this mechanical work to electrical energy using a generator involves further energy losses. On the other hand, if its electrodes were nearly perfect, and its electrolyte had a negligible resistance, a fuel cell could operate at an efficiency of nearly 100%.

The fuel cell has also been proposed as a part of a system for storing electrical energy. Hydrogen and oxygen gases produced by the electrolysis of water could be stored in high-pressure gas cylinders and allowed to react in a fuel cell

PROFILES IN CHEMISTRY

"Animal Electricity" and the Voltaic Pile One of the most important discoveries in electrical science was made by the Italian physician and professor of anatomy and surgery, Luigi Galvani (1737–1798). Late in the eighteenth century Galvani became interested in electrical effects on living tissue, and so he obtained an electrostatic generator and a Leyden jar, a simple device for storing up static charge. His famous observation was that contacting the muscle of a dead frog with a metal probe in the presence of his electrostatic generator caused the muscles of the frog to twitch. Galvani made the further important discovery that glass or bone probes would not produce the same effect, thus demonstrating that metal could conduct electricity and that glass or bone could not. Unfortunately, Galvani belonged to a time when people believed in innate forces and principles contained within substances and organisms. One example of this was phlogiston, the principle in a substance that was believed to cause combustion. Therefore, he ascribed his results with the frog as being something contained within the frog, which he called "animal electricity."

The interpretation of these results was questioned by the Italian professor of physics Alessandro Volta (1745–1827). He repeated some of Galvani's experiments on the frog and observed that the muscle twitch could occur without the electrostatic generator if the frog's tissue was probed by two dissimilar metals. Volta concluded that the twitch was the result of the electrical stimulus applied to the frog, not the result of some vital force already present in the animal. Volta extended his investigations to a variety of combinations of dissimilar metals and compiled a list indicating the positive and negative metals in each combination. Further experimentation by Volta showed that an electric potential could result from these metal combinations in the presence of saltwater. To maximize the effect, he constructed a "voltaic pile" consisting of pairs of silver and zinc plates in contact but separated from each neighboring pair by a pad moistened with saltwater. Strong shocks could be felt by touching the plates at opposite ends of this pile. Each of these pairs of plates is what we now call a galvanic (or voltaic) cell, and Volta's arrangement placed these cells in series so that their potentials would add together. Thus, a source of dependable electric current became available for the experiments by others, such as Humphry Davy and Michael Faraday, which would move society into the electric age.

Question

Write the half-cell reactions and the cell reaction and determine E° for the cells in Volta's pile.

when the energy was needed. Calculations have been presented that show this method of energy storage to be far more compact than the rechargeable batteries currently in use.

A hydrogen–oxygen fuel cell is shown in Figure 14.8. The two electrodes are porous nickel; the two gases are forced through them, hydrogen through the anode and oxygen through the cathode. In order to speed up the reaction, the cathode is coated with an NiO catalyst. As we shall see in the next chapter, a **catalyst** is a substance that changes the rate of a reaction without itself being used up. The electrolyte is concentrated KOH. Here is the cell chemistry, which is equivalent to the combustion of hydrogen:

$$2 \times (H_2(g) + 2OH^-(aq) \longrightarrow 2H_2O(\ell) + 2e^-) \qquad \text{anode reaction}$$

$$O_2(g) + 2H_2O + 4e^- \longrightarrow 4OH^-(aq) \qquad \text{cathode reaction}$$

$$2H_2(g) + O_2(g) \longrightarrow 2H_2O(\ell) \qquad \text{net reaction}$$

Although the hydrogen-oxygen fuel cell is relatively expensive, it has already found important uses in small-scale applications such as spacecraft.

Cells using a 3% solution of hydrazine (N_2H_4) as fuel in 25% KOH function rapidly and efficiently and have been used in propelling small test vehicles.

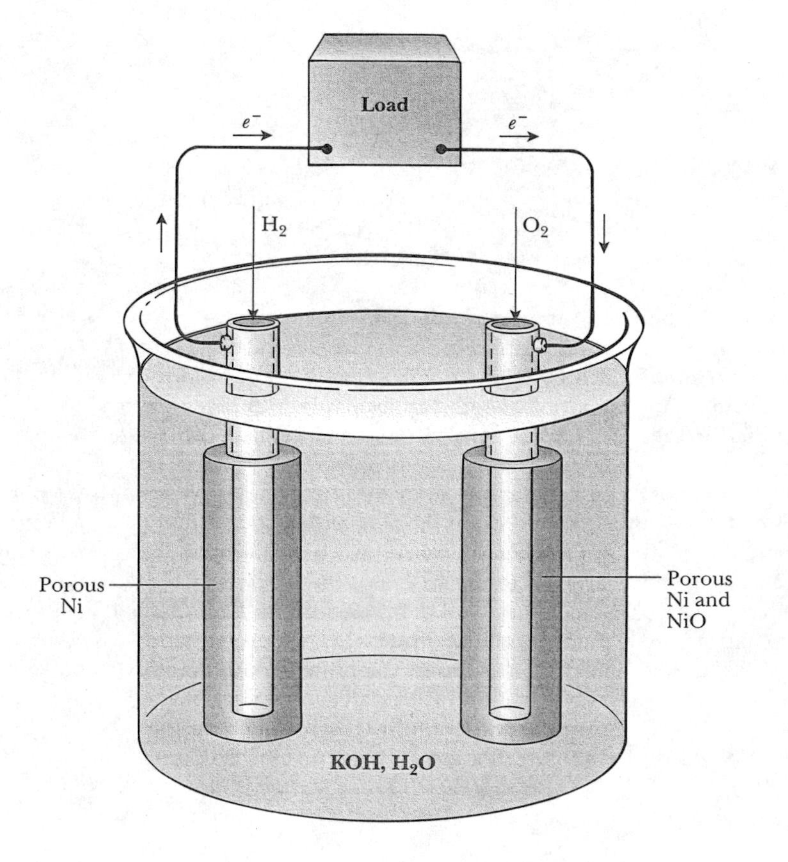

FIGURE 14.8 Hydrogen–oxygen fuel cell. The schematic drawing depicts the design of a typical hydrogen–oxygen fuel cell.

As in the hydrogen cell, porous nickel sheets can be used as the electrodes and catalyst. Disadvantages of such a system are the high cost of hydrazine and the fact that the pure substance is capable of exploding.

Biochemical fuel cells have reactions at one or both electrodes speeded by enzymes, the highly efficient catalysts of biochemical processes. Such cells could utilize as fuel the vast quantities of plant material that are available. They could also serve to consume organic waste in closed environments such as spacecraft. Problems that are encountered in this work are the susceptibility of enzymes to heat and to "poisoning" or deactivation by even traces of substances containing elements such as arsenic, mercury, and silver.

Research on fuel cell designs is aimed, particularly, at improving the materials used in the electrodes that catalyze the processes in order to advance fuel-cell efficiency toward the theoretical limit. Fuels vary considerably in cost. At the moment, fuel cells with relatively inexpensive fuels require expensive catalysts, whereas those with relatively inexpensive catalysts require expensive fuels. Obviously, it is desirable that a fuel cell require neither expensive, highly purified fuels nor precious metal or other expensive catalysts.

EXAMPLE 14.14

What standard cell potential can be expected from a fuel cell that consumes hydrogen and oxygen gases and produces water vapor?

PROFILES IN CHEMISTRY

The Invention of the Fuel Cell Was Earlier than You Might Think We usually associate the fuel cell, particularly the hydrogen–oxygen fuel cell, with the space age because it is so useful in spacecraft and holds so much promise for an era of pollution-free energy. It comes as a surprise to most people that the fuel cell, specifically the hydrogen–oxygen fuel cell, was discovered in December 1838 by Sir William Grove (1811–1896) in England. Grove was trained as a lawyer and received a knighthood for his services as a judge, but during a period of ill health he reduced his activities in the courtroom and turned his attention to science. His accomplishments in science clearly outweigh those in law, and his fame as a scientist should be much greater than it is.

The electrodes in Grove's first fuel cells consisted of glass tubes that were sealed at the top, with their open bottoms immersed in an electrolyte of dilute sulfuric acid. One of the tubes in each cell pair was filled with hydrogen gas, and the other was filled with oxygen gas. Each tube contained an electrode, which was platinum foil. The foil had been specially treated to increase its surface area and was connected to a wire that passed through the glass at the sealed end of each tube. To intensify the effect of his fuel cell, he connected 50 of them in series, thus getting 50 times the potential of a single cell. Grove reported that such a battery of his fuel cells could produce "a shock which could be felt by five persons joining hands, and which when taken by a single person was painful," and the battery could produce "a brilliant spark visible in broad daylight." Now the extraordinary promise of the fuel cell is being realized, some 160 years after Advocate Grove's invention.

Question

Write the half-reactions and the cell reaction for Grove's fuel cell.

COMMENT Obtain the value of $\Delta G°$ from Table 13.2 for the overall reaction, then proceed to solve for the cell potential using the equation $\Delta G° = -nFE°_{cell}$:

$$H_2(g) + \frac{1}{2}O_2(g) \longrightarrow H_2O(g) \qquad \Delta G° = -228.6 \text{ kJ/mol}$$

SOLUTION

$$E°_{cell} = -\frac{\Delta G°}{nF} = \frac{(228.6 \text{ kJ/mol})(1000 \text{ J/kJ})}{2(96,500 \text{ C/mol})} = 1.18 \text{ V}$$

EXERCISE 14.14

What would be the cell potential for a fuel cell that converted methane and oxygen to carbon dioxide and water vapor with the partial pressures of all gases at 1 atm? (*Hint:* First write half-reactions to show that $n = 8$ for one mole of methane.)

ANSWER 1.04 V

pH Determination

A galvanic cell in which differences in concentration produce a potential is the basis for some interesting and especially important applications involving the precise determination of cell concentrations and pH. Consider two cells, one of which contains an electrolyte of known concentration, whereas the other, which has a hydrogen gas electrode, contains an electrolyte solution of unknown hydrogen ion concentration. The pH of this solution can be determined by taking advantage of the Nernst equation, as in the following example.

EXAMPLE 14.15

In this example, a galvanic cell is constructed of a Cu electrode dipped into a 1 M solution of Cu^{2+} and a hydrogen gas electrode ($p_{H_2} = 1$ atm) in a solution of unknown pH. The Cu electrode is positive, and the potential measured is 0.573 V. The problem is to find the pH of the electrolyte solution in the hydrogen electrode compartment.

COMMENT Because the Cu electrode is positive and the cell is galvanic, its half-reaction must be using up electrons. The hydrogen ion concentration can be calculated using the Nernst equation:

SOLUTION The two electrode reactions are, therefore,

$$Cu^{2+}(aq) + 2e^- \longrightarrow Cu(s) \qquad E^\circ = 0.337 \text{ V for the cathode reaction}$$

$$H_2(g) \longrightarrow 2H^+(aq) + 2e^- \qquad E^\circ = 0 \text{ V for the anode reaction}$$

$$Cu^{2+}(aq) + H_2(g) \longrightarrow Cu(s) + 2H^+(aq)$$

$$E^\circ_{cell} = 0.337 \text{ V for the cell reaction}$$

Substituting into the Nernst equation gives

$$E_{cell} = 0.573 \text{ V} = E^\circ_{cell} - \frac{0.0591}{n} \log \frac{[H^+]^2}{[Cu^{2+}]p_{H_2}}$$

$$0.573 \text{ V} = 0.337 \text{ V} - \frac{0.0591}{2} \log \frac{[H^+]^2}{(1)(1)}$$

And the pH is obtained by solving for $-\log [H^+]$:

$$-\log [H^+]^2 = -2 \log [H^+] = \frac{2(0.573 - 0.337)}{0.0591} = 7.99$$

$$-\log[H^+] = 4.00 = \text{pH}$$

EXERCISE 14.15

A galvanic cell with a measured potential of 0.98 V consists of one compartment containing a positive Pt electrode dipping into a solution that is 0.100 M in Fe^{3+} and 0.0100 M in Fe^{2+}. The other compartment contains a hydrogen gas electrode [$p_{H_2} = 1$ atm] in a solution of unknown pH. Determine the unknown pH.

ANSWER pH = 2.6

Actual determination of pH with the half-cells just described would be very cumbersome. In the early 1930s Arnold Beckman invented a pH meter that was convenient and easily portable. His invention enjoyed immense success, and a company to produce these meters was formed. The schematics of the pH meter are shown in Figure 14.9. Each of the half-cells is contained in an electrode that dips into the solution of unknown pH. The glass electrode is sensitive to the pH of the solution into which it is dipping. The calomel electrode is simply a standard electrode that provides a reference half-cell potential at a constant temperature.

Determination of Solubility Products

Because cell potentials are very sensitive to changes in concentrations, it is possible to determine equilibrium constants that are difficult to achieve in other ways. The determination of K_{sp} values—solubility products of sparingly soluble salts—where the concentrations of dissolved ions are very low is an important example.

EXAMPLE 14.16

Here are the results of an experiment from which it is possible to determine the solubility product of AgCl. A galvanic cell has one compartment with a Zn electrode dipping into a solution of 1.00 M Zn^{2+} ions. In the other compartment there is an Ag electrode. At the bottom of the vessel is AgCl(s), which is in equilibrium with a solution of 1.00 M Cl^- ions. The Zn electrode is negative, and the measured potential is 1.00 V. Calculate K_{sp} for AgCl.

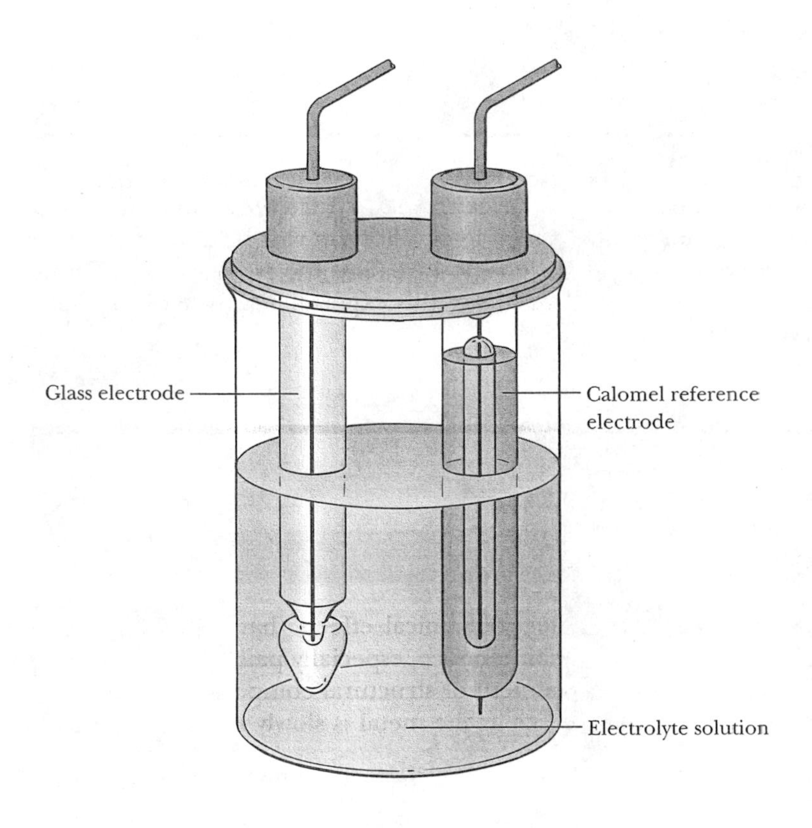

FIGURE 14.9 A glass electrode and a calomel reference electrode in an electrolyte solution. The glass electrode consists of a specially constructed thin-walled glass bulb containing a 0.10 M HCl solution about a silver wire/silver chloride electrode. The calomel electrode consists of a Pt wire that bridges the surrounding electrolyte solution and a pool of mercury that is layered with paste of Hg_2Cl_2 (calomel) and a saturated KCl solution.

COMMENT Once again, the Nernst equation provides the basis for solving the problem, but first you will have to find the correct value for the standard cell potential E°_{cell}. Then you can calculate the Ag^+ concentration in equilibrium with 1.00 M Cl^- and then the K_{sp}.

SOLUTION Because the Zn electrode is negative, here is the cell chemistry:

$$Zn(s) \longrightarrow Zn^{2+}(aq) + 2e^- \qquad E^\circ = 0.763 \text{ V anode reaction}$$

$$Ag^+(aq) + e^- \longrightarrow Ag(s) \qquad E^\circ = 0.799 \text{ V cathode reaction}$$

$$Zn(s) + 2Ag^+(aq) \longrightarrow Zn^{2+}(aq) + 2Ag(s) \qquad E^\circ = 1.562 \text{ V cell reaction}$$

Because $[Zn^{2+}] = 1.00$ M, substitution into the Nernst equation gives

$$1.00 \text{ V} = 1.562 \text{ V} - \frac{0.0591}{2} \log \frac{1}{[Ag^+]^2}$$

Rearranging to solve for $[Ag^+]$ gives

$$1.00 \text{ V} = 1.562 \text{ V} + \frac{0.0591}{2} \log [Ag^+]^2$$

$$\log [Ag^+]^2 = \frac{2(1.00 \text{ V} - 1.562 \text{ V})}{0.0591}$$

$$\log [Ag^+] = \frac{(1.00 \text{ V} - 1.562 \text{ V})}{0.0591} = -9.51$$

$$[Ag^+] = 3.1 \times 10^{-10} \text{ M}$$

Because $[Cl^-]$ is 1.00 M,

$$K_{sp} = [Ag^+][Cl^-] = (3.1 \times 10^{-10})(1.00) = 3.1 \times 10^{-10}$$

EXERCISE 14.16

A cell has one compartment with a Cu electrode in a 1.00-M solution of Cu^{2+}. The other compartment contains AgBr(s) in equilibrium with 0.500 M Br^- ions and an Ag electrode. The Cu electrode is positive, and the potential measured for the cell is 0.244 V. Based on the results of this experiment, what is the solubility product of AgBr?

ANSWER 5.65×10^{-13}

14.5 CORROSION OF METALS

Rust and corrosion of metals are electrochemical effects that cost Americans more than $10 billion per year for maintenance, especially painting, protecting exposed surfaces, and massive replacement of structural components. In corrosion processes, oxidation is taking place as the metal is slowly eaten away:

$$M(s) \longrightarrow M^{n+}(aq) + ne^- \qquad \text{oxidation of metal atoms}$$

The electrons produced by this half-reaction have to be taken up by a reduction reaction that occurs at the same time. In some cases this can take place right at the point where the corrosion is occurring as in the reaction with an acid:

$$2H^+(aq) + 2e^- \longrightarrow H_2(g) \qquad \text{reduction of protons}$$

This is the simple case in which an acid dissolves a metal. In other cases, the corrosion process is part of an electrochemical cell that has been created locally between an oxidation half-cell, where the metal is oxidized, and a reduction half-cell, where some substance such as dissolved oxygen or the ions of *some other metal* are reduced:

$$O_2(aq) + 2H_2O(\ell) + 4e^- \longrightarrow 4OH^-(aq) \qquad \text{reduction of dissolved oxygen}$$

$$M(s) \longrightarrow M^{n+}(aq) + ne^- \qquad \text{oxidation of another metal}$$

Thus, corrosion can take place in the presence of water that contains dissolved oxygen or dissolved electrolytes such as acids or salts. Corrosion can be expected to be particularly serious in environments where acid rain is a problem or in coastal regions where there is "salt" air and high humidity.

If corrosion is to be avoided, or at least minimized, the judicious choice of metals in the design of materials is important. The standard reduction potentials in Table 14.1 are the best starting point. Metals that have large negative reduction potentials such as calcium, vanadium, and zinc have high oxidation potentials and are easily oxidized. As a first approximation, we expect such metals to corrode easily. On the other hand, metals with positive reduction potentials such as copper, silver, platinum, and gold will have low oxidation potentials and can be expected to be corrosion resistant.

Some metals such as aluminum and chromium, which do have high oxidation potentials, are protected by a passivated surface. A **passivated surface** is the result of the metal being immediately oxidized upon exposure to the air, leaving a thin, tough oxide coating, which adheres very tightly to the surface, preventing additional corrosion from taking place.

$$4Al(s) + 3O_2(g) \longrightarrow 2Al_2O_3(s) \qquad \text{formation of passivated surface on Al}$$

Thus, chromium can be used as a protective coating on steel, and aluminum can be used in many otherwise surprising applications such as containers for concentrated nitric acid. Of course, other metals also form oxide coatings, most notably iron in the form of rust. However, many of these oxide coatings flake off and give no special protection to the metal.

Metals with reduction potentials more positive than that of H^+ are not corroded by pure nonoxidizing acids of 1 M concentration at 298K. Such metals as copper, silver, gold, and platinum require stronger oxidizing conditions for corrosion. Copper will dissolve in hydrochloric acid if dissolved oxygen is present:

$$2Cu(s) + 4H^+(aq) + O_2(aq) \longrightarrow 2Cu^{2+}(aq) + 2H_2O(\ell)$$

Copper and silver dissolve in nitric acid:

$$3Ag(s) + NO_3^-(aq) + 4H^+(aq) \longrightarrow 3Ag^+(aq) + NO(g) + 2H_2O(\ell)$$

Gold and platinum can be dissolved in aqua regia, a mixture of HCl and HNO_3:

$$Au(s) + 5H^+(aq) + 4Cl^-(aq) + NO_3^-(aq) \longrightarrow HAuCl_4(aq) + NO(g) + 2H_2O(\ell)$$

Corrosion of a metal can occur upon exposure to the ions of a less reactive metal. Thus, Cu^{2+} ions will cause iron to corrode:

$$Fe(s) + Cu^{2+}(aq) \longrightarrow Fe^{2+}(aq) + Cu(s)$$

Under standard conditions, the cell potential for this reaction is 0.78 V.

Practical Examples of Corrosion

The junction of two dissimilar metals provides the makings of a galvanic cell if an electrolyte is present. An example of this is a steel bolt in a copper plate with exposure to saltwater. The steel bolt is the anode:

$$Fe(s) \longrightarrow Fe^{2+}(aq) + 2e^-$$

The copper is the cathode:

$$Cu^{2+}(aq) + 2e^- \longrightarrow Cu(s)$$

We would expect that this second half-reaction would be severely limited by the low concentration of Cu^{2+} available. However, if the area of the steel anode is very small, as we would expect for a bolt, and the copper plate cathode has a relatively large area, a very small deposit of Cu^{2+} per unit area of copper ion on the cathode can result in considerable corrosion per unit area of the iron anode. In general, corrosion will be most serious when the anode has a small area and the cathode has a large area.

An example of corrosion resulting from dissolved oxygen is shown in Figure 14.10. A water drop on the surface of a piece of iron serves as the medium. The anode of the cell consists of pits in the iron surface where iron is being oxi-

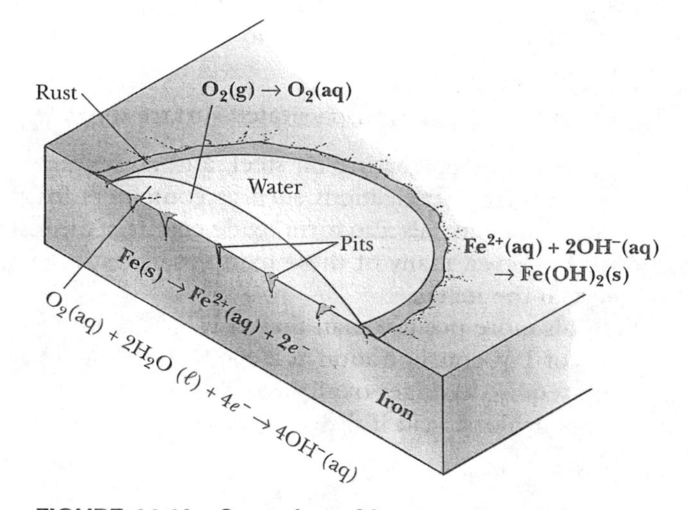

FIGURE 14.10 Corrosion of iron at a water drop. The process begins as dissolved oxygen comes in close contact with metal, usually in pits and cavities in the surface and particularly along the edge of the water drop where the oxygen concentration is the greatest. The iron itself serves as the electron conductor in this cell. Iron ions diffuse to the edge of the drop where they come in contact with hydroxide ions and there combine to form Fe^{2+}, and, subsequently Fe^{3+} hydroxides. Rust is the name given hydrated Fe(III) oxides and hydroxides.

dized. The cathode is at the edge of the water droplet where dissolved oxygen is being reduced. Electrons flow through the iron itself to complete the circuit. Corrosion will take place particularly at pits that are close to the edge of the water drop, minimizing the distance of electron conduction and, therefore, the distance. A line of rust can often be seen at the water level of a steel tank partially filled with water, and pitting of the steel in the tank will be most severe close to the rust line.

Pits in a metal surface are particularly susceptible to continued corrosion and eventual failure. The pit can start at the site of an impurity or a grinding mark. As the pit grows, positively charged metal ions accumulate in it. These attract chloride ions into the pit if they are available, forming a metal chloride. A number of metal chlorides such as iron(II) chloride can react with water to form hydrochloric acid:

$$FeCl_2(aq) + 2H_2O(\ell) \longrightarrow Fe(OH)_2(s) + 2HCl(aq)$$

Formation of the solid iron(II) hydroxide drives this reaction to the right while the hydrochloric acid can attack the iron, resulting in further corrosion.

Preventing Corrosion

There are a number of ways of preventing or limiting corrosion. A steel sample can be coated with the more active metal zinc. The zinc slowly corrodes, making any exposed steel surface the cathode and thus preventing oxidation of the iron. In this case the large area of the zinc anode compared with the small area of the exposed steel means that the corrosion per unit area of the zinc surface will be very small.

Some metal surfaces such as those of buried pipes, bridge abutments, piers, and the hulls of ships are continually exposed to electrolyte solutions. In order to protect against their constant susceptibility to corrosion, such structures are attached by a conductor to a sacrificial anode. A **sacrificial anode** is a metal more easily oxidized than the metal to be protected (Fig. 14.11). A complete cell is formed, with the electrolyte environment acting as the salt bridge and with the connection between the metals completing the circuit. The following are the important reduction half-reactions when magnesium is used to protect iron:

$$Mg^{2+}(aq) + 2e^- \longrightarrow Mg(s) \qquad E° = -2.36\ V$$

$$Fe^{2+}(aq) + 2e^- \longrightarrow Fe(s) \qquad E° = -0.44\ V$$

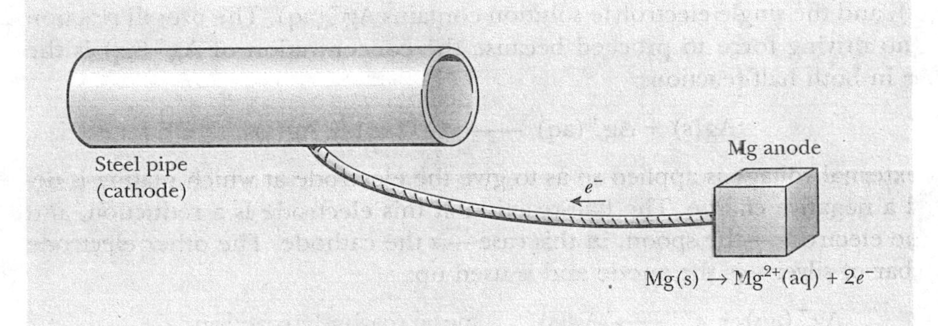

FIGURE 14.11 A sacrificial anode. Because magnesium is more easily oxidized than iron, the magnesium is the anode of this cell, and the iron does not corrode because it is the cathode. Replacement of the sacrificial anode from time to time is far easier than replacing the steel structure.

The spontaneous cell reaction for any reasonable range of concentrations is

$$Mg(s) + Fe^{2+}(aq) \longrightarrow Mg^{2+}(aq) + Fe(s)$$

Thus, the Mg anode is slowly consumed, and the Fe won't corrode. Periodic replacement of the sacrificial anode is far less expensive than replacement of the pipe, hull, or bridge structure.

14.6 ELECTROLYSIS

Electrolytic Cells and Electroplating

All of the electrochemical reactions we have discussed so far are galvanic. That is, they take place spontaneously. However, because such reactions produce a flow of electrons in the conductor connecting the electrodes, it is possible to oppose this flow by applying an external voltage source, stopping the reaction and forcing it in the opposite direction. The voltage source could be another electrochemical cell, a battery of such cells, or a direct current power supply. Thus, at standard conditions, the reaction

$$2Ag^{+}(aq) + Cu(s) \longrightarrow 2Ag(s) + Cu^{2+}(aq)$$

is spontaneous with $E^\circ_{cell} = 0.462$ V. If this process is carried out in a galvanic cell, the Cu electrode is negatively charged and is the anode. The Ag electrode is positively charged and is the cathode. The electron flow in the external circuit is from Cu to Ag.

However, if we insert a voltage source into the external circuit so as to oppose the spontaneous electron flow (Fig. 14.12), the reaction will stop if the imposed potential is 0.462 V and will reverse if the imposed potential is greater than 0.462 V. A cell in which an external voltage source forces a reaction in the nonspontaneous direction is called an electrolytic cell, and the process taking place is called **electrolysis**. The electrolysis reaction is the galvanic reaction reversed:

$$2Ag(s) + Cu^{2+}(aq) \longrightarrow 2Ag^{+}(aq) + Cu(s)$$

Now the Cu electrode is the cathode because reduction is taking place at its surface. It is still negatively charged. The Ag electrode is still positively charged, but it has become the anode.

One important use of electrolysis is **electroplating**, such as in the silver plating of tableware (Fig. 14.13). One electrode is the bar of silver, and the other is the surface to be plated. The surface of each electrode can be considered to be Ag(s), and the single electrolyte solution contains Ag^{+}(aq). The overall reaction has no driving force to proceed because the concentration of Ag^{+}(aq) is the same in both half-reactions:

$$Ag(s) + Ag^{+}(aq) \longrightarrow Ag^{+}(aq) + Ag(s)$$

An external voltage is applied so as to give the electrode at which plating is desired a negative charge. The half-reaction at this electrode is a reduction, and so the electrode—the spoon, in this case—is the cathode. The other electrode —a bar of silver—is the anode and is used up:

$$Ag^{+}(aq) + e^{-} \longrightarrow Ag(s) \qquad \text{spoon/cathode reaction}$$

$$Ag(s) \longrightarrow Ag^{+}(aq) + e^{-} \qquad \text{bar of silver/anode reaction}$$

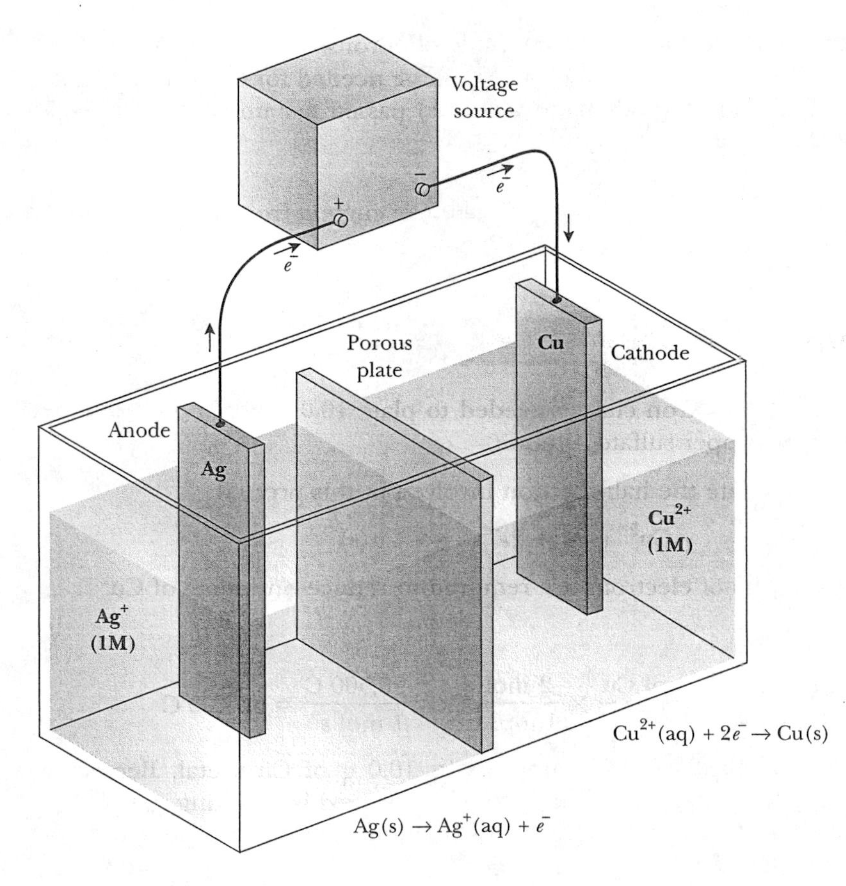

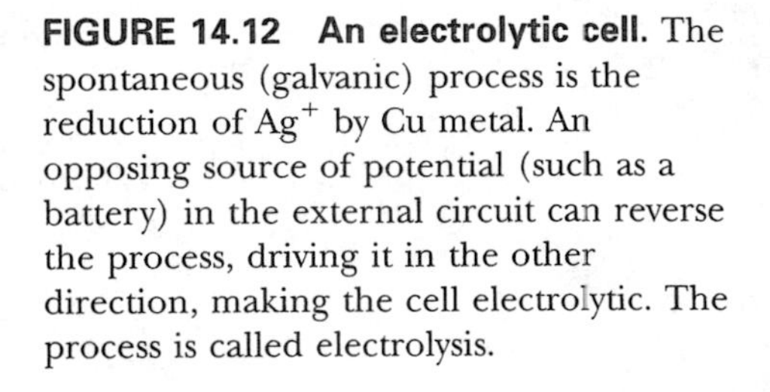

FIGURE 14.12 An electrolytic cell. The spontaneous (galvanic) process is the reduction of Ag^+ by Cu metal. An opposing source of potential (such as a battery) in the external circuit can reverse the process, driving it in the other direction, making the cell electrolytic. The process is called electrolysis.

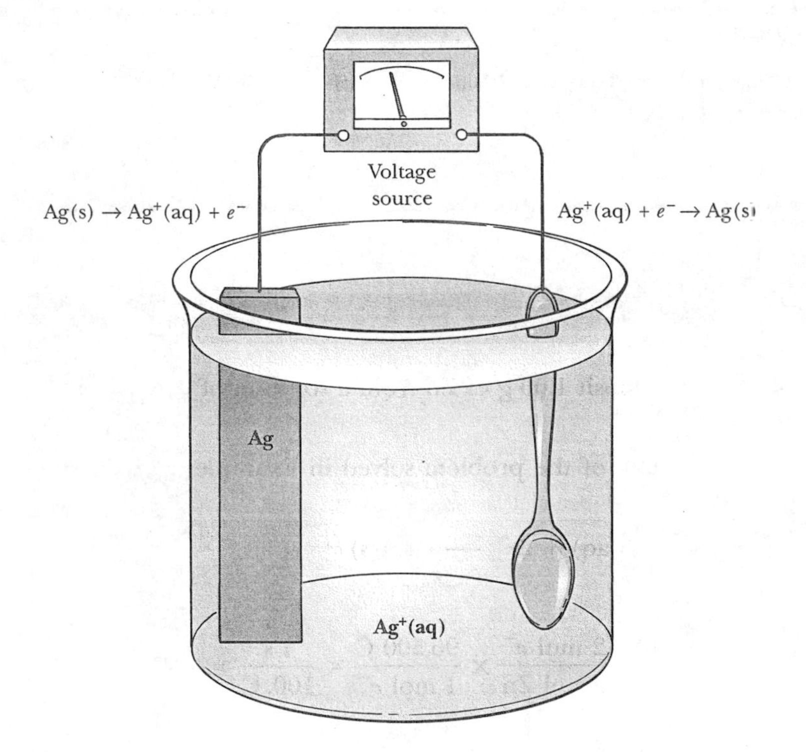

FIGURE 14.13 Silver-plated tableware. In the process, the bar of silver is consumed and the spoon is silver plated.

Because the total charge on one mole of electrons is 96,500 coulombs C or one faraday F of charge, the total electron charge needed for electrolysis can be calculated. Remember that charge (coulombs) passed per unit time (seconds) is the current, given in amperes:

$$\text{Amperes (A)} = \frac{\text{coulombs (C)}}{\text{seconds (s)}} \qquad \text{calculation of current from charge and time}$$

EXAMPLE 14.17

Calculate the total electron charge needed to plate 10.0 g of Cu onto a metal electrode from a copper sulfate solution.

COMMENT First state the half-reaction involved in this process:

$$Cu^{2+}(aq) + 2e^- \longrightarrow Cu(s)$$

Note that two moles of electrons are required to reduce one mole of Cu^{2+}.

SOLUTION

$$(10.0 \text{ g Cu}) \frac{1 \text{ mol Cu}}{63.5 \text{ g Cu}} \times \frac{2 \text{ mol } e^-}{1 \text{ mol Cu}} \times \frac{96{,}500 \text{ C}}{1 \text{ mol } e^-} = 30{,}400 \text{ C}$$

The passage of 30,400 C of charge deposits 10.0 g of Cu metal. Because a coulomb is an ampere–second, that could be arranged by allowing a 10 A current to flow for 3040 s or any other combination of amperes and seconds with a product of 30,400 C.

EXERCISE 14.17

Calculate the mass of Ni(s) that can be plated out of a solution of $NiSO_4$ by 1.00×10^5 C of charge.

ANSWER 30.4 g

EXAMPLE 14.18

How long would it take to deposit 1.00 g of Zn from a solution of $ZnCl_2$ by applying a current of 100. A?

COMMENT This is a variation of the problem solved in Example 14.17. Here is the half-reaction:

$$Zn^{2+}(aq) + 2e^- \longrightarrow Zn(s)$$

SOLUTION

$$(1.00 \text{ g Zn}) \frac{1 \text{ mol Zn}}{65.4 \text{ g Zn}} \times \frac{2 \text{ mol } e^-}{1 \text{ mol Zn}} \times \frac{96{,}500 \text{ C}}{1 \text{ mol } e^-} \times \frac{1 \text{ s}}{100. \text{ C}} = 29.5 \text{ s}$$

EXERCISE 14.18

What current is needed to deposit 2.00 g of Cr(s) from a solution of $Cr(NO_3)_3$ in 1.00 minute?

ANSWER 186 A

Electrolysis provides a means of reducing ores of reactive metals to the pure elements. For example, sodium metal is prepared by electrolysis of molten NaCl,

$$2NaCl(\ell) \longrightarrow 2Na(\ell) + Cl_2(g)$$

The cell must be constructed so as to keep the Na and Cl_2 separate. Otherwise, they would react instantly to reform the NaCl. Electrolysis reactions for the production of metals such as aluminum are of enormous commercial importance and will be described in Chapter 20.

Electrolysis of Aqueous Solutions

Electrolysis of aqueous solutions is of considerable importance. In these cases it is necessary to consult the table of electrode potentials to see what will be oxidized and what will be reduced. For example, suppose we pass a direct current through an aqueous solution of Na_2SO_4 (Fig. 14.14):

$$2H_2O(\ell) \longrightarrow O_2(g) + 4H^+(aq) + 4e^- \qquad \text{anode reaction}$$

$$2H^+(aq) + 2e^- \longrightarrow H_2(g) \qquad \text{cathode reaction}$$

The $H^+(aq)$ comes from the self-ionization of water,

$$H_2O(\ell) \longrightarrow H^+(aq) + OH^-(aq),$$

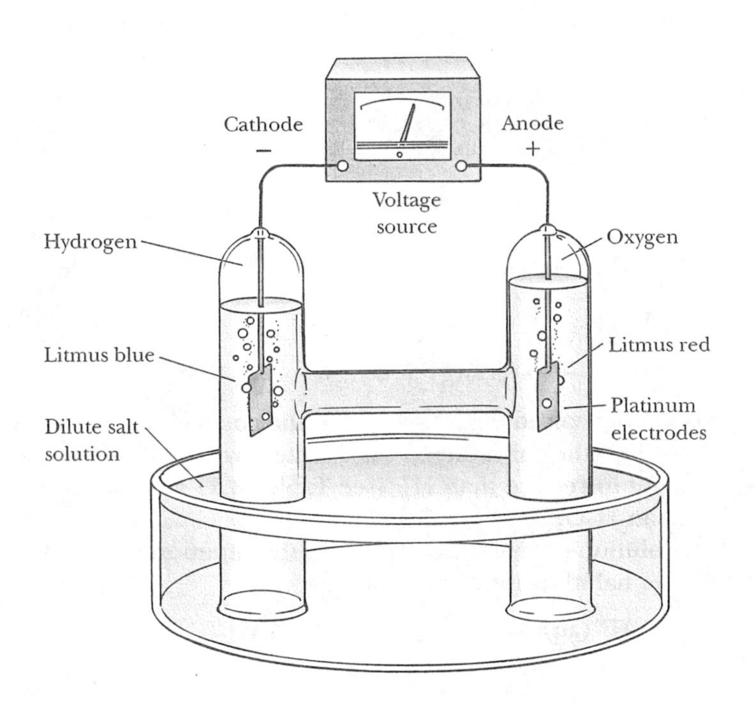

FIGURE 14.14 The electrolysis of water. Two inert electrodes are immersed in an electrolyte solution, typically a dilute, aqueous sodium sulfate solution. Water is oxidized at the anode, producing oxygen and hydrogen (hydronium) ions; and water is reduced at the cathode, producing hydrogen and hydroxide ions. Note that just as the balanced chemical equation predicts, twice as much hydrogen is produced in the left tube as oxygen in the right. In addition to hydrogen, hydroxide ion forms (on the left), producing an alkaline reaction that can be tested for with an indicator; hydronium ions are produced on the right.

PROCESSES IN CHEMISTRY

Chlorine and Industrial Electrochemistry. Chlorine is one of the most important of the chemical elements. It is only slightly soluble in water, and neither the gas nor the liquid is explosive or flammable. Chlorine gas has a greenish-yellow color and a characteristic, penetrating odor. It is about 2.5 times heavier than air. Liquid chlorine is clear amber in color and is about 1.5 times as heavy as water. At atmospheric pressure, it boils at about −100°C. One volume of liquid chlorine, when vaporized, will yield about 460 volumes of chlorine gas, which is an important piece of information because chlorine is generally transported throughout the world in the liquid state in specially constructed tank trucks and railroad cars. It is a very reactive material, entering into a wide variety of chemical reactions with organic and inorganic substances and materials. Dissolved in water, it is an oxidizing agent of moderate strength.

Listed as a hazardous substance, chlorine is subject to regulation under the Clean Water Act and more generally under regulations governing Best Management Practices and the Occupational Health and Safety Act and the Toxic Substances Control Act. It is an article of commerce that is manufactured and used on a scale of millions of pounds per year for everything from plant and animal drugs to cleaning fluids and the polymers of all manner of plastics materials used for protection, construction, and containing. Chlorine is manufactured today, as it has been for a century, by variations of the Hooker Chloralkali Process.

The electrochemical cells used for manufacturing chlorine by the Hooker Chloralkali Process are special cells known as diaphragm cells, consisting of three essential, identifiable parts:

1. Anodes in an electrolyte chamber
2. Cathodes in an electrolyte chamber
3. A diaphragm separating the chambers

A purified brine solution is fed into the anode compartment, and an electrical current is applied. The principal products are chlorine, hydrogen, and a cell liquor (solution), which is a mixture of NaCl and NaOH. For the passage of two faradays of electricity, the following chemical equation can be written:

$$2NaCl(aq) + 2H_2O(\ell) \longrightarrow Cl_2(g) + H_2(g) + 2NaOH(aq)$$

The brine solution that feeds into the operating diaphragm cell is essentially a saturated sodium chloride solution, about 23% by mass NaCl, though it may contain some sodium sulfate, sodium carbonate , and sodium hydroxide. The electrochemical reactions that occur are principally the oxidation of chloride ions and the electrolysis of water to produce hydrogen gas and hydroxide ions:

$$2Cl^-(aq) \longrightarrow Cl_2(g) + 2e^-$$ oxidation reaction consumes chloride ions

$$2H_2O(\ell) + 2e^- \longrightarrow H_2(g) + 2OH^-(aq)$$ reduction reaction from electrolysis of water

and the overall cell reaction is

$$2H_2O(\ell) \longrightarrow 2H_2(g) + O_2(g)$$

The Na^+ and SO_4^{2-} ions of the dissolved Na_2SO_4 increase the conductivity of the water, aiding the electrolysis. Neither of these ions is affected by the electrolysis because Na^+ is more difficult to reduce than H^+ (see Table 14.1) and SO_4^{2-} is more difficult to oxidize than H_2O.

When a concentrated solution of NaCl is electrolyzed, the situation becomes more complex. The possible half-reactions are as follows:

$$2H_2O(\ell) \longrightarrow O_2(g) + 4H^+(aq) + 4e^- \qquad E° = -1.23 \text{ V} \qquad \text{anode}$$

At first, the chlorine molecules formed dissolve in the electrolyte solution in the anode compartment, but as it becomes saturated, bubbles of chlorine gas emerge and are carried away. Hydrogen and NaOH solution are removed separately.

The voltage to drive the reaction is the thermodynamic potential, often referred to as the decomposition potential, plus the voltage required to overcome cell resistance and electrode overvoltage—overpotential—which is the additional voltage needed to drive the reaction. Overvoltage arises because of the nature of the electrode surface and is especially a factor in the electrolysis of water. Oxygen has a higher overvoltage than chlorine.

Question

Would you expect to find it easier to oxidize water, producing oxygen, than chloride ions, producing chlorine?

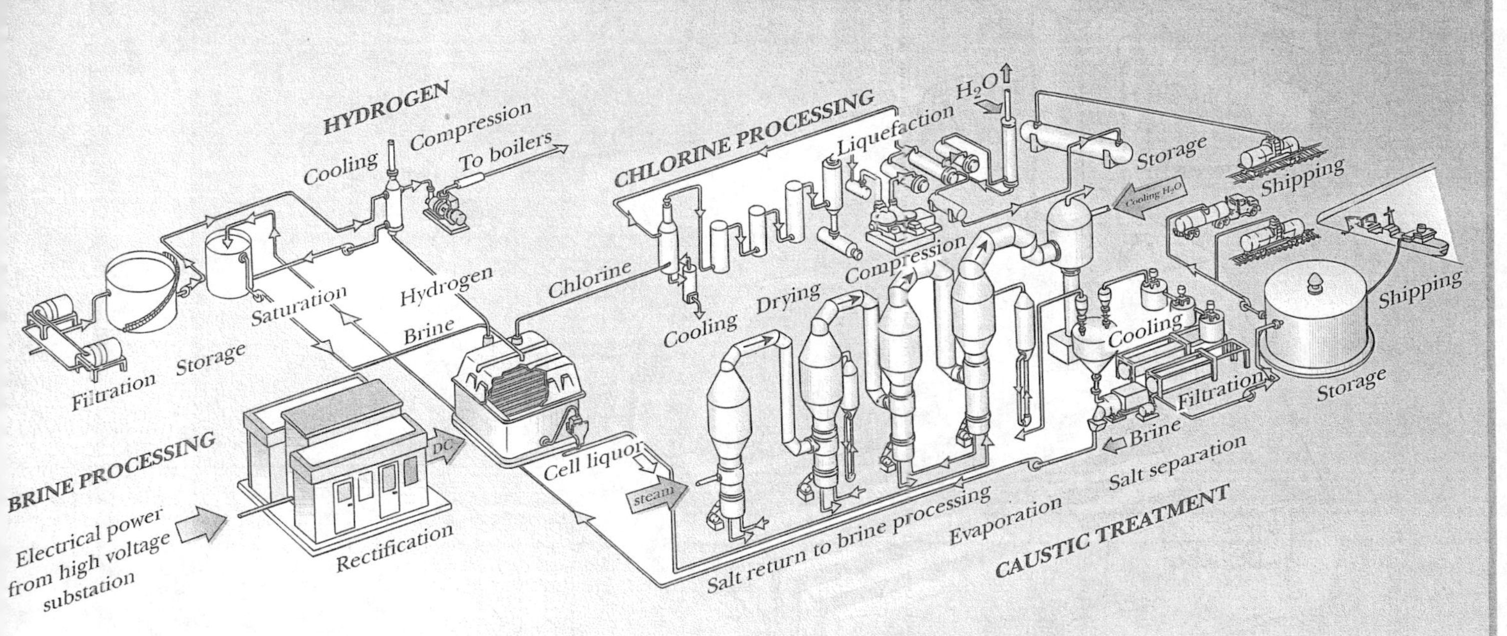

The Hooker Chloralkali Process. Of great importance industrially for the manufacture of chlorine and caustic (sodium hydroxide), the chemistry is essentially the electrolysis of a nearly saturated salt solution. By-product hydrogen is typically used on-site as an energy source for powering the process.

$2Cl^-(aq) \longrightarrow Cl_2(g) + 2e^-$	$E° = -1.36$ V	anode
$2H^+(aq) + 2e^- \longrightarrow H_2(g)$	$E° = 0.000$ V	cathode
$Na^+(aq) + e^- \longrightarrow Na(s)$	$E° = -2.714$ V	cathode

Under standard conditions this cell should be driven by the lowest necessary potential of 1.23 V to give hydrogen and oxygen gases. However, the conditions are not standard, and chlorine can be produced exclusive of any oxygen in the commercially important Hooker Chloralkali Process, in which graphite is the common anode material. This process is described in the accompanying box.

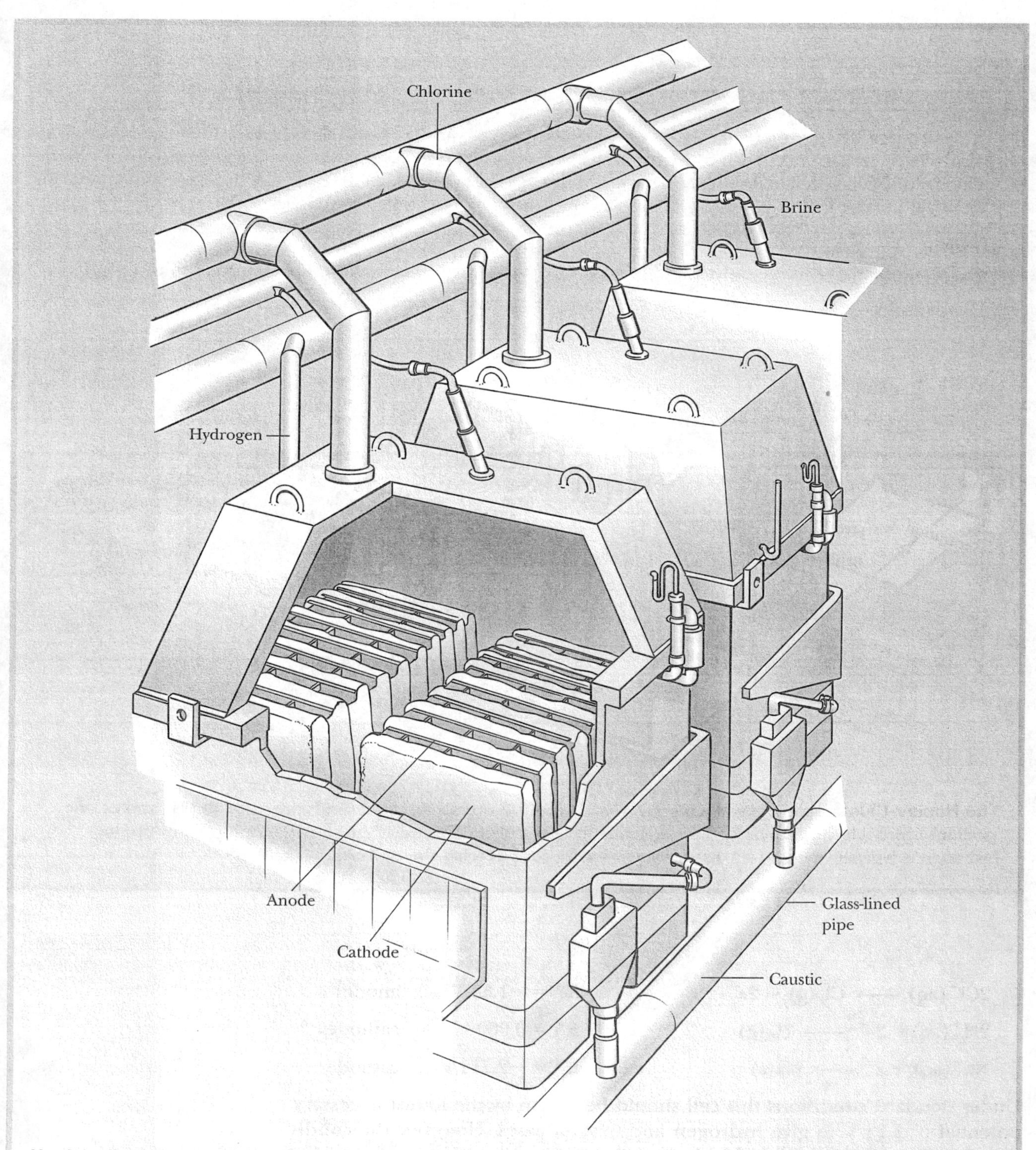

Hooker diaphragm cell for manufacturing chlorine. The cells are about 6 × 6 × 6 feet and made of poured concrete. The electrodes are placed in alternating rows. The electrolyte level is maintained, and chlorine, sodium hydroxide (caustic), and hydrogen are continuously drawn off.

SUMMARY

Electrochemical reactions involve the flow of electrons and ions through a conducting chemical system. This requires an oxidation–reduction, or redox, reaction in which electrons are transferred from one species to another. An oxidation–reduction reaction requires a change of oxidation states. A substance is oxidized if it loses electrons and its oxidation state increases. It is reduced if it gains electrons and its oxidation state decreases. In a redox reaction, the species that is reduced is called the oxidizing agent or oxidant, and the species that is oxidized is called the reducing agent or reductant.

Oxidation–reduction equations are often difficult to balance by inspection. In the half-reaction method an accounting is made of the electrons transferred in the balancing process. The actual chemical equation depends on whether the reaction is taking place in acidic or basic solution.

The half-reactions in an oxidation–reduction reaction can be separated in an electrochemical cell so that the electrons are transferred between the two halves by an electrical conductor such as a metal wire. A galvanic cell proceeds spontaneously and can produce a useful electric current. A cell upon which a current is imposed so as to produce a reaction that would otherwise not be spontaneous is called an electrolytic cell.

In any cell the anode is the electrode at which oxidation takes place, and the cathode is the electrode at which reduction takes place. The measured potential (E_{cell}) of a galvanic cell is the sum of the electrode potentials of the two half-cells. These are standard potentials (E°_{cell}) if concentrations of dissolved species are 1 M and partial pressures of gases are 1 atm. The sign convention for cell potentials is such that a spontaneous process has a positive potential. Half-cell potentials are defined so that the hydrogen electrode has a standard electrode potential (E°) of zero. Half-cells are normally tabulated as reductions. Changing the direction of a reaction or half-reaction simply changes the sign of its potential. Multiplying a reaction or a half-reaction by a constant has no effect on its potential.

Potentials for cells or half-cells that are not operating under standard-state conditions can be determined from the standard potential by means of the Nernst equation. The standard potential of a reaction is a measure of its spontaneity and is thus related to the standard free energy change. Because ΔG° is related to the equilibrium constant, there is also a relationship between E° and K.

Electrolysis cells can be used to carry out a number of useful reactions. The quantitative relationships in electrolysis depend on the number of moles of charge that pass through the cell during the process. The total charge passed in an electrochemical reaction can be determined by knowing that one mole of electrons carries a total charge of 96,500 coulombs.

TERMS

Electrochemistry (14.1)
Oxidation–reduction reactions, Redox reactions (14.1)
Oxidation states (14.1)
Oxidation (14.1)
Reduction (14.1)
Reducing agent, Reductant (14.1)
Oxidizing agent, Oxidant (14.1)
Half-reaction (14.1)
Disproportionation (14.1)
Electrochemical cell (14.2)
Galvanic cell (14.2)
Electrolytic cell (14.2)
Negative charge (14.2)
Positive charge (14.2)
Faraday (14.2)
Current (14.2)
Electrical potential (14.2)
Electrode (14.2)
Electrolyte (14.2)
Salt bridge (14.2)
Anode (14.2)
Cathode (14.2)
Standard cell potential (14.2)
Standard reduction potential (14.2)
Nernst equation (14.3)
Concentration cell (14.3)
Primary cell (14.4)
Secondary cell (14.4)
Specific energy (14.4)
Fuel cell (14.4)
Catalyst (14.4)
Passivated surface (14.5)
Sacrificial anode (14.5)
Electrolysis (14.6)
Electroplating (14.6)

IMPORTANT EQUATIONS

$\Delta G^\circ = -RT \ln K$	Relationship between ΔG° and the equilibrium constant K
$\Delta G^\circ = -nFE^\circ_{cell}$	Standard free energy change related to standard cell potential
$E^\circ_{cell} = \frac{RT}{nF} \ln K$	Standard potential and the equilibrium constant
$E^\circ_{cell} = \frac{0.0591}{n} \log K$	Alternative statement of relationship between E° and K for 25°C
$K = e^{nFE^\circ_{cell}/RT}$	Equilibrium constant calculated from standard potential
$E = E^\circ - \frac{RT}{nF} \ln Q$	Nernst equation for cell or half-cell potentials at other than standard-state conditions
$E = E^\circ - \frac{0.0591}{n} \log Q$	Alternative statement of Nernst equation for 25°C

QUESTIONS

Conceptual questions are denoted by a square screen. Extra-credit questions are denoted by a circular screen.

1. What is the difference between an oxidation-reduction (redox) reaction and an electrochemical reaction?
2. What is the difference between an electrolytic cell and a galvanic cell? Give an example of each.
3. How would you define the following terms?
   (a) oxidation (b) reduction
   (c) electrode (d) anode
   (e) cathode (f) electrolyte
   (g) oxidizing agent
4. What is the function of the salt bridge in a galvanic cell?
5. (a) What is meant by the term "standard electrode potential"?

(b) How can you relate the oxidation and reduction potentials of the same electrode?

6. How are standard electrode potentials used for measuring each of the following?
(a) Standard free energy change ($\Delta G°$) for a chemical process
(b) Equilibrium constant for a chemical reaction
(c) Direction of spontaneous change

7. How would you distinguish between each of the following pairs of terms?
(a) Oxidation and reduction
(b) Anode and cathode
(c) Electrolytic and galvanic cells
(d) Faradays and coulombs
(e) $E° > 0$ and $E° < 0$

8. Under what conditions does the equation $\Delta G° = -nFE°$ hold true?

9. In what way(s) can the Nernst equation be viewed as an example of Le Chatelier's principle?

10. Why would the sodium-sulfur cell be unsatisfactory for powering the starter of the family automobile? What possible uses can you imagine for this storage cell, which has a very high specific energy?

11. What are the advantages of using a fuel cell to produce electric energy compared with burning the fuel in a conventional power plant?

12. (a) Explain why copper piping might be preferable to iron piping for the standard application of carrying domestic water.
(b) Explain the basis for the plumbing problems that arise in copper piping when well water containing significant Fe^{3+} ion is used.

13. Certain types of brass are rapidly corroded by seawater as zinc dissolves from the alloy leaving behind a spongy mass of very nearly pure copper. Why is zinc more easily attacked than copper? Why does the corrosion process proceed rapidly once started?

14. Do you think anything of a chemical nature will occur to a pure gold coin dropped into a 1 M iron(II) sulfate [$FeSO_4$] solution? Why or why not? What about a copper wire placed in a 1 M silver nitrate solution? Write electrochemical equations for both cases.

15. The lead-acid storage battery is composed of galvanic cells that do not contain salt bridges. How is this possible?

PROBLEMS

Problems marked with a bullet (•) are answered in Appendix A, in the back of the text.

Oxidation–Reduction Reactions [1–10]

•1. For each of the following, give the apparent oxidation state for
(a) the chlorine atom in Cl_2O, ClO_2, and Cl_2O_7
(b) the sulfur atom in HS^-, HSO_3^-, $HS_2O_8^-$, $HS_4O_6^-$, and HSO_4^-
(c) the phosphorus atom in P_4, PH_3, P_2H_4, H_3PO_2, H_3PO_3, H_3PO_4, and $(NH_4)_4P_2O_7$

2. Give the oxidation state for each of the following:
(a) the uranium atom in UF_6, UO_2Cl_2, and UO_3
(b) the manganese atom in MnO, MnO_2, Mn_2O_3, Mn_3O_4, MnO_4^{2-}, MnO_4^-, and Mn_2O_7
(c) the chromium atom in $CrCl_2$, CrF_3, CrO_2, CrO_3, CrO_3Cl^-, CrO_4^{2-}, $Cr_2O_7^{2-}$

•3. For each of the following processes, identify the oxidizing agent and the reducing agent:
(a) $2Na(s) + Cl_2(g) \longrightarrow 2NaCl(s)$
(b) $Zn(s) + H_2SO_4(aq) \longrightarrow ZnSO_4(aq) + H_2(g)$
(c) $2Fe(s) + 3Cl_2(g) \longrightarrow 2FeCl_3(s)$
(d) $S(s) + 2F_2(g) \longrightarrow SF_4(g)$

4. Identify the oxidizing agent and the reducing agent in each of the following processes.
(a) $Br_2(aq) + 2I^-(aq) \longrightarrow I_2(s) + 2Br^-(aq)$
(b) $2KClO_3(s) \longrightarrow KClO_2(s) + KClO_4(s)$
(c) $6Li(s) + N_2(g) \longrightarrow 2Li_3N(s)$
(d) $2HF(aq) + Ca(s) \longrightarrow H_2(g) + CaF_2(s)$

•5. Which of the following correspond to oxidations? Which are reductions?
(a) An iron(II) salt is converted to an iron(III) salt: $Fe^{2+} \longrightarrow Fe^{3+}$
(b) A sulfide is converted to sulfur: $S^{2-} \longrightarrow S$
(c) A sulfite is converted to a sulfate: $SO_3^{2-} \longrightarrow SO_4^{2-}$
(d) A persulfate is converted to a sulfite: $S_2O_8^{2-} \longrightarrow SO_3^{2-}$

6. Tell what is oxidized and what is reduced in the following reactions.
(a) $Hg_2^{2+}(aq) + Pt(s) + 4Cl^-(aq) \longrightarrow 2Hg(\ell) + PtCl_4^{2-}(aq)$
(b) $5MnO_2(s) + 4H^+(aq) \longrightarrow 2MnO_4^-(aq) + 3Mn^{2+}(aq) + 2H_2O(\ell)$

•7. Complete and balance the following oxidation–reduction reactions, which take place in acid solution.
(a) $Cr_2O_7^{2-}(aq) + Cl^-(aq) \longrightarrow Cr^{3+}(aq) + Cl_2(g)$
(b) $MnO_2(s) + Hg(\ell) + Cl^-(aq) \longrightarrow Mn^{2+}(aq) + Hg_2Cl_2(s)$
(c) $Ag(s) + NO_3^-(aq) \longrightarrow Ag^+(aq) + NO(g)$
(d) $H_3AsO_4(aq) + Zn(s) \longrightarrow AsH_3(g) + Zn^{2+}(aq)$
(e) $Au^{3+}(aq) + I_2(s) \longrightarrow Au(s) + IO_3^-(aq)$
(f) $IO_3^-(aq) + I^-(aq) \longrightarrow I_3^-(aq)$
(g) $HS_2O_3^-(aq) \longrightarrow S(s) + HSO_4^-(aq)$
(h) $O_2^{2-}(aq) \longrightarrow O_2(g) + H_2O(\ell)$

8. Balance the following equations for the reactions in aqueous acid solution:
(a) $Cr_2O_7^{2-}(aq) + I_2(aq) \longrightarrow Cr^{3+}(aq) + IO_3^-(aq)$
(b) $S_2O_3^{2-} + I_2(aq) \longrightarrow S_4O_6^{2-}(aq) + I^-(aq)$
(c) $MnO_4^-(aq) + H_2O_2(aq) \longrightarrow Mn^{2+} + O_2(g)$
(d) $Hg_2Cl_2(s) + NO_3^-(aq) \longrightarrow Hg^{2+}(aq) + NO(g)$
(e) $MnO_4^{2-}(aq) \longrightarrow MnO_2(s) + MnO_4^-(aq)$
(f) $Pb(s) + PbO_2(s) + SO_4^{2-}(aq) \longrightarrow PbSO_4(s)$

•9. Complete and balance the following equations for reactions, which take place in basic solution:
(a) $Co(OH)_3(s) + Sn(s) \longrightarrow Co(OH)_2(s) + HSnO_2^-(aq)$
(b) $ClO_4^-(aq) + I^-(aq) \longrightarrow ClO_3^-(aq) + IO_3^-(aq)$
(c) $PbO_2(s) + Cl^-(aq) \longrightarrow ClO^-(aq) + Pb(OH)_3^-(aq)$
(d) $NO_2^-(aq) + Al(s) \longrightarrow NH_3(g) + AlO_2^-(aq)$
(e) $ClO^-(aq) \longrightarrow Cl^-(aq) + O_2(g)$
(f) $HXeO_4^-(aq) + Pb(s) \longrightarrow Xe(g) + HPbO_2^-(aq)$
(g) $Ag_2S(s) + CN^-(aq) + O_2(g) \longrightarrow S(s) + Ag(CN)_2(aq)$
(h) $MnO_4^-(aq) + S^{2-}(aq) \longrightarrow MnS(s) + S(s)$
(i) $Cl_2(g) \longrightarrow ClO^-(aq) + Cl^-(aq)$

10. Balance the following equations for the reactions in aqueous basic solution:
(a) $MnO_4^-(aq) + H_2O_2(aq) \longrightarrow MnO_2(s) + O_2(g)$
(b) $ClO_2(aq) \longrightarrow ClO_2^-(aq) + ClO_3^-(aq)$
(c) $CrO_4^{2-}(aq) + N_2H_4(aq) \longrightarrow Cr^{3+}(aq) + N_2(g)$
(d) $Ag(s) + CN^-(aq) + O_2(g) \longrightarrow Ag(CN)_2^-(aq) + OH^-(aq)$
(e) $Co(s) + ClO^-(aq) \longrightarrow Co^{2+}(aq) + Cl^-(aq)$
(f) $Cd(s) + H_2O(\ell) + Ni_2O_3(s) \longrightarrow Cd(OH)_2(s) + NiO(s)$

Standard Potentials [11–16]

•11. Using the data in Table 14.1, arrange the following substances as reducing agents in order of weakest to strongest: Al, Co, Ni, Ag, H_2, Na.

12. List the following in order of increasing strength as oxidants: Fe^{3+}, F_2, Pb^{2+}, I_2, Sn^{2+}, O_2.

•13. Using the standard reduction potentials listed in Table 14.1, pick an oxidizing agent

that could cause the following to happen spontaneously at standard conditions:
(a) $Fe(s) \longrightarrow Fe^{3+}(aq)$
(b) $Fe(s) \longrightarrow Fe^{2+}(aq)$
(c) $Fe^{2+}(aq) \longrightarrow Fe^{3+}(aq)$
(d) $2F^-(aq) \longrightarrow F_2(g)$

14. Making use of Table 14.1, select a reagent to perform the indicated task at standard conditions:
(a) $Zn(s) \longrightarrow Zn^{2+}(aq)$
(b) $F_2(g) \longrightarrow 2F^-(aq)$
(c) $Mn^{2+}(aq) \longrightarrow MnO_4^-(aq)$
(d) $H_2O_2(\ell) \longrightarrow H_2O(\ell)$
(e) $H_2O_2(\ell) \longrightarrow O_2(g)$
(f) $2I^-(aq) \longrightarrow I_2(s)$
(g) $Cr^{3+}(aq) \longrightarrow Cr_2O_7^{2-}(aq)$

•**15.** For each of the following unbalanced redox reactions below, calculate the standard cell potential and determine whether the reaction is spontaneous for 1 M solutions and 1 atm partial pressures, with solids in their standard states, all at 298K:
(a) $Fe^{2+}(aq) + MnO_4^-(aq) + H^+(aq) \longrightarrow Fe^{3+}(aq) + Mn^{2+}(aq) + H_2O(\ell)$
(b) $Ag(s) + H^+(aq) + Cl^-(aq) + O_2(g) \longrightarrow AgCl(s) + H_2O(\ell)$
(c) $Sn(s) + Hg_2Cl_2(s) \longrightarrow Sn^{2+}(aq) + Hg(\ell) + Cl^-(aq)$
(d) $HAsO_2(aq) + MnO_4^-(aq) \longrightarrow H_3AsO_4(aq) + MnO_4^{2-}(aq) + H^+(aq)$

16. Determine which of the following reactions would take place spontaneously as written under standard-state conditions. (The equations have not been balanced.)
(a) $Hg_2^{2+}(aq) \longrightarrow Hg(\ell) + Hg^{2+}(aq)$
(b) $Sn(s) + Fe^{2+}(aq) \longrightarrow Sn^{2+}(aq) + Fe(s)$
(c) $Ce^{4+}(aq) + 2Cl^- \longrightarrow 2Ce^{3+}(aq) + Cl_2(g)$
(d) $2Fe(CN)_6^{4-}(aq) + 2H^+(aq) \longrightarrow 2Fe(CN)_6^{3-}(aq) + H_2(g)$

Galvanic Cells [17–22]

•**17.** Draw a complete galvanic cell in which the two electrodes are a Ni bar in a 1.00 M Ni^{2+} solution and a Mn bar in a 1.00 M Mn^{2+} solution. Label all parts of the cell including the anode and the cathode. Give the cell notation, the half-reaction taking place at each electrode, and the overall cell reaction. Show the direction of electron flow and calculate the cell potential.

18. A galvanic cell consists of a Co electrode in a 1.00 M Co^{2+} solution and a Pt electrode in a solution containing Fe^{3+} and Fe^{2+} each at 1.00 M. The two compartments are connected by a salt bridge. Draw the cell and label all components. What is the notation for this cell? Indicate the anode and the cathode, the positive and negative electrodes, and the direction of electron flow. Give the half-reaction taking place at each electrode and the overall spontaneous cell reaction. Calculate the potential of the cell.

•**19.** Devise a cell with all concentrations at 1.00 M in which one electrode is a Cu bar in a Cu^{2+} solution. Choose the other half-cell so that the standard cell potential is about 0.4 V. Label the cell completely. Give the cell notation.

20. Draw a cell containing a Cu^{2+}/Cu^+ half-cell that has a standard cell potential of close to 1.20 V. Is the Cu^{2+}/Cu^+ electrode negative? Describe fully, giving the cell notation.

•**21.** A miniaturized battery designed for use in hearing aids is composed of Zn and a paste of KOH, water, mercury(II) oxide, and mercury. Write the separate half-reactions and the overall cell process if zinc and potassium hydroxide are consumed, mercury is deposited, and potassium zincate (K_2ZnO_2) is formed. What is the notation for this cell?

22. Devise a cell for the essential reaction in the corrosion of steel, considered as the conversion of Fe to Fe_2O_3 in an aqueous electrolyte environment. Give the cell notation. Write a complete equation for the process and draw a diagram for the cell.

Free Energy and Equilibrium Constants [23–32]

23. Calculate $\Delta G°$ for an oxidation–reduction reaction in which one electron is transferred and for which $E°_{cell}$ is +1.27 V.

24. Calculate E°_{cell} for an oxidation–reduction reaction in which two electrons are transferred and ΔG° is $-1300.$ kJ/mol.

•**25.** For the following reaction at 298K, determine the equilibrium constant and the standard free energy change ΔG°:

$$Sn(s) + Pb^{2+}(aq) \longrightarrow Sn^{2+}(aq) + Pb(s)$$

26. Use the table of standard reduction potentials to determine the equilibrium constant of the reaction,

$$2Hg^{2+}(aq) + 2Br^-(aq) \longrightarrow Hg_2^{2+}(aq) + Br_2(\ell)$$

•**27.** If 1.00 mol of silver is oxidized to Ag^+ by a stoichiometric quantity of $Br_2(\ell)$, what is the maximum electrical work that can be done by this process?

28. How much electrical work can be performed if 10.0 g of zinc metal is oxidized to Zn^{2+} by a sufficient amount of Ni^{2+}?

•**29.** A galvanic cell is constructed of two half-cells connected by a salt bridge. The first consists of an inert Pt electrode dipping into a solution that is 1 M in Fe^{2+} and 1 M in Fe^{3+}. In the other half-cell, a Zn electrode dips into a 1 M Zn^{2+} ion solution.

(a) Write the cell reaction for the spontaneous process.

(b) Which is the negative electrode? The positive electrode?

(c) Determine the cell voltage and the equilibrium constant.

(d) How will the cell voltage change if $[Fe^{3+}]$ increases?

30. A galvanic cell consists of a platinum wire dipping into a solution that is 1.0 M in Ce^{3+} and Ce^{4+}. The other electrode is silver, dipping into 1.0 M silver nitrate. Making use of Table 14.1, determine each of the following:

(a) the cell polarity

(b) the anode and the cathode

(c) the standard cell potential

(d) the equilibrium constant

31. A cell consists of Zn^{2+}/Zn and Cr^{3+}/Cr half-cells. The concentration of Zn^{2+} is 0.132 M at equilibrium. What is the equilibrium concentration of Cr^{3+}?

32. The chlorate ion, ClO_3^-, can disproportionate in basic solution according to the following reaction:

$$2ClO_3^-(aq) \longrightarrow ClO_2^-(aq) + ClO_4^-(aq)$$

What are the equilibrium concentrations of ions resulting from a solution initially 0.100 M in ClO_3^-?

Nernst Equation [33–38]

33. What would be the potential resulting from the following reaction if the concentration of Mn^{2+} ion is 0.10 M and the concentration of Cr^{3+} is 0.010 M?

$$3Mn(s) + 2Cr^{3+}(aq) \longrightarrow 3Mn^{2+}(aq) + 2Cr(s)$$

34. If the concentration of Fe^{2+} is 0.050 M, the partial pressure of O_2 is 0.30 atm, and the pH is 4.00, calculate the potential that would be developed by a cell with the reaction,

$$O_2(g) + 4H^+(aq) + Fe(s) \longrightarrow 2H_2O(\ell) + Fe^{2+}(aq)$$

35. Calculate the potential of a galvanic cell containing a Pt electrode dipping into a solution 0.00235 M in both Hg^{2+} and Hg_2^{2+} and a Pb electrode dipping into a 0.0936 M solution of Pb^{2+}. Which is the positive electrode?

36. A galvanic cell with a measured potential of 0.11 V contains a Pt electrode in a solution 0.0135 M in Cr^{2+} and 0.000216 M in Cr^{3+}. In the second compartment a Ni electrode dips into a solution of Ni^{2+} of unknown concentration. The Ni electrode is positive. Determine $[Ni^{2+}]$.

37. Use the following data to calculate the solubility product of PbF_2. A galvanic cell consists of a hydrogen gas electrode ($p_{H_2} = 1$ atm, $[H^+] = 1$ M) and a Pb electrode in a 1 M F^- solution in equilibrium with $PbF_2(s)$. The cell potential is 0.562 V and the hydrogen gas electrode is positive.

38. A galvanic cell contains one Pb electrode in a 1 M Pb^{2+} solution and the other Pb electrode in a saturated solution of $PbSO_4$. The cell potential is 0.235 V and the electrode in the saturated $PbSO_4$ is negative. What is the solubility product of $PbSO_4$ calculated from these data?

Concentration Cells [39–42]

•**39.** A concentration cell contains Pb electrodes in equal volume solutions of Pb^{2+} of 0.432 and 0.000149 M concentrations. Draw the cell and label all components. Write the spontaneous half-reactions at each electrode and the overall cell process. What will be the cell potential when the cell is first connected? What will happen to the cell potential as current is allowed to flow? What will be the final concentration of Pb^{2+} in each compartment?

40. The following half-cells are coupled together to form a concentration cell:

$$H_2(1\text{ atm}) \longrightarrow 2H^+(0.10\text{ M}) + 2e^-$$

$$H_2(1\text{ atm}) \longrightarrow 2H^+(1.0\text{ M}) + 2e^-$$

(a) Sketch the cell diagram, showing anode, cathode, flow of electrons, movement of ions, and the net cell reaction for the spontaneous process.
(b) Calculate the maximum cell voltage and comment on the source of the cell's driving force.
(c) Determine the change in the free energy and the electrical work the cell is capable of performing.

•**41.** A galvanic cell contains two hydrogen gas electrodes. One is in a 1.00 M H^+ solution and is positive. The other is in a solution of unknown pH. The cell potential is 0.0251 V. What is the unknown pH?

42. A galvanic cell has a Cu electrode in a 1 M Cu^{2+} solution and a hydrogen electrode (partial pressure of H_2 = 1 atm) in a solution of unknown pH. The cell potential is 0.7205 V. Why do these data allow for calculation of two possible pH values for the unknown solution? What are the two possible pH values? Which can be ruled out as unreasonable?

Electrolytic Cells [43–58]

43. What is the average current if 10,500 coulombs of charge flow in 1.00 hour?

44. How many coulombs of charge flow in 1.00 minute at a steady current of 4.50 A?

•**45.** How many grams of Cr(s) would be deposited in an electrolytic cell from $Cr^{3+}(aq)$ by 150,000 coulombs of charge?

46. What mass of Pb would be deposited by electrolysis of Pb^{2+} using a current of 100. A for 8.0 hours?

•**47.** What current would be necessary to plate 10.0 g of Co(s) from $Co^{2+}(aq)$ in 24.0 hours?

48. How long would it take to plate a 0.100 mm coating of Cr(s) on an automobile bumper having a total surface area of 1.00 m^2 from a solution of $Cr^{3+}(aq)$ employing a current of 100. A? The density of Cr(s) is 7.20 g/cm^3.

•**49.** Calculate the total energy in joules needed to produce 1 kg of Cl_2 by electrolysis from a 1 M solution of NaCl.

50. Calculate the maximum number of grams of cadmium deposited at the cathode by passage of 0.450 faraday of charge when a $CdCl_2$ solution is electrolyzed in a cell using inert electrodes.

•**51.** In each of the following cases, calculate the maximum mass in grams of vanadium that can be deposited on the cathode when a solution of $VO(NO_3)_3$ is electrolyzed with:
(a) a 1.00×10^4 C of electricity.
(b) a 1.00 A current flowing for 60.0 minutes.
(c) a 1.00 faraday of electricity.

52. A 0.750 A current passes for 1.00 hr through a cell that contains an aqueous sodium sulfate solution and nickel electrodes. At the cathode, water decomposes to hydrogen gas and hydroxide ions. At the anode H^+ ions are produced and nickel (II) oxide is deposited. Calculate each of the following after writing out the electrode reactions:
(a) the number of moles of H^+ and OH^- produced
(b) the number of moles of NiO deposited

•**53.** (a) An aqueous silver nitrate solution is electrolyzed using a 2.50 A current for a period of 1.00 h. Calculate the volume of oxygen gas collected at 1.00 atm and 273K. The anode reaction is as follows:

$$2H_2O(\ell) \longrightarrow O_2(g) + 4H^+(aq) + 4e^-$$

(b) In a second experiment, the cathode reaction resulted in an increase in mass of the silver electrode of 0.523 g. How long had current flowed through the cell during the second experiment?

54. A liter of a 1.0 M aqueous permanganate ion (MnO_4^-) solution is reduced at the cathode of an electrolytic cell. Determine how many faradays of electricity would be required to bring about the formation of each of the following:

(a) a 0.010 M solution of manganate ion (MnO_4^{2-}) in a total volume of 1.0 L
(b) 1.00 g of MnO_2
(c) 1.00 g of Mn

If the current in the cell was 10.0 A, how much time would be required to bring about each of these transformations?

•**55.** The coulometer, invented by Michael Faraday, is a device for measuring the total charge passing through a circuit. This is done simply by placing an electrolytic cell in series with the circuit and measuring the chemical result of the electrolysis. Faraday's coulometer measured the amount of $H_2(g)$ produced from the electrolysis of water. How many coulombs of electric charge would have passed through a circuit producing 3.62 L of $H_2(g)$ at 1.00 atm pressure at a temperature of 35°C?

56. A battery was used to drive an electrolytic cell. The current, as read on a meter in the external circuit, was believed to be exactly 0.450 A. The electrolytic cell was based on the electrolysis of a copper sulfate solution. During the 30.0 minutes that current was allowed to flow, a total of 0.3000 g of copper metal was deposited at the cathode. Determine the extent to which the meter was inaccurate.

•**57.** Four electrolytic cells are connected in series, that is, in a circuit with each negative electrode connected to the positive electrode of the next cell. In the first cell, silver ions (Ag^+) are reduced to metallic silver. In the second, copper is oxidized to copper(II) ions. In the third, a chromium(III) nitrate solution deposits chromium. In the fourth, water is electrolyzed. If 1.000 g of silver was deposited in the first cell, what are the masses of copper dissolved and chromium deposited? Determine the volumes of hydrogen and oxygen gas evolved at 1.00 atm and 273K.

58. Six different cells are connected in series so that they can be electrolyzed between inert electrodes, producing pure metal at each of the respective cathodes. In the first of the six cells, a 2.15 mL pool of mercury metal (d = 13.59 g/mL) was obtained from $Hg_2(NO_3)_2(aq)$. Calculate how many grams of Fe, Au, Co, or Cr could be deposited from solutions of the following:

(a) $K_4Fe(CN)_6$ (b) $K_3Fe(CN)_6$
(c) $Au(NO_3)_3$ (d) $Co(NO_3)_3$
(e) $Cr_2(SO_4)_3$

Additional Problems [59–70]

•**59.** For each of the following half-reactions, determine the number of electrons transferred and the direction of the transfer, and complete the balancing of the half-reaction equation for stoichiometry and charge.

(a) $VO^{2+}(aq) \longrightarrow VO_3^-(aq)$ (in basic solution)
(b) $Cr^{3+}(aq) \longrightarrow Cr_2O_7^{2-}(aq)$ (in acid solution)
(c) $Mn^{2+}(aq) \longrightarrow MnO_2(s)$ (in acid solution)
(d) $NO(g) \longrightarrow NO_3^-(aq)$ (in basic solution)
(e) $Fe^{3+}(aq) \longrightarrow Fe^{2+}(aq)$ (in acid solution)

60. From among the following half-reactions, and with the aid of the reduction potentials listed in Table 14.1, determine the strongest oxidant and the strongest reductant:

(a) $Na^+(aq) + e^- \longrightarrow Na(s)$
(b) $Br_2(\ell) + 2e^- \longrightarrow 2Br^-(aq)$
(c) $O_2(g) + 2H^+(aq) + 2e^- \longrightarrow H_2O_2(\ell)$

(d) $Ce^{4+}(aq) + e^- \longrightarrow Ce^{3+}(aq)$
(e) $Sn^{2+}(aq) + 2e^- \longrightarrow Sn(s)$

61. Devise a cell with all standard concentrations and gas pressures (1 M or 1 atm) that has a standard cell potential as close as possible to 1.40 V. Describe the cell fully.

62. Calculate the standard cell potential and the value for $\Delta G°$ for a galvanic cell composed of a silver electrode immersed in a 1.0 M silver nitrate solution and an aluminum electrode in a 1.0-M aluminum nitrate solution.

•63. Using standard reduction potentials from Table 14.1, determine the electrode potential for each of the following half-reactions at the concentrations given:
(a) $Fe^{2+}(aq) \longrightarrow Fe^{3+}(aq) + e^-$
$[Fe^{2+}] = 0.10$ M, $[Fe^{3+}] = 0.20$ M
(b) $Ni(s) \longrightarrow Ni^{2+}(aq) + 2e^-$
$[Ni^{2+}] = 0.0010$ M
(c) $MnO_4^-(aq) + 8H^+(aq) + 5e^- \longrightarrow Mn^{2+}(aq) + 4H_2O(\ell)$
$[Mn^{2+}] = 0.10$ M, $[MnO_4^-] = 1.00$ M, pH = 1.00

64. The solubility product for $PbSO_4$ can be calculated using the half-reactions

$$PbSO_4(s) + 2e^- \longrightarrow Pb(s) + SO_4^{2-}(aq) \qquad E° = -0.359 \text{ V}$$

$$Pb^{2+}(aq) + 2e^- \longrightarrow Pb(s) \qquad E° = -0.126 \text{ V}$$

Calculate K_{sp} for $PbSO_4(s)$ using these data.

•65. From data in Table 14.1, calculate the equilibrium constant for the following reaction:

$$2Ag^+(aq) + 2Hg(\ell) \rightleftharpoons Hg_2^{2+}(aq) + 2Ag(s)$$

With all substances in their standard states, determine the direction in which the cell would operate spontaneously.

66. A galvanic cell composed of a Zn electrode in a 1.00-M zinc sulfate solution and a hydrogen electrode $[H_2(1 \text{ atm})/H^+(1 \text{ M})]$ is connected in series to a second cell. The second cell is composed of a silver electrode dipping into a 1.00-M silver nitrate solution coupled to a hydrogen gas electrode $[H_2(1 \text{ atm})/H^+(1 \text{ M})]$. Galvanic cells connected in series constitute a battery and the cell potentials are added together.
(a) Write the half-cell reactions occurring at each electrode.
(b) Write the equation for the net reaction and calculate the standard cell potential.
(c) Sketch the entire battery, being sure to include the anodes, cathodes, electron flow, migration of ions, and all other significant features.
(d) If the two cells in series are now operated as an electrolytic cell rather than a galvanic cell by placing an opposing voltage of 1.50 V against the cell voltage, what will happen? If the opposing cell voltage is increased to 1.60 V, what happens?

•67. (a) For the cell process described by the following equation below, find the standard cell potential, the actual cell potential, and then find the concentration ratio $[Fe^{2+}]/[Co^{2+}]$ at which the potential generated by the cell is zero:

$$Fe(s) + Co^{2+}(0.50 \text{ M}) \longrightarrow Fe^{2+}(1.0 \text{ M}) + Co(s)$$

(b) Using the Nernst equation and the data in Table 14.1, calculate the cell potential for the following unbalanced reaction:

$$Fe(s) + MnO_4^-(0.010 \text{ M}) + H^+(0.10 \text{ M}) \longrightarrow MnO_2(s) + Fe^{3+}(0.010 \text{ M}) + H_2O(\ell)$$

68. The electrodes in two half-cells are both constructed of an inert platinum gauze supporting a paste of MnO_2. The electrolyte in each half-cell is 0.01 M MnO_4^- and they both operate according to the half-cell reaction,

$$MnO_4^-(aq) + 4H^+(aq) + 3e^- \longrightarrow MnO_2(s) + 2H_2O(\ell)$$

The difference between the two half-cells is that the pH in one is 1.0 whereas the pH in the other is 2.0.
(a) Write the net equation for the spontaneous process.
(b) What is the cell potential?
(c) What is the equilibrium constant?

69. Draw a simple schematic diagram of a cell for the electrolytic decomposition of water. What gases will be found and in what relative proportions? At which electrode will each collect? Write the half-cell reactions for each electrode processes and note the direction in which ions and electrons are flowing.

70. Exactly 1000. mL of a 0.100 M NaCl solution was electrolyzed for 100. minutes with a 1.0 A current. Calculate the pH of the solution in the cell due to the hydroxide ion concentration that accumulates there. (See *Processes in Chemistry: Chlorine and Industrial Electrochemistry.*)

Multiple Principles [71–73]

71. When XeO_3 is reduced, the product is Xe gas. A 5.00-g sample containing some Na_2SO_3 is treated with XeO_3. Under these conditions the Na_2SO_3 is oxidized to Na_2SO_4. The Xe gas produced is collected and measured and is found to occupy 175 mL at 35.0°C and 0.950 atm. What is the percentage of Na_2SO_3 in the sample?

72. A concentration cell has Zn electrodes. The electrolyte in each of the half-cells is a solution of $ZnCl_2$ dissolved in water. One has a freezing point of −2.00°C. The other has a freezing point of −0.90°C. What is the potential of the cell at 25°C?

73. A galvanic cell consists of one half-cell with a Zn electrode in a 0.100 M Zn^{2+} solution and a second half-cell with a Pt electrode in a 0.100 M solution of HIO_3. Write the equation for the spontaneous cell reaction and calculate the cell potential. The acid ionization constant (K_a) for HIO_3 is 1.9×10^{-1}.

Applied Problems [75–79]

74. Calculate the approximate specific energies of the nickel-cadmium and sodium-sulfur cells considering only the reacting species in the overall reactions.

75. Aluminum metal can be won from alumina (Al_2O_3) in an electrolytic process. What is the maximum number of grams of pure metal that can be obtained from 1.0 metric ton (1000 kg) of crude ore containing 1.5% alumina?

76. How much energy is required to electroplate 1.00 g of silver metal onto a metal surface from a 1.00 M solution of $AgNO_3$? How does the amount of energy required change if the $AgNO_3$ solution is 2.00 M?

77. Chlorine is one of the products of the electrolysis of molten sodium chloride. Use the van der Waals equation to calculate the pressure at 35°C that would result if the chlorine produced with a yield of 96.0% from 10.00 kg of sodium chloride were compressed into a tank with a volume of 75.0 L.

78. Calculate the standard cell potential for a fuel cell in which ammonia and oxygen gas react to produce nitrogen gas and water vapor.

79. If the fuel cell in Problem 79 operates at an efficiency of 40.0%, how many coulombs of charge will be produced if 100. g of ammonia are consumed?

ESTIMATES AND APPROXIMATIONS [80–83]

80. A balloon is being filled with hydrogen produced by the electrolysis of an aqueous solution of an acid. How long will it take to generate enough hydrogen to lift 1.50 kg using a current of 8.5 A?

81. A galvanic cell consists of a Cu electrode in 0.500 L of 1.00 M $CuSO_4$ and a Zn electrode in 0.500 L of 1.00 M $ZnCl_2$. A steady current of 1.00 A is drawn from the cell. Plot the cell potential as a function of time at 1-hour intervals from 0 to 10 hours and use this graph to estimate the power output during this time.

82. What do you think is the chemical basis for the corrosion resistance of stainless steel? Briefly explain.

83. How many coulombs of electricity are consumed in the average U.S. household each day?

FOR COOPERATIVE STUDY [84–86]

84. A lead storage battery is allowed to discharge until 23.92 g of PbO_2 have been reduced at the cathode. Determine each of the following:

(a) the numbers of grams of $PbSO_4$ formed

(b) the time required to recharge the battery to its original state using a current of 3.0 A.

85. Two different types of batteries discharge according to the following reactions:

$$Fe(s) + Ni_2O_3(s) + 3H_2O(\ell) \longrightarrow Fe(OH)_2(s) + 2Ni(OH)_2(s)$$

$$Cd(s) + Ni_2O_3(s) + 3H_2O(\ell) \longrightarrow Cd(OH)_2(s) + 2Ni(OH)_2(s)$$

Beginning with exactly 100.0 g of pure Fe and Cd, determine each of the following:
(a) Which battery will have undergone the greatest chemical change per unit time assuming equal currents?
(b) Assuming both batteries generate 1.40 V, which converted the greater quantity of chemical energy to electrical energy?
(c) If a 10 A recharging current is used, which is restored to its original condition faster?

86. Derive the equation for the value of E for the reaction in the lead-acid storage battery as a function of pH if all other concentrations are held at 1 M.

WRITING ABOUT CHEMISTRY [87–88]

87. The standard potential ($E°$), the standard free energy change ($\Delta G°$), and the equilibrium constant (K) are all indicators of the spontaneity of a reaction. Which of these three indicators gives you the best feeling of the degree of spontaneity of a reaction? Explain your answer carefully, and include a discussion of the nature of each of the three indicators and the nature of chemical reactions.

88. Electrochemical cells were the earliest method of producing an electric current. Luigi Galvani demonstrated in 1786 that two dissimilar metals (the electrodes) caused a twitch in a frog's muscle (the voltmeter). Alessandro Volta used zinc and silver electrodes to construct the first battery in 1800. However, Michael Faraday and Joseph Henry showed in the 1820s that an electric current could be produced by moving a magnet near a circuit and the electric generator was born. In succeeding years the generator eclipsed the battery as a source of electric current, a situation that exists today. Do you think the battery will come back and rival the generator? Give your reasons and answer this question from as broad a point of view as possible.

chapter fifteen

15

CHEMICAL KINETICS

15.1 HOW CHEMISTRY HAPPENS

Understanding Chemical Reactions

In this chapter we undertake the study of the second of two fundamental approaches to understanding how and why chemistry happens. The first was through thermodynamics, an approach concerned with the equilibrium conditions describing initial and final states associated with chemical change. Chapters 10–13 were devoted to aspects of chemical thermodynamics and chemical equilibrium. Thermodynamics offers a means of measuring the driving force for a system to change from one state to another. It lets us determine which direction is spontaneous, but provides no information about the rate of change. The second approach is through kinetics, which deals with the *rate of change* from initial to final states, under nonequilibrium conditions. Thus, the two approaches are complementary. You recognize the thermodynamic driving force to be the change in free energy of the reaction: The tendency to resist chemical change is known as the activation energy and is usually independent of the driving force. A very spontaneous reaction in the thermodynamic sense is not necessarily rapid, nor does a rapid process necessarily proceed to completion.

The details of a chemical reaction are of great interest for practical reasons. Suppose a reaction that we want to happen is known to be thermodynamically favorable—that is, has a negative value of ΔG. Such a reaction may not be useful if it is either too slow or too fast. If the reaction takes place too slowly, it may be uneconomical. As a manufacturer, for example, you will not profit, or as a researcher your time will be poorly spent. Although the very slow conversion of vegetation into petroleum and coal has indeed proved useful, the study of reactions that take place over geologic time frames is not often practical. On the other hand, very fast, sometimes explosive reactions can also be of limited value because they can prove difficult to control. But many chemical reactions are both very fast and very important. For example, reactions in the human body involving enzymes and neurons are very rapid and highly selective, and there are modern techniques available for their careful and effective study.

Understanding kinetic factors can lead to control of a chemical reaction by minimizing competing reactions and formation of undesirable side products and by accelerating slower reactions or moderating faster ones. Corrosion is an example of a slow process that can be made slower still if we first understand the details of the process—an obviously desirable result! The "pinging" sound sometimes evident as you push on the accelerator while driving up a steep incline in a car powered by an internal combustion engine is the result of very fast processes that can be largely controlled in the modern high-compression engine. Once again, highly desirable! By learning what we can about the kinetics of combustion processes, better designs for the explosion chamber of the automobile engine and the combustion characteristics of the fuel are both possible.

Thermodynamic Versus Kinetic Control of Chemical Reactions

Under normal conditions, precipitation of a sparingly soluble salt (Chapter 10) is governed by purely thermodynamic considerations. One example is the precipitate that forms when a small amount of sulfide ion (S^{2-}) is added to an aqueous solution of Zn^{2+} and Fe^{2+} ions:

$$Zn^{2+}(aq) + S^{2-}(aq) \rightleftharpoons ZnS(s) \qquad K_{eq} = 1/K_{sp} = 2.2 \times 10^{23}$$

$$Fe^{2+}(aq) + S^{2-}(aq) \rightleftharpoons FeS(s) \qquad K_{eq} = 1/K_{sp} = 2.7 \times 10^{18}$$

It is the larger equilibrium constant for the less soluble ZnS that governs this competition between Fe^{2+} and Zn^{2+} for S^{2-} ions, and the reaction that occurs first as S^{2-} ions are slowly added is selective precipitation of zinc sulfide. The reaction is said to be thermodynamically controlled. The result of a **thermodynamically controlled** reaction depends on the equilibrium constant. As a practical matter, the reaction affords us an efficient way of separating Zn^{2+} from Fe^{2+} ions by selective precipitation of their sparingly soluble sulfides.

Now consider two gas-phase reactions that outwardly appear to be similar but with puzzling differences:

$$2NO(g) + O_2(g) \longrightarrow 2NO_2(g)$$

$$2CO(g) + O_2(g) \longrightarrow 2CO_2(g)$$

Although the equilibrium constants for both reactions are large numbers and both reactions are spontaneous, the atmospheric oxidation of NO is very fast, and the atmospheric oxidation of CO is very slow. Different reaction pathways are available to CO and NO, and we speak of the two oxidations as being kinetically controlled. The result of a **kinetically controlled** reaction depends on the rate or speed of the reaction. There are practical implications. The sickly yellowish cast to the air during temperature inversions in urban environments is due to the rapid conversion of colorless NO to obnoxious, brownish fumes of NO_2. And have you ever experienced headache and nausea in urban traffic jams? These are symptomatic of carbon monoxide poisoning from steadily increasing concentrations of long-lived CO molecules due to the unfavorable *kinetics*, the slow rate of the reaction in which carbon monoxide is oxidized to carbon dioxide.

An elegant example of thermodynamic versus kinetic control is the reaction of ethanol with acid. At 140°C in *dilute* sulfuric acid solution, the principal reaction is dehydration of two moles of ethanol, yielding one mole of diethyl ether and one mole of water. At 180°C in *concentrated* sulfuric acid, the principal reaction is dehydration of one mole of ethanol, yielding one mole of ethene (ethylene) and one mole of water.

At 140°C in H_2SO_4 solution:

$$\underset{\text{2 molecules of ethanol}}{C_2H_5OH + HOC_2H_5} \rightleftharpoons \underset{\text{diethyl ether}}{C_2H_5{-}O{-}C_2H_5} + H_2O$$

At 180°C in concentrated H_2SO_4:

$$\underset{\text{1 molecule of ethanol}}{C_2H_5OH} \rightleftharpoons \underset{\text{ethene}}{C_2H_4} + H_2O$$

Success in the competition between the two reactions is not due to differences in equilibrium constants—thermodynamics—but rather because the rate of each of the reactions is governed by the particular conditions. This is kinetic control. The products of the faster reaction predominate in each case.

From these examples, it should be abundantly clear that the rates of reactions have important consequences. We will now consider what rates of reactions are and how they are measured.

15.2 REACTION RATES AND RATE LAWS

Rate of a Reaction

During the course of a chemical reaction, reactants, or starting materials, are converted into products. Chemistry is happening and things are changing. The speed of the reaction is measured as either the amount of product produced per unit of time or the amount of reactant consumed per unit of time. A classic lecture demonstration and laboratory example is the reaction of zinc with hydrochloric acid. Following the progress of the reaction is simplified by the fact that one of the products is a gas:

$$Zn(s) + 2H^+(aq) \longrightarrow Zn^{2+}(aq) + H_2(g)$$

The reaction begins immediately on addition of the metal to 1 M aqueous hydrochloric acid at 20°C. We can track the progress of the reaction by observing the volume of hydrogen gas produced as measured by the piston rising in the

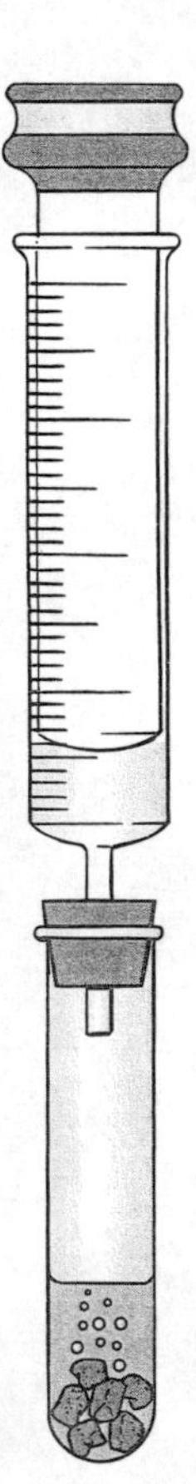

syringe (Fig. 15.1). The reaction is fast enough at that temperature and acid concentration that observations have to be made at relatively short intervals of no more than twenty seconds each. Data are recorded, organized conveniently in tabular form, and finally plotted as the total volume of H_2 collected as a function of time. The rate of the reaction, as measured in mL of H_2 produced per second, is the slope of the curve in Figure 15.2. Initially, hydrogen is produced at a speed of 15 mL/20 s—that is, 0.75 mL/s—after which the speed noticeably decreases, eventually falling to zero as the zinc supply is depleted. Note that the slope of the line, and, therefore, the rate of the reaction, changes as the reaction proceeds.

Repeating the experiment without altering any essential feature except the concentration of hydrochloric acid gives the results shown in Figure 15.3. When 2 M HCl is used, hydrogen is produced during the early part of the reaction at a more rapid rate, about 30 mL/20 s rather than the 15 mL/20 s produced at the lower concentration. The full effect is easily explored by adding zinc to several different concentrations of acid, from which you can conclude that the rate of the reaction increases with increasing concentration of the HCl reactant. Furthermore, the dependence of reaction rate on concentrations of reactants is observed in most chemical reactions. For reactions with gaseous reactants, the reaction rate depends on the partial pressures of reacting species. Now let us consider the nature of reaction rates and their dependence on reactant concentration or partial pressure more closely.

FIGURE 15.1 Following the course of a chemical reaction when one product is a gas. In this simple method, the test tube attached to the gas syringe contains zinc and hydrochloric acid:

$$Zn(s) + 2HCl(aq) \longrightarrow ZnCl_2(aq) + H_2(g)$$

Variation of Rate with Time

It is difficult to arrive at a quantitative statement for the rate of reaction, because the rate is different every time you look. The quantity measured—concentration, partial pressure, or gas volume—at first typically changes rapidly, then slows with time, and eventually approaches to completion of the reaction or reaches equilibrium. As the concentration decreases, so does the slope of the line (Fig. 15.2), which represents the rate of change in the concentration with respect to time.

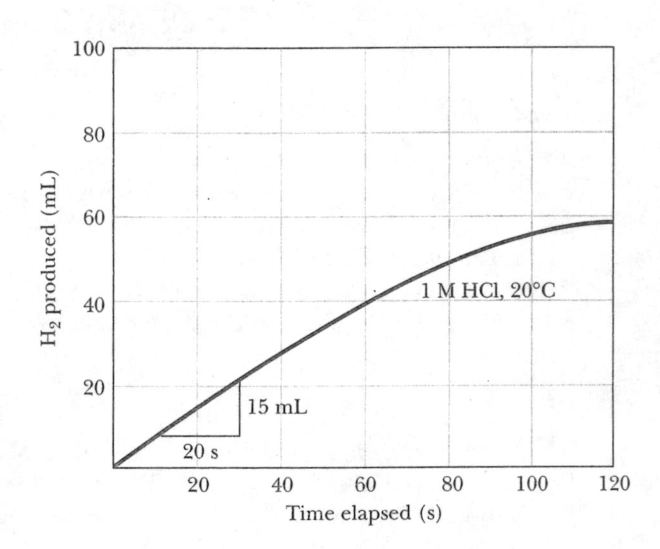

FIGURE 15.2 Production of hydrogen gas versus time elapsed. The slope of the line represents the time rate of change in concentration. In this example, the acid is 1 M and the evolution of hydrogen gas is measured as milliliters generated per second of time elapsed.

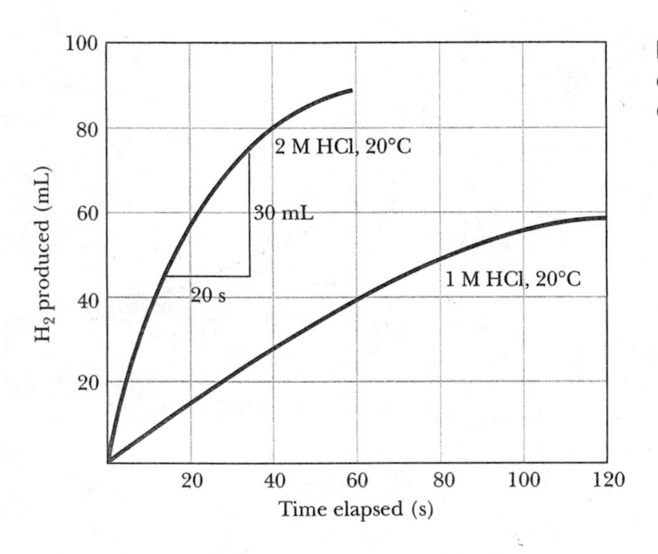

FIGURE 15.3 Concentration effects. Note the vast differences in the slopes of the two curves, one at double the concentration of the other at constant temperature.

If a change in concentration Δc occurs in a small time interval Δt, then the rate of change in concentration during that time interval is the ratio $\Delta c/\Delta t$. When the time interval Δt is made sufficiently small, $\Delta c/\Delta t$ can be regarded as the change in concentration per unit time at that particular instant and at that concentration. As an example, sucrose, common table sugar, can be split or *hydrolyzed* by water into two simpler sugars, glucose and fructose, which we depict simply as a word equation:

$$\text{Sucrose} + \text{water} \longrightarrow \text{glucose} + \text{fructose}$$

The rate of the reaction can be stated in terms of the appearance of either product, glucose or fructose, or the disappearance of the starting material, sucrose. Such reactions are normally run in a large excess of water, and so disappearance of water would not be monitored. Stating the rate of reaction in terms of glucose produced, we write the following:

$$\text{Rate} = \frac{\text{change in glucose concentration}}{\text{time elapsed}} = \frac{c_2 - c_1}{t_2 - t_1} = \frac{\Delta c_{\text{glucose}}}{\Delta t}$$

In this equation, c_2 is the concentration of glucose at time t_2, the end of the elapsed time, and c_1 is the concentration of glucose at time t_1, the beginning of the experiment.

If the rate of this reaction is expressed in terms of the disappearance of sucrose, the expression is similar. However, because c_2 will be less than c_1 as the sucrose is being used up, the expression involving the concentration change must be given a negative sign so that the rate will have a positive value:

$$\text{Rate} = -\frac{\text{change in sucrose concentration}}{\text{time elapsed}} = -\frac{c_2 - c_1}{t_2 - t_1} = -\frac{\Delta c_{\text{sucrose}}}{\Delta t}$$

The experimental data described in Figure 15.4 indicate that as sucrose hydrolysis proceeds, it slows. The plot of c versus t is not linear, and the actual reaction rate—that is, the slope—changes with time. As a consequence, $\Delta c/\Delta t$ represents the average rate during the particular time interval Δt and does not accurately represent the rate of change at a single instant. Strictly speaking, Δt

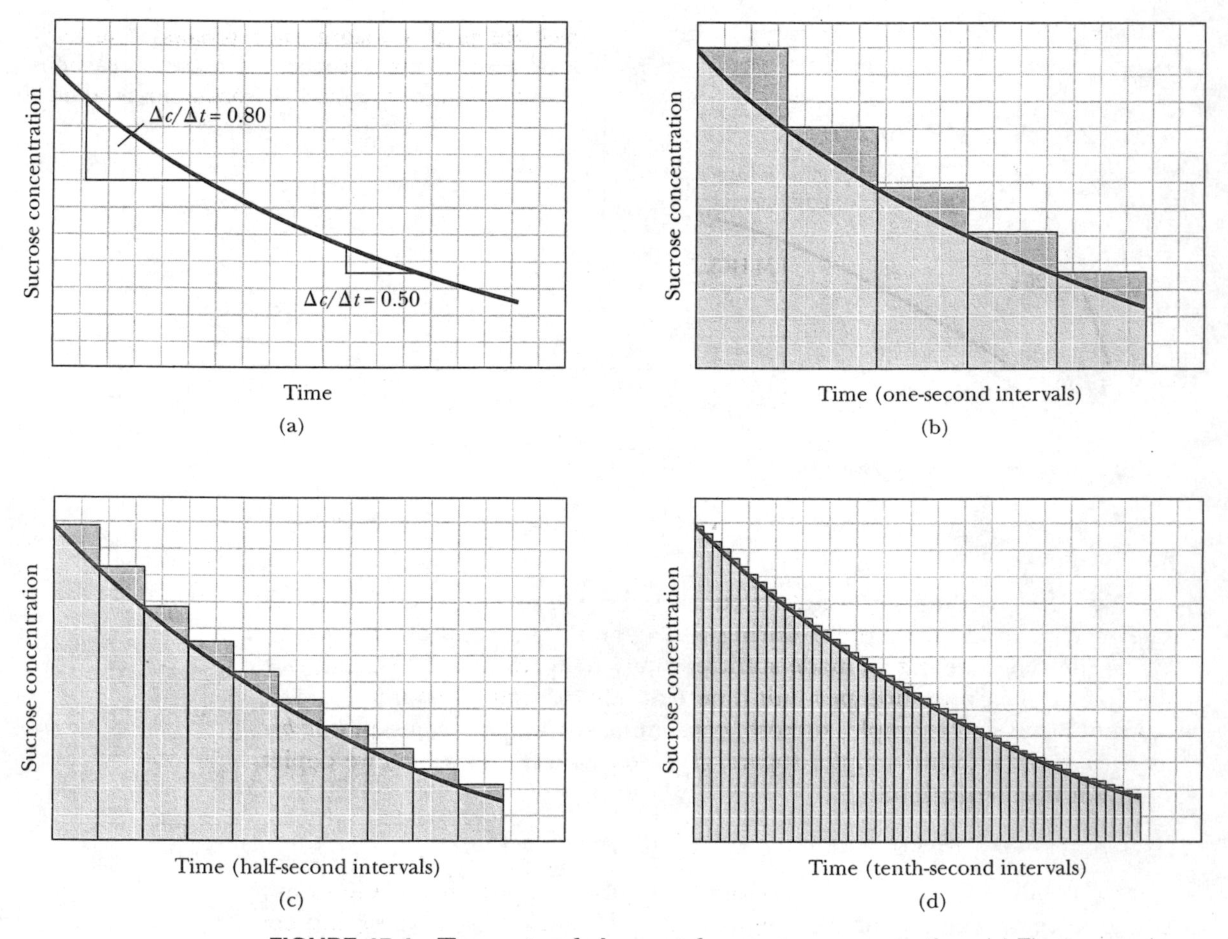

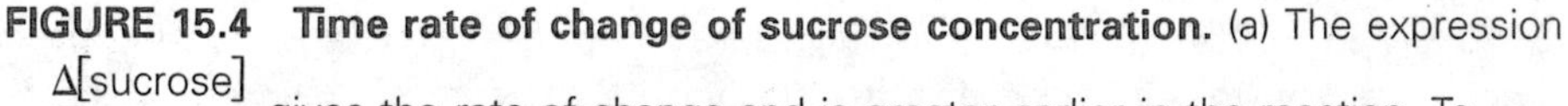
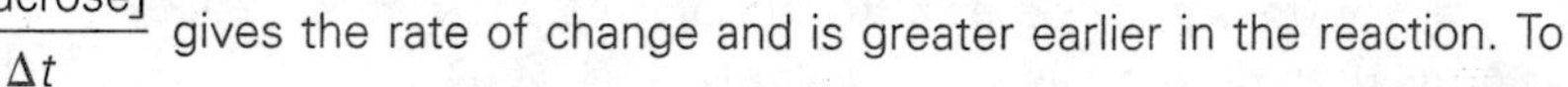

FIGURE 15.4 Time rate of change of sucrose concentration. (a) The expression $-\dfrac{\Delta[\text{sucrose}]}{\Delta t}$ gives the rate of change and is greater earlier in the reaction. To find the instantaneous change in the concentration with time, the intervals are made smaller: (b) five one-second intervals; (c) ten half-second intervals; (d) many more still smaller intervals. When the intervals are infinitesimally small, points on the curve represent the instantaneous rate of change.

should be infinitesimally small if we want to express the rate properly as the instantaneous change of concentration with time. Rates of reactions are always changing because they depend on concentrations, which are always changing.

As in the hydrolysis of sucrose, any reactant or product can be followed to determine the reaction rate. The particular choice depends on what is most easily and accurately measured. We will now consider how the dependence of the rate of the reaction on concentrations is expressed.

Rate Laws and Rate Constants

You can determine the rate of a chemical reaction by measuring the concentration of one component as a function of time. The **rate of a chemical reaction** is related to the rate at which products are formed or reactants disappear. But to learn more about the reaction you will also need to know how this rate of change depends on the concentrations of the other substances in the balanced chemical equation. Then there are other substances such as H_3O^+ and OH^- ions that may not appear in the balanced chemical equation but affect the rate and therefore need to be included in this discussion. What is needed is a *general* expression relating the rate of a reaction to the concentrations of *all* species involved. This general expression is called a rate law. A **rate law** is an algebraic expression that shows the dependence of rate on the concentrations of all species involved in the reaction.

As an example, imagine a chemical reaction between reactants A and B leading to product P.

$$A + B \longrightarrow P$$

If we measure the reaction rate before a significant amount of P is formed, we can be fairly certain that the rate depends only on the concentrations of A and B, because the concentration of P will still be so low. The rate would be $\Delta[P]/\Delta t$, the change in the concentration of P per unit time t, and the general expression takes the following form:

$$\text{Rate} = \frac{\Delta[P]}{\Delta t} = k[A]^m[B]^n$$

Rate law: For the general reaction between A and B yielding product *P*. The mathematical equation relates the reaction rate to the concentrations of the reactants, [A] and [B], raised to the powers *m* and *n*. If the rate is directly proportional to the concentration of a reactant, the exponent for that reactant is 1 and is not written. The values of *m*, *n*, and *k* are all determined by experiment, as we will discuss later.

In this rate-law expression, the values of the exponents m and n are generally small whole numbers that state the dependency of the rate on the concentrations of A and B. The proportionality constant k is the specific rate constant. The **specific rate constant** is a measure of the general rate of the reaction: Fast reactions have high values of k and slow reactions have low values of k. For a particular reaction at a fixed temperature, k has a *specific* value.

The rate in the previous law is given in terms of the change of the concentration of the product P. We could also write the reaction rate law in terms of either of the two reactants, A or B.

$$\text{Rate} = -\frac{\Delta[A]}{\Delta t} = -\frac{\Delta[B]}{\Delta t} = k[A]^m[B]^n$$

A negative sign must be used if the rate is written in terms of the disappearance of a reactant since the change in concentration will be negative and reaction rates are given as positive numbers.

Experimental results may simplify the rate law. If, for example, we determine that varying the concentration of B has no effect on the reaction rate, then the term $[B]^n$ can be dropped from the equation. That is, the value of n is zero and the rate law becomes a simpler expression:

$$\text{Rate} = \frac{\Delta[P]}{\Delta t} = k[A]^m$$

These are the basic principles of writing the equation for a reaction rate. Now consider a few specific examples and how the rate law for a reaction leads to a concept called the order of a reaction.

The Order of a Reaction

The individual values of m and n, the exponents in the general rate law expression, are referred to as the **order of the reaction** *with respect to that component.* In the general reaction given previously, the order of the reaction is m with respect to component A and n with respect to component B. The sum of the exponents in the rate law is called the **overall order of the reaction.** The overall order of the generalized reaction in the previous example would be $m + n$. Now let us consider two specific examples. The first is the gas phase decomposition of dinitrogen pentoxide:

$$N_2O_5(g) \longrightarrow N_2O_4(g) + \frac{1}{2}O_2(g)$$

The rate law has been experimentally determined to be the following:

$$\text{Rate} = -\frac{\Delta[N_2O_5]}{\Delta t} = k[N_2O_5]$$

The unwritten exponent for $[N_2O_5]$ in the rate law is 1. That means that the rate of reaction is directly proportional to the concentration of N_2O_5. We say, "This reaction is first order in N_2O_5" and "first order overall."

In our second example, nitrogen dioxide and fluorine gas react in an experiment known to be second order overall:

$$2NO_2(g) + F_2(g) \longrightarrow 2NO_2F(g)$$

The rate law is

$$\text{Rate} = k[NO_2][F_2]$$

and the sum of the exponents for $[NO_2]$ and $[F_2]$ is $1 + 1 = 2$, and so the reaction is first order in NO_2, first order in F_2, and second order overall. The reaction rate can be expressed three different ways: as production of NO_2F and as disappearance of either NO_2 or F_2:

$$\text{Rate} = -\frac{1}{2}\frac{\Delta[NO_2]}{\Delta t} = -\frac{\Delta[F_2]}{\Delta t} = \frac{1}{2}\frac{\Delta[NO_2F]}{\Delta t} = k[NO_2][F_2]$$

Note that NO_2 is used up at the same rate as NO_2F is produced but twice as fast as F_2 is used up. Therefore, rates expressed in terms of either NO_2 or NO_2F must be divided by 2 to be equal to a rate expressed in terms of F_2. In general, the coefficient of a substance in a chemical equation shows up in the denominator of the rate expressed in terms of that substance. Once again, note that a negative sign is necessary for a rate expressed in terms of the disappearance of a reactant.

EXAMPLE 15.1

Consider the reaction of dichromate ion with iron(II):

$$Cr_2O_7^{2-}(aq) + 6Fe^{2+}(aq) + 14H^+(aq) \longrightarrow 2Cr^{3+}(aq) + 6Fe^{3+}(aq) + 7H_2O(\ell)$$

The reaction obeys a third-order rate law and is first order with respect to each reactant. Write the rate law.

COMMENT Recall that the order for each species is the exponent of the concentration term for that species and that the overall order is the sum of the exponents.

SOLUTION $\text{Rate} = k[Cr_2O_7^{2-}][Fe^{2+}][H^+]$

EXERCISE 15.1(A)

For a reaction that obeys a stoichiometric equation of the type

$$A + 2B \longrightarrow 3C$$

relate the rate of formation of component C to the rate of the disappearance of components A and B, respectively.

ANSWER

$$\text{Rate} = \frac{1}{3}\frac{\Delta[C]}{\Delta t} = -\frac{\Delta[A]}{\Delta t} = -\frac{1}{2}\frac{\Delta[B]}{\Delta t}$$

EXERCISE 15.1(B)

Determine the relationships that exist between the different ways of expressing the rate of reaction for the following chemical equation performed in ether solution:

$$30CH_3OH + B_{10}H_{14} \longrightarrow 10B(OCH_3)_3 + 22H_2$$

ANSWER Even reactions that display unusual stoichiometry are governed by these same rules. Therefore,

$$\text{Rate} = -\frac{\Delta[B_{10}H_{14}]}{\Delta t} = -\frac{1}{30}\frac{\Delta[CH_3OH]}{\Delta t} = \frac{1}{10}\frac{\Delta[B(OCH_3)_3]}{\Delta t} = \frac{1}{22}\frac{\Delta[H_2]}{\Delta t}$$

Rate laws are not derived from theory but from experimental data. So how are the rate laws for chemical reactions determined? It is an important question to which we turn our attention in the next section.

Experimental Determination of Rate Laws

To answer the question we have posed—namely, how to determine rate laws experimentally—we once again consider the general case

$$A + B \longrightarrow P$$

$$\text{Rate} = k[A]^m[B]^n$$

We need to know what values to assign to m, n, and k. If we prepare a solution of the reactants and determine the initial rate right after the reaction starts, then the concentration of the product P is so small that it will not affect the reaction rate and the rate law. We can first determine the value of m by running a series of experiments in which the concentration of A is varied while maintaining the concentration of B at some fixed value. Because our experiments contain only one variable, the initial concentration of A, the computed initial rates should be directly proportional to $[A]^m$. If the rate of reaction doubles on doubling the concentration of A, then the rate depends on the first power of the concentration of A, m equals 1, and we say the reaction is first order in A. Should the reaction rate increase by a factor of four on doubling the concentration of A, then exponent m must equal two, and the reaction is said to be second order in A. That is, the reaction rate is proportional to the concentration of A squared. If doubling the concentration of A has no effect, then the reaction is zero order in A, because $2^0 = 1$. In this case, A would not appear in the rate law.

Once the order has been found with respect to A, more experiments are run, but now the concentration of A is fixed and the concentration of B is varied. That allows us to deduce the order of the reaction with respect to B. Finally, knowing the order with respect to each reactant, we can calculate the overall order of the reaction as the sum of the exponents and calculate the specific rate constant k using the concentration and rate data from any experiment:

$$\text{Rate} = -\frac{\Delta[\text{A}]}{\Delta t} = k[\text{A}]^m[\text{B}]^n$$

$$k = \frac{\text{Rate}}{[\text{A}]^m[\text{B}]^n}$$

If the data are good and the rate law is correct, the calculated rate constants for all experiments should be close in value and can be averaged.

EXAMPLE 15.2

Based on the experimental data provided in the following table, determine the overall order, the general rate law, and the specific rate constant for the following reaction (concentrations are in units of molarity, mol/L):

$$\text{A} + \text{B} \longrightarrow \text{products}$$

Experiment	Concentration of A, (M)	Concentration of B, (M)	Rate (M/min)
1	0.01	0.01	0.05
2	0.01	0.02	0.10
3	0.01	0.03	0.15
4	0.02	0.01	0.20
5	0.03	0.01	0.45

COMMENT Observe what happens to the reaction rate in those experiments for which the concentration of B is changed and the concentration of A is held constant. This will give the exponent in the rate law for [B]. Then do the same thing for A. After that, you should be able to write the rate law and calculate the value of the specific rate constant using data from any of the experiments.

SOLUTION

1. Compare experiments 1, 2, and 3. Because the concentration of A does not change, the observed twofold and threefold increases in the rate are directly due to the twofold and threefold increases in the concentration of B. We conclude that the reaction is first order in component B.
2. Now compare experiments 1, 4, and 5. Because the concentration of B does not change, the observed fourfold and ninefold increases in the reaction rate are due to the corresponding twofold and threefold increases in the concentration of A, leading to the conclusion that the rate of the reaction is directly proportional to the square of the concentration of A. Therefore, the reaction is second order in component A.
3. The reaction would be described as third order overall, and the rate-law expression is written as

$$\text{Rate} = k[\text{A}]^2[\text{B}]$$

4. To obtain the specific rate constant, use a set of concentration data for A and B from any of the five experiments. Choosing experiment 1, for example, gives the following:

$$k = \frac{\text{rate}}{[\text{A}]^2[\text{B}]} = \frac{5 \times 10^{-2}\ \text{M/min}}{[0.01]^2[0.01]} = 5 \times 10^4\ \text{M}^{-2}\ \text{min}^{-1}$$

Using the data from experiment 2 gives the same result:

$$k = \frac{\text{rate}}{[\text{A}]^2[\text{B}]} = \frac{0.1\ \text{M/min}}{[0.01]^2[0.02]} = 5 \times 10^4\ \text{M}^{-2}\ \text{min}^{-1}$$

EXERCISE 15.2(A)

Evaluate the overall order of the reaction from the following data:

$$\text{A} + \text{B} + \text{C} \longrightarrow \text{products}$$

Experiment	[A]	[B]	[C]	Rate (M/min)
1	1.0	1.0	1.0	1.0
2	2.0	1.0	1.0	2.0
3	1.0	2.0	1.0	2.0
4	1.0	1.0	2.0	1.0

ANSWER Rate = $k[\text{A}][\text{B}]$; overall second order

EXERCISE 15.2(B)

Again, consider a reaction of A, B, and C to give products, only this time it is found that the rate doubles when either the concentration of A or B doubles but the rate is halved when the concentration of C doubles. Write down a rate law that is consistent with these data.

ANSWER

$$\text{Rate} = k\,\frac{[\text{A}][\text{B}]}{[\text{C}]} = k[\text{A}][\text{B}][\text{C}]^{-1}$$

EXERCISE 15.2(C)

Show the relationships between the reaction-rate constant and the rates of the reaction as expressed in terms of disappearance of A and of B, and the appearance of C, respectively, for a reaction

$$3\text{A} + \text{B} \longrightarrow 2\text{C}$$

known to be first order in A and first order overall.

ANSWER

$$\text{Rate} = -\frac{1}{3}\frac{\Delta[\text{A}]}{\Delta t} = -\frac{\Delta[\text{B}]}{\Delta t} = \frac{1}{2}\frac{\Delta[\text{C}]}{\Delta t} = k[\text{A}]$$

Now that you have an idea of how rate laws can be determined for chemical reactions, we can discuss in the next section the kind of information that a rate law can provide about how chemical reactions proceed. We will limit our discussion to first-order, simple second-order, and zeroth-order reactions. In a zeroth-order reaction, the rate is independent of the concentrations of any of the reactants or products.

15.3 INFORMATION FROM RATE LAWS

Rate Law of a First-Order Reaction

We have already noted that the decomposition of dinitrogen pentoxide in the gas phase is a **first-order reaction** for which we can write the experimental rate law as follows:

$$N_2O_5(g) \longrightarrow N_2O_4(g) + \frac{1}{2}O_2(g)$$

$$\text{Rate} = -\frac{\Delta[N_2O_5]}{\Delta t} = k[N_2O_5]$$

If a reaction is first order, then the rate depends on the first power of the concentration, and the rate law is

$$\text{Rate} = -\frac{\Delta[\text{A}]}{\Delta t} = k[\text{A}]$$

Using calculus, this expression can be converted to an equation that relates the initial concentration of the reactant $[A]_0$, the concentration $[A]$ when time t has passed, and the rate constant.

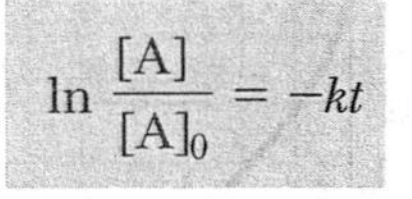

$$\ln \frac{[A]}{[A]_0} = -kt$$

Concentration as a function of time for a first-order reaction: In this expression, $[A]_0$ is the initial concentration of A at time $t = 0$, and $[A]$ is the concentration of A some time t later. The equation is of the form, $y = b + mx$, where b is zero, and so it plots as a straight line that intersects the origin.

A plot of $\ln([A]/[A]_0)$ versus t gives a straight line with negative slope that is equal to $-k$. This provides a graphical method for obtaining k and a pictorial statement of the order of the reaction. If plotting concentration and time in the form of this equation gives a straight line through the origin, then the reaction *may be first order.* And if plotting the data according to this equation does not give a straight line, then the reaction *cannot be first order.*

The plots in the following figures represent typical sets of kinetic data for a first-order chemical reaction describing the decomposition of A to products. Figure 15.5a (concentration versus time) shows concentration decreasing rapidly with time from some initial concentration but with the rate slowing all the while, and finally approaching zero at completion.

Rearranging the equation for a first-order reaction gives another way to plot the data and obtain k.

$$\ln \frac{[A]}{[A]_0} = -kt$$

$$\ln [A] - \ln [A]_0 = -kt$$

$$\ln [A] = \ln [A]_0 - kt$$

This is also the equation of a straight line ($y = b + mx$) and has an intercept of $\ln [A]_0$. Plotting $\ln [A]$ versus time gives a straight line with negative slope of $-k$, as shown in Figure 15.5b, providing another method of determining the specific rate constant. Taking the antilog of both sides of this equation yields

$$[A] = [A]_0 e^{-kt}$$

which shows that the initial concentration $[A]_0$ decays exponentially with time. Exponential decay is a unique feature of first-order rate processes.

Half-Life for a First-Order Reaction

A useful measure of reaction rate is the half-life ($t_{1/2}$) for a chemical reaction. The **half-life** is the time required for half the reactant initially present to decompose. A fast reaction has a short half-life and a slow reaction has a long

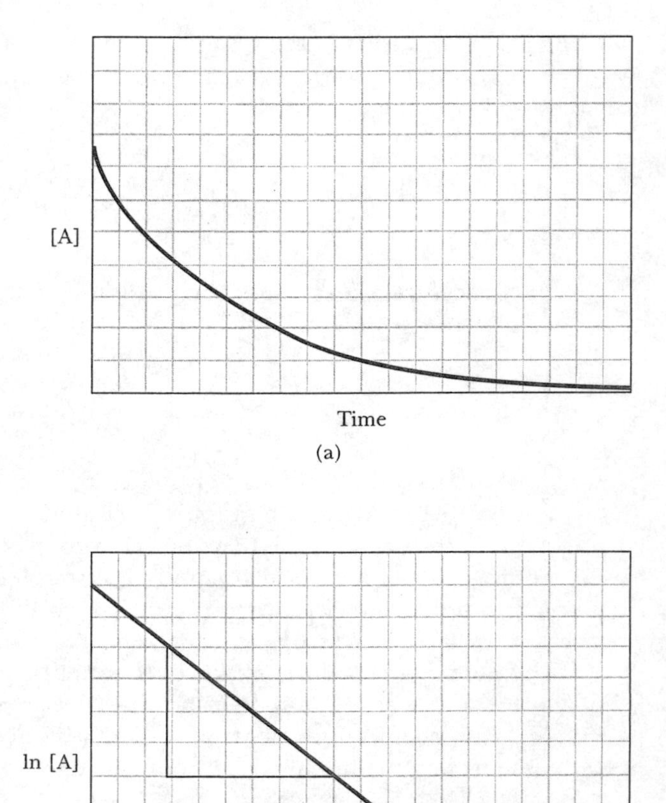

FIGURE 15.5 The order of a chemical reaction. (a) Plot of concentration versus time for a typical chemical reaction; (b) for a first-order reaction, a plot of ln [A] versus time is a straight line with a slope of $-k$.

half-life. To determine the equation for half-life, we start with the equation for a first-order reaction

$$\ln \frac{[A]}{[A]_0} = -kt$$

and recognize that $t_{1/2}$ is the special case in which half of A is decomposed, so that the concentration of A is

$$[A] = 0.5[A]_0$$

This leads to the simple but important relationship

$$\ln \frac{0.5[A]_0}{[A]_0} = \ln 0.5 = -0.693 = -kt_{1/2}$$

On rearrangement,

$$kt_{1/2} = 0.693$$

$$t_{1/2} = \frac{0.693}{k}$$

Half-life for a first-order reaction: *k* is the specific reaction rate constant. Note that the half-life is independent of concentration.

No concentration terms appear in the equation for $t_{1/2}$. No matter how concentrated the starting material is, it takes the same time for half of it to decompose. Concentration independence of the half-life is the unique signature of a first-order reaction.

To illustrate what we mean by half-life, consider the fate of a million molecules undergoing a first-order decomposition with $t_{1/2}$ equal to a minute (Fig. 15.6). During the first minute, 500,000 molecules disappear, leaving an equal number of molecules that have not reacted. During the next one-minute interval, half the remaining molecules react, leaving 250,000 that have not reacted. After three minutes, or three half-life intervals, 125,000 remain; then 62,500; 31,250; and so forth.

EXAMPLE 15.3

Sulfuryl chloride spontaneously decomposes at 600K according to the equation

$$SO_2Cl_2(g) \longrightarrow SO_2(g) + Cl_2(g)$$

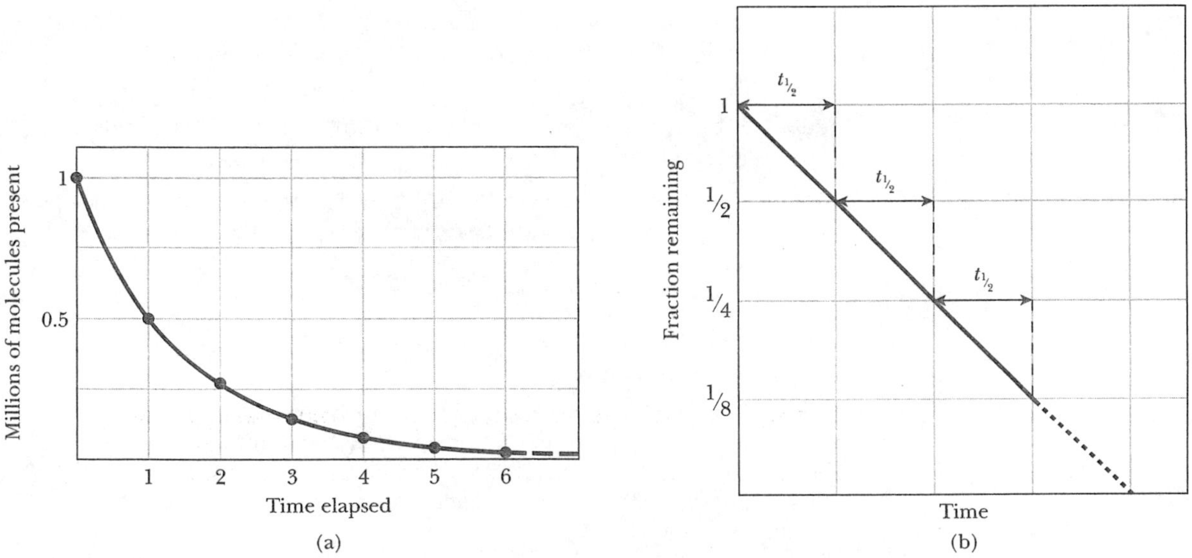

Millions of molecules present
1
0.5
1 2 3 4 5 6
Time elapsed
(a)

Fraction remaining
1
1/2
1/4
1/8
$t_{1/2}$
Time
(b)

FIGURE 15.6 The half-life of a chemical reaction: (a) During each interval, half the remaining molecules decay, decompose, or otherwise react (b) Here, three successive half-lives are marked off.

The first-order rate constant is 1.3×10^{-3} min^{-1}. What fraction of the original sulfuryl chloride sample remains after heating a sample for 30. minutes at 600K? How long will it take 50.% of the molecules in the original sample to decompose to products under these same conditions?

COMMENT Use the equation $\ln([A]/[A]_0) = -kt$ and note that $[A]/[A]_0$ is the fraction of the original sample that remains. Use $t_{1/2} = 0.693/k$ to determine the half-life.

SOLUTION

$$\ln\frac{[A]}{[A]_0} = -kt = -(1.3 \times 10^{-3}\ \text{min}^{-1})(30\ \text{min}) = -3.9 \times 10^{-2} = -0.039$$

$$[A]/[A]_0 = e^{-0.039} = 0.96 = \text{fraction remaining after 30 min}$$

$$t_{1/2} = \frac{0.693}{k} = \frac{0.693}{1.3 \times 10^{-3}\ \text{min}^{-1}} = 530\ \text{min} = \text{time for 50\% to decompose}$$

EXERCISE 15.3

If $t_{1/2}$ for a first-order reaction is 600. seconds, approximately how long will it take for 94% of the reactant molecules to decompose to products?

ANSWER About 2400 s

Rate Law of a Simple Second-Order Reaction

The following rate law represents an important class of **second-order reactions**. The rate law for a reaction with a rate that depends on the concentration of only one substance, but to the second power, leads to the equation

$\text{Rate} = k[A]^2$ Second-order rate law depending on one reactant

Once again, as with the rate law for a first-order reaction, calculus can be used to derive a useful form of the rate law for a second-order reaction:

$$\frac{1}{[A]} = \frac{1}{[A]_0} + kt$$

Concentration as a function of time for a simple second-order reaction: This gives the relationship between the reactant concentration and time in which the second-order rate law depends on only one reactant. $[A]$ is the existing concentration, and $[A]_0$ is the initial concentration.

Note that once again we have the equation for a straight line in the form

$$y = b + mx$$

If $1/[A]$ is plotted as a function of t for a second-order reaction, the result is a straight line with positive slope equal to the specific rate constant k. The intercept of the line at $t = 0$ is $1/[A]_0$. This intercept can be used to find the initial concentration. Plotting $1/[A]$ versus t and obtaining a straight line with a positive slope is the signature of a second-order reaction.

The half-life can be obtained for a second-order process by rearranging this equation:

$$\frac{1}{[A]} - \frac{1}{[A]_0} = kt$$

Just as was done for a first-order process, we set $[A] = [A]_0/2$ at $t = t_{1/2}$. The result is

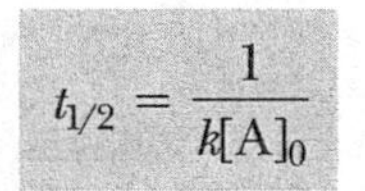

$$t_{1/2} = \frac{1}{k[A]_0}$$

Half-life for a second-order reaction: The half-life is inversely proportional to concentration.

Thus, the half-life for a second-order process is *concentration dependent:* It becomes progressively longer as the starting concentration decreases. If an initial half-life is 10 minutes, the *next* one will be twice as long (20 min), and the third one will be 40 minutes, and so on. This behavior is in marked contrast to first-order half-lives, which are independent of the starting point.

Rate Law of a Zeroth-Order Reaction

If the rate of a chemical reaction is independent of the concentration of any component, the reaction is described as being **zero order**. Consider the reaction

$$A \longrightarrow P$$

If the reaction is zero order, then the rate law must be

$$-\frac{\Delta[A]}{\Delta t} = k$$

from which calculus can again be called on to derive the following general relationship:

$$[A] = [A]_0 - kt$$

Zeroth-order reaction: The reaction is zero order if a plot of concentration versus t yields a straight line. The slope of the line is $-k$.

This is typical of many important industrial processes such as the addition of hydrogen to the double bond in ethylene on a platinum surface. We will have an

PROFILES IN CHEMISTRY

He Was Called "Kisty" George B. Kistiakowsky, known to his students and colleagues as "Kisty," was born in Kiev, Russia, in 1900. It was a time of revolution, and as a young man he fought in the White Russian Army. Kistiakowsky studied chemistry in Germany, and obtained his doctorate in 1926, the year he came to the United States. For more than 40 years afterward, he was professor of chemistry at Harvard University, receiving many awards, including the Priestley medal for his research in chemical kinetics, and was recognized for his work by three presidents of the United States (Truman, Eisenhower, and Kennedy). Kistiakowsky was Eisenhower's assistant for Science and Technology and was chair of the Science Advisory Committee from 1957–1963 under Eisenhower and Kennedy. Besides his benchmark studies on rates of chemical reactions, he was an authority on explosives, and as a key member of the Manhattan Project designed the triggering mechanism for the first atomic bomb, which was tested at Alamagordo, New Mexico, in July 1945. As director of the Los Alamos Laboratory he ranked high among those who developed the bomb, but he almost immediately sensed the extraordinary potential of atomic weapons for destruction and became an advocate for severely limiting, and then preventing, their use. He resigned from the Pentagon in 1967 and returned to full-time teaching and research at Harvard. Summarizing his opinions on nuclear weapons in the last year of his life, 1982, he wrote that the world could well explode unless an unprecedented mass movement for peace could be created to halt the threat of nuclear annihilation.

Among his classic studies on chemical kinetics in the 1930s is the dimerization, or coupling, of butadiene (C_4H_6), an important example of a second-order process involving one reacting species:

$$\underset{\text{monomer}}{2C_4H_6(g)} \longrightarrow \underset{\text{dimer}}{C_8H_{12}(g)}$$

Introducing butadiene into an empty flask at 600K, dimerization begins at time $t = 0$, and the progress of the coupling reaction can be followed by observing the pressure, which decreases as each pair of molecules of butadiene *monomer* becomes one molecule of *dimer*. Kistiakowsky's data support the following rate law:

$$\text{Rate} = -\frac{1}{2}\frac{\Delta p_{C_4H_6}}{\Delta t} = kp^2_{C_4H_6}$$

In the reaction, partial pressures are recorded rather than concentrations. Therefore, the equation for the second-order process takes the form, $1/p = kt + 1/p_0$, where p_0 and p refer to the partial pressure of C_4H_6. For a value of elapsed time, the partial pressure p of C_4H_6 is calculated using the reaction stoichiometry and the observed total pressure. By substituting that partial pressure of C_4H_6 at time t and the initial pressure of C_4H_6 into the equation for the second-order process, the value of the specific rate constant can be calculated. The rate law is consistent with the data, because the same value of the specific rate constant is calculated for each value of t within reasonable limits of error.

opportunity to describe such a reaction later in this chapter when we deal with the subject of catalysis.

Earlier in this chapter we discussed determining the rate of the reaction of HCl with Zn. The concentration dependence of the rate of this reaction can be demonstrated by reacting pieces of zinc with solutions of hydrochloric acid of different strengths and observing the rate of the hydrogen gas being evolved. As the concentration of the acid increased, so too did the rate of evolution of hydrogen gas. In the next section we will describe and discuss three other factors that affect reaction rates: changing the temperature of the zinc–hydrochloric acid reaction, changing the total surface area of the zinc, and adding an additional compound, in this case sodium chloride, to the reaction mixture.

Here is how the partial pressure of C_4H_6 is calculated from the total pressure: Letting $x = p_{C_8H_{12}}$,

$$2C_4H_6 \longrightarrow C_8H_{12}$$

Initial	p_0	0
Final	$p_0 - 2x$	x

Thus,

$$P_{total} = (p_0 - 2x) + x = p_0 - x$$

$$x = p_0 - P_{total}$$

$$p_{C_4H_6} = p_0 - 2x = p_0 - 2(p_0 - P_{Total}) = 2P_{Total} - p_0$$

If $p_0 = 632.0$ torr and $P_{total} = 606.6$ torr at $t = 367$ s, then

$$p_{C_4H_6} = 2P_{total} - p_0 = 2 \times 606.6 \text{ torr} - 632.0 \text{ torr} = 581.2 \text{ torr}$$

The values of t and p_0 and p for C_4H_6 are substituted into the equation

$$\frac{1}{p} - \frac{1}{p_0} = kt$$

$$\frac{1}{581.2} - \frac{1}{632.0} = k(367)$$

$$k = 3.77 \times 10^{-7} \text{ torr}^{-1} \text{ s}^{-1}$$

Similar computations for the other times yield consistent values of k and confirm that the reaction is second order.

Questions

Explain in your own words how Kistiakowsky determined the partial pressure of C_4H_6 as he followed his dimerization reaction.

Name one or more scientists of prominence who have served a visible and important political role.

Calculate $p_{C_4H_6}$ and k at 731 s, if $P_{total} = 584.2$ torr, and at 1038 s, if $P_{total} = 567.3$ torr ($p_0 = 632.0$ torr). Are the k values constant? Calculate $t_{1/2}$ for the reaction.

15.4 FACTORS INFLUENCING REACTION RATES

There are a number of factors that control the rate at which reactions proceed, and by varying these, rates can be controlled.

CONCENTRATION As we have already seen, reaction rates typically depend on the concentrations of reactants (and sometimes products) raised to various powers. Recall the reaction of zinc with hydrochloric acid from our earlier discussion. Hydrogen gas was generated and the rate of its evolution used to follow the progress of the reaction (Figs. 15.1, 15.2, and 15.3):

$$Zn(s) + 2H^+(aq) \longrightarrow Zn^{2+}(aq) + H_2(g)$$

FIGURE 15.7 Temperature effects on the rate of a chemical reaction. Data plot demonstrating temperature effects at 20°C and 35°C for 1 M HCl in the reaction with zinc. Note the rate has increased with temperature.

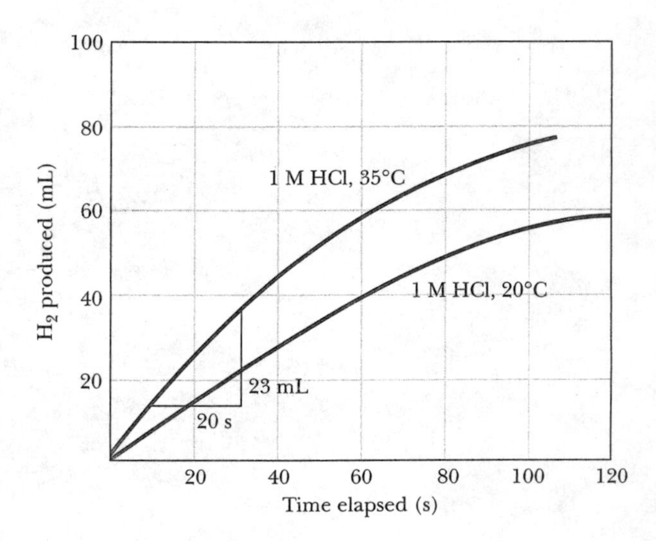

Repeating the experiment without altering any essential feature except the concentration of hydrochloric acid confirmed the concentration dependence of the reaction rate.

TEMPERATURE As pieces of zinc reacted with 1 M hydrochloric acid at 20°C (see Figure 15.2), hydrogen gas was produced at a rate of 15 mL/20 s. Figure 15.7 shows what happened when all experimental factors were kept the same except the temperature, which was raised to 35°C. Once again, hydrogen is produced, but the rate of evolution at the beginning of the reaction has increased to 23 mL/20 s. This particular reaction clearly speeds up at elevated temperature. In fact, the rates of most reactions increase when the temperature is increased. We will say more about the effect of temperature on reaction rates in the next section.

SURFACE AREA Again we can repeat the initial experiment with one essential difference. Instead of having several small pieces of zinc adding up to our 1-gram sample, we have chosen a single piece of zinc of nearly identical mass. The effect is to reduce the surface area compared to the original zinc samples.

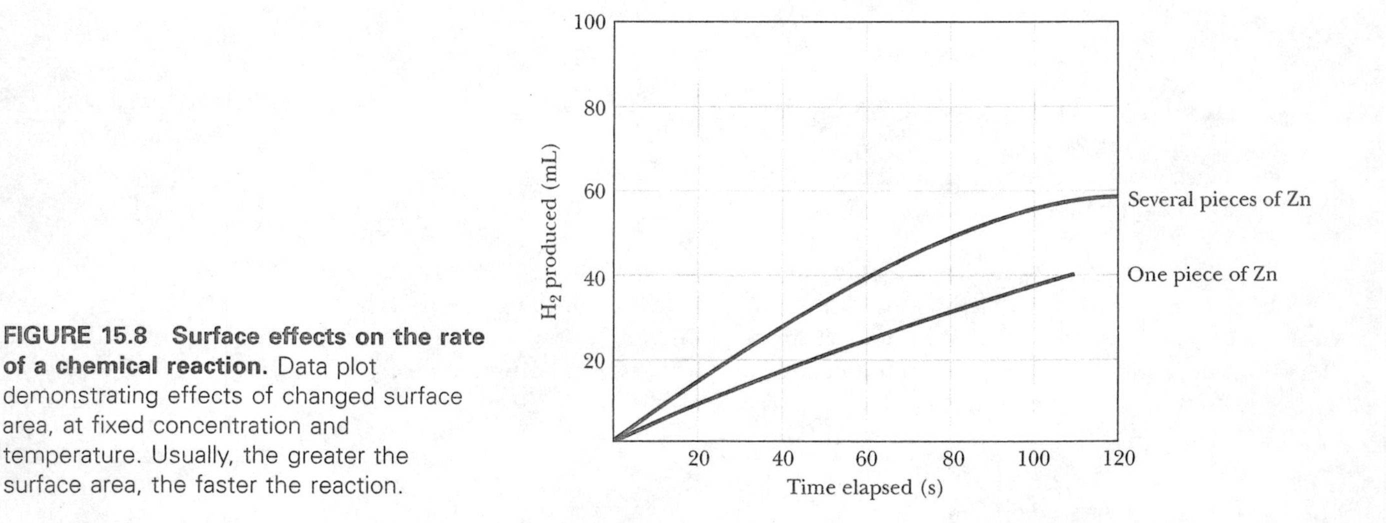

FIGURE 15.8 Surface effects on the rate of a chemical reaction. Data plot demonstrating effects of changed surface area, at fixed concentration and temperature. Usually, the greater the surface area, the faster the reaction.

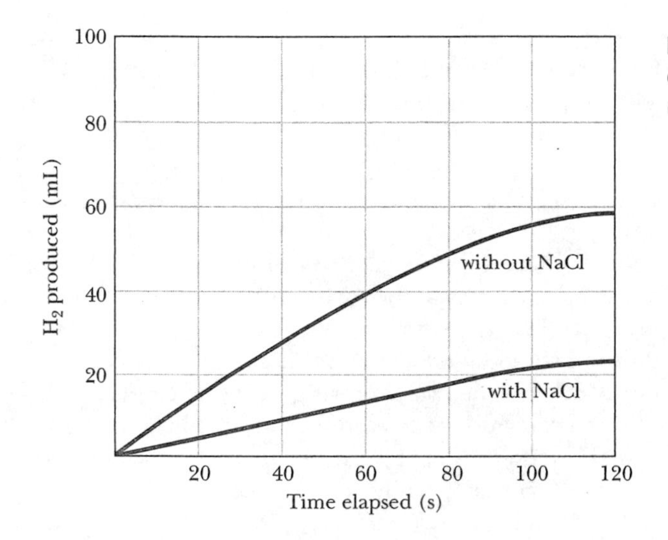

FIGURE 15.9 Data plot demonstrating the inhibiting effect of added NaCl. Salt clearly diminishes the rate of the reaction.

Hydrogen is once again produced smoothly during the course of the reaction, though the rate is considerably diminished (Fig. 15.8). On the other hand, if the metal had been finely powdered, the reaction might have been too vigorous to carry out safely. Thus, the reaction rate increases as the surface area of the metal is increased. This is generally observed in reactions of solids with liquids or gases and the rate is defined in terms of moles of reactant or product per unit area of surface.

CATALYSTS AND INHIBITORS The rates of many reactions are altered by the presence of catalysts or inhibitors. **Catalysts** are substances that participate in the events and accelerate chemical change without undergoing permanent change themselves. **Inhibitors** slow reaction rates and, in common with catalysts, do so without being part of the stoichiometry of the overall reaction. In the case of the Zn–HCl reaction, the presence of dissolved salts such as sodium chloride inhibits the rate of the reaction. You can see that by going back to our original experiment once more. Comparing it to one in which the hydrochloric acid has been saturated with NaCl demonstrates the inhibiting effect, reducing the rate significantly (Fig. 15.9). Sodium chloride inhibition requires that its ions be present in high concentration to be significantly effective. However, many catalysts and inhibitors produce considerable change in reaction rates when present in only very small or even trace quantities, which makes them particularly important. The subject is of such importance as to warrant our full attention later in this chapter. Now we take a closer look at the effect of temperature on the rate of a reaction.

15.5 TEMPERATURE DEPENDENCE OF REACTION RATE

Most chemical reactions proceed more rapidly if the reactants are heated. It is well-known among chemists and photographers, for example, that in the vicinity of room temperature an increase of 10°C approximately doubles the rate of reaction. The time needed to develop a photographic film with a particular developing bath is approximately halved if the temperature is raised from 20°C to 30°C. For a particular chemical reaction, the fact that the rate increases when the temperature is increased means that the value of its specific rate constant (k) increases.

Svante Arrhenius was the first person to explain the effect of temperature on reaction rate in precise terms. He was able to establish a mathematical relationship describing how the speed of a chemical reaction increases with temperature. This formula could be understood by assuming that a reaction between two molecules requires a collision. However, when molecules do collide, they usually part again without anything actually happening. But if the collisions are sufficiently violent—that is, the molecules are moving with sufficient speed—then the molecules might disintegrate and their atoms recombine into new molecules. We know that molecular speed increases with temperature. So a higher temperature will produce more of the necessarily violent collisions. However, it was also realized when a reaction rate could be measured that only a tiny fraction of the collisions involved really resulted in a reaction.

Arrhenius suggested that for a reaction to take place it was necessary for the reacting molecules to pass over an energy barrier separating reactants and products. He referred to the height of this energy barrier as the **activation energy**, E_a. Consider the potential energy diagram for a reaction given in Figure 15.10. The potential energy levels of the reactants and products are shown at the right and left sides of the diagram, and the overall change in energy ΔE for the reaction is simply equal to the difference between these. However, the indicated reaction does not go *directly* from reactants to products. The x-axis of the graph is a measure of the progress of the reaction from reactants to products and is labeled "reaction coordinate." The value of the **reaction coordinate** increases as the reaction proceeds from reactants to products. The activation energy for the forward reaction is the difference between the energy of the reactants and the energy maximum reached along the reaction coordinate. It is this energy maximum, or barrier, that keeps the reactants from being *instantly* transformed to

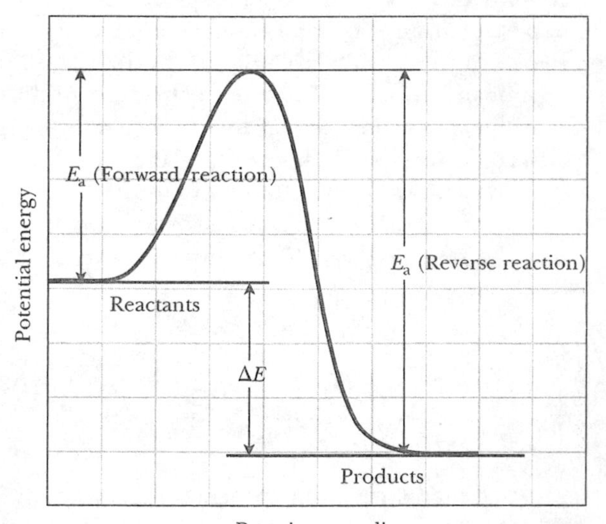

FIGURE 15.10 Progress of an exothermic reaction. The energy maximum corresponds to the activation energy, E_a, at approximately mid-reaction along the coordinate. Note the relationships between E_a for the exothermic forward reaction and the endothermic reverse reaction, as well as the net change in internal energy, ΔE.

products. The higher the activation energy barrier is, the slower the reaction. In a manner of speaking, the activation energy is a measure of the resistance to chemical change, opposing the driving force for chemical change. Whereas the driving force or change in free energy ΔG is determined by the methods described in Chapter 13, the activation energy is usually independent of ΔG and is measured by the methods of kinetics.

When Arrhenius plotted the logarithm of k against $1/T$ (k is the specific rate constant and T is the absolute temperature), he obtained a straight line with a negative slope. From this he produced a simple, general expression relating the specific rate constant, the absolute temperature, and the activation energy, which is widely known as the **Arrhenius equation**:

$$\ln k = -\frac{E_a}{RT} + \ln A$$

Logarithmic form of the Arrhenius equation: In this equation A is called the frequency factor, R is the ideal gas constant, and E_a is the activation energy. The temperature T is in kelvin units.

Because the equation takes the general form of a linear equation, a plot of ln k versus $1/T$ yields a straight line, the slope and the intercept of which yield the parameters E_a and A, respectively:

$$\text{Slope} = -\frac{E_a}{R}$$

$$\text{Intercept} = \ln A$$

Both terms are important, but the interpretation of E_a has special significance, as we will explain in the next section. Such a graph is known as an **Arrhenius plot**.

Taking the antilogarithm of both sides of the logarithmic form of the Arrhenius equation leads to the exponential form of the Arrhenius equation:

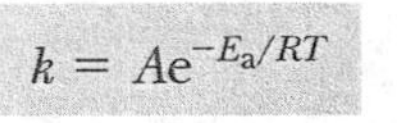

$$k = Ae^{-E_a/RT}$$

Exponential form of Arrhenius equation: Arrhenius equation in convenient form for calculating the rate constant k or the frequency factor A.

The activation energy E_a can also be determined algebraically using the logarithmic form of the Arrhenius equation and values for the rate constants, k_1 and k_2, at two temperatures, T_1 and T_2. In one experiment at temperature T_1,

$$\ln k_1 = \ln A - \frac{E_a}{RT_1}$$

In the second experiment, at some other temperature T_2,

$$\ln k_2 = \ln A - \frac{E_a}{RT_2}$$

Subtracting the expression for the first experiment from that for the second gives the following:

$$\ln k_2 - \ln k_1 = \left(\ln A - \frac{E_a}{RT_2}\right) - \left(\ln A - \frac{E_a}{RT_1}\right)$$

from which the following result can be obtained by rearranging and combining terms:

$$\ln \frac{k_2}{k_1} = -\frac{E_a}{R}\left(\frac{1}{T_2} - \frac{1}{T_1}\right) = -\frac{E_a}{R}\left(\frac{T_1 - T_2}{T_1 T_2}\right)$$

To illustrate how the Arrhenius equation is used algebraically, consider a chemical reaction for which rate constants were determined at two different temperatures:

$$\text{at } T_1 = 288\text{K it was found that } k_1 = 9.67 \times 10^{-6}\ \text{min}^{-1}$$

$$\text{at } T_2 = 333\text{K it was found that } k_2 = 6.54 \times 10^{-4}\ \text{min}^{-1}$$

First, solve the appropriate form of the Arrhenius equation for the activation energy E_a by rearranging the terms in the original equation:

$$\ln \frac{k_2}{k_1} = -\frac{E_a}{R}\left(\frac{1}{T_2} - \frac{1}{T_1}\right) = -\frac{E_a}{R}\left(\frac{T_1 - T_2}{T_2 T_1}\right)$$

$$E_a = -R\left(\frac{T_2 T_1}{T_1 - T_2}\right)\ln \frac{k_2}{k_1}$$

Then substitute the reaction rate and temperature data, use R with proper units of energy to give the answer in joules:

$$E_a = -8.314\ \text{J/mol·K}\left(\frac{333\text{K}\cdot 288\text{K}}{288\text{K} - 333\text{K}}\right)\ln \frac{6.54 \times 10^{-4}}{9.67 \times 10^{-6}} = 74{,}700\ \text{J/mol}$$

EXAMPLE 15.4

The rate of decomposition of N_2O_5 was measured at several different temperatures:

$$2N_2O_5(g) \longrightarrow 4NO_2(g) + O_2(g)$$

Temperature (K)	273	298	318	338
$k \times 10^5$ (min^{-1})	0.0787	3.46	49.8	487

Making use of an Arrhenius plot of log k versus $1/T$ for these data (Fig. 15.11), evaluate the slope and obtain E_a.

COMMENT Here we are plotting log k rather than ln k; ln k = 2.303 log k, and so the slope of the line is $-E_a/2.303R$.

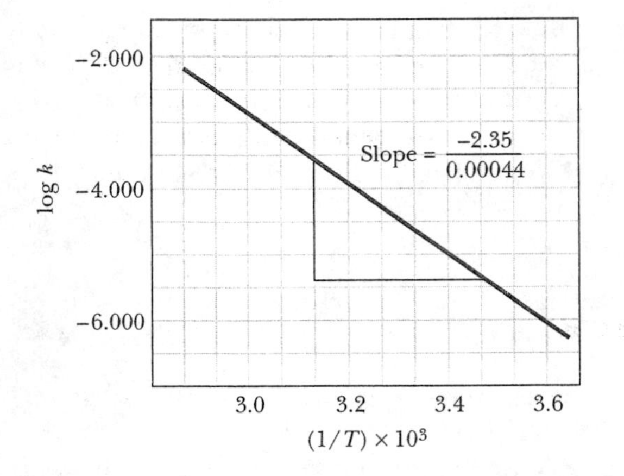

FIGURE 15.11 Arrhenius plot. Log k is plotted against $1/T$ for the decomposition of N_2O_5 in the vapor state.

SOLUTION

$$\text{Slope} = -E_a/2.303R = -5.3 \times 10^3\text{K}$$

$$E_a = 1.0 \times 10^5 \text{ J/mol}$$

EXERCISE 15.4

Use the data plotted in Figure 15.12 to determine E_a by a graphical method. Then taking the following rate constants from the same data plot, algebraically determine the value of the rate constant at 318K:

$$k_1 = 4.5 \times 10^{-5} \text{ min}^{-1} \quad \text{at 298K}$$

$$k_2 = 1.0 \times 10^{-3} \text{ min}^{-1} \quad \text{at 328K}$$

ANSWER $E_a = 8.3 \times 10^4$ J; $k = 4.1 \times 10^{-4}$ min^{-1}

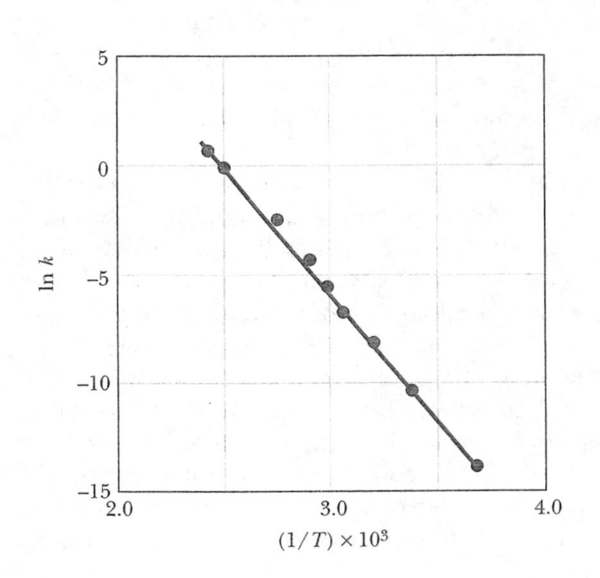

FIGURE 15.12 Arrhenius plot.

In this section we have shown how to calculate the activation energy and how the activation energy of a reaction is related to the temperature dependence of the reaction rate. It remains for us to give the concept of activation energy a physical reality in terms of what actually happens when atoms and molecules react. That reality takes the form of what we shall call collision theory.

15.6 COLLISION THEORY

So far, our discussions of chemical kinetics have been directed at the operational aspects of reaction rates. These have included measuring the rate of a reaction, establishing the experimental rate law, and determining the order of the reaction. However, the ultimate goal of chemical kinetics is to learn about the actual mechanism of the chemical reaction, the series of *chemical* steps that occurs. The **reaction mechanism** is the pathway the reaction follows in going from reactants to products described in microscopic terms—that is, in terms of atoms and molecules.

Information about the mechanism of a reaction must be contained in its rate law. Unfortunately, the amount of information in a rate law is limited, whereas the number of possible mechanisms is great. Therefore, a rate law cannot be definitive proof of a proposed mechanism, although an experimental rate law can be used to disprove a particular reaction mechanism or pathway. Because of these uncertainties, chemical kinetics is subject to interpretation more than most topics covered in general chemistry. We will now consider the possible mechanisms of two apparently simple gas-phase reactions that appear to be similar.

As we mentioned in the opening section of this chapter, the stoichiometries of the reactions of CO and NO with oxygen look alike:

$$2CO(g) + O_2(g) \longrightarrow 2CO_2(g)$$

$$2NO(g) + O_2(g) \longrightarrow 2NO_2(g)$$

However, even though the equilibrium constants of each of these reactions is very large, the reaction of CO with oxygen proceeds very slowly, whereas the reaction of NO with oxygen is very fast. We now need to establish some basic ideas that will allow us to understand how two such superficially similar reactions can have reaction rates that are so different.

To begin with, it is reasonable to assume that when two molecules react, they must have come in contact, or at least have suffered a near collision. Recall our model of the kinetic theory of gases (Chapter 4) in which a gas particle moves in a straight line through space until it collides with another gas particle or a wall of the container. Under similar conditions of temperature and pressure, we expect the number of particle collisions for either of the two reactions to be about the same. Because the reaction of NO with oxygen is so much faster, far more of the collisions must result in reaction. But even the faster NO oxidation is really a very slow reaction compared to the rate for a hypothetical model in which every collision leads to reaction. As a consequence, the number of *effective* collisions, those leading to reaction, must be very small compared to the total number of collisions that occurs. We need to explain why so few collisions are effective and why the fraction of collisions that is effective is greater for some reactions than for others. To explain these, let us go back to the concept of activation energy E_a.

Arrhenius viewed the activation energy for a chemical reaction as an energy barrier that had to be overcome for reactants to become products, a kind of resistance to chemical change. On the molecular level, this means that when two molecules collide with sufficient kinetic energy, they might be expected to react. Molecules that collide without sufficient energy to react simply bounce off each other. The kinetic energy of a gas particle is $mv^2/2$, where m is the mass of the particle and v is its speed. As we have mentioned (also Chapter 4) the particles of a gas are moving at different speeds, which can be described by a Maxwell–Boltzmann distribution (Fig. 15.13). Those particles with sufficiently high speeds, to the right of the line on the graph, have enough kinetic energy to react. If the temperature is increased, the fraction of particles with enough kinetic energy to react increases and the rate of the reaction increases accordingly.

However, reaction rates are much slower than would be predicted by the number of collisions of sufficient energy to lead to products. Recall the exponential form of the Arrhenius equation, $k = Ae^{-E_a/RT}$. The exponential term $e^{-E_a/RT}$ is simply the fraction of collisions that have sufficient energy to react. This fraction goes up when T is increased because of the negative sign in the exponential term. However, most of the collisions calculated by $e^{-E_a/RT}$ do not lead to products, and so the frequency factor A is the fraction of sufficiently energetic collisions that actually lead to reaction.

Why do so many sufficiently energetic collisions fail to lead to reaction products? Current collision theory postulates that the high point on the reaction coordinate curve is a **transition state** that must be reached for reactants to proceed to products. The transition state is a particular geometric combination of the reactants from which the products can be formed. The frequency factor reflects the fact that most colliding particles that have sufficient kinetic energy do not actually form the transition state. For instance, the formation of the transition state may depend on the molecules colliding in a particular orientation.

Now that we have established the ideas that chemical reaction involves collisions and that a certain fraction of these collisions leads to products, we can discuss some examples of mechanisms for specific reactions.

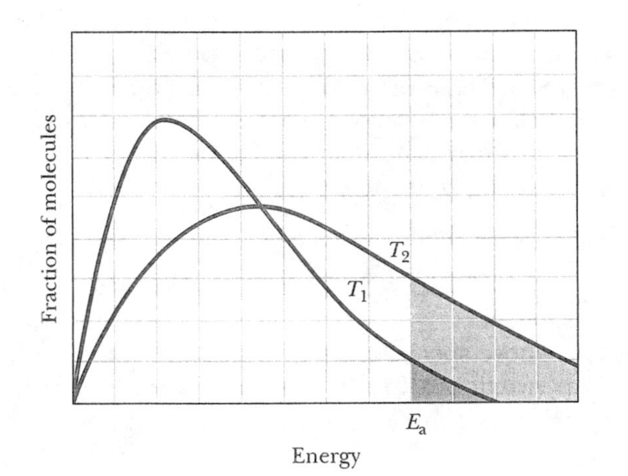

FIGURE 15.13 Distributions of energies. Sets of reacting molecules at different temperatures show different distributions where $T_2 > T_1$. E_a is the activation energy. At the higher temperature, a larger fraction of the molecules present is activated.

PROFILES IN CHEMISTRY

"The Theory of Rate Processes" Until the beginning of the twentieth century, what was known about chemical kinetics all came from the results of experiments. There were no accepted theories tying experiments together. Little was understood about rates of reactions in terms of the behavior of molecules and chemical bonds. Arrhenius had published his monumental paper on activation energies in 1889. This paper contained the idea of an energy barrier impeding or blocking the progress of a chemical reaction as it proceeded from reactants to products, but it was hardly given the attention it deserved. However, with the half-century long career of Henry Eyring, that all changed.

The year 1901 was remarkable for chemistry: It was the birth year of Linus Pauling (Chapter 6), Enrico Fermi (Chapter 22), and Henry Eyring, who will be discussed next. Born in Mexico of German and English descent, Eyring was trained and educated in engineering—as a mining engineer at the University of Arizona—and received a master's degree in metallurgy in 1924. On graduation he worked briefly for the Inspiration Copper Company, but became quickly disillusioned with the industry because of the poor safety practices at that time. During this short period of time, Eyring witnessed the accidental deaths of three miners. Returning to university to study chemistry, he received his doctorate from the University of California (Berkeley) in 1927. As a young instructor at the University of Wisconsin, Eyring came under the influence of Farrington Daniels, stimulating research interests in chemical kinetics that lasted to the end of his long and successful career. In 1946 Eyring moved to the University of Utah as professor of Chemistry and dean of the graduate school and remained there until his death in 1981. He was a devout Mormon and an elder of the Mormon Church.

With the publication of his 1934 paper, "The Activated Complex in Chemical Reactions," Eyring firmly established a line of research that profoundly influenced the direction of theoretical chemistry to the present day, connecting the new quantum theory with emerging ideas on the nature of the chemical bond and the understanding of the rates of chemical reactions. His several books and many scientific papers are testaments to his success as a teacher, especially "The Theory of Rate Processes," published in 1941, in which he presented a systematic account of transition state theory, which is still in use today. His last book on rate processes in medicine and biology was published posthumously in 1984. To many, the surprise of his career is the absence of the Nobel prize in Chemistry.

Questions

Draw a time line of Henry Eyring's career.

Do you think that awards like the Nobel prize should be given posthumously? Why or why not?

15.7 REACTION MECHANISMS

There is often a big difference between the mechanism for a reaction and the balanced chemical equation, representing the stoichiometric relationships that exist between reactants and products. More often than not, reactions involve multistep pathways. Consider the following reaction. Under no circumstances can it occur in a single step, because it is inconceivable to imagine the 21 reactant particles coming together and reacting at the same instant:

$$6Fe^{2+}(aq) + Cr_2O_7^{2-}(aq) + 14H^+(aq) \longrightarrow 6Fe^{3+}(aq) + 2Cr^{3+}(aq) + 7H_2O(\ell)$$

The reaction must take place by a series of steps, of which the balanced equation is only an expression of the net result.

An **elementary process** is one of those special cases in which the chemical equation truly represents a one-step reaction in which reactants go directly to products. Elementary processes are one-step reactions that occur all at once and are known to involve the reaction of one, two, or occasionally three molecules, atoms, or ions. A classic example of an elementary process is the proton-transfer reaction between hydronium and hydroxide ions:

$$H_3O^+(aq) + OH^-(aq) \longrightarrow 2H_2O(\ell)$$

Apparent simplicity in the equation is no guarantee of an elementary process. For example, consider the seemingly simple formation of hydrogen chloride from its elements:

$$H_2(g) + Cl_2(g) \longrightarrow 2HCl(g)$$

The equation suggests that HCl forms on collision of one hydrogen molecule and one chlorine molecule, but detailed kinetics studies demonstrate otherwise. Underlying this simple chemical equation is much greater complexity, and that is often the case.

How do we relate the mechanism of a reaction to the rate law and the concept of collisions leading to reactions? We generally begin the analysis by considering possible reactions that are elementary processes, those proceeding in only one step. In an elementary process, the reactants collide and then those collisions that get through the transition state go *directly* to the final products of the reaction. After we have considered the elementary processes, we will consider multistep reactions.

Elementary Processes

If a chemical reaction is an elementary process, then the rate law for the reaction is a simple statement of the number and kinds of particles that must collide for the reaction to take place. For example, if A and B collide in an elementary process to form products, $A + B \longrightarrow C$, then the rate law must be the following,

$$\text{Rate} = k[A][B]$$

This says that the reaction rate is proportional to the number of collisions between A and B because the number of A–B collisions is proportional to the concentration of A molecules and also to the concentration of B molecules. However, as we discussed in the previous section, only a fraction of the collisions are of sufficient energy to lead to a possible reaction, and then only a fraction of these proceed through the necessary transition state and on to products. The specific rate constant k is the overall proportionality constant linking [A][B] to the actual reaction rate.

An elementary process depending on the collision of two particles such as A and B is, of course, a second-order process. However, it is important to realize that knowing a rate law does not prove that we have an elementary process. Many reactions that proceed by multiple steps are described by simple first- and second-order rate laws.

It is not unusual for a second-order elementary process to involve only a single reactant, in this case the combination of two iodine atoms to form an iodine molecule:

$$2I(g) \longrightarrow I_2(g)$$

It is an elementary process, the number of collisions is proportional to the second power of the iodine atom concentration, and the rate law is simply stated as follows:

$$\text{Rate} = k[\text{I}]^2$$

Though rare, elementary processes are known that involve the simultaneous collision of three particles. In such cases, a third-order rate law is observed. One industrially and environmentally important example is the oxidation of nitrogen monoxide:

$$2NO(g) + O_2(g) \longrightarrow 2NO_2(g)$$

Once again, the rate law is derived directly from the elementary process that says that the reaction is second order in NO and first order in O_2—that is, the rate of the reaction is directly proportional to the square of the concentration of NO and directly proportional to the concentration of O_2:

$$\text{Rate} = k[\text{NO}]^2[\text{O}_2]$$

However, the simultaneous collision of more than three particles would truly be an extraordinary event, and elementary processes higher than third order are unknown.

First-order elementary processes are well-known. One example is the isomerization, or rearrangement, of cyclopropane into propene (propylene):

$$\underset{\text{cyclopropane}}{\begin{matrix} CH_2{-}CH_2 \\ \diagdown\ \diagup \\ CH_2 \end{matrix}} \longrightarrow \underset{\text{propene}}{CH_3{-}CH{=}CH_2}$$

The rate of this elementary process, being first order, is directly proportional to the cyclopropane concentration:

$$\text{Rate} = k[\text{cyclopropane}]$$

One reasonable interpretation of a first order elementary process is that molecules can acquire energy from heat or from absorption of photons of light. If enough energy is acquired by the molecule, it can spontaneously decompose by a first-order process.

A significant simplification of any determination of rate laws for a reaction is introduced when rate data are obtained at the start of the reaction, before significant concentrations of products have accumulated. Under such circumstances, we can safely ignore the effects of product concentrations on the reaction rate. However, as product concentrations build, a point is reached at which the reverse reaction has to be considered, and eventually equilibrium is established. There is a direct relationship between the equilibrium constant and the specific rate constants for the elementary forward and reverse reactions, which we will now establish.

Equilibrium Reactions

When the equilibrium condition is established for a chemical reaction (Chapter 10), the concentrations of reactants and products remain the same because the forward and reverse reactions are proceeding at the same rate. If these reactions are elementary, we can write rate laws for the forward and reverse reactions and show that the equilibrium constant K is the ratio of the rate constants k_1, for the forward reaction, and k_{-1}, for the reverse reaction:

$$K = \frac{k_1}{k_{-1}}$$

We can show how this works for a specific reaction, the equilibrium established between dinitrogen tetroxide and nitrogen dioxide:

$$N_2O_4(g) \underset{k_{-1}}{\overset{k_1}{\rightleftharpoons}} 2NO_2(g)$$

The forward reaction (involving k_1) is an elementary first-order decomposition process for which the following rate law can be written

$$\text{Rate} = k_1[N_2O_4]$$

In the reverse reaction, an elementary recombination process takes place according to the following rate law:

$$\text{Rate} = k_{-1}[NO_2]^2$$

Because the rates of the forward and reverse reactions are equal when equilibrium is established, we can set these two rate equations equal to each other:

$$k_1[N_2O_4] = k_{-1}[NO_2]^2$$

Rearranging terms to show the ratio of the rate constants gives the equilibrium constant:

$$\frac{k_1}{k_{-1}} = \frac{[NO_2]^2}{[N_2O_4]} = K$$

Of course, most chemical reactions are more complex, multistep processes. Now let us turn our attention to some aspects of the mechanisms of multistep chemical reactions.

Multistep Reactions and the Rate-Determining Step

Many reactions get you from starting materials to final products through a series of elementary chemical processes, each of which has its own rate law and specific rate constant, and the rates of these separate chemical steps are all different. You can be sure that a reaction is *not* an elementary process if (1) the exponents in the rate law are not the same as the coefficients in the balanced equation, (2) if a chemical species in the rate law is not a reactant, or (3) if the overall order of the rate law is greater than three.

The simplest concept for understanding the rates of multistep reactions is the rate-determining step. The **rate-determining step** is the one step in a mechanism that is sufficiently slower than all of the other steps so as to limit the overall rate of the reaction. Not all reactions have a step slower enough than the others to be a rate-determining step. In cases in which there is such a step, the overall rate is very close to the rate of that slowest step. It is analogous to a situation on a highway in which the number of cars per hour that can pass along the highway is about the same as the number of cars per hour that can pass through the slowest construction area.

An example of a reaction with a rate-determining step is the decomposition of hydrogen peroxide (H_2O_2) in the presence of iodide ion (I^-) to form water and oxygen gas. Here is the overall reaction:

$$2H_2O_2(aq) \longrightarrow 2H_2O(\ell) + O_2(g)$$

The *experimental* rate law for this reaction is based on the fact that the rate is directly proportional to the hydrogen peroxide concentration and also directly proportional to the iodide ion concentration:

$$\text{Rate} = k[H_2O_2][I^-]$$

This reaction cannot be considered an elementary process because the exponents of the rate law are not the same as the coefficients in the balanced equation, and the rate law includes I^- ion, which is not one of the reactants shown in the overall equation.

Further investigation into the nature of this reaction has led to the hypothesis that it takes place in two steps, a slow step followed by a fast step:

$$H_2O_2 + I^- \longrightarrow H_2O + OI^- \qquad \text{slow first step}$$

$$OI^- + H_2O_2 \longrightarrow H_2O + O_2 + I^- \qquad \text{fast second step}$$

The overall reaction is the sum of these two steps. The first step goes slowly and is considered to be rate determining. That is, as soon as an OI^- ion is produced in the slow first step, it quickly reacts with a molecule of H_2O_2 in the second step to form the final products. The rate of the reaction is thus the rate of the first step.

$$\text{Rate} = k[H_2O_2][I^-]$$

This agrees with the experimental rate law given previously. However, it would be possible to write other mechanisms that would be consistent with the experimental rate law. Thus, the rate law does not prove the mechanism. However, the rate law can rule out other suggested mechanisms. In this example, the experimental rate law rules out the possibility that H_2O_2 decomposition is an elementary process under these conditions.

The concepts of chemical kinetics that we have studied to this point are generally applicable to the study of catalysts. They play an important role in both the chemical process and products industries and will be considered in the next section.

15.8 CATALYSTS

Catalysts and Catalysis

Early in the nineteenth century, Jöns Jacob Berzelius described a number of laboratory situations in which chemical reactions were subject to the effects of certain substances that themselves remained unchanged. He described these reactions as "*catalyzed.*" In the modern world, industrial manufacture of just about everything from fuels to drugs is pushed along by **catalysts** (also called **accelerators**) that make a reaction go faster. Other agents, called inhibitors, are added to slow the spoilage of foods and reduce the rates of a host of important reac-

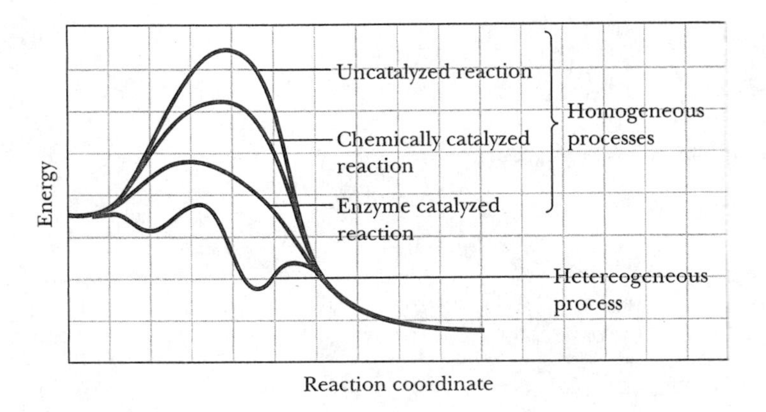

FIGURE 15.14 Catalyzed chemical reactions. The uncatalysed reaction has the highest "barrier."

tions to acceptable, manageable levels. **Inhibitors** make a reaction go more slowly. Catalysts and inhibitors are substances that change the rate of a chemical reaction without being consumed in the process. If a substance is to be a true catalyst, it should show up at the end of the reaction in the same chemical form and quantity as at the beginning. Catalysts exhibit identical effects on both the forward and reverse reactions.

Some 50 years after Berzelius, Wilhelm Ostwald was able to demonstrate conclusively for the first time that catalysts do not increase the yields in chemical reactions. He recognized that a given chemical reaction must be subject to the usual equilibrium conditions and constraints. No catalyst could operate beyond the theoretical limits set by the second law of thermodynamics. However, the approach to the state of equilibrium might be so slow as to make realization of the theoretical yield of reaction products all but unattainable under normal circumstances. If a substance is a catalyst, it speeds the approach to the equilibrium state. Inhibitors slow the rate of approach to the equilibrium condition. In catalysis lies the possibility that thermodynamically feasible but kinetically unfavorable reactions may be realized, often to our advantage but also disadvantageously on occasion. The spoilage of wine is greatly accelerated by a catalyst provided by the air-borne microorganism *acetobacter.*

Catalysts open alternative pathways with different activation energies. A useful way to think about how catalysts work is to consider the energy history of a chemical reaction depicted in Figure 15.14. The curve at the top represents the uncatalyzed reaction with the energy barrier E_a, which has to be overcome and acts to retard the reaction. The lower curves represent the reaction with a catalyst added to accelerate the reaction by providing a different pathway, one with a lower activation energy. Because E_a is lower, the value of the specific rate constant k will be larger according to the Arrhenius equation.

Inhibitors can block the normal path of a reaction and force it to proceed by a path with a higher value of E_a and therefore a lower value of the specific rate constant.

Catalyzed and Uncatalyzed Reactions

In addition to increasing the rate of a given chemical reaction, catalysts are often used in situations in which several reactions are possible. The catalyst favors

PROFILES IN CHEMISTRY

Jöns Jacob Berzelius Out of the Stockholm laboratory of Jöns Berzelius (Swedish, 1779–1848), one of the early nineteenth-century giants of chemistry, came significant discoveries and contributions to inorganic and organic chemistry, chemical analysis, and theoretical chemistry such as atomic masses, chemical combination, molecular geometry, and catalysis. His laboratory could count among its disciples the great Friedrich Wöhler, who in turn trained and educated the generation of chemists that contributed so much to the rise of chemistry in America in the second half of the nineteenth century. Berzelius was also a master of literary style and classroom scholarship. His *Lehrbuch der Chemie* (*Textbook of Chemistry*) contained the entire body of chemical knowledge known at that time in six volumes and made up of more than 4000 pages.

Questions

Explain the nature of a catalyst.

Name a scientist other than Berzelius whose textbook of chemistry was a great influence on its generation of readers. (See Chapter 1.)

one reaction, effectively excluding others. For example, depending on the nature of the catalyst, entirely different products can be obtained from the reaction of carbon monoxide and hydrogen (Table 15.1). Such selectivity obviously has great economic as well as scientific significance.

The essential feature of the catalyzed reaction is the entry of the catalyst into the process followed by its regeneration at the end of the reaction cycle. As a consequence, each catalyst unit can cause many reactant molecules to be converted to products. Consider a slow reaction in which reactant molecules (A) are converted into product molecules (P). On addition of catalyst molecules (X), an alternative, faster reaction pathway is available:

without the catalyst $A \longrightarrow P$

with the catalyst $A + X \longrightarrow B$

$B \longrightarrow X + P$

The entry of X into the catalyzed reaction scheme causes large numbers of reactant (A) molecules to be converted to product (P) molecules. The net result is still the same as the uncatalyzed process. What makes the overall process catalytic is the continuous regeneration of the X molecules.

TABLE 15.1 The CO–H_2 Reaction

Conditions	Reaction Products
Ni catalyst, 100°C, 1 atm	$CH_4 + H_2O$
$Zn(CrO_2)_2$, 400°C, 500 atm	$CH_3OH + H_2O$
ThO_2, 400°C, 200 atm	branched hydrocarbons suitable for use in unleaded gasolines

Homogeneous Catalysis

Catalysts fall into two major categories. If the catalyst is in the same phase as the other components of the reaction, it is a homogeneous catalyst. **Homogeneous catalysts** are normally dissolved in a liquid solution with the reactants. If the catalyst is in a different phase than the reactants, it is a heterogeneous catalyst. With a **heterogeneous catalyst** the reaction takes place at the interface between two distinct phases.

A classic example of homogeneous catalysis is the electron–transfer reaction between Ce^{4+} and Tl^{+} ions:

$$2Ce^{4+}(aq) + Tl^{+}(aq) \rightleftharpoons 2Ce^{3+}(aq) + Tl^{3+}(aq)$$

It is generally believed that the reaction is slow because Tl^{+} ions lose two electrons in going to product ions and each Ce^{4+} ion undergoes a one-electron transfer:

$$Tl^{+} \longrightarrow Tl^{3+} + 2e^{-} \qquad \text{oxidation half-reaction}$$

$$[Ce^{4+} + e^{-} \longrightarrow Ce^{3+}] \times 2 \qquad \text{reduction half-reaction}$$

$$2Ce^{4+} + Tl^{+} \longrightarrow 2Ce^{3+} + Tl^{3+} \qquad \text{net reaction}$$

Mn^{2+} successfully catalyzes the reaction, perhaps, because it separates the one-electron and two-electron transfers from each other:

$$Ce^{4+} + Mn^{2+} \longrightarrow Ce^{3+} + Mn^{3+}$$

$$Ce^{4+} + Mn^{3+} \longrightarrow Ce^{3+} + Mn^{4+}$$

$$Tl^{+} + Mn^{4+} \longrightarrow Tl^{3+} + Mn^{2+}$$

$$2Ce^{4+} + Tl^{+} \longrightarrow 2Ce^{3+} + Tl^{3+}$$

The net result is the same, catalyzed or uncatalyzed. In the catalyzed reaction, Mn^{2+} ions are consumed and regenerated.

The fact that you can purchase aqueous solutions of hydrogen peroxide of various concentrations with a substantial shelf life is a reflection of just how slow the decomposition of H_2O_2 is at 25°C:

$$2H_2O_2(\ell) \longrightarrow 2H_2O(\ell) + O_2(g)$$

But just a drop or two of a dilute solution of Fe^{3+} ions (Table 15.2) is sufficient to initiate the catalyzed peroxide decomposition reaction.

Nature's catalysts are enzymes, which are very large, high-molecular mass protein molecules containing a few active sites with catalytic activity. These giant organic molecules are highly effective and uniquely specialized catalyst materials. For example, a single molecule of the iron-containing enzyme known as catalase brings about the decomposition of 5 million molecules of H_2O_2 per minute at 273K. Because hydrogen peroxide, even in dilute solutions, is an acutely toxic substance, and because it is produced as a by-product of a number of other enzyme reactions, catalase plays the role of protector, destroying any hydrogen peroxide that is present.

Heterogeneous Catalysis

When the catalyst in a chemical reaction is in a different phase than the reactants, the process is described as heterogeneous catalysis. Heterogeneous catalysts

TABLE 15.2 Activation Energies for Some Catalyzed Reactions

Reaction	Catalyst	E_a (kJ/mol)
Homogeneous		
decomposition of H_2O_2	uncatalyzed	75
	iodide ion	58
	Fe^{3+}	42
	catalase (enzyme)	4
hydrolysis of grape sugar	H_3O^+	109
	sucrase (enzyme)	46
hydrolysis of urea	H_3O^+	105
	urease (enzyme)	54
Heterogeneous		
hydrogenation of ethylene	uncatalyzed	188
	Pt (on carbon)	146
	Ni (finely divided)	84

are commonly solids with the catalysis occurring on the solid surface. Many millions of tons of fertilizer ammonia are produced each year by the heterogeneously catalysed Haber process, which has gaseous reactants and a solid catalyst:

$$N_2(g) + 3H_2(g) \longrightarrow 2NH_3(g)$$

The process uses a catalyst consisting of an iron powder to which is added a few percent K_2O and Al_2O_3 to improve the overall performance of the primary iron catalyst.

Another important example of heterogeneous catalysis is the hydrogenation of ethene on the surface of a finely divided nickel or palladium catalyst:

$$CH_2{=}CH_2(g) + H_2(g) \xrightarrow{Ni(s)} CH_3CH_3(g)$$

The heterogeneous catalytic process can be broken into four distinct steps:

1. **Reactants diffuse to the catalyst surface.** Because ethylene ($CH_2{=}CH_2$) and hydrogen (H_2) are both gases at the reaction conditions, they will diffuse through the reaction vessel to the surface of the nickel catalyst. The greater the surface area of the catalyst, the more molecules can be adsorbed, and the faster the catalysis can occur.
2. **Reactants adsorb onto the catalyst surface.** Interactions between the reactant molecules and the surface of the catalyst lead to chemical bonding. For hydrogen this is postulated to involve the breaking of the H—H bond and the formation of Ni—H bonds. The weakening or breaking of bonds within the reactant molecules resulting from the formation of bonding interactions to the catalyst lowers the activation energy necessary to get the desired reaction to occur.

PROCESSES IN CHEMISTRY

Inexpensive Catalytic Hydrogenation Paul Sabatier (French, 1854–1941) spent his professional lifetime studying catalysts, for which he was eventually awarded the Nobel prize. But it was a failed experiment that he conducted in 1897 at the University of Toulouse that got him started on that line of research. Sabatier was fascinated by the fact that nickel could form a volatile compound, named nickel tetracarbonyl, on reaction with carbon monoxide:

$$Ni(s) + 4CO(g) \longrightarrow Ni(CO)_4(g) \qquad \text{preparation of nickel tetracarbonyl}$$

What so interested him was the extraordinary fact that a metal could form stable compounds that vaporized at relatively low temperatures. He reasoned that because ethylene has a multiple bond, as has carbon monoxide, perhaps it too could form a volatile compound with nickel. But when ethylene was added to the carbon monoxide gas stream being passed over his heated nickel metal, no volatile compound of nickel and ethylene could be found. Fortunately, Sabatier and his young graduate student saved what gases did form, and to their surprise, they found ethane. Nickel had catalyzed the addition of hydrogen to ethylene, forming ethane:

$$CH_2{=}CH_2(g) + H_2(g) \xrightarrow{Ni} CH_3CH_3(g) \qquad \text{nickel-catalyzed hydrogenation}$$

Sabatier, of course, in investigating his failed experiment, asked himself where the hydrogen had come from that added to the double bond in ethylene, bringing about the catalyzed hydrogenation reaction. He realized that the carbon monoxide had been prepared by his student from the reaction of methane and steam and used directly:

$$CH_4(g) + H_2O(\text{steam}) \rightleftharpoons CO(g) + 3H_2(g)$$

Stoichiometric quantities of hydrogen were present. And as they say, "The rest is history." Until the work of Sabatier and his students, catalytic hydrogenation required very expensive materials such as platinum and palladium, precluding widespread, economical use on an industrial scale. Nickel catalysis made it possible to perform reactions such as producing solid fats such as margarine and shortenings from vegetable oils economically.

Questions

Catalytic hydrogenation is used intensively for a large number of processes, from refining petroleum to making peanut butter. What was Sabatier's contribution to this technology?

Where did the hydrogen that added to the ethylene molecule, producing ethane, come from in Sabatier's classic "failed" experiment?

Suggest a possible reaction for preparing hydrogen-free carbon monoxide.

3. **Adsorbed reactants are converted to adsorbed products.** Lowering E_a results in an increased value for k according to the Arrhenius equation, and the reaction leading to the product (CH_3CH_3) is accelerated.
4. **Adsorbed products diffuse away.** This makes way for more reactants to diffuse to the catalyst surface, become adsorbed, and react.

Free Radical Processes

The sharing of electrons to form chemical bonds and stable molecules is one of chemistry's essential themes. However, there are chemical species called **free radicals** that are characterized by the presence of one or more unshared electrons. As you might expect of free radicals, they are relatively less stable because of the driving force to form a shared electron pair—a covalent chemical bond.

Free radicals are often produced by breaking a bonding electron pair between atoms so that one electron stays with each fragment. For example, light of certain wavelengths can break chlorine molecules into atoms:

$$Cl_2(g) \longrightarrow 2Cl\cdot(g)$$

The unpaired electron on each of the chlorine-free radicals is denoted by the dot. This kind of bond-breaking that splits the bond evenly, giving each fragment half of the electrons that had constituted the bond, is called **homolytic cleavage**. It does not produce ionic products.

Important early research on free radicals was performed by Max Bodenstein (German, 1871–1942) at the Kaiser Wilhelm Institute in Berlin early in the twentieth century. About the time he succeeded Walther Nernst as the head of the Institute for Physical Chemistry, he was studying the kinetics of apparently simple reactions such as the decomposition of hydrogen iodide or the combination of hydrogen and chlorine in the gas phase. But these reactions did not turn out to be kinetically simple. In particular, Bodenstein was the first to suggest a possible chain reaction in which the products of a molecular reaction serve to bring about the same kind of chemical change in another set of reacting molecules. In a **chain reaction**, the products of one cycle of a reaction start the next cycle. Introduction of a few chlorine atoms to a mixture of hydrogen and chlorine molecules starts an explosively rapid reaction in which a chain of steps is repeated many times. Here is the net reaction:

$$H_2(g) + Cl_2(g) \longrightarrow 2HCl(g)$$

Many millions of product molecules are produced from a few Cl atoms as each is used over and over again. Here are the steps in the free-radical chain reaction that turns hydrogen and chlorine molecules into hydrogen chloride molecules in the gas phase:

1. **Initiation step**: To begin the reaction, we need to generate a few chlorine atoms. To do this, we expose molecular chlorine to light of certain wavelengths. Sunlight will do. The resulting homolytic cleavage of the Cl—Cl bond splits the shared pair of electrons in the bond evenly between the two atoms and produces a pair of neutral atoms:

   $$Cl_2(g) + h\nu \longrightarrow 2Cl\cdot(g)$$

   This reaction starts the chain process and is called the **initiation step**.
2. **Propagation step:** The Cl· free radicals created in the initiation step are highly reactive, attacking the bonds of H_2 molecules. This reaction creates an HCl molecule and an H· radical, which in turn can attack a Cl_2 molecule:

   $$Cl\cdot + H_2 \longrightarrow HCl + H\cdot$$

   $$H\cdot + Cl_2 \longrightarrow HCl + Cl\cdot$$

   The radicals resulting from these steps can go on to attack a second molecule, and a third, and so on. These are referred to as propagation steps. The **propagation steps** continue the chain reaction. It would not be unusual for a light-induced chain reaction initiated by the pair of atoms from a single Cl_2 molecule to produce a million HCl molecules.

3. **Termination step:** The chain reaction does not go on forever, or even to the point at which one of the reactants is used up. At some point, by chance, the chain carriers—the chlorine and hydrogen atoms—enter into termination reactions. The **termination step** ends the chain. There are three possibilities. In each of them the unpaired electrons required to continue the chain reactions are consumed without replacement.

   Combination of chlorine atoms:

   $$Cl\cdot + Cl\cdot \longrightarrow Cl_2$$

   Combination of hydrogen atoms:

   $$H\cdot + H\cdot \longrightarrow H_2$$

   Combination of hydrogen and chlorine atoms:

   $$H\cdot + Cl\cdot \longrightarrow HCl$$

   In addition, molecules of other compounds present can react with the chain carriers to produce much less reactive species, thus disrupting the chain.

Among the interesting experimental features of chain reactions is auto-inhibition. Auto-inhibition is the slowing of the reaction by product molecules, HCl in this case. This occurs because of the increased rates of reactions that are the reverse of the propagation steps:

$$H\cdot + HCl \longrightarrow H_2 + Cl\cdot$$

$$Cl\cdot + HCl \longrightarrow Cl_2 + H\cdot$$

This presents us with a test for the presence of a free-radical chain process. Increase the concentration of the product, and see what happens to the rate of the reaction. If the reaction is free radical in nature, it should be inhibited.

EXAMPLE 15.5

Summarize the steps involved in a typical chain reaction by writing out the initiation, propagation, and termination steps for the synthesis of HBr from H_2 and Br_2. Because it is known to be a free-radical-initiated process, include an inhibition step, too.

COMMENT An initiation step in a free-radical chain reaction breaks a bond, forming two free radicals, which then can propagate the chain. The deeply colored Br_2 absorbs visible light and is more easily split into free radicals than H_2. Propagation steps will produce the desired product and another free radical to propagate the chain. Termination steps will involve combination of two free radicals.

SOLUTION

$$Br_2 + h\nu \longrightarrow 2\ Br\cdot \quad \text{initiation}$$

$$Br\cdot + H_2 \longrightarrow HBr + H\cdot \quad \text{propagation reactions}$$

$$H\cdot + Br_2 \longrightarrow HBr + Br\cdot$$

$$H\cdot + Br\cdot \longrightarrow HBr \quad \text{termination reactions}$$

$$H\cdot + H\cdot \longrightarrow H_2$$

$$Br\cdot + Br\cdot \longrightarrow Br_2$$

$$H\cdot + HBr \longrightarrow H_2 + Br\cdot \quad \text{inhibition reactions}$$

$$Br\cdot + HBr \longrightarrow Br_2 + H\cdot$$

EXERCISE 15.5

Summarize the steps involved in the free-radical-initiated chain reaction synthesis of chloromethane (CH_3Cl) by the photochlorination of methane (CH_4).

$$CH_4(g) + Cl_2(g) \longrightarrow CH_3Cl(g) + HCl(g)$$

There are many important chain reactions among industrial, commercial, and biological processes. Some are desirable and others are not. Among the desirable ones is the large-scale preparation of synthetic polymers. Undesirable free-radical chain reactions include the pinging or knocking of the internal combustion engine and the spoiling of foods. Certain aspects of the aging process in humans are also believed to be promoted by free radicals. Perhaps the most famous chain reaction is the nuclear chain reaction that is the essential process in nuclear fission.

Undesired chain reactions can often be suppressed by introducing substances that interrupt the chain. Very small amounts of lead and manganese compounds in motor fuels prevent engine knocking. Trace quantities of antioxidants prevent spoilage of foods caused by air oxidation. In some investigations, vitamin E appears to interfere with free radical oxidation reactions associated with aging processes in humans.

ADDITIVES IN FOODS A wide range of substances is added to many foods to retard the rate of breakdown of fats and oils from reaction with oxygen. By inhibiting free-radical chain reactions, they delay the onset and limit the degree of rancidity and the accompanying objectionable tastes and odors. In addition, they minimize the oxidative destruction of certain vitamins and essential fatty acids. Their action is due to their ability to react with oxygen-free radicals more readily than can the oxidizable substances in foods. For example, potato chips containing an antioxidant will remain fresh at 65°F for up to a month. Without the antioxidant, rancidity becomes evident within a few days at room temperature.

KNOCK INHIBITORS IN MOTOR FUELS When uncontrolled burning occurs in a spark-ignition internal combustion engine, there is a sudden increase in pressure; this is accompanied by sound waves that can be heard as pinging or knocking. Over thousands of miles of driving, this phenomenon can damage the engine. It is due to free-radical reactions that take place during the heating and compression of the fuel–air mixture prior to ignition. Certain gasoline-soluble organometallic compounds such as tetraethyl lead, called **knock inhibitors**, inhibit the reactions that cause knocking. However, the widespread use of tetraethyl lead in motor fuels for more than half a century caused unacceptable levels of lead pollution in the environment. Ecological concern in the United States and elsewhere resulted in legislation phasing out leaded motor fuels, and gasolines are almost all lead-free today. Chemical research has shown that the same knock inhibition can be accomplished by altering the structure and composition of the fuel itself by a process called reforming. The cost of this reforming process is much greater than that of the lead antiknock additive, but the environmental benefits are greater still.

SEEING THINGS

Free Radical Chemistry in a Flash The fact that light produces chemical reactions has been known for a long time. Light fades colors and causes changes in silver salts that makes it possible for us to create photographic images. For light to produce a chemical reaction, it has to be absorbed by a molecule that then gets so excited that it reacts. In chemistry, "excited" means "raised to a higher and less stable level of potential energy." The excited molecule must dissociate and then recombine into a new arrangement. The time intervals required for light-induced or photochemical excitation are very short. For the very fastest photochemical reactions, the times required are on the order of ten-thousandths of a millionth of a second, speeds approaching picoseconds. To estimate what that means, think of that small time interval in terms of seconds and years. One ten-thousandth of a millionth of a second is the same order-of-magnitude compared to one second as one second is to hundreds of years. It is fast! Until very recently, immeasurably fast. It was the exothermic synthesis of hydrogen chloride from its elements that first led to the conclusion that the primary reaction indeed involved the photolysis of the reactant molecules into atoms, followed by the well-known chain reaction:

$$Cl_2 + h\nu \longrightarrow Cl\cdot + \cdot Cl \qquad \text{dissociation into atoms}$$

$$H_2 + Cl_2 \longrightarrow 2HCl + \text{heat} \qquad \text{net result of the chain reaction}$$

The early work was done by giants of the period—Bodenstein, Warburg, Nernst, Polanyi, and no less a chemist than Einstein.

It is possible to demonstrate the elegant simplicity of the photochemical process and, at the same time, the dramatic qualities of fast reactions involving unpaired electrons brought about by propagating chain mechanisms in a now-classic lecture demonstration. A heavy-walled, clear glass bottle from a carbonated beverage will do just fine. Fill it 50–50% by volume with hydrogen and chlorine, stopper with a well-fitted cork, mount on a ring stand behind a polycarbonate shield with the stopper aimed to the ceiling, and directly expose the bottle (through the shield) to an intense pulse of light from a powerful flash as is easily obtained from a strobe-like camera-flash attachment. The photoflash causes enough chlorine molecules to dissociate to initiate the chain reaction—and blow the stopper sky high, perhaps 30 to 60 feet up.

Questions

Why does the mixture of H_2 and Cl_2 fail to react until it is exposed to an intense light?

Show how HCl is formed by chain-propagation reactions.

Would you expect the photobromination of hydrogen to follow a chain-reaction mechanism? How about the photochlorination of methane?

APPLICATIONS OF CHEMISTRY

The Catalytic Converter It is well known that automobile exhaust is a major air pollutant, containing a variety of incomplete fuel combustion products such as carbon monoxide and various lower-molar-mass hydrocarbons as well as combustion by-products such as assorted oxides of nitrogen and sulfur. In the United States and the rest of the industrialized world, catalytic converters are fitted onto automobile exhaust systems to convert combustion products to carbon dioxide and water and the nitrogen oxides back to nitrogen and oxygen. Converters that can handle both those tasks have been installed as original equipment on automobiles since 1980 in the United States. Manufacturing these converters is a billion-dollar-a-year business. Designs that meet the stringent 1991 Clean Air Act requirements have electrical preheaters, because the greatest amount of pollutants is produced during the first ten minutes of engine operation when it is essentially running cold.

Question

What is the fate of CO and unburned hydrocarbons in the oxidizing-reactor phase of an automotive catalytic converter?

SUMMARY

Chemical kinetics is the study of the rates of chemical reactions. Rates of some reactions are explosively fast, and others occur so slowly that one cannot be certain whether anything is happening. In between the very slow and the very fast lies the range of reactions whose rates most interest us. The study of chemical reactions gives information of the actual steps by which a reaction proceeds, and understanding chemical kinetics allows us to increase or decrease the rates of important reactions when it is desirable.

The rate of a chemical reaction is measured by the rate of change of the concentration or partial pressure of either a product or a reactant. Because rates are constantly changing, it is desirable to measure the change in concentration in as short a time period as possible. The rate of a reaction is given by a rate law, which is determined by experiment and has the general form

$$\text{Rate} = k[\text{A}]^m[\text{B}]^n$$

The proportionality constant k is called the specific rate constant. The exponent for a particular chemical species is the order of the reaction with respect to that species, and the sum of the exponents is the overall order of the reaction. Rate laws are determined by varying the concentration of each reactant independently and measuring the reaction rate.

Algebraic equations can be derived from the rate laws for each order of reaction describing the concentration of reactants as a function of time. This is particularly useful for first-order, simple second-order, and zeroth-order reactions. The half-life of a reaction is the time required for half of a given amount of reacting material to react.

Reaction rates are influenced by other factors besides concentrations including temperature, surface area, and the presence of catalysts or inhibitors. Temperature is an important variable and is related to the activation energy E_a. The activation energy is an energy barrier to the progress of the reaction that must be overcome. At higher temperatures, more molecules have sufficient energy to override the energy barrier and the reaction rate increases. This information is contained in the Arrhenius equation.

Chemical reactions are the result of collisions between molecules. However, not all collisions lead to the formation of products. Molecules must have sufficient energy to override the energy barrier, but they must also collide in such a way as to form the transition state, an intermediate structure, which is necessary if reactants are to proceed to products.

Some reactions are elementary processes, consisting of a single step. The rate law for such a reaction has exponents that are the same as the coefficients in the balanced equation. No elementary processes are observed above third order, because the simultaneous collision of more than three particles is so unlikely. The equilibrium constant for elementary processes is simply the specific rate constant for the forward reaction divided by the specific rate constant for the reverse reaction. Multistep processes proceed by a series of two or more elementary processes. If one step in the process is sufficiently slow, the rate of the overall reaction will be that of this rate-determining step.

Catalysts increase the rate of a reaction without being consumed by it. The function of a catalyst is to provide an alternative path with a lower activation energy for the process. Homogeneous catalysts exist in the same phase as the reactants and heterogeneous catalysts exist in a different phase. Many heterogeneous catalysts are solid surfaces. Free radicals, species containing an unpaired electron, play an important role in chemical kinetics. Reactions of free radicals characteristically proceed by chain reactions in which one free radical is regenerated over and over.

TERMS

Thermodynamic control (15.1)
Kinetic control (15.1)
Rate of a chemical reaction (15.2)
Rate law (15.2)
Specific rate constant (15.2)
Order (15.2)
Overall order (15.2)
First-order reaction (15.3)
Half-life (15.3)
Second-order reaction (15.3)
Zeroth-order reaction (15.3)
Catalysts (15.4)
Inhibitors (15.4)
Activation energy (15.5)
Reaction coordinate (15.5)
Arrhenius equation (15.5)
Arrhenius plot (15.5)
Reaction mechanism (15.6)
Transition state (15.6)
Elementary process (15.7)
Rate-determining step (15.7)
Accelerator (15.8)
Homogeneous catalyst (15.8)
Heterogeneous catalyst (15.8)
Free radical (15.8)
Homolytic cleavage (15.8)
Chain reaction (15.8)
Initiation step (15.8)
Propagation step (15.8)
Termination step (15.8)
Knock inhibitors (15.8)

IMPORTANT EQUATIONS

$\text{Rate} = \dfrac{\Delta[P]}{\Delta t} = k[A]^m[B]^n$	Rate law for a general reaction
$\ln [A] = \ln [A]_0 - kt$	Concentration as a function of time; first-order reaction
$\ln \dfrac{[A]}{[A]_0} = -kt$	Alternative equation; first-order reaction
$[A] = [A]_0 e^{-kt}$	Exponential form; first-order reaction
$t_{1/2} = \dfrac{0.693}{k}$	Half-life; first-order reaction
$\dfrac{1}{[A]} = \dfrac{1}{[A]_0} + kt$	Concentration as a function of time; second-order reaction
$t_{1/2} = \dfrac{1}{k[A]_0}$	Half-life; second-order reaction
$[A] = [A]_0 - kt$	Concentration as a function of time; zeroth-order reaction
$\ln k = \ln A - \dfrac{E_a}{RT}$	Logarithmic form of the Arrhenius equation
$k = Ae^{-E_a/RT}$	Exponential form of the Arrhenius equation
$K = \dfrac{k_1}{k_{-1}}$	Equilibrium constant and specific rate constants for elementary processes

QUESTIONS

Conceptual questions are denoted by a square screen. Extra-credit questions are denoted by a circular screen.

1. What do we mean by the terms "kinetic control" and "thermodynamic control" of chemical reactions, and why is it essential to understand the difference?
2. Why does a pile of flour burn slowly, but the same quantity suspended in the air as a dust can explode violently?
3. What is the rate expression for the appearance of product in the following reactions?
   (a) $X + Y + Z \longrightarrow XYZ$
   (b) $X_2 + Y_2 \longrightarrow 2XY$
4. If a reaction follows second-order kinetics, is it necessarily a second-order elementary process? Explain.
5. Why are third-order elementary processes considered to be extraordinary, and reactions of higher orders to be impossible?

6. Why would you be correct in assuming the following reaction does not proceed by the rate law that would come directly from the balanced equation?

$$4HCl + O_2 \longrightarrow 2H_2O + 2Cl_2$$

7. How do you account for the differences between the stoichiometry of the balanced chemical equation and the concentration dependencies expressed in the corresponding rate law that follows:

$$HCrO_4^- + 3Fe^{2+} + 7H^+ \longrightarrow Cr^{3+} + 3Fe^{3+} + 4H_2O$$

$$\text{Rate} = k[HCrO_4^-][Fe^{2+}][H^+]$$

8. What is your understanding of the expression "mechanism of a reaction?"

9. The following reaction is known to be an elementary process. What is the rate law?

$$CH_3I(g) + HI(g) \longrightarrow CH_4(g) + I_2(g)$$

10. The following reaction proceeds by an elementary process. What is its rate law?

$$2NO(g) + Cl_2(g) \longrightarrow 2NOCl(g)$$

11. What graphical methods could you use to test whether a reaction in which *A* goes to products is first or second order?

12. What is meant by the term "activation energy"?

13. Why do reaction rates increase with temperature?

14. What's wrong with the following logic? Reaction 1 is exothermic and reaction 2 is endothermic. Therefore, reaction 1 has the smaller activation energy and is the faster of the two reactions.

15. What is the essential difference between the members of each of the following pairs?
   (a) An elementary process and a stoichiometric reaction
   (b) The rate constant and the rate law
   (c) A transition state and a reaction intermediate
   (d) A reaction that is first-order in A and one that is second order in A

16. Are the following statements best described as true or false? Explain the reason for your choice.
   (a) The order of a chemical reaction is determined by inspection of the stoichiometry of the process.
   (b) An elementary process involving two molecules follows a second-order rate law.
   (c) An overall first-order reaction is described by a change in concentration that is exponential with respect to time.
   (d) The rate of a reaction is proportional to the accompanying change in free energy.
   (e) The rate constant for an elementary unimolecular reaction has units of concentration divided by time.
   (f) The slow step among the two or more that constitute a reaction mechanism is always considered to be rate-determining.
   (g) The Arrhenius theory suggests that the rate of reaction is a function of the number of collisions between reacting molecules.

17. Are the following statements best described as true or false? Explain the reason for your choice.
   (a) Catalysts are substances that can improve the yield of an otherwise poor reaction by favorably shifting the equilibrium without being altered in the overall process.
   (b) A given reaction mechanism is a true statement of the pathway from reactants to products.
   (c) Isolation of the transition state is necessary if one is ultimately to be able to verify proposed mechanisms involving such species.
   (d) The sum of the exponents for the concentration terms in the experimental rate law gives the overall order of the reaction.
   (e) The value of the activation energy for an endothermic reaction is ΔE.

18. For each of the following, answer the question from the selections of possible answers:
   (a) Consider the accompanying graph and determine whether it is essentially (cor-

rect, incorrect), as drawn, for an exothermic process.

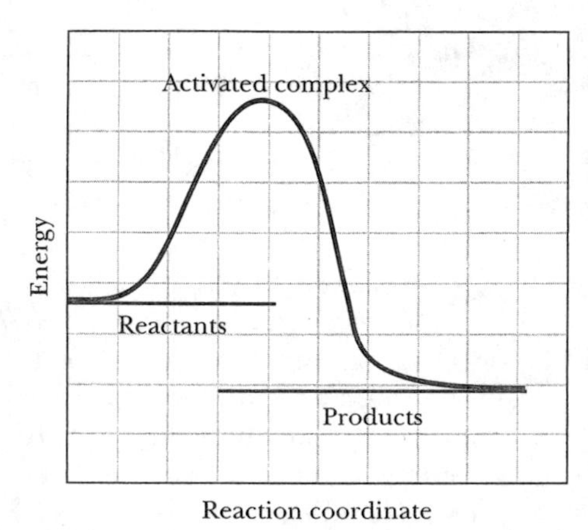

(b) An increase in temperature will usually lead to a (greater, smaller, unchanged) rate of reaction.

(c) An increase in E_a will usually lead to a (greater, smaller, unchanged) rate of reaction.

(d) An inhibitor for a given chemical reaction can provide a reaction pathway leading to a (more, same, less) favorable equilibrium state.

(e) The half-life of a chemical reaction is the time required for the reaction to proceed half way to completion. (True or false?)

(f) For a reaction with activation energy E_a at temperature T, the fraction of the total number of molecules reacting to produce products is given by $e^{-E_a/RT}$. (True or false?)

(g) The heterogeneous gas-phase hydrogenation of ethylene over a finely divided platinum catalyst would likely be described overall as (zero, first, second, indeterminate) order.

PROBLEMS

Problems marked with a bullet (•) are answered in Appendix A, in the back of the text.

Rates and Rate Laws [1–10]

•1. Which of the following reactions are *not* likely to be elementary processes?
(a) $A + B \longrightarrow AB$
(b) $2A_2 + 2B \longrightarrow 2A_2B$
(c) $A_2 + B + 2C \longrightarrow AC + ABC$

2. For the following reactions, write the rate in terms of the disappearance of each reactant and the appearance of product.
(a) $A + 2B \longrightarrow AB_2$
(b) $A_2 + 2B \longrightarrow A_2B_2$

•3. For each reaction, determine the relationship between the rate of reaction stated in terms of the reactant(s) and the rate stated in terms of the product—that is,

$$\frac{\Delta[\text{reactant}]}{\Delta t} \quad \text{and} \quad \frac{\Delta[\text{product}]}{\Delta t}$$

(a) $A_2 + B_2 \longrightarrow 2AB$
(b) $A_2 + B_2 \longrightarrow A_2B_2$

4. What are the rates of these reactions stated in terms of $\frac{\Delta[\text{reactant}]}{\Delta t}$ and $\frac{\Delta[\text{product}]}{\Delta t}$?
(a) $X + Y_2 \longrightarrow XY + Y$
(b) $X_2 + 2Y \longrightarrow X_2Y_2$

5. Consider the combination of nitrogen monoxide and chlorine:

$$2NO(g) + Cl_2(g) \longrightarrow 2NOCl(g)$$

Express the rate of formation of NOCl in terms of the rate of disappearance of Cl_2.

6. Consider the following reaction, which takes place under mildly acidic conditions:

$$4H_3O^+(aq) + MnO_4^-(aq) + 2Sb(OH)_3(aq) \longrightarrow Mn^{3+}(aq) + 2H_3SbO_4(aq) + 6H_2O(\ell)$$

It can be shown experimentally that the rate of permanganate disappearance doubles when either the permanganate concentration or the hydronium ion concentration doubles, but halves when the $Sb(OH)_3$ concentration is doubled. Write the experimental rate law for the reaction.

•7. Consider the following chemical reaction and the corresponding kinetic data showing the initial reaction rate as a function of the initial concentrations of the reactants, and establish the correct experimental rate law:

$$H_3AsO_4(aq) + 2H_3O^+(aq) + 3I^-(aq) \longrightarrow HAsO_2(aq) + I_3^-(aq) + 4H_2O(\ell)$$

Initial Rate	Molarity (mol/L)		
(M/s)	$[H_3AsO_2]$	$[H_3O^+]$	$[I^-]$
3.70×10^5	0.001	0.010	0.10
7.40×10^5	0.001	0.010	0.20
7.40×10^5	0.002	0.010	0.10
3.70×10^5	0.002	0.005	0.20

8. Using the data in the following table, determine the rate law for the reaction:

$$2A + 3B \longrightarrow 2C$$

Experiment	Molarity (mol/L)		Rate (M/min)
	[A]	[B]	
1	1.0	1.0	0.20
2	1.0	2.0	0.40
3	2.0	1.0	0.80

•9. Based on the accompanying experimental data, determine each of the following:
(a) the overall order of the reaction
(b) the order with respect to each component
(c) the specific rate constant

Experiment	Molarity (mol/L)		Observed Rate (M/s)
	[A]	[B]	
1	0.500	0.050	5×10^{-4}
2	0.500	0.100	2.0×10^{-3}
3	1.000	0.050	1.0×10^{-3}

10. The following data show the effect of concentration on reaction rate for a given reaction:

Molarity (mol/L)			Rate (M/s)
[A]	[B]	[C]	
0.01	0.20	0.10	2.8
0.01	0.40	0.10	5.6
0.01	0.80	0.05	5.6
0.02	0.20	0.10	2.8

(a) Write the indicated rate-law equation.
(b) Determine the specific rate constant.

Half-Life [11–18]

11. What is the half-life for a reaction that follows first-order kinetics and for which the specific rate constant is $3.5 \times 10^{-5}\ s^{-1}$?

12. A reaction known to be a first-order elementary process has a half-life of 250. s. What is its specific rate constant?

13. If $t_{1/2}$ for a first-order reaction is 10. minutes, approximately how long will it take for the reaction to be 99% complete?

14. The half-life for a first-order reaction is known to be 600. seconds. What is the specific rate constant, and what percentage of the reacting material remains after one hour of reaction time?

•15. A particular reacting substance is shown to be disappearing at a rate of 1.0% per minute at a given temperature. If the reaction is known to be first-order, what are the specific rate constant and the half-life for the reactant at the given temperature?

16. Consider the first-order decomposition of benzoyl peroxide in ether. The reaction is 60.% complete in 220. seconds at 60°C. Calculate the specific rate constant for the reaction, the half-life for the reaction, and the time required for the reaction to proceed 75% to completion.

•17. Consider a hypothetical reaction in which A is observed to decompose to products according to the following equation:

$$A \longrightarrow B + C$$

(a) Write the rate law for such a first-order reaction in terms of the disappearance of A.
(b) Develop a general expression for the time required for 90.% of the original quantity of A present to be used up.

18. A certain organic peroxide thermally decomposes in solution via first-order kinetics. The reaction was 72.5% complete after 600. seconds had elapsed.
(a) Calculate the specific rate constant.
(b) Determine $t_{1/2}$ for the reaction.

Activation Energy [19–24]

19. The frequency factor is 1.3×10^{11} s^{-1} for a reaction that has an activation energy of 5.5×10^4 J/mol at 298K. What is the specific rate constant for this reaction?

20. The specific rate constant is 5.03×10^{-2} M^{-1} s^{-1} for the following reaction at 289K. (M^{-1} s^{-1} is equivalent to 1/M·s.)

$$C_2H_5I(aq) + OH^-(aq) \longrightarrow C_2H_5OH(aq) + I^-(aq)$$

The activation energy is 8.87×10^4 J/mol. What is the value of the frequency factor A?

•21. Here is a set of experimental rate constants for a particular first-order decomposition:

Temperature (°C)	20	40
$k \times 10^5$, s^{-1}	47.5	576

Determine the activation energy for the reaction in this temperature range and calculate the half-life for the reaction at 30°C.

22. The elementary process:

$$2I\cdot \longrightarrow I_2$$

has a rate constant of 4.0×10^9 M^{-1} s^{-1} at 25°C.
(a) In a solution that is 1.0×10^{-4} M in I, what is the initial rate of the reaction?
(b) At 35°C, the reaction rate constant is 6.1×10^9 M^{-1} s^{-1}. What is the activation energy of the reaction?

•23. For a first-order reaction, the following experimental data were recorded:

Temperature (K)	273	293	313
$k \times 10^5$ (min^{-1})	2.45	45.0	575

Calculate the activation energy and the rate constant at 303K.

24. The data in Figure 15.11 give the temperature dependence of the rate constant for the reaction

$$2N_2O_5 \longrightarrow 4NO_2 + O_2$$

By graphical analysis of the data,
(a) Calculate the activation energy.
(b) Determine the frequency factor.

Mechanisms [25–28]

•25. Consider a bimolecular reaction involving A and B molecules slowly forming an intermediate complex C, which then rapidly breaks down to form product P.
(a) Write equations for the elementary processes that might illustrate the mechanisms of the reaction.
(b) From the information at hand, write the likely experimental rate law.

26. The following data were collected for the decomposition of hydrogen peroxide into water and molecular oxygen in the presence of iodide ion:

Time (min)	Fraction of Peroxide Decomposed	Iodide Ion Concentration, M
0	0	0.020
5	0.130	0.020
15	0.339	0.020
25	0.497	0.020
45	0.712	0.020
65	0.835	0.020

The rate of the reaction is proportional to the hydrogen peroxide concentration, and it is also proportional to the iodide ion concentration.
(a) Determine the experimental rate law.
(b) Determine the specific rate constant, and show that the data fit your rate law.

(c) Suggest a plausible mechanism that is consistent with all the information provided.

•**27.** It has been observed that NO_2 will react with molecular fluorine according to the following reaction:

$$2NO_2(g) + F_2(g) \longrightarrow 2NO_2F(g)$$

The experimental data fit a second-order rate law:

$$\text{Rate} = k[NO_2][F_2]$$

It appears that atomic fluorine may be present during the reaction. Propose a reasonable mechanism indicating the rate-controlling step based on this information. Can you suggest any obvious complicating factors?

28. Explain why the following statements or illustrations are best described as true or false:

(a) The following reaction is known to be a second-order elementary process:

$$CH_3CHO(g) + I_2(g) \longrightarrow CH_3I(g) + HI(g) + CO(g)$$

Therefore, rate $= k[CH_3CHO][I_2]$.

(b) The following sequence illustrates a catalyzed chemical reaction:

$$CH_3CHO + I_2 \longrightarrow CH_3I + HI + CO \quad \text{(slow)} \quad (1)$$

$$CH_3I + HI \longrightarrow CH_4 + I_2 \quad \text{(fast)} \quad (2)$$

$$CH_3CHO \longrightarrow CH_4 + CO \quad \text{(overall)} \quad (3)$$

(c) A compound C decomposes to products by a first-order reaction:

$$\text{Rate} = \frac{\Delta[C]}{\Delta t} = k[C]$$

Additional Problems [29–39]

29. The reaction

$$2Ti^{3+} \longrightarrow Ti^{2+} + Ti^{4+}$$

proceeds by a single elementary second-order step in aqueous solution. At 25°C, k is 1.0×10^2 when expressed in the usual units. A solution is initially 0.015 M in Ti^{3+}.

(a) What are the "usual units" for k?

(b) What is the initial rate of the reaction in the solution?

(c) What is the initial rate of change of the concentration of Ti^{3+} in the solution?

(d) How long is required for the concentration of Ti^{3+} to drop to 10.% of its initial value?

30. A sample of alcohol was dehydrated to a mixture of diethyl ether and water at 140°C in dilute sulfuric acid. One particular set of experiments produced the following data:

Time (min)	Fraction of Alcohol Reacted
60	0.197
93	0.290
143	0.409
295	0.672
590	0.889

Determine the probable experimental order of the reaction based on these data.

•**31.** Consider the kinetic experiment described by the following graph of log concentration versus time for N_2O_5 decomposition to $NO_2(g)$ and $O_2(g)$. [C] is the concentration of N_2O_5.

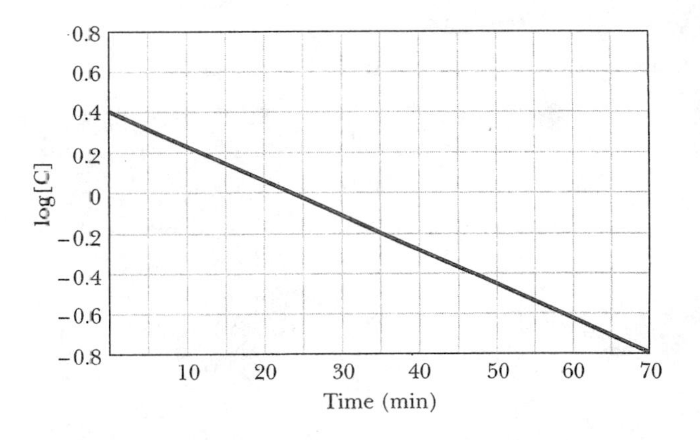

(a) Determine the order of the reaction.

(b) Evaluate the specific rate constant.

(c) Determine the initial concentration of the reacting material.

32. For a second-order reaction in which

$$2A \longrightarrow \text{products}$$

show clearly that $t_{1/2} = 1/k[A]_0$.

•**33.** At 415°C, ethylene oxide undergoes a first-order gas phase decomposition producing methane and carbon monoxide as the only products:

$$C_2H_4O(g) \longrightarrow CH_4(g) + CO(g)$$

The specific rate constant was found to be $0.0123\ \text{min}^{-1}$.

(a) After 60. min at 415°C, what percentage of any given ethylene oxide sample will have decomposed to products?

(b) How much time will be required for three quarters of the starting material to react?

34. A compound named dimethyldiimide ($CH_3N{=}NCH_3$) is known to decompose according to the following reaction in the gas phase:

$$CH_3N{=}NCH_3(g) \longrightarrow CH_3CH_3(g) + N_2(g)$$

After 20. min, the total pressure in the reaction flask is 50.% greater than the initial pressure. Assuming that nothing but $CH_3N{=}NCH_3$ molecules were initially present and that the decomposition is first order, what is the specific rate constant?

•**35.** (a) Briefly explain why the following reaction is not likely to be an "elementary" reaction:

$$A + B + C + D \longrightarrow X$$

(b) If the reaction were elementary, state the order with respect to A, the overall order of the reaction, and the rate law.

36. For an endothermic reaction, state the minimum value that the activation energy can have. For an exothermic reaction, state the minimum value that the activation energy can have.

37. Draw and label a diagram, including ΔE, E_a, and a possible transition state for any of the propagation steps leading to the principal product in the free-radical chlorination of methane to methyl chloride:

$$2CH_4(g) + Cl_2(g) \longrightarrow 2CH_3Cl(g) + H_2(g)$$

38. For a reaction for which

$$-\frac{\Delta[A]}{\Delta t} = \frac{\Delta[B]}{\Delta t} = k$$

what is the overall order of the reaction?

39. For the oxidation of V^{3+} by Fe^{3+}, the following mechanistic pathway has been suggested:

$$Fe^{3+} + Mn^{2+} \longrightarrow Fe^{2+} + Mn^{3+}$$

$$Fe^{3+} + Mn^{3+} \longrightarrow Fe^{2+} + Mn^{4+}$$

$$V^{3+} + Mn^{4+} \longrightarrow V^{5+} + Mn^{2+}$$

Which ion is the catalyst? What would be the rate law if the first step is slow?

Cumulative Problems [40–41]

40. For the imaginary reaction that follows, the rate constant k_1 at 298K for the forward reaction is $5.03 \times 10^{-2}\ s^{-1}\ M^{-1}$, and the rate constant at 298K for the reverse reaction k_{-1} is $7.35 \times 10^{-5}\ s^{-1}$. Calculate $\Delta G°$ of the reaction at 298K.

$$2A(g) \longrightarrow B(g)$$

41. Rate constants, forward and reverse, for the following hypothetical reaction at 298K are taken to be $k_1 = 2.44 \times 10^{-7}\ s^{-1}\ M^{-1}$ and k_{-1} is $3.19 \times 10^{-1}\ M^{-1}\ s^{-1}$. Calculate the equilibrium partial pressures starting with $p_A = 0.10$ atm and $p_B = 0.15$ atm.

$$A(g) + B(g) \longrightarrow C(g) + D(g)$$

Applied Principles [42]

42. The growth, multiplication, and expansion of colonies or cultures of bacteria are classic kinetic processes. When a growing bacterium reaches a critical size, it divides in two; then into four, eight, sixteen, and the multiplication process continues. In one growth experiment, the following data were obtained ([B] = number of bacteria per L):

Time (min)	0.0	60.0	120	180	240	300
$[B] \times 10^{-3}$	3.50	7.00	14.0	28.0	56.0	112

(a) How many minutes must pass before the population of the colony doubles?

(b) What is the bacteria concentration at 6 hr?
(c) When the bacteria concentration has reached 22.4×10^4/L, how much time has elapsed?
(d) Calculate the growth constant, the equivalent of the specific rate constant, for this multiplication process.

ESTIMATES AND APPROXIMATIONS [43]

43. Consider the following reaction, which is known to be an elementary process,

$$CH_3I(g) + HI(g) \longrightarrow CH_4(g) + I_2(g)$$

Postulate a possible transition state for the reaction mechanism. Then using this transition state, a table of bond enthalpies, and lots of approximations, estimate the energy of activation for the reaction.

FOR COOPERATIVE STUDY [44–45]

44. Data describing the first-order decomposition of N_2O_5 in CCl_4 solution at 45°C are given in Table 15.3.

$$2N_2O_5(g) \longrightarrow 4NO_2(g) + O_2(g)$$

The analytical procedure depends on the fact that the N_2O_5 and NO_2 are soluble in carbon tetrachloride (CCl_4), whereas O_2 is not. These data have been used to obtain the plots shown in the following figures. [A] = [N_2O_5]; x = fraction of N_2O_5 reacted.

TABLE 15.3 Decomposition of N_2O_5 in CCl_4 at 45°C*

Time Elapsed (s)	N_2O_5 Concentration (mol/liter)
184	2.08
319	1.91
526	1.67
867	1.36
1198	1.11
1877	0.72
2315	0.55
3144	0.34

*Recorded in a classic kinetics paper by Eyring and Daniels in 1933.

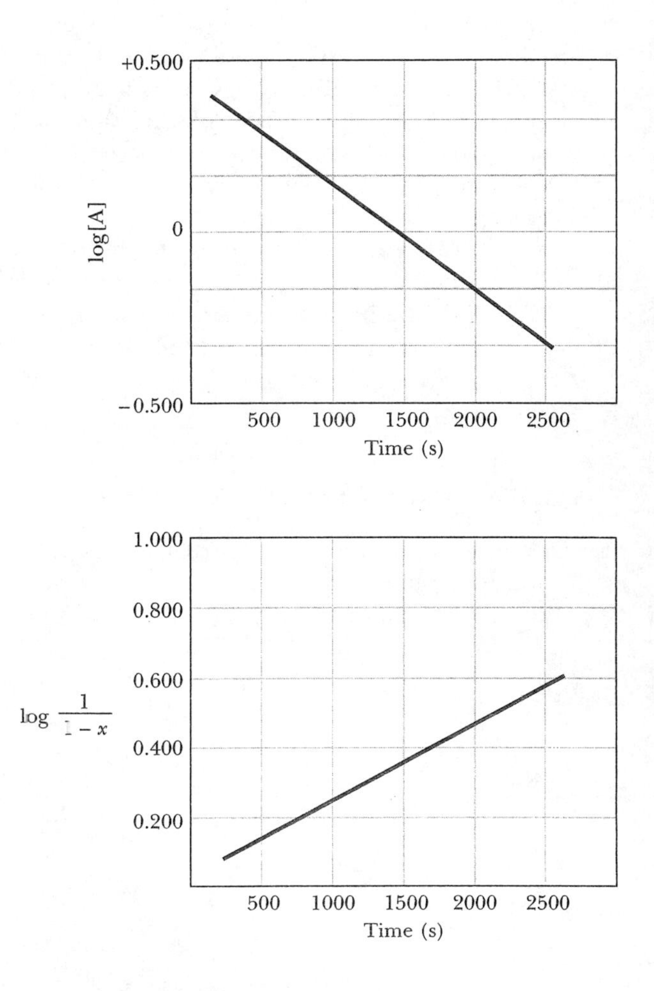

(a) How might you "follow" the progress of the reaction if you were conducting the experiment?
(b) Explain how these data points confirm the statement that the reaction is first order.
(c) Using both data plots, calculate the specific rate constant, and the initial N_2O_5 concentration.
(d) Determine the N_2O_5 concentration at the point at which the reaction has diminished to half the initial rate.
(e) What is the half-life of N_2O_5?
(f) Take any two consecutive sets of numbers from the data table and determine

the rate constant algebraically (in contrast to the graphical determination employed earlier in this problem).

45. Taking a cue from "Kisty" (*Profiles in Chemistry*), consider the thermal decomposition of dinitrogen oxide (N_2O) to N_2 and O_2 at 650.°C. With the initial N_2O pressure recorded at 300. torr, it was found that 50.% reacted in 250. seconds. When the initial N_2O pressure was 350. torr, it took only 214 seconds for 50.% to react.
(a) Is the reaction first or second order? Explain.
(b) With the initial N_2O pressure set at 1 atm, how much time would be required to achieve the same 50% decomposition?

WRITING ABOUT CHEMISTRY [46–48]

46. Knowing how the rate of a reaction depends on the concentrations of the reactants can give information on the mechanism of the reaction. Explain carefully why that is. Why is it that knowing the equilibrium constant for a reaction tells nothing about the mechanism of the reaction?

47. Describe a possible macroscopic model for the activation energy theory of chemical reactions. Explain why your model is reasonable, but also explain its limitations for duplicating the microscopic phenomena involved in the reactions of molecules.

48. A catalyst increases the rate of a chemical reaction. An inhibitor slows it down. Explain how catalysts and inhibitors work, including the theory of the activation energy in your explanation. Is there a fundamental difference between the functioning of catalysts and inhibitors? Also explain why catalysts and inhibitors should or should not have the same effect for the reverse reaction.

chapter sixteen

16

SOLIDS

16.1 CRYSTALLINE SOLIDS AND THE AMORPHOUS STATE

Solids are the building blocks of the world. They give it shape and form. Included among the solids are all of the structural metals, rocks and minerals, gems such as diamond and emerald, and ceramics such as bricks and porcelain. Solids range from the semiconductors, at the heart of all electronics, to our bones and teeth. States of matter are central issues for the applied science of chemistry, and the solid state is of enormous practical importance.

The words *solids, liquids,* and *gases* convey general meanings to most of us that need not be explained. For scientists and technologists, however, the words have precise meanings, which are universally accepted and understood. We recognize these three states or forms as those in which all the materials of the universe can exist. The distinguishing features are due to the differences in the energies of the particles (atoms, ions, or molecules). Particles in the gaseous state are most energetic and widely separated. The less energetic particles that characterize liquids try to collect themselves together but are still so mobile that they can easily slide by each other. Compared to gases, there is not much space between particles in liquids, but there is enough to allow for a characteristic lack of organization and for the ability to flow. Particles in solids have relatively little energy, compared to their gaseous and liquid-state counterparts. As a result, they organize themselves into highly ordered crystalline structures having characteristic patterns.

For our purposes, the words "solid" and "crystal" (or crystalline) carry the same meaning. A **crystal** has a structure in which the particles are arranged in a regularly repeating pattern. This pattern extends in three dimensions over the entire crystal, extremely long distances on the atomic scale. To put it another way, on solidification, the particles of a crystalline solid find themselves in an orderly arrangement with respect to each other. Generally, these crystalline arrays include an enormous number of particles. Imagine counting the atoms lined up along an edge of a large copper crystal that has a volume of 1 cm^3. Something on the order of 10^7 copper atoms line the edge from one corner to the other, and in the perfect copper crystal, the orderliness of the atoms at the beginning is the same as at the end or anywhere along the line.

We find the origins of the word crystal in the Greek word *krystallos,* which in turn comes from *krystos,* meaning icy cold. To the ancient Greeks, quartz was ice that had frozen so hard it never thawed. Over time, the name came to be used not only for quartz but for many other crystalline substances, and even for certain kinds of glasses used for goblets, decanters, and other decorative tableware . . . which are not crystalline at all. Such objects are glasses and are actually liquids. They become glasses when their component particles (molecules and ions) suddenly find themselves reduced to the energy of a solid before they had a chance to get organized.

Hard, apparently solid, materials such as glasses are described as amorphous rather than crystalline. **Amorphous** or noncrystalline substances can display short-range or local organization, but long-range order is characteristically absent. You might picture the amorphous state as an instantaneous "snapshot" of the liquid state in which all the motion of molecules from one location to another has been momentarily stopped. Accidental local organization might be evident here and there, but when you examine large arrays of particles over long distances, you find them to be characteristically random, as though they had been moving about with the freedom of a fluid rather than a solid. A **glass** is actually a supercooled liquid that has such a high viscosity that it will hold its shape for a very long time.

A useful criterion for distinguishing between a crystalline substance and a glass is the fact that crystalline solids have definite melting points. Unless a pure compound decomposes on heating, it can be characterized by its normal melting point, which is uniquely and precisely defined. A glass, on the other hand, does not pass through a clearly defined transition between the solid and liquid states. Instead, its viscosity simply decreases with increasing temperature until it visibly softens and finally flows, ending up as a puddle. Liquids are characteristically less viscous at higher temperatures. Glasses have viscosity behavior similar to other liquids across the entire temperature range.

Polymers or plastics are collections of giant molecules composed of much smaller molecules that have been snapped together into chains of extraordinary length. Because long, flexible chains tend to get tangled up in each other, polymers are characteristically amorphous rather than crystalline. However, polymers do show some degree of crystallinity due to local organization. Polymeric materials deserve separate attention, and we will say no more about them here, deferring that discussion to the two chapters on materials, which follow.

16.2 CRYSTAL STRUCTURE

The Internal and Surface Structure of Crystals

Crystals are composed of atoms arranged in an ordered fashion in three dimensions. The science of crystallography is concerned with the orientation of the atoms in space and the geometric shapes that result. The following is a summary of five important properties characteristic of crystalline substances:

- All crystals of the same substance, when prepared in the same way, have the same shape whether the crystals are large or small.
- Crystals are characterized by their faces and edges. Each face normally has another face parallel to it, and each edge normally has other edges parallel to it. The edges intersect at fixed angles.
- Crystalline substances have a high degree of symmetry. **Symmetry** makes an object look even and balanced. In more technical terms, symmetry is a property of an object that allows operations such as rotations and reflections to be performed on it without an apparent change in its orientation. Thus, a five-pointed star is symmetrical because it can be rotated by one fifth of a complete circle and look as though no operation has been done.
- When most crystals shatter, they tend to fracture along planes parallel to the faces of the original crystal, producing smaller crystals shaped just like the larger crystals from which they broke away. The planes along which fracture occurs are called **cleavage planes**.
- Because cleavage occurs along specific planes, the properties of a crystal are not the same in all directions. That is, crystals are **anisotropic**, which means different in different directions. On the other hand, amorphous materials are **isotropic**; they have the same properties in all directions. Thus, when glass shatters, splintering occurs in all directions from the point of impact, producing shards—sharp and irregularly shaped pieces.

Because crystals have characteristic external shapes, the particles that make up the crystal must be fixed in specific positions. Furthermore, because small crystals of a substance have the same shape as large crystals of the same substance, we can safely conclude that the basic arrangement within the crystal is repeated over and over in a regular way as the size of the crystal increases. Accordingly, a simple two-dimensional model of a crystal might look like Figure 16.1. The lines suggest possible directions of cleavage in this hypothetical model.

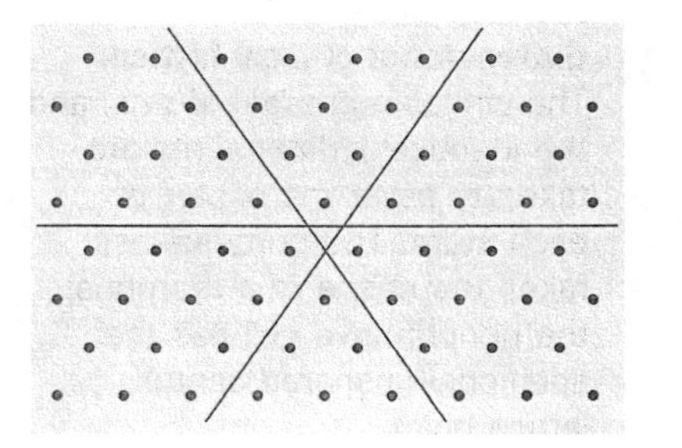

FIGURE 16.1 Model of a hypothetical two-dimensional crystal. Favored directions of cleavage are shown by lines.

FIGURE 16.2 Atoms cover a surface like so many soap bubbles in a single layer. The aggregate of bubbles illustrates the structure of metals on the atomic scale. Note the boundaries and the defects but especially the large areas of regularity where atoms (spheres) have lined up in regular rows.

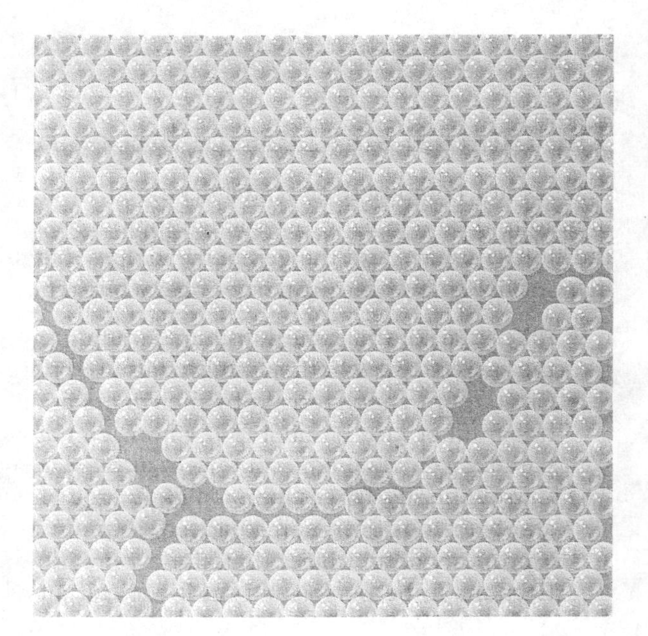

Lattice Structure and Unit Cells

To simplify our introduction to crystal structure and the solid state, let us change from the three-dimensional perspective of the world around us to an imaginary two-dimensional world. Crystals in two dimensions are analogous to the layers of atoms adsorbed onto a surface at high density or the single layer of soap bubbles in Figure 16.2. A tiny region of a crystal where circles represent identical atoms can be shown in Figure 16.3. The repeating arrangement of atoms in a crystal can be represented by a crystal lattice. The **crystal lattice** for a structure is a collection of all points that have identical environments. We can identify all the points in the centers of the circles as the complete lattice.

The **unit cell** is a very small part of the lattice structure, formed by connecting lattice points in such a way that the entire structure can be generated from the unit cell simply by translation of the unit cell. **Translation** is a simple shift of position involving no other movements, such as rotations. The unit cell in a structure has as many dimensions as the structure itself. In a three-dimensional lattice the unit cell has three dimensions, whereas in our two-dimensional

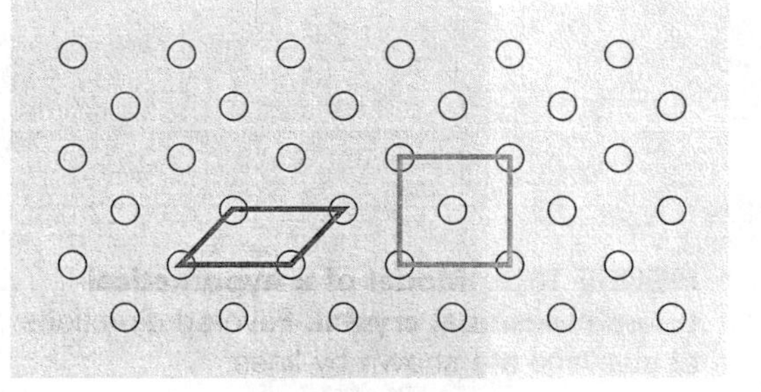

FIGURE 16.3 Two-dimensional crystal lattice. The circles represent atoms, and the identical lattice points are taken to be at the center of each atom. The primitive cell takes the shape of a rhombus; the nonprimitive cell has the additional unshared center lattice point.

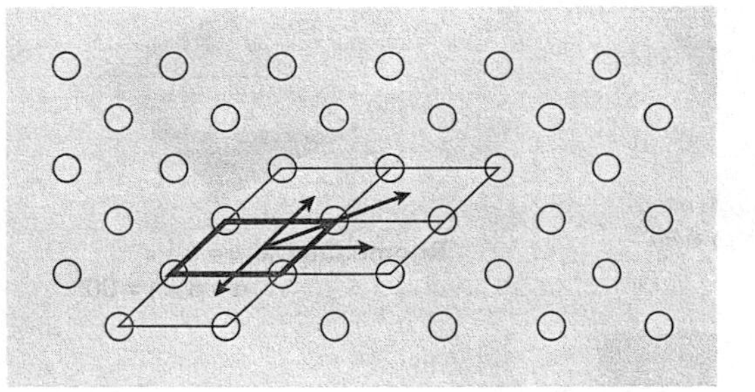

FIGURE 16.4 Generating the crystal lattice. Movement and translation of the primitive cell along the directions indicated by the arrows shows how the entire crystal lattice can be generated.

structure, the unit cell has two dimensions. The smallest possible unit cell for a given lattice is a **primitive cell**. A primitive cell for a lattice is outlined in Figure 16.3 and takes the shape of a parallelogram. A primitive cell is sometimes called a **simple cell**. Movement of this cell in the directions shown in Figure 16.4 generates the entire structure. The geometry of the unit cell in two dimensions is specified by the lengths of the two intersecting sides and the angle between them.

The cell shown as a square in Figure 16.3 is a nonprimitive cell. Any cell larger than the primitive cell is a **nonprimitive cell**. Nonprimitive cells are normally chosen if they have more symmetry than the primitive cell. Primitive and nonprimitive unit cells may be distinguished from each other by determining the total number of lattice points per unit cell. For the primitive cell in two dimensions, there is a lattice point at each corner. However, each corner is shared by four unit cells, and, therefore, only one fourth of a lattice point is within each of these four cells. The total number of lattice points within a planar primitive cell is then

$$\frac{4 \text{ corners}}{\text{cell}} \times \frac{1 \text{ lattice point}}{4 \text{ corners}} = 1\frac{\text{lattice point}}{\text{cell}}$$

All primitive two- or three-dimensional unit cells contain only one lattice point. Because the nonprimitive cell has the additional unshared lattice point at its center, the number of lattice points within the nonprimitive cell is 2. All nonprimitive cells contain more than one lattice point.

For three-dimensional lattices and unit cells, we need only extend these principles. The unit cell will have a lattice point at each corner, and a primitive cell will have the smallest possible volume. Three lengths and three angles are now necessary to specify the geometry of the unit cell. They are, respectively, the three intersecting edges a, b, and c and the three angles between these edges, shown in Figure 16.5. The angles are represented by α, β, and γ. Angle α is the angle between edges b and c, β is the angle between a and c, and γ is the angle between a and b. The possible geometries for three-dimensional primitive cells are given in Figure 16.6.

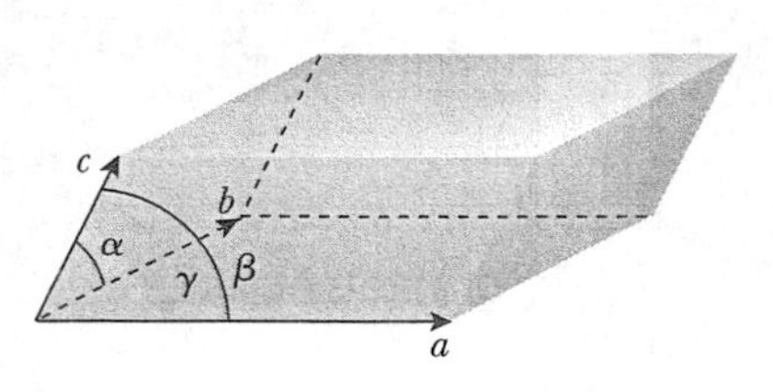

FIGURE 16.5 The dimensions of a three-dimensional unit cell.

FIGURE 16.6 The geometries for three-dimensional primitive cells. Seven are possible.

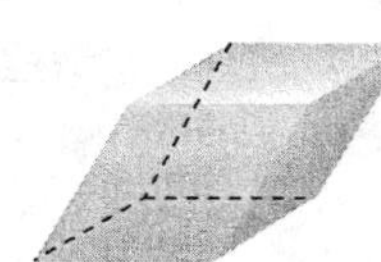

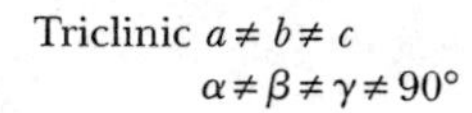

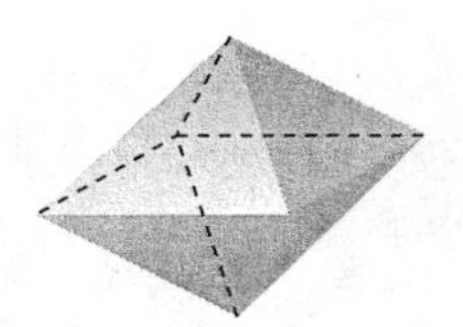

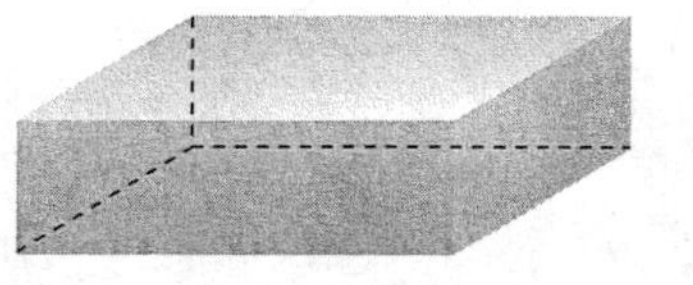

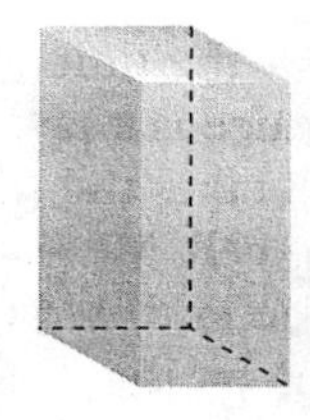

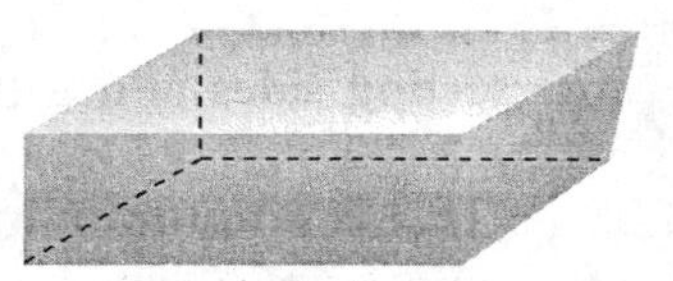

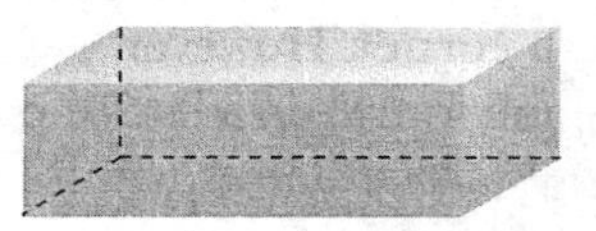

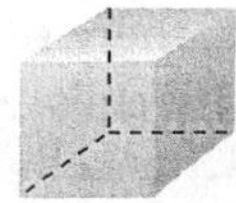

Crystal Lattices Based on the Cube

Of the possible crystal lattices, three have a cubic unit cell (Fig. 16.7). One is the simple or primitive cubic cell, with lattice points only at the corners. There are two nonprimitive cubic lattices: **Face-centered cubic (FCC)** has an extra lattice point in the center of every face, and **body-centered cubic (BCC)** has an extra lattice point in the center of the cell. These lattices are very important in understanding the structures of metals and salts. Consider the number of lat-

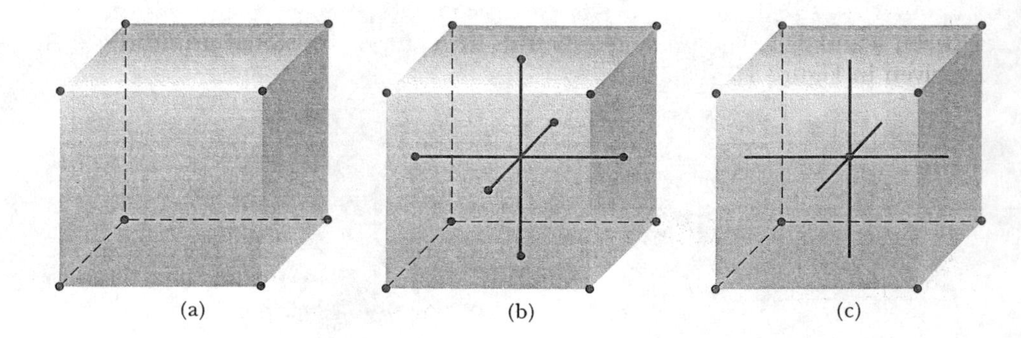

FIGURE 16.7 Cubic unit cells. (a) The primitive or simple cubic unit cell; (b) the face-centered cubic unit cell; (c) the body-centered cubic unit cell.

PROFILES IN CHEMISTRY

"Tout Est Trouvé!"—All Is Discovered! The branch of science relating to the structure and properties of crystals and crystalline substances is known as crystallography. The essence of the science of crystallography lies in the relationship between the shape you see and the internal regularity that exists in the great variety of nature's crystals. Crystallography originates in the scientific works of the Abbé René Haüy (French, 1743–1822), canon of the Cathedral of Notre Dame and Professor of Mineralogy at the University of Paris Museum of Natural History. To his great credit, Haüy was able to show that all known crystalline forms can be reduced to a few elementary shapes.

It is rare that seminal discoveries of a scientific nature are accidental. Rather, they are made by observers who are prepared to interpret more accurately than others what they have experienced. Legend has it that while visiting a close friend with an exceptional mineral collection, the good Abbé Haüy accidentally let a large crystal of calcite (calcium carbonate) he was examining slip to the floor, where it broke upon impact into many pieces. While gathering up the ruins, Haüy's imagination was caught by the fact that the fragments—regardless of size—were all quite similar and had the same shape. Then (presumably) to the chagrin of his host, he proceeded to break some of the fragments into still smaller pieces. Each of these proved to be identical in shape to the others as well as to the original fragments from which they were fractured, an observation that could hardly be considered coincidental. With an exclamation of "*Tout est trouvé!*"—"All is discovered!"—he returned to his own mineral collection and proceeded to systematically demolish all the calcite crystals he could find, further confirming his discovery. No matter what the size or shape of the parent crystal from his collection, the fragments all showed faces at the same angles.

What was Haüy's discovery? What did this pattern of destruction suggest about the nature of calcite and other crystalline materials? Here was Haüy's great achievement, the idea of internal order. The precise angles between adjoining faces of a crystal were determined by a regular stacking of identical units, a concept that ultimately provided a means of identifying and classifying crystals.

Question

What determines the precise angles between adjoining faces of a crystal, and what does that have to do with the classification of crystals?

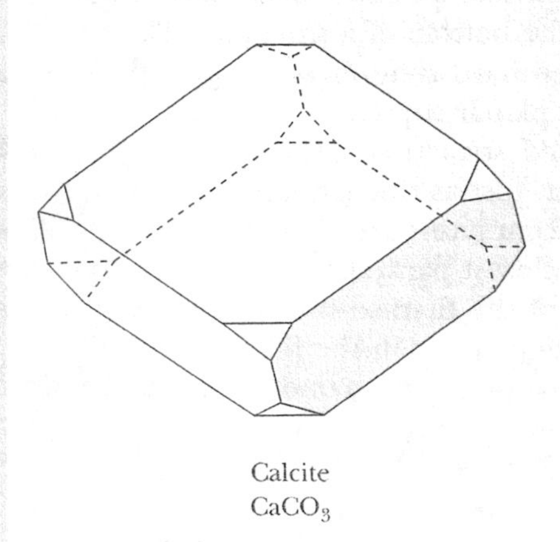

Calcite
$CaCO_3$

Calcite, a form of calcium carbonate. Crystals cleave into rhombohedral units of different sizes.

tice points in the unit cell for the three cubic lattices, just as we did for the planar or two-dimensional lattice. Lattice points located at unit cell corners are shared by eight unit cells. In the body-centered cubic lattice, the extra lattice point located within the cube is unshared. In the face-centered cubic lattice, the extra lattice points in the face centers are shared by two unit cells. Thus, the number of lattice points for each kind of unit cell are as follows:

- Primitive (simple) cubic:

$$\left[\frac{8\ \text{corners}}{\text{cell}} \times \frac{1\ \text{lattice point}}{8\ \text{corners}}\right] = 1\,\frac{\text{lattice point}}{\text{cell}}$$

- Body-centered cubic:

$$\left[\frac{8\ \text{corners}}{\text{cell}} \times \frac{1\ \text{lattice point}}{8\ \text{corners}}\right] + 1\ \text{lattice point} = 2\,\frac{\text{lattice points}}{\text{cell}}$$

- Face-centered cubic:

$$\left[\frac{8\ \text{corners}}{\text{cell}} \times \frac{1\ \text{lattice point}}{8\ \text{corners}}\right] + \left[\frac{6\ \text{faces}}{\text{cell}} \times \frac{1\ \text{lattice point}}{2\ \text{faces}}\right] = 4\,\frac{\text{lattice points}}{\text{cell}}$$

16.3 METALLIC CRYSTALS

Closest Packing of Spheres

In metals and their alloys, the solid state is characterized by long-range three-dimensional order, a fact that is strongly reflected in their properties and behavior, as we shall describe in this and succeeding sections. In many ways, elemental metallic solids are the simplest of structures because all the atoms are identical, regularly arranged, and separated by fixed distances. If the atoms are closest packed, the situation is very much like so many balls racked up for a game of pool, or ball bearings covering the bottom of a shoe box (Fig. 16.8). A **closest packed** layer forms naturally when hard spheres of equal diameter are placed as close together as possible on a planar surface.

In a three-dimensional closest packed structure, layers of closest packed spheres are stacked on top of each other. Just as the spheres within the layers are as close together as possible, the adjacent layers are also as close together as possible. This results in what is called a closest packed three-dimensional crystal structure. A second layer sits on top of the first, so that spheres in the second layer are positioned directly over indentations in the first layer (Fig. 16.9). A third layer of spheres sits over indentations in the second layer, and so forth. Two principal crystal structures arise from this model: hexagonal closest packing (HCP) and cubic closest packing (CCP).

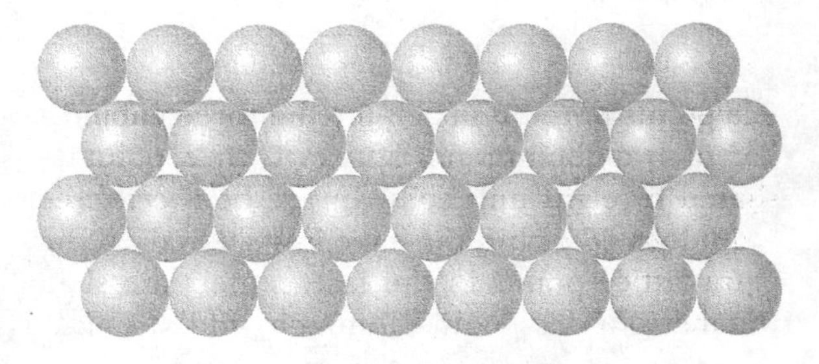

FIGURE 16.8 One layer of closest packed spheres.

FIGURE 16.9 Two layers of closest packed spheres. The second layer is shown sitting over half the indentations between the spheres in the first layer.

If the third layer of spheres is positioned directly over the first layer in an *ababab* . . . arrangement, then **hexagonal closest packing (HCP)** is produced, as shown in Figure. 16.10a. Alternatively, if the third layer is aligned so that it is directly above neither layer one nor layer two and the fourth layer is over layer one (Fig. 16.10b) in an *abcabcabc* . . . arrangement, then **cubic closest packing (CCP)** is produced. Examination of models reveals that the actual structure within the cubic closest packed structure is that of the face-centered cubic lattice, with a metal atom centered at every lattice point. Thus, the CCP structure is identical to the face-centered cubic, or FCC, structure and is adopted by a number of metals, including silver, aluminum, calcium, lead, and copper. The HCP arrangement is observed for magnesium, osmium, ruthenium, and others.

A third simple metallic structure that is observed has metal atoms centered on the lattice points of a body-centered cubic (BCC) lattice, with an atom at every corner of the cubic lattice and an atom buried in the body of the cube (Fig. 16.11). The BCC structure is observed for iron, potassium, tungsten, and several other metals. Note that BCC is not closest packed.

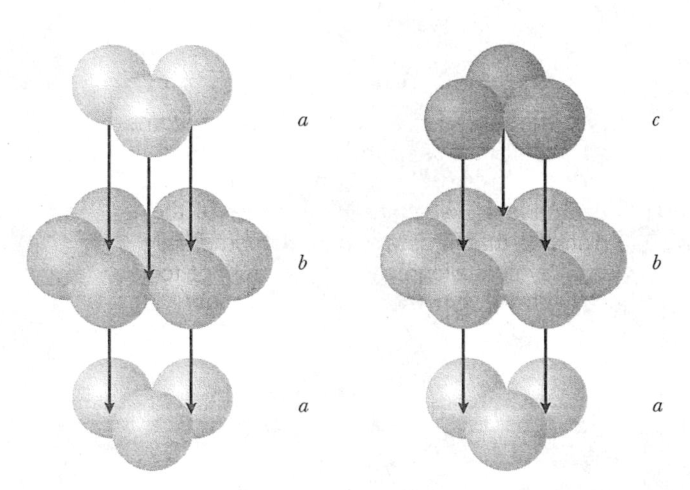

FIGURE 16.10 Three layers of closest packed spheres. First, in (a) is an *abab*... hexagonal closest packed arrangement, and (b) is an *abcabc*... cubic closest packed arrangement.

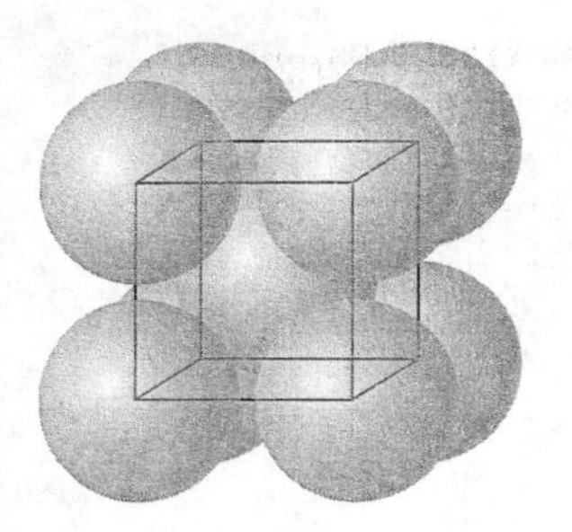

Body-centered

FIGURE 16.11 Packing of spheres. Here is the body-centered arrangement of atoms (spheres), one of three cubic systems.

Because the unit cell is the building block from which the entire crystal lattice is generated, the density of the metal must be the same as the density of its unit cell,

$$d_{\text{entire sample}} = d_{\text{unit cell}}$$

Therefore, the density of a solid depends on its crystal structure. The density of the unit cell can be calculated if we know the volume of the unit cell and the number of atoms contained in it. For structures in which atoms are centered on lattice points, the number of atoms per unit cell is the same as the number of lattice points per unit cell. Beginning with the general definition of density as mass per unit volume, the density of the unit cell can be defined as follows:

$$d_{\text{unit cell}} = \frac{m_{\text{unit cell}}}{V_{\text{unit cell}}}$$

where m is the total mass of the atoms in the unit cell and V is the volume of the unit cell. The mass of the atoms in the unit cell in grams is

$$m_{\text{unit cell}} = \frac{ZM}{N_A}$$

where Z is the number of atoms per unit cell, M is the molar mass, and N_A is Avogadro's number. This mass can be substituted into the equation for $d_{\text{unit cell}}$,

$$d = \frac{ZM}{N_A V}$$

For a cubic unit cell where a is the length of the edge of the unit cell and $V = a^3$,

$$d = \frac{ZM}{N_A a^3}$$

Density of a cubic crystal: In the equation, the number of atoms per unit cell is Z, M is the molar mass of the atom in g/mol, and N_A is Avogadro's number, 6.022×10^{23} atoms per mole. For a unit cell that is a cube, the volume is a^3, where a is the length of the edge of the unit cell, a distance that is determined experimentally by a technique known as X-ray crystallography. Be careful not to confuse Z in this equation with the symbol for atomic number.

EXAMPLE 16.1

X-ray diffraction experiments show that copper crystallizes in an FCC unit cell 3.608 Å along an edge. Density measurements give the value 8.92 g/cm^3. What is the molar mass of copper?

COMMENT Solve for M by rearranging the preceding equation for calculating density,

$$M = \frac{dN_A a^3}{Z}$$

Because the atoms sit on lattice points, the number of atoms per unit cell Z is determined in exactly the same manner as the number of lattice points per unit cell:

$$Z = \left[\frac{8 \text{ corners}}{\text{cell}} \times \frac{1 \text{ atom}}{8 \text{ corners}}\right] + \left[\frac{6 \text{ faces}}{\text{cell}} \times \frac{1 \text{ atom}}{2 \text{ faces}}\right] = 4\,\frac{\text{atoms}}{\text{cell}}$$

Now convert the length of unit cell edge from Å to centimeters using the relationship 1 Å $= 1 \times 10^{-8}$ cm.

SOLUTION

$$M = \frac{dN_A a^3}{Z}$$

$$= \frac{(8.92 \text{ g/cm}^3)(6.022 \times 10^{23} \text{ atoms/mole})(3.608 \times 10^{-8} \text{cm})^3}{4 \text{ atoms}}$$

$$= 63.1 \text{ g/mol}$$

EXERCISE 16.1

Tantalum crystallizes in a BCC arrangement with a unit cell edge of 3.281 Å. The density of tantalum has been measured to be 16.6 g/cm^3. Calculate the molar mass of tantalum.

ANSWER 176 g/mol

PROCESSES IN CHEMISTRY

Polymorphic Modifications and Processing Steel Certain elements—notably iron—are capable of existing in different crystalline forms, known as polymorphic modifications, which change with temperature. For example, at temperatures below 1185K, the crystal structure of iron is called α-iron and is BCC. A polymorphic transformation into the FCC β-iron structure takes place at that temperature, and the resulting FCC crystal structure remains stable up to 1667K, at which point it changes to γ-iron and reverts back to a BCC structure, which is the stable crystal structure up to the melting point. These differences in crystal structure are reflected in significant differences in the densities of the forms, and this is very important in steel manufacturing and processing because steel is an iron alloy. As a result of the transformations into these polymorphic modifications (different crystal structures), steel exhibits the annoying habit of contracting on heating and expanding on cooling through the transformation temperatures.

Question

To make iron into steel, small amounts of carbon need to be dissolved in it. What would you expect to happen to the solubility of carbon in iron with increasing temperature? Why?

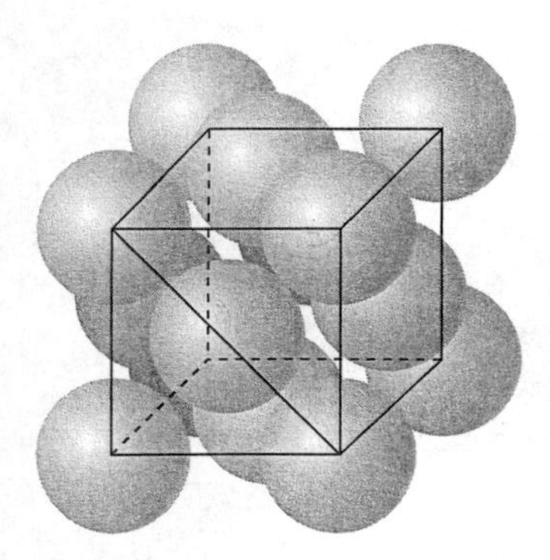

FIGURE 16.12 Face-centered cubic lattice (FCC) of closest packed spheres. The face diagonal is four times the radius of the spheres.

Using a model for the structure of a metal of hard spheres in direct contact, one can calculate the relationship between the size of the spheres and the size of the unit cell. The radius r of the sphere corresponds to the atomic radius r of the atom, and the size of the cubic unit cell is determined by its atomic radius. Consider the FCC unit cell of a metal using a diagram in which the atoms are shown at their full space-filling size (Fig. 16.12). Corner atoms do not touch each other. However, atoms on the corners do touch atoms in the centers of faces, and the face diagonal fd is equal to four atomic radii. By the Pythagorean theorem

$$(fd)^2 = a^2 + a^2 = 2a^2$$
$$fd = a\sqrt{2} = 4r$$

Thus, in a metal that crystallizes in an FCC lattice, the atomic radius r can be calculated:

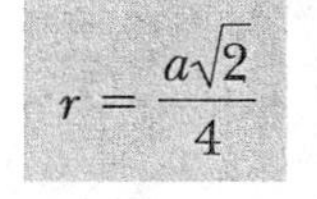

$$r = \frac{a\sqrt{2}}{4}$$

Atomic radius for the FCC crystal structure: This concept is based on the assumption that hard-sphere atoms, centered on the lattice points, are in contact. Spheres touch along the face diagonal *fd* of the unit cell. The distance *a* separates lattice points along the edges of the cube, that is, the length of the unit-cell edge.

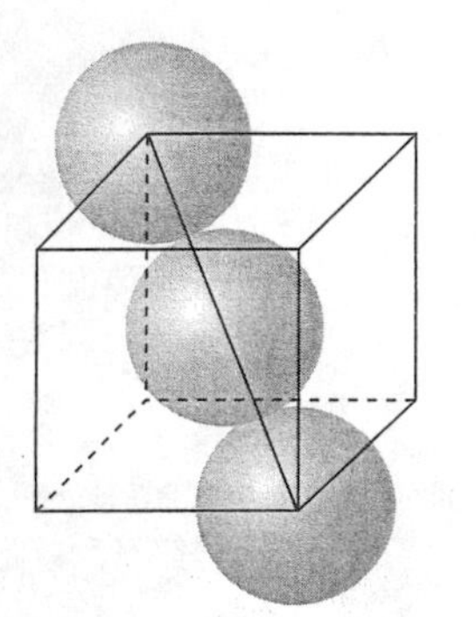

FIGURE 16.13 Body-centered cubic (BCC) lattice. The body diagonal is four times the radius of the spheres.

In the BCC structure of a metal (Fig. 16.13), the corner atoms again do not touch. However, the corner atoms do touch the atom in the center of the unit cell, and in this case the body diagonal bd is equal to four atomic radii. The body diagonal bd is the hypotenuse of a right triangle, the other two sides of which are a and fd. Thus, according to the Pythagorean theorem,

$$(bd)^2 = (fd)^2 + a^2$$
$$(bd)^2 = \left(a\sqrt{2}\right)^2 + a^2 = 2a^2 + a^2 = 3a^2$$
$$bd = a\sqrt{3} = 4r$$

from which the atomic radius r can be calculated:

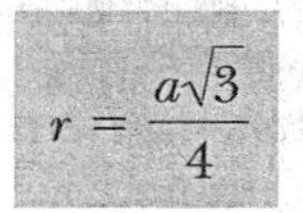

$$r = \frac{a\sqrt{3}}{4}$$

Atomic radius for the BCC crystal structure: This, too, is based on the assumption that hard-sphere atoms, centered on the lattice points, are in contact. The spheres touch along the body diagonal *bd* of the unit cell. The distance *a* is the distance between lattice points along the edges of the cube.

EXAMPLE 16.2

Copper crystallizes in an FCC lattice with a unit cell 3.608 Å along an edge. Calculate the atomic radius of copper.

COMMENT Remember that equations for the radius r are different for FCC and BCC structures, so substitute the value of the distance a along an edge into the proper equation.

SOLUTION

$$r = \frac{(3.608\ \text{Å})\sqrt{2}}{4} = 1.276\ \text{Å}$$

EXERCISE 16.2

Potassium crystallizes in a BCC structure. The unit cell edge length is 5.333 Å. Calculate the atomic radius of potassium.

ANSWER 2.309 Å.

Bonding and Conduction in Metals

Having discussed the general features of the structures of metals, we now consider how the atoms in these structures are bonded together. In turn, this can provide some insight into the characteristic properties of metals, such as conductivity of electricity and heat, malleability, and ductility, all of which depend on the presence of mobile electrons.

The metallic elements occupy the left side and lower portion of the periodic table of the elements. Compared to the nonmetallic elements on the right side and upper portion, ionization energies and numbers of electrons in the valence shell are typically low for metals. Low ionization energies indicate weakly held valence shell electrons as well as a small tendency to add electrons to their own valence shells. As a result, not only do atoms of metallic elements fail to form anions, but formation of covalent bonds by sharing electron pairs provides little stabilization. Therefore, species such as Na_2 and Pb_2 are not observed except under extraordinary conditions. The molecular orbital diagram for the ground-state diatomic sodium molecule (Fig. 16.14a) shows the two $3s$ orbital electrons in one molecular orbital. That should be favorable to bonding, but the bonding molecular orbital is only slightly stabilized relative to the atomic orbitals, and the covalent bond is weak.

Metals exist principally as crystalline solids. In the solid state, each atom is surrounded by many nearest neighbors. The number of nearest neighbor atoms is called the **coordination number,** and for metals it is quite high, usually eight or twelve. Common structures are those in which the atoms are packed together as closely as mutual repulsion of their electronic and nuclear charges permits. However, metals have relatively small numbers of valence electrons. For example, sodium has only one valence shell electron and lead has only four. No metal comes even close to having the eight or twelve valence electrons needed for ordinary covalent bonding to all its nearest neighbors. Clearly, then, the available valence electrons are not localized in fixed position, but must be deployed in a delocalized way to hold the structure together. Thus, the electrons must be moving among the pairs of atoms to bond the structure, in marked contrast to ordinary covalent bonding where the pairs of electrons are fixed in definite bonds.

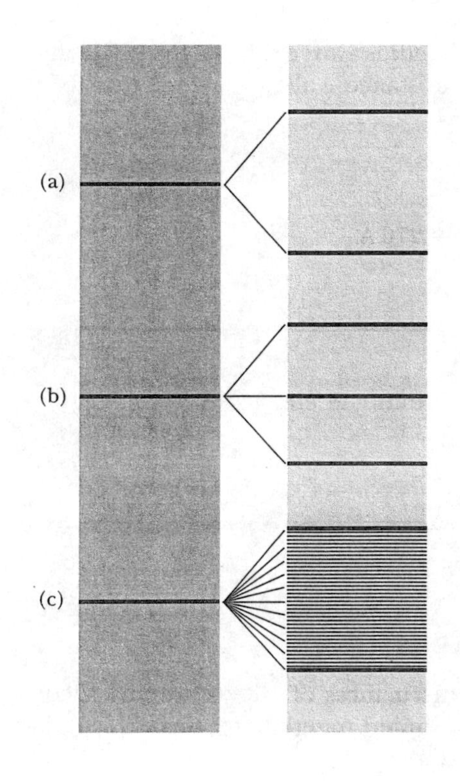

FIGURE 16.14 Molecular orbital diagram. Bonding of (a) two, (b) three, and (c) many sodium atoms.

By adding more atoms to the molecular orbital diagram of diatomic sodium, the situation that exists in the structure of the solid can be approached. Suppose we bond together three atoms, as in Figure 16.14b. Three atomic orbitals form three molecular orbitals, and two of the valence electrons occupy the lowest molecular orbital, while the third electron occupies the middle or nonbonding molecular orbital. As the number of atoms increases, the number of molecular orbitals produced increases accordingly. For n atoms—where n is a very large number—there are n molecular orbitals at very close intervals (Fig. 16.14c). The valence electrons occupy the lower portion of this band of orbitals. Movement of electrons through the metal for conduction of electricity or heat is easy because the electrons near the top of the band can jump to a slightly higher level that is empty and move about the structure. Conductivity in metals, then, is the result of the movement of these electrons, and a good **conductor** allows an easy passage of an electric current. Applying an electric potential to a piece of metal results in a flow of these electrons, that is, an electric current. Metals are also characteristically good conductors of heat as a result of electrons that are free to move.

The mobility of the electrons in metals such as aluminum, copper, gold, and lead also helps to explain their malleability and ductility. In each case, the metal maintains its strength and integrity even after severe deformation such as hammering it flat or drawing it into a long narrow wire. Under such stress, a nonmetal such as sulfur would simply crumble. However, the highly mobile electrons in a metal readily form new bonds as the atoms are moved about, and so the structure continues to be held firmly together in spite of remarkable changes in shape.

APPLICATIONS OF CHEMISTRY

Palladium and Gold Lattices Both palladium and gold crystallize in the face-centered cubic arrangement. In general, unit cell lengths for metals are on the order of 3 to 6 Å, and both palladium and gold have similar lengths. Thus, the number of atoms in cubes of equal volume, and therefore length along an edge, must be about the same for both metals. However, the atomic mass of palladium is about 60% that of gold. This results in a large density difference between the two metals, which turns out to have very practical implications.

Palladium and gold have been widely used in the microelectronics industry for electroplating what are commonly known as *interconnects.* All microelectronics devices contain interconnects that enable an electrical signal, and hence a message, to be transmitted from one component to another. Whereas copper and silver are excellent interconnects because of their conductive characteristics, they corrode too easily. On the other hand, although gold is not as conductive as silver or copper, it is conductive enough and extremely resistant to corrosion. However, since 1980, gold has traded in the range of $300 per troy ounce. At a U.S. consumption level of 22.5 metric tons or 724,000 troy ounces, that is some $220 million worth of gold for microelectronics devices, about 60% of which was used in connector contacts. Another 25% was used in printed circuit wiring board fingers. Palladium is also suitable for interconnects because it resists corrosion. At a recent price of $200 per troy ounce for palladium, which, furthermore, has a lower density than gold, the economic advantages of palladium over gold are obvious.

Question

If both palladium and gold crystallize in the FCC arrangement, and if both are characterized by similar distances between atoms along an edge of the unit cell, what accounts for the important, radically different densities of the two metals?

This *band structure* of the molecular orbitals of a metal also accounts for the fact that metals are opaque. Having a wide range of very closely spaced molecular orbitals means that a wide range of electronic transitions is possible. Thus, light of most frequencies is absorbed, and little passes through.

16.4 SALTS

Ionic Crystals

The first crystal structure to be determined experimentally, by X-ray crystallographic techniques, was that of an ionic compound. Common table salt—plain old NaCl—was found to consist of an array of Na^+ and Cl^- ions regularly arranged so that each ion was surrounded by six ions of unlike charge (Fig. 16.15). Contrary to the expectations of the time, there was no evidence for pairing of ions into anything resembling a NaCl molecule. In general, the crystals of ionic compounds are composed of lattices consisting of three-dimensional arrays of positive and negative ions.

Ionic compounds are characterized by very high melting points (Table 16.1), a property associated with unusually high thermal stability resulting from the attraction between ions of unlike charge. These forces are called **coulombic forces,** and according to Coulomb's law, the coulombic force F is directly proportional

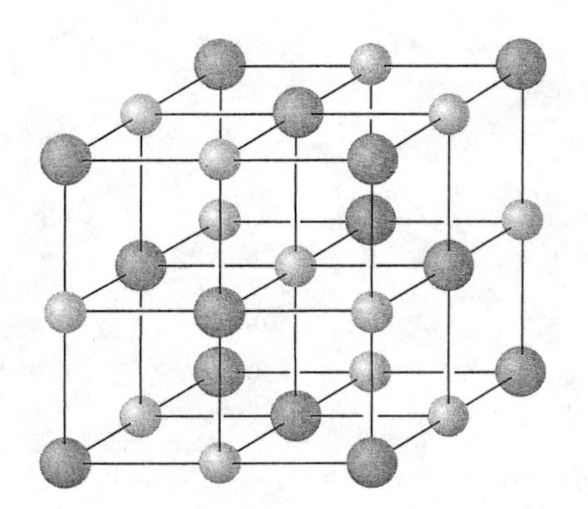

FIGURE 16.15 NaCl structure. In this model, the spheres at the corners and face centers represent Cl^- ions; the sphere at the center and the spheres at the mid-points of the edges represent Na^+ ions.

to the magnitude of the two charges q_1 and q_2, and is inversely proportional to the square of the distance d separating them. The force is an attraction if the charges are unlike, and a repulsion if the charges are like. Stating that algebraically,

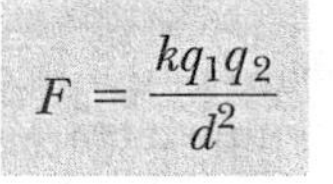

$$F = \frac{kq_1q_2}{d^2}$$

Coulomb's law: Determines the force F of attraction or repulsion between separated charges—attractive forces between unlike charges and repulsive forces between like charges. The magnitudes of the charges are q_1 and q_2, d is the distance between them, and k is a proportionality constant. It is especially important to note that in this law the forces between ions increase or decrease as the reciprocal of the square of the distance between the charges. Therefore, the forces will drop off very rapidly as the ions are separated.

Because ionic crystals contain both like and unlike charges, there must be repulsions as well as attractions within the structure. But attractions must be favored because the structures hold together. The key to understanding this lies in the reciprocal relationship between force and distance—the square of the

TABLE 16.1 Melting Points of Some Salts

Salt	Melting Point (°C)
AgCl	455
$CaCl_2$	772
LiF	870
NaCl	801
NaI	651
KCl	776

distance separating them. With each ion surrounded by ions of unlike charge, the greatest attractive forces are between adjacent ions, whereas repulsions result from ions at greater distances. Because the distance is squared, the forces fall off rapidly as the distance is increased. There are huge numbers of attractions and repulsions in an ionic crystal, but, overall, the attractions considerably outweigh the repulsions.

THE CESIUM CHLORIDE STRUCTURE CsCl has the simplest ionic crystal structure. It is based on the primitive cubic lattice and has lattice points only at the corners of a cubic unit cell. Each corner has an identical environment. For CsCl there is a Cl^- ion at each unit cell corner. Cubic symmetry is preserved by placing a Cs^+ ion at the center of the cube, as depicted in Figure 16.16. The number of Cl^- ions per unit cell is determined as follows:

$$\frac{8 \text{ corners}}{\text{cell}} \times \frac{1 \text{ Cl}^-}{8 \text{ corners}} = 1\frac{\text{Cl}^-}{\text{cell}}$$

Because the Cs^+ in the center of the unit cell is not shared with any other unit cell, each unit cell contains one Cs^+ ion and one Cl^- ion, and the ratio of the number of Cs^+ ions to Cl^- ions is therefore one-to-one (1:1) throughout the crystal.

The CsCl structure can accommodate only ions of equal charges, resulting in a 1:1 ion ratio. In fact, the few salts that crystallize in the CsCl structure all have cations with a charge of +1 and anions with a charge of −1. The same structure is observed in CsBr, NH_4Cl, and TlBr. It is also found in a few of the so-called *stoichiometric alloys* that contain strict one-to-one molar ratios of two metals, such as CuZn, AuZn, and MgHg.

If we extend the CsCl lattice to several adjacent unit cells, the structure could be viewed just as well as unit cells, with Cs^+ ions at the corners and Cl^- ions at the unit cell centers. Either viewpoint is as good as the other for determining the coordination number of the ions, the number of nearest neighbors that a particular type of ion has. Thus, a unit cell with the Cs^+ ion in the center has eight nearest neighbor Cl^- ions, giving the Cs^+ ion a coordination number of eight. Similarly, if we choose to draw the unit cell with the Cl^- atom in the center, the Cl^- can be seen to have eight nearest neighbor Cs^+ ions and a coordination number of eight.

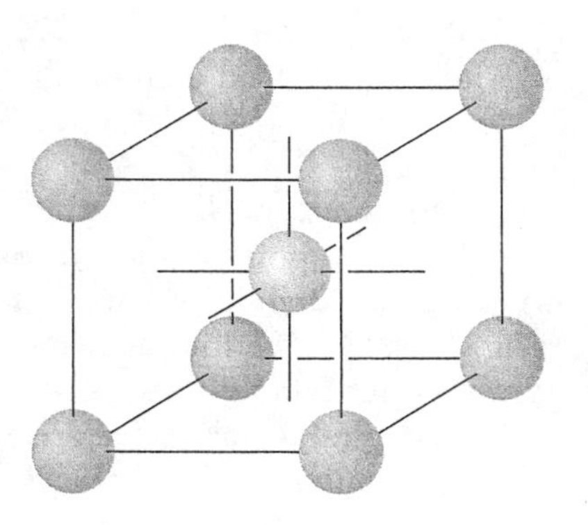

FIGURE 16.16 Cesium chloride structure. Cubic symmetry is preserved with corners of the unit cell being occupied by Cl^- ions; the Cs^+ ion is placed at the center of the cube.

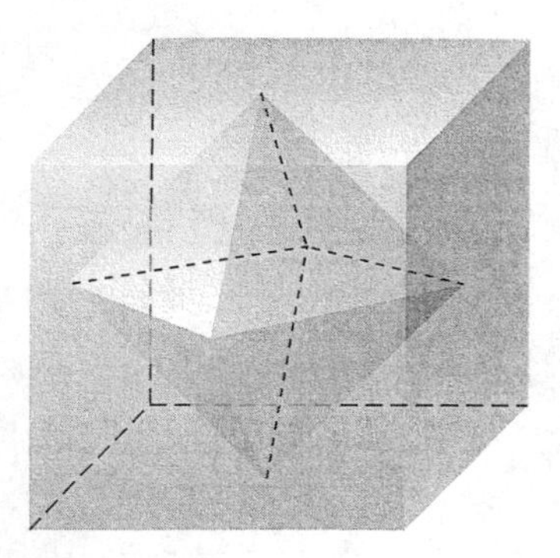

FIGURE 16.17 Octahedral hole. Embedded in the center of the face-centered (FCC) lattice of the cubic crystallographic system is an octahedral hole.

THE ROCK SALT STRUCTURE The NaCl structure (see Fig. 16.15) is based on an FCC lattice, but also uses spaces or holes in the FCC structure. A **hole** is a location in the structure in which there is enough room to accommodate an atom or ion. Chloride ions in the NaCl structure are located on the FCC lattice points. The Na^+ ions are situated in holes between the Cl^- ions that are located in the center of the unit cell and halfway along each edge. All of these holes are equivalent in the overall structure, but we will consider the hole in the center of the FCC lattice first. As can be seen in Figure 16.17, an ion located in this hole has six nearest neighbors and a coordination number of six. Furthermore, these six nearest neighbor ions can be thought of as placed in space at the vertices of an octahedron, one of the few perfectly regular solid geometries. The octahedron has eight equal faces, each of which is an equilateral triangle, twelve equal edges, and six equal vertices. The hole thus described is called an **octahedral hole**. The Na^+ ions located halfway along the unit cell edges are also in octahedral holes, but it is necessary to consider neighboring unit cells in order to see the octahedra. In the NaCl structure, all of the octahedral holes in the lattice are occupied by Na^+ ions.

Just as the CsCl structure could be drawn having either Cs^+ or Cl^- ions at the unit cell corners, the NaCl lattice can be viewed in an alternative way, with Na^+ ions in an FCC lattice and Cl^- ions in the octahedral holes. Thus, each ion is surrounded by an octahedron of the opposite ions and each ion has a coordination number of six. The ratio of Na^+ to Cl^- ions can be verified by considering the number of each in a unit cell. If we draw the cell with Cl^- ions at the lattice points, the number of Cl^- ions is determined as follows:

$$\left[\frac{8 \text{ corners}}{\text{cell}} \times \frac{1 \text{ Cl}^- \text{ ion}}{8 \text{ corners}}\right] + \left[\frac{6 \text{ faces}}{\text{cell}} \times \frac{1 \text{ Cl}^- \text{ ion}}{2 \text{ faces}}\right] = 4 \text{ Cl}^- \frac{\text{ions}}{\text{cell}}$$

In order to count the Na^+ ions properly we must consider that an ion on the unit cell edge is shared by four adjacent unit cells.

$$\left[\frac{12 \text{ edges}}{\text{cell}} \times \frac{1 \text{ Na}^+ \text{ ion}}{4 \text{ edges}}\right] + \frac{1 \text{ center unshared Na}^+ \text{ ion}}{\text{cell}} = 4 \text{ Na}^+ \frac{\text{ions}}{\text{cell}}$$

Accordingly, the cell contains ions in the ratio of 4 : 4, which reduces to 1 : 1, indicating that the numbers of Na^+ ions and Cl^- ions in a sample of sodium chloride of any size are equal.

The rock salt NaCl structure is very common for 1 : 1 salts. Among the many solids that adopt this structure are AgCl, KCl, KBr, LiCl, and NH_4I with +1 cations and −1 anions; BaO, BaS, and MgO with +2 cations and −2 anions; BN with ions that are formally +3 and −3; and ZrC with a cation and anion that are formally +4 and −4.

THE ZINC BLENDE STRUCTURE Zinc sulfide (ZnS) also crystallizes into an FCC lattice. In this case the S^{2-} ions are located at the lattice points of an FCC lattice. Each Zn^{2+} ion is located in a hole, but not an octahedral hole. Instead, it is in a hole that has four nearest neighbors (Fig. 16.18a). The four nearest neighbors can be considered to be at the vertices of another regular polyhedron, the tetrahedron (Fig. 16.18b), making it a **tetrahedral hole**. There are eight tetrahedral holes in the FCC unit cell and an alternating one half of them are occupied by Zn^{2+} ions in the zinc blende structure. The coordination num-

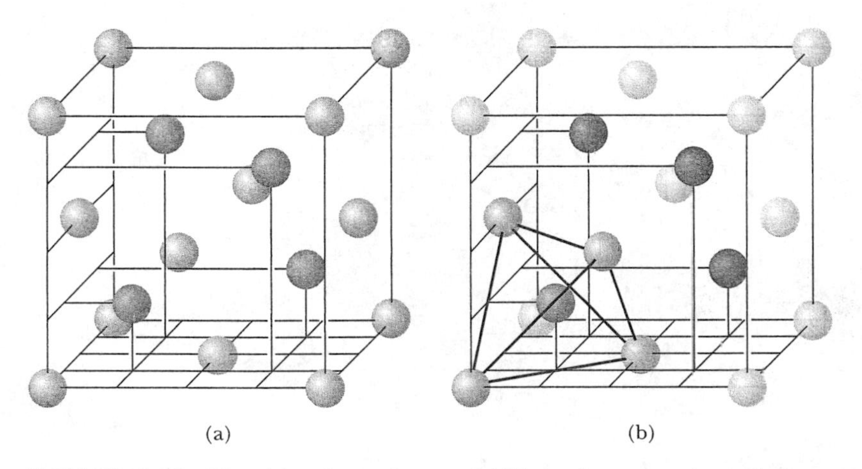

FIGURE 16.18 Zinc blende structure. (a) The spheres on the cell corners and face centers (FCC positions) represent S^{2-} ions. The spheres in the tetrahedral holes represent Zn^{2+} ions. (b) The tetrahedral hole, front lower left; Zn^{2+} lies within a tetrahedron formed by four S^{2-} ions.

PROFILES IN CHEMISTRY

X-Rays and Crystallography How is it that we know so much about crystalline structures and the packing of atoms in solids when the particles we wish to know about are so small, with diameters on the order of 10^{-10} m? The first studies in crystallography depended on visual examination of crystalline substances. Crystals of directly observable size had to occur naturally or be grown in the laboratory. Development of microscopic and other optical techniques improved our understanding by orders-of-magnitude. But by far the most powerful technique for structural studies of crystalline solids has been X-ray crystallography.

X-rays were discovered in 1895 by Wilhelm Röntgen, a physics professor at the University of Würzburg. By smashing electrons onto a heavy metal target, electromagnetic radiation of very short wavelengths, on the order of 10^{-10} m, was released. Such short distances are similar to the spacings between atoms or ions in substances, which leads to the possibility of the diffraction of the X-rays. (You are familiar with the diffraction of light from an oil slick on water where the thickness of the oil layer is comparable to the wavelength of visible light.) In 1912, Max von Laue found that a beam of X-rays directed onto a salt crystal produced a series of uniformly placed spots on a photographic plate. The spots fell in a pattern that was determined by the geometric arrangement of the ions in the solid. W. H. Bragg and W. L. Bragg then showed how to analyze the spots. Röntgen, Laue, and the Braggs all won Nobel prizes for their work; the Braggs were one of only two father-and-son combinations ever to have been so recognized.

In what may prove to be the ultimate X-ray crystallographic experiment, the double helix model of James Watson and Francis Crick (Cambridge University) for the structure of the genetic material known as DNA was made possible by analysis of the X-ray patterns of Rosalind Franklin and Maurice Wilkins (University College, London). Except for Franklin, who died prematurely at age 37, all of these also received Nobel prizes for their work. Unfortunately, such recognition is never given posthumously.

Questions

How much of the microscopic structure of a solid do you think can be deduced from the macroscopic structure of single crystals?

Who were the other father and son Nobel laureates?

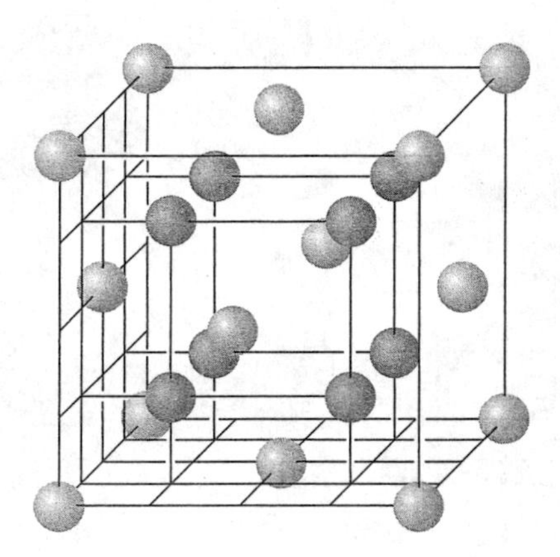

FIGURE 16.19 Fluorite or calcium fluoride (CaF_2) structure. The spheres in the FCC positions represent Ca^{2+} ions, and the eight spheres in tetrahedral holes represent F^- ions.

ber of each ion in the zinc blende structure is four. Each unit cell of the structure contains four S^{2-} ions on the FCC lattice points, just as the unit cell of the NaCl structure contains four Cl^- ions. The four Zn^{2+} ions are not shared with adjacent unit cells, so the ratio of Zn^{2+} and S^{2-} ions is 4:4, or 1:1. Other ionic compounds that adopt this crystal structure are BeS, HgS, CuCl, CuBr, and AgI.

THE FLUORITE STRUCTURE In the fluorite (CaF_2) structure (Fig. 16.19) the unit cell can be drawn so that Ca^{2+} ions are situated on FCC lattice points with the F^- ions in all of the eight tetrahedral holes. This gives a ratio of Ca^{2+} to F^- ions of 4:8, which reduces to 1:2, as in CaF_2. The coordination numbers for the ions in fluorite are eight for the Ca^{2+} and four for the F^-. Other 1:2 compounds that adopt this structure are BaF_2, $SrCl_2$, ZrO_2, and HfO_2.

Ionic Radii

Calculating ionic radii from crystal structures of ionic compounds follows principles similar to those for calculating atomic radii from crystal structures of metals. Once again we assume that the lattice points are occupied by ions represented by spheres of fixed radius that are in contact with nearest neighbor ions of unlike charge. However, the fact that two radii are involved in each structure prevents a unique solution. For example, in the CsCl structure (Fig. 16.20) we assume that unlike ions touch along the body diagonal of the cube. Thus,

$$bd = 2r_{Cl^-} + 2r_{Cs^+}$$

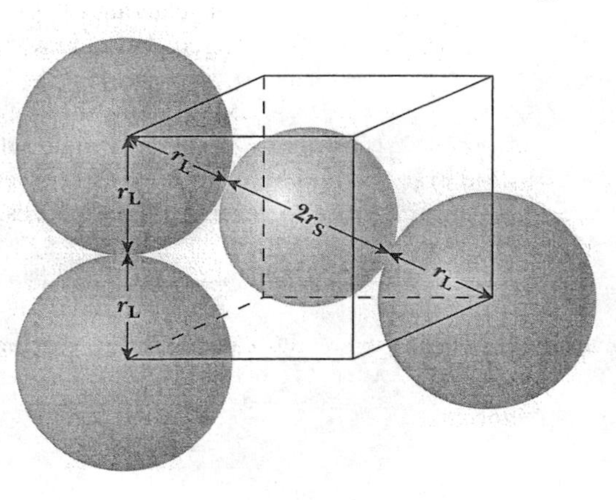

FIGURE 16.20 CsCl structure. Part of the structure showing the relationship between the radii of the large (r_L) and small (r_S) ions. The large ions touch only at the minimum of r_S.

where r_{Cl^-} and r_{Cs^+} are the ionic radii of Cl^- and Cs^+. We also know that

$$bd = a\sqrt{3}$$

where a is the length of the unit cell edge. Thus

$$2r_{Cl^-} + 2r_{Cs^+} = 2(r_{Cl^-} + r_{Cs^+}) = a\sqrt{3}$$

If the value of a, the unit cell length, is known, we can calculate $r_{Cl^-} + r_{Cs^+}$, but we cannot calculate a unique value for either one. To compile tables of ionic radii such as Table 16.2 requires that a self-consistent set of ionic radii be constructed from data for many crystals.

EXAMPLE 16.3

CdO has the NaCl structure and a unit cell edge length of 4.689 Å. Calculate the ionic radius of Cd^{2+} using these crystal data and ionic radii given in Table 16.2.

COMMENT Examination of the NaCl structure (see Fig. 16.15) shows that unlike ions touch along a unit cell edge, so that for CdO

$$2r_{O^{2-}} + 2r_{Cd^{2+}} = a = 4.689\ \text{Å}$$

and

$$r_{O^{2-}} + r_{Cd^{2+}} = \frac{4.689\ \text{Å}}{2} = 2.345\ \text{Å}$$

Table 16.2 gives $r_{O^{2-}}$ as 1.45 Å.

SOLUTION

$$r_{Cd^{2+}} = 2.345\ \text{Å} - 1.45\ \text{Å} = 0.90\ \text{Å}$$

TABLE 16.2 Ionic Radii, (Å), for Some Common Monatomic Ions

Li^+ 0.68	Be^{2+} 0.30			O^{2-} 1.45	F^- 1.33
Na^+ 0.98	Mg^{2+} 0.65		Al^{3+} 0.45	S^{2-} 1.90	Cl^- 1.81
K^+ 1.33	Ca^{2+} 0.94	Zn^{2+} 0.74	Ga^{3+} 0.60	Se^{2-} 2.02	Br^- 1.96
Rb^+ 1.48	Sr^{2+} 1.10		In^{3+} 0.81	Te^{2-} 2.22	I^- 2.19
Cs^+ 1.67	Ba^{2+} 1.29		Tl^{3+} 0.91		

EXERCISE 16.3

FeO has a NaCl structure and a unit cell edge length of 4.294 Å. Use the value of the ionic radius of O^{2-} given in Table 16.2 to calculate the ionic radius of Fe^{2+}.

ANSWER 0.70 Å

16.5 NETWORK CRYSTALS, NONCONDUCTORS, AND SEMICONDUCTORS

Covalent Network Crystals

We now know there are three allotropes of pure elemental carbon: the classic solids—graphite and diamond—and a new class of discrete carbon molecules called buckminsterfullerenes, or just fullerenes. Here we will study the two classic solid-state allotropic forms with their strikingly different properties. Graphite (Fig. 16.21) is made up of flat sheets or planes of carbon atoms in a network of six-membered rings, much like chicken wire. The graphite crystal consists of layered stacks of these sheets, separated by a fixed interplanar distance of 3.35 Å. The bonding within the sheets is very strong, whereas the forces of attraction between separate sheets is very weak. Thus, the sheets slide easily over each other, giving graphite its slippery, greasy feel. This structure makes graphite useful as a lubricant and in pencils, where layers of graphite slide off the surface and adhere to the paper. Because of its very different structure in different directions, graphite is strongly anisotropic; that is, it has different behavior in different directions. It is an electrical conductor in the direction of the sheets and an insulator perpendicular to them. Molybdenum disulfide also has a layered structure; it has been used as a lubricant additive in motor oil, especially because it is a solid and the lubricant shows no appreciable change of viscosity with temperature.

Diamond (Fig. 16.22) is an example of a purely covalent crystal with a three-dimensional network of atoms. It adopts the four-coordinate zinc blende struc-

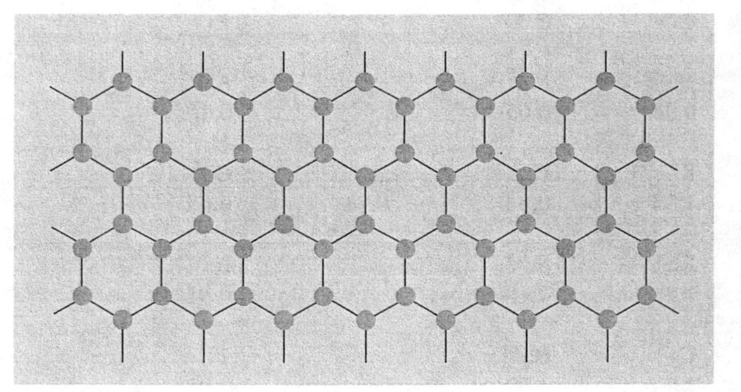

FIGURE 16.21 Graphite. Part of a layer of the crystal structure of one of the two common allotropic forms of the element carbon.

SEEING THINGS

The Vigorous Synthesis of Molybdenum Disulfide The layered compounds molybdenum disulfide (MoS_2) and tungsten disulfide (WS_2) are interesting because of potential applications as lubricants, battery cathodes, and catalysts. Unfortunately, traditional synthetic methods are difficult and time consuming, involving intermittently heating and grinding the elements together for several days at temperatures in excess of 900°C, or running reactions that lead to complex and difficult-to-separate mixtures. Recently, solid-state chemists working at the University of California in Los Angeles successfully carried out syntheses in which the transition metal halides and alkali metal sulfides react rapidly to give the desired products:

$$2MoCl_5(s) + 5Na_2S(s) \longrightarrow 2MoS_2(s) + 10NaCl(s) + S(s)$$

These reactions are vigorous, highly exothermic, and visually dramatic. The starting materials either ignite spontaneously or start with the touch of the tip of a hot metal rod or filament and produce an intense flash of light and molten sodium chloride.

Question

Graphite and molybdenum disulfide (MoS_2) have in common the interesting property that they are solid lubricants. What characteristic structural feature(s) do they share that accounts for their lubricity?

ture (see Fig. 16.18), except that every location of a Zn^{2+} or S^{2-} ion now contains a carbon atom. Each carbon atom is covalently bonded by a shared pair of electrons to four neighboring carbon atoms in a tetrahedral arrangement. This produces a network extended by a huge number of unit cells in each direction, forming a covalent network crystal that is resistant to deformation, and therefore extremely hard. In fact, diamond is the supreme abrasive, being harder than any other material. Natural and synthetic diamonds are widely used for grinding, cutting, and boring metals, rock, and other substances. In the age of the phonograph record, diamond was the material of choice for the tip of the stylus of a player because of its wear resistance. Being a brittle material, diamond can be readily crushed into a powdery grit that is widely used in industry as an abrasive. Small single crystals are used for turning and boring tools in oil-well drilling and mining operations, for dies for drawing wire, for dressing tools for grinding wheels, and for single point cutting tools. Diamonds of near flawless quality have been prized as gem stones for thousands of years.

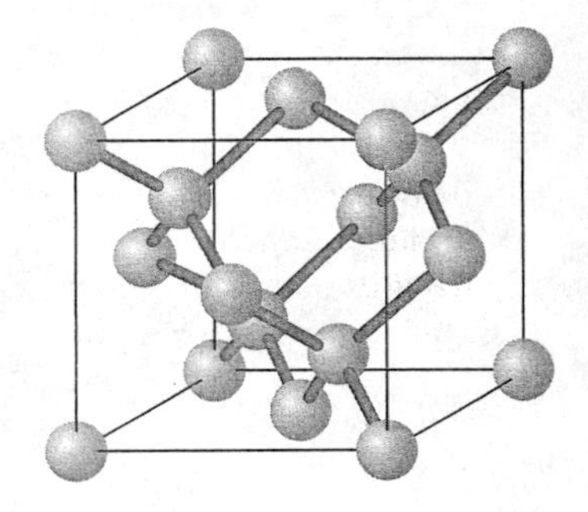

FIGURE 16.22 Diamond. One unit cell in the crystal lattice of the other common allotropic form of the element carbon.

Diamond and Graphite Compared

	Diamond	Graphite
Density	3.5 g/cm^2	2.0 g/cm^2
Hardness	Hardest material known	Very soft
Conductivity	Nonconductor	Conductor
Color	Transparent	Black, opaque
Cost	$10,000/lb	$0.35/lb
Use	Abrasive	Lubricant

PROFILES IN CHEMISTRY

Diamonds for Everything and Forever The first synthetic diamonds were prepared in 1955 by scientists at the General Electric Company. This completed a task that had stimulated scientific thought and engineering practice since the last years of the eighteenth century, when it first became known that diamonds were indeed elemental carbon. Graphite is the stable form of carbon at standard atmospheric pressure and temperature, so in the General Electric process, a charge of graphite and a catalyst metal are heated to melt temperatures at a pressure high enough for diamond to be stable. The graphite dissolves in the metal, and diamond is produced from it. Iron was the first molten metal used, but many are now known to serve as effective catalysts: Cr, Mn, Co, and Ni, for example. This is also an engineering problem, requiring a specially designed "belt" apparatus to produce the required temperature and pressure, in the range of 1400 to 2400°C and 55,000 to 130,000 atm.

The General Electric scientists identified those first, small, dark, synthetic diamond crystals by the well-known characteristics of natural diamond: (1) they readily scratched silicon carbide, which is, next to diamond, the hardest abrasive in common use, and they scratched tungsten carbide, too; (2) they burned cleanly in pure oxygen without leaving a residue; (3) they had a density of 3.52 g/cm^3 at 25°C; and (4) they had the same X-ray diffraction pattern as pure diamond.

In 1970, General Electric announced the creation of carat-sized diamonds of gemstone quality. The process starts with tiny seed crystals, usually synthetic diamonds no bigger than the period at the end of this sentence. This seed crystal, along with a metal catalyst and a powder charge of synthetic diamond, is subjected to high pressures and temperatures in a special press. At the high temperatures of the molten metal, the diamond powder dissolves, but the end of the tube containing the seed crystal is kept cool enough to prevent the seed from dissolving. By carefully controlling the pressures and temperatures, the carbon atoms from the diamond powder can be made to migrate through the molten metal catalyst, finally redepositing themselves upon the seed crystal and eventually building to a large diamond.

In 1986, scientists at Pennsylvania State University reported a way of coating objects with thin films of synthetic diamond, which promises to revolutionize major segments of electronics, optics, machine tools, chemical processing, and military technology. Diamond film is likely to become a commonplace material, and although such materials may have the appearance of a lacquer coating, they promise to radically improve the quality of the objects they coat in ability to resist abrasion and corrosion.

Question

Would you expect tungsten carbide crystals to scratch a silicon carbide surface? Why (or why not)?

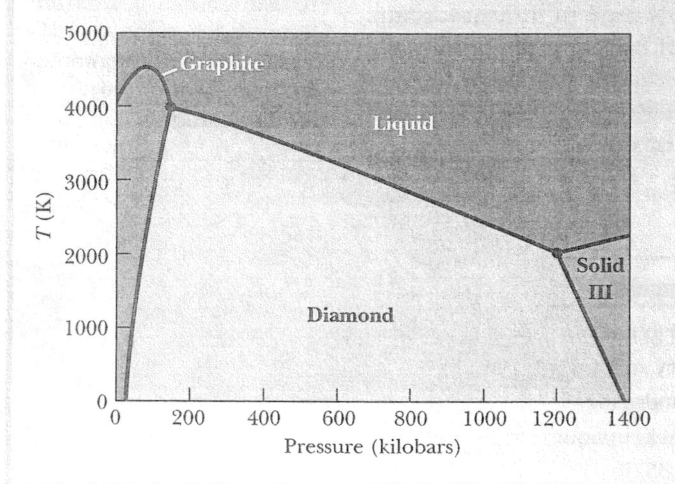

Phase diagram for carbon. Identify the point on the diagram corresponding to 25°C and 1 atm. Note that graphite and not diamond is the standard form of the element.

The diamond structure is also found in germanium, silicon, and the gray form of tin. However, because the bonds between atoms in these elements are weaker than those between carbon atoms in diamond, they are all softer. A number of other elements and compounds form crystals consisting of a covalent network, including a variety of important minerals containing silicon. A few of these generally complex structures will be considered in the next chapter.

Boron nitride (BN) is composed of equal numbers of boron and nitrogen atoms and has the same number of electrons as pure carbon. This is because boron has one less electron than carbon, but nitrogen has one more. Remember that substances with the same numbers of electrons are isoelectronic, and because the bonding is determined by electrons, isoelectronic compounds such as this often have very similar structures. Boron nitride forms structures analogous to diamond and graphite. The diamond form can be described as having the zinc blende structure, and the bonds in this structure are principally covalent and about as strong as in diamond. This gives a covalent network structure with hardness comparable to diamond.

One would, of course, prefer to have abrasives that are much less expensive than diamond or boron nitride. One in particular that has been commercially successful is silicon carbide (SiC), or carborundum. Silicon carbide crystallizes in a number of modifications. Among these are the zinc blende structure and several closely related structures. The bonds are somewhat weaker than those in diamond, and so silicon carbide is not quite as hard. However, it is relatively inexpensively made from sand and coal.

Insulators and Semiconductors

Diamond has properties very different from those of a metal such as copper. As a nonmetallic solid, diamond is a very poor conductor of electricity, and it cannot be seriously stressed and still maintain its basic structure. Copper, on the other hand, is typically metallic and is what diamond is not, namely a good conductor of electricity and capable of being hammered into sheets and drawn into wires. As we mentioned before, a conductor of electricity is characterized by the flow of electrons from atom to atom.

In the diamond structure, each carbon atom has four valence orbitals. There are also four valence electrons on each carbon atom with the electron configuration $2s^2 2p^2$. An assembly of n carbon atoms in a diamond structure (see Fig. 16.22) has a total of $4n$ molecular orbitals, of which $2n$ are bonding and $2n$ are antibonding. There are $4n$ valence electrons, and these just fit in the $2n$ bonding molecular orbitals, two to an orbital. Thus, the bonding orbitals are completely full and the antibonding orbitals are empty. Because the ionization energy of carbon is high, bonding results in considerable stabilization and very strong bonds. These strong bonds in the rigid, three-dimensionally linked structure account for diamond's extreme hardness as well as a considerable energy difference, or band gap, between the two sets of orbitals (Fig. 16.23a). The **band gap** is the difference in energy between the highest energy molecular orbitals, which are occupied with electrons, and the lowest energy molecular orbitals, which are not. It represents the minimum energy necessary to excite an electron into a higher energy state.

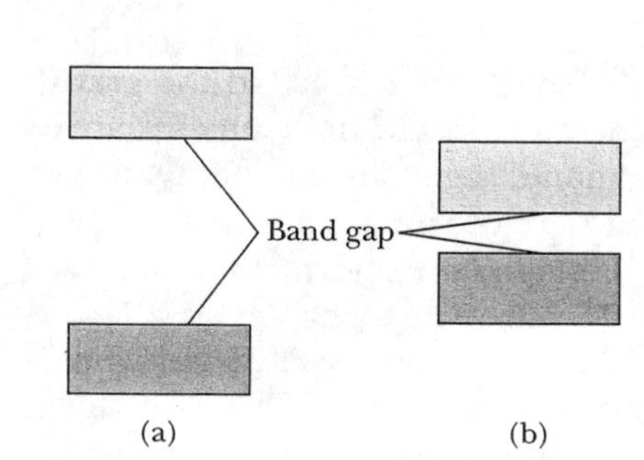

FIGURE 16.23 Band gap. The energy difference between molecular orbitals. Shown here are valence (lower) and conduction (upper) bands for (a) an insulator and (b) a semiconductor.

The valence electrons in the diamond structure are all held tightly in bonding orbitals, with no opportunity for movement from one atom to another. In order for this substance to conduct electricity, an electron in the valence band, the bonding orbital band, would have to jump up to the conduction band, the antibonding orbital band, where it would be free to move. At higher temperatures the number of electrons excited to the point of making this jump increases. However, because of the large band gap in diamond, the temperature would have to be extremely high to excite a significant number of electrons, so diamond is effectively an **insulator.** In an electrical insulator, no electrons flow from atom to atom. If the band gap were narrower, however, conduction of electricity would be a more likely possibility at reasonable temperatures. This is observed with silicon ($3s^23p^2$) and germanium ($4s^24p^2$) for which the ionization energies are not so high. Substances such as these have the diamond structure but smaller band gaps (Fig. 16.23b), so electrons can jump to the conduction band and move through the structure. Pure substances that exhibit conductive behavior like that of silicon and germanium are called **semiconductors**.

Once an electron has jumped, other electrons are free to move in the valence band as well. The electron jumping out of the valence band leaves behind a hole, a missing electron location. The movement of other electrons within the valence band that is now possible causes the hole to appear to be moving. Each time an electron fills an available hole it leaves a new hole behind it. Conduction is possible as soon as an electron leaves the valence band. At higher temperatures, more electrons make the jump, and, as a consequence, substances such as silicon and germanium are better conductors at higher temperatures. In contrast, metals become poorer conductors at higher temperatures because the increased motion of the metal atoms impedes the movement of the electrons.

As a practical matter, the conductivities of pure silicon and germanium are too low at room temperature to be of much use as semiconductor materials. In order to increase the number of electrons or the number of holes that are free to move, the pure substances are carefully **doped**, or impregnated with other elements. Silicon can be doped with phosphorus atoms, for which the valence electron configuration is $3s^23p^3$. Each phosphorus atom occupies a location in the structure equivalent to a silicon atom. However, each phosphorus atom brings along an extra electron that occupies an energy level just below the conduction band. The electron easily jumps into the conduction band and moves about the crystal. Because the current is carried by the negatively charged electrons, phosphorus-doped silicon is called an ***n*-type semiconductor**.

On the other hand, when silicon is doped with aluminum, for which the valence electron configuration is $3s^2 3p^1$, the foreign atoms enter the silicon structure. But there is now one electron less per aluminum atom. Holes have been introduced into the valence band. These holes are free to move in the direction opposite the direction of electron flow. Thus, the holes can be regarded as positively charged current carriers, and aluminum-doped silicon is a ***p*-type semiconductor.** Very pure silicon is needed as a starting material for doping, because the number of conductor electrons or holes must be very carefully controlled. Purities up to 99.9999999%, by far the purest substances ever produced to date, have been realized.

Germanium, with the diamond structure, is also widely used in semiconductor devices. Additional semiconductors with the same structure and number of valence electrons as silicon or germanium can be prepared from equal moles of an element with one fewer valence electron and an element with one more valence electron. These compounds are known as the III–V compounds because they consist of an element from Group IIIA and an element from Group VA. Examples are aluminum phosphide and gallium arsenide.

Transistors, Solar Cells, and Integrated Circuits

The microelectronics revolution that has so altered the way we live in the late twentieth century depends on *n*-type and *p*-type semiconductors. Perhaps the simplest application of these semiconductors is in **current rectification**, which is changing alternating current to direct current by allowing current to flow in only one direction in a circuit. In Figure 16.24 an *n*-type and a *p*-type semiconductor are in contact. This produces a *p–n* rectifier, each side of which can be connected to the terminal of a battery. If the negative terminal of the battery is connected to the *p*-type semiconductor, and the positive terminal is connected to the *n*-type semiconductor, two things happen: Electrons from the battery fill all of the holes in the *p*-type semiconductor, and the extra electrons in the *n*-type semiconductor drain away to the positive terminal of the battery. Thus, each side of the transistor becomes an insulator, and the flow of current ceases.

If the battery is connected in the opposite manner, electrons from the negative terminal of the battery can flow into the *n*-type semiconductor. The excess electrons on this side of the transistor can now flow across the *p–n* junction into the holes in the *p*-type semiconductor; these in turn can flow into the positive

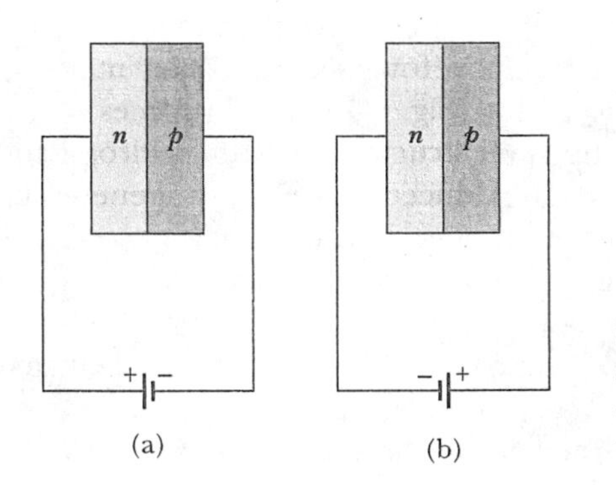

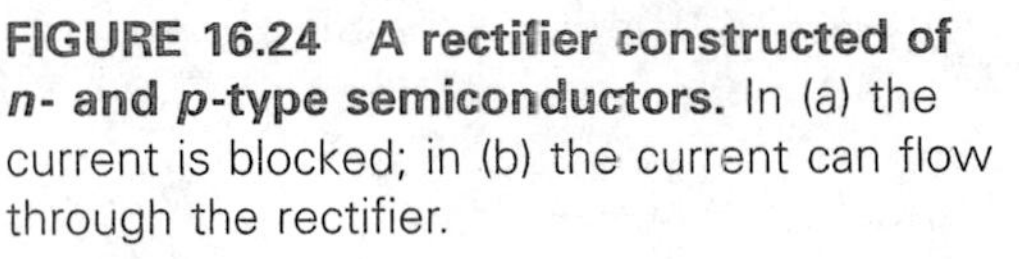

FIGURE 16.24 A rectifier constructed of *n*- and *p*-type semiconductors. In (a) the current is blocked; in (b) the current can flow through the rectifier.

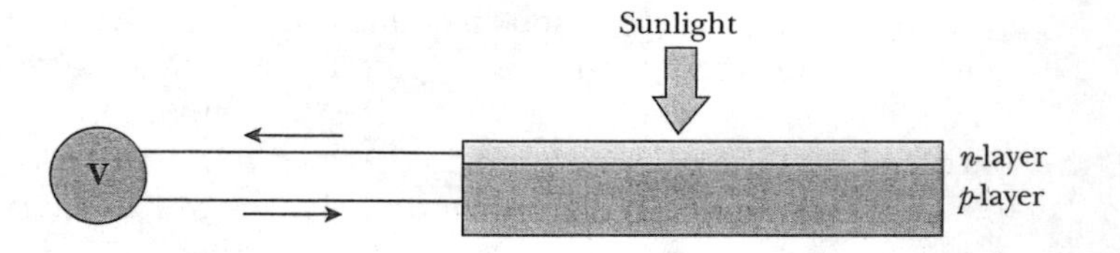

FIGURE 16.25 A solar cell employing a *p–n* junction. Sunlight striking the *n*-layer causes a flow of electrons through the voltmeter as indicated.

terminal of the battery, and a continuously flowing current is established. If the rectifier is connected to an alternating current source, current flow in one direction will be stopped, giving a half-wave rectified or direct current.

Another use of the *p–n* junction is in solar cells (Fig. 16.25). A thin *n*-layer is placed on top of a *p*-layer by the controlled diffusion of phosphorus or another element that supplies electrons. Electrical contact strips are connected to each layer. Sunlight shining on the *n*-layer causes some electrons to jump to the conduction band, leaving holes in the valence band in that layer. The electrons in the conduction band can flow into the conductor, leaving the *n*-layer with a positive charge. This is neutralized by the holes in the *n*-layer flowing across the junction and through the *p*-layer, where they meet up with electrons coming through the outside circuit. The overall result is a flow of current through the circuit.

Gallium arsenide semiconductors have very fast responses and are revolutionizing the design of semiconductor devices. Integrated circuits based on gallium arsenide have already achieved operating speeds up to five times that of the fastest silicon chips. Solar cells based on the photovoltaic properties of alternating thin films of GaAs and GaAlAs have already achieved a surprising milestone at Sandia National Laboratory in New Mexico, where scientists demonstrated a solar cell that converted more than 30% of the light that struck it into electricity.

16.6 MOLECULAR CRYSTALS AND LIQUID CRYSTALS

Crystal Structures Made of Molecules

None of the structures that we have discussed so far involve compounds consisting of discrete molecules. Compounds that are made of molecules can also form crystals, although generally at much lower temperatures than metals, ionic compounds, and covalent network structures. In ice, for example, water molecules are arranged in a regular repeating pattern (Fig. 16.26). The forces between the molecules that hold them in the crystal structure include hydrogen bonds, other dipole–dipole interactions, and induced dipoles. In general, stronger intermolecular forces lead to higher melting points. However, these intermolecular forces are much weaker than the coulombic attractions of unlike charges in ionic compounds, or covalent bonds in network structures like diamond and boron nitride. Therefore, molecular crystals tend to be broken up more easily by increasing temperature and have relatively low melting points. In addition, they are relatively soft and generally do not have clearly defined cleavage planes.

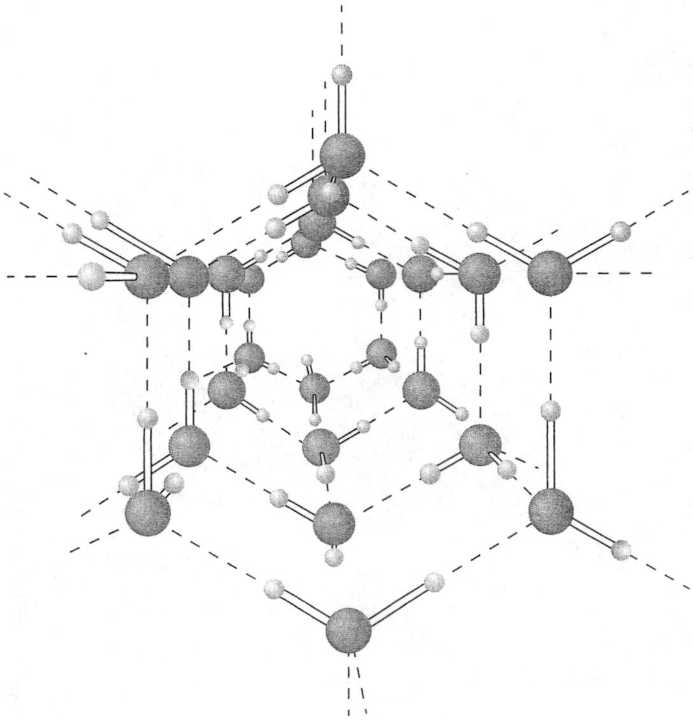

FIGURE 16.26 The regular, hexagonal arrangement of the water molecules in ice crystals. This microscopic structure appears in the macroscopic structure of snowflakes. Molecules are held in place within the crystal structure mainly by hydrogen bonds and weaker dipole–dipole interactions and dispersion forces. The presence of large, open channels in the ice structure results in ice having a lower density than liquid water.

Orderliness in Liquids

Liquid crystal is neither a contradiction—*liquid* and *crystal*—nor an oxymoron. **Liquid crystals** are an intermediate form of matter, displaying the flow properties of liquids and the light-scattering properties of solids. That combination of properties has taken on great scientific and technological significance in the last thirty years in such diverse fields as computer electronics and automotive products. Liquid crystals have also been a part of biomedical research into the nature of the synapse mechanism of nerve transmission and the treatment of arteriosclerosis. Orderliness approaching that seen in crystalline solids is possible in the liquid state under certain circumstances.

The liquid crystal state depends on the presence of molecules with rigid, rod-like structures. At higher temperatures, these rods are oriented in a random fashion, resulting in typical properties characteristic of a disordered liquid (Fig. 16.27a). However, as the compound cools, some ordering of the random structure is possible. Ordering is observed because the molecules have preferred orientations. The simplest of these is the **nematic** phase (Fig. 16.27b), in which the molecules are all parallel to each other, but the centers of the molecules are distributed at random. Shown in Figure 16.27c is the **smectic** phase, which is more ordered than the nematic phase because the molecules not only have the same preferred direction, but their centers are located on planes that run through the structure. Finally, the full ordering of a molecular crystal is shown in Figure 16.27d, completing the transition from liquid to crystalline solid.

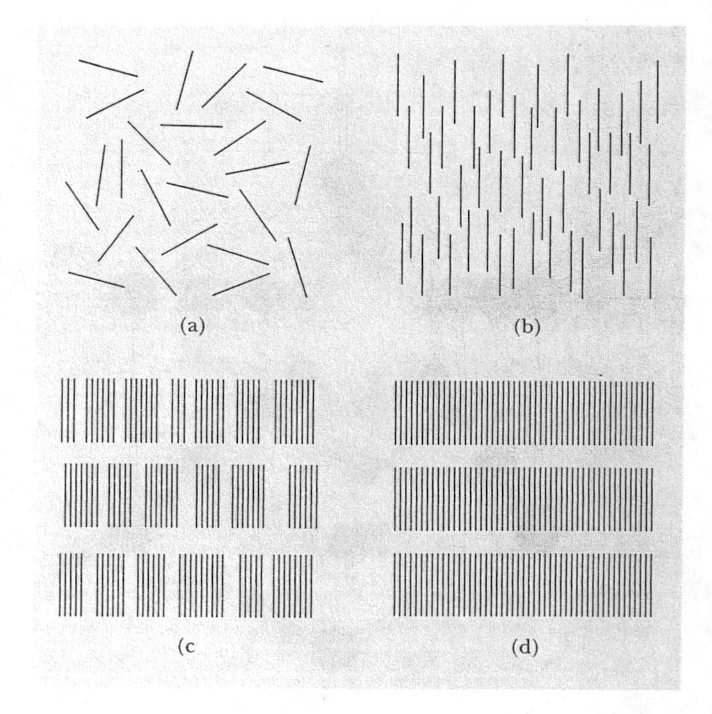

FIGURE 16.27 Four degrees of orientation of rod-shaped molecules. (a) *Liquid:* characterized by a typically disordered, random arrangement, or only short-range order. (b) *Nematic phase:* simplest liquid crystal phase exhibits a preferred orientation while still retaining the general random behavior of ordinary liquids. (c) *Smectic phase:* moving toward higher degrees of organization typical of molecular crystals. Note the layered structure. (d) *Molecular crystals:* well-organized and orderly.

The liquid crystal substance known as cholesteryl myristate is a good example.

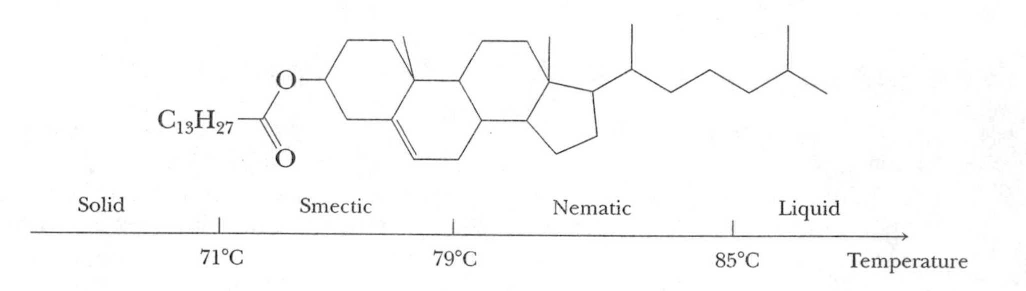

Phase transitions for cholesteryl myristate. The liquid crystal has both a nematic and a smectic phase. The structure of the molecule suggests a rigid, rod-like structure centered around the cholesteryl backbone and long hydrocarbon chains at the ends. (In this structure shorthand, carbon atoms are located at angles, intersections, and ends of lines where element symbols are not shown.)

Cholesteryl myristate has a rigid rod-like structure. It is solid at room temperature and at all temperatures below 70°C, which is approximately where it melts into a cloudy liquid. That is extraordinary because most crystalline solids melt into clear liquid phases. If we continue to heat the substance, it does finally clear above 85°C and behaves like a normal liquid thereafter. It is in the range between these two temperatures that we have a uniform liquid crystal phase, not solid crystals mixed into the liquid phase.

The orientation of the molecules in a liquid crystal is very sensitive to electric and magnetic fields. This is useful in liquid crystal devices, particularly liquid crystal displays, or LCDs. Liquid crystal displays can vary the intensity of the light emerging from different parts of the display. Figure 16.28 shows the schematic of a liquid crystal display as used in a digital watch or a calculator. A slight change in the electric field in a part of the display causes that part—perhaps in the shape of a letter or a number—to be darkened.

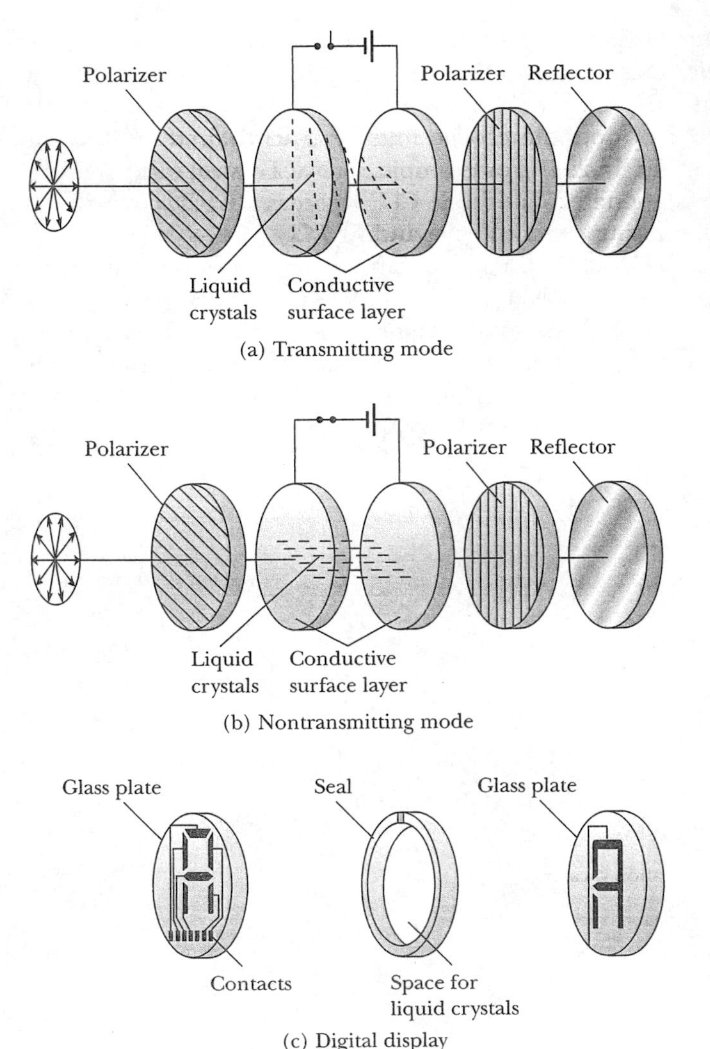

FIGURE 16.28 Simplified diagram of a liquid crystal display device. (a) After passing through a polarizing filter, light from a source is turned by the liquid crystal molecules so it can pass through the second polarizing filter and be returned by the reflecting mirror to give a bright display. (b) If the liquid crystal molecules are rotated by the application of a potential difference applied to the display, then the light is blocked from reaching the reflecting mirror and that part of the display appears black. (c) The digital display.

16.7 IMPERFECT CRYSTALS AND NONSTOICHIOMETRIC COMPOUNDS

Defects and Irregularities in Crystals

Even though the most noticeable aspect of crystals is the regularity and perfection in their design, they do have their imperfections, or defects. A **defect** is any variation from the strictly repeating structure of the crystal. Remember that the number of unit cells in a crystal is huge. Thus, if only a small percentage is *defective*, there still will be a considerable number of defects in the crystal. For example, a cubic crystal of NaCl that is 0.1mm on a side contains about 10^{17} unit cells. Even if there are defects in only 0.01% of these unit cells, there are still 10^{13} defects in the crystal. There are several kinds of crystal defects. We will consider only **point defects**, defects that involve only one or a very few lattice sites.

The simplest point defect in a crystal is nothing more than a vacancy—a hole where a missing atom or ion should be. If the missing species is an ion, electrical neutrality and the proper stoichiometry can be maintained by also having a nearby ion or ions of opposite charge missing. Such paired defects are called **Schottky defects** (Fig. 16.29a). For example, an NaCl crystal with a Schottky defect would be missing one Na^+ and one Cl^-, whereas a $CaCl_2$ crystal with such a defect would be missing one Ca^{2+} ion and two Cl^- ions.

If an atom or ion is out of its usual position, occupying a normally vacant hole, we have a **Frenkel defect** (Fig. 16.29b). Frenkel defects are most common when the smaller ion, normally the cation, is much smaller than the anion. For example, Ag^+ is a very small cation, and Frenkel defects in the silver halides (AgCl, AgBr, and AgI) are so common that Ag^+ ions can diffuse quite freely through the crystals. Both Frenkel and Schottky defects preserve the regular stoichiometry of the crystal compound.

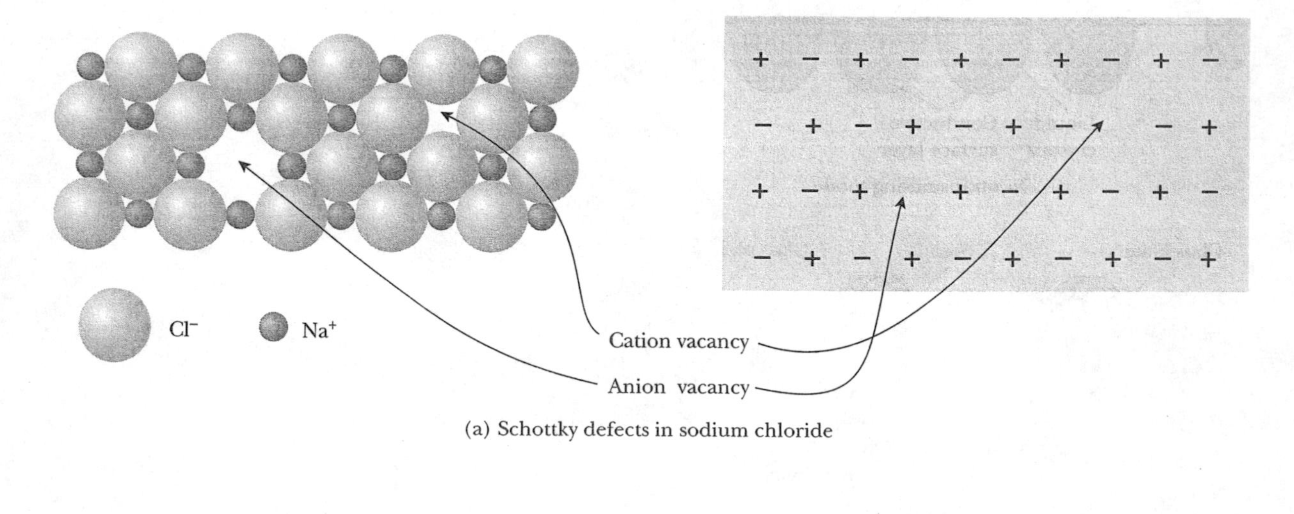

(a) Schottky defects in sodium chloride

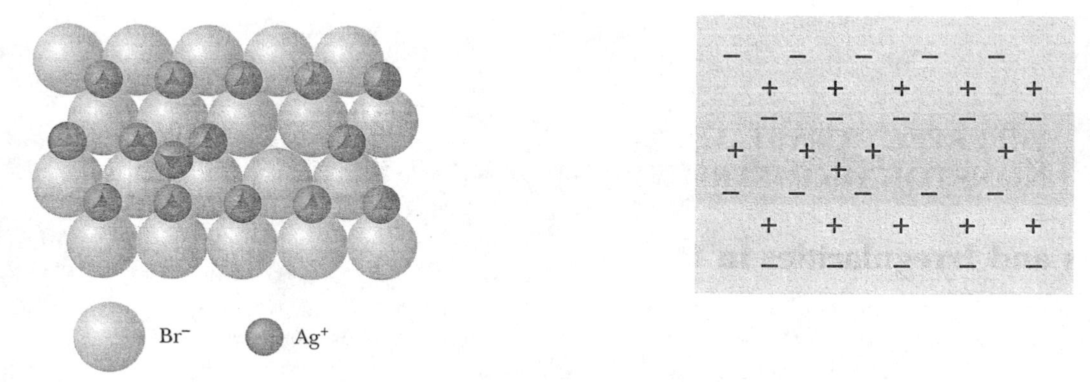

(b) Frenkel defects in silver bromide

FIGURE 16.29 Point defects in crystals. In pure crystalline substances, we find holes, or vacancies, because atoms are missing from lattice sites; and we find interstitials, atoms in unexpected places in the lattice. (a) Schottky defects come in pairs: a hole where a missing ion ought to be, and nearby, another from a lost ion of opposite charge. (b) Frenkel defects are caused by ions out of position, occupying a normally vacant hole and leaving a hole behind.

Some point defects do not maintain the regular stoichiometry of the compound, although they do maintain the required neutrality. In these cases an impurity of some type is either substituted for a regular constituent atom or ion or occupies a vacant hole in the crystal lattice. Substitutions involving ions of the same charge and similar size are quite common. For example, Ba^{2+}, and Sr^{2+} have similar ionic radii. Therefore, precipitation of $BaSO_4$ in the presence of Sr^{2+} ions will result in some Sr^{2+} ions occupying Ba^{2+} sites in the $BaSO_4$ lattice.

Sometimes an electron will be substituted for an anion in a crystal lattice. This situation can be caused by irradiating a salt such as NaCl with ultraviolet light. The energy of the ultraviolet light can remove an electron from a Cl^- ion, forming a Cl atom.

$$Cl^- \longrightarrow Cl + e^-$$

The uncharged Cl atom is much smaller than the Cl^- ion and is no longer held by coulombic forces in the crystal structure, so it migrates to the surface and is lost. The electron, with its negative charge, is held in a vacant anion site by the neighboring positive ions. The salt is now nonstoichiometric, with more Na^+ ions than Cl^- ions. However, electrical neutrality is maintained, as it must be because an electron replaces each Cl^- ion that was removed. The presence of the trapped electron causes a blue color in the NaCl crystal, and the defect is called an F-center, from the German word *farbe* for color.

Adding small amounts of impurities to metal structures introduces defects that modify properties. For example, metals are weakened by the presence of defects in their crystal structures in which lines of atoms are missing, allowing the structure to bend too easily or break. However, metal impurities that have smaller atoms than the principal metal tend to fill these holes and help hold the structure together. Thus, 10% zinc increases the strength of copper by 30%.

Nonstoichiometric Compounds

Some compounds commonly have **nonstoichiometric** ratios of their elements as a result of crystal defects. That is, their formulas differ from the whole-number relationships of ordinary stoichiometry compounds. This is especially true for materials composed of ionic compounds in which two oxidation states are possible for one of the ions. For example, iron oxides are known in which iron is present in both Fe^{2+} and Fe^{3+} states. The mineral siderite, which is iron(II) carbonate, can be heated to produce an oxide that is mainly iron(II) oxide (FeO) but has some iron(III) oxide (Fe_2O_3) from reaction with the air:

$$FeCO_3 \longrightarrow FeO + CO_2 \qquad \text{heating iron(II) carbonate}$$

$$4FeO + O_2 \longrightarrow 2Fe_2O_3 \qquad \text{in air, some iron(III) oxide forms}$$

The ratio of Fe^{2+} to Fe^{3+} depends on temperature and the ambient oxygen pressure. Variations in stoichiometric composition of the respective oxides range from $Fe_{0.95}O_{1.00}$ to $Fe_{1.00}O_{1.00}$. The lattice in this case is the NaCl structure, but formation of Fe^{3+} upsets electrical neutrality by introducing an extra +1 charge, which must then be offset by some type of defect. Imagine removing one Fe^{2+} for every two Fe^{3+} ions introduced (Fig. 16.30). That preserves electrical neutrality but alters the stoichiometry because there are now fewer iron ions than oxygen ions.

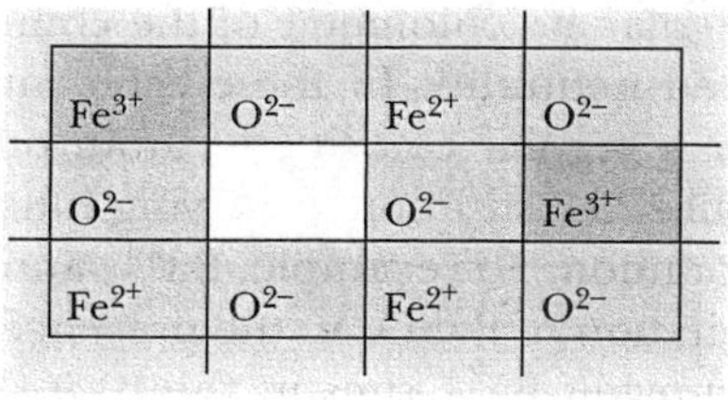

Fe^{3+}	O^{2-}	Fe^{2+}	O^{2-}
O^{2-}		O^{2-}	Fe^{3+}
Fe^{2+}	O^{2-}	Fe^{2+}	O^{2-}

FIGURE 16.30 Defects allow for electrical neutrality but result in nonstoichiometry. Here, insertion of two Fe^{3+} ions provides charge compensation for loss of a single Fe^{2+} ion. However, note that the ratio of iron-to-oxygen ions drops below 1 : 1 due to the missing ion.

NiO also has the NaCl structure, and its composition can vary from $Ni_{0.97}O_{1.0}$ to $Ni_{1.0}O_{1.0}$. The nonstoichiometric form, with the low nickel content and some missing Ni^{2+} ions, can be prepared in an excess of oxygen. In this case Ni^{3+} ions replace some Ni^{2+} ions to preserve the electrical neutrality. Transfer of an electron converts a Ni^{3+} to a Ni^{2+},

$$Ni^{3+} + e^- \longrightarrow Ni^{2+}$$

In nonstoichiometric NiO, electrons can jump through the structure from Ni^{2+} to Ni^{3+}, making it a modest conductor—actually a semiconductor—whereas the stoichiometric compound has no ready mechanism for electron transfer and is an insulator.

SUMMARY

Solids are composed of crystals that consist of a lattice of particles—atoms, molecules, or ions—that regularly repeat themselves in a space-filling pattern throughout the structure. The amorphous, noncrystalline state of a material exists when long-range order is absent. Sometimes referred to as the glassy state, such a substance is best understood as a liquid that is frozen or trapped in an unstable energy state.

Lattice points are equivalent points of a crystalline array. A unit cell is formed by connecting lattice points in such a way that the resulting cell can generate the entire crystal by simple translations in three dimensions. The smallest possible unit cell is called the primitive or simple cell. However, a nonprimitive unit cell is sometimes useful in order to incorporate all of the symmetry features of the crystal lattice. The focus of the discussions in the chapter is the highly symmetrical cubic unit cell in its three variations: primitive or simple, body-centered (BCC), and face-centered (FCC).

Metals normally have closest packed structures, in which the atoms can be approximated as hard spheres in contact with each other. Two regular closest packed structures are possible, cubic closest packing (CCP) and hexagonal closest packing (HCP). Cubic closest packing is identical to the FCC face-centered cubic lattice, with an atom at every lattice point. Metals also crystallize in the BCC lattice. Atomic radii of metals can be calculated from their crystal structures and unit cell dimensions.

Ionic compounds have structures in which the lattice sites are occupied by the positive and negative ions. Electrostatic interactions are the principal bonding forces. Crystals containing networks of atoms connected by covalent bonds are called covalent network crystals. Diamond and graphite, the two classic crystalline allotropic forms of carbon, have very different covalent network structures and very different properties. The electronic structures of some network substances such as silicon and germanium result in semiconductor properties. These properties can be enhanced by doping with other atoms. Molecules also form crystals; these usually have low melting points as a result of their being held together by relatively weak intermolecular interactions.

Liquid crystals are an important intermediate state of matter bridging the normal crystalline and liquid states of certain natural and synthetic substances. They behave like liquids in their flow properties, but there is sufficient organization of the molecules, so they also behave like crystalline solids, particularly with respect to light scattering and generation of color. LCDs are among the many modern applications of liquid crystals.

All crystals can be expected to have some defects. Three important defects in crystals are those resulting from vacancies in the crystalline lattice (Schottky defects), from atoms being out of their normal positions (Frenkel defects), and from different atoms substituting for the normal atoms in the lattice. Defects in crystals can account for the formation of nonstoichiometric compounds.

TERMS

Crystal (16.1)
Amorphous (16.1)
Glass (16.1)
Symmetry (16.2)
Cleavage plane (16.2)
Anisotropic (16.2)
Isotropic (16.2)
Crystal lattice (16.2)
Unit cell (16.2)
Translation (16.2)
Primitive cell (16.2)
Simple cell (16.2)
Nonprimitive cell (16.2)
Face-centered cubic (16.2)
Body-centered cubic (16.2)
Closest packed (16.3)
Hexagonal closest packing (16.3)
Cubic closest packing (16.3)
Coordination number (16.3)
Conductor (16.3)
Coulombic forces (16.4)
Hole (16.4)
Octahedral hole (16.4)
Tetrahedral hole (16.4)
Band gap (16.5)
Insulator (16.5)
Semiconductor (16.5)
Doping (16.5)
n-type semiconductor (16.5)
p-type semiconductor (16.5)
Current rectification (16.5)
Liquid crystal (16.6)
Nematic (16.6)
Smectic (16.6)
Defects (16.7)
Point defects (16.7)
Schottky defect (16.7)
Frenkel defect (16.7)
Nonstoichiometric (16.7)

IMPORTANT EQUATIONS

$$d = \frac{ZM}{N_A V}$$ Density of crystal

$$r = \frac{a\sqrt{2}}{4}$$ Atomic radius for face-centered cubic

$$r = \frac{a\sqrt{3}}{4}$$ Atomic radius for body-centered cubic

$$F = \frac{kq_1q_2}{d^2}$$ Coulomb's law, force between charges

QUESTIONS

Conceptual questions are denoted by a square screen. Extra-credit questions are denoted by a circular screen.

1. How can you most easily distinguish between a glass and a crystalline solid?
2. Can you determine the atomic mass of an unknown metal if you know its density and the dimensions of its unit cell? Why or why not?
3. Look up the melting points of water, ethanol, dimethyl ether, and methane. Considering that ethanol and dimethyl ether have the same molecular mass, and water and methane nearly so, what can you say about the intermolecular forces between these molecules?
4. How does the bonding theory of metals explain why most metals can be bent without breaking?
5. What kind of semiconductor results if germanium is doped with gallium? If gallium arsenide is doped with phosphorus? If aluminum phosphide is synthesized with a slight excess of aluminum?
6. Explain why a semiconductor device consisting of a sandwich of two *n*-type semiconductors on either side of a *p*-type semiconductor would pass electric current in neither direction.
7. Distinguish between the following pairs of terms:
   (a) isotropic and anisotropic solids
   (b) hexagonal close packing and cubic close packing of spheres
   (c) crystal lattice and unit cell
8. Explain the differences between the properties of diamond and graphite on the basis of their fundamentally different crystal structures.
9. Using the familiar properties of solid carbon dioxide and table salt, justify the contention that the former is a molecular crystal and the latter is an ionic crystal.
10. Why you would expect high-energy radiation that causes the formation of ions to damage a covalent layer such as a cell membrane but have no effect on a metallic film or layer?
11. What are the principal reasons for the similarities and differences between metallic and ionic crystals?
12. Why is diamond hard and insulating in contrast to graphite, which is soft and conducting?
13. Explain each of the following:
    (a) A *p(ositive)*-type semiconductor can be formed by substituting an aluminum atom for a silicon atom.
    (b) A *n(egative)*-type semiconductor can be formed by substituting an arsenic atom for a silicon atom.
    (c) What kind of semiconductor is obtained if silicon is doped with indium? With antimony?
14. Explain why gallium arsenide is referred to as a III–V semiconductor.
15. What is the effect of doping silicon crystals with phosphorus?
16. Offer a structural explanation for the differences between conductors, semiconductors, and insulators.
17. How does the change in the electrical resistance of copper compare with that of silicon when the temperature increases? Explain.
18. Explain why silicon is an insulator—not a semiconductor—at very low temperatures.
19. Show how the sodium chloride lattice can be generated by filling octahedral holes in a face-centered cube.
20. What is an LCD, and how does it work?

PROBLEMS

Problems marked with a bullet (•) are answered in Appendix A, in the back of the text.

Crystal Density and Atomic Radii [1–14]

•1. By what factor would the density of a metallic crystal change if the atomic radius of the metal were doubled and all other factors were kept the same?

2. What would be the ratio of densities of two metals A and B, with the same structures and all other factors the same except that metal A had an atomic mass 1.50 times as great as metal B?

•3. Use the following data to calculate the value of the Avogadro number N_A. Tungsten crystallizes in a body-centered cubic structure with a unit cell edge of $a = 3.1583$ Å. The density of tungsten metal is 19.3 g/cm^3 and its molar mass is 183.85 g/mol.

4. Calculate Avogadro's number from the following data for iridium, which has a face-centered cubic structure with a unit-cell edge of 3.823 Å and a density is 22.42 g/cm^3.

•5. Silver crystallizes in an FCC lattice with a unit cell edge of $a = 4.0776$ Å. The density is experimentally determined to be 10.5 g/cm^3. Calculate the molar mass of silver.

6. Vanadium has a body-centered cubic unit cell with a unit cell edge of 3.011 Å. Its density is 5.96 g/cm^3. Calculate its molar mass.

•7. Copper crystallizes into a face-centered cubic structure with a unit cell edge of 3.61 Å. Calculate the density of copper.

8. Chromium has a body-centered cubic arrangement for its unit cell, a length along an edge of 2.89 Å and a measured density of 7.0 g/cm^3. What would be the calculated density if the unit cell were mistakenly assumed to be a face-centered cube?

•9. Calcium has a face-centered cubic lattice with a lattice constant $a = 5.56$ Å. Determine the atomic radius of calcium.

10. Iron has a body-centered cubic lattice and a density of 7.86 g/cm^3. Calculate the atomic radius of iron.

•11. Copper has a face-centered cubic structure with a density of 8.92 g/cm^3. Calculate its atomic radius.

12. Calculate the atomic radius of niobium using its molar mass, its density of 8.55 g/cm^3, and the fact that it crystallizes in a body-centered cubic structure.

•13. Calculate the fraction of empty space in an FCC structure of a metal, assuming the metal atoms are hard spheres that just touch along the face diagonal.

14. Calculate the fraction of free space in a BCC structure of a metal. Compare your results with the previous problem. Tell why your results seem reasonable or why they do not.

Ionic Radii [15–20]

15. Use the ionic radii in Table 16.2 to calculate the distance between the Ba^{2+} and Cl^- ions in $BaCl_2$(s).

16. Magnesium oxide has the NaCl (rock salt) structure. Use the ionic radii in Table 16.2 to calculate the length of a unit cell edge in a crystal of MgO.

•17. Thallium(I) bromide has the CsCl structure with a unit-cell edge length of 3.97 Å. The ionic radius of Br^- is 1.96 Å. Calculate the ionic radius of Tl^+.

18. Iron(II) oxide has the NaCl structure with a unit cell edge of 4.294 Å. The oxide ion has a radius of 1.45 Å. Calculate the ionic radius of Fe^{2+} in iron(II) oxide.

•19. A sample of pure sodium chloride has a density of 2.167 g/cm^3. The average distance between adjacent sodium and chloride ions in the crystal lattice is 2.184 Å. From these data, calculate Avogadro's number.

20. Cesium bromide has the CsCl structure. Use the ionic radii in Table 16.2 to calculate the closest distance between unlike ions, then use this to calculate the density of CsBr.

Additional Problems [21–25]

•**21.** Silicon has the diamond structure. How many silicon atoms are there in one unit cell?

22. Tantalum metal has a body-centered cubic structure, with a unit cell edge of 3.281 Å. Its molar mass is 180.9. Calculate its density.

23. Barium sulfide has the NaCl structure. Use the ionic radii in Table 16.2 to calculate the length of the edge of the unit cell of BaS.

24. Determine the average distance between K^+ and Cl^- ions in crystalline potassium chloride, given the density of 1.984 g/cm^3. KCl crystallizes in the NaCl structure.

•**25.** Given the interionic distance between Li^+ and Cl^- ions in LiCl to be 0.257 nm, determine the density of lithium chloride. LiCl has the NaCl structure.

Multiple Principles [26–29]

26. A compound of mercury and chlorine has a crystal structure with two formula units per unit cell. The unit cell edges are 4.47, 4.47, and 10.89 Å, and the unit cell angles are all 90°. The density of the substance is 7.15 g/cm^3. What is the formula of the compound?

•**27.** An unknown substance that is a gas at room temperature can be condensed to a solid at −80°C. X-ray analysis of its crystals can then be performed. As a solid, it is found to have a cubic unit cell, 5.15 Å on each side, containing 4 molecules. The density of the solid is 0.73 g/cm^3. What is the density of the substance as a gas at 25°C and a pressure of 1.00 atm?

28. A compound of cobalt and sulfur has a cubic structure containing four formula units per unit cell with a unit cell edge of 5.212 Å. Its density is 4.269 g/cm^3. What mass of sulfur would be required to produce 1.50 kg of this compound?

•**29.** The parallel planes of carbon atoms in graphite are 3.35 Å apart. Each plane is composed of six-membered rings of carbon atoms connected together (Fig. 16.21). The density of graphite is 2.25 g/cm^3. Use these data to calculate the carbon–carbon distance within the carbon planes. What does this result tell you about the bonding between the carbon planes?

Applied Principles [30–32]

30. Semiconducting, nonstoichiometric nickel(II) oxide can be prepared by reaction of the stoichiometric compound with oxygen gas. What volume of $O_2(g)$ at 25°C and 2.50 atm would be consumed in converting 1.00 kg of NiO to the appropriate amount of $Ni_{0.97}O_{1.00}$? What mass of the semiconductor would be produced?

•**31.** Preparation of silicon for semiconductor uses requires the purification of the silicon up to 99.9999999%. At this level of purity, approximately how many silicon atoms per gram have been replaced by some other element?

32. What is the relative cost advantage to plate a surface to a given thickness with palladium ($200/troy ounce) rather than gold ($375/troy ounce)? The densities at 25°C are Au = 19.3 g/cm^3 and Pd = 12.0 g/cm^3.

ESTIMATES AND APPROXIMATIONS [33–37]

33. Complete the following reaction, balance the equation, and offer an explanation of what it might be that drives this vigorous and dramatic reaction:

$$WCl_6(s) + Na_2S(s) \longrightarrow$$

34. What kind of crystalline solid structure is found for dry ice? Why?

35. Given the lattice constant for C_{60} of 10.04 Å , make careful estimates of the density of solid buckminsterfullerene and the fraction of "free space" available.

36. Predict the relative conductivity and the electron band gap for solid C_{60}, buckminsterfullerene, compared with other substances discussed in this chapter.

37. Computers crash if the temperature gets too high. You can put that statement on a scientific basis by answering this question: Why do semiconductors lose their effectiveness with increasing temperature?

FOR COOPERATIVE STUDY [38–43]

38. Determine the number of lattice points per unit cell in each of the following lattices. See Figure 16.6 for cell characteristics.
(a) primitive tetragonal
(b) body-centered tetragonal
(c) end-centered orthorhombic (extra lattice points in two opposite faces)

•39. How many lattice points are there in one unit cell of each of the following lattices?
(a) face-centered orthorhombic
(b) primitive triclinic
(c) body-centered monoclinic

40. The tetragonal unit cell contains a fourfold rotation axis. This means that the unit cell can be rotated by 360°/4 or 90° about this axis and appear as though no operation has taken place. Locate this axis. Now place an extra lattice point in the center of each of the two faces that are perpendicular to this fourfold axis. This would be a nonprimitive "end-centered" tetragonal unit cell. However, this cell is equivalent to the primitive tetragonal cell. Explain why.

41. A sample is found to have the formula $Fe_{0.97}O_{1.00}$. What fraction of the metal ions are Fe^{3+}?

42. An "FeO" sample is found to have the actual formula $Fe_{0.95}O_{1.00}$. What fraction of the metal ions are Fe^{2+}?

43. Calculate the fraction of nickel ions that are Ni^{2+} in a sample of "nickel(II) oxide" having the formula $Ni_{0.97}O_{1.00}$.

WRITING ABOUT CHEMISTRY [44–45]

44. What do you find difficult about studying crystals? Explain.

45. Two important concepts used in describing the structure of a crystal were the crystal lattice and the unit cell. Explain why a crystal does or does not actually have a crystal lattice and a unit cell. Offer any evidence you can to support your point of view on this. Why do we use these questionable concepts in describing the structure of a crystal? Could we describe crystal structures without these concepts?

17

chapter seventeen

MATERIALS

17.1 INTRODUCTION TO MATERIALS SCIENCE

Glass shatters! Wood splinters! Steel is tough! A rubber band is elastic and stretches easily. The floor does not usually collapse under our weight and the roof is not likely to fall in on us during a snowstorm. We do not think about the possibility of the bridge falling as we ride over it or a plane's wings coming off in flight. Yet floors, roofs, bridges, and wings *do* fail, and on occasion catastrophically. Why does this happen? Are our materials what we expect them to be? Can we improve on existing materials, making them tougher, stronger, and lighter? Are the properties of materials at least measurable, if not always easily or rationally understood? Our goal in this chapter is to shed some light on a world of materials and their properties, a subject that is too often taken for granted. Imagine how different our lives would be without tonnage steel, reinforced concrete, glass and ceramics, newsprint paper, synthetic textile fibers, rubber tires and elastic bands, transistors and semiconductors, Saran wrap and Styrofoam, aluminum cans and plastic beverage bottles, pressure-sensitive tape and stick-on notes, O-rings and all manner of adhesives, caulks and sealants—all these testify to the accomplishments of materials science.

As a discipline, materials science brings together knowledge gained from physics, from a variety of areas of theoretical and practical chemistry, from important areas of engineering practice—including electrical, thermal, mechanical, and chemical engineering—and from fabrication and processing technology. Thus, a materials scientist tends to be something of a renaissance person, knowledgeable in many related fields that are combined to produce interesting and sometimes astonishing results. Modern materials are often required to perform in unfriendly environments, characterized by very high or very low temperatures, in which they are mercilessly bent, stretched, and compressed, testing their flexibility and rigidity, and in some circumstances leading to fatigue—and ultimately failure. This is the age of stress and strain, and our materials need to be able to withstand the deforming forces we put on them. When planes crash or bridges collapse or spacecraft explode, questions are always raised about materials failure. What caused the accident? Did the brakes fail? Did the seal fail, as in the *Challenger* explosion? Did a cable fail? Did a gear fail? We begin by surveying the range of useful materials and saying something of their properties.

17.2 MATERIALS

Planet Earth has been blessed with a bounty of raw materials. For most of the history of humanity these materials have been used as they were found: stone implements, wood and forest products, a rare mineral or two, and air and water. As civilization has progressed, humans have introduced new uses for existing materials and have discovered new materials. Advances of extraordinary magnitude have been coupled to refining, alloying, and fabricating of metals. The metallurgy of iron, copper, tin, and zinc and the discovery of bronze and brass alloys allowed the production of the first advanced weapons and other material objects.

The great natural division into metallic and nonmetallic materials arises because the properties of metals and nonmetals are, for the most part, so different. It is only within the past century and a half that people began to synthesize, formulate, and fabricate secondary materials using the Earth's primary resources in ways we recognize today as modern. What may be the most important discovery of the beginning of this period was the availability of cheaper and tougher iron and steel. This made contemporary technology characteristically metallic. With important new discoveries in ceramics and plastics, metals have lost their supremacy, and we can expect the technology of the twenty-first century to be more evenly represented by metals and nonmetals. Useful materials for engineering purposes can be grouped into several major categories that include both natural and synthetic materials.

Metallic Materials

As we discussed in the previous chapter, metal structures are composed of atoms, not molecules. However, it is just not possible to produce a pure metal, one consisting of atoms of only one kind. Impurities are always present. Most of the more familiar engineering metals are alloys. **Alloys** are solutions of two or more metals in each other. The presence of small amounts of other elements, metallic as well as nonmetallic, sometimes profoundly affects the characteristic properties of a metal. Of all the metal alloys, the most prominent is structural steel, an alloy containing iron as its principal component, along with a little carbon, usually less than 1%. Structural steel can be readily fabricated into useful shapes and has outstanding qualities of strength combined with flexibility. It can be formed into girders strong enough to support modern skyscrapers yet capable

PROFILES IN CHEMISTRY

Art, Science, and Technology The antecedents of modern materials science can be found in the decorative arts, from which one can conclude that creative discovery combines the love of beauty with the workings of human intellect. Few can doubt the influence of artistic motivation in discovering the properties of materials, which led to scientifically and technologically important ways of making and using matter. Consider the following examples:

- The first uses of ceramics and metals in decorative objects, such as fire-hardened figures of clay found in many Middle Eastern archaeological sites long before any clay pottery
- Evidence of copper dress ornaments and beads dating from about 700 B.C. recovered from archaeological digs in Anatolia; copper showed up some time before in axes, swords, and other weapons
- Jewelry from the royal graves at Ur, suggesting that 5000 years ago early metallurgists anticipated modern use of alloying elements to strengthen metals and to lower their melting points—cold-working metals to harden them and annealing metals to soften them
- The casting of monumental bronze doors and great church bells, undertaken nearly 500 years ago

In more contemporary examples of the confluence of art, science, and technology, consider the art of printing and the knowledge needed to produce type metal, inks, and dyes. Or think about the remarkable work of silversmiths and porcelain makers; artists who work in oil, watercolors, charcoal, crayon, acrylic, and an almost endless stream of materials. Photography and filmmaking have evolved over a century of innovation, invention, and discovery, from early daguerreotypes to Polaroids and color printing and processing. The history of clothing and costume reflects not only the social history of times past and present but the emergence of rayon, nylon, and polyester fibers, and enhanced cotton and wool fibers with extraordinary and unexpected properties. The list is endless!

Questions

What can be the scientific benefit of producing works of art? Give several examples.

What additional subjects depicting the confluence of art, science, and technology can you describe?

of deforming when subjected to sudden forces. Thus, in an earthquake such a building can give a little, absorbing the shock wave, instead of crumbling as it would if it were constructed of a brittle material that was unable to bend.

Steel shares the qualities generally found in other metals. It is opaque, lustrous, and a good conductor of heat and electricity, but unfortunately it is subject to corrosion by acids and other oxidizing compounds that convert atoms to cations and diminish the favorable structural properties of the material. An example of the strength of steel is the George Washington Bridge connecting New York and New Jersey. When completed in 1931, its central span of 3500 feet was twice as long as any bridge previously constructed. It is held up by four cables, two on each side, each consisting of 26,474 zinc-coated steel wires, each of which is 0.196 inches in diameter and compacted into a cable diameter of about 36 inches. Each wire is strong enough to support a mass of 7000 pounds, and together the wires can hold aloft 184 million pounds. The cable has a strength of more than 240,000 pounds per square inch. A few of the huge number of alloys are listed in Table 17.1.

Glass and Ceramic Materials

Glasses and **ceramics** are substances composed of molecules or ions, and very often contain silicon and oxygen. They are generally hard, resistant to heat, and often brittle. Simple examples of ceramics are materials fabricated from silica

TABLE 17.1 Compositions and Uses of Some Alloys

Alloy	Composition	Uses
Iron Alloys		
Sinimax	54% Fe, 43% Ni, 3% Si	High-frequency coils
ACI Type HH	0.25% C, 2% Mn, 24–28% Cr, 11–14% Ni, 56–63% Fe	Heat-resistant cast steel
Type 9 Ni	0.1% C, 8–10% Ni, 0.5% Mn, 0.2% Si, 89–91% Fe	Low-temperature applications
Type 410 Cb	0.15% C, 12.5% Cr, 87% Fe	Stainless steel
Gray cast iron	3% C, 0.5% Mn, 2% Si, 94% Fe	Castings
Aluminum Alloys		
AZ91A	9% Al, 0.7% Zn, 0.1% Mn, 90% Mg	Castings, auto parts, etc.
Alnico I	12% Al, 20–22% Ni, 5% Co, 41–43%, 18–22% Fe	Permanent magnets
Copper Alloys		
Gilding metal	95% Cu, 5% Zn	Ammunition, coinage
Free-cutting brass	62% Cu, 35.5% Zn, 2.5% Pb	Screws, gears, keys
Tin Alloys		
Pewter	91% Sn, 7% Sb, 2% Cu	Ornamental items
Solder 50–50	50% Sn, 50% Pb	General-purpose solder
Silver Alloys		
Denture amalgam	33% Ag, 52% Hg, 12.5% Sn, 2% Cu, 0.5% Zn	Dental fillings
Sterling silver	92.5% Ag, 7.5% Cu	Tableware
Nickel Alloy		
Monel	72% Ni, 25% Cu, 3% Fe	Corrosion-resistant labware

(SiO_2, silicon dioxide), magnesia (MgO, magnesium oxide) and alumina (Al_2O_3, aluminum oxide). These oxides are all **refractory** substances, meaning that they maintain their useful properties at high temperatures. Refractory substances have high melting points and also include silicon carbide (SiC), silicon nitride (Si_3N_4), boron carbide (B_4C), and tungsten carbide (WC). The refractory oxides, carbides, and nitrides are all extremely hard substances with very high melting points, because the interatomic forces are generally greater than in metals. That makes them ideal for fabrication into materials that can perform at high temperatures—for example, the refractory linings for furnaces. Oxygen-containing ceramics are not susceptible to reaction with oxygen or with compounds such as nitric and sulfuric acids—the oxidizing acids—so they are highly resistant to corrosion.

We mentioned amorphous substances, particularly glass, at the beginning of the last chapter. Such materials are characterized by a lack of the long-range order of crystalline solids and therefore the absence of a definite melting point. When heated, these substances soften slowly as the viscosity continuously decreases. When sufficiently soft, glass can be worked easily into forms such as sheets, bottles, tubes, and fibers in modern, high-speed, low-cost processes.

The principal difference between ceramics and glasses is that ceramics have a regularly repeating, crystalline structure. Bricks and tiles are traditional ceramic materials. Fine "whiteware" dinnerware products are ceramics, as are porcelains, cements, and abrasives. Important new materials called glass ceram-

TABLE 17.2 Compositions and Uses of Some Ceramics

Ceramic	Formula	Properties/Uses
Alumina	Al_2O_3	Strong, inert, refractory/cutting tools, insulators
Beryllia	BeO	High-thermal conductivity, strong/electronic components
Cordierite	$2MgO \cdot 2Al_2O_3 \cdot 5SiO_2$	Thermal-shock resistant, low dielectric losses/high-frequency insulators
Mullite	$3Al_2O_3 \cdot 2SiO_2$	Thermal-shock resistant/refractory furnace parts, laboratory ware
Silicon carbide	SiC	Stable at very high temperatures, very hard/refractory furnace parts, abrasives
Silicon nitride	Si_3N_4	Resistant to wear, corrosion, high temperatures/furnace components
Zirconia	ZrO_2	High melting point, chemically inert, low thermal conductivity/refractory uses, glazes, enamels

ics start with a glass that is then crystallized in a network of microscopic crystals. The starting glass in these cases often contains lithium and aluminum oxides along with SiO_2. The desired object is shaped in the glassy state. Then a heat treatment converts the glassy material into a form with very fine-grained crystals and no porosity. The resulting products have exceptional mechanical strength and heat resistance and are excellent for cooking ware. Unusually break-resistant tableware that will not shatter when dropped on a ceramic floor and ceramic floor tiles that will not crack when struck by a falling object have been introduced by DuPont and Dow Corning.

In marked contrast to metals and metal alloys, the main shortcoming of ceramics and glasses is their brittleness, which limits their use in applications such as engines, in which their excellent heat resistance would be extremely useful. Efforts are being made to find ceramics with properties that will permit such uses in spite of their inherent brittleness. They cannot be easily modified to bend rather than break in response to deforming forces. However, structural ceramics, compositions that include silicon nitride, are being tested for high-temperature automobile engines. In our discussions on thermodynamics in Chapter 13, we noted that the higher the temperature at which an engine operates, the more efficient it is. Structural ceramics open the way to greater efficiency by allowing construction of engines that can operate at high temperatures for protracted periods of time. Tables 17.2 and 17.3 give the composition and some properties of selected ceramics and glasses.

TABLE 17.3 Essential Compositions and Uses of Some Glasses

Glass	Composition	Uses
Soda–lime	72% SiO_2, 14% Na_2O, 10% CaO, 3% MgO	Bottles, windows
Borosilicate	81% SiO_2, 4.5% Na_2O, 12.5% B_2O_3, 2% Al_2O_3	Laboratory and cookware
Cer-vit C-101	66.4% SiO_2, 21.4% Al_2O_3	Telescope mirrors
Corning 4602	18.5% SiO_2, 14.1% Al_2O_3, 57.3% P_4O_{10}	Heat absorbing
Owens-Corning SF	59.5% SiO_2, 5% Al_2O_3, 7% B_2O_3, 14.5% Na_2O, 8% TiO_2, 4% ZrO_2	Insulation
Float	72.9% SiO_2, 13.9% NaO_2, 4% MgO, 8.6% CaO, 0.3% SO_3	Windshields

PROFILES IN CHEMISTRY

Glass–Ceramic Cooking Surfaces Electric heating for domestic cooking was first introduced in the last decade of the nineteenth century with the electrification of major cities. It caught on rapidly. Early electric stove tops held cast-iron plates with resistance wires underneath that transferred heat by conduction through the plate. Others contained wire spirals that transferred heat by thermal radiation. But by the 1960s, these early prototypes were replaced by flat coils in which a nickel-chromium resistance wire was embedded in a tube of Inconel, a corrosion-resistant alloy of nickel and chromium. This tube was insulated from the resistance wire by a dense, magnesium oxide-powder packing. Such burners heated fairly rapidly, were long-lasting and efficient, and inexpensive to operate. But cleaning up spills was as much of a problem for the 1960s chef as it was for predecessors half a century earlier.

The answer to the spill and clean-up problem emerged through materials science research at the Corning Glass Company with the development of a special family of glass–ceramic cooktops for electric stoves that became commercially available in the 1980s. These stove tops came fitted with a glass–ceramic sheet separating and insulating the heating coils from the pots and pans while providing a smooth, attractive, and easy-to-clean surface. What are the requirements for such a surface? What kind of material will meet those demands? How can such a material be fabricated? To begin with, heat passes to the pots and pans by conduction, and if the surface is transparent, by radiation. However, the surface cannot simply be a terrific thermal conductor because you do not want the entire surface to get hot—only the space under the pot or pan placed above the heating coils. After all, a hot surface any other place will waste heat and be dangerous to the chef. To clean the surface easily, it must be smooth and stay so throughout its lifetime—which means it must be resistant to scratches and abrasion. At the same time, it must be able to withstand the thermal shock of rapid cooling when cold water is accidentally dropped on a very hot surface or the mechanical shock caused by a dropped object striking the surface. Of course, the top surface must be able to withstand the 600°C punishment to its bottom side from the underneath heating coils throughout a long lifetime.

So this must be a truly tough material. Yet you may be well aware of how un-tough a material glass is, once scratched. A glass cutter can score a glass with a diamond edge of some kind and then the glass can be broken uniformly along the scratch. For stove-top applications, this property could be a catastrophic problem, because the glass surface is bound to be scratched in normal service. The solution to this problem in materials science? A crystalline ceramic of very fine grains—single crystals with a regular arrangement of atoms within—that effectively limits the ability of a crack to propagate through the material. In simple terms, the propagating crack runs into crystals that stop the crack from moving further. At the same time, the glass must have a very small coefficient of thermal expansion to maintain its toughness.

A ceramic based on lithium aluminum silicate, which crystallizes in a fine grain structure as hexagonal crystals and is therefore anisotropic, expands thermally across the hexagons differently than along the hexagons. The practical result is that as the atoms spread sideways, they pack down. So there is little thermal expansion, and cracks will not spread easily.

Manufacturers of such stove tops begin by melting more than a dozen oxides into a homogeneous glass mixture at about 1650°C: SiO_2 (68%), Al_2O_3 (19%), Li_2O (4%), MgO (2%), ZnO (2%), and smaller percentages of several others, including oxides of titanium, zirconium, sodium, and potassium. The titanium and zirconium oxides are especially important because they act as nucleating agents, leading to crystallization. It is the controlled crystallization that is the key to the process and the resulting properties. From the melt, the glass–ceramic composition is formed into continuous sheets by a system of rollers at a temperature of about 1500°C and is then cut to size. Crystallization occurs while the slabs are aging at temperatures ranging from 1000 to 1500°C, and by the time the day-long aging process is complete, the glass–ceramic material may be more than 90% crystalline.

Polymers and Plastic Materials

Plastic materials are composed principally of long, chain-like molecules called **polymers**. Plastic milk bottles are composed of polyethylene molecules, the polymer with the simplest structure. The basic building block or repeating unit in the long chain is ethylene, the **monomer**, or small molecule repeating units of which the polymer is made.

Questions

Consider the different types of materials as candidates for a flat cooking surface on a stove. Explain how each type would function for this application.

What is your recommendation for the coefficient of thermal expansion for a ceramic stove top? Should it be high? Low? Why?

A sealed glass ampoule is easily and cleanly decapitated by scoring the glass at the neck and snapping it off. Why is this property of glass a problem for stove tops? How is the problem resolved?

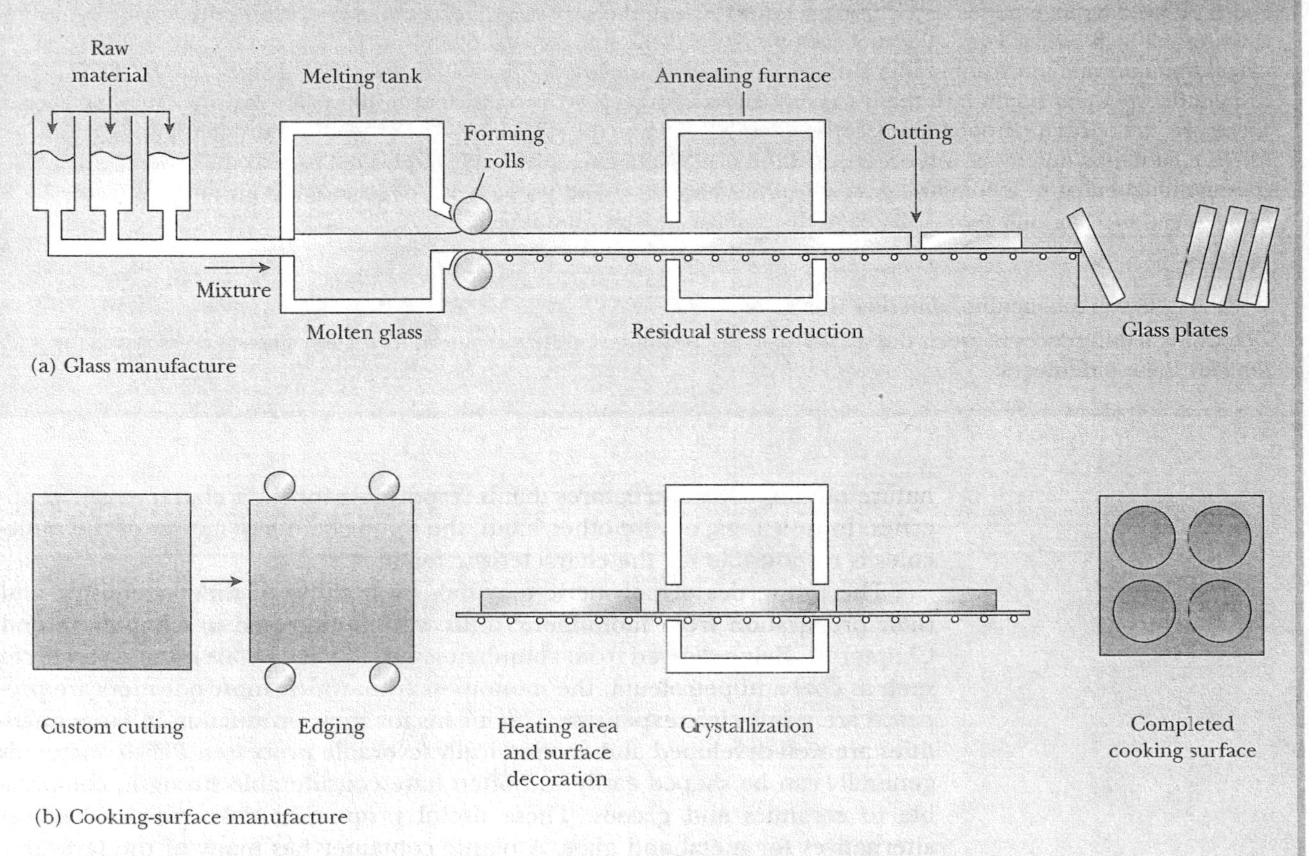

Ceramic stove tops. The schematic diagram outlines the Corning glass process, beginning with the introduction of the raw materials mixture into the melting furnace and ending with the finished slab ready to be installed on the stove top.

$$nCH_2{=}CH_2 \longrightarrow -(CH_2CH_2)_n-$$

ethylene (monomer) → polyethylene (polymer); n is the number of times the basic monomer unit repeats

The subscript letter n in the polyethylene formula is the number of repeating ethylene monomer units in the polymer, which can easily range into the tens of thousands, leading to huge molecular masses. For metals and ceramics, it is the

SEEING THINGS

Melting Glass **(This demonstration should not be performed without supervision and proper safety precautions.)** Soft soda–lime–silicate glass, the glass of windowpanes and laboratory tubing, has a much higher coefficient of thermal expansion than Pyrex (borosilicate) glass, a composition that is much less sensitive to thermal shock and that is the glass of all manner of oven ware. In this demonstration, a piece of soft glass tubing is heated with a propane torch to its melting point and allowed to drip from its end into a beaker of cold water. As it strikes the water, it freezes suddenly, then craze-cracks and shatters violently—visibly and audibly. In a similar experiment with a piece of Pyrex glass rod, the resulting teardrop that drips into the water not only does not shatter but has become extremely strong.

As an amorphous material—a frozen liquid—it is not surprising to find that both these glasses soften, then melt and drip, holding together as water droplets do because of the very strong forces that exist within the liquid. But because of the high coefficient of thermal expansion, when the soft glass droplet hits the water the molten inside keeps expanding into the now frozen outer surface, causing it to shatter. No such shattering of the outer skin of the Pyrex droplet occurs when it falls into the beaker of cold water, because its coefficient of thermal expansion is lower. Furthermore, the resulting droplet has become very hard because the still molten glass within the droplet slowly expands into the hardened outer wall, a process called thermal toughening. Although only heated to near the softening point, thermal toughening of automotive glass effectively hardens it and prevents it from shattering into the characteristically long, sharp, and dangerous shards of the familiar broken windowpane.

Questions

What is "thermal toughening" and how is it used?

Describe the differences between the shatter patterns of window glass and automotive safety glass and discuss the reasons for these differences.

nature of their crystal structures that is responsible for their characteristic properties. In polymers, on the other hand, the long, chain-like nature of the molecules is responsible for the characteristic features.

The properties of polymeric materials, something of their chemistry, and their preparation from monomeric units will be discussed in Chapter 18 and Chapter 21. Being derived from abundant and readily available natural resources such as coal and petroleum, the monomers from which most polymers are prepared are relatively inexpensive. The means for their production in large quantities are well-developed and economically favorable processes. Plastic materials generally can be shaped easily and often have considerable strength, comparable to ceramics and glasses. These useful properties make plastics low-cost alternatives for metal and glass. A plastic container has many of the favorable properties of glass, but it will not break when dropped and its shipping mass is much less. A plastic frame for a backpack has all the strength of its metal counterpart, yet it is much lighter in weight. The shortcomings of plastics are generally poor resistance to heat and their reactivity in the presence of a variety of chemicals, including some common solvents such as dry cleaning fluids and gasolines.

Composite Polymeric Materials

Composites are combinations of a polymeric plastic base impregnated with reinforcing fibers or sheet materials such as Fiberglas and Kraft paper. Such compositions are increasingly important and offer an attractive way to enhance useful properties; the reinforcing fibers bear the load, and the polymer base serves as the matrix, holding the fibers in place. The flexibility and lightness of the polymer matrix is enhanced by the high strength of the reinforcing fibers or sheets, producing a structural material used in boat hulls, auto bodies, fuel-storage tanks, and many other applications. Carbon fibers offer exceptional strength when used to reinforce polymer composites. Wood is a naturally occurring composite material that is reinforced by fibers. Its unusually wide range of uses and applications results directly from its lightness, flexibility, and overall strength.

Selecting Materials

What is the basis for selecting a material for a particular application? Bear in mind that cost is almost always a factor and often the deciding one. In some cases, low-cost materials are absolutely necessary because an established market exists and a competitive product that is only marginally better, or a new product introduced at a higher price, may not be given a chance to establish itself. Substitutes for wood in construction or window glass in automotive applications must be competitively priced. On the other hand, applications requiring extremely high-performance standards, such as space vehicles, race car engines, or microelectronics can sometimes tolerate the use of more costly materials. Iron and aluminum are modestly priced metals, whereas tungsten, iridium, and platinum are very costly. Ceramics range from low-cost fired clays used in flowerpots and dishes to high-cost silicon nitride and related specialty ceramic materials suitable for fabricating blades for gas turbines and jet engines. Polymers are generally less expensive materials, but do range from low-cost polymers used in commodity-plastics applications such as transparent food wrap to very expensive polymeric fibers such as those used in weaving fabric for bulletproof vests and flameproof firefighting clothing.

Bonding in Materials

The bonds found in materials are, of course, the same kinds of bonds that we have already discussed. Ionic and metallic bonding were covered in Chapter 16 as part of our discussion of solids. The properties of covalent bonds were covered in Chapter 6, and the theories of these bonds were discussed in Chapter 7. The weaker intermolecular bonds, often called van der Waals forces and including hydrogen bonds, were also introduced and discussed in Chapter 6. It is useful to summarize these different bonds in the context of the study of materials.

The ionic bond is a nondirectional attraction—that is, the magnitude of the bond is the same in all directions surrounding an ion. Ionic bonds are a consequence of the coulombic forces of attraction between ions of opposite charge that make up the structure. The net energies of attraction for crystalline substances such as sodium chloride are on the order of 10^3 kJ/mol. When compared with examples of other types of bonds, such energies turn out to be among the very strongest. Materials that are characteristically ionic tend to be hard and brittle and electrically and thermally insulating. The ionic bond is the principal bond in many ceramic materials.

Metallic bonds are also nondirectional. The structures of metals and their alloys are characterized by high-coordination numbers, which means high numbers of nearest neighbors around a given metal atom. Metal atoms are held together by their valence electrons, which usually number no more than three but are highly delocalized into large numbers of overlapping bonding orbitals. We have described the valence electrons in metallic materials as free to drift through the entire structure. Because of the freely moving electrons, metals are characteristically strong, malleable, ductile, and conducting materials. Consider, for example, that it is possible to pound copper into sheets for roofing leaders and gutters, pull it into electrically conducting wires, and extrude it into piping for plumbing.

In marked contrast to ionic and metallic bonds, covalent bonds are strongly directional. They involve sharing pairs of electrons between atoms and creating definite bond angles with respect to shared pairs between a central atom and any other two to which it is covalently bonded. Bond angles are tabulated as average values (Table 7.1) and the strengths of covalent bonds are taken as average bond enthalpies or energies (Table 6.2). Many materials are held together by covalent bonds and on the whole such bonds prove to be reasonably strong, and in some cases comparable to ionic bonds. However, covalent materials such as diamond, with amazing strength and high melting points, are exceptional.

Van der Waals forces are weaker attractions between molecules, with typical energies of 10 kJ/mol or less. They exist to a greater or lesser extent between all atoms or molecules, although they often cannot be measured because of the overpowering presence of the three primary bonds—ionic, metallic, and covalent. Van der Waals forces do not involve sharing or transfer of electrons between atoms or into bonding orbitals. Instead they result from attractions between permanent and induced dipole moments. They are directional, but not to the same extent as covalent bonds. The hydrogen bond is the best known and generally the strongest of the bonds of this group. The molecules of ammonia and water in the solid and liquid phases are held together by hydrogen bonds. Hydrogen bonds are the result of the relatively powerful attraction between

TABLE 17.4 Melting Points of Some Substances Related to Type of Bonding

Substance	Bonding	Melting Point (°C)
NaCl	Ionic	800
CaO	Ionic	2580
Fe	Metallic	1535
W	Metallic	3370
C (diamond)	Covalent	~3500
B	Covalent	2300
Br_2	Covalent (intramolecular) van der Waals (intermolecular)	−7.2
NH_3	Covalent (intramolecular) Hydrogen bonding (intermolecular)	−77.7
$—(CH_2—CH_2)_n—$	Covalent (intramolecular) van der Waals (intermolecular)	~120 (softens)

dipoles containing hydrogen and one of the very electronegative elements: nitrogen, oxygen, or fluorine.

A measure of the strength of the forces that hold a substance together is the melting point. The higher the melting point, the more energy is required to break the bond, and therefore the stronger the bond. If a material has more than one kind of bond, the weakest break first, and this will be noted by melting or softening of the material. Table 17.4 gives the types of bonding present in some substances and their melting points.

17.3 PERFORMANCE CHARACTERISTICS OF MATERIALS

Stress and Strain

Most human beings, having successfully progressed beyond childhood, are clever enough to not skate on ice that is too thin or lean against a glass window of less than adequate thickness. After enough years, our experience serves to keep us out of harm's way. We seem to know our own strength and the strength of materials in a general way. But words such as "tough," "hard," "strong," and "rigid," though commonly used in everyday speech, are too vague to use in answering serious scientific and engineering questions about strengths of materials. The specific details describing a material in terms of its limitations for use are called the performance characteristics. **Performance characteristics** are actual measured values rather than qualitative statements such as "tough," "hard," and "rigid." Because standard tests of stress and strain are the most widely performed for quantifying and comparing strengths of materials, they are the performance characteristics we will describe and use in some detail.

The words "stress" and "strain" have particular meanings. **Stress** (τ) is the load, or force, imposed on a test specimen divided by the original cross-sectional area of the specimen:

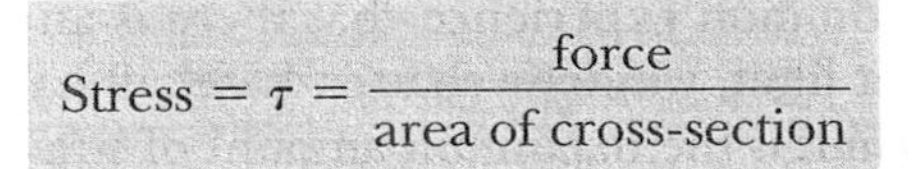

$$\text{Stress} = \tau = \frac{\text{force}}{\text{area of cross-section}}$$

If you suspend a mass of 150.00 kg on a wire with a cross-sectional area of 0.4000 mm^2—that is, $4.000 \times 10^{-7}\ \text{m}^2$—you are applying a stress to the wire that can be calculated by converting mass to force in Newtons using $F = mg$ and then dividing by the cross-sectional area:

$$\text{Stress} = \frac{150.00\ \text{kg} \times 9.8066\ \text{m/s}^2}{4.000 \times 10^{-7}\ \text{m}^2} = 3.677 \times 10^9\ \text{N/m}^2$$

The stress results in a strain that in this case is tensile elongation, or stretching of the wire. **Strain** (ε) is the deformation, or increase, in the length of the specimen as a result of the stress, divided by the original length:

$$\text{Strain} = \varepsilon = \frac{\text{increase in length}}{\text{original length}}$$

If the length of the wire under this load increases from 2.000 m to 2.012 m, the increase in length is 0.012 m, and the strain can be calculated as follows:

$$\text{Strain} = \frac{0.012 \text{ m}}{2.000 \text{ m}} = 0.0060 \text{ m/m}$$

Strain is sometimes reported without units, but it is best to keep the units here to remind us that we are dealing with a relative change in length. If the stress were a compressive load, then the material would contract.

On removing the stress, materials exhibit two kinds of behavior. If the material is permanently strained, then the strain behavior is termed **plastic deformation**. Wet sand and freshly poured concrete, both of which retain an imprint, are good examples of plastic deformation. Stretching the plastic rings that hold together a soda six-pack is an example of plastic deformation, because the material is permanently deformed. If it were not for plastic deformation, it would not be possible to crimp the lid of a can or stamp sheet metal into shapes for auto-body parts. Malleability and ductility are characteristics that permit plastic deformation. Aluminum, tin, lead, and other metals can be drawn into wires and hammered and rolled into sheets. Sodium can be forced through a slit in a die, producing ribbons of the metal.

On the other hand, if the material reverts to its original size and shape when the deforming force is removed, then the strain behavior is termed **elastic deformation**. For example, we speak of a rubber band as having a "memory," because it knows to return to its original condition after the stretching force is removed. Rubber bands are made of materials that completely and quickly recover from stresses that extend them dimensionally. These elastic, rubbery materials are called elastomers. An elastomer is a highly elastic polymer.

Trees bend in the wind, and the wings of jet aircraft flex with changing lift and drag on their surfaces. Trudging across the lawn leaves footprints in the grass that only disappear slowly, but eventually do disappear. We depend for our very lives on the elasticity of arteries that expand and recover with the pulsating rhythm of the heart as it pumps blood through the human body. Less obvious is the perfectly elastic nature that can be observed in glass, wood, and bone.

It is a matter of common experience that there is an elastic limit for these materials. Beyond that limit, windows shatter, baseball bats splinter, and bones fracture. The **elastic limit** is the maximum amount of strain a material can withstand and still return to its original shape. An evergreen tree in winter that finds itself bent over with a heavy, wet snow may never recover fully, as evidenced in springtime by a permanent curvature in the trunk. When an aneurysm develops in an arterial wall, it becomes more brittle as it becomes larger, a characteristic of elastic rubber materials, and can fail catastrophically. Of the two million deaths that occur annually in the United States, 1% occur from exploded aneurysms. The British writer Robert Louis Stevenson died at age 44 of an exploded aneurysm; more recently, Joseph Biden, the senior senator from Delaware, was fortunate to be able to have a ballooned artery surgically repaired before failure. In the early history of commercial jet aircraft, no story of exceeding the elastic limit of a material is more catastrophic than the three crashes that ended the career of the British Comet jet passenger plane. In the aftermath of these tragic events, it became clear that the pressurized fuselage had failed to withstand the stresses imposed on it.

Creep is long-term plastic deformation, a familiar effect. Metals, wood, paper products, and plastics such as polyethylene creep by sagging when subjected to stress under relatively light loads for long periods of time. Although few metals creep at room temperature, creep is a cause of failure of metal parts under severe or prolonged conditions of operation.

To mold or form a material into an object of desired size and shape, it must be plastic under fabricating conditions. Traditionally, fenders, quarter panels, hood, roof, and trunk material for an automobile have been sheet steel. However, fender parts for the Cadillac Seville are being made from flat plastic composite sheets, and the Pontiac Fierro and Buick Reatta bodies are molded entirely in plastic. The Mazda Miata, a popular sports car on the road today, has a plastic body. Both the steel sheets and the plastic sheets must be able to undergo plastic deformation during fabrication. However, once the fender is formed and is installed on the frame of the car, only elastic deformations are desired. Of course, that has rarely been achieved with steel fenders, where even a small stone, spun off the road by another car, is enough to leave a permanent dent. External plastic parts, having a much greater elastic range, can withstand minor accidents without being permanently deformed.

Modulus of Elasticity

Structural materials come in all shapes, sizes, degrees of complexity in design, and composition. To understand and anticipate the performances of these structures in response to forces people and nature impose on them, they are usually analyzed by the ways in which they are treated—bent, twisted, compressed, and so forth. A solid can respond to an applied force with elastic or plastic deformation. A graph of stress versus strain for a material under tension (being pulled) is called the material's tensile history. Figure 17.1(a) is a tensile history of a structural material that is perfectly elastic, with the deformation being a directly proportional response to the applied load. Removing the stressing force restores the original size and shape of the structural material. Elastic behavior gives a

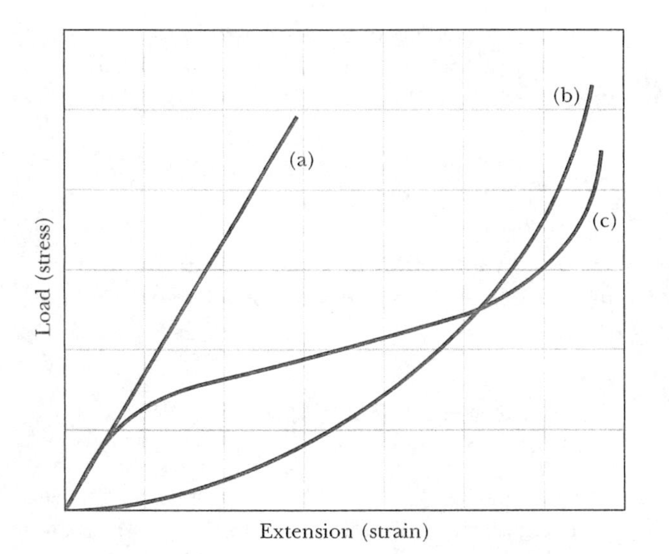

FIGURE 17.1 Modulus of elasticity. For a solid to resist an applied force, it must yield to it—that is, contract under a compressive load or stretch under a tensile load. Here are tensile histories of some elastic materials. (a) The straight line indicates the directly proportional relationship between stress (load) and strain (deformation), showing perfect elasticity. (b) A J-shaped curve is characteristic of the tensile behavior of many biological materials. (c) An S-shaped curve is characteristic of many rubber elastic materials in tension.

straight line in the tensile history and is said to obey Hooke's law: Stress and strain are directly proportional. The J-shaped curve in Figure 17.1(b) is characteristic of many biological materials—a pierced earlobe with a heavy earring, for example. Figure 17.1(c) shows an S-shaped curve describing a typical rubber–elastic material such as a balloon.

The **modulus of elasticity** (E) is the ratio of stress (τ) to strain (ε). It is a measure of elasticity or stiffness of a material and is a characteristic of the material in the elastic region:

$$\text{Modulus of elasticity } E = \frac{\text{stress}}{\text{strain}} = \frac{\tau}{\varepsilon}$$

Modulus of elasticity: Ratio of the stress or applied load, as measured by the force per unit area, to the strain, which is the elongation (in tension) divided by the original length. Because strain is units of length divided by units of length, it is a dimensionless number, and therefore the stress, typically reported as N/m^2 or psi ($lb/in.^2$), determines the units of the modulus. In other words, both stress and modulus have the same units. Note that the psi is a traditional unit of measure still widely used: 1 N/m^2 (Pa) = 10 $dyne/cm^2$ = 1.45×10^{-4} psi.

Stiffness is resistance to deformation. High values for the modulus of elasticity mean high stiffness, because the stress or load the material can absorb will be very large with respect to the strain or resulting deformation. Axles for cars, wings for airplanes, and many other applications require materials with high stiffness and only a small amount of elasticity to absorb strong shocks. Low values of the modulus indicate high elasticity. A rubber band may have a modulus of elasticity on the order of only a few hundred pounds per square inch; metals have moduli several orders-of-magnitude greater (Table 17.5). The maximum a material under stress can withstand before permanent deformation occurs at the elastic limit, the point where the curve leaves the straight line of a perfectly elastic material (Fig. 17.1c). Beyond this point, the material deforms to the point of rupture, the break point.

A plot of stress versus strain for a steel piece is shown in Figure 17.2. Note that the strain is linear up to 42,000 psi, which is the elastic limit and the end of the elastic range. Calculating the modulus of elasticity at a stress of 30,000 psi (lb/in^2) gives

$$E = \frac{30,000 \text{ lb/in.}^2}{0.001 \text{ in./in.}} = 3 \times 10^7 \text{psi}$$

The piece ruptured at 63,000 psi at an elongation of 14%, ending the test. Although not brittle, this steel piece is not especially ductile. A more ductile metal might stretch as much as 50%.

EXAMPLE 17.1

A mass of 100.00 kg is suspended by a wire that has a circular cross-section with a radius of 0.500 mm. Before the mass was attached, the length of the wire was 100.00 cm. With the mass in place the length of the wire was 100.62 cm. When

TABLE 17.5 Modulus of Elasticity: Values of Some Typical Materials

Material	Modulus* in Pa (N/m^2) $\times 10^{-8}$
	Metals
aluminum	820
copper	1100
gold	830
iron	2100
nickel	2100
silver	760
steel, carbon	2000
steel, stainless	1700
	Ceramics
glass	690
Al_2O_3	3400
$3Al_2O_3 \cdot 2SiO_2$	1400
MgO	2000
SiC	4100
	Polymers
epoxy resin (cast)	33
nylon-66 fiber	6.9
polycarbonate	23
polyethylene	2.0
polytetrafluoroethylene	2.7
	Composites
concrete	210
wood	130

*1 N/m^2 (Pa) = 10 $dyne/cm^2$ = 1.45×10^{-4} psi ($lb/in.^2$)

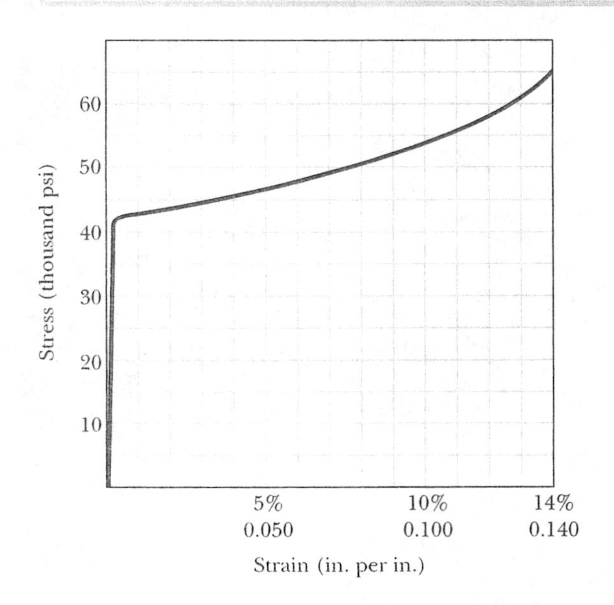

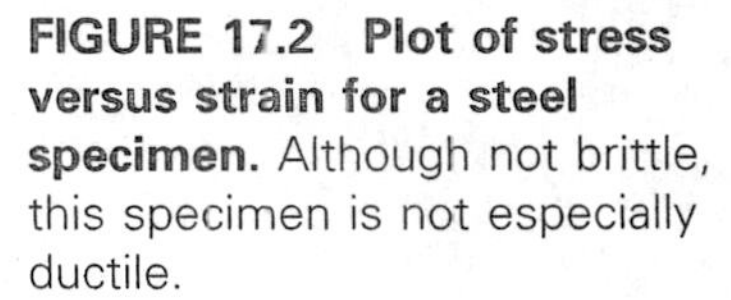

FIGURE 17.2 Plot of stress versus strain for a steel specimen. Although not brittle, this specimen is not especially ductile.

APPLICATIONS OF CHEMISTRY

Measuring the Modulus of Elasticity Actual measurements of the modulus of elasticity are made on a standardized test piece mounted in a specially designed hydraulic system capable of loading materials in tension or compression. In a tension, or tensile, test of a polymeric material, a parallel-sided strip is held by two clamps that are separated by the hydraulic system at constant speed, and the force needed to carry this out is recorded as a function of clamp separation. The test strips are usually dog-bone-shaped to promote deformation between the clamps and deter flow in the clamped section; the two-inch center of the test strip, called the gauge length, is the active-material region in the test.

The modulus of elasticity is computed from data in the linear region before the elastic limit at which the stress-to-strain ratio is constant—point b on the curve. At the top of the curve, in the vicinity of the stress maximum cd and beyond the proportional limit, the material begins to neck down, or narrow. The necked region stabilizes at a particular reduced diameter as deformation continues at more or less constant stress, until the neck has propagated across the entire gauge length de to the point of rupture e. Note how the cross-section of the necking portion of the test piece in the figure decreases with increasing extension.

The necking process is also known as "yielding," "drawing," "cold-drawing," or "cold-flow." It is an essential element in the orientation processes used to strengthen synthetic fibers such as nylon. An undrawn nylon fiber can be

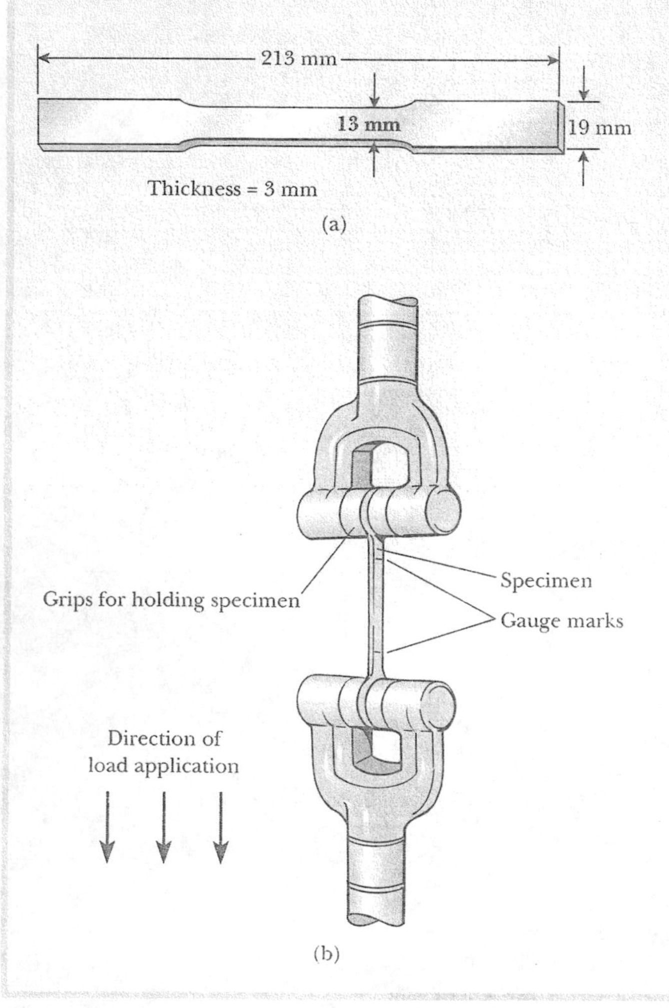

Tensile strength. Perhaps the most important single measure of strength in a material, the force necessary to pull the specimen apart, is determined, along with how much the material stretches before breaking. Both ends of a standard specimen (a) are clamped in the jaws of (b) a testing machine. The jaws then move apart at controlled rates, pulling the sample from both ends.

elongated to several times its original length. If in the process it is also twisted or warped, it takes on the familiar strength typical of fishing line. Undrawn nylon can be pulled apart or necked down with two pairs of pliers. You can perform a simpler demonstration test on a polyethylene sample by grasping two close points in a six-pack ring and pulling it steadily apart. It stretches out some 80%. To do the experiment put two pencil marks a few centimeters apart along a segment of one of the rings. Then stretch it slowly with your hands while holding tight just beyond the two marks. At first, the plastic material will resist the pull but then will give suddenly and stretch easily. Carefully note the manner in which the material seems to change its dimensions as it flows from two points, or necks. Note as well any other changes, especially the increased strength of the material in the necked region, which is now much stronger than the original material.

Questions

Perform the suggested experiment of stretching a segment of a six-pack top. Describe what you observed and explain it as well as possible.

Perform the suggested experiment with a rubber band instead. Compare your observations. Offer a brief explanation for the differences in behavior of the materials.

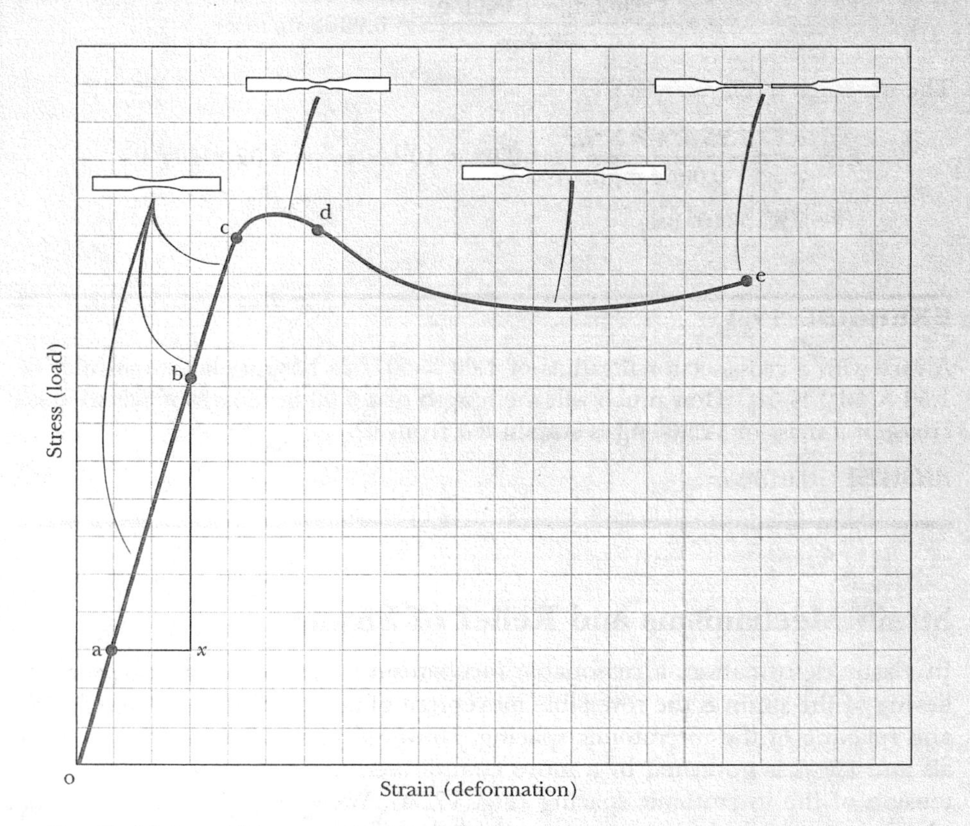

Tensile history. This test specimen has been deformed by "cold-drawing." Along the segment 0abc, Hooke's law holds: Stress is proportional to strain. This is the elastic (linear) region of the curve and the slope, bx/ax, is the modulus of elasticity. Point c is the yield point, the highest stress the material can withstand and still "recover." The segment cde, beyond the yield point, is the region of plastic (permanent) deformation; and finally, the break point is observed at e.

the mass was removed, the wire returned to its original length. Calculate the stress and strain of the wire and its modulus of elasticity.

COMMENT The stress τ is the force divided by the wire's cross-sectional area; the strain is the increase in length divided by the original length; and the modulus of elasticity is the ratio of stress to strain.

SOLUTION Begin with the force or load on the wire:

$$F = mg = (100.00 \text{ kg})(9.807 \text{ m/s}^2)$$
$$= 9.807 \times 10^2 \text{ kg·m/s}^2 = 9.807 \times 10^2 \text{ N}$$

The stress τ is this force divided by the wire's cross-sectional area (πr^2)

$$\tau = \frac{9.807 \times 10^2 \text{ N}}{\pi(5 \times 10^{-4} \text{ m})^2} = 1.25 \times 10^9 \text{ N/m}^2$$

The strain ε is

$$\varepsilon = \frac{1.0062 \text{ m} - 1.0000 \text{ m}}{1.0000 \text{ m}} = 0.0062 \text{ m/m}$$

The modulus of elasticity E is

$$E = \frac{\tau}{\varepsilon} = \frac{1.25 \times 10^9 \text{ N/m}^2}{0.0062 \text{ m/m}} = 2.02 \times 10^{11} \text{ N/m}^2 = 2.02 \times 10^{11} \text{ Pa}$$
$$(= 2.93 \times 10^8 \text{ psi})$$

EXERCISE 17.1

A wire with a cross-sectional radius of 1.00×10^{-3} m has an elastic modulus of 1.50×10^{11} N/m^2. How much will the length of a 5.00-m length of this wire increase if a mass of 125.00 kg is suspended from it?

ANSWER 0.0130 m

Strain Mechanisms and Relief of Strain

In elastic deformation, a reasonable mechanism of strain and subsequent relieving of the strain is the reversible movement of atoms through the stretching and relaxing of the interatomic spacing. However, plastic deformation in metals and alloys is governed by a more complicated mechanism than simple extension of the interatomic spacing (Fig. 17.3a). We say **slippage** has occurred when atoms move a unit, or interatomic distance, along a plane that we call the slip plane. When the stress is removed, there is no driving force to return atoms that lie below the slip plane to their original position. Slippage occurs along planes of high atomic or molecular density. These planes are shown for face-centered cubic (FCC), body-centered cubic (BCC), and hexagonal close-packed (HCP) structures in metals in Figure 17.3b. Because the FCC lattice has the greatest number of orientations of its slip planes, twelve in all, such metals are

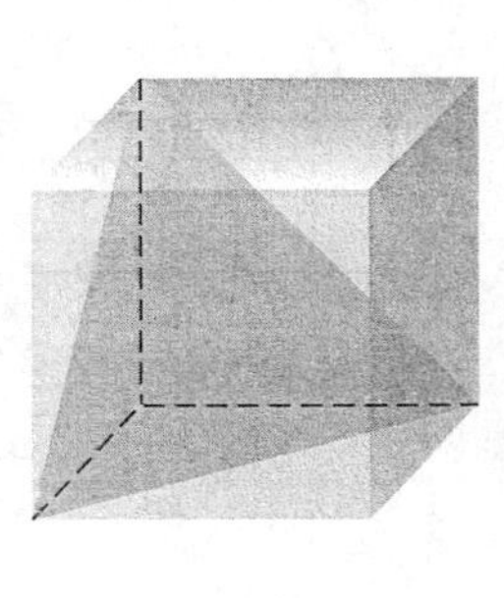

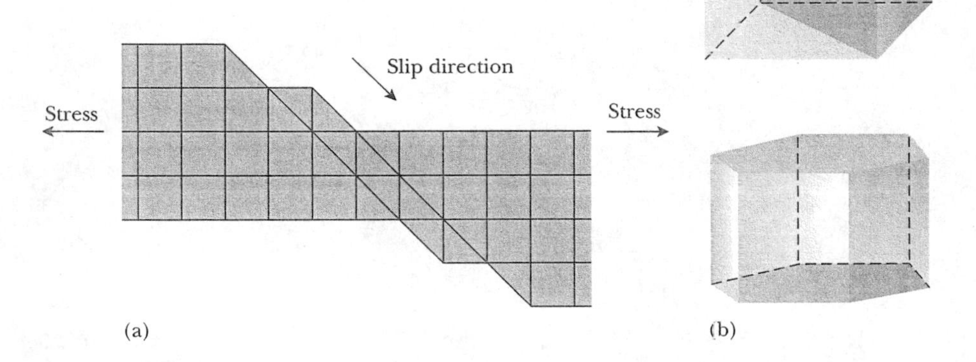

FIGURE 17.3 Plastic deformation in metals and alloys. (a) Schematic diagram of a single-crystal metal sample oriented for plastic deformation with the least applied stress. (b) Slip planes in the FCC, BCC, and HCP metals. These planes are those of greatest atomic density.

the most ductile. Copper and aluminum are examples. In contrast, titanium with an HCP lattice is brittle.

In principle, if metals crystallized perfectly from the melt without defects, all the bonding interactions between the atoms that lie across the slip plane would have to break simultaneously for slippage to occur. Because of that, metals would be hundreds of times stronger than any ever tested. In practice, the strengths of metals are limited by defects. Slippage occurs as the defects move bond-by-bond along the slip plane (Fig. 17.4). Eventually, the defect is consumed and the crystal becomes more perfect, with fewer defects and greater strength. It becomes **work hardened**, or strain hardened. This sequence ends when no more defects are available to slip in response to a stress because they have all run off the edge of the crystal or into each other. The stress required to slip any further equals the stress required to rupture the structure, and the structure in turn fails.

To understand the brittle character of ceramic materials, consider the magnesium oxide structure and the idea of slippage along a slip plane. Slippage cannot occur (Fig. 17.5) because shifting one interatomic unit along a plane would force ions with like charges to align next to each other, destroying the ionic arrangement that holds the crystal together. Imposing a large enough stress to force such a distortion results in shattering the structure. This does not happen in a metal because all the atoms are the same.

FIGURE 17.4 Plastic permanent deformation of a crystal. (a) Results from sliding one plane of atoms past an adjacent plane. In contrast to a perfect crystal, in which high stress is required to slip the planes, the presence of a defect lowers the required slip stress. (b) In the simplest sense, the defect slips like the movement of a caterpillar.

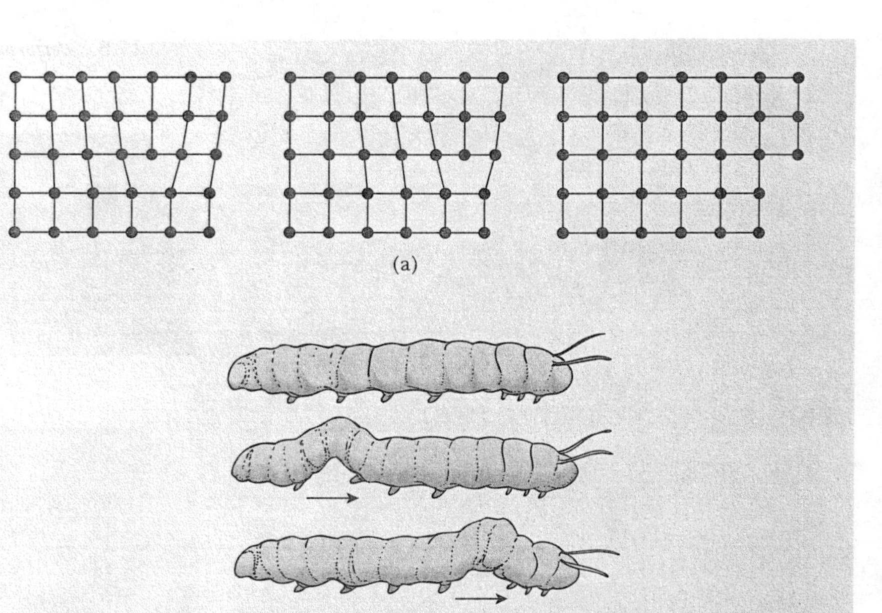

PROFILES IN CHEMISTRY

Robert Hooke, Curmudgeon of Science Robert Hooke (British, 1635–1703), the son of an English clergyman, was educated at Oxford University, where he supported himself by waiting on tables in the dining hall. While at Oxford, he attracted the attention of Robert Boyle (Chapter 4), who took him on as an assistant. Because of his mechanical aptitude and skills, Hooke was able to help make a success of Boyle's designs for the air pump used to study the pressure–volume relationship that led to Boyle's law.

Hooke's accomplishments in astronomy, geology, and mathematics were arguably among the greatest discoveries of his age. We are most interested in his accomplishments in physical science and his studies of the behavior of springs. In 1678 he discovered what we now call Hooke's law—that the force tending to restore a spring (or any elastic material, for that matter) to its equilibrium position is proportional to the distance by which it is displaced from that position. His earlier observations of the expansion and contraction of spiral springs led to what we now call the hairspring that made small and accurate timepieces possible and, by eliminating the cumbersome and awkward pendulum then used for the design of clocks, eventually led to the wristwatch.

But for all his ingenious experiments, discoveries, and inventions, he was well-known to his seventeenth-century colleagues as a cantankerous and sometimes nasty fellow who seemed to take perverse pleasure in controversy. It is said that he drove Isaac Newton to the brink of a nervous breakdown on more than one occasion. One wag of the day referred to Hooke as the curmudgeon of his age.

Questions

How could you test Hooke's law?

The mechanical behavior of both the smallest molecules and the largest can be modeled after Hooke's law. Consider the H_2 molecule and a rubber band, and briefly explain their behavior in light of Hooke's law.

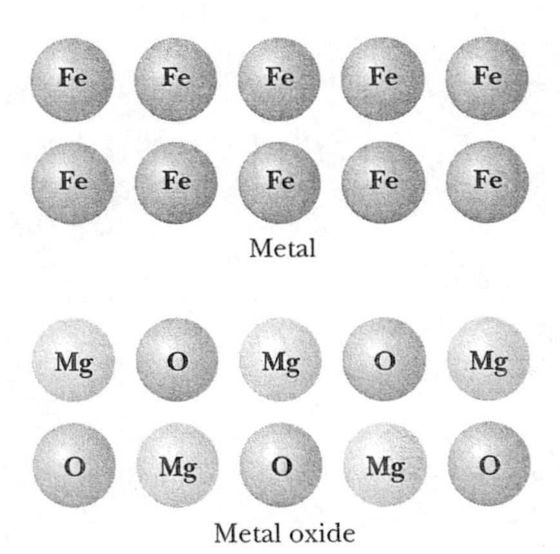

FIGURE 17.5 Metal oxides such as MgO are brittle, and metals such as Fe are not. The fact that a metal oxide is brittle and a metal is not can be understood by comparing their respective cubic crystal structures.

17.4 METALS AND ALLOYS

Metallic Structures

Recall from the previous chapter that our model of metals considered the atoms to be hard spheres that are in contact with each other. The two closest packed arrangements, face-centered cubic (FCC) and hexagonal-closest packed (HCP), occur frequently in metal structures. Remember that the CCP structure is identical with the face-centered cubic (FCC) structure. Each of the closest packed structures—FCC and HCP—contains closest packed layers of metal atoms. In the FCC structure, the layers are superimposed over each other in an *abcabc* arrangement that repeats at every third layer. In the HCP structure the arrangement is *ababab*, repeating every other layer.

The other common metal structure is the body-centered cubic (BCC) structure. This structure has metal atoms at the cube corners and in the body center. It is not a closest packed structure. This can be shown by calculating the atomic packing factor. The **atomic packing factor** is the fraction of the unit cell occupied by actual atoms. The atomic packing factor is 0.68 for the BCC structure, whereas it is 0.74 for the closest packed structures.

EXAMPLE 17.2

Show that the atomic packing factor for BCC is 0.68.

COMMENT From Chapter 16, recall Figure 16.13 and the geometric arguments that gave rise to the equation $r = \frac{a\sqrt{3}}{4}$ for the BCC structure, where r is the atomic radius and a is the length of the unit-cell edge. Also recall that the BCC unit cell contains a total of two atoms. Then calculate the volume of the atoms in terms of a and compare it with the volume of the cube.

SOLUTION Proceed as follows:

$$\text{Volume of atoms} = 2\left(\frac{4}{3}\pi r^3\right) = 2\left(\frac{4}{3\pi}\left[\frac{a\sqrt{3}}{4}\right]^3\right)$$

$$= 2\left(\frac{4}{3}\pi \cdot \frac{3\sqrt{3}}{64}a^3\right)$$

$$= 0.68a^3$$

$$\text{Volume of the cell} = a^3$$

$$\text{Atomic packing factor} = 0.68a^3/a^3 = 0.68$$

EXERCISE 17.2

Calculate the atomic packing factor for a simple cubic metal structure. Why do you think that this structure is not actually observed in metals?

ANSWER Atomic packing factor = 0.52; too little of the space in the unit cell is used.

Crystals and Grains

So far, we have considered the structure within the crystal, taking the position that the crystal contains countless millions of atoms that seem to extend infinitely from any atom under consideration. However, we also know that crystals have finite sizes and they are, in fact, usually quite small. Except in the semiconductor industry, which depends on single crystals, crystalline materials are composed of many small crystals joined together in a polycrystalline structure.

Consider, for a moment, the surface of a crystal. Atoms on the surfaces of crystals do not have exactly the same properties as atoms within the bulk of the structure because their coordination numbers are different—that is, surface atoms have empty bonding sites available. Furthermore, even though the face of a crystal may appear extremely flat, we do not really expect to find an ab-

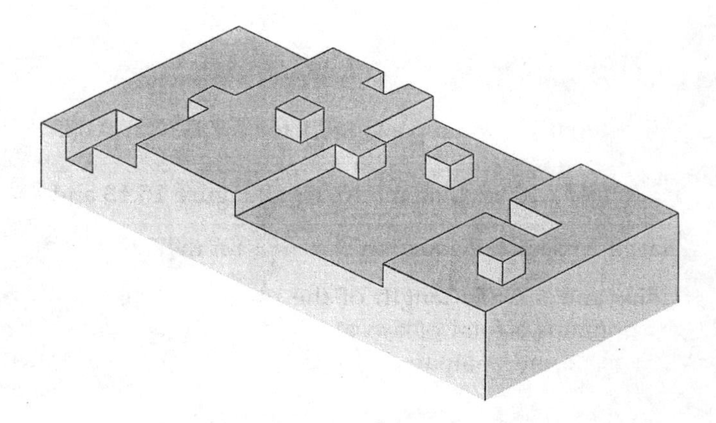

FIGURE 17.6 The Hirth–Pound model of the surface of a crystal. In reality, grains are much more irregularly shaped than the figure would suggest. Yet it remains a useful model.

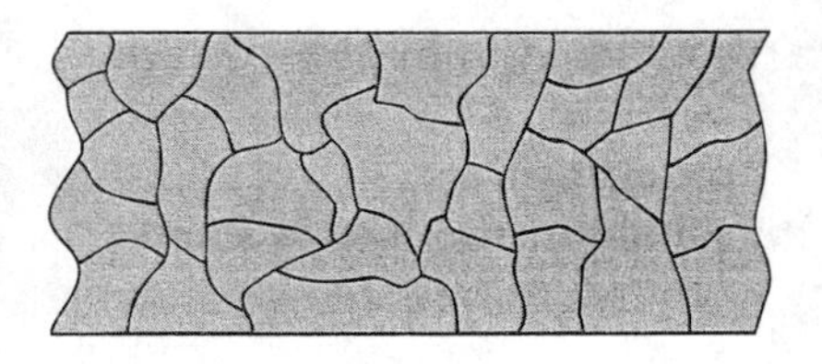

FIGURE 17.7 Model of grains in a metal. Grain boundaries typically exist between the grains.

solutely flat surface of a perfect plane of atoms. Rather, the surface has irregularities in the form of steps or terraces as layers of atoms start or stop. An attractive visualization of this is the Hirth–Pound model of a crystal surface, shown in Figure 17.6.

Before we can discuss polycrystalline materials, however, we need to answer the question, "What occurs at the boundary between one crystal and another?" There is a defect in which the regular arrangement of one crystal ends and a new regular arrangement, oriented in a different direction, begins. Polycrystalline materials are considered to be composed of grains. Each **grain** is a single crystal with a regular arrangement of atoms within it. However, because the grains pack tightly together, the crystals are not able to form characteristic flat faces because of the limited space to grow. Thus, the shapes of grains are irregular, much more irregular than the Hirth–Pound model. Grain boundaries exist between the grains as shown in the schematic diagram in Figure 17.7.

Solid Solutions

Many metals dissolve in each other when they are melted, resulting in **solid solutions** called alloys when the melt cools. In some cases of solid solutions, atoms of one metal simply fit into sites that would otherwise be occupied by the other. This is called a **substitutional solid solution**. Nickel, for example, dissolves in copper, forming a substitutional solid solution in which some copper atoms have been replaced by nickel atoms. One plane of the FCC structure of this solution is shown in Figure 17.8.

Only some combinations of metals form solid solutions in which the mole fraction of the lesser component is 10% or more. A set of three rules called the **Hume–Rothery rules** is useful in predicting which metals dissolve in each other to at least this extent:

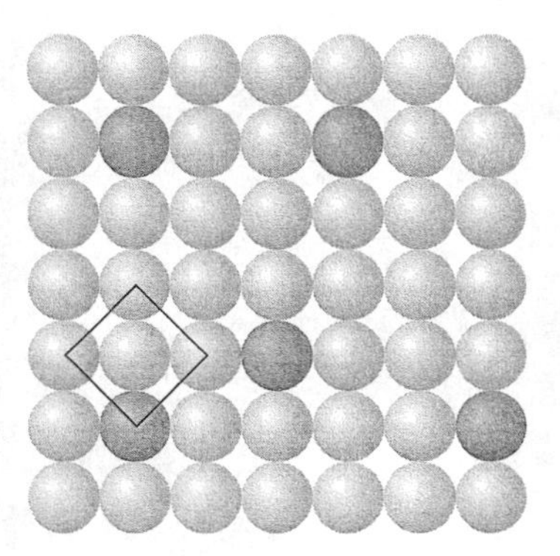

FIGURE 17.8 One layer of a substitutional solid solution in the FCC (CCP) metal structure.

Rule 1. The atomic radii of the two metals should be within 15% of each other.
Rule 2. The metals should have similar electronegativities.
Rule 3. To form a complete range of solid solutions, the metals should have the same crystal structures.

Compliance with the second rule can be checked in a table of electronegativities or by simply choosing metals close to each other on the periodic table. The crystal structures and unit-cell lengths of some important metals are given in Table 17.6. The relationship between unit-cell length and atomic radius was discussed in Chapter 16. Note that for any two metals with the same crystal structure, the ratio of their atomic radii and unit cell lengths is identical.

EXAMPLE 17.3

Which metals from Table 17.6 would you expect to form substitutional solid solution with iron to the extent of 10% or more?

TABLE 17.6 The Crystal Structures and Lattice Parameters of the Stable Forms of Some Important Metals

Metal	Structure	Unit Cell Length(s) (Å)
Ag	CCP	4.078
Al	CCP	4.041
Au	CCP	4.070
Be	HCP	2.283, 3.607*
Cd	HCP	2.973, 5.606
Co	HCP	2.514, 4.105
Cr	BCC	2.878
Cu	CCP	3.608
Fe	BCC	2.861
K	BCC	5.333
Mg	HCP	3.203, 5.196
Na	BCC	4.24
Ni	CCP	3.514
Pb	CCP	4.941
W	BCC	3.158
Zn	HCP	2.658, 4.934

*The hexagonal closest packed structure has two different unit-cell lengths, because the cell is not a cube. If they are within 15% of each other, the Hume–Rothery requirement is met.

COMMENT Table 17.6 shows that Fe has a BCC structure in common with Cr, K, Na, and W. Once again, going back to Chapter 16, the atomic radius in the BCC case is

$$r = \frac{a\sqrt{3}}{4}$$

and thus the atomic radii of these metals are all in the same proportion as their unit cell lengths, *a*.

SOLUTION The percentage differences between the unit cell dimensions of Fe and the BCC metals are

$$\text{Cr:}\quad 100\% \times \frac{|2.861 - 2.878|}{2.861} = 0.59\%$$

$$\text{K:}\quad 100\% \times \frac{|2.861 - 5.333|}{2.861} = 86.4\%$$

$$\text{Na:}\quad 100\% \times \frac{|2.861 - 4.24|}{2.861} = 48.2\%$$

$$\text{W:}\quad 100\% \times \frac{|2.861 - 3.158|}{2.861} = 10.4\%$$

W and especially Cr fall within the atomic radius requirement. The Pauling electronegativities of Fe, Cr, and W are 1.8, 1.6, and 1.7, respectively, all of which are relatively close. Therefore, we would expect both Cr and W to form solid solutions with iron quite freely, which, in fact, they do.

EXERCISE 17.3

Which metals of Table 17.6 would you expect to be capable of forming substitutional solid solutions with Cu to the extent of 10% or more?

ANSWER Ag, Al, Au, and Ni all satisfy the atomic radius requirement. Of these, all but Al satisfy the electronegativity requirement.

Small atoms such as hydrogen, boron, carbon, and nitrogen are able to form interstitial solid solutions with metals. **Interstitial solid solutions** are solutions in which atoms are inserted into holes or interstices of the metal structure. Such solutions are rare because of size constraints, but they can have dramatic consequences when present. The various carbon steels are examples of interstitial solid solutions in which carbon atoms are inserted into tetrahedral holes of the FCC form of iron, as shown in Figure 17.9.

Properties of Simple Metal Structures

So far our discussion has been limited to solid solutions of pure metals and their alloys. The solid solution is composed of crystals of the same kind and often is referred to as a **single-phase alloy**. Recall that a phase is a homogeneous and physically distinct region of matter, and so a single-phase alloy is a homogeneous

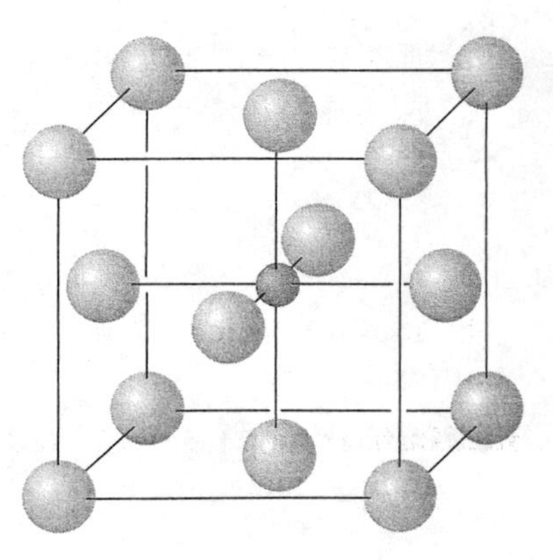

FIGURE 17.9 Interstitial solid solutions. An interstitial solution of carbon, the small solid central sphere, in the FCC (CCP) form of iron, the larger spheres. Equivalent locations for the carbon atoms are at the centers of the unit-cell edges. The carbon atoms are too large to fit into interstitial sites without causing some strain in the structure and so the carbon content is limited to about 2%, distributed at random.

mixture. In this section we will consider a few properties of pure metals and single-phase alloys, all homogeneous materials, which we will group as **simple metals.** Substitution of atoms of one pure metal for those of another metal normally increases the strength and hardness of a metal while decreasing its ductility. There are only a few exceptions. The general reason for these effects is that when atoms in a structure have been replaced by atoms of a somewhat different size, the planes of atoms in the structure are no longer as flat and cannot slide across each other so easily. Interstitial atoms tend to have the same effect, locking the atomic network into a more rigid structure.

When stress is applied to a metal it generally deforms. However, the elasticity of a simple metal crystal depends on the direction in which the force is applied. This is a consequence of the fact that crystals are anisotropic. Anisotropic means having different properties in different directions. For example, a single crystal of iron (BCC) is more than twice as stiff—half as elastic—along the body-diagonal direction as along the direction parallel to the unit-cell edge. In Figure 17.10 observe that in the BCC structure of iron the density of atoms is less in the direction paralleling the unit-cell edge, allowing for more movement of atoms and accounting for the lesser stiffness. However, an ordinary polycrystalline sample of iron displays an intermediate degree of stiffness because the grain structure is characterized by grains oriented in random directions, giving the sample an averaged stiffness that is less than the maximum.

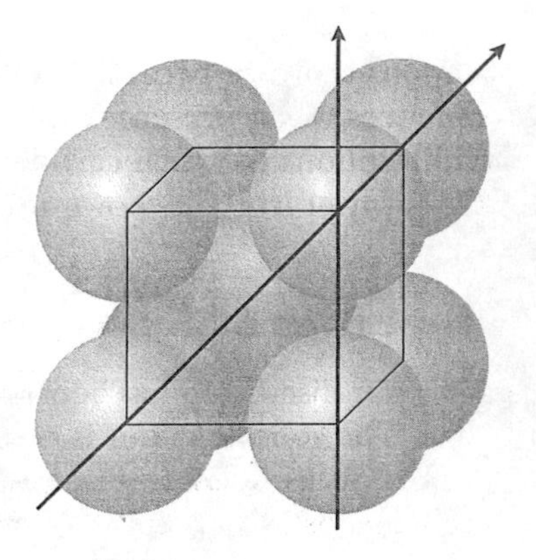

FIGURE 17.10 Body-centered cubic structure. Shows the effect of strain along the edge and the diagonal of the cell.

Polyphase Structures and Phase Diagrams

In contrast to the subject of the previous discussion in which all of the grains were of the same phase, many important alloys contain grains of different phases. The existence of these different crystals in the material has enormous effects on the properties of an alloy. For example, an alloy of copper and tin can contain two different kinds of crystals. The principal crystals that compose a copper-rich solid solution of Cu and Sn are rounded. However, needle-shaped tin-rich crystals are also present. These long crystals act like reinforcing rods in concrete and give the alloy increased strength.

Carbon is involved in the metallurgy of iron and steel. The number of phases in cast irons and carbon steels is large, including the FCC and BCC forms of the pure metal, the compound Fe_3C, graphite, and a variety of solid solutions of these components. The hard and brittle material called cast iron has a high carbon content and includes large, coarse crystals of Fe_3C that determine the material's properties. Low carbon steels contain the two phases, BCC iron and a solid solution of BCC iron and Fe_3C.

A **polyphase system** contains a number of different phases such as the different phases of steel and cast iron. The phase diagram depicted by Figure 17.11 for a two-component system composed of copper and silver aids in understanding a polyphase system. A **phase diagram** is a graph on which it is possible to locate the state of the system in terms of the temperature along the y-axis and the mass percentage of each component along the x-axis. The nature of the phases that exist for a particular composition and temperature can be determined by locating the appropriate point on the diagram. Thus, different areas of the diagram are marked out for different phases. Equilibrium between two phases exists along the line on the diagram between the two phases. A particu-

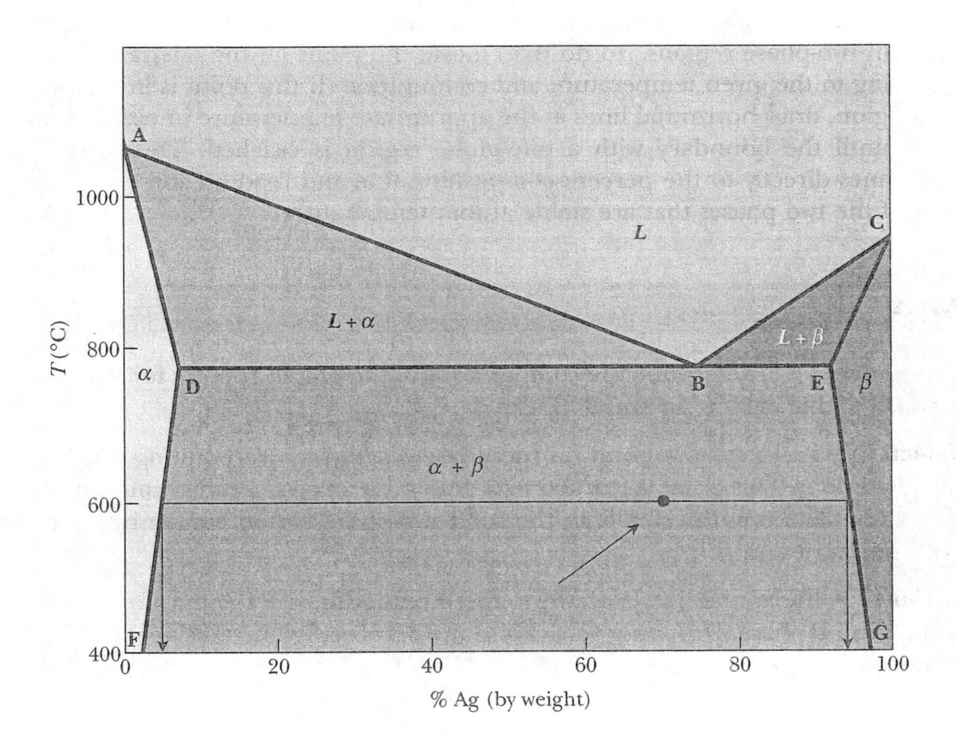

FIGURE 17.11 The phase diagram for the copper–silver system. Dark lines mark the boundaries of regions occupied by particular phases. The dot at 70% Ag and 600°C is in the region containing both the α and the β phases. The composition of each of these phases for this composition and temperature is determined by extending a horizontal line to the region boundary on either side and then dropping directly to the percent Ag axis.

larly important equilibrium exists between liquid and solid phases, and along this line you will find the melting point as a function of composition. The minimum of the melting point occurs at a point known as the eutectic. The **eutectic** is the composition that has the lowest melting point.

In the phase diagram for copper and silver in Figure 17.11, L is a liquid phase, α is a copper-rich solid solution, and β is a silver-rich solution. Above the line ABC, the sample exists entirely as a liquid, L. The line ABC represents the melting point as a function of composition. The melting point of pure copper occurs at point A and the melting point of pure silver occurs at point C. As described in an earlier discussion of colligative properties (Chapter 9), the melting points of the pure substances are lowered by the addition of the dissolved solute. Point B is the eutectic, corresponding to the composition of minimum melting point, which occurs at about 75% silver.

Below the line ABC, except at the three points A, B, and C, a mixture of two phases exists. Each region contains a liquid solution and a solid solution in a heterogeneous mixture. Thus, if a liquid with any composition except pure copper, pure silver, or the eutectic mixture is cooled, it will start to solidify when it crosses the ABC line but will not be fully solid until the ADBEC line is crossed. Pure solid solutions exist at high-copper or high-silver compositions. These are denoted by the α and β regions, where α is rich in copper and β is rich in silver. For most compositions, however, the solid consists of a two-phase mixture of α and β. This is the region below the FDEG line. Within this region, the alloy contains two kinds of grains, which strongly affect the properties of the alloy.

Phase diagrams provide useful information about what phases are present for any composition and temperature. From Figure 17.11 we can tell that at 70% Ag and 600°C two phases are present, copper-rich solid solution α and silver-rich solid solution β. The compositions of the two phases are dependent on temperature, and the phase diagram allows us to find the compositions of the phases present in two-phase regions. To do that, locate the point on the diagram corresponding to the given temperature and composition. If this point is in a two-phase region, draw horizontal lines at the appropriate temperature in either direction until the boundary with a one-phase region is reached. Then drop vertical lines directly to the percent-composition line and read off the compositions of the two phases that are stable at that temperature.

EXAMPLE 17.4

Using Figure 17.11, determine the compositions of the phases present for a sample at 600°C that is 70% Ag and 30% Cu.

COMMENT First locate the point on the phase diagram corresponding to the given conditions. This point is marked by a dot on the diagram and pointed out by an arrow. The point is clearly in the solid two-phase region containing the solid solutions α and β.

SOLUTION A horizontal line is drawn to the boundaries with the pure α phase and the pure β phase. Dropping verticals to the percent Ag line tells us that the composition of the α phase is about 5% Ag and 95% Cu, and the composition of the β phase is about 95% Ag and 5% Cu.

EXERCISE 17.4

Determine the compositions of the phases present for a composition of silver and copper at 800°C that is 40% Ag and 60% Cu, using Figure 17.11.

ANSWER A copper-rich solid solution containing about 8% Ag and a liquid solution containing about 71% Ag

Rapid cooling of certain compositions of alloys can result in nonequilibrium conditions that persist indefinitely. On cooling a silver–copper alloy with 5% Ag at 800°C, it should convert from a single-phase solid solution α to a two-phase mixture of the solid solutions α and β. However, rearrangement of atoms in a solid is slow, especially at lower temperatures. Thus, if the alloy is cooled fairly rapidly, it is possible to a have pure solid solution α at room temperature containing 5% Ag, and it can persist for a long time. Obtaining nonequilibrium alloys is an important aspect of metallurgy.

17.5 GLASS AND CERAMIC MATERIALS

Ceramics are among our oldest materials and they are also among the most durable. Archaeologists have found pottery specimens that are thousands of years old and still in excellent condition. Glassmaking probably dates back to 2500 B.C., and the use of crystalline ceramics is older still, with commercial pottery dating back to about 4000 B.C. Some simple earthenware vessels date from 15,000 B.C. Concrete and cement are ceramics as are bricks and tile.

Glass and ceramic materials have always been important, but perhaps never more than today. Several centuries ago, the introduction of the glass windowpane revolutionized the way we protect ourselves from often unfriendly environments. And who can doubt the historic contributions of eyeglasses. Today, glass has revolutionized the way we communicate through the introduction of optical fibers that carry orders-of-magnitude more information than any conventional copper cable. Now, scientists are looking forward to being able to transmit and store electrical energy without loss through technology from new superconducting metal-oxide ceramics.

Glass and ceramics are generally inorganic and nonmetallic materials with exceptional properties obtained through high-temperature processing. Glasses are amorphous materials; ceramics are crystalline materials. Modern ceramic materials are indispensable to an industrialized society, particularly refractory materials that tolerate the higher processing temperatures often required in basic manufacturing industries. Beyond that, ceramics encompass the range of materials from the single ruby crystal in a laser to dense polycrystalline materials, pigments in paints, and porcelain enamel finishes. Compared to metals as a class of materials, ceramics are better thermal and electrical insulators and are chemically more stable. Furthermore, ceramic materials are appreciably more stress-resistant in compression and display greater stiffness and thermal stability than ordinary carbon-based polymeric materials. However, in contrast to metals, ceramics tend to be brittle.

Crystalline Ceramics

Ceramics are held together by ionic and covalent bonds and by relatively weaker van der Waals forces that are the first to break, if they are present, when a ceramic is fractured. The corundum crystal structure for Al_2O_3 consists of a hexagonal close-packed arrangement of O^{2-} ions, with the Al^{3+} ions occupying octahedral holes. Magnesium oxide has the sodium chloride structure. The bonding in diamond is purely covalent. Quartz, a crystalline form of silica (SiO_2), tungsten carbide (WC), and many other ceramics have bonding that is intermediate between purely ionic and purely covalent. Ceramics from fired clays contain ionic and covalent bonds and van der Waals forces.

The most common ceramics are the silicates, which include fired clays and Portland cement. A tetrahedron of four oxygen atoms surrounding a central silicon atom is the basic building block of the structure. A single such unit has the formula SiO_4^{4-} and is called orthosilicate. Its structure and those of several other silicate ions are given in Figure 17.12 and Figure 17.13. The Lewis structure of SiO_4^{4-} and two other representations of this ion are given in Figure 17.12(a). Note that the Lewis structure of SiO_4^{4-} requires 32 electrons, whereas only 28 electrons are available in the valence orbitals of the elements. Hence, the ion must have the charge of -4. In general, for each terminal oxygen atom—an oxygen atom not shared by another silicon—a charge of -1 is necessary. The reason for this is that the oxygen needs to be surrounded by eight electrons for a stable configuration. Six of these electrons come from the oxygen, one comes from the silicon it is bonded to, and one is extra and accounts for the -1 charge.

For the other silicate ions in Figures 17.12 and 17.13, the SiO_4 structural unit is simply shown as a tetrahedron. An oxygen atom sits at each of the four

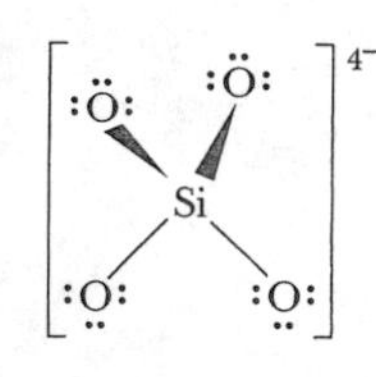

=

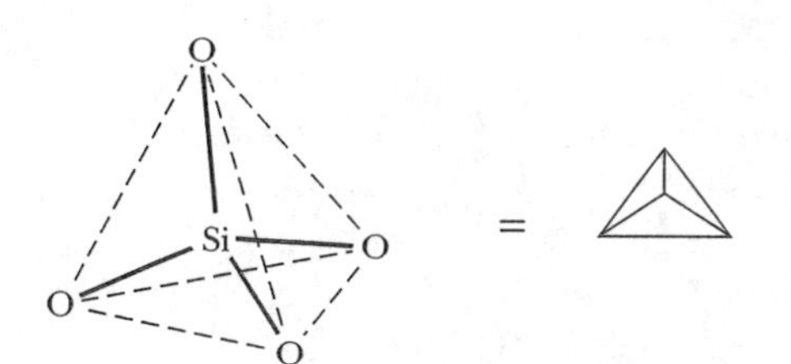

(a)

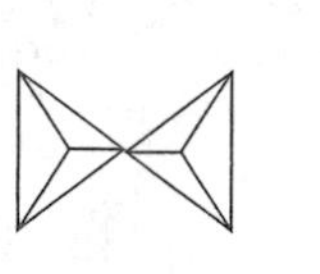

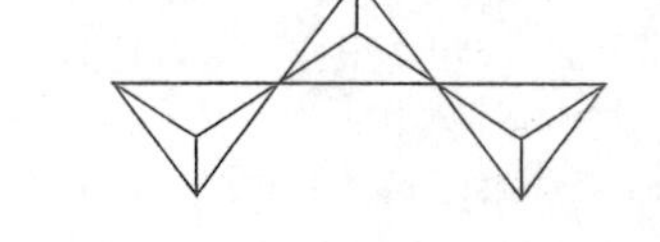

(b)

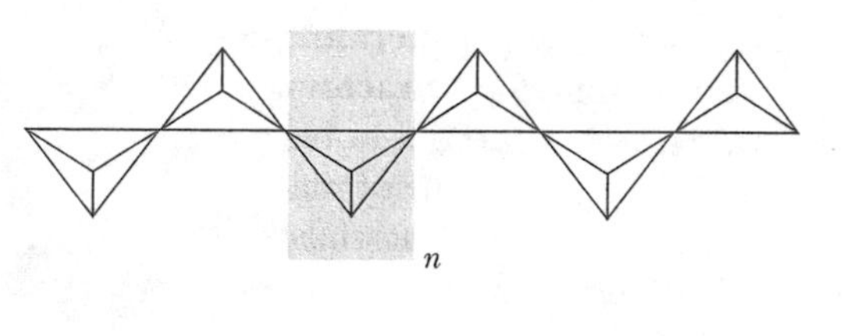

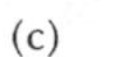

(c)

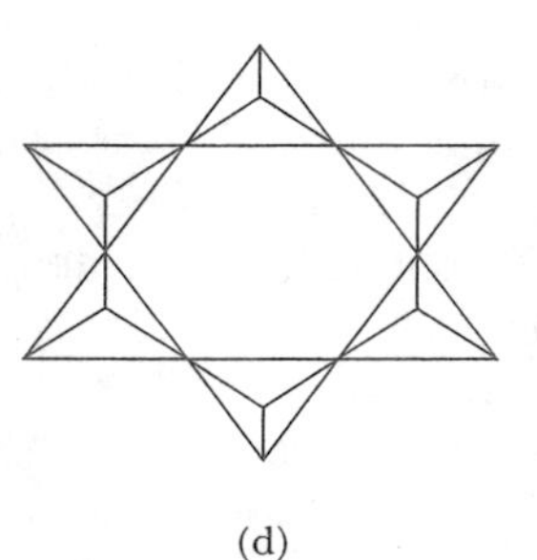

(d)

FIGURE 17.12 Silicate ions. (a) Three equivalent representations of SiO_4^{4-}, the orthosilicate ion; (b) disilicate, $Si_2O_7^{6-}$, and trisilicate, $Si_3O_{10}^{8-}$; (c) an "infinite" single-chain silicate, $(SiO_3^{2-})_n$; (d) a ring silicate $Si_6O_{18}^{12-}$.

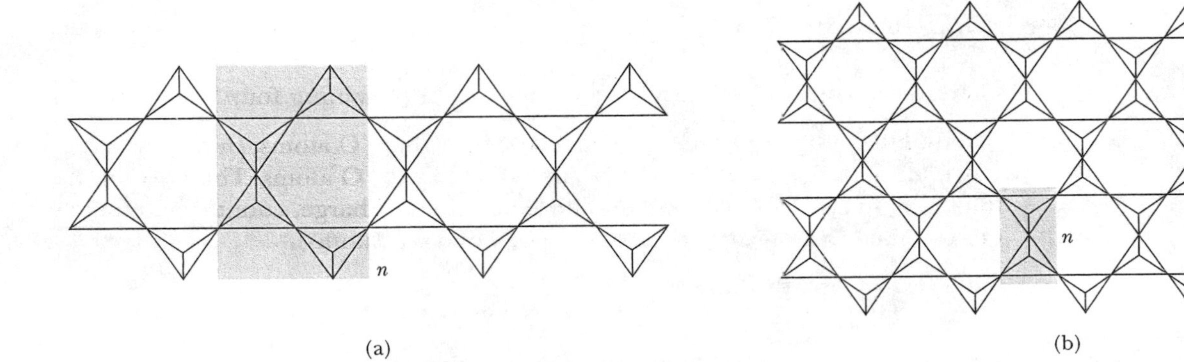

FIGURE 17.13 Complex silicate ions. (a) An "infinite" double-chain silicate, $(Si_4O_{11}^{6-})_n$; (b) an "infinite" sheet-silicate ion $(Si_2O_5^{2-})_n$.

corners of the tetrahedron with a silicon atom at the center. Connecting tetrahedra by sharing corners leads to a large number of structures. As we go through these structures, count up the numbers of silicon atoms, oxygen atoms, and negative charges. The disilicate ion has two silicon atoms, one in the center of each tetrahedron. There are six terminal oxygen atoms and one shared between the two tetrahedra, for a total of seven. The six terminal oxygen atoms establish a charge of −6. The disilicate ion's formula is thus $Si_2O_7^{6-}$. You should be able to draw the Lewis structure of this ion showing a correct charge of −6. Similar reasoning leads to an $Si_3O_{10}^{8-}$ formula for trisilicate ion with three silicon atoms.

Longer and longer single-chain silicate ions can be formed. To determine the formula of a polymeric silicate, it is necessary to locate the building-block unit that, when repeated over and over, generates the whole structure. In Figure 17.12(c), a building block has been marked off, and n is a large number that gives the total molar mass of the ion. Within the building block there is a single silicon atom plus two terminal oxygen atoms and two oxygen atoms shared with neighboring building blocks. Any atom shared between two neighboring blocks can be counted only as one-half in each block. Thus, there are three oxygen atoms in each block and the formula is $(SiO_3^{2-})_n$.

Figure 17.12(d) shows a ring-silicate ion with the formula $Si_6O_{18}^{12-}$. Note that the proportional numbers of atoms and charges are the same as for the infinite-chain silicate in Figure 17.12(c). A double-chain silicate is shown in Figure 17.13(a). Counting the atoms in the building block that has been marked gives the formula $(Si_4O_{11}^{6-})_n$. This structure occurs in some important minerals called amphiboles. In these minerals, the very long ions are separated by cations such as K^+ and some water molecules. The bonding that holds parallel silicate ions to each other is much weaker than the bonds within the ion itself. Thus, the mineral cleaves between the ions, giving mineral fibers such as are found in asbestos.

An infinite sheet-silicate ion with the formula $(Si_2O_5^{2-})_n$ is shown in Figure 17.13(b). Such ions occur in talcs. In these minerals the sheets are separated by cations and water molecules and cleavage takes place between the sheets, yielding flat platelets.

EXAMPLE 17.5

Give the formula for tetrasilicate, a single chain containing four SiO_4 groups.

COMMENT Link four SiO_4 groups together by shared O atoms. Draw the structure. Be sure that each Si atom is surrounded by four O atoms. Then count up the total number of O atoms, and, to determine the charge, count the number of terminal O atoms (those not bridging between Si atoms).

SOLUTION The formula is $Si_4O_{13}^{10-}$.

EXERCISE 17.5

Give the formulas for the following silicate ions.
(a) A ring silicate containing eight tetrahedra
(b) An infinite double-chain silicate based on rings of four tetrahedra each.

ANSWERS $Si_8O_{24}^{16-}$, $(Si_2O_5^{2-})_n$

So far, our silicate ions have included isolated ions and polymers that extend in one dimension (chains) and two dimensions (sheets). Is it possible to form a silicate ion that extends in three dimensions? Various ways of attaching the SiO_4 tetrahedra to form a three-dimensional network can be imagined. However, extending the network in three dimensions requires that all four of the oxygen atoms of each tetrahedron be shared with another tetrahedron. Recall that to form a sheet extended in two dimensions, three oxygen atoms per tetrahedron had to be shared (Fig. 17.13b). If all four oxygen atoms are shared, there is no excess charge and the silicate is no longer an ion but rather a neutral species. Furthermore, because each silicon atom will be surrounded by four oxygen atoms, each of which is shared, the formula must be SiO_2, the same as silica in quartz or sand.

However, it is possible to construct an ionic, three-dimensional silicate structure by replacing some of the silicon atoms with aluminum atoms, a reasonable substitution because aluminum is similar to silicon in size and electronegativity. However, the aluminum atom has only three valence electrons compared to four for silicon, so for every silicon atom that is replaced by an aluminum atom, an extra electron and a charge of -1 is also gained. The general formula of these aluminum-substituted silicas—aluminosilicates—is $Si_mAl_nO_{2(m+n)}^{n-}$. Aluminosilicates with structures that extend in all three directions include feldspars and zeolites.

Aluminosilicates also form layered structures. A structure can be formed of two sheets that share the formerly unshared oxygen atom at the apex of each tetrahedron, as shown in Figure 17.13(b). Without replacing silicon atoms by aluminum, this would be a neutral species with the formula SiO_2. Replacing some silicon atoms by aluminum atoms leads to an ionic-layer structure. Such layer structures appear in micas and in clays, the all-important precursors of ceramics. When micas are subjected to rapid heating, the water between the sheets is vaporized quickly, puffing it up and giving the familiar packing material called vermiculite. Because of their charged-layer structures, clays can absorb large amounts of water. This and the sheet-like cleavage are what give clays their greasy feel. The wet mass is easily shaped to the desired configuration and then it will dry to a rigid structure. The product, at this point called greenware, is weak and brittle and has no practical uses. Slow heating to a high temperature or "firing" is necessary to produce the desired ceramic properties. During the firing process the ceramic shrinks and is strengthened as a number of chemical changes take place:

- Water is driven off. This includes water that was merely adsorbed as well as water that is the product of chemical reactions.
- Calcium and magnesium carbonates decompose. For example,

$$CaCO_3(s) \longrightarrow CaO(s) + CO_2(g)$$

- The metal oxides produced react with silica to produce silicates. For example,

$$mCaO + nSiO_2 \longrightarrow Ca_mSi_nO_{m+2n}$$

- Matter in the clay from plants is decomposed and oxidized.
- Quartz is transformed to a different structure called crystabolite.
- A material called mullite ($3Al_2O_3 \cdot 2SiO_2$) is formed with long needle-like crystals that reinforce the structure.

Portland cement is a ceramic of exceptional importance and utility. It hardens slowly on reaction with water, forming a hard, durable solid. The fact that it hardens even when submerged in water makes it useful for bridge abutments and similar applications where such conditions are encountered. Properly reinforced, Portland cement can be used as a substitute for steel beams and even for the hulls of boats.

Portland cement is manufactured by firing a mixture of clay and ground limestone ($CaCO_3$) in a cement kiln. The limestone is converted to calcium oxide, and then a number of reactions finally lead mainly to $2CaO \cdot SiO_2$ (dicalcium silicate) and $3CaO \cdot SiO_2$ (tricalcium silicate). This mixture is then cooled and mixed with gypsum ($CaSO_4 \cdot 2H_2O$), which regulates the hardening rate, and the mixture is ground to a fine powder. A number of slow reactions leading to a complex and extensive network of silicon and oxygen atoms take place when the powder is mixed with water. These continue as the cement slowly hardens over more than a year.

Other important ceramic materials include alumina (aluminum oxide, Al_2O_3) and magnesia (magnesium oxide, MgO). Alumina melts at more than 2000°C, and so it is an important refractory. In practice, grains of alumina are pressed together or sintered at high temperatures with a small amount of a flux, a lower melting compound such as silica, SiO_2. The flux melts and also lowers the melting point of the alumina at the surface of the grains so that they stick together, soften further, and can be formed into the desired shape. Alumina tubes are used to contain the sodium in sodium vapor lamps because the material is resistant to sodium. It is translucent when formed with a microstructure that is almost pore free. Magnesia has an unusual combination of properties, being an electrical insulator and yet a good conductor of heat. It is used to insulate the heating elements of electric stoves.

Very hard, high-melting ceramic materials can be formed from the carbon compounds of tungsten, silicon, and boron—tungsten carbide (WC), silicon carbide (SiC) or Carborundum, and boron carbide (B_4C). Tungsten carbide, formed by the direct combination of the elements at high temperature (1500°C), is said to be one of the strongest, if not the strongest, of all structural materials:

$$W(s) + C(s) \xrightarrow{1500°C} WC(s)$$

Silicon carbide also competes for strongest materials honors and is used as an abrasive for cutting and grinding and in fibrous form for manufacturing high-strength, heat-resistant composites. To prepare silicon carbide, graphite and silica are brought together in an electric furnace at 2000°C:

$$3C(s) + SiO_2(\ell) \xrightarrow{2000°C} SiC(s) + 2CO_2(g)$$

Hard black crystals of boron carbide are produced when boron oxide is heated with graphite in an electric furnace:

$$2B_2O_3(s) + 7C(s) \longrightarrow B_4C(s) + 6CO(g)$$

Among its special applications are control rods for nuclear reactors (Chapter 22) and reinforcing agents in composites for military aircraft.

Amorphous Ceramics

Glasses are amorphous ceramics. In contrast to metals and crystalline ceramics, their most characteristic structural feature is the absence of long-range order in their structures. This difference is clearly evident on heating. Crystalline substances have definite melting points. Below their melting points they are solids with regular crystalline structures having the same bonding throughout. The melting point is the temperature at which the kinetic energy is sufficient to break enough bonds to collapse the structure. In principle, this collapse proceeds without change in temperature until all the solid is liquid. Beyond the melting point the resulting free-flowing fluids display the typical random or disordered structure of the liquid state.

The behavior of glasses on heating or cooling is more complex. They are characterized by orderliness only over very short-range distances. There is no discrete melting point, because the bonds holding the structure together are not uniform. However, as the temperature of a glass is raised, it is possible to identify a number of points for which a significant change in the properties can be detected.

Silica is the basic ingredient for most glasses. A glass can be formed by simply melting quartz, which is the familiar beach sand form of SiO_2, and then cooling it sufficiently rapidly so that it does not have time to crystallize. Such glass is chemically resistant and generally useful at elevated temperatures because of its high softening range. However, it is expensive to fabricate because of the high temperatures needed to produce it, and it is difficult to shape. Adding the oxides of sodium and calcium, Na_2O and CaO, lowers the melting range and gives the familiar soda–lime–silicate glasses found in window panes and bottles. Other oxides can be added to impart special properties. Borosilicate glasses contain B_2O_3, which improves the thermal shock resistance of the glass: These products carry famous trade names such as Pyrex. Lead oxide makes the glass softer and heavier and increases the index of refraction, giving the glass more sparkle. Such glass is used for fine decanters and tableware.

Ordinary soda–lime–silicate glass typically consists of a uniformly dispersed mixture of 75% sand (silica, SiO_2), 20% soda ash (anhydrous sodium carbonate, Na_2CO_3), and 5% lime (calcium oxide, CaO, Fig. 17.14). The compounds added to the SiO_2 disrupt the orderly structure of quartz and lower the melting point to a reasonable working temperature. For such a glass at a low temperature, no permanent or plastic deformation is possible without fracture. That is, bending of the glass without breakage is limited to elastic strain from which it can spring back to its original shape. This behavior is observed at temperatures up to the strain point. Above the **strain point** small amounts of permanent deformation are possible. When the temperature is increased further, the anneal point is reached. At the **anneal point** any stresses that exist in the glass are readily removed by movement of the atoms in the structure to new locations. Newly fabricated glass objects are normally held at the anneal point long enough to remove all stresses in the glass. Otherwise the glass object would be excessively susceptible to mechanical or thermal shock. If the glass is heated still more, the softening point is reached. At the **softening point** the glass will deform under its own weight. Finally, if the glass is heated still further, the working point is reached. Glass at the **working point** can be readily drawn, blown, or pressed into the desired shape. The fact that glass can be heated to the point of easy working makes it especially valuable for fabricating items of all conceivable shapes.

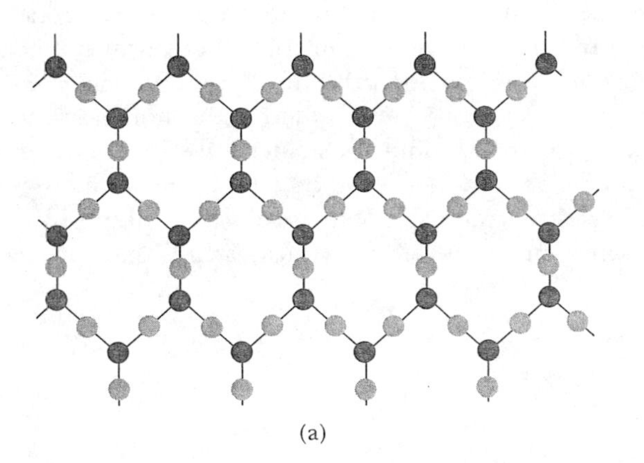

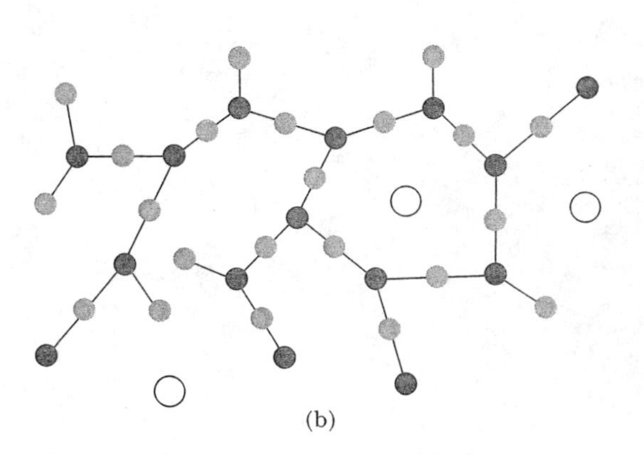

FIGURE 17.14 Two-dimensional representations of the silica structure. (a) The crystalline, highly ordered hexagonal arrangement. (b) An amorphous arrangement, made that way because of the presence of foreign substances. Silicon atoms are shown in black, oxygen atoms in gray, and other substances as larger empty circles.

Characteristic Temperatures for Soda–Lime–Silicate Windowpane Glass	
Strain point	550°C
Anneal point	590°C
Softening point	750°C
Working point	1050°C

Silicate glasses are unquestionably the most important commercially because of their excellent transparency, extraordinary chemical resistance, and the fact that they are made from inexpensive ingredients. Where silicate glasses fail as materials is in brittle fracture. Flaws easily propagate to result in material failure when thermally, mechanically, or chemically stressed. The reason for this is their continuous and rigid three-dimensional structure composed of SiO_4 tetrahedra with oxygen atoms occupying all four corners (Fig. 17.15). Although short-range order is clearly defined by every silicon atom connected to four oxygen atoms and every oxygen connected to two silicon atoms, there is no evidence of long-range order. When a fracture begins, it can propagate throughout the material (Fig. 17.16) in spite of the strength of the silicon-to-oxygen bond. The sharp crack at the point of failure has virtually no plastic flow, resulting in enormous pressures—or amplification of stress.

Fiber optics is an important modern mode of communications made possible in part by glass technology. In **fiber optics** a beam of light is transmitted along a thin glass fiber with very little loss through the fiber walls. Special glass is used for the fiber, and the walls of the fiber are coated with another glass, which keeps reflecting the light back into the fiber instead of allowing it to escape. Bunches of these fine fibers are now being used by telephone companies to transmit signals and messages without the use of electrical wires. Fiber optics has an advantage in the enormously increased number of messages that can be propagated simultaneously through one carrier.

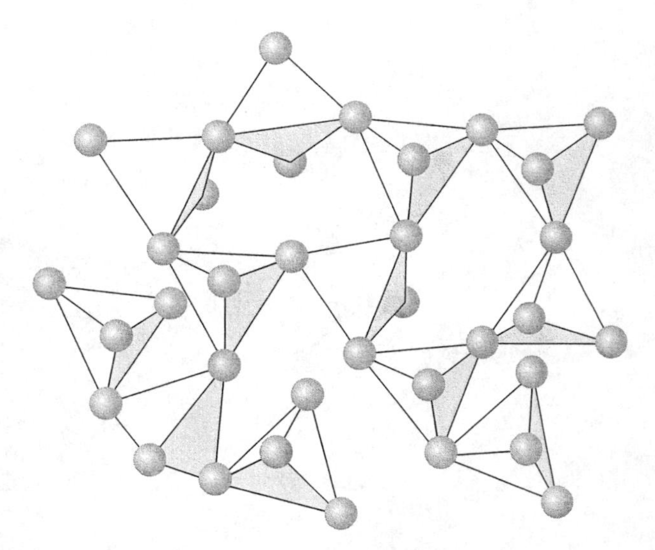

FIGURE 17.15 Silicate glass structures. SiO_4 tetrahedra are shown with oxygen atoms sharing the corners.

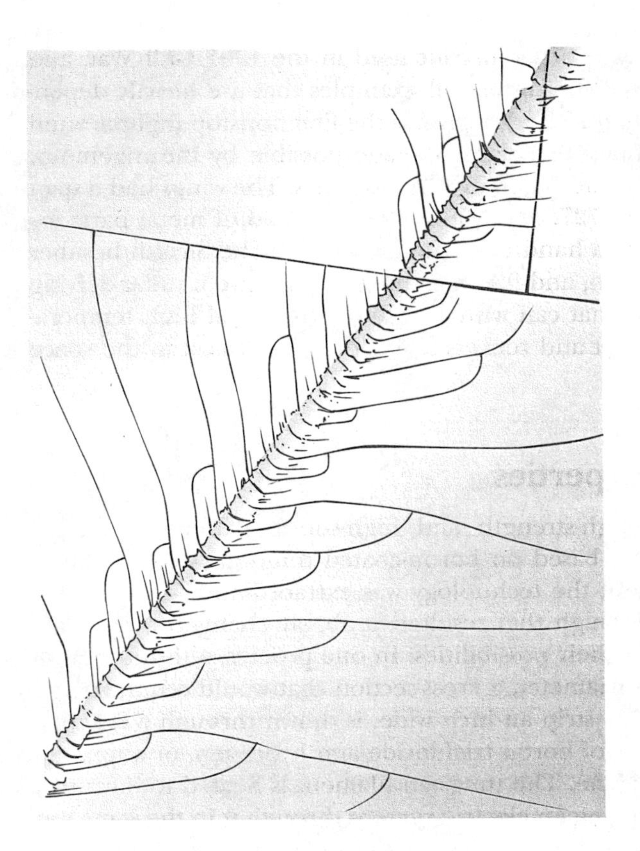

FIGURE 17.16 Crack propagation in silicate glass leading to failure. Result from a deliberate scrape by a needle point on the surface of a glass microscope slide. Direction of the scratch s upper right to lower left. Magnification is about 700 times.

17.6 COMPOSITE POLYMERIC MATERIALS

As we noted in our introductory comments earlier in this chapter, composite materials are composed of reinforcing particles, fibers, or sheet materials embedded in a firm base or matrix. These materials serve in applications as diverse as the hulls of most sailboats and powerboats, growing numbers of automotive body parts, kitchen countertops and cookware, popular sporting equipment such as tennis rackets and skis, and the skin of the frames on many modern commercial and military aircraft.

Early boat builders in the Middle East 7000 years ago used pitch as a binder for reeds. Many cultures have used straw to strengthen bricks made of mud and clay, producing some of the early composite ceramic materials Laminated woods based on shellac have been known for 3000 years. We know composites based on materials such as glue and egg have been in use for 2000 years. Papier-mâché has been used as a material of construction for centuries in much the same form as schoolchildren use it today, soaking pieces of paper in a glue made from starch and water and pasting the pieces into a mold, or simply molding them by hand. Artisans have known for a long time that plaster of paris is much improved in strength by adding horse or cow hairs to the composition.

Bakelite polymers impregnated with paper and cloth appeared in many products after 1910, but the real revolution in engineering properties created by composite materials has taken place only in the past 30 years with the introduction of boron, carbon, and Kevlar fibers. The 1986 flight of the *Voyager* aircraft, the

Stealth bomber, the generation of jet aircraft used in the 1991 Gulf War, and the materials-intensive space program are all examples that are heavily dependent on composites. The flight of the *Voyager* was the first nonstop flight around the world without refueling and was largely made possible by the ingenious, featherweight polymer-composite skin and wing supports. The wings had a span greater than that of a Boeing 727, and the body was devoid of metal parts except for the twin engines and a handful of nuts and bolts. The Stealth bomber requires composite airfoils, skin, and frame components to give it radar-defying qualities. Special composites that can withstand high stress and high temperatures are needed for spacecraft and rockets and the vehicles used in the space shuttle program.

Construction and Properties

The first of the modern high-strength and high-stiffness fiber composites emerged about 1960 and was based on boron-coated tungsten-wire filaments. The chemistry was fascinating; the technology was extraordinary; the cost was prohibitive; and the breakthrough that resulted probably changed our way of thinking about materials and their possibilities. In one process, a thin tungsten filament about 0.013 mm in diameter, a cross-section that would require more than 1000 filaments to cover a strip an inch wide, is drawn through a reaction chamber filled with a mixture of boron trichloride and hydrogen, or a mixture of diborane (B_2H_6) and chlorine. The tungsten filament is heated to incandescence—about 1300K—by passing an electric current through it in the same way a lightbulb is lighted, causing an 0.05-mm layer of boron to deposit on the surface:

$$2BCl_3(g) + 3H_2(g) \longrightarrow 2B(s) + 6HCl(g) \qquad \text{boron from boron trichloride}$$

$$B_2H_6(g) + 3Cl_2(g) \longrightarrow 2B(s) + 6HCl(g) \qquad \text{boron from diborane}$$

At the end of the process, a boron-coated tungsten fiber is wound on a drum and is ready for use in polymeric or aluminum-based metal composites. These are largely for military and aerospace applications, because the costs—which are about \$250/lb—would not be competitive in any other marketplace.

Carbon fibers date back more than a century to Thomas Edison and his search for suitable lightbulb filaments. One of Edison's most successful early materials was carbon obtained by heating bamboo. Construction of a typical carbon fiber-based polymeric composite begins with a filament of polyacrylonitrile, a fiber that is a close relative of the kind of synthetic fiber used to weave sweaters (Acrilan) and carpets (Crestlan). These very fine fibers are stretched and twisted, and then roasted at elevated temperatures in the absence of air or oxygen, leaving behind relatively stiff and brittle fibers that are thinner than human hairs but with greater strength. Fabricated into tapes, a 1-cm-wide strip might consist of 20,000 fibers. Woven into sheets that look like cloth or simply chopped directly into microfibers of a millimeter or so in length, these carbon residues are ready to be embedded into a polymer base. Great strength is imparted to the resulting composite material. The base acts as a binder, or matrix, for the fibers and at the same time stabilizes and protects them from environmental degradation—thermal, electrical, and chemical, for example—and the net result is a set of mechanical properties beyond that of either the polymer base or the embedding fibers. The carbon filament composites used to fabricate the air surfaces of the Stealth bomber consist of carbon filaments embedded in an epoxy-resin base. The resin is similar to the two-tube adhesives commonly used for bonding ceramics, glasses, metals, and other difficult-to-mend materials.

Kevlar is rather different in that it is produced in much the same way as nylon and many other synthetic textile fibers, in a direct process that does not require the extreme conditions needed for preparing boron and carbon fibers. It is important to note that Kevlar is much less costly and, although not quite as strong or stiff as the other two, it can be used in many commercial applications.

The greatest benefits realized from composite polymeric materials are the high strength-to-mass ratio, the high stiffness-to-mass ratio, and the relative ease of fabrication into complex parts that serve many functions. Parts fabricated from carbon fiber composites, for example, can be as strong as steel yet weigh only a fraction as much. At the same time, very stiff materials can withstand considerable loads without deforming. The ability to mold one single part that may be the equivalent of 10 or 20 parts bolted, welded, or otherwise fused together is important in both the strength and the economic advantages that can be realized. But there are disadvantages, too. In spite of their apparent strengths, composite polymeric materials can be damaged, especially when drilling holes through them, which causes splintering, sometimes delaminating or peeling apart, and costs can be an order-of-magnitude greater than metal, wood, and glass or ceramic parts.

During the past 30 years, a great deal of scientific research and engineering development effort has been expended on high-stiffness composite materials for the benefits of saving weight in critical structural applications. The fuel efficiencies achieved in the automobile over the past 20 years, for example, would not have been possible without downsizing the vehicle and further reducing mass by replacing some metal components with molded composite polymeric materials. Such considerations are even more important in aerospace engineering and design, where delivering still greater payloads—more passengers or equipment, under greater stress—are essential design requirements.

Reducing mass while maintaining the necessary stiffness or resistance to deforming of structural members is a key concept in aircraft design. The common materials—steel, aluminum, magnesium, and even wood and glass—deliver about the same amount of stiffness for the same mass of material. However, boron, carbon, and Kevlar fibers deliver much more resistance to deforming for the same mass as the more common materials, and one can expect major breakthroughs will be achieved through the use of these new materials.

SUMMARY

Engineering materials can be divided into four important categories: metallic materials, glass and ceramic materials, polymers and plastic materials, and composite materials. Each of these categories is defined by a distinct range of structures that result in their unique properties. Ionic, metallic, covalent, and van der Waals bonds are included in materials, and the strength of a material and its performance characteristics depend on the bonding. Polymers and plastics are briefly introduced in this chapter but will be discussed mainly in Chapter 18. Semiconductors, another important class of materials, were introduced in Chapter 16.

In a general way, materials made of metal, wood, glass, and plastics have particular responses to mechanical, thermal, environmental, and electrical demands, which in turn dictate the choice of material for a particular application. Stress and strain are widely measured performance characteristics. Stress is the force imposed on a test specimen divided by the original cross-sectional area of the specimen. Strain in tension is the increase in length of the specimen divided by the original length. Strain can result in permanent or plastic deformation, but if the material recovers completely, then strain behavior is said to be elastic. Creep is long-term plastic deformation. Comparing such performance characteristics among metals and plastics is particularly important, and plastics have successfully replaced metals in many applications.

Metals are normally made up of crystals or grains that are frequently solid solutions of metals. The Hume–Rothery rules allow us to predict which metals will dissolve significantly in each other. Alloy composition and crystal defects are important in determining the strength of a metal. Many important alloys have polyphase structures with different kinds of crystal grains present. These affect the properties of the metal significantly. Phase diagrams are used to help understand polyphase systems.

Ceramics are crystalline structures held together by ionic and covalent bonds and the weaker van der Waals forces. The most common ceramics are the silicates, based on SiO_4 units. Substitution of aluminum for silicon atoms leads to aluminosilicate ions. Such ions in sheet-like forms are the basis of clays, many of which are shaped into the desired configurations and then fired. Portland cement is a ceramic of exceptional importance. Glasses are amorphous ceramics commonly based on silica.

Composite polymeric materials are composed of reinforcing fibers embedded in a base material that acts as a matrix for the fibers. Great benefits can be realized from composite polymeric materials reinforced with boron, carbon, or Kevlar. All have characteristically high strength-to-mass ratios and high stiffness-to-mass ratios, both of which are important in designing for demanding applications, including automotive and aerospace industries.

TERMS

Alloys (17.2)
Glass (17.2)
Ceramics (17.2)
Refractories (17.2)
Plastic materials (17.2)
Polymers (17.2)
Monomers (17.2)
Composites (17.2)
Performance characteristics (17.3)
Stress (17.3)
Strain (17.3)
Plastic deformation (plasticity) (17.3)
Elastic deformation (elasticity) (17.3)
Elastomer (17.3)
Elastic limit (17.3)
Creep (17.3)
Modulus of elasticity (17.3)
Stiffness (17.3)
Slippage (17.3)
Work hardened (17.3)
Atomic packing factor (17.4)
Grain (17.4)
Solid solution (17.4)
Substitutional solid solution (17.4)
Hume–Rothery rules (17.4)
Interstitial solid solution (17.4)
Single-phase alloy (17.4)
Simple metal (17.4)
Polyphase system (17.4)
Phase diagram (17.4)
Eutectic (17.4)
Portland cement (17.5)
Strain point (17.5)
Anneal point (17.5)
Softening point (17.5)
Working point (17.5)
Fiber optics (17.5)

IMPORTANT EQUATIONS

Stress (load) $= \tau = \dfrac{\text{force}}{\text{area of cross-section}}$	Stress on a sample
Strain (deformation) $= \varepsilon = \dfrac{\text{change in length}}{\text{original length}}$	Strain exhibited by a sample
Modulus of elasticity $= E = \dfrac{\text{stress}}{\text{strain}} = \dfrac{\tau}{\varepsilon}$	Elasticity or stiffness

QUESTIONS

Conceptual questions are denoted by a square screen. Extra-credit questions are denoted by a circle screen.

1. To which one of the four major materials categories would you assign each of the following?
   (a) Kevlar (b) asphalt used for roads
   (c) paper (d) rope (e) blackboard chalk
2. Why do pure metals find limited use whereas applications for alloys abound?
3. Why are metals corroded by oxidizing agents such as acids and oxygen, whereas ceramics are resistant?
4. What properties allow polymers to replace metals for some applications? In what general aspects are polymers inferior to metals? In what respects are they superior?
5. What does it mean to say that a bond is directional? Which kinds of bonds are directional and which are not?
6. How can you show whether the crystal structure of a metal has a closest packed lattice structure?
7. What is the difference between a crystal and a grain? Why do grains grow in the shapes they do?
8. The Hume–Rothery rules say that to form a solid solution, metals should have about the same electronegativity. What is the reason for this? If, say, potassium and tin were compatible in other respects, what effect would the difference in their electronegativities have?
9. What distinguishes substitutional and interstitial solid solutions?
10. What principal characteristics distinguish metals, glasses and ceramics, and plastics?
11. What is the effect on the properties of a metal of substituting some different atoms into the structure?
12. What is the difference between elastic and plastic strain? Explain this in terms of macroscopic properties as well as in atomic terms.
13. Why are tensile strengths for metals only about one-tenth of the theoretical values?
14. Why does work hardening increase the tensile strengths of metals? What other properties are also affected by cold working?
15. Under what conditions can a polyphase structure increase the strength of a metal?
16. How and why does the behavior of a glass differ from the behavior of a metal or a crystalline ceramic on heating?
17. Why is fused (melted and then crystallized) silica of limited use as a glass? How is it modified to make it more usable?
18. In what essential ways can the properties of silicates and aluminosilicates be explained by their structures?
19. Why is clay able to hold so much water?

20. How could you distinguish between crystalline and amorphous ceramics?

21. Offer a brief explanation why windowpane glass is transparent to visible light.

22. What is a eutectic?

23. In what ways have the elastic- and plastic-performance characteristics of steel established its universal value?

24. Imagine a material that is elastic in the solid phase under all conditions. How might you fabricate it into a useful product? (Suggest more than one way.)

25. From your experience, name several materials—metals, plastics, and wood products—that creep at room temperature over a long time.

26. What essential features do carbon and Kevlar-fiber composites share? What features distinguish them?

27. In aeronautical engineering, designing parts in aluminum, which has a much lower density and can be as strong as steel, usually provides little weight saving when substituted for steel. Why?

28. Silly Putty creeps when left to stand or when very slowly stretched, but fractures like glass when suddenly shocked (stressed). Try it! Roll some into a rod the size of a piece of blackboard chalk. Then, grasping the ends, pull it slowly apart. Repeat the experiment by pulling suddenly and sharply. Note the result and briefly explain the difference in behavior.

PROBLEMS

Problems marked with a bullet (•) are answered in Appendix A, in the back of the text.

Types of Materials [1–2]

•1. For each of the following applications list the properties required and suggest which category or categories of materials would be applicable. Do not hesitate to suggest possibilities that are not currently in use.
(a) Paperclip (b) Compact disc
(c) Raincoat (d) Chair
(e) Insulator for carrying a high-voltage conductor

2. List the properties required and suggest which categories of materials would be applicable for each of the following applications.
(a) Stock for a rifle (b) Fender of a car
(c) Bumper of a car (d) Drive shaft of a car
(e) Cover of a golf ball

Bonding in Materials [3–6]

•3. Which of the following bonds in polyethylene break first when the material is heated?
(a) Covalent bonds along the "backbone" of the polymeric molecule
(b) van der Waals forces between the polymeric molecules

4. Arrange the following substances in order of increasing melting or softening point:
(a) Copper (b) Diamond (c) Polyethylene polymer

•5. Predict approximate melting points for the following materials:
(a) SiC (b) MgO (c) Cu

6. Give your best estimate of the melting point of each of the following substances:
(a) I_2 (b) Polypropylene (c) LiF

Stress and Strain [7–8]

7. (a) What is the stress placed on a nylon thread with a diameter of 1.00 mm if it is holding up a load of 50.0 kg?
(b) What is the strain on the nylon thread if its length increases from 50.0 cm to 51.3 cm?

8. A rubber band with a cross-sectional area of 1.00 mm^2 increases in length from 4.00 cm to 10.55 cm when a mass of 350. g is suspended from it. Calculate the stress and strain on the rubber band under these conditions and determine the modulus of elasticity.

Solid Solutions [9–10]

•**9.** What metals would you expect to form solid solutions with beryllium?

10. What metals would you expect to form solid solutions with nickel?

Phase Diagrams [11–12]

•**11.** Using Figure 17.11, tell what phases would be present in the copper–silver system and what the composition of each phase would be for each of the following sets of conditions:
(a) 70% Ag; 900°C (b) 90% Ag; 800°C

12. Use the copper–silver phase diagram in Figure 17.11 to tell what phases would be present and what would be the composition of each phase for each of the following temperatures and mass percentages:
(a) 98% Cu; 800°C
(b) 10% Cu; 500°C

Ceramic Materials [13–18]

13. What is the formula of the anion containing a single chain of five SiO_4 units?

14. What is the formula of a silicate anion containing a ring of ten SiO_4 units?

15. Draw silicate ions that are consistent with each of the following formulas:
(a) $Si_8O_{24}^{16-}$ (b) $(Si_2O_5^{2-})_n$

Suggest a structure other than the one we have mentioned.

16. Draw silicate ions that would fit the following formulas:
(a) $Si_5O_{16}^{12-}$ (b) $Si_4O_{12}^{4-}$

•**17.** Calculate the correct value for x and y in each of the following aluminosilicate ions:
(a) $Si_3AlO_x^{y-}$ (b) $Si_2Al_xO_8^{y-}$

18. Determine x and y for the following ions:
(a) $Si_xAl_4O_{12}^{y-}$ (b) $Si_4Al_xO_{12}^{y-}$

Additional Problems [19–22]

•**19.** Calculate the atomic-packing factor for the face-centered cubic structure.

20. Which metals do you think would be able to form substitutional solid solutions with chromium?

•**21.** Write the balanced equation for the complete reaction of polyethylene $(—CH_2—CH_2—)_n$, with air.

22. Complete and balance the following equations:
(a) $BCl_3(g) + H_2(g) \longrightarrow$
(b) $B_2H_6(g) + Cl_2(g) \longrightarrow$

Multiple Principles [23]

•**23.** A sample that contains 55.0% by mass of Ag_2O and 45.0% by mass of CuO reacts with hydrogen to produce the metals. These are then heated to form a liquid–liquid solution and cooled. At what temperature will the solution start to solidify? How many phases will be present and what will be their compositions when the temperature has dropped to 700°C?

Cumulative Problems [24–28]

24. Calculate the stress on a wire with a cross-sectional radius of 0.10 mm if a mass of 10.0 kg is supported by it.

•**25.** Calculate the strain for the wire in problem 24 if it increases elastically in length by 4.35% when stressed by the mass.

26. Lead is a soft material at room temperature. Why?

•**27.** Copper is malleable and can be "hammered" flat at room temperature. What is the effect on the number of defects and dislocations in the crystal structure?

28. For the purpose of suspending a heavy load from a cable, does it make a difference whether you use a cable of pure iron or one containing a small percentage of carbon. Why? If cables of equal diameter are used, which will support the greater load before failure?

ESTIMATES AND APPROXIMATIONS [29–33]

29. Select the category of materials that you think would be best for the construction of a propeller blade for a small boat to be used in saltwater. List all the properties necessary for this application and explain which kinds of materials are satisfactory for each property and why. Then come up with the overall best material and explain your choice.

30. Estimate the relative melting or softening points of polyethylene (a polymer used for food storage films and milk bottles), lithium chloride, potassium chloride, chlorine, and ethylene (C_2H_4).

31. Estimate the melting or softening points of the following materials. Then check in a chemistry handbook to see how close you came.
(a) Si (b) KBr (c) Polyvinyl chloride (another useful polymer found in phonograph records and raincoats) (d) Cr (e) BCl_3

32. Comment on the possibilities of building a Boeing 727 out of wood.

33. Comment on the possibilities of materials for constructing a corrosion-resistant water main.

FOR COOPERATIVE STUDY [34–38]

34. A 3.500-m length of a wire that has a modulus of elasticity of 1.2×10^{12} dynes/cm^2 increases elastically to a length of 3.582 m when a load of 50.00 kg is suspended by it. What is the diameter of the wire?

35. Consider the following applications. For each, list the desired characteristics and select a possible material from the four major categories. In some cases, there may be more than one choice. When multiple possibilities exist, suggest advantages and disadvantages of each choice.
(a) The hull of a canoe
(b) A tank for a gas under pressure
(c) The heating elements of an electric stove
(d) A pipe to carry molten iron

36. Stretching an elastic band is clearly an energy-releasing event. Try it! Stretch a fairly broad rubber band (say, $\frac{1}{8}$ to $\frac{1}{4}$ in. in diameter) across your lower lip (which is more sensitive to heat than your fingers). Note the sensation. Then let the elongated rubber band recover and note the sensation. Offer a thermodynamic argument in explanation.

37. Explain why the top of a tin can that has been opened but not completely removed becomes brittle on successive bending and eventually breaks (fails), whereas a polypropylene hinge of a case for eyeglasses can be bent many thousands of times without breaking off (failing).

38. Try developing an argument to explain why, in complete contrast to most other materials, the modulus of elasticity of rubbery materials often increases with temperature.

WRITING ABOUT CHEMISTRY [39–40]

39. Analyze how chemistry is arranged and taught in this chapter compared to previous chapters, giving as many details as you can. Then explain how this works for you.

40. The concepts of stress and strain, elasticity and stiffness, and modulus of elasticity could be applied to human relations as well as to performance characteristics of materials. Explain what stress and strain would be in human relations. Tell what kinds of people exhibit the characteristics of elasticity and stiffness and explain with a few examples. Explore whether modulus of elasticity could be applied to people in the terms that you have developed.

18

chapter eighteen

PROPERTIES OF POLYMERS

18.1 GIANT MOLECULES

Origins of Polymers

Polymeric materials are composed of high-mass molecules that themselves are made up of smaller, repeating units. The giant molecules are called **polymers**; the smaller, repeating units are called **monomers**. Monomers, polymers, and polymeric materials are the subject of this chapter. Our earlier brief introduction to the nature and properties of polymeric materials in the preceding chapter was necessary to provide a more complete picture of materials science. Here, our purpose is to focus on the general nature and properties of polymers, leaving their synthesis and chemical reactivity to Chapter 21.

As has so often been the case, craft tradition and technology led the way in the discovery of methods for preparing and identifying polymeric materials with useful properties. Anecdotal history has it that Christopher Columbus found natives of the West Indies playing games with rubber balls. Rubber artifacts have been found among Mayan ruins in Mexico's Yucatan Peninsula. By the beginning of the nineteenth century, it was well known to workers on South American and Southeast Asian plantations that rubber could be mechanically worked, or *masticated,* to make it suitable for coating cloth. Charles Goodyear was awarded U.S. patents in the 1840s for curing rubber. His process came to be known as **vulcanization**, which involves incorporating sulfur to eliminate the stickiness of the natural product, which led to the commercial production of rubber.

For thousands of years, we have taken advantage of the wondrous properties of cotton and wool, fibrous polymers found in nature. Cellulose nitrate, a modified plant product, was used by John Wesley Hyatt as early as 1869. Hyatt was seeking a substitute for ivory and found a product that was much more generally useful. It came to be known as *celluloid* and showed up in everything from men's shirt collars to the first motion-picture film material. But it was Leo Baekelund's discovery of Bakelite that launched the Age of Plastics. **Plastics** are polymeric materials manufactured industrially. Polymers in their unprocessed state before being made into useful articles are called **resins**. Bakelite was the first hugely successful plastic. The second, discovered 30 years later, was nylon.

Despite all these important historical events, polymer chemistry has only recently become widely recognized as a science in its own right. And despite our extreme familiarity with polymers in the second half of the twentieth century, most people still do not have a clear idea what they are. In part, this is because materials like metals and ceramics have been around for so much longer. Artisans have been skilled at smelting and glassblowing for centuries, whereas practical polymer chemistry is little more than a century old.

The range of properties displayed by many of the polymeric materials we regularly encounter is extraordinarily broad. A polymer can be a flexible shower curtain or a rigid piece of plumbing pipe; a compact disk or a bullet-proof protective shield; a release film or an adhesive; an insulator or a conductor. Polymers are so diverse and adaptable that in only a few decades this class of materials has developed into an almost $100 billion industry in the United States alone.

PROFILES IN CHEMISTRY

Baekelund and Bakelite Leo Baekelund (1863–1944), a young Belgian émigré chemistry professor, inventor, and entrepreneur, was in search of a synthetic substitute for shellac. While working in his garage laboratory in the Bronx, New York, he discovered an acid-catalyzed condensation reaction of phenol with formaldehyde. The amber-colored, viscous product quickly turned to a hard, intractable solid that could be cut or machined, was an excellent electrical insulator, and was both water and organic-solvent resistant. It could also be placed on surfaces as a protective and decorative coating. Quickly (and correctly) assessing the potential commercial applications for such a material, Baekelund refined his process and modified his products to maximize the unique potential of their properties.

The original 1907 patent application cites potential uses ranging from linings for pipes, valve seats, billiard balls, phonograph records for that fledgling industry, and numerous parts for electrical appliances. Baekelund also noted that his product is thermally set, or *thermoset*. This means that heat is the agent used to bring about permanent chemical changes that set the phenol-formaldehyde mixture into the shape of the vessel containing it. Once cured in this way, the formed polymer no longer goes through a melting transition (fusion) on heating.

Beginning in 1909, Baekelund marketed his invention under the name Bakelite (after himself). It was not the first commercial plastic material—celluloid owns that title—but it was the first of major industrial importance. Nearly a century later, millions of pounds are still being manufactured worldwide each year. Baekelund's Bakelite launched polymer chemistry and the plastics industry.

Questions

What items do you believe to be made of Bakelite?

In what sense is a billiard ball made of Baekelund's polymer one giant molecule?

Monomers and Polymers

Length and mass account for the special properties of the covalent molecules called polymers and the polymeric materials that we encounter. By the 1930s it had become increasingly clear that the molecules of natural rubber and many other such polymeric materials were composed of high-molecular-mass molecules consisting of long chains of carbon atoms connected by normal covalent bonds. It is true that the waxy petroleum products used in making candles (called paraffins, or paraffin wax) are composed of chains of carbon atoms, but the chains are not long enough (perhaps only 30 to 40 atoms in length) and their molecular masses are too low (only about 500 amu) for them to exhibit polymeric properties. Only when the length of the chain of atoms results in molecular masses on the order of 5,000 to 50,000 amu do we begin to see polymeric properties.

Polyethylene is the best known and most widely used polymeric material. It is constructed from the gas ethylene, $CH_2{=}CH_2$. This low-molecular-mass monomer is the molecular building block that becomes strung together chemically into the giant molecule we call a polymer—in this case, polyethylene:

$$n\mathrm{CH_2{=}CH_2} \longrightarrow -(\mathrm{CH_2{-}CH_2})_n-$$

Note that the double bond in the monomer has been replaced by a single bond in the polymer. If the value of n is only about 10 and the chain is terminated at each end with hydrogen atoms, the substance is a soft, waxy material that melts at just about body temperature and is quite unlike the gaseous monomer from which it was constructed (Fig. 18.1).

$H{-}(CH_2{-}CH_2)_{10}{-}H$
Molecular mass = 282
m.p. = 37°C
$CH_2{=}CH_2$
Molecular mass = 28
b.p. = −104°C

Snapping molecules together to form very large molecules is called **polymerization**. If the polymerization process is extended to the point where the molecular mass is in the range of 5,000 to 20,000 amu, a form of polyethylene suitable for coatings and adhesives is produced. This form is widely used in waterproofing paper products. Further polymerization produces still higher molecular masses. At 200,000 amu, a polyethylene polymer with much greater mechanical strength and flexibility, which is suitable for the manufacture of films and bottles, is produced. Molecular masses can be in the millions, with properties varying accordingly.

If you write the formula of polyethylene to show the repeating monomer units as follows,

$$-(\mathrm{CH_2{-}CH_2})_n-$$

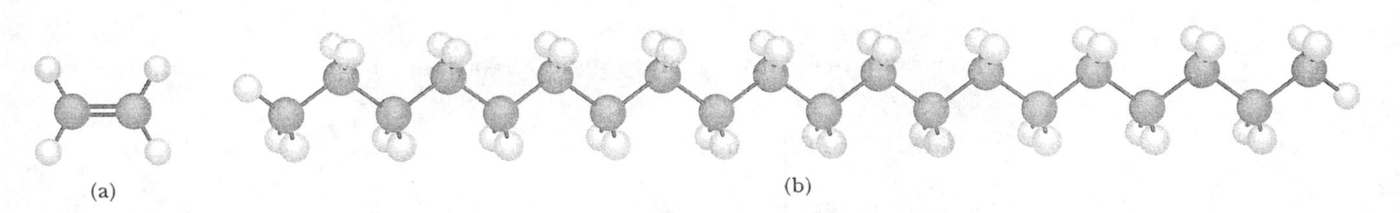

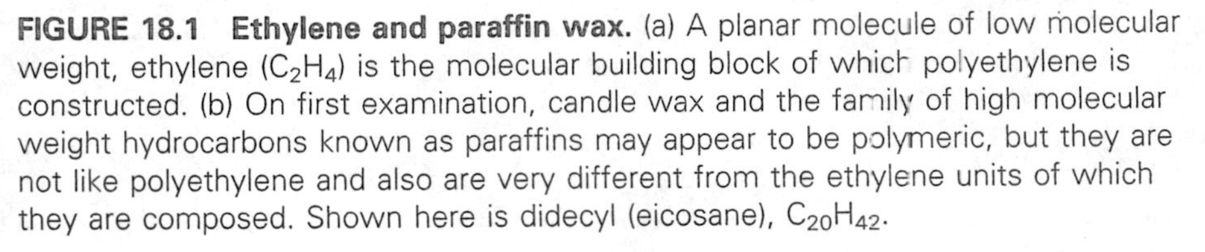

FIGURE 18.1 Ethylene and paraffin wax. (a) A planar molecule of low molecular weight, ethylene (C_2H_4) is the molecular building block of which polyethylene is constructed. (b) On first examination, candle wax and the family of high molecular weight hydrocarbons known as paraffins may appear to be polymeric, but they are not like polyethylene and also are very different from the ethylene units of which they are composed. Shown here is didecyl (eicosane), $C_{20}H_{42}$.

the subscript n in the formula is the degree of polymerization. The **degree of polymerization (DP)** refers to the number of repeating units in the polymer chain. It is a measure of the chain length. The molecular mass of the ethylene repeating unit is 28 amu. Thus, if n is 1000, then the molecular mass must be 28,000 amu:

$$\text{Molecular mass} = 1000 \times 28 \text{ amu} = 28{,}000 \text{ amu}$$

If the molecular mass is 56,000 amu, then the degree of polymerization must be 2000:

$$\text{DP} = \frac{56{,}000 \text{ amu}}{28 \text{ amu}} = 2.0 \times 10^3$$

EXAMPLE 18.1A

Polytetrafluoroethylene can be represented by the formula $—(CF_2—CF_2)_n—$. If the degree of polymerization is 10,000, what is the molecular mass?

COMMENT The subscript n is the degree of polymerization and the multiplier. The $—CF_2—CF_2—$ tetrafluoroethylene repeating unit has a molecular mass of 100. amu.

SOLUTION Molecular mass = 10,000 × 100. amu = 1,000,000 amu.

EXAMPLE 18.1B

If the molecular mass of a polytetrafluoroethylene molecule is 3 million, what is the degree of polymerization (DP)?

COMMENT This is the previous problem in reverse. Simply divide the polymer's molecular mass (3 million amu) by the molecular mass of the polymer's repeating unit (100. amu) to obtain the value for the degree of polymerization.

SOLUTION

$$\text{DP} = \frac{3 \times 10^6 \text{ amu}}{100. \text{ amu}} = 3 \times 10^4$$

EXERCISE 18.1(A)

If the degree of polymerization of a sample of polyethylene oxide is 1400, what is the molecular mass? Polyethylene oxide can be represented by the formula $—(CH_2CH_2O)_n—$.

ANSWER 62,000 amu

EXERCISE 18.1(B)

If the molecular mass of a polyethylene oxide molecule is 840,000 amu, what is the degree of polymerization?

ANSWER 19,000

There is no marker on the scale of molecular dimensions that allows one to say, "Here polymer chemistry begins." Rather, it has to do with transitions that occur over broad ranges. Typically, somewhere around 10,000 amu, we can say we are dealing with a high-molecular-mass, *polymeric* material. At that point and beyond, the characteristically sharp melting temperature of most small molecules has broadened into a *softening range.* The crisp change from solid to liquid that is characteristic of crystalline solids no longer occurs. Another way of looking at molecular dimensions on the scale of polymers is to consider certain mechanical properties such as stiffness—that is, high modulus of elasticity (Chapter 17). A characteristic of polymers is that adding one more monomer unit to the polymer chain does not significantly alter these properties of the polymeric material.

Homopolymers and Copolymers

Polymers are broadly distinguished by the makeup of the repeating units along the chain. For purposes of simple illustration, let us consider the repeating groups as different geometric shapes such as circles, squares, and diamonds. In a **homopolymer** the repeating units—the blocks along the polymer chain—are all the same. A very clear example of this is polyethylene of the type used for milk bottles, in which the repeating unit is just —(CH_2—CH_2)—. Less obvious homopolymers result from two different monomers that can line up in the polymer only in a strictly alternating fashion. In this case the repeating unit is the same throughout the polymer molecule, even though each repeating group is the result of two monomer units. Polyethylene terphthalate (PET), which is used in the manufacture of soft drink bottles, is made from two kinds of molecules; however, it is considered a homopolymer because once made, the repeating units along the chain are identical (Fig. 18.2).

Copolymers have polymer chains that do not consist simply of one unit repeating over and over. Rather, they are composed of at least two different kinds of units. A **random copolymer** has repeating units that are randomly distributed along the polymer chain. One of the best examples is the copolymer of vinyl acetate and vinyl chloride, which was used in huge amounts in the manufacture

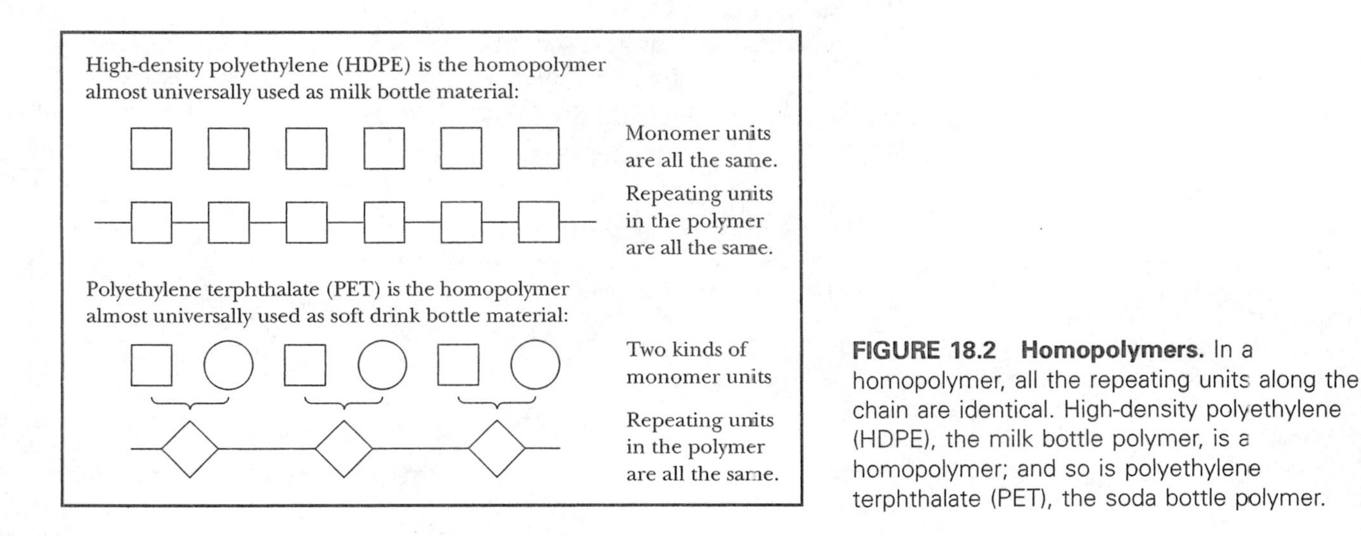

FIGURE 18.2 Homopolymers. In a homopolymer, all the repeating units along the chain are identical. High-density polyethylene (HDPE), the milk bottle polymer, is a homopolymer; and so is polyethylene terphthalate (PET), the soda bottle polymer.

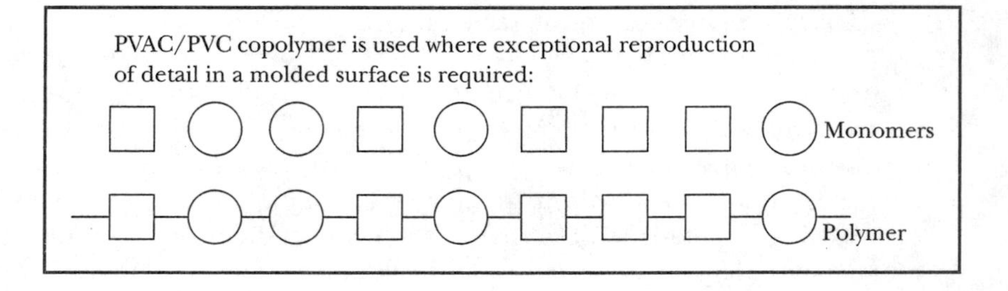

PVAC/PVC copolymer is used where exceptional reproduction of detail in a molded surface is required:
Monomers
Polymer

FIGURE 18.3 Random copolymer. Here the polymer has a random arrangement of two or more units that repeat along the chain. In this example, the copolymer consists of polyvinyl acetate (PVAC) and polyvinyl chloride (PVC), a material used for half a century in huge amounts for manufacturing phonograph records.

of phonograph records until records were replaced by magnetic tapes and, more recently, by laser disks. The added 5 to 25% vinyl acetate makes the copolymer less brittle than the homopolymer polyvinyl chloride (Fig. 18.3).

Some copolymers have repeating units that, although not strictly alternating, come close to that kind of regular arrangement (Fig. 18.4). These are called **alternating copolymers**. Another kind of copolymer has branches of one kind of monomer grafted onto already formed polymers of another. These are called **graft copolymers.** It is possible to improve the ability of Orlon to take up dye by grafting water-attracting cellulosic branches such as cotton and rayon fibers onto the homopolymer chain without diminishing the basic strength of the polymer (Fig. 18.5).

Block copolymers have long segments of different monomer residues repeating along the chain. DuPont's once-revolutionary Spandex fibers are block copolymers composed of polymer chains in which stiff segments are linked to soft segments. The soft segments permit the polymer to uncoil and stretch to three or four times its original length, while the links provide hydrogen bonds that are strong enough to prevent permanent distortion (Fig. 18.6).

Because of their length and flexibility, the properties of polymeric materials depend on the degree of bending, twisting, and folding of the homopolymer and copolymer chains. Furthermore, as on a railroad, **branches** may extend off the main line, and their presence affects the properties of the polymeric material by interfering with the ability of the chains to come together or slide by each other. Finally, the presence of natural or artificial **cross-links**, which are chemical bridges between adjacent chains, contributes to properties of polymeric materials by creating networks of molecules. For example, if long-chain molecules are closely packed, uninterrupted by branching, and contain few connections between chains, then the molecules can slide readily by one another; as the temperature increases, the polymeric material is characterized by viscous

Low-molecular-mass SAN polymers are widely used as dispersing agents for pigments in coloring paints:

Monomers

Polymer

FIGURE 18.4 Alternating copolymer. Polymers such as these are characterized by an alternating arrangement of repeating units along the chain. SAN is a copolymer of styrene (S) and acrylonitrile (AN). Applications include textile and carpet fibers.

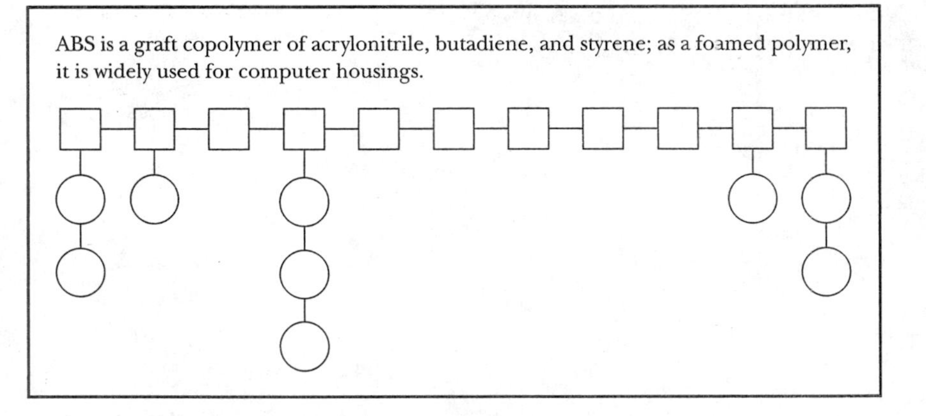

ABS is a graft copolymer of acrylonitrile, butadiene, and styrene; as a foamed polymer, it is widely used for computer housings.

FIGURE 18.5 Graft copolymer. Copolymers can be produced by grafting smaller segments along the main chain, as indicated in the simple drawing. For example, ABS is a graft copolymer formed from acrylonitrile (A), butadiene (B), and styrene (S), which is widely used as an engineering material for office machines, television casings, and computer frames.

flow. If branching leads to entanglement of the chains, then its constituent molecules cannot as easily move by each other, and the flow characteristics of the material are considerably altered. Some degree of cross-linking leads first to elastic properties because segments of these polymeric molecules find themselves tied to their original positions and return there when a stretching force (as can be produced in a rubber band) is removed (Chapter 17). But when the number of cross-links becomes so large as to radically reduce the segments of polymeric molecules that are unconnected, the polymeric material loses elasticity and becomes rigid and brittle. It becomes set.

Polymer Synthesis and Polymerization Reactions

The snapping together of monomer units to make polymers is a highly specialized aspect of chemical synthesis to which we cannot give much attention until we have treated organic molecules in Chapter 21. But a few words and illustrations here will hold us over. Polymer synthesis can be divided into chain-growth and step-growth polymerization. **Chain-growth polymerization** is characteristic of polymer syntheses in which the small molecular building blocks that we call monomers are *added* to each other, and in which no by-products result. Typically, a catalyst, high temperature, and high pressure are required. A chemical reaction attaching two molecules without by-product formation is called an *addition.* Consequently, chain-growth polymerization is often called **addition poly-**

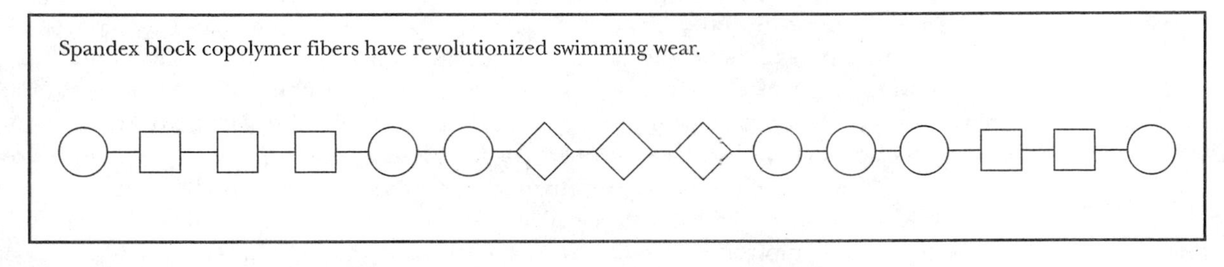

Spandex block copolymer fibers have revolutionized swimming wear.

FIGURE 18.6 Block copolymer. These fibers consist of blocks that impart stiffness alternating with softer segments that make material elastic.

Polymer models. Homo- and copolymers; branched and cross-linked polymers.

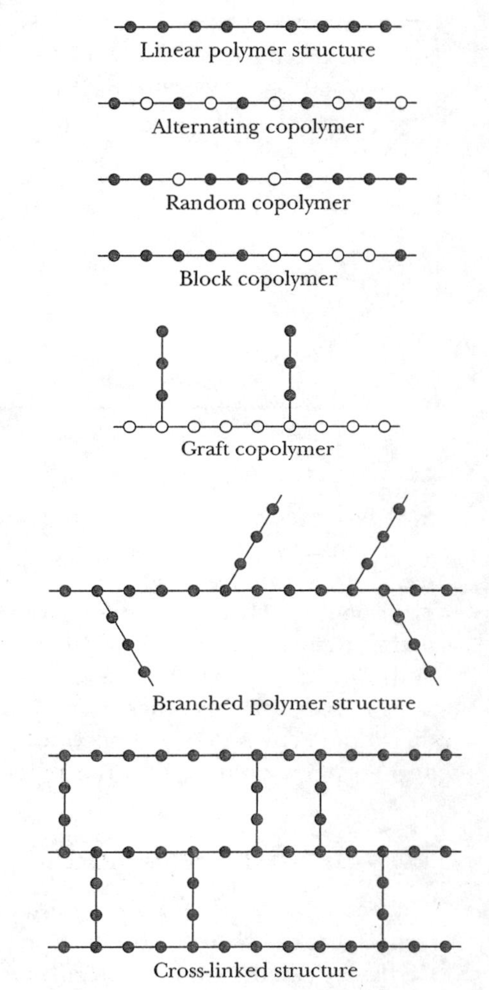

merization (Fig. 18.7). The polymerization of ethylene and its derivatives demonstrate chain-growth or addition polymerization. The polymerization reaction involves the double bond, which is expended in the process, but it leaves all the atoms of the monomers present in the polymer:

$$n\mathrm{CH_2{=}CH_2} \longrightarrow -(\mathrm{CH_2{-}CH_2})_n-$$

Step-growth polymerization differs in that double bonds are not typically involved and in that the "snapping together" process releases small molecules, often water; as a result, the monomer residues in the repeating unit are left with fewer atoms. Because such reactions are known in chemistry as *condensations,* step-growth polymerization is often referred to as **condensation polymerization**. A simple example of step-growth polymerization occurs in the preparation of the chains of silicon–oxygen polymers. These polymers with alternating silicon and oxygen atoms are often called siloxane polymers (polysiloxanes) or Silicones. The monomer is $\mathrm{HO{-}Si(CH_3)_2{-}OH}$ (dimethyldihydroxysilane), which is produced by allowing $\mathrm{Cl{-}(CH_3)_2Si{-}Cl}$ (dimethyldichlorosilane) to react with water:

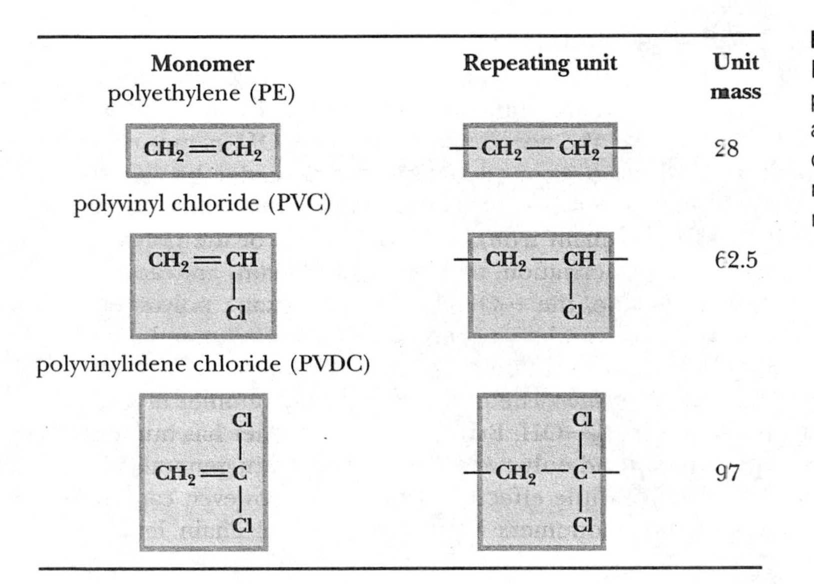

FIGURE 18.7 Chain-growth polymerization. Polyethylene, polyvinyl chloride, and polyvinylidene chloride are all formed by additional polymerization reactions involving the double bonds. Note that the monomer and the repeating unit in the polymer have the same number of atoms.

$$\mathrm{Cl{-}(CH_3)_2Si{-}Cl + 2H_2O \longrightarrow HO{-}Si(CH_3)_2{-}OH + 2HCl}$$

Building of the siloxane chain can then begin. Two units of silane combine by losing one molecule of water:

$$\mathrm{HO{-}Si(CH_3)_2{-}OH + HO{-}Si(CH_3)_2{-}OH \longrightarrow HO{-}Si(CH_3)_2{-}O{-}Si(CH_3)_2{-}OH + H_2O}$$

As more and more monomer units condense in this manner, the chain grows to polymeric dimensions:

$$\mathrm{H{-}[{-}O{-}Si(CH_3)_2]_n{-}OH}$$

One molecule of water is produced for every link formed between monomer units. Step-growth polymerization always results in the expulsion of one small molecule for every link formed between the reacting monomers (Fig. 18.8).

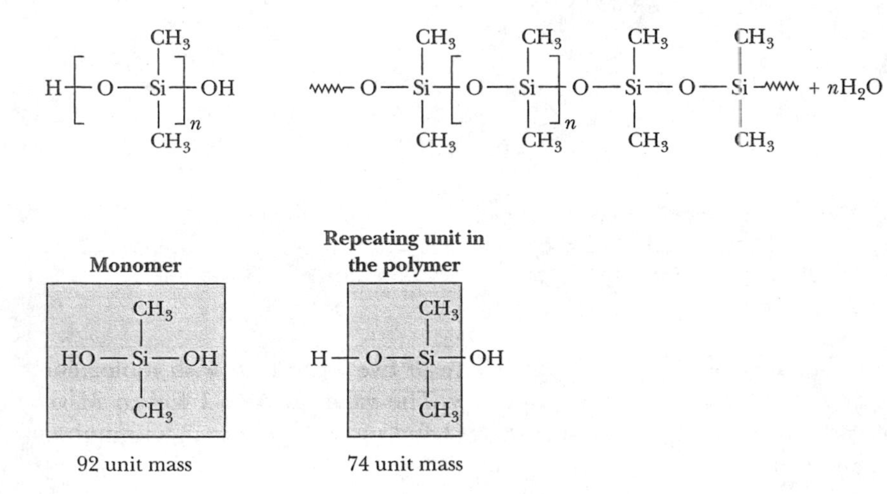

FIGURE 18.8 Step-growth polymerization. Polysiloxane polymers are formed by condensing, or snapping together repeating units while releasing small molecules in the process. In this case, the small molecule released is H_2O and the unit molecular mass is accordingly reduced by 18.

A critical aspect of step-growth polymerization is that the monomers must have two sites available for reaction. That is, each monomer unit must be able to connect on both ends like the couplers on either end of a railroad car. In the polymerization reaction above the two sites are the two —OH groups on the Si atom. Compounds that have two groups that can react are said to be *difunctional*.

The ends of polymer molecules are special places. They are obviously going to be very different from the main trunk or the branches of the polymer. For example, in step-growth polymerization, the ends of the chains are capped with unreacted groups, for example, the —OH groups in siloxane polymers. That means that they may be reactive when no further reaction is wanted. One way around this is to add a very small amount of *monofunctional* monomer to act as a chain stopper or chain *terminator*. The monofunctional monomer in the siloxane polymer would be $Si(CH_3)_3$—OH. But if a typical polymer has hundreds to thousands of repeating units and only two end groups, we can generally assume that the end groups will have little effect on properties. However, capping the chain with monofunctional monomers is a way to control chain length and molecular mass.

Molecular Masses of Polymers

Defining the molecular mass of a polymer is not as straightforward as for a small molecule. First, it is impossible to get anything close to a mole of monomer molecules to polymerize and stop at the same length. A pot full of polyethylene contains many strands of nearly the same—but not exactly the same—chain length. This is in contrast to a typical pure compound composed of small molecules in which every molecule is exactly the same as every other, and the molar mass in grams per mole is the mass of a mole of these identical small molecules. With polymers, however, one has to live with average molecular masses—except in the case of certain biological molecules. Second, the values obtained for the masses of polymer molecules are dependent on the method of measurement.

Methods that depend on colligative properties—usually osmotic pressure measurements—give **number-average molecular masses**, $\overline{M}_n$, the simple arithmetic average of the molecular masses of all molecules. Colligative properties measure numbers of particles regardless of size, and number-average molecular masses depend on the number of molecules of each mass:

$$\overline{M}_n = \frac{W}{\Sigma N_i} = \frac{\Sigma N_i M_i}{\Sigma N_i}$$

Number-average molecular mass, $\overline{M}_n$: This is the simple numerical average, which is the total mass W of the molecules divided by their total number ΣN_i, where N_i is the number of molecules of each mass M_i.

For example, consider an imaginary mixture of five molecules with molecular masses of 2, 3, 3, 3, and 4 amu, respectively. The value of N_i is 1 for an M_i of 2 amu, N_i is 3 for an M_i of 3 amu, and N_i is 1 for an M_i of 4 amu. The number average molecular mass $\overline{M}_n$ is calculated as follows:

$$\overline{M}_n = \frac{\text{total mass } W}{\text{number of molecules}} = \frac{\Sigma N_i M_i}{\Sigma N_i}$$
$$= \frac{[1 \times 2] + [3 \times 3] + [1 \times 4]}{1 + 3 + 1} = \frac{15}{5} = 3 \text{ amu}$$

Mass-average molecular masses, $\overline{M}_w$, are based on experimental measurements that are affected by the sizes of the molecules. These include light-scattering techniques, which are more sensitive to larger masses. In contrast, if one measures a property that depends on the number of particles, each molecule contributes equally, regardless of its mass. But with light-scattering techniques, the larger molecules have a greater impact on the average molecular mass because they scatter light more. Thus, the average molecular mass that will be obtained by such techniques will emphasize the higher values and thus be higher than the number-average. An average that emphasizes the higher values can be obtained if the individual values are squared:

$$\overline{M}_w = \frac{\Sigma N_i M_i^2}{\Sigma N_i M_i}$$

Mass-average molecular mass $\overline{M}_w$: The mass-average molecular mass is computed by dividing the sum of the squares of the individual molecular masses by the sum of the molecular masses.

Once again, consider our imaginary mixture of 5 molecules of molecular masses 2, 3, 3, 3, and 4 amu, respectively. This time our calculation is the average of squared values, which emphasizes size as well as number,

$$\overline{M}_w = \frac{\Sigma N_i M_i^2}{\Sigma N_i M_i} = \frac{[1 \times 2^2] + [3 \times 3^2] + [1 \times 4^2]}{[1 \times 2] + [3 \times 3] + [1 \times 4]} = \frac{47}{15} = 3.1 \text{ amu}$$

Note that the result gives a larger value for the mass-average than the number-average, which is always the case, unless, of course, all the molecules have the same chain length—which is not observed except in some biological molecules.

If all molecules have the same molecular mass M, then the equation for M_w reduces to

$$\overline{M}_w = \frac{\Sigma NM^2}{\Sigma NM} = \frac{NM^2}{NM} = \frac{M^2}{M} = M$$

because there is only one value of M and, therefore, N is the total number of molecules. This same result M is obtained with the number-average molecular mass if all molecules are identical. Thus, under these conditions,

$$\overline{M}_w = \overline{M}_n$$

EXAMPLE 18.2

Calculate the number- ($\overline{M}_n$) and mass- ($\overline{M}_w$) average molecular masses for an imaginary mixture of polymers containing four strands whose molecular masses are 2000, 4000, 6000, and 8000 amu, respectively.

COMMENT The total mass is the sum of 2000, 4000, 6000, and 8000 amu, and the number of polymer strands is 4.

SOLUTION

Number-average molecular mass M_n

$$\overline{M}_n = \frac{\text{total mass } W}{\text{number of molecules } N} = \frac{\Sigma N_i M_i}{\Sigma N_i}$$

$$= \frac{(1 \times 2000) + (1 \times 4000) + (1 \times 6000) + (1 \times 8000)}{1+1+1+1} = \frac{20{,}000}{4} = 5000 \text{ amu}$$

Mass-average molecular mass $\overline{M}_w$

$$\overline{M}_w = \frac{\Sigma N_i M_i^2}{\Sigma N_i M_i}$$

$$= \frac{(1 \times 2000^2) + (1 \times 4000^2) + (1 \times 6000^2) + (1 \times 8000^2)}{(1 \times 2000) + (1 \times 4000) + (1 \times 6000) + (1 \times 8000)}$$

$$= \frac{1.20 \times 10^8}{20{,}000} = 6000 \text{ amu}$$

Once again, note that $\overline{M}_w$ is greater than $\overline{M}_n$, which is as it should be.

EXERCISE 18.2

Calculate the number and mass-average molecular masses for a mixture of three polymer strands whose molecular masses are 1.10×10^4, 1.20×10^4, and 1.30×10^4 amu, respectively, and compare their values.

ANSWER

$$\overline{M}_n = 12{,}000 = 1.20 \times 10^4 \text{ amu}$$

$$\overline{M}_w = 12{,}056 = 1.21 \times 10^4 \text{ amu}$$

and $\overline{M}_w$, as expected, is larger than $\overline{M}_n$

Another way of illustrating what we mean by number-average and mass-average molecular masses is given by the graph in Figure 18.9a, which shows a plot of N_i, the number of molecules at each molecular mass, as a function of molecular mass M_i. The curve takes the general form of a distribution function, reflecting the fact that there are many chains of differing length. Some molecules have short chain lengths and low molecular masses. Other molecules have long chain lengths and high molecular masses. The number-average molecular mass $\overline{M}_n$ can be found just above the maximum in the distribution function, whereas the mass-average molecular mass $\overline{M}_w$ must have a higher value because it emphasizes the molecules with high molecular masses. When all the chain lengths are the same, we say that the system is **monodisperse**, and $\overline{M}_n$ must equal $\overline{M}_w$. Some biopolymers such as β-keratin and insulin are monodisperse. All other

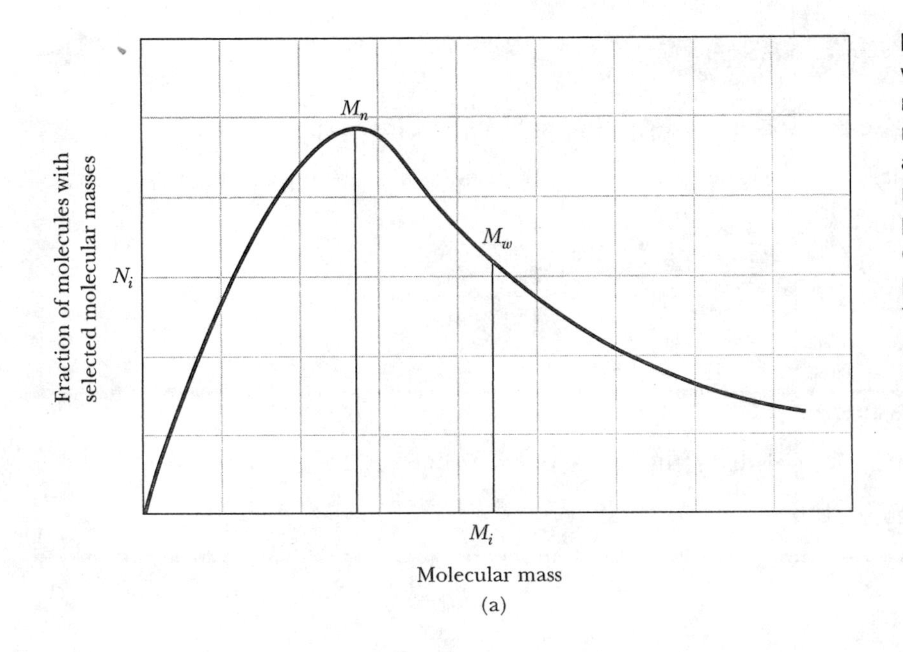

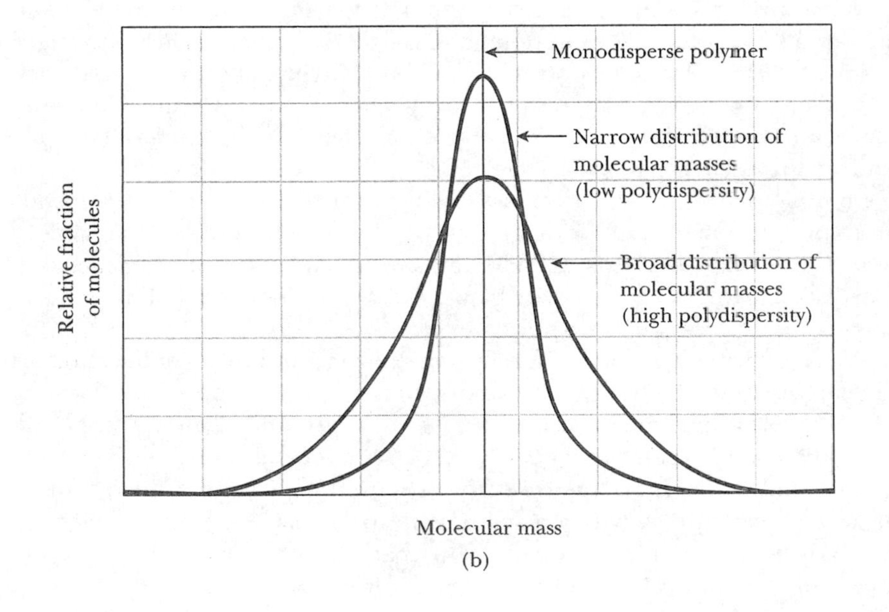

FIGURE 18.9 Polymer molecular weights. (a) Graph of number of molecules with selected mass (N_i) versus molecular mass (M_i) showing number-average (M_n) and mass-average (M_w) molecular masses. Many commercial polymers such as nylon have distributions close to this. (b) The polydispersity gives the distribution of the chain lengths and molecular masses. For a monodisperse system, all the molecules are identical.

polymer systems have at least a small range of chain lengths, and are said to be polydisperse. **Polydisperse** means that the molecules are not all the same. Because $\overline{M}_n$ and $\overline{M}_w$ differ more if the spread of masses is greater, the ratio $\overline{M}_w/\overline{M}_n$ can be used as a measure of the **polydispersity**, the extent to which molecules are of different masses (Fig. 18.9b).

EXAMPLE 18.3

Calculate the polydispersity of the polymer discussed in Example 18.2.

COMMENT The ratio $\overline{M}_w/\overline{M}_n$ can be used as a measure of the polydispersity.

SOLUTION

$$\frac{\overline{M}_w}{\overline{M}_n} = \frac{6000 \text{ amu}}{5000 \text{ amu}} = 1.2$$

EXERCISE 18.3

Calculate the polydispersity of the polymer in Exercise 18.2 above.

ANSWER 1.01

18.2 MECHANICAL PROPERTIES OF POLYMERS

Besides molecular mass, other factors affect the properties of a polymer. Polyethylene has a relatively low tensile strength. Polyvinyl chloride, with a Cl atom in place of an H atom in the monomer ($CH_2{=}CHCl$), has a tensile strength three times as great. Imagine polyethylene and polyvinyl chloride chains each being stressed by pulling. In order for the polymer sample to elongate, the long molecules must move with respect to each other. The chlorine atom is the negative end of the carbon–chlorine dipole $C{+}\!\!\longrightarrow Cl$, whereas a hydrogen is the positive end of the much weaker carbon–hydrogen dipole $H{+}\!\!\longrightarrow C$. The Cl ends of the dipoles in polyvinyl chloride attract the H ends; this establishes stronger van der Waals forces than exist in polyethylene and increases the resistance of the material to stretching (Fig. 18.10a,b). These van der Waals forces make polyvinyl chloride stronger than polyethylene.

In the remarkable fluorocarbon polymer known as Teflon, all of the H atoms have been replaced, this time by F atoms; hence, the monomer is $CF_2{=}CF_2$ (Fig. 18.10c). Each fluorine is at the strongly negative end of the $C{+}\!\!\longrightarrow F$ dipole. Negative dipole ends repel each other, and as a result the material has little strength and deforms or creeps readily at room temperature. These properties make Teflon very difficult to machine, and parts made of this polymer are usually made by casting or molding. But this opaque, soft, waxy, white crystalline polymer is important for its other properties; it is virtually immune to chemical attack, has marvelous electrical resistivity, is especially stable at high temperatures, and has one of the lowest coefficients of friction of any known material. It is used for cable insulation, electrical components, bushings and bearings, and nonstick products such as cookware, food processing equipment, and conveyor equipment. Unfortunately, Teflon is especially difficult to process. Typically, the granular polymer resin is compressed and then fused at high temperatures using techniques developed for processing ceramics and glass. The polymer has been successfully modified to retain some of the properties by removing some

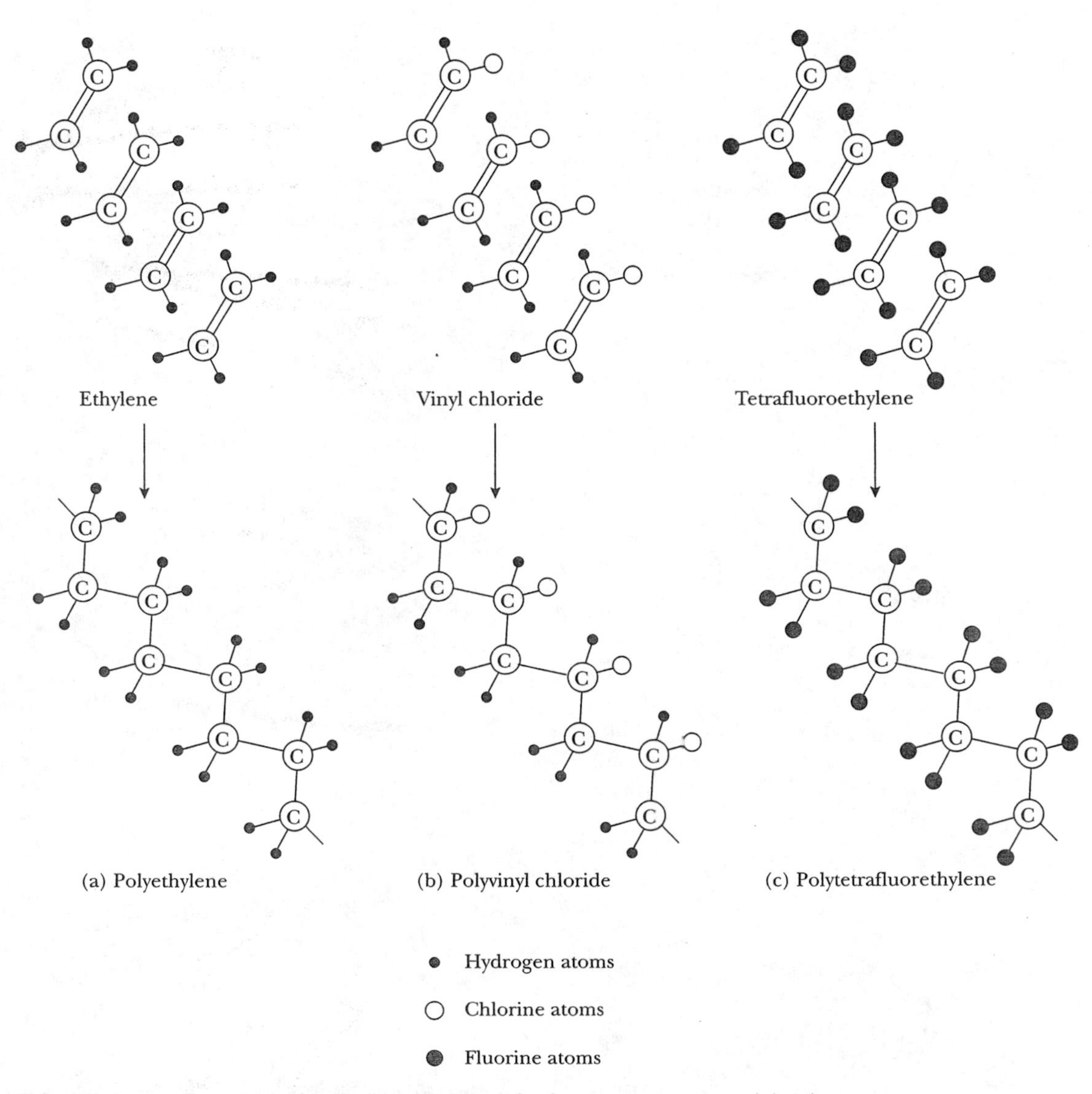

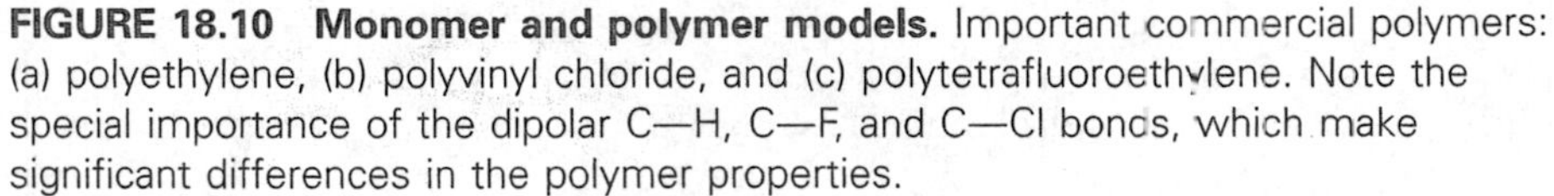

FIGURE 18.10 Monomer and polymer models. Important commercial polymers: (a) polyethylene, (b) polyvinyl chloride, and (c) polytetrafluoroethylene. Note the special importance of the dipolar C—H, C—F, and C—Cl bonds, which make significant differences in the polymer properties.

but not all of the fluorine atoms and replacing them with hydrogens: for example, $CHF{=}CF_2$ and $CH_2{=}CF_2$ are both used as monomers for that purpose.

Thermoplastics and Thermosets

We mentioned earlier in this chapter that where cross-links connect adjacent long chains, the polymer is strengthened and its thermal stability increased. Rigid **thermoset polymers** are highly cross-linked. **Elastomers**, polymers that can be stretched and then return to their original shape, are lightly cross-linked materials, and **thermoplastic polymers** are not cross-linked at all (Fig. 18.11).

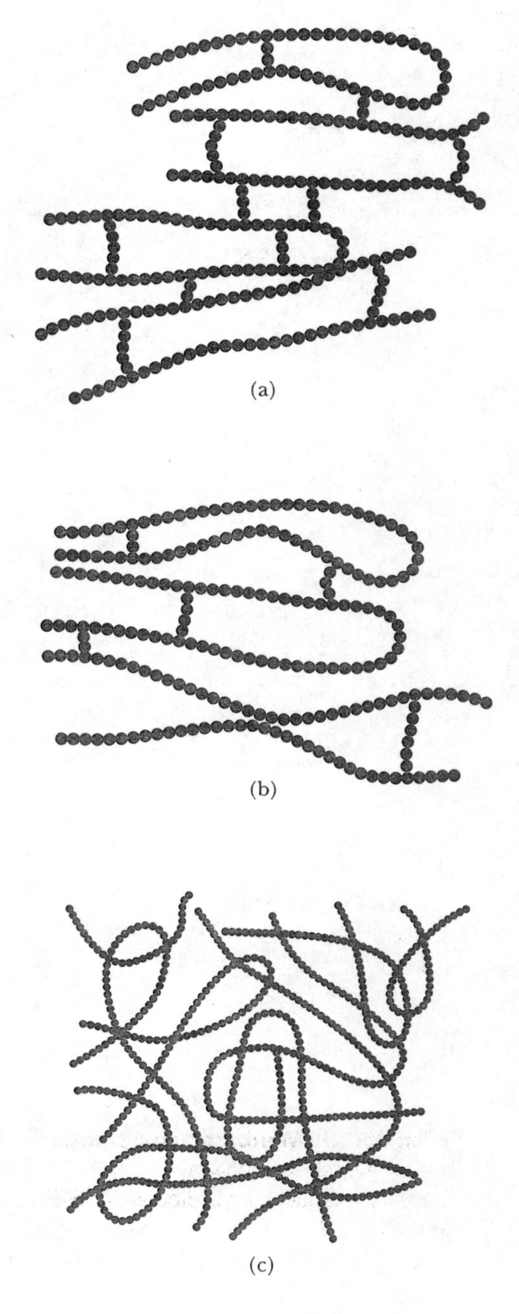

FIGURE 18.11 Polymer properties. (a) Thermoset polymeric materials are highly cross-linked, leading to rigidity and inflexibility. (b) Elastomeric materials are lightly cross-linked so the chains can be stretched out but snap back when the stressing force is removed. (c) Thermoplastic materials are not cross-linked and the chains can slide by each other.

- **Thermoset polymers** harden, or cure, becoming fixed in a given shape, by a chemical reaction that produces a highly cross-linked, three-dimensional network. Heat and pressure are generally used to kick off the reaction. Once formed, the cured polymer is generally insoluble in all solvents and cannot be heated to deform it without causing chemical decomposition. Typical examples are Bakelite and Formica, epoxy resins used to glue things permanently, and a composite material called sheet molding compound, which is used in the fabrication of large auto parts such as hoods and fenders.

- **Thermoplastic polymers** soften and can be made to flow when heated. On cooling, they harden and retain the shape established by molding or extrusion at the higher temperature, and the heating and cooling cycle can be repeated. A few of the more familiar examples are polyethylene, polystyrene, and polyvinyl chloride.

The different properties characteristic of thermoplastic and thermoset materials can be understood in terms of the model polymer structures shown in Figure 18.11. Whether the individual molecules are linear or branched, van der Waals forces loosely hold one polymer chain to another. In all cases except the cross-linked structure, simply heating the polymer provides enough energy to break up these weak interactions, allowing the chains to begin to separate from each other. This is thermoplastic behavior and it increases with temperature, increasing the plasticity of the material—that is, the ease with which it can be shaped. Most typically, no definite melting point is observed, but there is an approximate softening point, such as that of a glass.

Our model structure for a thermosetting resin (Fig. 18.11a) shows long-chain polymer molecules held together by covalent bonds formed by chemical reactions usually initiated by heating. The cross-links have covalent chemical bonds between molecules and give the material additional strength and rigidity, preventing the softening and melting that occur on heating thermoplastic resins. Thermoset polymeric materials do not melt or flow on heating, and they cannot be reshaped once cross-linking has taken place—they can be molded only once. Thermosets also tend to be insoluble and chemically resistant. Furthermore, if you accept the idea that each molecular chain is linked to every other by the cross-links, then a thermoset material must be one giant molecule. Can you see a "single" molecule? All by itself? Without a microscope? Next time you are in a sporting goods store or a bowling alley, just pick up a bowling ball. You will be holding a single, giant molecule in which all the chains are bridged, or cross-linked. Table 18.1 compares several characteristics of thermoplastic and thermoset materials.

Crystalline and Amorphous Polymers

If a long-chain molecule does not have bulky appendages—branches—off the backbone, it may crystallize, or form a **crystalline polymer**—that is, a polymeric material possessing a significant degree of molecular organization or *orderliness.* We cannot, however, expect the high degree of order that we find in covalent network structures like diamond, ionic crystals like table salt (NaCl), or crystals of familiar small molecules like cane sugar ($C_{12}H_{22}O_{11}$). Crystalline polymers are really only semi-crystalline at best, with small crystalline regions called **crystallites** dispersed through amorphous regions that have no characteristic long-range order (Fig. 18.12). Crystalline polyethylene, for example, can be grown from solution as the chains fold back and forth upon each other. The failure to grow perfect crystals is a result of the twisting and curling that occur because of the great lengths of the polymer chains. Polyethylene can exist in either a crystalline form or in a disordered amorphous form. The amorphous form is characterized by transparency, which is a desirable property for food wrap. The crystalline form is translucent because light scattering occurs between the crystallites and the amorphous regions in which they are embedded.

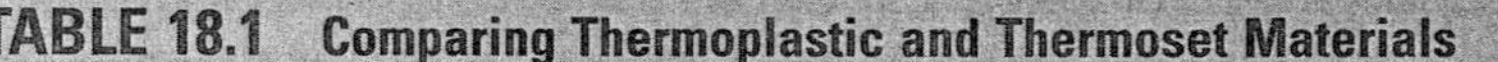

TABLE 18.1 Comparing Thermoplastic and Thermoset Materials

Characteristic	Thermoplastic	Thermoset
Cross-linked chains	No	Yes
Resoftenable on heating	Yes	No
Reformable parts	Yes	No
Recycle scrap	Yes	No
Strength	Moderate	Higher
Impact resistance	Moderate to good	Poor
Modulus of elasticity	Low	Moderate
Chemical resistance	Moderate to good	Good
Colorability	Good to excellent	Poor
Thermal stability	Fair	Good
Cost of material	Moderate-to-high	Low-to-moderate
Examples	**Thermoplastic**	**Thermoset**
	Polyethylene	Phenol-formaldehyde
	Polypropylene	Urea-formaldehyde
	Polystyrene	Polyurethane
	Polyfluoroethylene	Polysilane
	Polyester	Epoxy resins
	And many, many more	

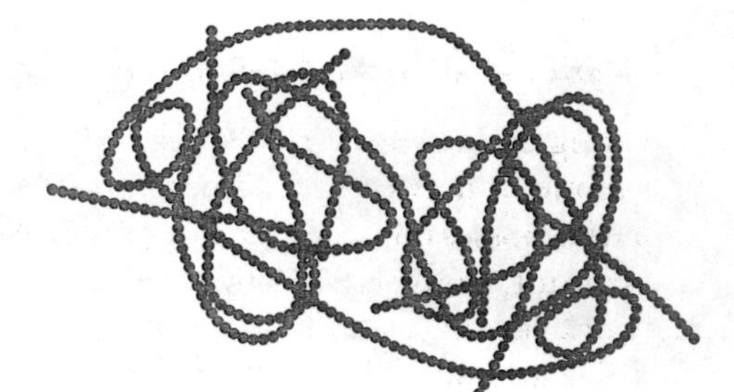

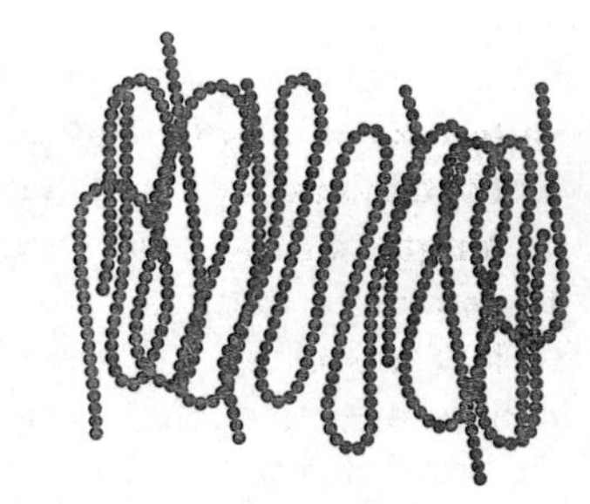

FIGURE 18.12 Amorphous and crystalline polymers. Ordered crystallites (bottom) are areas where the chains come together in some kind of alignment and regularity. These crystalline regions are generally found dispersed through amorphous regions of a polymer.

TABLE 18.2 Comparing Amorphous and Crystalline Polymers

Characteristic	Amorphous	Crystalline
Molecular structure	Random and loose	Partially ordered
Thermoplastic	Yes	Yes
Appearance	Clear	Opaque
Chemical resistance	Poor	Good
Forming parts	Slow	Fast
Change in shape on forming parts	Low shrinkage	High shrinkage
Basic processing	Physical change	Chemical change
Examples	**Amorphous**	**Crystalline**
	Polystyrene	Polyethylene
	Polycarbonate	Polypropylene
	Polyphenylene oxide	Polyester
	Natural rubber	Nylon-6,6
	And many, many more	

Amorphous polymers are disordered chains, much like the strands in a bowl of cooked spaghetti; we will use this analogy in a later section of this chapter to explain their properties. Table 18.2 compares some characteristics of amorphous and crystalline polymers. The behavior of amorphous and crystalline polymers as a function of temperature is particularly important, and we will describe this next.

Polymeric Materials and Temperature Effects

Heating provides an important means of distinguishing crystalline solids and amorphous substances (Chapter 16). A crystalline solid, for example, has a sharp melting point and goes directly from a solid to a liquid when this temperature is reached. On the other hand, an amorphous substance such as a glass simply softens and becomes less and less viscous as it is heated, eventually reaching a temperature where it flows freely.

In common with other amorphous substances, amorphous polymers have no definite melting points and generally become softer as the temperature is increased. There is, however, one temperature of considerable importance for amorphous materials, and this is called the glass transition temperature. The **glass transition temperature T_g** is the temperature at which the behavior of the material changes from being similar to that of a rigid glass to being similar to that of a liquid. It is sometimes also called the "brittle point," and below T_g the polymer is, in fact, brittle. One can demonstrate this by submerging a length of flexible plastic hose into liquid nitrogen and then shattering the hose with a hammer. Somewhere between room temperature and liquid nitrogen temperature, the glass transition temperature of the polymer has been crossed. An amorphous polymer must have a value of T_g well above room temperature if it is to hold its rigid shape under ordinary conditions.

The temperature behavior of a crystalline polymer is more complex. Such a polymer will also have a T_g value because it contains considerable amorphous regions. However, the ordered regions, or crystallites, will behave as a crystalline solid, and there will be a melting point T_m that is above T_g. Below T_g, the crystallites of a crystalline polymer will be buried in a glassy matrix of amorphous material, and the polymer will be brittle. Between T_g and T_m, the crystallites will be embedded in a flexible matrix, and the material will be flexible, yet tough. This is the best temperature range for *working* the polymer into a particular shape. For example, polyethylene bottles are blow molded within that temperature range as the polymer flows out to the walls of a bottle mold in response to the stress caused by a sudden burst of hot air. Above T_m, the amorphous regions will be highly flexible and the crystallites will have melted, and the polymer takes on the properties of a free-flowing liquid.

A graph of volume as a function of temperature differentiates amorphous and crystalline polymers (Fig. 18.13a). The behavior of an amorphous polymer follows the upper curve, with an increase in volume at T_g where the molecules have separated sufficiently to be free to move around. A crystalline polymer will show a slight volume increase at T_g as the amorphous regions become fluid, and then a greater increase in volume at T_m when the polymer becomes liquid. The effect of temperature on another mechanical property, stiffness or elastic modulus, is shown in Figure 18.13(b).

Amorphous Polymers and the Spaghetti Principle

An amorphous polymer can be compared to a bowl of well-cooked spaghetti. The strands are twisted and tangled in every direction, and there is no semblance of order. In contrast, the package of uncooked spaghetti with its straight strands all lined up in parallel rows shows considerable order. In fact, the disorder in a bowl of cooked spaghetti is less than the disorder in an amorphous polymer for two reasons. First, the polymer chains are much longer than the spaghetti strands in proportion to their width. Thus, you cannot pull a single molecule out of a polymer the way you can pull a single strand out of a bowl of spaghetti. Second, the polymer chain is not a smooth tube but rather a tight spiral coil that then is twisted and tangled. In a microscopic view, an amorphous polymer would look like a real mess (Fig. 18.14).

The disorder of an amorphous polymer can be increased by the addition of certain small molecules as impurities. These added compounds, called **plasticizers**, can occupy random sites in the polymer structure, thus decreasing the order even more. A plasticizer lowers the glass transition temperature and makes the polymer more flexible at a given temperature. The polymer polyvinyl chloride (popularly called *vinyl*) is used to make raincoats and curtains as well as the old LP records. A plasticizer is added to give the flexibility necessary in raincoats and curtains.

The tangled strands in a dish of spaghetti could be ordered somewhat by pulling the mass in opposite directions and largely orienting the strands in one direction. The same thing can be done with the chains in a polymer, which will tend to line up when they are pulled. Further stretching of the polymer will start to uncoil the spiral of the polymer chain. Two possible types of behavior are possible after such stretching and subsequent release of the stress. If the polymer is at least semi-crystalline, it will tend to stay in its new stretched shape. Such a

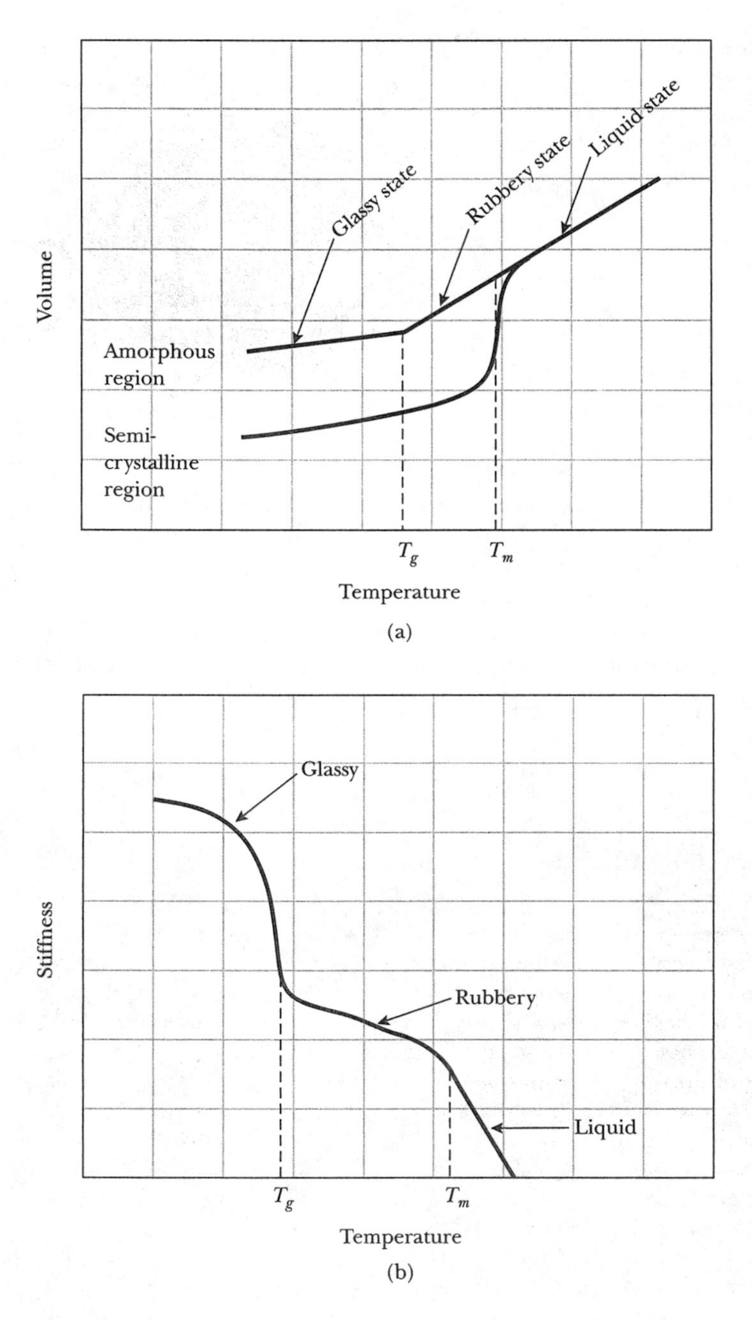

FIGURE 18.13 Distinguishing amorphous and crystalline polymers. (a) Change from the glassy to the liquid state with increasing temperature for amorphous and crystalline polymers; (b) change in stiffness (elastic modulus) with temperature for an amorphous polymer.

drawn polymer is a good candidate for a film or fiber. On the other hand, if the polymer is amorphous, the stretched polymer will likely exhibit elastic behavior and recoil quickly when the stress is released. As mentioned earlier, amorphous structure is normally associated with chain branching. Cross-linking also has an important effect on elastic behavior.

If the strands, or chains, are cross-linked, the structure becomes less flexible because movement is restricted. Properties of cross-linked polymers range from rigid to rubbery, depending on the degree of cross-linking. A rubber band has elasticity and is called an *elastomer* because it is mildly cross-linked, whereas

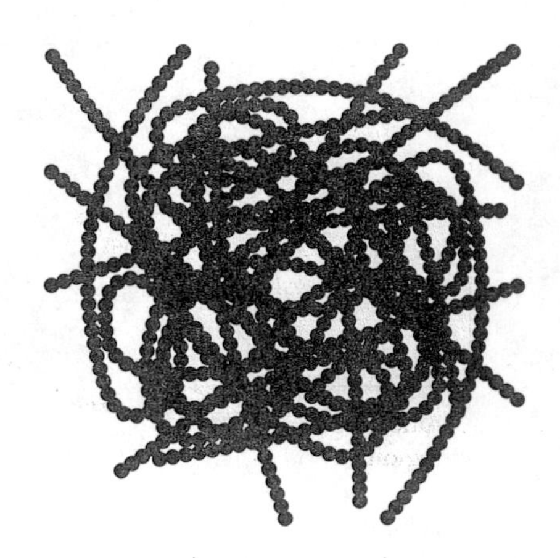

FIGURE 18.14 Spaghetti principle. Disorganization in amorphous polymers can be thought of as the random arrangement of floppy chains in a bowl of cooked spaghetti.

rubber tires have been vulcanized into a more highly structured three-dimensional arrangement. Natural rubber (polyisoprene) is vulcanized by adding sulfur, which produces disulfide (—S—S—) cross-links that tie the strands together (Fig. 18.15). These ties prevent the polyisoprene strands from separating from each other, thereby greatly increasing the mechanical strength of the material. For example, once vulcanized by adding about 5% sulfur and heating, the tensile strength of polyisoprene increases by a factor of ten.

PROFILES IN CHEMISTRY

Vulcanite Charles Goodyear of New Haven, Connecticut, realized that if rubber could be stabilized, it might have wide-ranging and remarkable applications. He had been experimenting with this substance, known to South American Indians as caoutchouc, which means "the tree that weeps." Others had introduced this material in Europe and the United States for erasing, or rubbing out, pencil marks; as a result, the name "rubber," suggested by Joseph Priestley, the discoverer of oxygen, caught on. It had found its way into rain gear and tablecloths as a waterproof coating for cloth, but unfortunately, it was not stable. If it was too hot, it became sticky and degraded; if it was too cold, it became brittle and cracked. In 1839, after several years of unsuccessful experiments, Goodyear accidentally dropped a chunk of rubber mixed with sulfur on a hot stove. The astonishing result was that it did not melt or degrade but actually hardened while retaining a good deal of its elasticity. By 1858, when his first rubber patent was issued, he had refined his process to produce a darkly colored substance he called "Vulcanite." Of all the applications Goodyear thought of for Vulcanite, there were at least two that he failed to recognize: inflatable tires, developed by John Dunlop in Ireland and later by Harvey Firestone in the United States, and the first new material for dental prostheses (false teeth) in 500 years. This use was discovered by Thomas Evans, an American dentist practicing in Paris, and as a result Vulcanite displaced gold, silver, ivory, and animal bone. In 1868, Vulcanite cost $4 per pound; from one pound, a dentist could produce up to ten dentures, for each of which he might charge $15, generating $146 in profit. Or, to look at it another way, that $15 denture contained about $0.40 worth of this new polymeric material. Very profitable, indeed! And with the contemporary discovery of nitrous oxide as an anesthetic in 1844 by Horace Wells of Hartford, Connecticut, extractions and dentures became the mainstay of the profession, leading more than one dentist to complain that too little attention was being devoted to the proper care of real teeth.

Questions

How would an automobile tire made of unvulcanized rubber perform?

What problems can you imagine the early dental prosthesis material (Vulcanite) produced?

George Washington's false teeth were made of wood. Comment on the use of wood for such an application.

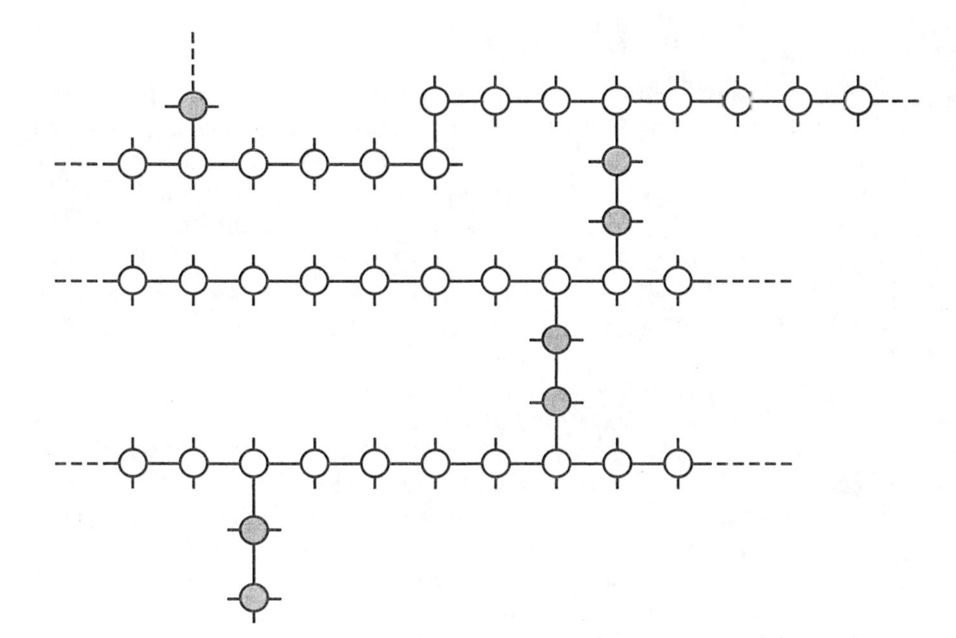

FIGURE 18.15 Vulcanization produces cross-linking in natural rubber. Natural rubber (polyisoprene) is vulcanized by adding sulfur, which produces disulfide (—S—S—) cross-links (indicated by shaded circles), tying chains together with chemical bonds.

PROCESSES IN CHEMISTRY

Latex Rubber Natural rubber, or latex, is a milky white, 35% aqueous suspension of the carbon–hydrogen polymer called *polyisoprene.* The most important source is the tropical tree *Hevea braziliensis,* which was the only source of rubber up to the time of World War II. After the water is removed, the crude polyisoprene rubber is a poor material with little strength and no elasticity because it has no cross-links. Vulcanization transforms it into a useful material. Heating the rubber with sulfur produces some disulfide links between C atoms in the polymer adjacent to the double bonds in the molecule. By adjusting the processing conditions, the degree of cross-linking can be altered and the properties controlled—the more cross-links, the stiffer and harder the rubber. For example, the vulcanized natural rubber in a typical car tire contains 3–5% (by mass) of sulfur and results in a cross-link for every 500 or so carbon atoms along the polymer chain. A very tough, hard rubber material that would be appropriate for the bumper on a truck or a train may contain up to 40% sulfur and is so extensively cross-linked that it is close to being a thermoset polymer. In fact, with time, rubber products often become harder and more brittle due to the cross-linking action of oxygen in the air.

Synthetic latexes such as styrene–butadiene copolymer or polyvinyl acetate are made by emulsion polymerization. Styrene–butadiene copolymer is widely used as a binder in exterior and interior paints, replacing the traditional drying oils. Polyvinyl acetate emulsions are found in well-known products such as Elmer's glue and binders and adhesives for wood and paper products.

Questions

What happens when natural rubber is vulcanized?

To what do you attribute the aging of automotive windshield wipers that have become brittle and ineffective over time, requiring that they be changed?

Effects of Stress on a Polymer

When subjected to a stretching force, a large number of thermoplastic polymeric materials exhibit the phenomenon of narrowing or *necking*, or *cold drawing*. At first, there is strong resistance to the stretching force. Then, the yield point is reached, and almost suddenly the material gives, stretching out and narrowing. Once necking has begun, the polymer is easier to stretch, and the narrowing begins to spread. At some point, it once again becomes more difficult to continue stretching, and finally, at the break point, it breaks. You can easily demonstrate this for yourself by taking a polyethylene carrying top for a six-pack of soda or beer and by following the procedure described in Chapter 17.

Necking is largely a consequence of the spaghetti principle. As you pull on the polymer, the molecules become aligned in the direction of the pull and stretch out. Molecules slide by each other, and thickness is sacrificed for length. More molecules in the necked region align, and as stretching continues, molecules are pulled away from the shoulders where the necking began. Finally, the strength of the material reaches the point where you need to pull hard enough to cause it to snap. Amorphous polymers can only be cold-drawn below their glass transition temperatures. If they are above T_g, they are in the rubbery state and simply stretch without necking, which is the case with Silly Putty.

SEEING THINGS

Silly Putty Without a doubt, one member of a family of polymers known as methyl borosiloxanes must be the most famous silicon compound ever. This polymer was accidentally discovered by scientists at the General Electric Central Research and Development Laboratories in Schenectady, New York, in 1943. What they were seeking was a very hard, tough material; what they got was extraordinary. It was a putty-like material that could be stretched like taffy, rolled up into a ball and bounced, and collapsed into a puddle if left on a bench top for any length of time. General Electric did nothing with the discovery except, we presume, play with it. But in 1949, an advertising copywriter by the name of Peter Hodgson heard about it. He obtained the rights to it, put a plug of it in a plastic egg, called it Silly Putty, and started marketing it as a novelty item through Nieman–Marcus and Doubleday Bookstores. Today, it is no longer the fad it was in the 1950s, when nearly 6 million pounds of the stuff were sold, but it is still around, sold by Binney & Smith (the manufacturer of Crayola crayons) in outrageous colors and even glow-in-the-dark versions.

At room temperature, Silly Putty is above its glass transition temperature, and because it is an amorphous polymer, it behaves in a characteristic way that you can easily demonstrate by rolling it between the palms of your hands into a rod-shaped piece about the size of a stick of chalk. If you now stretch it slowly, it will thin, sag (or creep) under its own mass, and eventually break apart. If you repeat the experiment, but this time stretch it abruptly, it fractures in what almost appears to be brittle failure, without stretching, with sharp edges being produced at the break. In the first experiment, the slow stretching allowed time for the molecules to pull by and through each other. But in the second experiment, the tensile stretching force was too great, and the bonds between polymer strands ruptured together.

Questions

Why does a ball of Silly Putty collapse flat if left alone but bounce off the floor if dropped?

What would be the consequences of the Silly Putty polymer's being below its glass transition temperature at room temperature?

What would be the consequences of the Silly Putty polymer's being crystalline rather than amorphous?

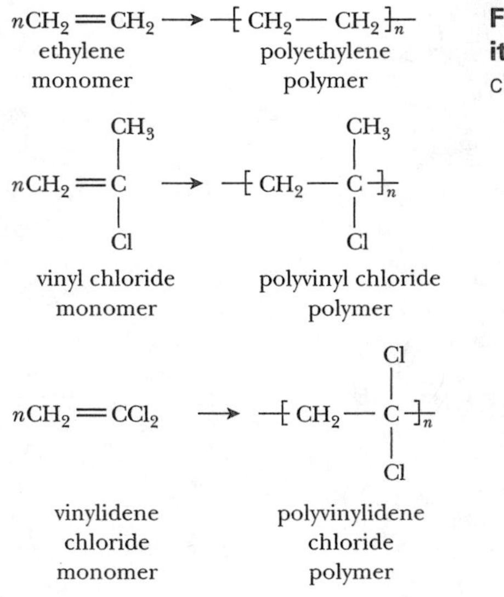

FIGURE 18.16 Polyethylene and two of its derivatives. Polyethylene, polyvinyl chloride, and polyvinylidene chloride.

Before turning our attention to the practical applications and processing of polymers, it will be helpful to consider the chemical structures of a few of the monomers and polymers that we have been discussing. Look at their chemical structures in terms of the general features of polymers that give them their characteristic properties. The details of the structure within the monomers will be easier to understand after we have covered organic chemistry (Chapter 21).

Figure 18.16 shows polyethylene and two of its derivatives. Attractions among strong dipolar C—Cl bonds account for polyvinyl chloride's and polyvinylidene chloride's having greater relative strength than polyethylene.

The repeating units of the polyester known as PET and the polyamide known as nylon-6,6 are shown in Figure 18.17 and Figure 18.18. These polymers are produced by step-growth (condensation processes). In each case, H_2O is the compound eliminated to connect the monomers. In both examples the monomers are difunctional. Proteins are polyamide polymers, and nylon polymers are synthetic polyamides. In both, "amide linkages" bridge the monomer

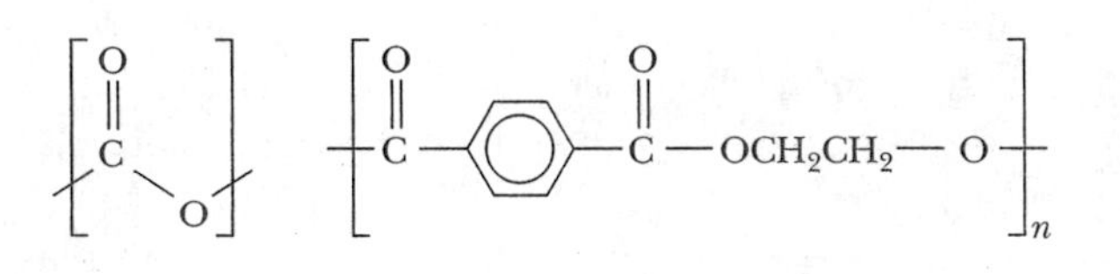

FIGURE 18.17 Polyethylene terphthalate. The polyester known more commonly as PET. The ester linkage is shown on the left.

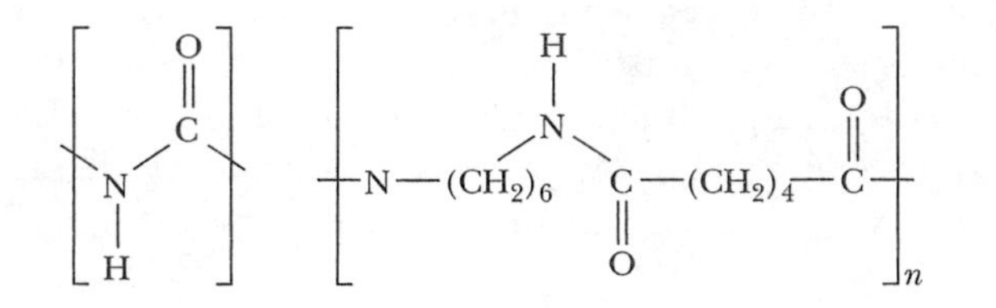

FIGURE 18.18 Nylon-6,6. The repeating unit in the first synthetic textile fiber, the one commonly referred to as simply "nylon." It s a polyamide, and the amide linkage is shown at the left.

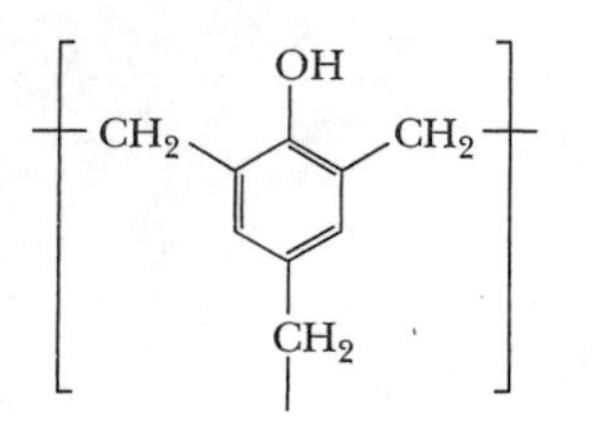

FIGURE 18.19 Bakelite. The repeating unit is characterized by the rigidity of the ring and extensive cross-linking.

units to form the macromolecules. The repeating unit at the right in Figure 18.18 is found in nylon-6,6, given that name because each monomer unit contains six carbon atoms.

Figure 18.19 shows the repeating unit in the thermoset resin Bakelite. The unit is trifunctional, permitting monomer units to link into a network. The result is a strong and complex three-dimensional arrangement. Once this polymerization has taken place, the product cannot be softened by heating and cannot be dissolved in any common solvent. It is said to be a thermoset.

18.3 POLYMERS AND PLASTICS

Commodity and Engineering Plastics

Commodity plastics are low cost, synthetic polymeric materials used for applications in which great demands are not made on their properties in manufacturing or use. They include the credit card plastics, the plastics of the shampoo bottle and food wrap, and the plastics of the "jewel" box in which compact disks come packaged (but not the disks themselves), the Styrofoam coffee cup, and *disposable* plates and kitchen utensils. For the most part, commodity plastics cost \$0.50 to \$1.00 per pound. Typical examples are thermoplastic materials such as polyethylene, polypropylene, polyvinyl chloride, polyisoprene, polystyrene, and polyvinyl acetate, as well as thermoset plastics such as phenolic polymers. You have probably heard of most of these materials.

Engineering plastics cost more than commodity plastics do. Typically, they sell for \$1.00 to \$5.00 per pound, and sometimes much more. These materials command higher prices because they possess enhanced thermal, mechanical, chemical, and electrical properties that allow them to be used in demanding engineering applications, especially in competition with metal, wood, and glass. You are also probably familiar with many of these. Compact disks are made of polycarbonate, as is the bulletproof plastic called Lexgard commonly found as the transparent barrier that separates the tellers from the public in banks. Bushings and bearings are often made of nylon, as are fishing tackle, ropes, and lines of all kinds and sizes. Kevlar is a woven material of polyamide fibers, and car bumpers are thermoplastic composites made of polyester and polycarbonate. Soda bottles are molded of thermoplastic polyester, which is also one of the most versatile polymers known to the textile and fibers industry. The multilayered ketchup bottle draws strength and character from polycarbonates, while other layers deflect oxygen in the air trying to enter and prevent moisture from leaving. Table 18.3 lists some important properties and applications of engineering plastics.

Of course, the line of demarcation between engineering and commodity plastics is not distinct. For example, polyethylene is widely used in the manufacture of milk bottle containers, an application that does not require great strength of material or significant chemical resistances and will not require a long service life. However, polyethylene is also almost exclusively the polymer of choice for pipelines distributing natural gas. Here, polyethylene must be able to resist the effects of various solvents in the gas stream, have enough elasticity to respond to mechanical deformation and internal pressure, and last for the better part of a century.

TABLE 18.3 Selected Properties and Applications for Engineering Plastics*

Field	Required Properties	Typical Applications
Electronics	Thermal stability; chemical and flame resistance; dimensional stability	Chip carriers; mountings; housings; storage disks; tapes; wire; cable
Packaging	Functionality; shelf life; permeation resistance; heat resistance	Films and containers
Automotive	Light weight; rigidity	Fenders; hood and trunk lids; frames; seats; glazing.
Aerospace	Thermal stability	Vertical panels; doors
Construction	Impact resistance; transparency; long-term strength	Glazing; plumbing fixtures Wall panels; ceilings, floors; roofing tiles; exterior shingles and boards.

*In 1997, packaging was far and away the leading market for engineering materials, followed by construction, electronics, transportation, and aerospace.

Processing Polymers

Much of the shaping and fabrication of solid polymers is similar to the ways in which metals are treated. They can be sliced, bent, creased and machined, molded and stamped, extruded and drawn. We will restrict our introduction to extrusion and molding, the two most common and important processing techniques. Keep in mind the major distinction between thermoplastic and thermosetting polymers. Thermosetting plastics may flow into a mold, but heat and pressure cause them to solidify quickly because of the chemical cross-linking that takes place. The shapes are permanent, and the scrap cannot be reworked. On the other hand, thermoplastics are processed by heating to the softening point at which they can be processed, and they are then cooled to hold their shape. The processing is reversible, and the finished parts and manufacturing scrap can be reworked repeatedly and, generally, inexpensively. But the processing of very large parts may be slowed by the time it takes to cool them to nearly room temperature, after which they can be removed from a mold.

Extrusion is an effective fabricating technique for thermoplastics and basically consists of the continuous driving of a liquid polymer through a die. The polymer is forced into the extruder as powder or pellets by the turn of a screw. Heat and pressure are applied, and the fluidized polymer emerges like so much toothpaste from a tube. A schematic diagram of a simple single-screw extruder is shown in Figure 18.20. Objects of many shapes can be made. Pipe and tubing are fabricated by forcing polymer through a ring-shaped die, films and sheets are produced, wires are coated, and fibers are "spun" by forcing the liquid polymer or a solution of the polymer through tiny pinholes.

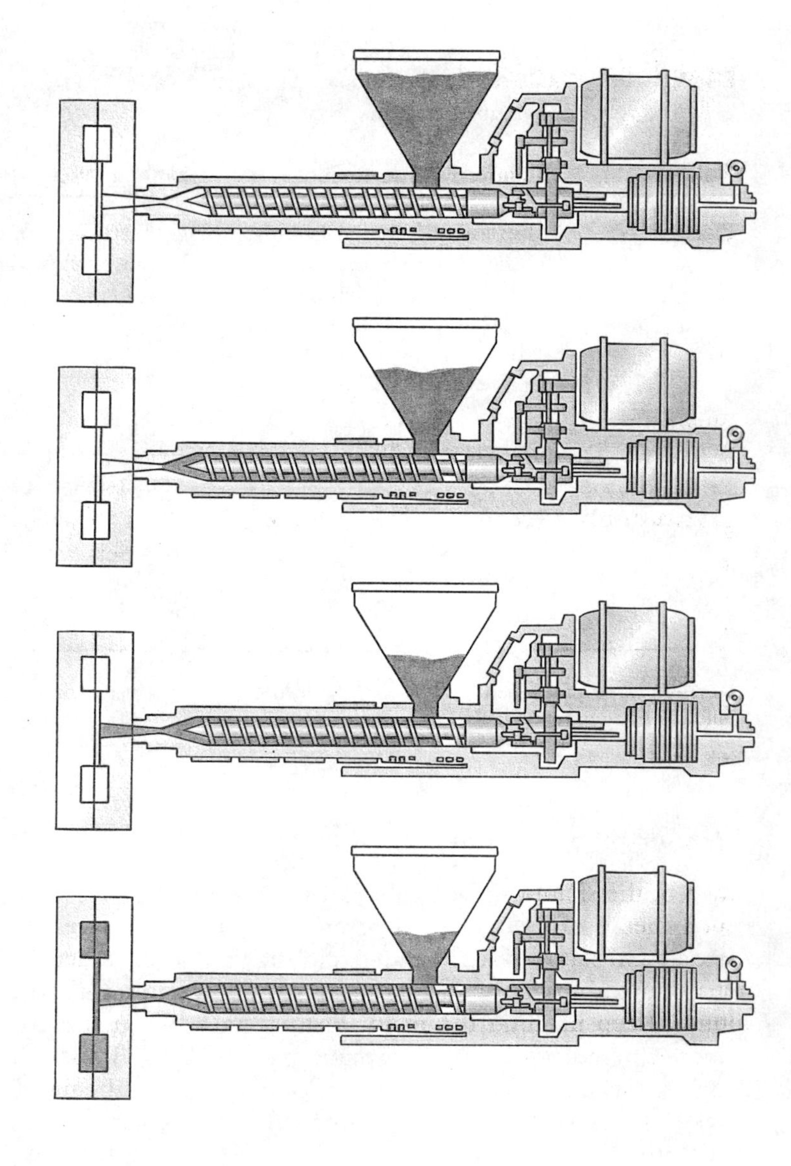

FIGURE 18.20 Fabricating plastic materials. Schematic diagram of a single screw extruder for molding objects. Successive stages for forcing a polymer into a mold are shown.

Molding involves filling a container called a *mold* with a material that can flow or be made to flow. The material hardens in the shape of the mold and is then removed. There are many different molding techniques. Compression molding involves placing a plug of thermosetting polymer in a mold and putting it under heat and pressure, which causes it to flow and fill the mold. Thermoplastic polymers are processed by injection molding: that is, they are literally shot into the mold from the extruder. Foamed plastics such as the egg carton and insulation board are produced by expansion molding, in which steam is used to expand polymer beads and fuse them together into the shape of the mold. In blow molding, the polymer fills the mold by being blown up like a balloon. This is the classic way in which bottles are produced.

18.4 POSTCONSUMER PLASTICS WASTE

Municipal Solid Waste (MSW)

About 200 million tons of municipal solid waste (MSW) is collected each year from households, institutions, and light industrial establishments. This amount represents about 4 pounds per day for every man, woman, and child in the United States. Approximately 13% is recycled; 14% is incinerated for energy recovery; and the remainder is put to rest in landfills (Fig. 18.21). As the population increases (if there is no per capita reduction), the total MSW will continue to rise. At the same time, our ability to rid ourselves of it in traditional ways is diminishing because the capacities of landfills are being exhausted, and incineration is becoming more expensive and more difficult because of government air pollution regulation and legislation. Major cities such as New York, Los Angeles, and Philadelphia are already experiencing difficulty, and the disposal problem is getting worse. Medical wastes and highly toxic heavy industrial wastes are part of the problem but are not addressed in this discussion. Our principal focus is on those components of MSW that are plastics.

Plastics Recycling

Approximately 18% by volume—8% by mass—of the MSW stream is plastics. Packaging materials, especially bags, bottles, and boxes, make up about half of the mass of the plastics. The most common plastic container being recycled today is the PET (polyethylene terphthalate) soft drink bottle. It is recycled at a level of about 20%, which is extraordinarily high. However, because of the fear of infectious contamination, PET bottles cannot be just ground up, melted down, and reused for bottles, even though the material is thermoplastic. For this reason, ground PET bottles have found their way into industries outside food packaging, such as the manufacture of carpet backing and Fiberfill insulation for

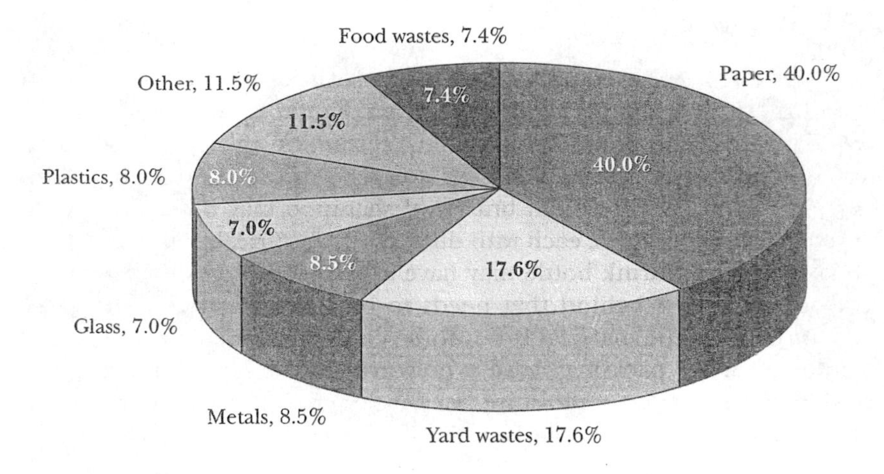

FIGURE 18.21 Municipal solid waste (MSW). Pie chart showing percent and distribution of major components of solid waste in landfills.

sleeping bags and ski jackets. Some PET from bottles is depolymerized, allowing recovery of the original monomers that then can be used as virgin chemical ingredients for many purposes, including refrigerator insulation, automobile bumpers, furniture, and food packaging.

HDPE (**h**igh-**d**ensity **p**oly**e**thylene) is widely used for milk. Like PET, ground HDPE cannot be reused in applications involving contact with food because of the risk of contamination. HDPE is easily identifiable and not mixed with other materials in these applications, and so HDPE is recycled for use in the manufacture of such products as trash cans, traffic cones, flower pots, and new base cups for soft drink bottles. On a small scale, ground HDPE is used for *plastic* lumber in applications such as railroad ties, decking for boat piers and docks, fencing, and park benches. It has advantages over wood for such applications because it will not splinter or chip, it does not need to be painted because the color can be included in the fabrication, and it will not rot.

One of the most visible forms of waste from plastic materials is polystyrene foam items such as fast-food containers, including *clamshell* burger boxes, *Styrofoam* coffee cups, disposable bowls and plates, and the *peanuts* used as a packing material. Although recycling programs for these items are being instituted in many regions, other areas are passing legislation banning polystyrene foam items even though institutional and fast food use makes it relatively convenient to recover and recycle these items. Unfortunately, the economics of polystyrene foam recycling is barely favorable at best because of the initial low cost of raw materials and manufacturing and of the relatively high cost of recovery and recycling. Typically, polystyrene foam items are collected in special trash containers in schools, fast-food restaurants, and institutional lunchrooms and cafeterias. They are cleaned and converted to pellets, which can then be mixed with other plastic materials to create second- and third-line items such as plastic lumber. The McDonald's burger box recycle program produced food trays for restaurant use and wall board and roofing panels for construction within the company. Unfortunately, the recycle program itself was trashed.

Other important recycle programs involving essentially unadulterated plastics include the recovery of plastic bags of LDPE (**l**ow-**d**ensity **p**oly**e**thylene), largely from supermarkets, and polypropylene cases from 12-volt automobile batteries.

Mixed Polymers

The greatest problems to be overcome in recycling plastics arise in cases in which multiple polymers are used in the same application. Examples are the complex multilayer barrier packages used for food items, cosmetics, and drugs, and the complicated multimaterial designs used for solving sophisticated automotive problems. For example, five different brands of shampoo may be packaged in five differently fabricated bottles, each with different caps and labels. Something as simple as the PET soft drink bottle may have a polypropylene or metal cap. Both of these leave a ring behind that needs to be removed after the cap is twisted off if the PET bottle material is not to be contaminated on recycling. A typical multilayer barrier pack may have a polycarbonate layer for strength, a polypropylene layer acting as a moisture barrier, a polyethylene vinyl alcohol copolymer serving as an oxygen barrier, the necessary tie layers of adhesives to glue them all together, as well as some ground polymer of uncertain origin as a between-layer filler. Try separating all of that! Or imagine trying to economically separate all the metal, glass, and plastic composites in an automobile. Advanced separation technologies are needed to cope with such problems.

Scrap tires are a large percentage of automotive plastic scrap. Well over 2 billion old tires are stockpiled around the country awaiting a sound economic solution to the problem of their recycling and reuse. To that outrageous number, 280 million more are added each year. Burying them is no longer an acceptable solution. One interesting idea is to grind them up and use them as an asphalt substitute. However, although such a substitute will last longer than the bituminous variety of pavement, it at present costs too much to process and prepare.

Biodegradability

A simple idea for dealing with MSW components such as newspapers, food, and many plastics materials is to let natural organisms bring about **biodegradation**. Unfortunately, biodegradation has not yet had a significant impact on MSW and the disposal problem in the United States. Biodegradation occurs when microorganisms such as fungi or bacteria produce enzymes that chemically degrade a polymeric material. Certain plastics such as the LDPE used in manufacturing grocery bags can be formulated with additives such as corn starch or vegetable oil that assist the biodegradation process in a landfill. Generally, these bags are not as durable and cost much more than the standard variety. Furthermore, the presence of these bags among standard LDPE bags may have the undesirable effect of drastically diminishing the properties of a recycled product.

An interesting and important problem is the choice of disposable plastic bags or disposable paper bags in the supermarket. Another is the choice between reusable cotton diapers and disposable plastic diapers. Neither choice is obvious. In a landfill, there seems to be little difference between paper and plastic because both deteriorate slowly and take up considerable space. Lots of trees get cut down to produce paper bags, and paper manufacturing pollutes huge amounts of water. On a pound-per-pound basis, the manufacture of most consumer plastics is much less energy-demanding and less polluting than that of paper and textile products, and a plastic bag weighs far less and carries far more than a paper one. It is true that the number of disposable diapers sent to landfills each year is gargantuan; however, washing cotton diapers is hugely energy-demanding and water-polluting.

Resource Recovery and Recycling

In the face of recent technological advances, it is possible and reasonable to expect that we will eventually be capable of recovering, separating, and recycling most polymers, plastics, and composite materials. This is a far cry from the world in which we presently live, where only 15% of our MSW is recycled, most of which is metal, glass, and paper—not plastics. Only about 2% of plastic materials are recovered and recycled at the present time. But it does not have to be that way!

The National Bottle Code

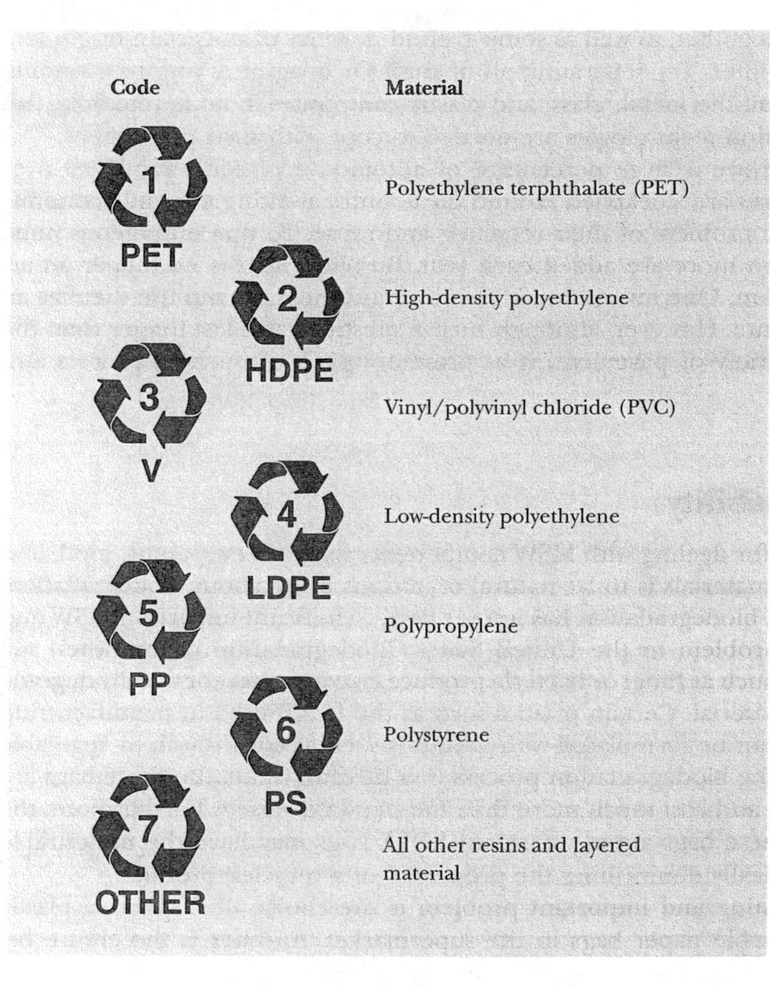

Code
Material
1
PET
Polyethylene terphthalate (PET)
2
HDPE
High-density polyethylene
3
V
Vinyl/polyvinyl chloride (PVC)
4
LDPE
Low-density polyethylene
5
PP
Polypropylene
6
PS
Polystyrene
7
OTHER
All other resins and layered material

In a typical recovery and recycle scenario, different types of plastics are placed in special compartmentalized trucks to keep materials separated. A code number is stamped on the bottoms of bottles to aid the process. The trucks empty their segregated contents at state-of-the-art municipal recycling facilities where they are shredded into strips or chunks and then into centimeter-sized flakes. Careful segregation and separation are important in order to guarantee (as much as possible) that an odd bottle of HDPE does not get into the PET compartment. This is important because of the huge differences in the softening points of these two materials; unwanted black specks can result when the softening point of one polymer is high enough to char the other. One HDPE bottle among 10,000 PET bottles is enough to ruin the clear, transparent appearance of the lot. With PVC (polyvinyl chloride) and PET, it requires only one bottle in 20,000. To effect separation, hydrocyclone tanks are filled with water, allowing HDPE to float free of PET, which sinks. Such density-based separations are possible with many plastic materials.

The sorted plastics flakes are then dried and transferred to pelletizers, which turn them into materials of different grades and colors. After being cooled and dried again, the pellets of recycled plastics are ready to be bagged and shipped to companies able to manufacture products designed within the property profiles of these recovered and recycled plastics.

Sound simple and straightforward? So why aren't more plastics materials being recycled? The answer to that question is complicated. Part of the answer is sociological, and part is psychological—our throwaway society needs to become the recovery-and-recycle society. Part of the answer is economic—products and markets need to be created where recycled plastics are used independently or are compatible with virgin plastics feedstock. In many cases, it is just cheaper to use virgin plastics instead of recycled materials. Part of the problem is governmental—federal, state, and local governments often act at cross-purposes and not in concert with industry and environmental groups. Part of the problem is technological—because there are so many classes and varieties of polymers, plastics and composites, the separation technologies required are often difficult and changing all the time. Recycling is a problem in need of a solution. Good progress is being made, but we are not there yet.

SUMMARY

Size accounts for the special properties of the covalent molecules called polymers. When the length of the chain results in molecular masses on the order of 5000 to 50,000 amu, properties become observably polymeric. Monomers are the small molecule building blocks that get chemically *strung* together into these giant molecules called polymers. Homopolymers are composed of identical repeating units; copolymers are made up of different repeating units.

There are two principal classifications of polymers that differ according to whether the polymers are formed by adding monomer units—chain-growth polymerization—or condensation of monomer units with loss of a small molecule in the process—step-growth polymerization. Polyethylene is typical of the chain-growth polymers with the double bonds providing the means for snapping units together. Siloxane polymers are made by the step-growth process.

The number of monomer repeating units that comprise the polymer chain is the degree of polymerization. Because the polymerization process produces a distribution of different length chains all forming at once, polymer molecular masses are always average values except for a few biological polymers. Molecular mass determinations depend on the way they are measured—number-average molecular masses, M_n and mass-average molecular masses, M_w. Mass-average molecular masses are the larger. Dividing M_w by M_n gives the polydispersity, a measure of the range of molecular masses in the polymer.

The extent to which a polymer is cross-linked determines whether the properties will be primarily thermoplastic or thermoset. Because thermoplastic polymers are not cross-linked, they soften on heating and can be made to flow. On cooling, they harden and retain the shape established by molding or extrusion at the higher temperature, and the heating and cooling cycle can be repeated. Thermoset polymers are chemically hardened, or cured, becoming fixed in the desired shape because of reactions that produce highly cross-linked, three-dimensional networks. Once formed, the cured polymer is generally insoluble and cannot be heated to deform it without causing chemical decomposition. As the degree of cross-linking of the polymer chains increases, the material becomes less flexible. The mechanical properties of natural rubber are greatly enhanced by vulcanization, a cross-linking process, which adds disulfide bonds between strands, turning a thermoplastic elastomer into a thermoset polymer.

All polymers are amorphous materials but many contain organized regions called crystallites resulting from alignment of segments of chains. Crystalline character is favored by lack of branches and cross-links. The temperature range across which the transformation from a flexible material to a brittle glass takes place is known as the glass transition temperature T_g. Below this temperature, the polymer exists as a glassy or semicrystalline substance. The polymer is flexible above this temperature. If the polymer is crystalline there is a melting temperature T_m where the polymer assumes the usual characteristics of a free flowing liquid. In the working temperature range between T_g and T_m, the polymer will flow like a liquid when subjected to stress.

Even though postconsumer plastic waste comprises only 8% by mass of the materials in landfills, it is a valuable resource worth recovering for commercial and environmental reasons.

TERMS

Polymeric materials (18.1)
Polymer (18.1)
Monomer (18.1)
Vulcanization (18.1)
Plastics (18.1)
Resins (18.1)
Polymerization (18.1)
Degree of polymerization (18.1)
Homopolymer (18.1)
Copolymer (18.1)
Random copolymer (18.1)
Alternating copolymer (18.1)
Graft copolymer (18.1)
Block copolymer (18.1)
Branches (18.1)
Cross-links (18.1)
Chain-growth polymerization (18.1)
Addition polymerization (18.1)
Step-growth polymerization (18.1)
Condensation polymerization (18.1)
Number-average molecular mass (18.1)
Mass-average molecular mass (18.1)
Monodisperse (18.1)
Polydisperse (18.1)
Polydispersity (18.1)
Thermoset polymer (18.2)
Elastomer (18.2)
Thermoplastic polymer (18.2)
Crystalline polymer (18.2)
Crystallite (18.2)
Amorphous polymer (18.2)
Glass transition temperature (18.2)
Plasticizer (18.2)
Commodity plastics (18.3)
Engineering plastics (18.3)
Extrusion (18.3)
Molding (18.3)
Biodegradation

IMPORTANT EQUATIONS

$$M_n = \frac{W}{\Sigma N_i} = \frac{\Sigma N_i M_i}{\Sigma N_i}$$ Number-average molecular mass

$$M_w = \frac{\Sigma N_i M_i^2}{\Sigma N_i M_i}$$ Mass-average molecular mass

QUESTIONS

Conceptual questions are denoted by a square screen. Extra-credit questions are denoted by a circular screen.

1. In what ways does size account for the differences in properties of small molecules and very large molecules? Consider $H—[CH_2—CH_2]_n—H$ where n is any number 1 through 10 as an example of a small molecule and $n = 1000$ as an example of a very large molecule.

2. As the basis for a brief discussion of the difference between homo- and copolymers, explain why a strictly alternating copolymer could just as well be described as a homopolymer.

3. Polytetrafluoroethylene, well known for its nonstick surface properties and marketed under the trade name Teflon, is a polymer with a backbone, or main chain, composed of regularly repeating $—CF_2—$ units. In order to modify the properties somewhat so materials have some degree of stickiness, a material known as polyvinylidene fluoride has been synthesized. It has a backbone made up of alternating $—CH_2—$ and $—CF_2—$ units. What are the chemical formulas for the respective monomers for each of these polymers?

4. In what way is the monomer unit in a homopolymer related to its molecular mass and degree of polymerization?

5. What are the general effects of branching on the properties of polymers?

6. What are the effects of cross-linking on the properties of polymers such as an elastic band and the sidewall of a tire?

7. Using your answers to Questions 5 and 6, distinguish between thermoplastics and thermosets.

8. What happens if you include a small percentage of monofunctional molecules among the difunctional molecules being used to prepare a homopolymer in a step-growth polymerization?

9. What would be the critical result of accidentally or purposely including a small percentage of trifunctional molecules among the difunctional molecules being used to prepare a homopolymer in a step-growth polymerization?

10. How could you tell whether a polymerization reaction follows a chain-growth or step-growth pathway to high molecular mass?

11. Why is M_w typically greater than M_n? What would you conclude about the odd result when they are the same?

12. How do you account for the increased tensile strength observed in polyvinyl chloride versus polyethylene, based on a difference of only one atom—substituting a chlorine (in the vinyl chloride monomer, $CH_2{=}CHCl$) for a hydrogen (in the ethylene monomer, $CH_2{=}CH_2$)?

13. What simple thermal observation can be used to distinguish between crystalline and amorphous polymers, and why is that the case?

14. The temperature range between T_m and T_g is significant for polymers. Why?

15. Why won't Silly Putty cold flow, or neck? Or to put it another way, where is the silicone polymer known as Silly Putty with respect to T_m and T_g? Briefly explain.

16. In a traditional landfill, little of the organic wastes—paper, food, and plastics—biodegrades in spite of the fact that a large percentage of these materials are biodegradable. Why is that so? What can be done about it?

PROBLEMS

Problems marked with a bullet (•) are answered in Appendix A, in the back of the text.

Polymers in Practice [1–8]

•1. Identify *three* woven items of apparel as being constructed of synthetic, partially synthetic, or natural textile fibers.

2. Identify *three* items of sporting goods that are molded of synthetic plastics or are made of composite materials.

•3. Identify *four* materials, products or applications found in the kitchen that are not made of wood, metal, glass, or ceramics.

4. Identify *four* materials, products or applications found in the bathroom that are not made of wood, metal, glass, or ceramics.

•5. Identify *five* materials, products, or applications found on the outside of a house that are not made of wood, metal, glass, or ceramics.

6. Identify *five* materials, products, or applications found on the outside of an automobile that are not made of wood, metal, glass, or ceramics.

•7. Identify *six* materials, products, or applications found within the passenger compartment of an automobile that are not made of wood, metal, glass, or ceramics.

8. Identify *six* materials, products, or applications found under the hood or body panels of an automobile that are not made of wood, metal, glass, or ceramics.

Molecular Masses of Polymers [9–20]

•9. Calculate the degree of polymerization of a polyphenylene oxide polymer having an average molecular mass of 13,800. The monomer unit in the polymer has a molecular mass of 92 amu.

10. Calculate the degree of polymerization of a polyethylene oxide polymer having an average molecular mass of 660,000. The monomer unit has a molecular mass of 44 amu.

•11. Calculate M_n and M_w values for a mixture of five polymer chains having the following molecular masses, respectively: 120,000, 135,000, 150,000, 175,000, and 190,000.

12. Calculate M_n and M_w values for a mixture of five polymer chains having the following molecular masses, respectively: 65,000, 68,000, 68,500, 69,000, and 70,000.

•13. Calculate the polydispersity for the polymer in Problem 11 above.

14. Calculate the polydispersity for the polymer in Problem 12 above.

•15. If the polydispersity of a particular polymer happens to be 1.0, what is the relationship between M_n and M_w? Why?

16. Would you use colligative properties or light scattering methods to determine M_w values? Why?

•17. The degree of polymerization of a polyphenylene sulfide polymer is 180, and the molecular mass of the monomer unit is 108. Calculate the average molecular mass of the polymer.

18. The degree of polymerization of a polycarbonate polymer is 245 with an average molecular mass of 62,200. Calculate the molecular mass of the monomer.

•19. Determine the degree of polymerization for the nylon polymer of molecular mass 113,000 and formula $—[NH—(CH_2)_6—CO]_n—$.

20. Determine the degree of polymerization for the polyester of molecular mass 86,400 in Figure 18.17.

Additional Problems [21–25]

21. Equal masses of two polymers are blended together. Given the following data, calculate M_n and M_w and determine the polydispersity of the mixture.

	Polymer A	Polymer B
M_n	25,000	25,000
M_w	75,000	225,000

22. Equal masses of two polymers are blended together. If the polymers are monodisperse and have molecular masses of 9000 and 45,000, respectively, what are the values of M_n and M_w?

•**23.** Polypropylene and polyisoprene are both hydrocarbon polymers, but polyisoprene contains double bonds and polypropylene does not. One is an excellent, fiber-forming thermoplastic material whereas the other is a thermoplastic elastomer. Which is which, and why the difference?

24. Subject a thermoplastic polymer—like the polyethylene carrying top for a soft drink or beer six-pack top—to a stress by pulling on it. At some point, the applied stress is sufficient to cause the material to yield (Chapter 17). Describe what happens at the molecular level (a) under stress but before the yield point and (b) after the yield point. Why is the polymer more transparent in the *stretched* region?

•**25.** You have three *alcohols*:

CH_3CH_2OH
ethanol, a monofunctional molecule

$HOCH_2CH_2OH$
ethylene glycol, a difunctional molecule;

$$\begin{array}{l} HOCH_2\underset{\displaystyle |}{C}HCH_2 \\ \quad\quad\;\; OH \end{array}$$
glycerol, a trifunctional molecule

Which would you choose to prepare a
(a) thermoplastic polymer?
(b) thermosetting polymer?

Cumulative Problems [26–29]

26. In the recycling of standard liter polyethylene terphthalate soft drink bottles, one process depolymerizes the polymer into monomers. After separating the paper labeling material, remaining cap material, base material, and base glue, 42.5 g remains. How many grams of each monomer could possibly be recovered? The polymer is a condensation homopolymer composed of two strictly alternating monomers with molecular masses of 62 and 122. One molecule of water is eliminated with the formation of each polymer link.

27. Briefly explain whether a plasticized polymer will have a higher or lower entropy than the pure polymer.

28. Polypropylene is a hydrocarbon polymer, yet gasoline tanks for automobiles are designed in polypropylene and you do not have to worry about the gasoline dissolving the tank. Why?

•**29.** It is well known that the heat capacities of polymeric materials and composites are universally higher than those of metals and their alloys. Why?

ESTIMATES AND APPROXIMATIONS [30–31]

30. It is said that a single PVC bottle among 20,000 PET bottles is enough of a contamination to prevent the successful reintroduction of recovered material in similar applications. Estimate the contamination in parts per million (ppm).

31. It has been known for a long time that polyisoprene (natural) rubber can be toughened and hardened without losing elasticity by adding sulfur and heating, a process Goodyear called "vulcanization." If cross-linking during vulcanization is assumed to occur at the expense of the remaining double bonds, estimate the fraction of monomers that are cross-linked in a rubber containing 5% sulfur. Assume the sulfur cross-links are disulfide (—S—S—) groups.

FOR COOPERATIVE STUDY [32–36]

32. Suggest an experiment to determine if stress-induced crystallization occurs when a thermoplastic elastomer like a rubber band is stretched.

33. One of the great controversies in the early years of polymer chemistry was whether these giant polymers were aggregates, or collections, of small molecules, held together perhaps as the particles in stable colloidal suspensions like the globules of fat in homogenized milk. That is not the case. The chemical bonds linking monomer units in polymers are *regular* covalent bonds. How can you tell?

34. Another result of the great polymer controversies of the 1920s was the notion that no polymer was a "pure" chemical compound in the usual sense of the word. Why are true polymers—with the exception of a some biopolymers—never "pure"?

35. What is the bond angle between successive carbon atoms along the backbone of a linear polyisoprene polymer? At the points where the polymer is branched, in natural rubber? At the points where the polymer is cross-linked, in vulcanized rubber?

36. In what way do melting point data provide characteristic differentiation between monomeric and polymeric substances?

WRITING ABOUT CHEMISTRY [37–38]

37. Before polymer science was even thought of there were many natural polymers such as cotton, wool, and rubber. Explain why the development of polymer science had such an important effect on the progress of society in spite of the fact that natural polymers were already in use. Could we now do completely without natural polymers? Are some natural polymers better for some uses than any synthetic polymer? Explain your answer for each of these questions.

38. Discuss the various ways of disposing of plastics waste including recycling. Which way of disposing of this waste do you think is the best? Explain why you feel this way.

chapter nineteen

TRANSITION METALS

19.1 PROPERTIES OF THE TRANSITION METALS

Practical properties such as electrical conductivity and malleability make metals among the most useful of all substances. We have already discussed the alkali metals (Group IA) and the alkaline-earth metals (Group IIA). These elements have quite typical metal properties, but they are probably not the elements that first come to mind when you think of metals. You are far more likely to identify iron, copper, gold, and silver as typically "metallic" elements. Other metals that easily come to mind are nickel, chromium, and platinum. All of these metallic elements are transition metals and all have important uses, ranging from metallurgical applications to chemotherapy. The **transition metals** include the 30 elements in the *d* block of the 4th, 5th, and 6th periods (Fig. 19.1), as well as the 14 actinides and 14 lanthanides. The lanthanides and actinides are often called inner-transition elements. In addition, the elements numbered 104 to 112 fill out the 7th period of the *d* block, giving us still more transition metals. Altogether the transition metals account for more than half the known elements.

Because they are metallic, the transition elements are generally lustrous, electricity and heat conducting, opaque, malleable, and ductile, but they also have some distinguishing characteristics of their own. For example, most exhibit multiple oxidation states in the compounds they form. A large fraction form colored compounds, and many of the elements and their compounds are attracted to a magnetic field. Collectively, these properties are the result of partially filled

FIGURE 19.1 The transition metals include the 30 elements in the *d* block. These are the metallic elements of the 4th, 5th, and 6th periods. There are also the 14 actinides and 14 lanthanides, often called inner-transition elements. The elements numbered 104 to 112 fill in the 7th period in the *d* block.

21 Sc 44.9559	22 Ti 47.88	23 V 50.9415	24 Cr 51.9961	25 Mn 54.9380	26 Fe 55.847	27 Co 58.9332	28 Ni 58.69	29 Cu 63.546	30 Zn 65.39
39 Y 88.9059	40 Zr 91.224	41 Nb 92.9064	42 Mo 95.94	43 Tc (98)	44 Ru 101.07	45 Rh 102.9055	46 Pd 106.42	47 Ag 107.8682	48 Cd 112.411
57 *La 138.9055	72 Hf 178.49	73 Ta 180.9479	74 W 183.85	75 Re 186.207	76 Os 190.2	77 Ir 192.22	78 Pt 195.08	79 Au 196.9665	80 Hg 200.59

d shells in the elements themselves or in their compounds or ions. A typical example is iron with its partially filled *d* shell and an electron configuration of $[Ar]4s^2 3d^6$. The *d* orbitals are completely filled in the ground-state copper atom, and there is a half-filled *s* orbital $[Ar]4s^1 3d^{10}$. However, the copper(II) ion does have the partially filled *d* shell $[Ar]3d^9$ and displays all the distinguishing properties of an ion of a transition metal.

The Group IIB elements zinc, cadmium, and mercury have filled-shell electron configurations. Zinc, for example, is $[Ar]4s^2 3d^{10}$. These elements and their compounds are not attracted to a magnetic field, and multiple oxidation states and colored compounds are rare. They are often not counted as transition metals, but we find them more conveniently treated as such.

One of the defining properties of transition-metal elements is the ease with which they form coordination complexes. **Coordination complexes** are stable arrangements of metal atoms or cations surrounded by atoms, ions, or molecules. Transition metals tend to form coordination complexes because their ions have relatively small sizes and high charges.

The key to understanding most of the distinguishing properties of the transition metals and their coordination complexes lies within their electronic configurations. In Chapter 5 we discussed the buildup of atoms according to the aufbau principle involving the addition of electrons, one by one, with accompanying increase in nuclear charge. As the nuclear charge increases, the energy of the orbitals drops in response to the increased attraction to the nucleus. However, orbitals with regions that are quite distant from the nucleus are **shielded** by the electrons in inner orbitals. Shielding reduces the nuclear charge felt by outer electrons and has the effect of reducing the amount of stabilization, raising the energy of the outer orbitals. Because the $3d$ orbitals are well-shielded by inner orbitals, their energy does not drop significantly as the nuclear charge is increased while the $1s$, $2s$, $2p$, $3s$, and $3p$ orbitals are being filled. Thus, when the filling of the $3p$ orbitals is completed, the next orbital to be filled is the $4s$ for potassium and calcium. The $4s$ orbital is sufficiently far from the nucleus that it does not completely shield the $3d$ orbitals, and these finally can drop in energy as the nuclear charge increases. After two electrons have been placed in the $4s$ orbital, the next ten electrons fill the five $3d$ orbitals, giving us the first transition series from scandium to zinc. Then the $4p$ orbitals fill for the elements gallium to krypton. In a similar way $4d$ fills after $5s$, and $5d$ starts to fill after $6s$.

The *f* orbitals are shielded even more effectively than *d* orbitals, and only when the $6s$ orbital is being filled do the $4f$ orbitals begin to show a significant drop in energy. At this point the $5d$ and $4f$ orbitals drop to competitive values. One electron goes into a $5d$ orbital to give lanthanum and then the seven $4f$ orbitals begin to fill, giving rise to the next fourteen elements. Because electrons in f orbitals do not contribute significantly to chemical properties, the fourteen elements are all similar to lanthanum and are called the lanthanide elements or rare earths. In the next period, the same explanation can be used to explain the similar properties of the actinide elements, although the competition between $6d$ and $5f$ orbitals is even keener and the filling is more irregular.

19.2 THE FIRST TRANSITION SERIES

Fourth-Period Elements from Scandium to Zinc

It is the $3d$ electrons that characterize the first transition series, elements Sc to Zn. Within this series, the properties change in a relatively smooth manner with increasing atomic number. Some properties of the first transition series are shown in Table 19.1. Note that from scandium through copper there is a steady increase in density, a periodic property that depends on decreasing size of the atoms as electrons are added to the d shell. This decreasing size is a result of the less effective shielding of the nucleus by electrons in the $3d$ shell that have previously been added. As the nuclear charge increases, the effective nuclear charge felt by these electrons increases, drawing them in closer to the nucleus. The **effective nuclear charge** for an electron is the nuclear charge exerted on the electron after the charge reduction due to shielding is considered.

The increasing values of the first ionization energy also reflect the increased attraction of the electrons for the nucleus as the effective nuclear charge they are subjected to increases. Zinc has a particularly high first-ionization energy, because its $3d$ and $4s$ subshells are complete.

TABLE 19.1 Properties of the First Transition Series

Metal	m.p. (°C)	b.p. (°C)	d (g/cm^3)	I_1 (eV)*	I_2 (eV)	I_3 (eV)
Sc	1200	2400	2.5	6.54	12.80	24.45
Ti	1677	3277	4.51	6.83	13.57	27.47
V	1919	3400	6.1	6.74	14.65	29.31
Cr	1903	2642	7.14	6.76	16.49	30.95
Mn	1244	2095	7.44	7.43	15.64	33.69
Fe	1535	3000	7.87	7.90	16.18	30.64
Co	1493	3100	8.90	7.86	17.05	33.49
Ni	1453	2732	8.91	7.63	18.15	35.16
Cu	1083	2595	8.95	7.72	20.29	36.83
Zn	420	908	7.14	9.39	17.96	39.7

*Ionization energies in electron volts. (1 eV for 1 mol of electrons = 96.49 kJ.) I_1, I_2, and I_3 are the energies necessary to remove the first, second, and third electrons.

TABLE 19.2 Ground-State Electron Configurations of the First Transition Series

Orbital	Number of Electrons									
	Sc	Ti	V	Cr	Mn	Fe	Co	Ni	Cu	Zn
4*s*	2	2	2	1	2	2	2	2	1	2
3*d*	1	2	3	5	5	6	7	8	10	10

Now let us look at the electron configurations for the elements of the first transition series (Table 19.2). Notice the irregularities that occur at Cr and Cu. This is the result of the stability of a half-filled or fully filled *d* shell, either 5 or 10 *d* electrons. Thus, Cr and Cu with $4s^13d^5$ and $4s^13d^{10}$ shell configurations are more stable arrangements than the $4s^23d^4$ and $4s^23d^9$ configurations. The electron configurations for the two series below the first transition series are even more irregular.

Removal of electrons from transition-metal atoms produces cations. With smaller numbers of electrons, the nuclear charge felt by the outer-shell electrons increases. Therefore, the energy of the 3*d* orbitals drops relative to that of the 4*s* orbitals, and experimental evidence shows that all first transition-series cations have empty 4*s* orbitals. In addition, one or more *d* electrons may also be missing. However, higher ion charges are increasingly less stable, going from Sc to Zn, and therefore the total number of 4*s* and 3*d* valence electrons that are removed is generally limited. For most of the metals, several different ions are known (Table 19.3).

Starting with the first of the fourth-period transition elements, Sc loses all three valence electrons to form the Sc^{3+} ion. Titanium can lose all four 4*s* and 3*d* electrons to form the Ti^{4+} ion. However, such a high charge cannot be stabilized in a pure ionic compound, and so Ti(IV) compounds such as $TiCl_4$ are largely covalent. The melting point of $TiCl_4$ is $-30°C$, much lower than that of

TABLE 19.3 Common Oxidation States of the First Transition Series

	+1	+2	+3	+4	+5	+6	+7
Sc			x				
Ti		x	x	x			
V		x	x	x	x		
Cr		x	x	x	x	x	
Mn		x	x	x	x	x	x
Fe		x	x				
Co		x	x				
Ni		x					
Cu	x	x					
Zn		x					

an ionic salt. More ionic compounds occur for Ti^{3+} and Ti^{2-}. Proceeding along the first transition series through Fe, successively higher oxidation states become possible, although actual cations are generally limited to the +2 and +3 states. Cobalt forms stable compounds only as Co^{2+} and Co^{3+}, Ni as Ni^{2+}, and Cu as Cu^{+} and Cu^{2+}. For Zn, formation of an ion greater than +2 would involve breaking into the very stable filled *d* shell, so Zn^{2+} is the only Zn ion. Oxidation states commonly observed for the first transition series are listed in Table 19.3. For transition metals with multiple oxidation states, ions that have high states can drop to lower states and thus tend to be good oxidizing agents, whereas those that have low states tend to be good reducing agents.

EXAMPLE 19.1

Determine the valence electron configuration of the Fe^{3+} ion.

COMMENT Note first that the valence electron configuration of Fe is $4s^2 3d^6$. Removal of the two *s* electrons gives the Fe^{2+} ion. It is necessary to remove one *d* electron to produce the Fe^{3+} ion.

SOLUTION The valence electron configuration is $3d^5$.

EXERCISE 19.1

Determine the electron configurations of the Cu^{2+}, Cr^{3+}, Mn^{2+}, and V^{4+} ions.

ANSWER

Cu^{2+} $[Ar]3d^9$
Cr^{3+} $[Ar]3d^3$
Mn^{2+} $[Ar]3d^5$
V^{4+} $[Ar]3d^1$

Metal Compounds of the First Transition Series

Compounds formed by these elements reflect oxidation states the metals are capable of attaining. Oxygen and fluorine are strongly oxidizing and tend to form stable compounds with the metals in their higher oxidation states. On the other hand, the lower oxidation states of the metals tend to be the most stable in the presence of an anion of a strongly reducing element such as I_2 or S—that is, I^- or S^{2-}. For example, copper forms Cu(I) and Cu(II) compounds. CuF_2 is the only copper fluoride. The chlorides and bromides both exist for Cu(I) and Cu(II), but CuI is the only copper iodide. Both oxides and sulfides exist. However, Cu_2S, which melts at 1100°C, is much more stable than CuS, which decomposes at 220°C.

For each metal the compounds at the higher oxidation states are more covalent. That is, in the higher oxidation states the ions are not stable as independent entities in the way they are when the oxidation states are lower. Mn^{2+} is a stable entity and MnO has the NaCl structure of a typical ionic compound. However, Mn^{7+} cannot exist in the same way. The charge of +7 is much too high for an independent ion, and such a species, if formed, would immediately seize electrons from any available source. The oxidation state of +7 can be realized only if Mn is covalently bonded. For example, Mn_2O_7 is a covalently bonded, explosive, green liquid.

Another way of looking at this is to consider the electronegativity of a metal as a function of its oxidation state. The higher its oxidation state, the greater its demand to draw electrons toward itself. This means that the electronegativity increases with oxidation state and metals in higher oxidation states behave more like nonmetals.

Another example of the increase in nonmetal character as the oxidation state of a metal increases is the reaction of the metal oxide with water. As we have mentioned before, metal oxides are basic when dissolved in water and nonmetal oxides are acidic under the same circumstances. The oxides of chromium are a good example:

- Chromium(II) oxide is basic and dissolves in aqueous acidic solutions:

$$CrO(s) + 2H_3O^+(aq) + 3H_2O(\ell) \longrightarrow [Cr(H_2O)_6]^{2+}(aq)$$

  Ions such as $[Cr(H_2O)_6]^{2+}$ are complex ions, which will be discussed in Section 19.4.
- Chromium(III) oxide is amphoteric, dissolving in aqueous acidic or basic solutions forming complex ions. **Amphoteric** means able to act as either an acid or a base.

$$Cr_2O_3(s) + 6H_3O^+(aq) + 3H_2O(\ell) \longrightarrow 2[Cr(H_2O)_6]^{3+}(aq)$$

$$Cr_2O_3(s) + 2OH^-(aq) + 7H_2O(\ell) \longrightarrow 2[Cr(OH)_4(H_2O)_2]^-(aq)$$

- Chromium(VI) oxide is strongly acidic. It will dissolve in neutral solutions to form chromic acid, H_2CrO_4, which dissociates to form protons (hydronium ions) and hydrogen chromate ($HCrO_4^-$) and chromate (CrO_4^{2-}) ions:

$$CrO_3(s) + H_2O(\ell) \longrightarrow H_2CrO_4$$

$$H_2CrO_4 + H_2O(\ell) \longrightarrow H_3O^+(aq) + HCrO_4^-(aq)$$

$$HCrO_4^-(aq) + H_2O(\ell) \rightleftharpoons H_3O^+(aq) + CrO_4^{2-}(aq)$$

19.3 THE SECOND AND THIRD TRANSITION SERIES

The metals of the second and third transition series are distinctly different from those of the first. Those of the second and third transition series tend to be hard, high melting, and fairly unreactive elements. They form compounds in a wide range of oxidation states and tend to have higher oxidation states than the first transition-series metals. The compounds they form generally have a high degree of covalent character, although there are a few simple ionic compounds such as WO_3. Typical covalent compounds such as WCl_5 and WCl_6 have low melting points of only 244°C and 280°C, respectively.

The similarities between adjacent family members in these two transition series are very strong. For example, molybdenum chemistry is similar to that of tungsten, the element just below it, but neither is very similar to chromium, the element above. Chromium forms very stable salts and complexes in the +2 and +3 oxidation states, whereas molybdenum and tungsten do not. Molybdenum and tungsten form a large series of complex oxygen-containing anions, whereas chromium does not.

The differences between the properties of pairs of first- and second-transition-series elements can be attributed mainly to the large differences in atomic and ionic radii. This is the result of the electrons being in orbitals with a higher value of the principal quantum number n for the second series. However, the radii for pairs of second- and third-series elements are very close and properties are similar in spite of the difference in n. This rather surprising fact is the result of the filling of the $4f$ orbitals of the lanthanides prior to the filling of the $5d$ orbitals of the third transition series. The incomplete shielding by the $4f$ electrons for each other results in a steady decrease in the radii of the lanthanides, which results in a decrease in the atomic and ionic radii of subsequent elements. This decrease almost exactly balances the increase in radii that would be expected from the increase in the principal quantum number n on going from the second to the third transition series. This effect is called the lanthanide contraction. Typical results are the ionic radii of V^{3+} (first series), Nb^{3+} (second series), and Ta^{3+} (third series), which are 0.780, 0.86, and 0.86 Å, respectively. With equal charges and ionic radii, we can expect Nb^{3+} and Ta^{3+} to behave very similarly.

19.4 COORDINATION COMPLEXES

One of the most important aspects of transition-metal chemistry is the formation of coordination complexes. A coordination complex can be a neutral species or a complex ion. A **complex ion** is an ion containing a number of different atoms. At the center of the coordination complex is a central cation or sometimes a neutral atom. Arranged around this center are a number of ligands, usually four or six. **Ligands** are normally either anions or polar molecules such as H_2O or NH_3, which are bonded in some way to the central cation or atom. Coordination complexes of the transition metals can be prepared from their salts. As an example, chlorine reacts with nickel to produce the yellow salt, nickel(II) chloride:

$$\mathrm{Ni(s)} + \mathrm{Cl_2(g)} \longrightarrow \mathrm{NiCl_2(s)}$$

If this salt is added to water, heat is given off, and a green solution forms. Evaporation of the water from this solution gives a green solid, $NiCl_2{\cdot}6H_2O(s)$. So the aqueous reaction is

$$\mathrm{NiCl_2(s)} + \mathrm{6H_2O}(\ell) \longrightarrow \mathrm{NiCl_2{\cdot}6H_2O(aq)}$$

If ammonia is added to the green solution, it turns purple. If this solution is evaporated to dryness, the compound that is isolated has the formula $NiCl_2{\cdot}6NH_3$. The aqueous reaction in this case is

$$\mathrm{NiCl_2{\cdot}6H_2O(aq)} + \mathrm{6NH_3(aq)} \longrightarrow \mathrm{NiCl_2{\cdot}6NH_3(aq)}$$

This chemistry is much more complicated than that of main-group metals such as sodium or calcium. The difference is that nickel, in common with other transition-metal ions, forms coordination complexes.

Discovering the Nature of Transition-Metal Complexes

The formation of coordination complexes by transition-metal ions has been known for a long time. Pioneering chemical studies were carried out by the German-Swiss chemist Alfred Werner (1866–1919), who prepared ammonia complexes, which were known as ammines. Ammines of Ni^{2+}, Cu^{2+}, Co^{3+}, Pt^{2+}, and Pt^{4+} were well-known, although poorly understood, at that time. Werner made careful and complete studies of families of complexes. For example, when water solutions of $PtCl_4$ reacted with different amounts of ammonia, five different ammines formed, with formulas ranging from $PtCl_4{\cdot}2NH_3$ to $PtCl_4{\cdot}6NH_3$. Werner prepared pure solutions of all five complexes and performed two key experiments on each:

- First, by treating the solutions of the five complexes with a silver nitrate solution and measuring the mass of precipitated silver chloride, he was able to calculate the number of "free" chloride ions in each compound.
- Second, by measuring the conductance of the ammine solutions, he was able to determine the total number of ions in each complex, as shown in the following table.

Complex	Formula	Free Cl^- Ions	Total Ions
1	$PtCl_4{\cdot}6NH_3$	4	5
2	$PtCl_4{\cdot}5NH_3$	3	4
3	$PtCl_4{\cdot}4NH_3$	2	3
4	$PtCl_4{\cdot}3NH_3$	1	2
5	$PtCl_4{\cdot}2NH_3$	0	0*

* This is nonionic.

Werner interpreted these results as follows: In complex 1, four of the five ions were free chloride ions and the remaining ion was a complex ion made up of all the other atoms, Pt and six NH_3 molecules. In each successive complex, one more Cl atom is incorporated into the complex ion and becomes unavailable for reaction with silver nitrate solution. He recognized the actual formulations as follows:

Complex	Formulation
1	$Pt(NH_3)_6^{4+}$, $4Cl^-$
2	$PtCl(NH_3)_5^{3+}$, $3Cl^-$
3	$PtCl_2(NH_3)_4^{2+}$, $2Cl^-$
4	$PtCl_3(NH_3)_3^{+}$, Cl^-
5	$PtCl_4(NH_3)_2$

PROFILES IN CHEMISTRY

Alfred Werner: Colorful Chemist Born the son of a factory foreman in the Alsace region of France, Alfred Werner was German-educated but as a result of the Franco–Prussian war retained strong French sympathies all his life. At age 20, after a year of compulsory German military service, he moved to Switzerland, where he became a Swiss citizen and remained the rest of his life. Werner earned his doctorate degree at the University of Zürich in 1890, with a thesis on the spatial arrangements of atoms in molecules containing nitrogen. He was a postdoctoral student with Marcelin Berthelot in Paris, where he extended the newly published works of Jacobus van't Hoff and Joseph LeBel on tetravalent carbon to other atoms. This study, done when he was 24 years old, led the way to Werner's greatest work on a coordination theory of molecular chemistry, for which he won the Nobel prize in chemistry in 1913.

Werner's coordination theory came to him (he recalled later in life) while he slept. Waking at 2:00 AM and arising immediately, he worked intensively until 5:00 AM, writing down the details of his ideas. The theory was based on relationships between atoms that were not limited to ordinary chemical bonds of the day that were ionic (after Arrhenius's concept of simple inorganic substances), or covalent (after Kekulé and van't Hoff's systems of organic molecules built around tetravalent carbon). In Werner's visionary theory, atoms or groups of atoms could be spatially arranged around a central atom according to geometric principles.

Among the extraordinary and important features of his work was the stereo-isomeric conclusion that coordination compounds of metal ions (or atoms) with complex radicals (ligands) of the form MeA_3B_3 do in fact occur in two isomeric forms. As Werner described it in his Nobel lecture of December 11, 1913,*

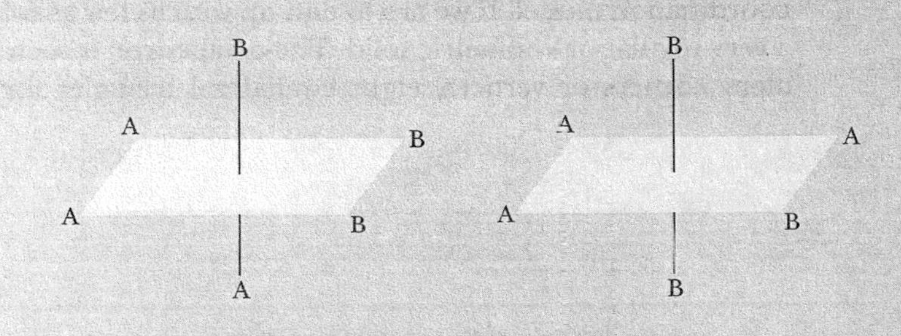

[T]he octahedral distribution of the six groups about the center atom means that compounds with complex radicals must occur in two stereo-isomeric forms, according to the arrangement of the three B groups in a plane or in a sectional plane of the octahedron (see the previous example). The compound $CoCl_3 \cdot (NH_3)_3$, for example, exists in two isomeric forms, of which one is indigo blue and the other violet.

The coordination theory immensely broadened chemical theories of the day to explain the existence and properties of these mysteriously colorful complexes of transition-metal atoms or ions. Often spoken of as "secondary valences," it was left to another generation of chemists led by Linus Pauling to tie together Werner's geometric constructions to ordinary valence bonds.

It is interesting to note in passing the extraordinary success Werner achieved while working and teaching under the most primitive conditions. Werner was an extraordinary and hugely popular lecturer who attracted many students to his research laboratories, founding an entire field of chemistry—coordination chemistry.

Questions

How did Werner expand the concept of chemical bonding beyond what was already accepted?

Werner was also able to show the existence of stereo-isomeric compounds of metal ions (or atoms) with complex radicals of the form MeA_4B_2. How many isomeric forms are possible?

*"On the constitution and configuration of higher-order compounds," Nobel Lecture, December 11, 1913, Amsterdam, Elsevier Publishing Co., 256–269.

Two more members of the series, $Pt(NH_3)Cl_5^-$ and $PtCl_6^{2-}$, were discovered later. Werner established that in each of these Pt(IV) complexes, there were six species bonded to the central ion. The number of these species, or ligands, about an atom or ion is called its **coordination number**. Platinum(IV) has a coordination number of six in these cases.

On the basis of results of similar sets of experiments, Werner was able to show that Co^{3+} and Cr^{3+} formed complexes with coordination number six, whereas Pt^{2+} and Pd^{2+} formed complexes with coordination number four. Some transition-metal ions such as Ni^{2+} and Cu^{2+} formed complexes with either coordination number four or coordination number six.

Werner was able to infer the geometry of the four- and six-coordinate complexes by counting the number of isomers. **Isomers** are compounds of identical composition but with different arrangements of atoms, resulting in different properties. For example, there are two isomers of the Co^{3+} complex ion $[Co(NH_3)_4Cl_2]^+$. Both have distinctly different properties. What does this tell us about the coordination geometry? Try to draw structures representing possible coordination geometries with coordination number six. For example, if the coordination about the central-metal ion were in the form of a planar hexagon, there would be three isomers of this complex ion (Fig. 19.2a). Because the planar geometry is not satisfactory, try a solid or three-dimensional figure for the coordination model. If we are to end up with as few as two isomers we must have a very regular or symmetric solid. The octahedron is such a solid. It has six equivalent corners or vertices, eight equilateral triangles for the faces, and twelve

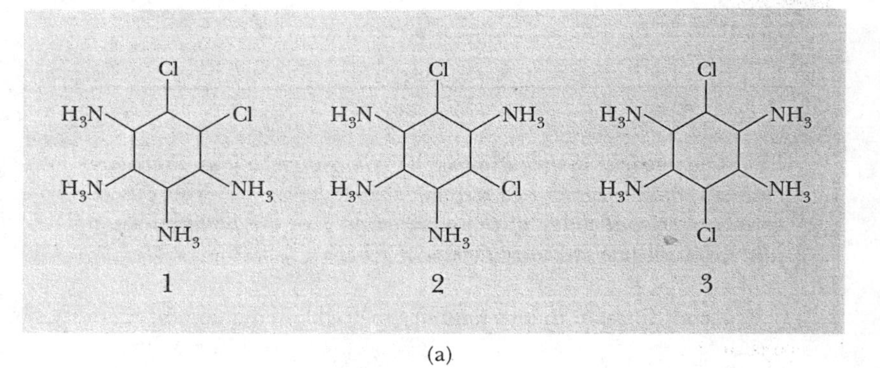

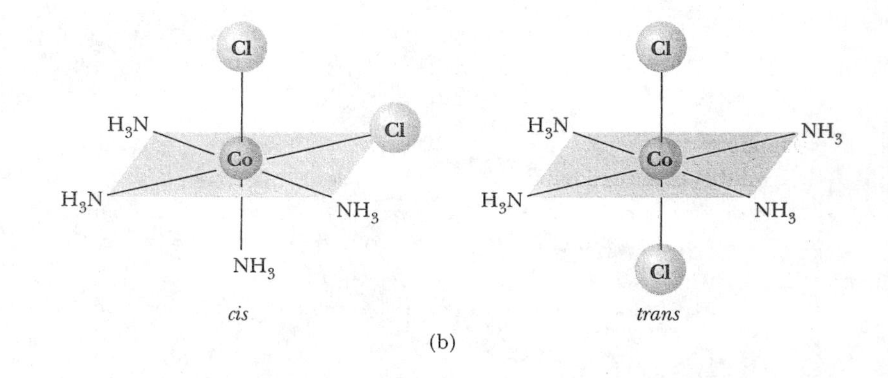

FIGURE 19.2 Isomers are compounds of identical composition but with different arrangements of atoms, resulting in different properties. (a) If the coordination about the central-metal ion takes on a planar, hexagonal arrangement, then three isomers of the $[Co(NH_3)_4Cl_2]^+$ ion would be possible. (b) Octahedral geometry leads to only two possible isomers.

APPLICATIONS OF CHEMISTRY

Cisplatin: An Effective Chemotherapeutic Agent Cisplatin is the *cis*-isomer of a coordination compound called diamminedichloroplatinum(II), with the formula $Pt(NH_3)_2Cl_2$, or simply *cis*-DDP. It is an effective chemical agent for fighting cancer. *cis*-DDP can successfully penetrate the nucleus of the cell, where it works to block the alarming and deadly rapid rate of reproduction of unhealthy, cancerous cells by building a bridge between adjacent rings called guanidine bases in the cell's DNA. *trans*-DDP does the same thing, but the bridges are much longer and not nearly as stable, allowing the cell to easily recognize and remove these blocking groups within a few hours of having formed. The cell is then permitted to go on about its business of out-of-control reproduction.

Like many chemotherapeutic agents, *cis*-DDP brings with it toxic side effects. But it has proved generally far more toxic to unhealthy than healthy cells and has been widely used for many years. Unraveling its mode of action as an antitumor agent provides the possibility of learning about the chemical and molecular nature of the disease and how to approach its ultimate cure. The chemotherapeutic action of the drug became well-known in the 1970s; its mode of action became understood in the 1980s; but the compound was first prepared by Alfred Werner a century ago.

Werner correctly concluded that the existence of two isomers, a yellow one and an orange one, could only mean the complex compounds were square planar and not tetrahedral. Later it was shown that the yellow isomer has no dipole moment, and so it must be the *trans*-isomer.

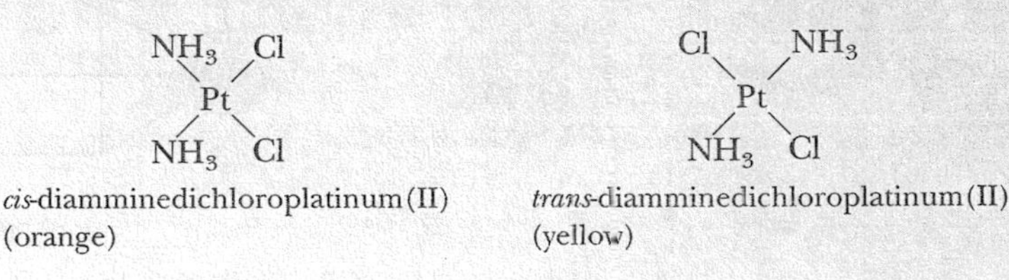

cis-diamminedichloroplatinum(II) (orange)

trans-diamminedichloroplatinum(II) (yellow)

Questions

Why does the existence of isomeric forms of $Pt(NH_3)_2Cl_2$ mean the structure is square planar and not tetrahedral?

Why is it important to know the actual geometry of the ligands around the central atom of a complex?

Why does the fact that the orange isomer of $Pt(NH_3)_2Cl_2$ is a polar molecule mean it must be the *cis* form?

edges, all of the same length. Because this geometry occurs fairly regularly in chemical structures, a skeleton notation is used that shows four vertices in a square plus one vertex above and one below. We will use this convenient notation, although it creates the illusion that two of the vertices are different from the other four when in fact all six are equivalent. Figure 19.2(b) shows the two isomers of the $[Co(NH_3)_4Cl_2]^+$ ion. The isomer with two adjacent chlorine atoms, 90° apart, is referred to as the ***cis*** form; the one with the two chlorines opposite or 180° apart, is the ***trans*** isomer. Keep in mind that "trans" means "across," and the chlorine atoms are across from each other.

Similar reasoning was used to deduce the coordination geometry about Pt^{2+} in $Pt(NH_3)_2Cl_2$. Two isomers exist, leading to a prediction of the square-planar geometry, as shown in Figure 19.3(a). But not all four-coordinate complexes are square planar. Some, particularly those of Co^{2+} and Zn^{2+}, exhibit four coordination with tetrahedral geometry (Fig. 19.3b). Note that if $Pt(NH_3)_2Cl_2$ had been tetrahedral instead of square planar, there would be only one isomer for the complex rather than the two that were found.

FIGURE 19.3 The coordination geometry about Pt^{2+} in $Pt(NH_3)_2Cl_2$. (a) The two isomers of the square-planar $Pt(NH_3)_2Cl_2$ coordination complex. If the geometry were that of the regular tetrahedron, shown in (b)—which is not the correct structure—then no isomers would be possible for $Pt(NH_3)_2Cl_2$.

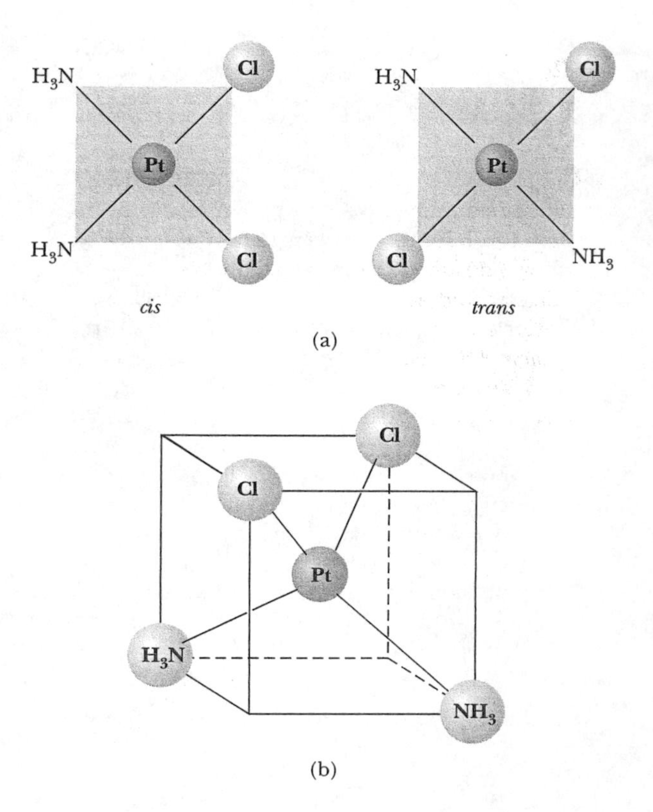

Geometries of Coordination Complexes

Although octahedral geometry occurs frequently in coordination complexes for the more common metals and has been the most thoroughly studied, many other geometries are observed (Fig. 19.4). Some of the most important are

- Linear, two-coordination (Fig. 19.4a); observed for Ag^+ and Au^+ complexes: $[Ag(NH_3)_2]^+$, $[Au(CN)_2]^-$
- Trigonal, three-coordination (Fig. 19.4b); rare, observed for Ag^+ and Hg^{2+}: $[HgI_3]^-$
- Tetrahedral, four-coordination (Fig. 19.4c); common, observed for Cu^+, Co^0, Co^{2+}, Ni^0, Ni^{2+}, Zn^{2+}, Cd^{2+}: $[Zn(CN)_4]^{2-}$, $[CoCl_4]^{2-}$
- Square planar, four-coordination (Fig. 19.4d); common, observed for Ni^{2+}, Pd^{2+}, Pt^{2+}, (d^8 electron configurations): $[Ni(CN)_4]^{2-}$, $[PtCl_4]^{2-}$
- Octahedral, six-coordination (Fig. 19.4e); common, observed for transition metals in most oxidation states: $[PdCl_6]^{2-}$, $[Fe(H_2O)_6]^{2+}$, $W(CO)_6$

Naming Coordination Complexes

Coordination complexes have a central-metal ion or atom surrounded by two or more ligands. The ligands are either anions or polar-neutral molecules. We will not consider the highly unusual cases of cationic ligands. In naming coordination complexes, the anionic ligands are identified by names ending in "o." The important examples are

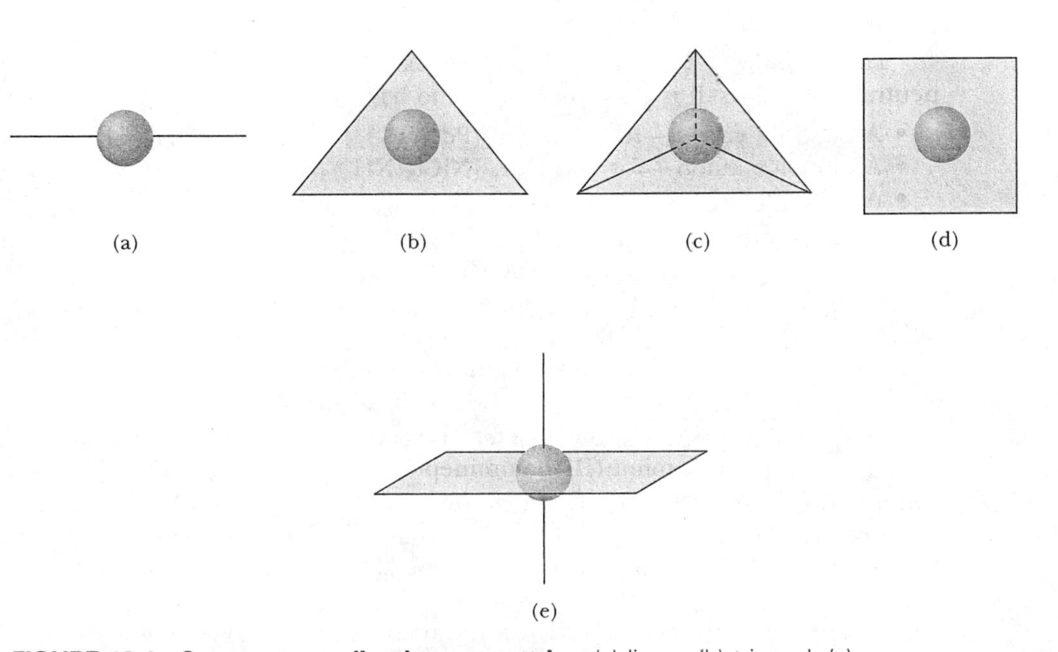

FIGURE 19.4 Common-coordination geometries: (a) linear, (b) trigonal, (c) tetrahedral, (d) square planar, and (e) octahedral.

F^-	fluoro	NO_2^-	nitro
Cl^-	chloro	$C_2O_4^{2-}$	oxalato
Br^-	bromo	SCN^-	thiocyanato
I^-	iodo	OH^-	hydroxo
CN^-	cyano		

The names of neutral compounds acting as ligands are not systematic, and the following three important examples should be learned.

H_2O	aqua
CO	carbonyl
NH_3	ammine

Standard ways of writing and naming coordination complexes that eliminate confusion have been universally adopted. We will first consider complexes that are neutral molecules, such as $Pt(NH_3)_2Cl_2$. The ligands are named before the central atom in alphabetical order, and the prefixes "di," "tri," "tetra," "penta," "hexa," and so on, are used if there is more than one of a particular ligand. Thus, $Pt(NH_3)_2Cl_2$ would be "diamminedichloroplatinum."

However, because the oxidation states of transition metals tend to be variable, it is necessary to show the oxidation state of the metal in roman numerals enclosed in parentheses after the name of the metal. The overall charge for a neutral complex calculated from the known charges of the ligands must be zero. For $Pt(NH_3)_2Cl_2$, the NH_3 ligands are neutral and the two Cl^- ligands each have a charge of −1. This requires that the Pt have an oxidation state of +2. Thus, the complete name of $Pt(NH_3)_2Cl_2$ is "diamminedichloroplatinum(II)."

The naming of salts containing complex ions is similar to the naming of neutral complexes. It is possible for a salt to have

- A complex cation—for example, $[Pd(NH_3)_6]^{4+}$ in $[Pd(NH_3)_6]Cl_4$
- A complex anion—for example, $[NiCl_3(NH_3)_3]^-$ in $Na[NiCl_3(NH_3)_3]$
- A complex cation and a complex anion—for example, $[Cr(H_2O)_4Br_2]^+$ and $[PtI_5(NH_3)]^-$ in $[Cr(H_2O)_4Br_2][PtI_5(NH_3)]$

The complex ions are enclosed in [square] brackets, and for these salts the cation precedes the anion in both the name and the formula. The oxidation states shown must be such that the overall charge on the salt is zero. Thus, $[Pd(NH_3)_6]Cl_4$ is hexaamminepalladium(IV) chloride. The suffix **-ate** is added to the name of the metal if it is in a complex anion. Thus, $Na[NiCl_3(NH_3)_3]$ is sodium triamminetrichloronickelate(II), and $[Cr(H_2O)_4Br_2][PtI_5(NH_3)]$ is tetraaquadibromochromium(III) amminepentaiodoplatinate(IV). The oxidation states of metals in salts with a complex cation and a complex anion can be ambiguous. Determining the oxidation state of Cr in this complex required knowing that Pt is normally in an oxidation state of +4 when it is coordinated to six ligands.

Alternatives are encountered in the naming of some very common metals such as iron. Just as $FeCl_2$, iron(II) chloride, is sometimes known as ferrous chloride, the salt, $Na_4[Fe(CN)_6]$, is sodium hexacyanoferrate(II).

EXAMPLE 19.2

Name the following coordination complexes:

$$K_3[Al(OH)_4] \quad \text{and} \quad [Cr(NH_3)_4(H_2O)Cl]Cl_2$$

COMMENT For the first complex, the one K^+ ion indicates that the complex ion has a charge of −1. Each of the four —OH groups supplies a −1 charge, so that the Al atom has a +3 oxidation state. In the second complex, two chloride ions outside the brackets indicate that the complex ion has a charge of +2. The chromium atom is surrounded by six ligands, five of them neutral species. The sixth ligand is a chloride ion, so the chromium atom must have an oxidation state of +3.

SOLUTION Potassium tetrahydroxoaluminate(III) and tetraammineaquachlorochromium(III) chloride.

EXERCISE 19.2

Give the formula for hexaamminechromium(III) hexachlorocobaltate(III).

ANSWER $[Cr(NH_3)_6][CoCl_6]$

19.5 BONDING IN COORDINATION COMPLEXES

The coordination complexes we are studying consist of a positively charged metal ion surrounded by a number of ligands that are either anions or polar molecules, such as H_2O or NH_3. If the ligand is a molecule, the complex forms with the negative end of the molecular dipole directed toward the metal cation. The tendency to form these complexes can be ascribed largely to electrostatic attraction. The charges of transition metal ions are normally +2 as a result of losing two valence shell *s* electrons, and exceed +2 if one or more *d* electrons are lost. Chromium with the $[Ar]4s^1 3d^5$ configuration and copper with $[Ar]4s^1 3d^{10}$ have the possibility of losing just the one *s* electron to form the +1 ion. The Cr^+ ion has no significant chemistry, and the Cu^+ ion is the only important +1 ion formed by the transition metals. In addition to charges of +2 or greater, the transition metal ions have relatively small sizes, and this results in a strong electrostatic attraction to the ligands and the formation of stable complexes. In contrast, the relatively large size of the Na^+ ion and its charge of only +1 results in a weak electrostatic attraction to ligands and no significant formation of complexes. The bonding model that we will use for coordination complexes will be simplified by the assumption that the bonding is purely the result of electrostatic attractions; it is called the crystal-field theory.

Crystal-Field Theory

The **crystal field theory** begins with three simplifying assumptions:

1. All bonding is assumed to be the result of electrostatic attractions—that is, the attraction between the opposite charges of the ligand and the metal ion.
2. All ligands are assumed to have a negative charge. For neutral but polar ligands, the negative end of the dipole is directed toward the positive ion and functions similarly to a free-negative charge.
3. All charges are assumed to be located at points.

The negative charges of the ligands are arranged around the cation according to the geometry of the complex. This arrangement of negative ions creates a static-electric field, which is called the crystal field. With this in mind, consider the most commonly encountered geometry in the first transition series, the six-coordinate octahedral geometry. The octahedral crystal field is shown in Figure 19.5. All transition-metal ions have lost their *s* valence shell electrons. Therefore, the only valence electrons and orbitals that we need to consider in this model are *d* electrons and orbitals. The principal question that needs to be answered is this: "What effect does the presence of the crystal field have on the relative energy levels of the *d* orbitals?"

The positions of the negative point charges of the octahedral crystal field with respect to the shapes of the five *d* orbitals are shown in Figure 19.6. In the

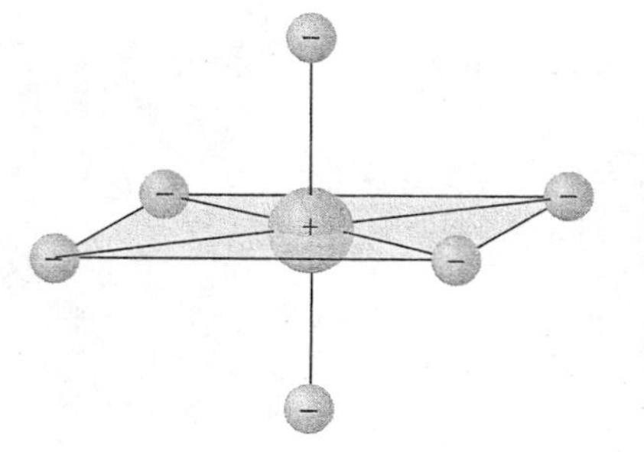

FIGURE 19.5 The crystal field. The six negative point charges of the ligands are arranged symmetrically about a positively charged central ion, creating a static-electric field called a crystal field. In this case, the arrangement of charges about the central ion is octahedral.

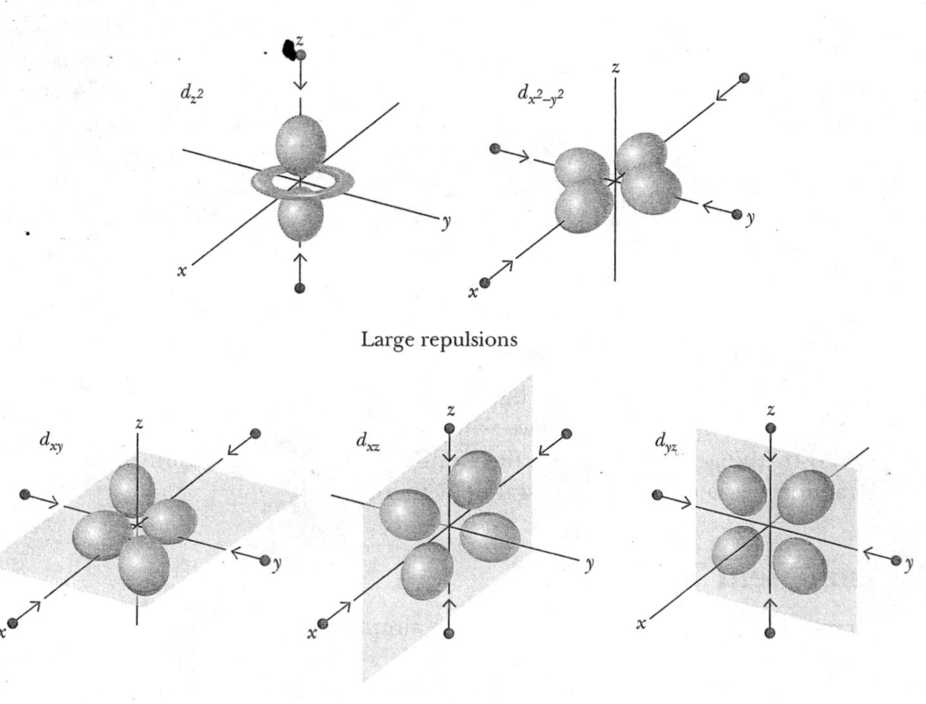

FIGURE 19.6 Octahedral-crystal-field splitting of the 3*d* orbital energies. The six negative point charges are located on the Cartesian axes at equal distances from the origin. As these external charges (associated with the ligands) approach the distributed orbitals, the largest repulsions arise in the orbitals that are pointed directly at the charges. The $d_{x^2-y^2}$ and d_{z^2} orbitals, pointing directly along the *x*-, *y*-, and *z*-axes, generate greater repulsions, whereas the d_{xy}, d_{xz}, and d_{yz} orbitals, pointing between these axes, generate smaller repulsions.

absence of any disturbing influences such as electric charges, all five are said to be **degenerate**, which means that they have the same energy. However, the presence of the crystal field breaks the degeneracy. For example, if an octahedral crystal field is superimposed on the *d*-orbital diagrams, two of the orbitals, $d_{x^2-y^2}$ and d_{z^2}, point directly at the negative charges. Repulsion between the point charges and electrons in these two orbitals will destabilize the orbitals somewhat and raise the orbital energy. The other three orbitals, d_{xy}, d_{xz}, and d_{yz}, are directed between the point charges. Their energies would not be raised as much by the presence of the negative charges. The degeneracy of the five orbitals has therefore been broken into two groups, a group of two at higher energy and a group of three at lower energy.

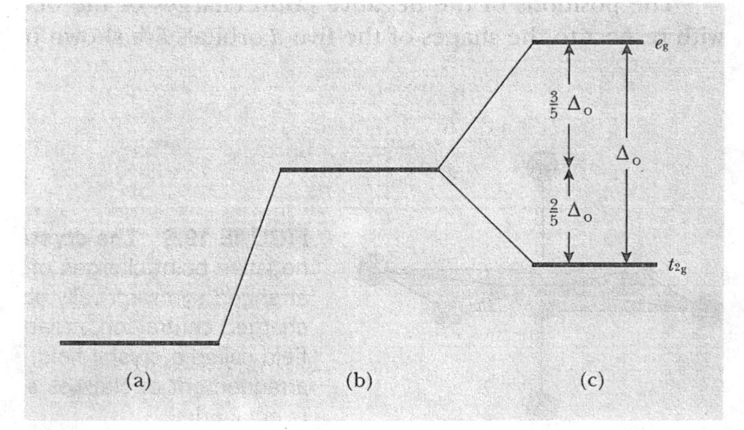

FIGURE 19.7 Energy-level diagram. Shows relative energies of *d* orbitals of a transition-metal ion. (a) The energy is lowest in the absence of ligand charges. (b) With ligand charges spread equally, as would be the case if they were spread over a concentric sphere surrounding the metal ion, the unsplit higher energy level would be observed. (c) Finally, in an octahedral crystal field the indicated splitting would be observed.

The two higher energy orbitals, $d_{x^2-y^2}$ and d_{z^2} are included in a group labeled e_g, where e is used to designate two orbitals of the same energy—that is, two degenerate orbitals. The lower energy d_{xy}, d_{xz}, and d_{yz} orbitals are assigned to a group labeled t_{2g}, where the t designates three degenerate orbitals. The energy difference or "splitting" between the two sets is called Δ_O, where the subscript stands for octahedral (Fig. 19.7).

Once the orbitals have been split in an octahedral field, it is necessary to populate them with electrons. We will add them one at a time (Fig. 19.8). The first three electrons enter the three lower energy t_{2g} orbitals. Each occupies a different orbital according to Hund's rule, and the spins are not paired. However, there are two options for the fourth electron:

1. Enter a t_{2g} orbital that already contains another electron and pair its spin with the second electron.
2. Enter an empty e_g orbital.

Each alternative requires more energy than was expended for any one of the first three electrons:

- If the fourth electron enters a t_{2g} orbital it must overcome the spin-pairing energy, including the electrostatic repulsion of the other electron.
- If the fourth electron enters an e_g orbital there is the added energy cost of Δ_O—that is, it must overcome the octahedral-splitting energy.

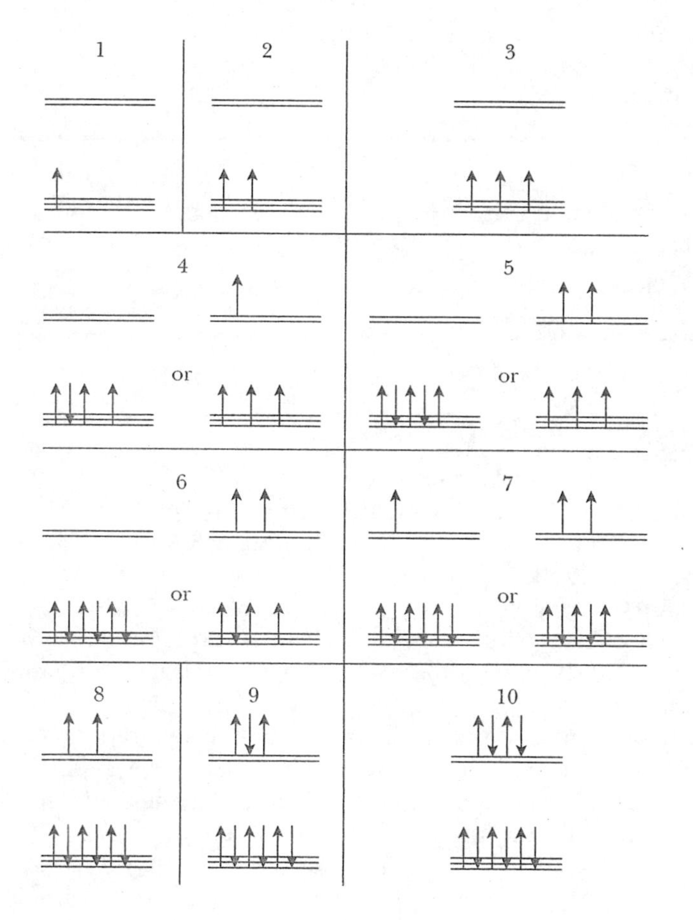

FIGURE 19.8 Electrons in the orbitals of a transition-metal ion in an octahedral crystal field. Possible electron configurations are shown for one or more *d* electrons, up to a full complement of ten.

The result depends on which energy is greater. The spin-pairing energy depends principally on the metal ion. The octahedral-splitting energy depends on both the metal ion and the ligand. Two alternatives also occur for five, six, and seven electrons. But for one, two, three, eight, nine, and ten electrons, there is only one possible arrangement.

Important properties, particularly regarding magnetism, depend on the number of unpaired electrons in a coordination complex. For a d^4 complex it is two unpaired electrons if the spin-pairing energy is greater than the octahedral splitting energy, and four unpaired electrons in the opposite situation. The configuration with fewer unpaired electrons is called the **low-spin configuration** and that with more unpaired electrons is called the **high-spin configuration**. Thus, in an octahedral crystal field, low- and high-spin configurations exist for four, five, six, and seven d electrons. Only one possible configuration exists for ions with one, two, three, eight, nine, and ten d electrons.

EXAMPLE 19.3

Determine the possible numbers of unpaired electrons for the Fe^{3+} ion in an octahedral crystal field.

COMMENT Note that Fe^{3+} has a d^5 electron configuration and consult Figure 19.8.

SOLUTION This can result in either a low-spin configuration with one unpaired electron or a high-spin configuration with five unpaired electrons.

EXERCISE 19.3

Determine the possible numbers of unpaired electrons for each of the following ions in an octahedral crystal field: Cr^{2+}, Co^{2+}, Cu^{+}, Fe^{2+}, Zn^{2+}, and Sc^{3+}.

ANSWER Four or two; three or one; zero; four or zero; zero; zero

Magnetic Properties of Complexes

Because transition-metal complexes and complex ions can have unpaired electrons, they can be distinguished by a characteristic magnetic property or by its absence. The magnetic properties, diamagnetism and paramagnetism, were discussed in Chapter 7, Section 7.6. Based on our earlier discussion on the structure of the atom in Chapter 5, two electrons can be accommodated in a single orbital as long as they have different values for the spin-quantum number, $m_s = \pm 1/2$. That is, one electron will have m_s equal to $+1/2$, and the other will have m_s equal to $-1/2$. These two values of m_s are equivalent to opposing electron spins, and we commonly say that such electrons have their spins paired. Using arrows to represent electrons, the two values of m_s are given by the directions in which the arrows point. Such spin-paired electrons are like bar magnets oriented in opposite directions, each canceling the effect of the other. Substances with all electron spins paired in this manner are called **diamagnetic** and are not attracted to a magnetic field. However, if a substance has at least one unpaired electron, it will have a net magnetic moment, and the substance is said to be **paramagnetic**. Paramagnetism in a substance depends on having one or more orbitals containing an unpaired electron, and paramagnetic substances are pulled into a magnetic field. Such situations are particularly common in transition-metal chemistry in which there are partially filled *d* orbitals.

Paramagnetism can be measured by determining the force by which a sample is drawn into a magnetic field. The more unpaired electrons in a given amount of sample, the stronger this effect will be. An instrument used for this purpose is the Gouy balance (Fig. 19.9). The difference in the measured mass of the sample with the electromagnet off and on is determined, from which one can calculate the number of unpaired electrons.

Spectral Properties of Complexes and the Spectrochemical Series

One of the characteristic properties of transition metals is the prevalence of colored ions. Another is the color change that often accompanies the addition of a reactant such as ammonia to a solution of a transition-metal complex. To explain the source of these optical effects, let us look at the Ti^{3+} ion with its one *d* electron. When Ti^{3+} is placed in water, it forms the $Ti(H_2O)_6^{3+}$ complex ion with an octahedral ligand environment. As shown in Figure 19.8, this one electron will reside in a t_{2g} orbital in the most stable state—the ground state. However, absorption of an amount of energy equal to Δ_O would result in the *d* electron jumping up to the e_g orbital, producing an excited state.

The exact amount of energy needed to excite the $Ti(H_2O)_6^{3+}$ complex is most conveniently delivered by electromagnetic radiation of the proper frequency, according to the Planck equation (Chapter 5):

$$E = h\nu$$

Quantum of radiation: Energy E is proportional to the frequency ν of the radiation. The constant of proportionality h (Planck's constant) is 6.626×10^{-34} J·s.

Because the energy desired is Δ_O, the frequency is related to the octahedral crystal field splitting in the following manner:

$$\Delta_O = h\nu$$

$$\nu = \frac{\Delta_O}{h}$$

The process for a d^1 system (one *d* electron) is shown in Figure 19.10a. The crystal-field splitting for $Ti(H_2O)_6^{3+}$ is such that the radiation absorbed is in the green and yellow region of the visible spectrum. If white light passes through a solution of this complex, the yellow and green will be absorbed and the transmitted violet light is the color you see. The light absorbed will eventually appear as heat. As a general rule, any colored solution absorbs light of a wavelength that is complementary to the color it transmits. For example, a yellow solution absorbs in the blue region of the visible spectrum and a violet solution absorbs in the yellow and green regions (Table 19.4, Figure 19.11).

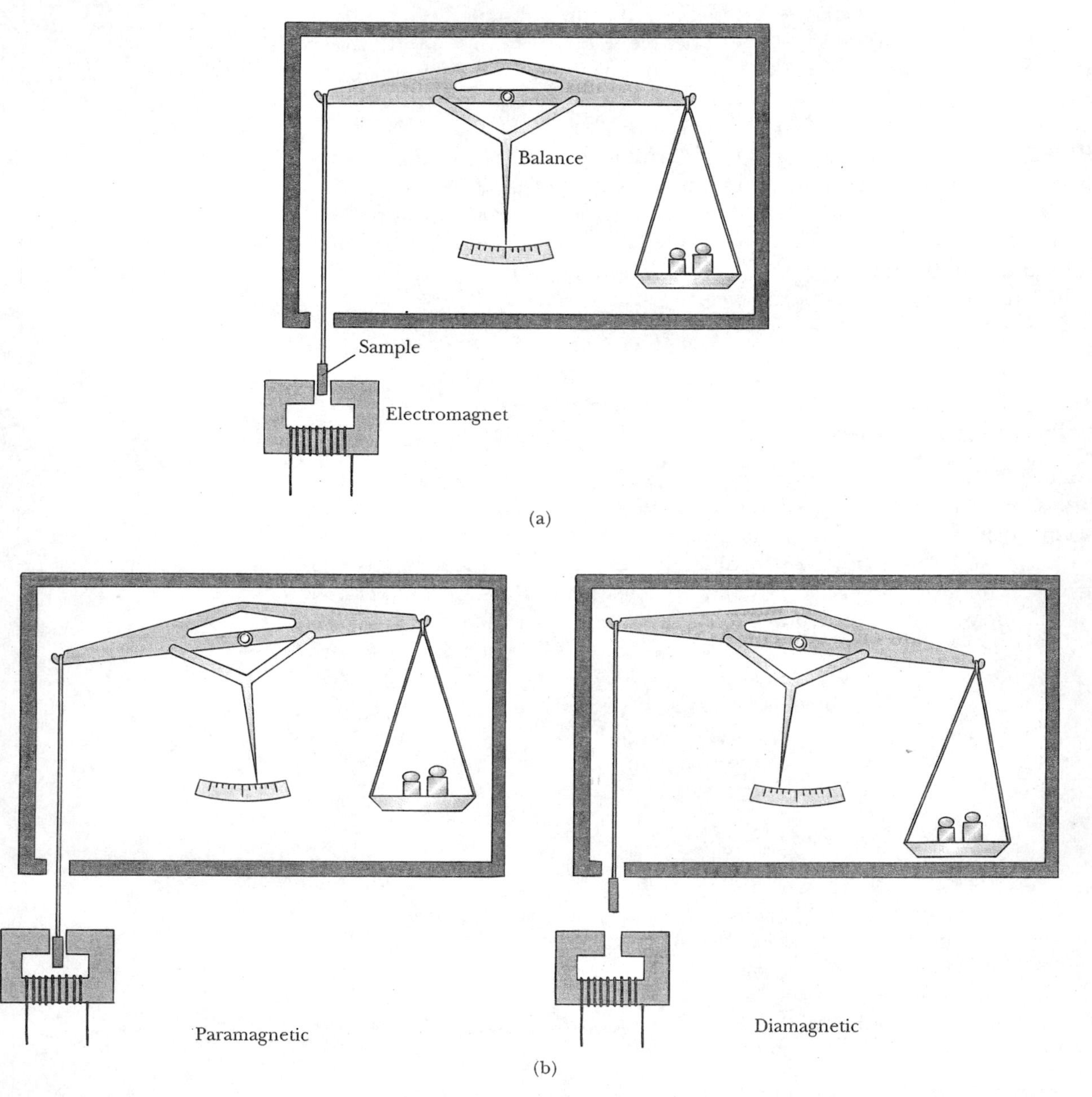

FIGURE 19.9 Measuring the magnetic susceptibility of a paramagnetic substance. Schematic representation of a Gouy balance for measuring magnetic susceptibility. (a) The mass of the sample is determined with the power to the electromagnet turned off. (b) The power is then turned on and the change apparent in the mass is used as a measure of the extent to which the sample is drawn into the magnetic field. If the sample is paramagnetic, the sample is drawn into the magnetic field and apparently weighs more than in the absence of the field. If the sample is diamagnetic, it is very weakly repelled and appears to weigh slightly less than with the magnet turned off.

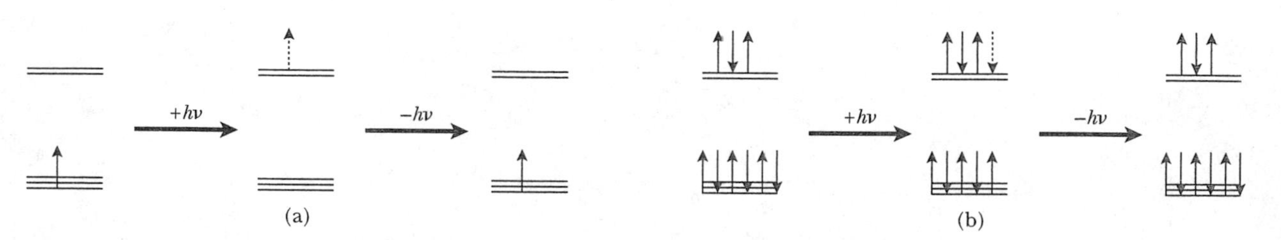

FIGURE 19.10 Absorption and emission of one quantum of energy. One electron is promoted to the higher (e_g) level and then falls back, in an octahedral crystal field: (a) $Ti(H_2O)_6^{3+}$ has one *d* electron; (b) $Cu(H_2O)_6^{2+}$ has nine *d* electrons.

TABLE 19.4 Relationships Between Light Absorbed and Color Transmitted

Wavelength Absorbed, nm	Color Absorbed	Color Transmitted
400–450	Violet	Yellow
450–480	Blue	Yellow
480–490	Green/blue	Orange
490–500	Blue/green	Red
500–560	Green	Purple
560–580	Yellow/green	Violet
580–590	Yellow	Blue
590–640	Orange	Green/blue
640–700	Red	Blue/green

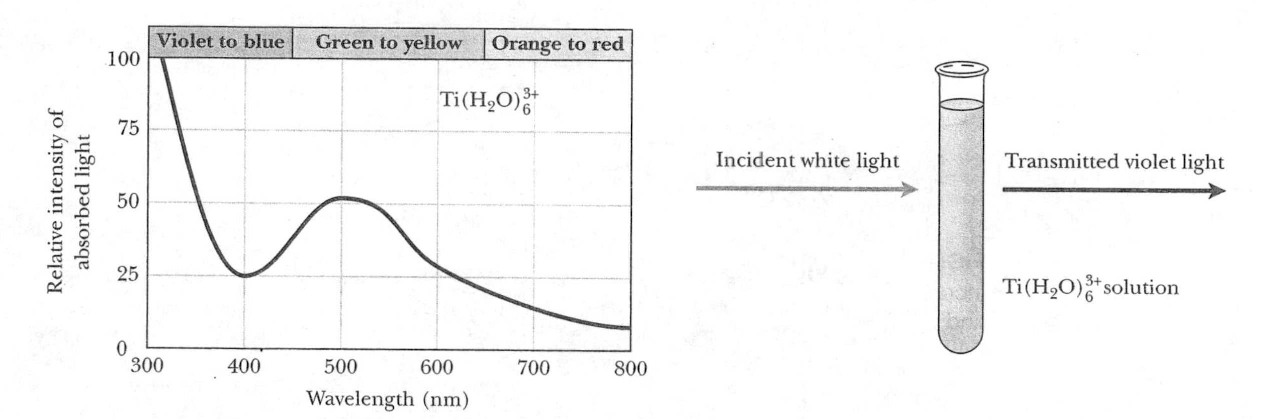

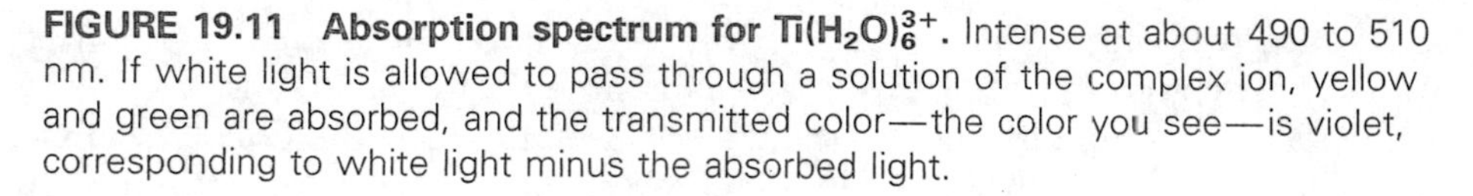

FIGURE 19.11 Absorption spectrum for $Ti(H_2O)_6^{3+}$. Intense at about 490 to 510 nm. If white light is allowed to pass through a solution of the complex ion, yellow and green are absorbed, and the transmitted color—the color you see—is violet, corresponding to white light minus the absorbed light.

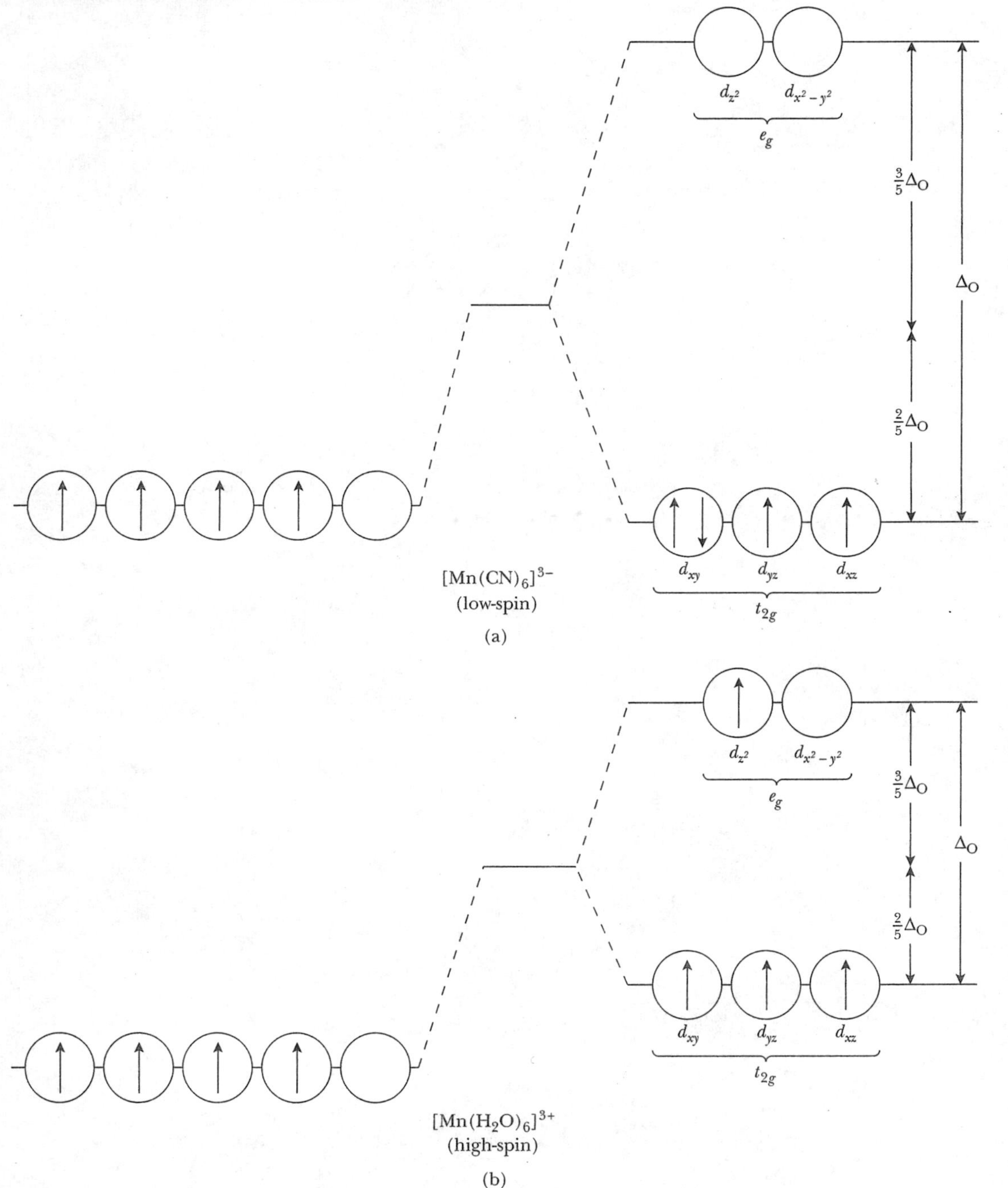

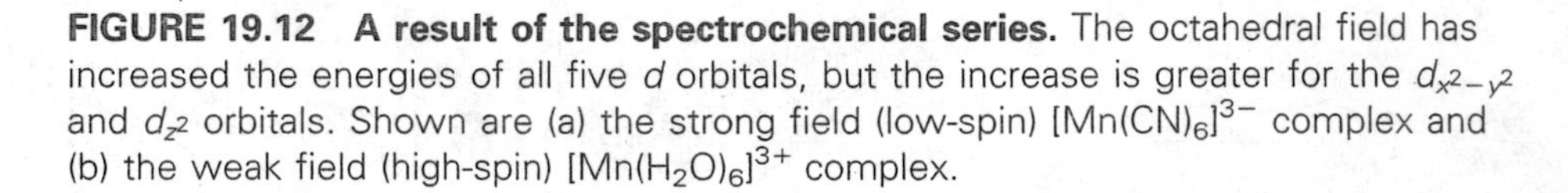

FIGURE 19.12 A result of the spectrochemical series. The octahedral field has increased the energies of all five *d* orbitals, but the increase is greater for the $d_{x^2-y^2}$ and d_{z^2} orbitals. Shown are (a) the strong field (low-spin) $[Mn(CN)_6]^{3-}$ complex and (b) the weak field (high-spin) $[Mn(H_2O)_6]^{3+}$ complex.

An equally simple situation occurs with d^9 configurations such as Cu^{2+} (Fig. 19.10b). The ground state has one space or "hole" in the e_g orbitals, and the excited state has a hole in the t_{2g} orbitals. It is useful to view the hole as jumping to the lower energy orbitals when radiation is absorbed. Spectra for the configurations d^2 to d^8 are more complex but follow the same general ideas.

It is well-known that addition of simple reactants can cause dramatic changes in the colors of complex ions in solution. The frequency of the absorbed light changes when the ligand is changed, and because

$$\nu = \frac{\Delta_O}{h}$$

the value of Δ_O must depend on the identity of the ligands.

The list of ligands in order of increasing Δ_O values for a given metal ion is called the **spectrochemical series**. The order of the more common ligands in the spectrochemical series is

$$I^- < Br^- < Cl^- < F^- < OH^- < C_2O_4^{2-} < H_2O < NH_3 < NO_2^- < CN^-$$

When the ligands of a transition metal complex are exchanged, the change in Δ_O will result in a change in color. For example, adding NH_3 to a water solution of Ni^{2+} ions exchanges NH_3 ligands for H_2O ligands. According to the spectrochemical series, this increases the value of Δ_O (Fig. 19.12). This is observed as a change in color, the transmitted light, from green to purple. To fully understand the reasons for the order of the spectrochemical series, one has to go beyond the crystal-field theory.

In addition to the effect of the ligand on the value of Δ_O giving the spectrochemical series, the value of Δ_O for a particular complex also depends on the identity of the metal ion and its charge. Two important observations follow:

- **Higher oxidation states result in higher values of Δ_O.** The higher the charge on the ion, the stronger the attraction between the metal and the ligands. This will pull the ligands in more closely so that the perturbation of the metal orbitals and the splitting of the energy levels by the ligands will be greater. Therefore, the occurrence of low-spin complexes will be greater for metals in their higher oxidation states.
- **The value of Δ_O increases on going from the first to the second to the third transition series.** The observed result of this effect is that practically all complexes of second- and third-transition series metals are low spin.

Geometries other than octahedral also have different values of Δ. For example, for tetrahedral coordination the value of Δ is about half that for octahedral coordination.

19.6 REACTIONS OF SOME COORDINATION COMPLEXES

Formation of Aqua Complexes and Ligand-Exchange Reactions

Transition-metal ions dissolved in water can be considered aqua complexes in cases in which water molecules are the ligands. A strong tendency toward formation of the aqua complex can be seen in the strongly exothermic reaction of anhydrous transition-metal salts with water. For example,

$$CrCl_3(s) + 6H_2O(\ell) \longrightarrow [Cr(H_2O)_6]^{3+}(aq) + 3Cl^-(aq)$$

TABLE 19.5 Enthalpies of Hydration, kJ/mol, of Some of the Gaseous Ions of the Metals, from Potassium to Zinc

$M^{n+}(g) + 6H_2O(\ell) \longrightarrow [M(H_2O)_6]^{n+}(aq)$	
K^+	−322
Ca^{2+}	−1577
Cr^{2+}	−1904
Mn^{2+}	−1841
Fe^{2+}	−1946
Co^{2+}	−1996
Ni^{2+}	−2105
Cu^{2+}	−2100
Zn^{2+}	−2046

Table 19.5 lists enthalpies of hydration for gaseous ions of some of the elements of the first transition series as well as potassium and calcium. The huge negative values suggest something of the ease with which aqua complexes are formed by transition-metal ions. Notice how low this value is for K^+, a large ion with a small charge.

Aqua complexes of metals are all acidic to some extent. You can see why by examining ionization reactions such as the following:

$$[Cr(H_2O)_6]^{3+} + H_2O(\ell) \rightleftharpoons [Cr(H_2O)_5(OH)]^{2+} + H_3O^+$$

The driving force for this reaction is the stronger coordination of the metal cation to the hydroxo ligand OH^- with its negative charge compared to the neutral H_2O ligand. The acidity of these ions is comparable to that of other weak acids. In this case, $K_a = 1.26 \times 10^{-4}$.

For all +2 and +3 ions of the first transition series, the aqua ions are octahedral, $M(H_2O)_6^{2+}$ or $M(H_2O)_6^{3+}$, although there are distortions from perfect octahedral geometry in some cases. When in water solution, the complexed water molecules are often not identified explicitly, and the complexed ion is usually given simply as $M^{n+}(aq)$.

The water molecules in aqua complexes can be replaced in a stepwise manner with other ligands, and an equilibrium-constant expression can be written for each step. Here is an example of the first step:

$$Cr(H_2O)_6^{2+}(aq) + NH_3(aq) \rightleftharpoons Cr(NH_3)(H_2O)_5^{2+}(aq) + H_2O(\ell)$$

$$K_1 = \frac{[Cr(NH_3)(H_2O)_5^{2+}]}{[Cr(H_2O)_6^{2+}][NH_3]}$$

The sum of the six stepwise reactions gives the net reaction for which the equilibrium constant for this reaction is the overall formation constant of the complex:

$$Cr(H_2O)_6^{2+}(aq) + 6NH_3(aq) \rightleftharpoons [Cr(NH_3)_6^{2+}](aq) + 6H_2O(\ell)$$

$$K_{overall} = \frac{[Cr(NH_3)_6^{2+}]}{[Cr(H_2O)_6^{2+}][NH_3]^6}$$

EXAMPLE 19.4

The overall formation constant for $Ni(NH_3)_6^{2+}$ is 4.07×10^8. Calculate the concentration of $Ni^{2+}(aq)$ when $[NH_3] = 0.120$ M and $[Ni(NH_3)_6^{2+}] = 0.501$ M.

COMMENT Substitute into the equilibrium expression for the net reaction.

SOLUTION

$$[Ni^{2+}] = \frac{[Ni(NH_3)_6^{2+}]}{K_{overall} \times [NH_3]^6}$$

$$= \frac{0.501}{(4.07 \times 10^8)(0.120)^6} = 4.12 \times 10^{-4}\ \text{M}$$

EXERCISE 19.4

Cu^{2+} forms the ammine complex, $Cu(NH_3)_4^{2+}$. The overall formation constant for this complex is 6.80×10^{12}. Calculate the concentration of $Cu^{2+}(aq)$ in equilibrium with $NH_3(aq)$ at 0.331 M and $Cu(NH_3)_4^{2+}$ at 0.413 M.

ANSWER $[Cu^{2+}] = 5.06 \times 10^{-12}$ M.

Complexes with Chelating Ligands

Ligands such as Cl^- or NH_3 have just one coordination site, either the negative charge or the negative end of a molecular dipole. Ligands such as Cl^- or NH_3 are called monodentate. A **monodentate** ligand is a ligand with only one coordinating site—"monodentate" actually means "having only one tooth." Ethylenediamine, abbreviated *en,* is similar to NH_3 but has two coordinating nitrogen sites:

$$NH_2—CH_2—CH_2—NH_2 \qquad \text{ethylenediamine (en)}$$

Ethylenediamine with two coordinating sites is a **bidentate** ligand, and "bidentate" means "having two teeth."

Because of the flexibility of the ethylenediamine molecule, it can bend so that both nitrogen atoms are able to complex the same metal ion, forming a five-membered ring. Thus, three ethylenediamine molecules can be substituted for six NH_3 molecules in $Ni(NH_3)_6^{2+}$ to form the complex ion shown in Figure 19.13. In addition to being bidentate, ethylenediamine is also a chelating ligand. The term "chelating" comes from the Greek word meaning "claw," because a chelating ligand can use two or more binding sites on the same metal ion and appears to grasp it. The magnitude of the overall formation constant increases surprisingly from the nonchelating, monodentate NH_3 ligand to the chelating

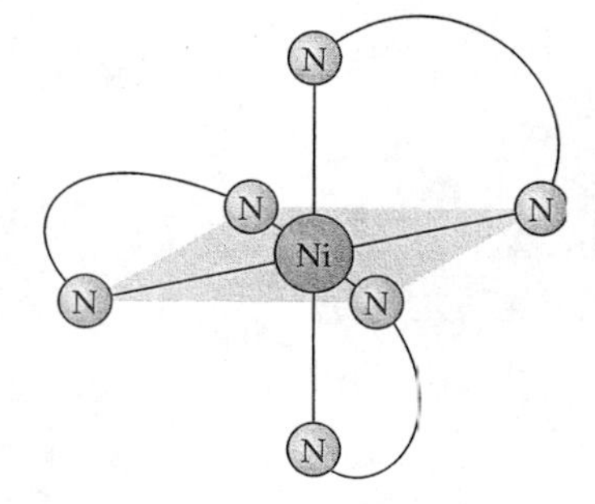

FIGURE 19.13 Chelate complex of Ni^{2+} with ethylenediamine (en): $[Ni(en)_3]^{2+}$. Hydrogen atoms have been omitted for simplicity of illustration, and the carbon atoms are located at the bends in the lines linking the nitrogen atoms. Complexing is made possible by the flexibility of ethylenediamine (en) = $H_2NCH_2CH_2NH_2$.

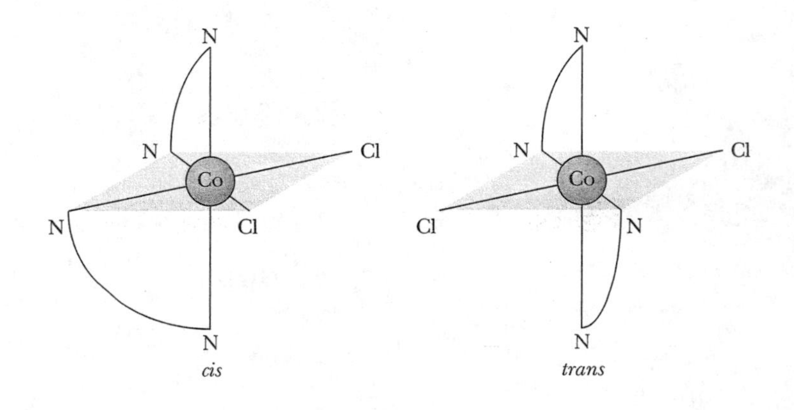

FIGURE 19.14 Geometric *cis*- and *trans*-isomers. On the left, the *cis*-$Co(en)_2Cl_2^+$ isomer; on the right, the *trans*-$Co(en)_2Cl_2^+$ isomer. Both cations have the same chemical composition but different spatial orientation of the chlorine atoms about the central metal ion.

bidentate $H_2NCH_2CH_2NH_2$ ligand. For Ni^{2+}, the increase is from 4.07×10^8 for $[Ni(NH_3)_6]^{2+}$ to 1.91×10^{18} for $[Ni(en)_3]^{2+}$. This tendency is known as the **chelate effect**: Chelating ligands are more strongly bound than otherwise-equivalent monodentate ligands.

Several factors contribute to the chelate effect. Consider the formation of $Ni(en)_3^{2+}$ from the aquated ion:

$$Ni(H_2O)_6^{2+}(aq) + 3en(aq) \longrightarrow Ni(en)_3^{2+}(aq) + 6H_2O(\ell)$$

- The reaction is favored by entropy because four reactant particles form seven product particles. The equivalent reaction of NH_3 has no entropy advantage because seven reactant molecules form seven product molecules.
- Only three ethylenediamine molecules need to find each metal ion to form the complex, compared with six NH_3 molecules.
- Once one end of the ethylenediamine molecule has attached to the metal ion, the other end is close by, and it is a relatively convenient and favorable matter to close the five-membered ring.

Another example of a six-coordinate bidentate ligand system is the $Co(en)_2Cl_2^+$ complex ion, which can exist as a *cis*- and *trans*-isomer (Fig. 19.14).

EXERCISE 19.5(A)

Pt^{2+} ions form a four-coordinate, square-planar complex. Draw a sketch of its ammonia and ethylenediamine complexes.

EXERCISE 19.5(B)

Oxalic acid (HOOC—COOH) is a diprotic acid that forms a divalent anion

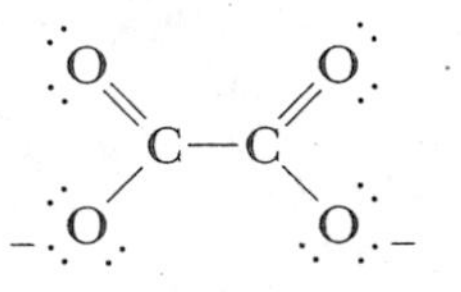

APPLICATIONS OF CHEMISTRY

EDTA and Heavy-Metal Poisoning Chelating ligands with more than two complexing sites show even greater effects. Ethylenediaminetetraacetate ion, widely known as EDTA, is a notable example:

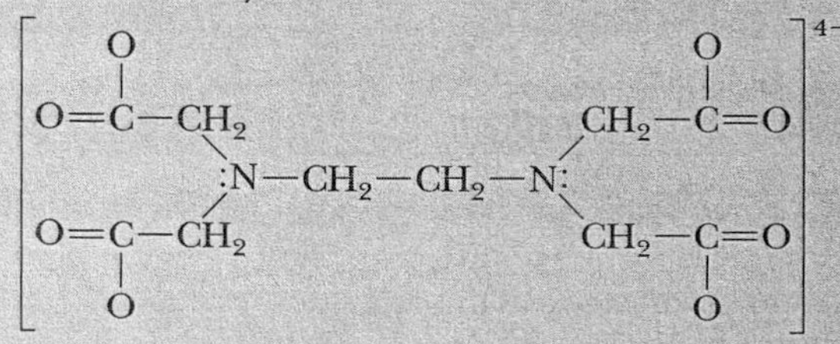

Both nitrogen atoms and the four single-bonded oxygen anions are capable of coordinating with a metal ion. Such a ligand is hexadentate, or six-toothed, and is a potent chelating ligand. In the case of Fe^{3+}, an octahedral complex is formed in which the flexible molecule wraps itself around the Fe^{3+} ion in such a way as to occupy all six octahedral sites of the coordination complex. Adding the sodium salt of EDTA to solutions containing Fe^{3+} ions results in the complete complexation of the Fe^{3+} ions. Once complexed, such ions can be maintained in the presence of other ions in solution that would normally have caused their precipitation, or the complexed ions can be removed from solution. Hydroxide ions, for example, normally precipitate Fe^{3+} ions, but the EDTA complex of Fe^{3+} is soluble in the presence of hydroxide.

As a pharmaceutical preparation known as Versine, the use of EDTA as a complexing agent in the treatment of heavy-metal poisoning has proved to be one of the most important practical applications of coordination chemistry. After EDTA is administered as an antidote in lead and mercury poisoning, the undesired ions are rapidly complexed. These complexed ions dissolve and are excreted by the body. This is essential because even small quantities of lead, from peeling paint or improperly glazed pottery, for example, can affect certain enzyme systems involved in the manufacture of hemoglobin needed for red blood cells. Minor symptoms such as headache and irritability may be followed by gastrointestinal symptoms—loss of appetite, constipation, nausea, vomiting, and abdominal pain. The central nervous system may be involved. However, the treatment is not without its side effects, because EDTA complexes so strongly that it also removes calcium ions from body tissue along with several essential trace metal ions, which must be replenished when EDTA is used therapeutically.

Questions

Explain the principle involved in using EDTA to treat lead or mercury poisoning.

Which of the following are bidentate ligands?

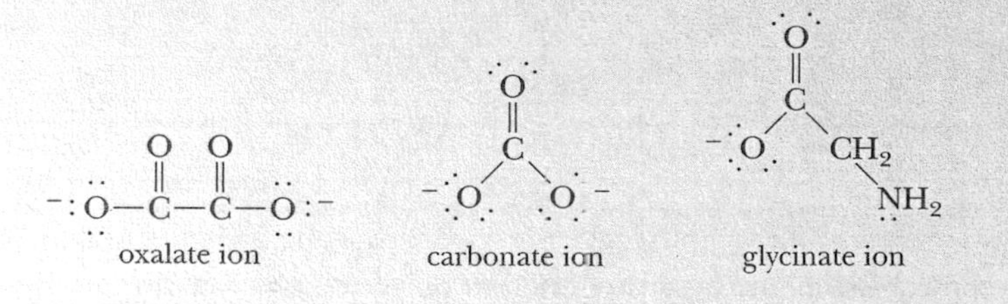

capable of acting as a bidentate chelating agent. Draw a sketch of the octahedral complex it forms with Cr^{3+} ions.

EXERCISE 19.5(C)

Sketch the octahedral complex formed between EDTA and Fe^{3+} ions. (See Applications of Chemistry box above.)

PROCESSES IN CHEMISTRY

Metallocene Catalysts for Polymer Synthesis Catalysts are essential for many reactions in which monomers are snapped together to form polymers. An unusual family of these catalysts is called metallocenes. These are complexes in which a transition metal is sandwiched between two rings of carbon atoms. They are very useful catalysts for the synthesis of polyethylene and polypropylene, for example.

Propylene has the structure $CH_2{=}CH{-}CH_3$ and it is the double bond in its structure that is sacrificed in building the polymer, polypropylene. Likewise the double bond in ethylene, $CH_2{=}CH_2$, allows units to add forming polyethylene. In both cases, long chains of monomer units are snapped together with the aid of a catalyst. However, in polypropylene each monomer unit has a CH_3 group attached to the double bond, which ends up as a side chain hanging off the polymer backbone. Different orientations of this CH_3 group are possible. If all of the CH_3 groups are hanging off on the same side of the polymer chain, the polymer is called isotactic—isotactic polypropylene:

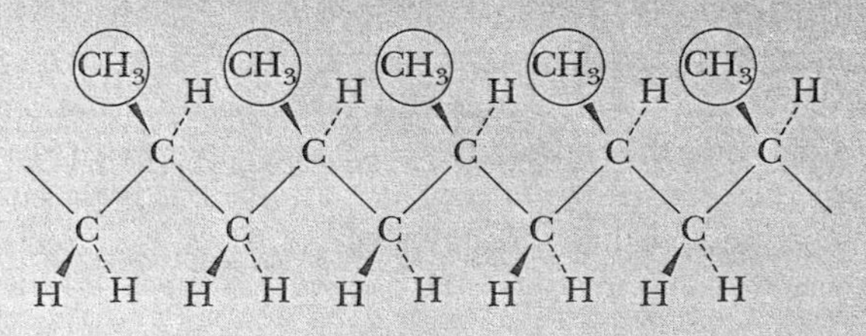

If the CH_3 groups are in a regular pattern on alternate sides, the polymer is called syndiotactic—syndiotactic polypropylene:

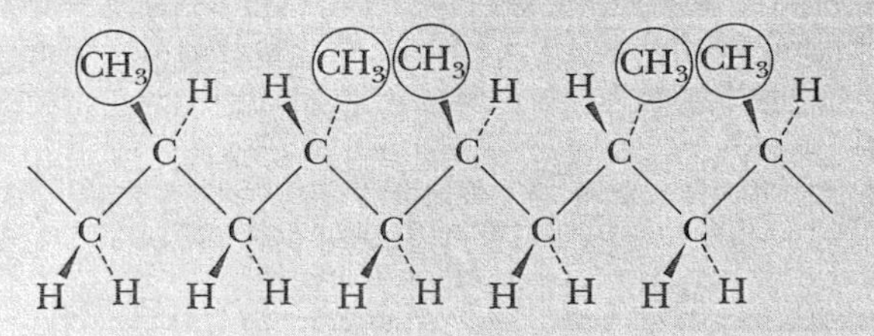

If the CH_3 groups are randomly situated on one side or the other, it is atactic—atactic polypropylene:

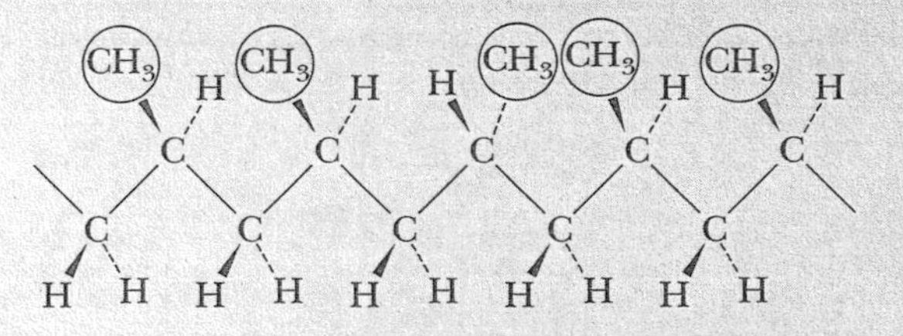

As traditionally produced, polypropylene is about 95% isotactic, a few percentages atactic, and an even lower percentage syndiotactic. The atactic is highly undesirable, but the syndiotactic form has potential uses. Using metallocene catalysts it is possible to control the amounts of the different polymer types that are produced.

Metallocene catalysts control the degree to which the polymer has isotactic, syndiotactic, or atactic geometry, and that determines the properties. Syndiotactic polypropylene could not be produced as a pure polymer before the discovery of metallocene catalysts. The material is significantly different from iostactic polypropylene in its physical properties—much softer but also tougher and clearer. An especially important property that strongly suggests use of products made of the material for medical applications is its stability to sterilizing gamma radiation.

Metallocene polymerization catalysts were discovered in the 1950s by Giulio Natta (Italy) and Ronald Breslow (United States), who first worked with these coordinated-metal compounds. However, these early studies on "sandwich" structures lapsed because no support for the work emerged at the time. Then, in the 1980s, German chemistry professors Walter Kaminsky (Hamburg) and Hans Brintzinger (Konstanz) were able to produce metallocene catalysts that did have real value. As a result, metallocene-based polymers, ranging from crystalline to elastic materials, have been commercially available since 1991.

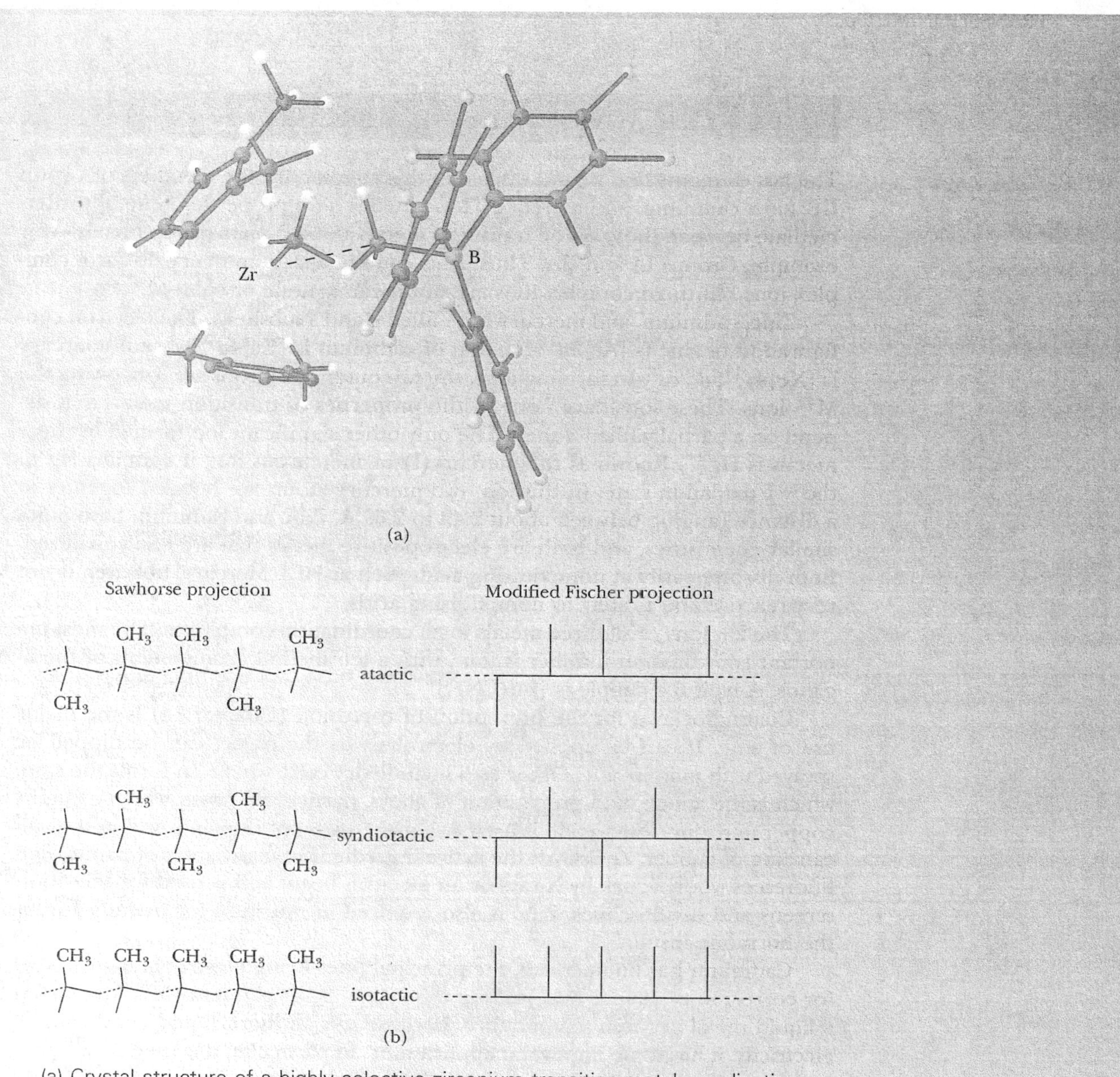

(a) Crystal structure of a highly selective zirconium transition-metal-coordination catalyst for making stereoregular polymers such as isotactic and syndiotactic polypropylene. (b) Atactic, syndiotactic, and isotactic polypropylene chain-length segments, shown in two different projections, indicating the organization of the side chains.

Questions

Why are there different forms of the polypropylene and only one form of polyethyene?

Would you expect "tacticity" and metallocene catalysts to be important issues in the preparation of polyvinyl chloride: $CH_2{=}CHCl$?

19.7 ZINC, CADMIUM, AND MERCURY

The last elements that we will discuss in this chapter are the members of Group IIB: zinc, cadmium, and mercury. These metals have properties generally intermediate between those of the transition metals and the main-group metals—for example, Groups IA and IIA. Thus, zinc, cadmium, and mercury do form complex ions but these complex ions are not paramagnetic or colored.

Zinc, cadmium, and mercury have filled *d* and *s* subshells. The electron configuration of zinc is $[Ar]3d^{10}4s^2$, that of cadmium is $[Kr]4d^{10}5s^2$, and mercury is $[Xe]4f^{14}5d^{10}6s^2$. In forming ions, the two outer *s*-electrons are lost, giving the M^{2+} ions. These ions have none of the properties of transition metals that depend on a partially filled *d* shell. The only other significant ion formed by these metals is Hg_2^{2+}. Known as the mercury(I) or mercurous ion, it contains Hg in the +1 oxidation state. In this ion, two mercury atoms are bonded together at a distance ranging between about 2.49 to 2.69 Å. Zinc and cadmium have quite similar chemistries, and both are electropositive metals that are easily oxidized. Both dissolve easily in nonoxidizing acids such as HCl. Mercury, however, is not very reactive and is inert to nonoxidizing acids.

The 2+ ions of all three metals form coordination complexes. The most important coordination number is four, with a tetrahedral arrangement of the ligands. A typical example is $[Zn(CN)_4]^{2-}$.

Coating of iron for the prevention of corrosion (Chapter 14) is the major use of zinc. It can be applied by electrolysis or the object can be dipped or sprayed with molten zinc. Other uses include dry cells, where Zn forms the case, which is the anode, and preparation of alloys, particularly brass, which contains copper and zinc. Zinc oxide is used as a white pigment in paints and in the vulcanizing of rubber. $ZnCrO_4$ is the active ingredient in many antirust paints. ZnS fluoresces when struck by X-rays or an electron beam and is used for television screens and oscilloscopes. Zinc is also required in nutrition for manufacturing the hormone insulin.

Cadmium has limited uses, the principal ones being electroplating on steel for corrosion resistance and making the yellow paint pigment CdS. Mercury is a liquid metal at room temperature. Because it is an inert, liquid conductor of electricity, it has some important applications. In particular, it is used in switches and as the flowing electrode in electrolysis cells, the most important being the chlor-alkali cell (Chapter 14) for the electrolysis of sodium chloride to sodium hydroxide and chlorine gas. Excited mercury atoms emit intensely in the green and ultraviolet regions of the electromagnetic spectrum. Therefore, mercury is used in mercury vapor lamps on streets and highways and in fluorescent tubes. In the fluorescent tubes, the ultraviolet radiation that is emitted by mercury causes the fluorescent coating on the tube to give off visible light.

SUMMARY

The transition elements include the *d*-block elements, the lanthanides, and the actinides. All are metals and characteristically exhibit variable oxidation states and colored compounds, many of which are paramagnetic. Much of their chemistry centers around the properties of coordination complexes. The electron configurations of the elements show a preference for half-filled and filled *d* shells.

Coordination complexes are composed of a central-metal ion and coordinating ligands. The nature of these complexes was determined by Alfred Werner, who studied the solution properties of families of complexes having different proportions of the same ligands complexed to the same metal ion. Of the several observed coordination numbers, four and six are very common.

Crystal-field theory assumes a model in which the positively charged metal ion is surrounded by ligands represented by negative point charges. These charges split the otherwise equal-energy *d* orbitals. For octahedral coordination, the *d* orbitals are split into two groups and magnetic and spectral properties can be explained accordingly. The list of ligands in order of degree of splitting is called the spectrochemical series.

Chemical equilibria in solution are demonstrated by these complexes when they exchange ligands. Particularly stable complexes are formed with chelating ligands, which have more than one site that can complex a single metal ion.

The elements zinc, cadmium, and mercury are often not considered to be transition metals because their *d* orbitals are always full. They resemble the transition metals in that they form coordination complexes. However, they form essentially no colored ions, and their only instance of multiple-oxidation states occurs with mercury, which has the ions Hg^{2+} and Hg_2^{2+}.

TERMS

Transition metals (19.1)
Coordination complexes (19.1)
Shielding (19.1)
Effective nuclear charge (19.2)
Amphoteric (19.2)
Complex ion (19.4)
Ligand (19.4)
Coordination number (19.4)
Isomer (19.4)
cis-(19.4)
trans-(19.4)
Crystal-field theory (9.5)
Degenerate (9.5)
Low-spin configuration (9.5)
High-spin configuration (9.5)
Diamagnetic (9.5)
Paramagnetic (9.5)
Spectrochemical series (9.5)
Monodentate (19.6)
Bidentate (19.6)
Chelating (19.6)
Chelate effect (19.6)

QUESTIONS

Conceptual questions are denoted by a square screen. Extra-credit questions are denoted by a circular screen.

1. Why do chromium and copper have electron configurations with only one *s* electron in the valence shell, whereas their neighboring elements have two?

2. What establishes the maximum oxidation state for a transition metal?

3. Why does the fact that $Pt(NH_3)_2Cl_2$ has two isomers eliminate the possibility that this particular complex might have the geometry of a tetrahedron?

4. Octahedrally coordinated Cr^{3+} is particularly stable. In terms of the crystal-field theory, why is this so?
5. Why do water solutions of many transition-metal ions change color when ammonia is added?
6. Why is the crystal-field theory uninformative regarding the actual colors of complexes?
7. Are water solutions of transition-metal ions basic or acidic? Explain.
8. Why do chelating ligands form complexes so much more readily than comparable nonchelating ligands?
9. Transition-metal ions can be titrated with solutions of ligands. The end point occurs when essentially all of the transition-metal ions have been converted to complexes of the titrating ligand. Why can EDTA be successfully used to titrate Cu^{2+} ions to a sharp end point and ammonia cannot?
10. Suppose a transition metal gives colored complexes with all ligands in the spectrochemical series. How will the colors of the complexes tend to change, proceeding from one end of the spectrochemical series to the other?

PROBLEMS

Problems marked with a bullet (•) are answered in Appendix A, in the back of the text.

Electron Configurations [1–4]

•1. Determine the electron configurations for the metallic elements calcium and scandium.

2. What are the electron configurations of the metallic elements titanium and vanadium?

•3. Determine the electron configurations for each of the following ions: Ni^{2+}, Cu^{+}, Zn^{2+}, and Ti^{2+}.

4. What is the electron configuration for each of the following ions: Cr^{3+}, Au^{3+}, Co^{3+}, and V^{3+}?

Coordination Complexes: Structures, Names [5–14]

•5. What would be the total number of ions per formula unit and the number of Cl^- ions that could be precipitated per formula unit from a solution of the salt $[Cr(H_2O)_4Cl_2]Cl$?

6. How many "free" (nonligand) Cl^- ions and total ions would there be per formula unit of the salt $K_2[CoCl_4]$?

•7. $PdCl_2$ in water solution will react with two, three, and four molecules of NH_3 to give three different complexes. Propose formulas for each of these compounds, state the number of "free" chloride ions each contains, and list the total number of ions in each formulation.

8. The Co^{3+} ion has a coordination number of six. Give the formulas for all possible combinations that would result from adding $NH_3(aq)$ to an aqueous solution of $CrCl_3$. For each formula predict the number of "free" Cl^- ions and the total number of ions per formula unit.

•9. How many isomers would you expect for the complex ion $[Ni(NH_3)_3Cl_3]^-$ if the coordination geometry is octahedral? Draw the isomers.

10. Draw the isomers that would be observed for $[Co(NH_3)_4Cl_2]Cl$ if the coordination had the geometry of a triangular prism (which it does not). Do the same for a pentagonal pyramid. How many isomers would there be in each case?

•11. Name the following compounds:

(a) $K_2[Ni(CN)_4]$
(b) $Na_3[Cr(C_2O_4)_3]$
(c) $[Pt(NH_3)_5Cl]Cl_3$
(d) $[Fe(NH_3)_4(NO_2)_2]_2SO_4$

(e) $[Co(H_2O)_6]I_3$
(f) $[Ru(NH_3)_5Cl]Br_2$
(g) $K_2[CoCl_4]$

12. Give the formulas of the following compounds:
(a) Pentaammineiodochromium(III) bromide
(b) Sodium tetrachloroferrate(III)
(c) Potassium tetracyanonickelate(II)
(d) Pentaaquachlorochromium(III) chloride
(e) Sodium tetrahydroxozincate(II)
(f) Hexaaquachromium(III) sulfate
(g) Potassium trichlorotrithiocyanatochromate(III)

•13. Determine both the coordination number and the oxidation state for the metal atom in each of the following complexes:
(a) $CuCl_4^{2-}$ (b) $Ag(CN)_2^-$
(c) $Zn(NH_3)_4^{2+}$

14. Give the coordination number and oxidation state of the metal in each of the following complexes:
(a) $Fe(CO)_5$ (b) $Cr(H_2O)_6^{3+}$
(c) $Fe(CN)_6^{3-}$

Coordination Complexes: Bonding, Magnetic Properties, and Spectra [15–24]

15. How many unpaired electrons will there be in a Ti^{3+} ion that is in an octahedral crystal field?

16. What would be the number of unpaired electrons if a Zn^{2+} ion were subjected to an octahedral crystal field?

•17. Determine the possible numbers of unpaired electrons in complexes of the following ions in an octahedral crystal field: Mn^{2+}, Co^{3+}, and Ni^{2+}.

18. Give the possible numbers of unpaired electrons in octahedral complexes of Cu^{2+} and Fe^{3+}.

•19. If an octahedral Fe^{2+} complex is paramagnetic, what do you know about its electron configuration?

20. Give the configuration of the 3*d* electrons (the number of electrons in the t_{2g} and e_g orbitals) in an octahedral complex of Co^{2+} that has one unpaired electron.

•21. Calculate the value of Δ_O for a d^9 octahedral complex with a strong visible absorption centered at 600. nm. What color would you expect a solution of this complex to be?

22. What is the value of Δ_O for an octahedral d^1 complex with an absorption centered at 550 nm? Predict the color of the solution.

•23. Which of the following species when aquated would you expect to exhibit paramagnetic behavior? Consider both high-spin and low-spin situations if they are possible.
(a) Cr^{3+} (b) Ni^{2+} (c) Co^{2+}
(d) Fe^{2+} (e) Fe^{3+} (f) Cu^{2+}

24. Which of the following ions would you expect to contain unpaired *d* electrons? Consider both high- and low-spin situations, if possible.
(a) $Cr(NH_3)_6^{3+}$ (b) $Fe(CN)_6^{3-}$
(c) $Fe(CN)_6^{4-}$ (d) $Cu(NH_3)_6^{2+}$

Reactions of Coordination Complexes [25–30]

25. Write the balanced chemical equation for the reaction of Cr^{3+} with water. This ion normally forms octahedral complexes.

26. What is the chemical equation for the reaction that occurs when Zn^{2+}, an ion that normally forms tetrahedral complexes, is added to water?

•27. Calculate the pH of the resulting solution if 10.0 g of $CrCl_3$ is dissolved in sufficient water to produce 1.00 L of solution. The first K_a of $[Cr(H_2O)_6]^{3+}$ is 1.26×10^{-4}.

28. The first acid-ionization constant of $Al(H_2O)_6^{3+}$ is 1.07×10^{-5}. Calculate the pH of a solution initially 0.100 M in $Al(H_2O)_6^{3+}$.

•29. The overall formation constant for $[Co(NH_3)_6]^{3+}$ is 5.0×10^{31}. What will be the concentration of $Co^{3+}(aq)$ at equilibrium if the concentrations of $[Co(NH_3)_6]^{3+}(aq)$ and $NH_3(aq)$ are, respectively, 0.100 and 0.120 M?

30. The overall formation constant for $[Cu(NH_3)_4]^{2+}$ is 6.8×10^{12}. What concentration of $NH_3(aq)$ would be present at equilibrium with $[Cu^{2+}]$ at 0.00100 M and $[Cu(NH_3)_4]^{2+}$ at a concentration of 0.633 M?

Additional Problems [31–32]

•**31.** Give the name of each of the following compounds and give the oxidation state of every metal in each compound.
(a) $[Cu(H_2O)_6]Cl_2$ (b) $Ni(NH_3)_4Cl_2$
(c) $Na_2[PtCl_4]$ (d) $K_4Fe(CN)_6$

32. Draw all possible isomers for the complex ion $[Cr(H_2O)_3(NH_3)_2Cl]^+$, assuming octahedral coordination. (Remember how symmetrical the octahedron is. There are probably fewer isomers than you might first think.)

Cumulative Problems [33–35]

•**33.** Magnetite, Fe_3O_4, has a density of 5.18 g/cm^3 and a cubic-unit cell with a cell edge of 8.37 Å. How many formula units of Fe_3O_4 are there per unit cell? How many Fe^{2+} ions and how many Fe^{3+} ions are there per unit cell? (You should begin by considering how many Fe^{2+} ions and how many Fe^{3+} ions there are in one formula unit of the compound.)

34. Determine the equilibrium constant for the reaction

$$[Ni(NH_3)_6]^{2+}(aq) + 6H_3O^+(aq) \rightleftharpoons 6NH_4^+(aq) + [Ni(H_2O)_6]^{2+}(aq)$$

The overall formation constant of $[Ni(NH_3)_6]^{2+}$ is 4.07×10^8.

•**35.** Determine $\Delta G°$ for the reaction

$$[Cr(H_2O)_6]^{3+}(aq) + CN^-(aq) \rightleftharpoons [Cr(H_2O)_5(OH)]^{2+}(aq) + HCN(aq)$$

K_a for $[Cr(H_2O]_6^{3+}$ is 1.26×10^{-4}.

Applied Problem [36]

36. Plating parts of automobile trim with chromium serves to protect as well as improve the appearance. Suppose you are electroplating a bumper with a total area of 1.50 m^2 in a bath containing Cr^{3+} in acid solution. You want a coating exactly 0.12 mm thick. Exactly how would you arrange the conditions so that this would happen? The density of chromium is 7.14 g/cm^3.

ESTIMATES AND APPROXIMATIONS [37–38]

37. Would you think $[CoF_6]^{3-}$ is a paramagnetic complex ion or not? Offer a rationale for your contention.

38. The absorbing species in aqueous solutions of Cu^{2+} is hydrated Cu(II), most likely $Cu(H_2O)_6^{2+}$, and the solutions are characteristically blue. Briefly explain how you might prove that the coordinated water molecules are critical to the color.

COOPERATIVE STUDY [39–43]

39. What might you reasonably infer from the fact that the synthesis of compounds in which metallic elements are in high oxidation states benefit from basic reaction conditions?

40. Silver forms the sparingly soluble salt AgCl ($K_{sp} = 1.6 \times 10^{-10}$) and the soluble complex ion $[Ag(NH_3)_2]^+$ ($K_{formation} = 1.5 \times 10^7$). Calculate the solubility of AgCl in 1.00 M NH_3.

41. The values of the first three stepwise-formation constants for $[Cu(NH_3)_4)]^{2+}$ are 1.7×10^4, 3.2×10^3, and 8.3×10^2. The overall formation constant is 6.8×10^{12}. What is the value of K_4?

42. A tetrahedral crystal field splits the *d* orbitals into two groups, a lower energy group of two and a higher energy group of three. Draw a diagram and determine which two *d* orbitals will end up in the lower energy group.

43. Ions with seven *d* electrons such as Co^{2+} tend to form some tetrahedral complexes. How do you think the *d* orbitals would be split in a tetrahedral field? How many unpaired electrons would these complexes have?

WRITING ABOUT CHEMISTRY [44–45]

44. Transition metals are necessary for life. Just look at the metals listed on the package of a container of multivitamin and mineral tablets. The amount of each transition metal in these tablets is small. In sufficiently large quantities transition metals are poisonous, the result being called heavy-metal poisoning. Write a short essay that draws an analogy between transition metals and something else in human life that is necessary but dangerous or fatal in large doses.

45. Some theories of chemistry seem to be based on almost unreasonable premises, and yet they lead to reasonable conclusions. An example is the assumption made in the kinetic theory of gases that gas particles are hard round spheres that have mass but no volume. Yet the kinetic theory of gases gives important results on the relationship of molecular speed to temperature and also to molar mass. Graham's laws of diffusion and effusion demonstrate this. Analyze the crystal-field theory. How reasonable are the premises? How good are the results? How could you make the assumptions more realistic, and what effect might that have on the results?

chapter twenty

20

METALLURGY

20.1 FROM MINERALS TO METALS

Natural Sources of Metals

One measure of civilization is the ability to discover, extract, and then use nature's metallic elements. It is with this in mind that we refer to the Bronze Age and the Iron Age. Although three quarters of nature's elements are metals, their abundance is remarkably low—less than one percent of the universe and a quarter of the Earth's crust. Furthermore, except for aluminum and iron, these elements are largely present in deposits that are not usable. For example, there is a lot of gold in the seas, but the current market price of gold is incompatible with the cost of its extraction. Fortunately for past civilizations, a few metals such as copper, gold, and silver occurred in the native or metallic form, and these were the first metals to be used. The metals that have proved to be the most important in advancing civilization have been concentrated in mineral deposits composed of much higher proportions of these metals than in the Earth's crust. However, the world's richest mineral deposits are fast disappearing due to the rapidly increasing rate of exploitation, especially throughout the twentieth century. For example, from less than 10 million tons of metals at the beginning of the century, annual production passed 100 million tons by 1940 and a billion tons by 1990. Metals that were of no interest in 1900, such as titanium and vanadium, are themselves produced at annual levels of tens of millions of tons today.

Metallurgy is the science and technology that leads from minerals to metals and alloys. Extraction metallurgy employs process chemistry to remove metals from their ores. An **ore** is a naturally occurring mineral from which it is practical to extract a metal. Classically, the major works of metallurgy have been associated with iron alloys, particularly steels, and have been of enormous technological importance. The metallurgies of metals such as aluminum and titanium, which are so important to today's technology, were developed much more recently than that of iron.

The metallurgy of a metallic element depends on the nature of its available ores. Because metals are electropositive, that is, lose electrons easily, we expect to find them in oxidized forms as the cations in salts and bound to electronegative elements such as oxygen and sulfur. Figure 20.1 shows the periodic table divided into regions approximately reflecting the kinds of compounds in which the elements occur. Figure 20.1(a) divides the elements of the Earth and the Earth's atmosphere according to traditional geochemical classifications: **lithophiles**, elements that appear as oxides and halides in the crust; **siderophiles**, metallic alloys in the core; **atmophiles**, as volatile gases of the atmosphere or in dissolved form in the oceans; and **chalcophiles**, elements found as sulfides in the crust.

The template in Figure 20.1(b) outlines the elements according to their reactivity, which, in turn, determines the likely methods of separation and isolation. Generally, their reactivity changes in decreasing order from Region A to Region D. Furthermore, because elements close to each other in the periodic table have similar chemical properties, they tend to occur together in nature.

Region A of Figure 20.1 includes the most reactive metals, those of Group IA followed by those of Group IIA. These elements are far too reactive to be found uncombined. Sodium and potassium are the most common elements in Group IA. Almost all of the compounds of sodium are water soluble, as are most of those of potassium. The most important ores of each of these metals are their water-soluble chlorides found in seawater, salt wells, and salt domes such as those in Louisiana and Michigan. The metals of Group IIA form insoluble sulfate and carbonate salts, except for $MgSO_4$, which is water soluble. These salts, especially limestone ($CaCO_3$) and gypsum ($CaSO_4$), are important minerals: limestone as marble for decoration and construction and gypsum for plaster and railroad chalk. Magnesium ions are present in seawater, an important source of the metal.

The metals of region B are also quite reactive. They tend to have high charges and a high affinity for oxygen, and they are found as oxides such as TiO_2 or Al_2O_3, oxo anions such as WO_4^{2-} and double oxides such as $FeO \cdot Cr_2O_3$.

The metals in region C generally occur associated with sulfur, including MoS_2, PbS, $PbSO_4$, and ZnS and more complex species such as CoAsS and $CuFeS_2$. Region D metals are the least reactive and occur either in the native state as free copper, silver, and gold, or in compounds that are easily decomposed to yield the free metal. Region E is exclusively nonmetals.

Extraction processes follow the general divisions of Figure 20.1. The simplest processes are mechanical separation and simple heat treatment, which can be used for the metals of region D, which are the least reactive. Chemical reductions are generally used for the metals of intermediate reactivity in regions B and C, but are not practical for the most reactive metals, those of region A. Electrolytic reduction is the common method for the separation of these metals from their ores. Furthermore, because these metals react with water, the electrolytic process must be carried out in nonaqueous media, specifically as molten salts.

Siderophile | Lithophile | Chalcophile | Atmophile

H																	He
Li	Be											B	C	N	O	F	Ne
Na	Mg											Al	Si	P	S	Cl	Ar
K	Ca	Sc	Ti	V	Cr	Mn	Fe	Co	Ni	Cu	Zn	Ga	Ge	As	Se	Br	Kr
Rb	Sr	Y	Zr	Nb	Mo	(Tc)	Ru	Rh	Pd	Ag	Cd	In	Sn	Sb	Te	I	Xe
Cs	Ba	La*	Hf	Ta	W	Re	Os	Ir	Pt	Au	Hg	Tl	Pb	Bi	(Po)	(At)	(Rn)
(Fr)	(Ra)	(Ac)†															

( ) Radioactive element of very low abundance
* Including lanthanides Ce–Lu
† Including actinides Th,U

(a)

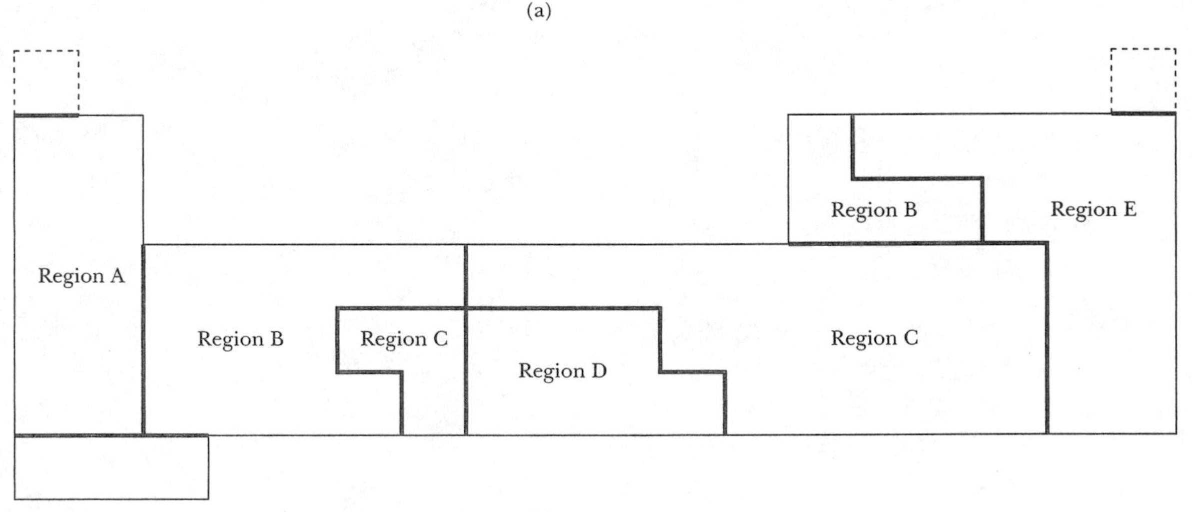

(b)

FIGURE 20.1 Classification and isolation of the elements. (a) Geochemical classification of the elements on Earth. *Lithophiles*, as oxides and halides in the crust; *siderophiles*, in the core as metal alloys; *atmophiles*, as gases in the atmosphere or in dissolved form in the oceans; *chalcophiles*, as sulfides in the crust. (b) Naturally occurring metal compounds and methods of extraction. *Region A:* Most reactive metals, generally found as water-soluble salts or insoluble sulfates and carbonates. Electrolysis of molten salts is the most common method of reduction. *Region B:* Reactive elements generally occurring in combination with oxygen. The ores are reduced by electrolytic or chemical means. *Region C:* Less-reactive metals generally occurring in combination with sulfur. The sulfides are usually roasted to give oxides, which are then chemically reduced. *Region D:* Least reactive metals, occurring in the free state or in ores that can be decomposed simply by heating. *Region E:* These are nonmetals, occurring freely in the atmosphere or as anions in salts.

PROFILES IN CHEMISTRY

Lead, Zinc, and the Badgers Wisconsin is known as the Badger State. However, the "badgers" that the state was named after were miners from Missouri and southern states who came to southwestern Wisconsin in the early 1800s to make their fortunes on the rich lead deposits that had been found there. These rough-and-ready men lived in holes that they dug close to their mines, which came to be known as "badger holes," and the miners became known as "badgers."

In those days lead was used for a variety of purposes including glassmaking and making paint pigments, but its most important use in that still untamed area was for making bullets. The primary ore of lead is PbS, commonly called galena. This ore always comes mixed with similar ores of other metals, particularly FeS_2, ZnS, and Ag_2S. The ores in southwestern Wisconsin had particularly large amounts of ZnS and some $ZnCO_3$, but because no important use was known for zinc at the time, the zinc-containing rocks were just piled up as waste.

Later in the nineteenth century a number of important uses were found for zinc including preparing paint pigments and, most importantly, for coating or galvanizing iron to prevent corrosion. The great piles of waste rocks from the lead mining were suddenly remembered, and the zinc was extracted at great profit. Zinc mining then took over from lead mining, and skilled miners from Cornwall, England, came to the area to apply modern deep-mining methods.

To free zinc economically from the sulfide requires a bit of creative chemistry because carbon cannot be used as the reducing agent. However, the oxide is reducible by carbon, so the trick in freeing the metal is to convert the sulfide into the oxide and then reduce this oxide:

(1) $2ZnS(s) + 3O_2 \longrightarrow 2ZnO(s) + 2SO_2(g)$ roasting of the sulfide at 850°C

(2) $ZnO(s) + C \longrightarrow Zn(s) + CO(g)$ reduction of the oxide in a blast furnace, at 600°C

Heating zinc carbonate is another source of zinc oxide:

(3) $ZnCO_3(s) \longrightarrow ZnO(s) + CO_2(g)$ at 600°C

If the roasting of ZnS is carried out at lower temperatures, the sulfate, not the oxide, is produced:

(4) $ZnS(s) + 2O_2(g) \longrightarrow ZnSO_4(s)$ roasting of the sulfide at 600°C

The sulfate is water soluble and its solutions can be electrolyzed if metal plating is the desired end result. But no such extraordinary means are necessary to free lead from its sulfide, which is directly convertible by roasting:

(5) $PbS(s) + O_2 \longrightarrow Pb(s) + SO_2(g)$ Roasting of the sulfide at 850°C

The mining era is now far in the past in this area but signs of the once intense mining activity can still be seen here and there . . . and Wisconsin remains the "Badger" State.

Questions

If you know of one mineral that can be found in a specific location, how can you predict what other metals may also occur there? Give a couple of examples.

Why is lead sulfide less stable than zinc sulfide, as evidenced by the direct conversion of the former to the metal (by roasting)?

How would you explain the difference in reactivity to carbon of ZnS and ZnO?

Mechanical Separation Processes

The most familiar form of mechanical separation is the gold panning method used by prospectors to separate out the very dense flakes of gold from the less dense grains of other minerals that are present. The pan is loaded with the ore and then repeatedly filled with water and swirled so that the less dense minerals spill out over the edge of the pan and the gold flakes remain behind. This is an application of selective settling in a liquid, which, in this case, is water. In **selective settling** the more dense granules settle more rapidly, and the less dense granules are preferentially washed away. Although such procedures usually do not lead directly to the pure metal, mechanical processes for the concentration of ores are used in most cases to separate the wanted mineral from the unwanted impurities or gangue. **Gangue** contains various minerals such as SiO_2.

Settling processes are common because, in general, the impurities such as SiO_2 present in ores are less dense than the metal compounds themselves. For example, SiO_2, which is tremendously abundant in the Earth's crust, has a density of 2.65 g/cm^3, whereas metallic oxides and sulfides generally have densities in the range of 3.5 g/cm^3 and up. One cm^3 of native gold weighs in at 14 to 19 g. Modern improvements on the gold prospector's pan involve flowing water over vibrating, inclined tables with roughened surfaces that trap the heavy particles. Settling methods have been developed to handle vast quantities of material in continuous-stream processes. If the liquid has a density in between those of the desired and undesired material, the separation can be based on what sinks and what floats. The density of about 3.00 g/cm^3 needed to float SiO_2 and other impurities is above that of any suitable liquid. However, liquid suspensions of tiny particles of very dense minerals or alloys are suitable for this purpose.

Magnetic separation can be used to separate magnetic compounds such as magnetite (Fe_3O_4) and ilmenite ($FeTiO_3$), from unwanted impurities. This technique, though useful, is limited. A number of important iron ores are not magnetic. However, conversion of at least some of the iron present to magnetite can sometimes be achieved by heating Fe_2O_3 in the presence of CO in what is called a magnetizing roast:

$$3Fe_2O_3(s) + CO(g) \longrightarrow 2Fe_3O_4(s) + CO_2(g)$$

The most widely used technique for mechanical separation is flotation (Fig. 20.2). **Flotation** is a separation method that depends on some particles being wetted by water and some by oil. The particles that are wetted by the oil are beaten into a froth and float to the surface in a tank of water. The method is extremely selective and can be used to concentrate ores with very high levels of impurities.

The key to flotation is the difference in the surfaces of the mineral particles and the gangue. This can result in a difference in the kinds of wetting that occur on each. Generally, however, the surfaces are not sufficiently different to give a clean separation. Therefore, a substance called a collector must be added; this attaches itself to the mineral particles. Metal mineral particles are primarily ionic or strongly polar substances because they contain an electropositive metal and normally a strongly electronegative element such as O or S. Therefore, they are normally *wetted* more effectively by the polar water molecules than the nonpolar oil molecules. However, the collector gives the mineral surface a nonpolar nature so that it is wetted by the oil. Thus, the collector molecule must have one end that is attracted to the ionic mineral surface and one end that will be attracted to the oil. The dodecyl (12 carbon atoms) ammonium ion is such a molecule:

$$CH_3{-}CH_2{-}CH_2{-}CH_2{-}CH_2{-}CH_2{-}CH_2{-}CH_2{-}CH_2{-}CH_2{-}CH_2{-}CH_2{-}NH_3^+$$

dodecylammonium ion

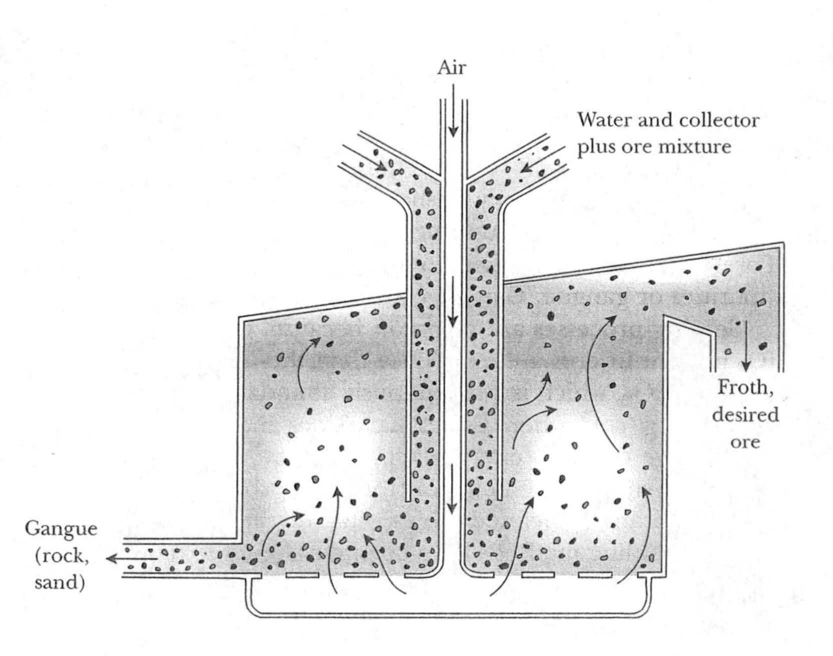

FIGURE 20.2 The flotation process for separating minerals from gangue. The mineral surface is treated with a collector, which allows it to be wet by oil and swept up to the top by the froth. The gangue is wet by water and settles to the bottom. Concentrated sulfide ores (ZnS and PbS) can be separated by such flotation techniques. The collector molecules are typically surface active agents (surfactants).

Here the $—NH_3^+$ group would be attracted to negative ions or the negative ends of bond dipoles on the mineral surface, and the long chain would dissolve in the oil to which it is very similar. Only a single layer of collector molecules on the surface of the mineral is necessary to cause the particles to be wetted completely by the oil. Impurities such as SiO_2, lacking ionic or strongly polar bonding, would not attract the collector molecules.

A foaming agent such as pine oil is normally added to the flotation mixture to help build a large amount of froth to carry off the oil-wetted particles.

20.2 REDUCING METAL OXIDES TO METALS

Chemical Reductions

As we mentioned in Section 20.1, metals are normally found in oxidized states because atoms of these elements easily lose electrons. Therefore, freeing the metal from its ore is normally a reduction process, and, consequently, some other species must be oxidized at the same time. In many cases the ore that is reduced is the metal oxide or sulfide.

We will consider a number of different ways to reduce metal oxides to the free metals, beginning with the use of another metal as the reducing agent. This direct chemical method can be a commercially viable process if the metal being produced is more valuable than the metal being sacrificed as the reducing agent. Let us begin with a general reaction in which metal oxide M′O is to be reduced with the aid of another metal M″ acting as the reducing agent. Here is the general reaction for this process where M′ and M″ both form oxides with the stoichiometry MO:

$$M'O + M'' \longrightarrow M' + M''O \qquad \text{reduction of a metal oxide with a metal}$$

Suitable coefficients have to be added to the equation if other stoichiometries exist. For example, the reduction of titanium(IV) oxide with aluminum would follow the equation

$$3TiO_2(s) + 4Al(s) \longrightarrow 2Al_2O_3(s) + 3Ti(s) \qquad \text{reduction of } TiO_2 \text{ with Al}$$

Going back to the general equation, the reaction can be broken into two parts, the sum of which is the net equation

$$M'O \longrightarrow M' + \tfrac{1}{2}O_2$$

$$M'' + \tfrac{1}{2}O_2 \longrightarrow M''O$$

$$M'O + M'' \longrightarrow M' + M''O$$

The value of $\Delta G°$ for the overall process will be the difference between $\Delta G_f°$ values for the respective oxides:

$$\Delta G° = \Delta G_f°(M''O) - \Delta G_f°(M'O)$$

Standard free energy change: This equation applies to the reaction of one metal with the oxide of another in 1:1 stoichiometry. $\Delta G_f°(M''O)$ is the standard free energy of formation of M″O, and the $\Delta G_f°(M'O)$ is the standard free energy of formation of M′O.

The overall process will have a negative value of $\Delta G°$ if $\Delta G_f°$ for M″O is more negative than $\Delta G_f°$ for M′O.

Our most useful equation for predicting the spontaneity of a chemical reaction is the thermodynamic statement for the change in free energy:

$$\Delta G° = \Delta H° - T\Delta S°$$

Free energy change at standard conditions: For simplicity, we will generally use the values of ΔH and ΔS at standard conditions, that is, $\Delta H°$ and $\Delta S°$. $\Delta H°$ can be calculated from a table of standard enthalpies of formation and $\Delta S°$ from a table of standard entropies.

We can calculate $\Delta G°$ at a variety of temperatures assuming that $\Delta H°$ and $\Delta S°$ do not change with temperature. This assumption, although only an approximation at best, greatly simplifies the calculations and our discussion of the principles involved. For a reaction of the type

$$M'' + \tfrac{1}{2}O_2 \longrightarrow M''O$$

we expect a negative value of $\Delta H°$ because of the formation of the strong metal–oxygen interaction, and we expect a negative value of $\Delta S°$ because of the loss of the gaseous reactant. Therefore, for this reaction, the value of $\Delta G°$ for any metal can be expected to become more positive with increasing temperature, as shown for the formation of a number of different metal oxides in Figure 20.3.

The general rule for interpreting Figure 20.3 is that a metal will reduce the oxides of all the metals above it on the graph because such a reaction will have

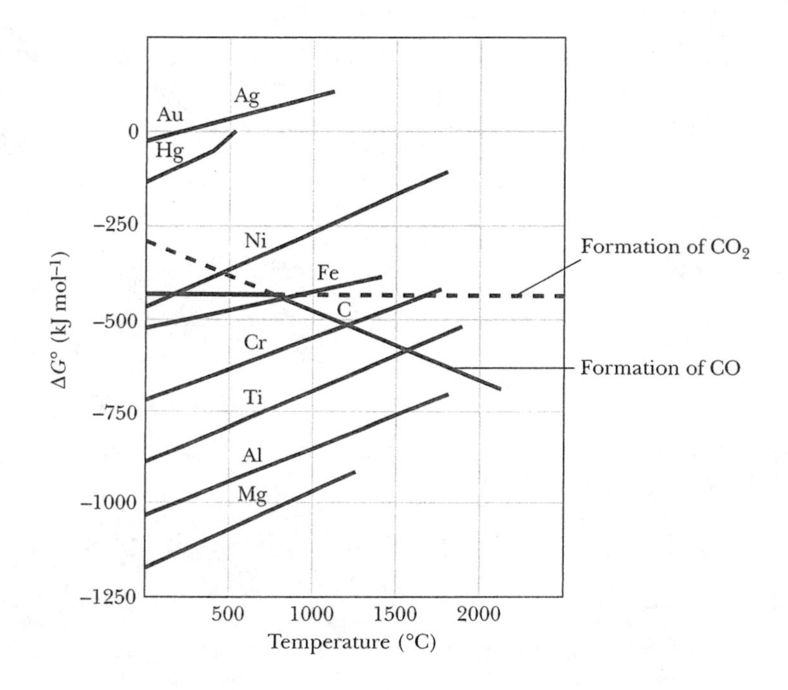

FIGURE 20.3 Free energies of formation of the most common oxides of carbon and some selected metals, as a function of temperature. For carbon the free energy of formation is shown for the favored oxide, CO_2 or CO, depending on the temperature.

a negative value of $\Delta G°$. Thus, titanium will reduce the oxide of chromium, and magnesium will reduce the oxides of all of the metals that are given. Because the slopes of all of the lines of $\Delta G°$ versus T are about the same, the ability of one metal to reduce the oxide of another is essentially independent of temperature.

EXAMPLE 20.1

Use data from Tables 12.4 and 13.1 to calculate $\Delta G°$, and find the value of K for the reduction of CuO with Fe at 300.K and 2000.K. Assume that the Fe is oxidized to Fe_2O_3.

COMMENT First, write the chemical equation. Because we will be calculating $\Delta G°$ at two temperatures, 300.K and 2000.K, we will need to know $\Delta H°$ and $\Delta S°$ of the reaction. Use standard enthalpies of formation and standard entropies to calculate $\Delta H°$ and $\Delta S°$ of the reaction. Then determine $\Delta G°$ using $\Delta G° = \Delta H° - T\Delta S°$, and, finally, calculate K using $\Delta G° = -RT \ln K$.

SOLUTION

$$3CuO(s) + 2Fe(s) \longrightarrow Fe_2O_3(s) + 3Cu(s)$$

$$\Delta H° = \Delta H_f°(Fe_2O_3) - 3[\Delta H_f°(CuO)] = -822 - 3(-155)\text{ kJ} = -357\text{ kJ}$$

Remember that $\Delta H_f°$ for an element in its standard state is zero.

$$\Delta S° = S°(Fe_2O_3) + 3S°(Cu) - 3S°(CuO) - 2S°(Fe) = 90.0 + 3(33.3) - 3(43.5) - 2(27.2)\text{ J/K} = 5.0\text{ J/K}$$

$$\Delta G° = \Delta H° - T\Delta S°$$

At 300.K,

$$\Delta G^\circ = -357\text{ kJ} - 300.\text{K} \times 5.0\text{ J/K} \times \frac{1\text{ kJ}}{1000\text{ J}} = -358\text{ kJ}$$

$$\Delta G^\circ = -RT\ln K$$

$$\ln K = -\frac{\Delta G^\circ}{RT}$$

$$= -\frac{-358\text{ kJ} \times 1000\text{ J/kJ}}{(8.31\text{ J/mol·K})(300.\text{K})} = 144$$

$$K = e^{144} = 3 \times 10^{62}$$

At 2000.K,

$$\Delta G^\circ = -357\text{ kJ} - 2000.\text{K} \times 5.0\text{ J/K} \times \frac{1\text{ kJ}}{1000\text{ J}} = -367\text{ kJ}$$

$$\ln K = -\frac{\Delta G^\circ}{RT}$$

$$= -\frac{-367\text{ kJ} \times 1000\text{ J/kJ}}{(8.31\text{ J/mol·K})(2000.\text{K})} = 22.1$$

$$K = e^{22.1} = 4 \times 10^9$$

EXERCISE 20.1

Calculate ΔG° and the value of K for the reduction of Ag_2O with Zn at 400.K and 1500.K. Use the data in Tables 12.4 and 13.1 and the additional value, $S^\circ(Ag_2O) = 121.7$ J/mol·K.

ANSWERS At 400.K, $\Delta G^\circ = -304.2$ kJ and $K = 5.29 \times 10^{39}$. At 1500.K, $\Delta G^\circ = -266.7$ kJ and $K = 1.94 \times 10^9$.

Note that for Ag and Hg, a temperature can easily be reached at which the reaction

$$M + \tfrac{1}{2}O_2 \longrightarrow MO$$

has a positive value of ΔG°. This means that the reaction can be made to run spontaneously in the reverse direction, and the metal can be separated from the oxide by simple heating. This made possible Priestley's classic preparation of oxygen by heating HgO,

$$HgO(s) \longrightarrow Hg(\ell) + \tfrac{1}{2}O_2(g)$$

For gold, the value of ΔG° for oxide formation is positive at room temperature, and the oxide decomposes readily.

EXAMPLE 20.2

Calculate the temperature at which mercury and HgO exist in equilibrium with $O_2(g)$ at a partial pressure of 1.00 atm. This temperature can be considered to be the minimum temperature for the decomposition of HgO to mercury metal. Thermodynamic data needed in addition to Tables 12.4 and 13.1 are as follows: $\Delta H_f^\circ(\text{HgO}) = -90.83$ kJ/mol, $S^\circ(\text{HgO}) = 70.29$ J/mol·K, $S^\circ(\text{Hg}) = 76.02$ J/mol·K.

COMMENT Write the chemical equation, and determine ΔH° and ΔS°. At equilibrium $\Delta G^\circ = 0$. Solve $\Delta G^\circ = \Delta H^\circ - T\Delta S^\circ$ for T when ΔG° is equal to zero.

SOLUTION

$$\text{HgO(s)} \longrightarrow \text{Hg}(\ell) + \tfrac{1}{2}\text{O}_2\text{(g)}$$

$$\Delta H^\circ = -\ (-90.83 \text{ kJ/mol}) = 90.83 \text{ kJ}$$

$$\Delta S^\circ = \tfrac{1}{2}(205) + 76.02 - 70.29 \text{ J/K} = 108 \text{ J/K}$$

At equilibrium,

$$\Delta G^\circ = \Delta H^\circ - T\Delta S^\circ = 0$$

$$T = \frac{\Delta H^\circ}{\Delta S^\circ} = \frac{90.83 \text{ kJ} \times 1000 \text{ J/kJ}}{108 \text{ J/K}} = 841\text{K} = 568°\text{C}$$

EXERCISE 20.2

Calculate the minimum temperatures for the decomposition of PbO and Al_2O_3 to the metals. Use thermodynamic data in Tables 12.4 and 13.1 as well as the following standard entropies in J/mol·K: Pb, 64.81; PbO, 68; Al, 28.33; Al_2O_3, 50.92.

ANSWERS 2180K for Pb and 5350K for Al.

The ΔG° values for the formation of CO and CO_2 have also been included on Figure 20.3. For the oxidation of carbon to carbon dioxide,

$$\text{C(s)} + \text{O}_2\text{(g)} \longrightarrow \text{CO}_2\text{(g)}$$

the value of ΔS° is about zero because the number of moles of gas is unchanged by the reaction. Thus, ΔG° is essentially constant with temperature. For

$$\text{C(s)} + \tfrac{1}{2}\text{O}_2\text{(g)} \longrightarrow \text{CO(g)}$$

ΔS° is positive because an extra half-mole of gas is formed in the reaction. Therefore, ΔG° for this reaction becomes more negative as the temperature increases. The two values of ΔG° cross at about 400°C. Therefore, below this temperature, CO_2 is the favored product of the reaction of carbon with oxygen, whereas above 400°C, CO is the favored product.

Carbon is a particularly useful reducing agent for metal oxides because it is inexpensive and readily available and because of the thermodynamic favorability of its oxidation. In order to reduce a metal with carbon by the reaction

$$\text{M}'\text{O(s)} + \text{C(s)} \longrightarrow \text{M}' + \text{CO(g)}$$

it is necessary to raise the temperature above the point where the ΔG° value for the formation of CO equals the ΔG° value for the formation of M′O. This means exceeding the temperature where the ΔG° versus T lines cross for the two processes on Figure 20.3. Thus, carbon could theoretically reduce any metal oxide. However, two factors interfere. One is the practical temperature limit of the reaction vessel and the other is the tendency for some metals, particularly reactive ones, to form compounds with carbon called carbides. Chromium and iron are reduced with carbon, whereas this method is impractical for titanium, aluminum, and magnesium because of the temperature required. The use of other reducing agents or electrochemical reductions is necessary in these cases.

EXAMPLE 20.3

Calculate the minimum temperature necessary to reduce ZnO to the metal with carbon (graphite form). Use thermodynamic data from Tables 12.4 and 13.1.

COMMENT Consider the minimum temperature for the reaction to be the temperature at which $\Delta G°$ reaches zero. Write the reaction and calculate $\Delta H°$ and $\Delta S°$. Remember carbon is oxidized to CO at high temperatures because the entropy change is more favorable than oxidation to CO_2.

SOLUTION $ZnO(s) + C(s) \longrightarrow Zn(s) + CO(g)$

$$\Delta H° = -110 - (-348.3)\text{ kJ} = 238\text{ kJ}$$

$$\Delta S° = (198 + 41.6 - 43.9 - 5.73)\text{J/K} = 190\text{ J/K}$$

$$T = \frac{\Delta H°}{\Delta S°} = \frac{238\text{ kJ} \times 1000\text{ J/kJ}}{190\text{ J/K}} = 1250\text{K} = 980°\text{C}$$

EXERCISE 20.3

Calculate the minimum temperature necessary to reduce Al_2O_3 to aluminum metal using the graphite form of carbon. The necessary standard entropies in J/mol·K are Al, 28.33 and Al_2O_3, 50.92.

ANSWER 2310K

Extraction of sulfur-containing ores presents different problems. Reduction of sulfides using carbon, that is,

$$2MS + C(s) \longrightarrow 2M + CS_2$$

is not useful because the reaction

$$C(s) + 2S(g) \longrightarrow CS_2(g)$$

has a value of $\Delta G°$ close to zero at 298K and a negative value of $\Delta S°$, which causes $\Delta G°$ to become more positive at increasing temperatures. Therefore, it is necessary to convert the sulfides to other compounds, usually oxides, by reaction with oxygen in the air in a process called roasting. For example,

$$2ZnS(s) + 3O_2(g) \longrightarrow 2ZnO(s) + 2SO_2(g)$$

The ZnO can then be reduced with carbon. Roasting of ores of metals like Zn, Cu, and Ni has been a serious environmental problem due to the SO_2 produced.

More effective methods of stack gas clean-up to recover the valuable SO_2 for sulfuric acid production are still being developed.

Other chemical reducing agents include H_2 and reactive metals. An example of the use of a reactive metal will be discussed when we consider the refining of titanium later. Using H_2 is limited by the fact that $\Delta G°$ for the reaction

$$H_2(g) + \tfrac{1}{2}O_2(g) \longrightarrow H_2O(g)$$

becomes less negative with increasing temperature because of the negative value of $\Delta S°$. Therefore, increasing the temperature does not increase the spontaneity of the reduction process or the number of different metal oxides that can be reduced. Hydrogen reduction does find uses, however, where reduction with carbon would result in a metal carbide, as with tungsten. We will discuss this example later.

Electrolytic Reductions

The oxides of metals near the bottom of Figure 20.3 have very large negative $\Delta G_f°$ values. Therefore, the free-energy change for extraction of the metal is very unfavorable, and the temperature required for chemical reduction may therefore be impractical. However, electrochemical reductions should require the application of only sufficient electrical potential to exceed the potential that would be generated if the reaction proceeded in the spontaneous direction, that is, producing the metal oxide rather than reducing it. If ΔG is the free-energy change of the spontaneous reaction, the minimum potential to run the reaction in the nonspontaneous direction can be calculated from the equation

$$\Delta G = -nFE$$

Relationship between free energy change and cell potential: In this equation *n* is the number of electrons transferred per atom reduced, *F* is the Faraday constant of 96,500 coulombs of charge per mole of electrons, and *E* is the applied potential.

In practice, the electrical potentials necessary for even the most unfavorable reductions are very modest. The reduction of Al_2O_3 to Al is performed commercially at an applied potential of about 6 V. The calculation of cell potentials for electrolysis cells is complicated by two factors. First, the conditions in the cells, for example fused salts, are far from the standard conditions of our thermodynamic and reduction potential tables. Second, potentials for electrolysis are always higher than would be calculated according to thermodynamics. The difference is called overvoltage. **Overvoltage** is simply defined as the voltage actually needed beyond that calculated to create spontaneous conditions. It results partly from the change of concentration that develops as the reaction proceeds in the region of the electrolyte that is directly in contact with the electrodes. Part of the overvoltage is necessary to get the reaction to run at a rate that is fast enough for significant formation of products.

The reactive metals for which electrochemical reductions are necessary cannot be produced from electrolysis in aqueous solutions because the metals would react with the water as they were formed. Nonaqueous electrolytes must be used

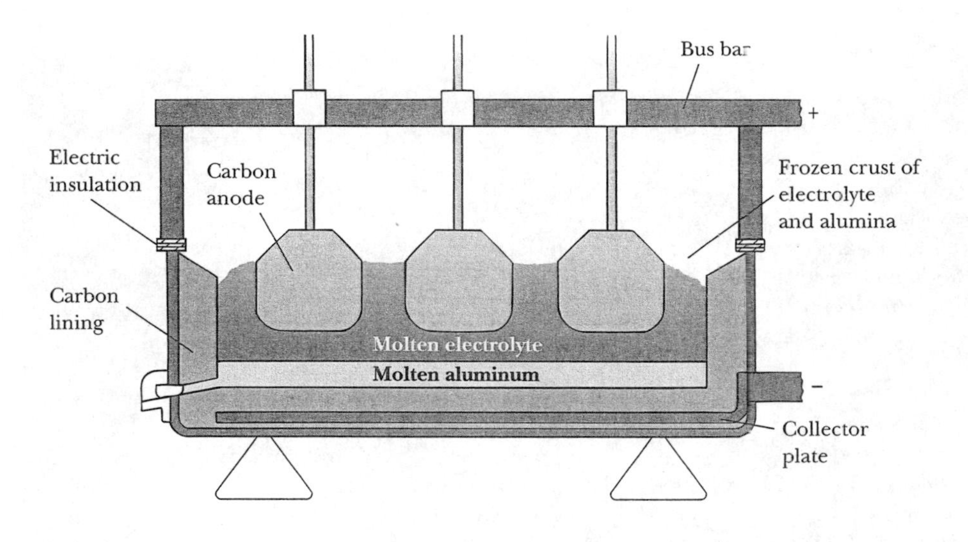

FIGURE 20.4 Production of aluminum from Al_2O_3 by electrolysis. The Hall–Heroult process depends on the electrolysis of bauxite ore (Al_2O_3) in a bath of cryolite (Na_3AlF_6) under carefully controlled conditions. Temperature control is critical, and the system usually operates at about 950°C, which is considerably below the normal melting point of bauxite.

and these are mostly fused salts. For sodium and potassium, the chlorides NaCl and KCl are simply melted and electrolyzed. For aluminum, the high melting point of alumina (Al_2O_3) from bauxite ore requires the use of molten salt cryolite (Na_3AlF_6) to dissolve the oxide (Fig. 20.4). For elements with low melting points, such as Na, K, Mg, and Al, the metal can be easily withdrawn from the electrolysis cell as a liquid. For high-melting metals such as titanium, electrolysis is less satisfactory because the metal forms as powdery crystals, which mix with the electrolyte. A chemical reduction with an active metal is more satisfactory in such cases.

The relationships discussed in Chapter 14 can be used to calculate the mass of metal deposited in an electrolysis. The mass in grams of metal deposited in t seconds is given by

$$m = \frac{iMt}{Fn}$$

Mass of metal produced by electrolysis: A metal compound in the $n+$ oxidation state will require n moles of electrons to produce one mole of metal of molar mass M. The Faraday constant F is 96,500 coulombs (C) per mole of electrons, and the current i is in amperes, that is, C/s.

For example, in the reduction of aluminum oxide to produce one mole of aluminum, n equals 3:

$$\tfrac{1}{2}Al_2O_3 + 3e^- \longrightarrow Al + \tfrac{3}{2}O^{2-}$$

EXAMPLE 20.4

For what length of time would a 50.0 A current need to be applied to produce 100. g of magnesium by electrolysis?

PROCESSES IN CHEMISTRY

Electrolytic Production of Aluminum Although aluminum is an abundant and widespread element, methods of producing it from common aluminum-containing resources such as clay have not been developed. The only practical ore of aluminum is bauxite, which is mostly Al_2O_3. Aluminum is a very reactive metal, so its preparation must be either by reaction with an even more reactive metal or by electrolysis. Bauxite melts at 2030°C, however, which is impractically high. Until a satisfactory method of electrolysis was discovered for aluminum it was prepared only through reduction by an active metal like sodium. Aluminum was so expensive and valuable in the nineteenth century that medals for the highest order of achievement were often struck from this metal. Napoleon used gold and silver tableware for most guests, reserving aluminum tableware for special occasions when royalty visited. In 1884, when the Washington Monument was completed, a 6-pound cone of aluminum was placed at the very top.

The satisfactory method of electrolysis was discovered independently by Paul Heroult of France and Charles Martin Hall of the United States in 1886. Hall had only recently graduated from Oberlin College, and it is reported that he did the necessary experimental work to develop the electrolysis in an old woodshed. The trick is to perform the electrolysis in a bath of molten cryolite, Na_3AlF_6, which melts at 1010°C, far below the temperature of molten bauxite. The addition of small amounts of other substances lowers the operating temperature still further to about 950°C. The anode of the cell is carbon, and a carbon lining of the cell serves as the cathode. The cathode is inert in this process, but the carbon anode actually takes part in the electrolytic reaction and has to be replaced from time to time. The principal reactions are as follows:

$$AlF_6^{3-} + 3e^- \longrightarrow Al(\ell) + 6F^- \qquad \text{cathode}$$

$$2Al_2O_3(\ell) + 24F^- + 3C(s) \longrightarrow 4AlF_6^{3-} + 3CO_2(g) + 12e^- \qquad \text{anode}$$

$$2Al_2O_3(\ell) + 3C(s) \longrightarrow 4Al(\ell) + 3CO_2(g) \qquad \text{net}$$

The aluminum produced by this process can be up to 99.9% pure.

After his brilliant chemical success, Charles Martin Hall abandoned science and turned his attention to business, specifically the aluminum business. His ability in business appears to have been as great as his ability in science. The enterprise he formed became the Aluminum Company of America, now officially called Alcoa, Inc., a widely known corporate enterprise. Charles Martin Hall remembered his *alma mater* well and contributed generously to Oberlin College. As a result of his and Heroult's work, aluminum could be used economically for an enormous number of applications, but now those prize aluminum medals that had been given in Europe didn't seem so special.

Questions

The extensive use of aluminum occurred only in the twentieth century, whereas many of our other important metals have been used for hundreds of years. Why did the use of aluminum come so late?

The production of aluminum requires enormous inputs of electricity, as can be seen from the anode reaction and its 12 moles of electrons in the balanced half-reaction. How might the anode reaction be altered to cut the demand in half, saving electricity and still producing the same amount of aluminum?

COMMENT Rearrange the equation for the mass of metal produced by electrolysis and solve for t. Note that Mg in its compounds exists in the oxidation state of 2+ and so n will be 2.

SOLUTION

$$t = \frac{mFn}{iM} = \frac{(100.\text{ g})(96{,}500\text{ C/mol})(2)}{(50.0\text{ C/s})(24.3\text{ g/mol})} = 15{,}900\text{ s} = 4.41\text{ hr}$$

EXERCISE 20.4

What current would be necessary to produce 10.0 g of sodium in 1.00 hr?

ANSWER 11.6 A

20.3 EXTRACTION OF SOME IMPORTANT METALS

The production and refining of the metals from their ores depends on a large number of chemical principles. We will consider five significant examples, drawn from gold, iron, copper, titanium, and tungsten production and refining.

Extraction of Gold

Although highly prized and generally considered rare, the amount of gold available in the ground appears to be large relative to the demand for it. South Africa and the former Soviet Union produce almost 70% of the world's gold, but there are productive deposits in 13 other countries. Thus, gold is quite widespread relative to some metals like cobalt, for which Congo holds almost all of the known reserves; niobium and tantalum, where almost 90% of the production is in Brazil; or tungsten, where 75% of reserves are in China. The lower the percentage of gold in a deposit, the more expensive the deposit is to work. Therefore, economic factors determine which gold deposits will be mined, and the amount of gold produced will increase when the price of gold rises.

Gold resists oxidation and therefore does not tarnish. The instability of gold oxides has been mentioned previously, and, in fact, gold occurs almost exclusively as the uncombined metal. The only exceptions are some mixtures of tellurium compounds of gold and silver. The prospector panning the gold-bearing sands in streams and riverbeds of California, the Yukon, New Zealand, and many other places was performing a selective settling process based on the density difference. For gold, this process not only enriched the ore, it produced the desired metal. Large-scale equipment working on a continuous basis now performs the density separation.

Chemical methods are used to ensure the extraction of all the gold that is present and, in addition, free it from impurities. Some of these impurities, such as silver, have considerable value of their own. The ore concentrate from the settling process is contacted with mercury coated on the surface of a rotating drum. Liquid mercury very readily forms alloys called **amalgams** with a number of metals, and the gold amalgam that is formed sticks to the drum surface. Amalgams are often formed directly on the contact of mercury with another metal and include the silver amalgam used in filling teeth and the copper amalgam formed when mercury is rubbed onto a copper coin. When sufficient gold amalgam has formed on the drum surface, it is scraped off. Some of the mercury can be removed simply by pressing it out, and the rest is distilled off and reused.

For very low-grade ores, the gold can be extracted by reaction with cyanide solutions, generally NaCN or KCN. The CN^- ion forms a very strong coordination complex with the Au^+ ion. This complex can be formed from gold metal and CN^- ion in the presence of the oxygen from the air,

$$4Au(s) + 8CN^-(aq) + O_2(g) + 2H_2O(\ell) \longrightarrow 4[Au(CN)_2]^-(aq) + 4OH^-(aq)$$

The tendency of gold to form coordination complexes is consistent with its being a transition metal. In this reaction, elemental gold is being oxidized to the +1 oxidation state. Although Au_2O_3 and the free Au^+ ion are not stable, gold in the +1 oxidation state is stable when complexed to the two CN^- ions. It is quite common for the reactivities of metal ions in oxidation–reduction reactions to change drastically when the metal ion is complexed or the ligands on the metal are exchanged.

The Au^+ in $[Au(CN)_2]^-$ can be reduced back to Au(s) with powdered zinc metal. The Zn is oxidized to Zn^{2+} ions, which can be complexed by the CN^-

Prospecting for Gold James Marshall (1810–1885), an itinerant carpenter, discovered gold in Sutter's Creek, setting off the legendary California gold rush, the largest of its kind in American history. Arriving in California in 1845 and settling at Sutter's Fort (Sacramento), he was employed by John Sutter to set up an undershot water sawmill 40 miles up the American River from the Fort. Marshall dug the races, erected the wheel, and, on January 24, 1848, released the waters into the system. But there was a problem with some unexpected earth wash, so he shut down the system and walked the tailrace looking for the blockage.

As Marshall walked along, the glitter of something lodged in a small crevice caught his eye. He picked up the substance and found it to be quite heavy for its size and of a peculiar color. It was unlike anything he had ever seen in the river before. He sent an Indian boy back to his camp to fetch a tin plate on which he collected several small cubical crystals and flakes that sparkled in the sunlight. Marshall knew what he had discovered from that very first moment. Stories of Colorado gold had preceded him to California. But Marshall was not one to be suckered into a discovery of "fool's gold," iron pyrites (FeS_2), as others before him, so he ran the few simple tests he knew to confirm his suspicions. He had the camp cook boil the flakes in the lye of her soap kettle and found next morning no tarnishing or discoloration at all. He then had the blacksmith beat some of the flakes on his anvil and was pleased to note "the crystals would flatten but not break."

Marshall, convinced he had struck gold, collected more of the plentiful samples from the races, runs, and creeks coming off the American River and returned to Sutter's Fort. Together, he and John Sutter tested the metallic nuggets with nitric acid and roughly determined their density compared with silver. Convinced their tests positively identified the substance as pure gold, they worked desperately for some months at mining gold and attempting to keep their discovery a secret. But everyone in the camp and at the Fort—nearly a hundred people—quickly learned of the discovery, and by the next spring 50,000 miners, adventurers, entrepreneurs, and assorted ne'er-do-wells descended on Sutter's Fort, the American River, and California. The greatest gold rush in history had begun.

Sadly, neither Marshall nor Sutter shared in the fortune beyond the few pounds of gold taken out that first year, which they shared with their campers and for which they received only a fraction of current fair market value. Sutter at least has become an icon of a kind for having his name associated with the 49ers and the place of discovery for all time to come. Few know of James Marshall and the practical chemistry that he applied to that monumental discovery.

Questions

How did Marshall's tests establish that his yellow crystals and flakes were likely to be elemental gold, not iron pyrites (FeS_2)?

How did Sutter's additional tests validate and confirm Marshall's suspicions of his discovery?

ligands, thus breaking up the Au complex and facilitating the reduction back to Au(s).

$$2[Au(CN)_2]^-(aq) + Zn(s) \longrightarrow [Zn(CN)_4]^{2-}(aq) + 2Au(s)$$

Copper and silver, which are in the same group as gold, have somewhat similar chemical reactions and exist in gold ores as impurities. These are separated from the gold by passing chlorine gas through the molten metal. Gold is not oxidized by the chlorine, but both copper and silver are, and they are removed as the chlorides:

$$2Ag(\ell) + Cl_2(g) \longrightarrow 2AgCl(\ell)$$

$$Cu(\ell) + Cl_2(g) \longrightarrow 2CuCl(\ell)$$

Refining of Iron

Iron is a soft, reactive metal that has essentially no uses in its pure state. Its alloys, particularly steels, however, are the mainstay of industrialized civilization. The principal ores of iron are

- **hematite** Fe_2O_3 and hydrated Fe^{3+} oxides such as $2Fe_2O_3{\cdot}3H_2O$, which are red.
- **magnetite** Fe_3O_4, which is brown to black and strongly magnetic. The apparent fractional oxidation state of Fe in magnetite is the result of a crystal structure containing both Fe^{2+} and Fe^{3+} ions.
- **siderite** $FeCO_3$, which is gray to black.
- **taconite**, which contains all of the other iron ores. It is commonly green, strongly magnetic, and relatively low grade because of the presence of significant quantities of silica, which must be removed at the mining site so that the ore will not be too expensive to ship.

Iron ore, normally containing considerable SiO_2 and other Si compounds as impurities, is charged into a blast furnace (Fig. 20.5) along with limestone ($CaCO_3$) and coke. Coke is produced from heating coal in the absence of air and is composed of carbon and the fused coal ash. This is the most important example of using carbon to reduce a metal oxide to the free metal. A blast of hot air at the bottom of the furnace results in temperatures up to 1300°C as the coke burns,

$$2C(s) + O_2(g) \longrightarrow 2CO(g)$$

The carbon monoxide progressively reduces the iron oxides to lower oxides and finally to the molten metal, which is drawn off at the bottom of the furnace:

$$3Fe_2O_3(s) + CO(g) \longrightarrow 2Fe_3O_4(s) + CO_2(g)$$

$$Fe_3O_4(s) + CO(g) \longrightarrow 3FeO(s) + CO_2(g)$$

$$FeO(s) + CO(g) \longrightarrow Fe(\ell) + CO_2(g)$$

Hot carbon in the coke can react with the carbon dioxide evolved from these reactions to regenerate carbon monoxide:

$$CO_2(g) + C(s) \longrightarrow 2CO(g)$$

FIGURE 20.5 Blast furnace for the reduction of iron ore (oxides) to iron. The chemistry is complex, but basically this is a carbon monoxide and carbon reduction of Fe_2O_3. The density difference between molten iron and the calcium silicate slag allows them to be drawn off at separate points at the bottom of the furnace. The pig iron that is produced has a high carbon content, which tends to make it brittle and limits its usefulness without further processing.

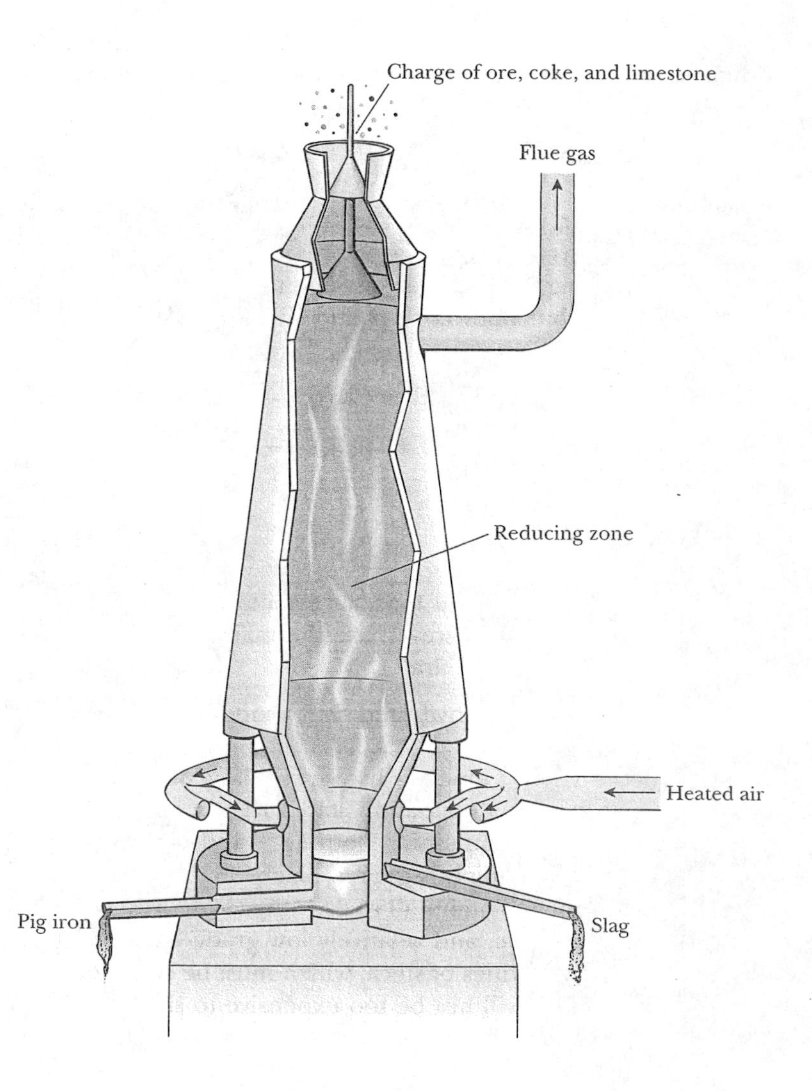

The heat of the furnace decomposes the limestone,

$$CaCO_3(s) \longrightarrow CaO(s) + CO_2(g)$$

and the calcium oxide helps to remove the silicon-containing impurities by forming calcium silicate *slag,* which is a liquid at the temperature involved:

$$CaO(s) + SiO_2(s) \longrightarrow CaSiO_3(\ell)$$

This slag is less dense than the molten iron and floats on top of it. From time to time it is drained off. Some of it is blown with air to make a fluffy, nonflammable insulation known as rock wool.

The iron produced by the blast furnace is called cast iron or pig iron. It contains considerable impurities: C, 2.0–4.5%; Si, 0.7–3.0%; S, 0.1–0.3%; P, 0–3.0%; Mn, 0.2–1.0%. As a result of these impurities, pig iron is brittle and suitable only for producing castings that will not be subjected to shock. Reduction of the carbon content to between 0.05 and 2.0% and removal of almost all of the other nonmetallic impurities leads to steel, a family of alloys of iron with more desirable qualities of flexibility, hardness, strength, and malleability. This is normally achieved in an open hearth furnace (Fig. 20.6), consisting of a shallow pool of molten iron heated by gas flames over the surface. The furnace is lined with magnesium oxide or mixed magnesium and calcium oxides.

It is necessary to oxidize the nonmetal impurities in the iron in order to remove them. This is done by adding a controlled amount of an iron oxide, which will be reduced to metallic iron in the process:

$$3C(s) + 2Fe_2O_3(s) \longrightarrow 3CO_2(g) + 4Fe(\ell)$$

$$3S(s) + 2Fe_2O_3(s) \longrightarrow 3SO_2(g) + 4Fe(\ell)$$

$$12P(s) + 10Fe_2O_3(s) \longrightarrow 3P_4O_{10}(g) + 20Fe(\ell)$$

$$3Si(s) + 2Fe_2O_3(s) \longrightarrow 3SiO_2(\ell) + 4Fe(\ell)$$

These nonmetal oxides are all acidic and will react with the basic magnesium oxide or calcium oxide–magnesium oxide lining of the furnace. For example,

$$P_4O_{10}(g) + 6MgO(s) \longrightarrow 2Mg_3(PO_4)_2(\ell)$$

$$SiO_2(\ell) + MgO(s) \longrightarrow MgSiO_3(\ell)$$

Sulfur (as FeS), another harmful impurity for steelmaking, can be removed by reaction with coke and lime:

$$FeS(s) + CaO(s) + C(s) \longrightarrow Fe(s) + CaS(s) + CO(g)$$

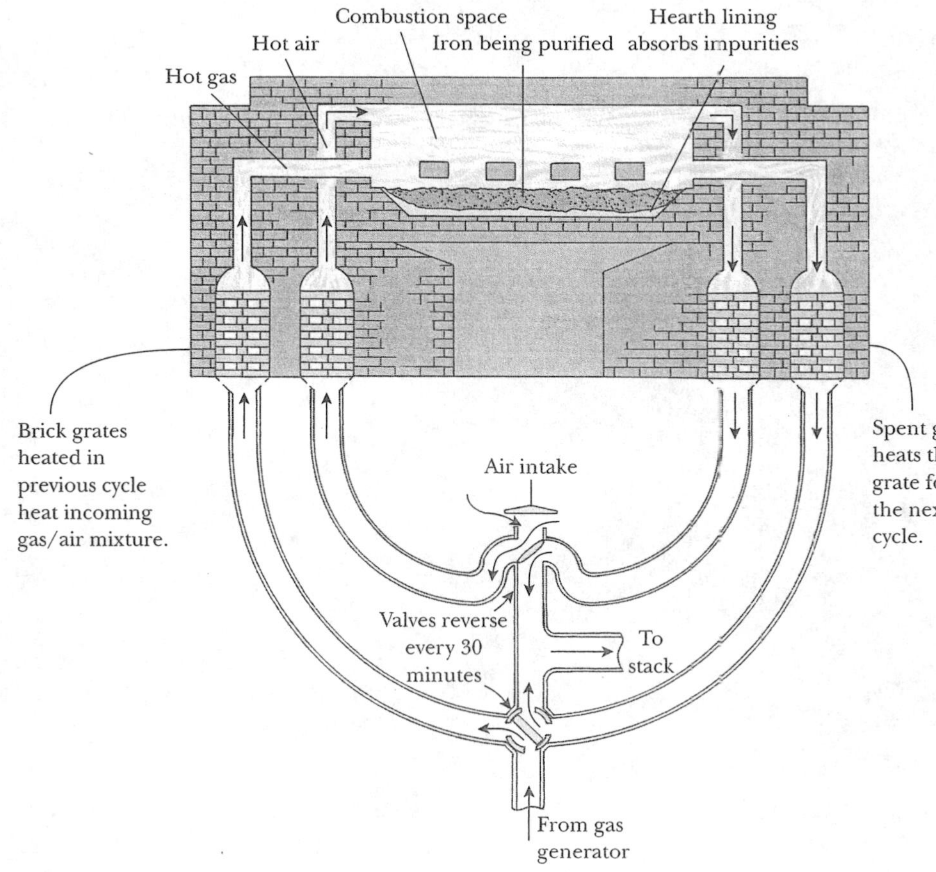

FIGURE 20.6 Open hearth steel. The open hearth process serves to lower the carbon content of pig iron and convert it to steel. In the process, the properties improve dramatically due to reduction of the carbon content. Hot fuel gases and air (sometimes enriched with oxygen) play directly over the surface of a molten pool of the metal. The basic charge is cast iron (pig iron), scrap steel, and hematite (Fe_2O_3).

PROFILES IN CHEMISTRY

Sir Henry Bessemer and His Process At the time of the Crimean War (which pitted England and France against Russia), the search was on for reasonably priced steel strong enough for high-power cannons. Here was the problem: Iron directly from the smelter was *cast iron,* rich in carbon and exceedingly hard, but very brittle. Removing the carbon to produce *wrought iron,* a soft but tough and not-brittle material that could be beaten into any shape, was a painstakingly expensive process. However, steel, with an intermediate carbon content between cast iron and wrought iron was just right—hard and tough. The trouble was that one had to go through the trouble of making wrought iron from cast iron, and then adding back the right amount of carbon.

Enter Henry Bessemer (1813–1898), a British mining engineer and metallurgist. He first considered the method being used for converting cast iron to wrought iron. Just the right amount of iron ore was added to cast iron to provide the oxygen to convert the carbon to carbon monoxide, which would burn off, leaving pure iron behind. Was there another way? Bessemer reasoned that a blast of air might provide the needed oxygen, but conventional wisdom suggested that might lower the temperature and shut down the entire process. Being a tinkerer at heart, he decided just to try it and see what happened. To his surprise, the blast of air not only burned off the carbon, but the heat of the reaction made the process self-sustaining with no further heating needed. By stopping the process at the right time, Bessemer found he had made steel without having to make wrought iron first or spend any money on fuel after starting up the process.

Bessemer opened his own steelworks at Sheffield, England, and began to sell steel on the world market at a tenth of the prevailing price, which very quickly made him a rich man. Almost as quickly, the world's steelmaking capacity converted to Bessemer process steel, opening the era of cheap steel and making possible the age of the giant ocean liner, the steel-skeletoned skyscraper, and the long suspension bridge, all of which are benchmarks of twentieth century civilization. Bessemer did not invent steel, but it was through his inventions and discoveries that it became widely available. In 1879, a grateful British nation recognized his achievements, and he was knighted. In that same year, he was elected to fellowship in the Royal Society, joining Boyle, Davy, Dalton, and Faraday.

Currently, steel is produced by a refinement of the Bessemer technique, the basic oxygen process, in which a jet of 99.5% oxygen at 10 atm pressure is forced beneath the surface of the molten charge. This process is computer controlled and has the advantages over the Bessemer process of increased speed and a purer, more uniform product.

Question

Write the chemical reactions involved in the Bessemer process for making steel. Then make a thermodynamic analysis of the process.

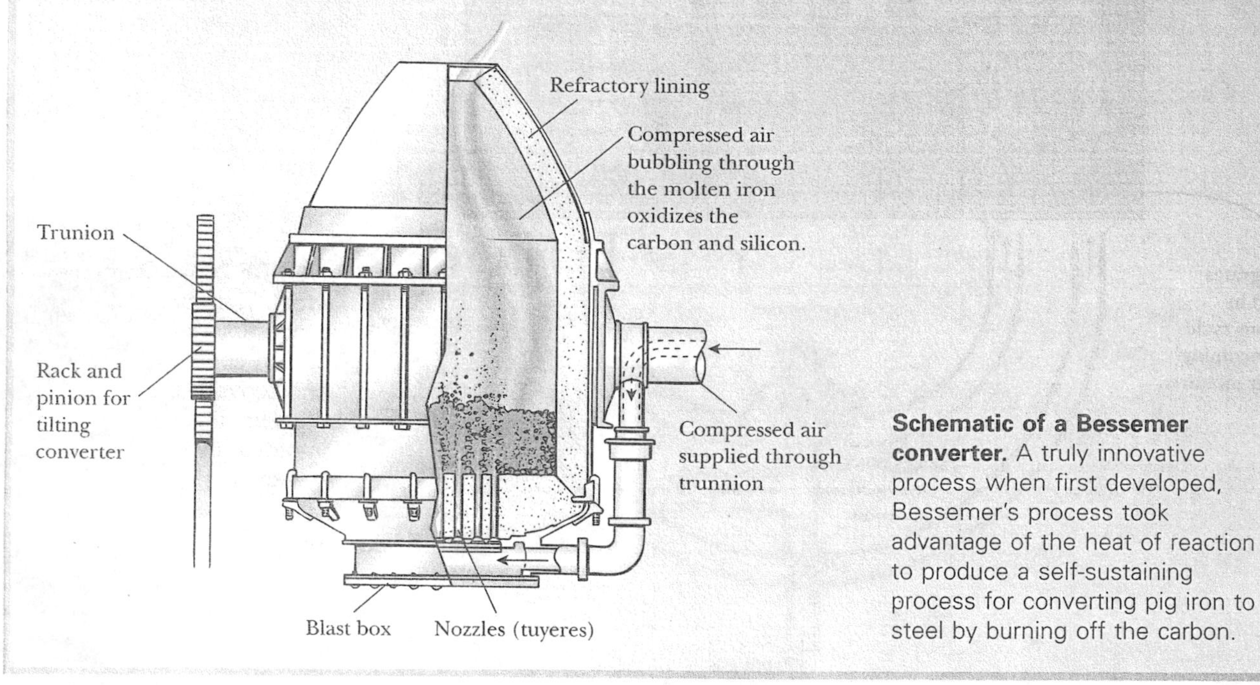

Schematic of a Bessemer converter. A truly innovative process when first developed, Bessemer's process took advantage of the heat of reaction to produce a self-sustaining process for converting pig iron to steel by burning off the carbon.

Refining of Copper

Copper is an attractive, reddish, durable metal, second only to silver in conductivity of heat and electricity. Bronze, brass, and other copper alloys have long been of enormous utility. Copper occurs in the two important classes of ores:

- Sulfide ores such as $CuFeS_2$, Cu_3FeS_3 and Cu_2S
- Oxidized ores such as CuO, $Cu_2(OH)_2CO_3$, and $Cu_3(OH)_2(CO_3)_2$

Copper ores now available are mainly low grade and contain a large proportion of sand and rock gangue. The ore is separated from the gangue by flotation in the following manner: First, the impure ore is ground with oil and water. Then, a collector is added, which attaches to the more polar ore particles, as described in Section 20.1. The ore particles are wetted by the oil, whereas the gangue is wetted by the water. This mass is then added to a larger amount of water containing a foaming agent, and the mixture is blown with air and beaten until a froth forms and rises to the top (Fig. 20.2). The ore particles stick to the surface of the bubbles of the froth and are carried off at the surface. Weighed down with the water, the gangue particles settle to the bottom. This method will remove 95% of the pure copper ore from an impure ore that is 98% waste.

The purified ore is then "roasted" at about 1600°C in the presence of air. This performs a number of useful functions including oxidizing arsenic and antimony impurities to volatile oxides that distill off and converting iron in sulfide ores to FeO:

$$2CuFeS_2(s) + 3O_2(g) \longrightarrow 2CuS(s) + 2FeO(s) + 2SO_2(g)$$

Sand (SiO_2) and limestone ($CaCO_3$) are then added to the mixture, which is heated in a reverberatory furnace (Fig. 20.7). The limestone and sand form calcium silicate, which is liquid at the furnace temperature and acts as a flux for removal of the FeO,

$$FeO(s) + SiO_2(s) \xrightarrow{CaSiO_3(\ell)} FeSiO_3(\ell)$$

The iron silicate produced in this reaction is a form of slag. Sulfur still present in the mixture reduces the CuS to Cu_2S, which is a liquid at this temperature and is called copper matte. The slag is less dense than the matte and floats on top of it.

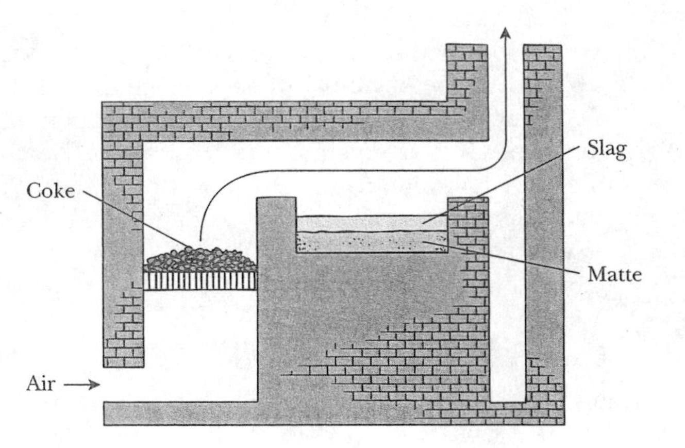

FIGURE 20.7 Reverberatory furnace for producing copper matte (Cu_2S). The principal separation of iron from copper as they are found in the ore ($CuFeS_2$) is accomplished by a roasting process. Then, in this kind of furnace a flame plays over the top of the ore, the iron oxide separates as a silicate slag, and the copper is left as the sulfide.

PROCESSES IN CHEMISTRY

***Kodo Zuroku*—"Copper Is the Handiwork of Heaven"** Half a century before Commodore Matthew Perry led a squadron of four steam-powered U.S. naval vessels into Edo Harbor on July 8, 1853, to negotiate a treaty opening Japanese ports to commerce and ending centuries of isolation, the Sumitomo Mining Company (Osaka) commissioned an artist and a poet to chronicle the classical activity of copper mining and extraction as it had been done in Japan for 4000 years. The *Kodo Zuroku*, as the book is called, is in two parts, the first containing 22 hand-colored woodblock prints showing details of stages in the production of copper, from mining through smelting and desilvering to casting forms for the market, together with drawings of various tools employed. The second part gives a detailed account of the operation and the Sumitomo family involvement in the copper industry. This name is carried today by the multinational Sumitomo Electric Company.

Questions

Judging from the text, how similar to modern methods does the *Kodo Zuroku* description of producing copper seem to be? Write reactions when possible.

Why not use inexpensive carbon reduction to free copper from its compounds?

Why is it necessary to purify the 98% copper that normal processing produces?

The matte is drawn off and placed in a copper converter, where air is blown through the molten Cu_2S, and the free metal is produced by the reaction

$$Cu_2S(\ell) + O_2(g) \longrightarrow 2Cu(\ell) + SO_2(g)$$

In this reaction copper and oxygen are both reduced as the S^{2-} is oxidized to SO_2. The product at this point is called blister copper and is about 99% pure.

Most of the copper produced is used for electrical transmission. The high electrical conductivity of copper is not realized unless it is very pure. Copper is refined electrolytically using a blister copper bar as the anode and a thin sheet of pure copper as the cathode. Both electrodes dip into the same solution containing Cu^{2+}, so the cell potential is zero. A potential is applied so that the bar of impure copper is positive, causing the following reactions:

$$Cu(s) \longrightarrow Cu^{2+}(aq) + 2e^- \qquad \text{impure Cu (anode)}$$

$$Cu^{2+}(aq) + 2e^- \longrightarrow Cu(s) \qquad \text{pure Cu (cathode)}$$

Careful control of the potential imposed on the cell results in the deposition of pure copper at the cathode. The impurities in copper include metals that are more difficult to oxidize than Cu, such as silver, gold, and platinum, and some that are easier to oxidize, such as iron, zinc, lead, and nickel. Keeping the imposed potential sufficiently low will ensure that the Ag, Au, and Pt will not be oxidized because their electrode potentials for oxidation are more negative than that for Cu. For example,

$$Cu(s) \longrightarrow Cu^{2+}(aq) + 2e^- \qquad E° = -0.337 \text{ V}$$

$$Ag(s) \longrightarrow Ag^{+}(aq) + e^- \qquad E° = -0.799 \text{ V}$$

Thus, as the anode disintegrates these valuable metals simply fall to the bottom of the cell for later reclaiming. The Ag, Au, and Pt recovered from this "anode mud" pay for the electrolytic purification process. The other impurities, Fe, Zn, Pb, and Ni, are oxidized along with the Cu because they are more easily oxidized. However, a sufficiently low imposed potential will ensure that they are not reduced back to the metal at the cathode because their electrode potentials for reduction are lower than copper. For example,

$$Zn^{2+}(aq) + 2e^- \longrightarrow Zn(s) \qquad E° = -0.763 \text{ V}$$

$$Ni^{2+}(aq) + 2e^- \longrightarrow Ni(s) \qquad E° = -0.250 \text{ V}$$

$$Cu^{2+}(aq) + 2e^- \longrightarrow Cu(s) \qquad E° = +0.337 \text{ V}$$

These metals remain in the electrolyte solution, which therefore has to be changed from time to time. Proper control of the cell can result in 99.95% pure copper at the cathode. The impurity at this point is mostly oxygen, and removal of this oxygen results in copper that is 99.999% pure, the purest substance produced from a large-scale chemical process.

Refining of Titanium

Titanium is abundant, tough, low in density, and corrosion resistant. It can maintain its strength at high temperatures. Unfortunately, its preparation is expensive because no chemical reduction of the oxide (TiO_2) gives satisfactory results. The metal finds use only in specialized aircraft engine and airframe applications, where relatively high cost can be tolerated. Ilmenite ($FeTiO_3$) the chief ore of titanium, can be separated magnetically from silica (SiO_2) with which it is combined. Then, the ore is heated with coke in the presence of chlorine,

$$FeTiO_3(s) + 3Cl_2(g) + 3C(s) \longrightarrow 3CO(g) + FeCl_2(s) + TiCl_4(g)$$

Because $TiCl_4$ is a volatile compound, it can be purified by fractional distillation. Pure $TiCl_4$ can then be reduced by reaction with an active metal. Magnesium is most commonly used for this purpose. The $TiCl_4$ vapor is passed over molten magnesium and the metallic titanium forms as a spongy mass:

$$TiCl_4(g) + 2Mg(\ell) \longrightarrow 2MgCl_2(\ell) + Ti(s)$$

Alternatively, molten sodium can be used as the reducing agent:

$$TiCl_4(g) + 4Na(\ell) \longrightarrow 4NaCl(s) + Ti(s)$$

In either case, the titanium is washed with water to remove the chloride salt. The metal can then be pressed into the shape of an electrode and melted and then cooled in this shape under high vacuum. This process prepares it for electrolytic purification.

The electrolyte in the electrolytic purification of titanium is molten sodium chloride containing $TiCl_4$. The solution is a liquid at the operating temperature of 850°C. The impure titanium is oxidized to Ti^{2+} at the anode and then redeposited as purified metal at the cathode. Cell potentials of 0.3 to 2.5 V are applied. The purity of the metal depends on a number of factors, including applied potential, cell design, rate of deposition as controlled by the current, and electrolyte composition.

The high cost of titanium results from the need to use magnesium or sodium in the preparation. This can be avoided by producing TiI_4 instead of $TiCl_4$ in the first step, by reaction with I_2 instead of Cl_2. The TiI_4 can be decomposed directly to titanium, making the use of magnesium or sodium unnecessary. In fact, very pure titanium crystals can be produced by decomposition of TiI_4 on an electric filament by what is known as the van Arkel process. Unfortunately, the relatively high cost of iodine (compared to chlorine) more than makes up for the savings on the magnesium or sodium.

Refining of Tungsten

Tungsten is a very dense metal (19.3 g/cm^3) with extremely high melting and boiling points (3370°C and 5900°C, respectively). Its very low vapor pressure at high temperatures makes it useful in electric light filaments, X-ray-tube targets, electrical contacts and arcing points, and furnaces. Its main use, however, is in making steel alloys where small amounts result in enormous increases in hardness and strength. Tungsten and chromium are used to make steels for high-speed cutting tools that remain hard at red heat.

The main ores of tungsten are tungstates: $FeWO_4$, $MnWO_4$, $CaWO_4$, and $PbWO_4$. After mechanical and magnetic separation, the ore is reacted with aqueous NaOH to produce water soluble Na_2WO_4, which ionizes in solution to the WO_4^{2-} ion; for example,

$$CaWO_4(s) + 2OH^-(aq) \longrightarrow WO_4^{2-}(aq) + Ca(OH)_2(s)$$

The aqueous tungstate ion is then reacted with acid to produce tungsten(VI) oxide (WO_3):

$$WO_4^{2-}(aq) + 2H_3O^+(aq) \longrightarrow WO_3(s) + 3H_2O(\ell)$$

The WO_3 is dried and then reduced to the metal with hydrogen gas:

$$WO_3(s) + 3H_2(g) \longrightarrow W(s) + 3H_2O(g)$$

Hydrogen is used for the reduction in order to obtain a high-purity product. In contrast, the use of carbon as the reducing agent would lead to the formation of tungsten carbide (WC). The temperature is held at 800°C during the reduction. The hydrogen is used as a flowing gas so that the water formed will be swept away without reacting with the metal.

Before it is exposed to air, the metal is cooled to room temperature under an atmosphere of hydrogen to prevent oxidation. The metal, a powder at this point, is compacted into porous bars at room temperature and very high pressure. It is then slowly heated to 1300°C under a hydrogen atmosphere, which again prevents oxidation. In this process, called sintering, the powder grains are stuck together by movements of the atoms at their surfaces. An electric current passed through the bar heats it to 3000°C. The sintering continues, and spaces between the original powder grains decrease until the full density of the metal is approached. The metal can then be worked mechanically at a temperature of about 900°C into the desired shape.

20.4 RECYCLING PROCESSES

Recycling is an obvious alternative to producing metals from their ores, considering (1) the current emphasis on recycling our resources and (2) the huge amount of metal that is scrapped in everything from automobiles to drink cans.

As we have discussed above, reducing metal ores to the pure metals requires a great deal of energy. Recycling a scrapped metal avoids using as much energy as reducing the metal ore because the scrapped metal is already in the reduced state. Unfortunately, recycling of scrap metal is greatly hampered by the fact that almost all metals are used in the form of alloys.

Alloys are mixtures or solid solutions of metals and were discussed in some detail in Chapter 17. For most uses, alloys are preferable to pure metals because they provide a greatly enlarged spectrum of possible metallic materials. We mentioned in Chapter 17 how alloying can increase the hardness and strength of a metal. In addition, alloying can be used to modify a large number of important properties. There are hardly any cases where unalloyed metals are useful materials, except copper for electrical conduction. Even trace impurities increase the resistance. Thus, the refining of copper is aimed at producing a high-purity product, and recycling of copper wires and other conductors is mainly a matter of melting and reshaping the metal.

Steels are among our most important alloys. These are alloys of iron with some carbon and usually one or more other metals. In fact, pure iron is a relatively soft and easily corroded metal, which has no uses as a material and which you may never have even seen. However, its alloys, the steels, have an enormous range of useful properties. Examples of steel alloys include the following.

- **Corrosion-resistant steels**. Steels that resist the familiar rusting of iron and attack by acids are generally known as **stainless steels**. Chromium is the most common metal alloyed with iron to produce stainless steel, and nickel is often added as well. Alloys called Hastelloys are used for surgical instruments and contain large amounts of nickel as well as other metals such as tungsten, molybdenum, and cobalt. The addition of about 15% silicon to iron gives an alloy called Duriron, which resists acids and bases and is used for drains in laboratories.
- **Spring steels**. Springs, including those used in automobiles and other vehicles, contain molybdenum, which gives them their characteristic properties.
- **Heat-resistant steels**. These alloys can be heated to red heat (600–1100°C) without oxidizing. An example is Incoloy, which is used in the heating elements of electric stoves and contains 46% Fe, 32% Ni, 20% Cr, and 2% Mn.
- **Magnetic steels**. The alloy chosen depends on whether the magnet is intended to be permanent, as in a stereo speaker, or able to be magnetized and demagnetized very quickly, as in an electric motor. Alnico alloys are used for permanent magnets. As the name suggests, they contain aluminum, nickel, and cobalt, and a typical composition is 51% Fe, 24% Co, 14% Ni, 3% Cu, and 8% Al. Silicon is added to steels that are to be used in the magnet cores of electric motors.
- **Electrical resistance alloys**. These appear in a variety of applications, such as stoves, hair dryers, and all other electrical heaters. Some of the alloys used for this purpose do not contain iron—Nichrome, for example, is 80% Ni and 20% Cr. A useful iron-containing alloy for electric heaters contains 72% Fe, 22% Cr, 5% Al, and 1% Co.
- **Deep-drawing steel**. This kind of alloy is used for such purposes as stamping out auto bodies and requires the opposite properties of spring steel. Lead is generally included in these alloys.

An automobile contains ten or more different steels plus other metals, such as copper in the wiring and often aluminum for the engine block. Thus, simply crushing, shredding, and melting junk cars results in a complex mixture of metals from which it will be very difficult to reproduce the precise alloys necessary to build other cars. Less-demanding uses of steel, such as reinforcing rods for concrete construction, provide a commercial outlet for recycled cars as well as steel cans. More demanding uses require sorting of the various parts, and separation techniques that are different from those applied in metallurgy of ores must be applied. Gravity devices using air or water separate metallic and nonmetallic scrap from automobiles. Crushing and screening techniques are used to separate different metals. Cryogenic (low-temperature) techniques can be used to facilitate this kind of separation because some metals become more brittle than others at lower temperatures. For example, zinc becomes brittle at $-50°C$ and will break into small pieces, whereas copper and aluminum remain malleable at this temperature.

A typical automobile contains about 10 kg of copper, and removal of this metal from a scrapped vehicle is essential. Copper is not a useful component in steel alloys and contributes to difficulties in hot-rolling as well as susceptibility to corrosion. Addition of Na_2SO_4 to molten, high-carbon steel at around 1500°C results in the removal of copper as Cu_2S and some of the carbon as CO. The reaction is

$$Na_2SO_4(\ell) + 2Cu(\ell) + 3C(s) \longrightarrow Cu_2S(\ell) + Na_2O(g) + 3CO(g)$$

Because of the great reactivity of aluminum, the energy cost of its production is extremely high. Recycling the aluminum from drink cans is a simpler but not trivial process. The cans are made of aluminum alloyed with some magnesium. The outside paint and inside protective resin coating must be burned off, and the cans must be thoroughly dried, because any water present would react violently with molten aluminum. The cans can then be melted. Because the can tops contain more magnesium than the can bodies, pure aluminum must be added to each batch to make an alloy suitable for the main part of the cans. Then pure magnesium can be added to that part of the melt that will be used for producing more can tops.

SUMMARY

This chapter concentrates on the preparation of metals from their naturally occurring ores. The chemical properties of a metal determine the nature of its ores, which determines the necessary refining processes. The periodic table provides useful generalizations about the nature of ores and the methods of refining that are necessary. Metals close to each other on the periodic table will have similar metallurgy.

Metal production from an ore generally requires mechanical separation and then reduction of the ore to the metal. Standard chemical reduction processes are used for most metals except those that are too reactive and for which active metal or electrolytic reductions are necessary. An understanding of chemical reductions is aided by a graph of free energies of formation of metal oxides as a function of temperature. From such a graph it is evident that a metal will reduce the oxide of a less-reactive metal irrespective of temperature. Carbon will reduce a larger number of metal oxides as the temperature is raised, but hydrogen gas will actually reduce fewer metal oxides as the temperature is increased.

Electrolytic reductions provide a means of preparing metals that are too reactive to be prepared chemically. Such metals must be electrolyzed from nonaqueous solutions—generally molten salts—because the metal produced would otherwise react with water.

The details of the preparations of five important metals were discussed. Gold occurs as the pure metal, and its refining involves only purification. Iron is the most important example of the carbon reduction of oxide ores. Copper occurs in sulfur-containing ores that are not reduced effectively by direct reaction with carbon. Titanium is reduced by more active metals, and tungsten is reduced with hydrogen gas.

The general problems with recycling metals were discussed. These processes are complicated, particularly by the fact that metals are almost always used as alloys.

TERMS

Metallurgy (20.1)
Ore (20.1)
Lithophiles (20.1)
Siderophiles (20.1)
Atmophiles (20.1)
Chalcophiles (20.1)
Selective settling (20.1)
Gangue (20.1)
Flotation (20.1)
Overvoltage (20.2)
Amalgam (20.3)
Stainless steel (20.4)

IMPORTANT EQUATIONS

Equation	Description
$\Delta G° = \Delta G_f°(M''O) - \Delta G_f°(M'O)$	Standard free energy of metal plus metal oxide reaction
$\Delta G° = \Delta H° - T\Delta S°$	Free energy change at standard conditions
$\Delta G = -nFE$	Relationship between free energy change and cell potential
$m = \frac{iMt}{Fn}$	Mass of metal produced by electrolysis

QUESTIONS

Conceptual questions are denoted by a square screen. Extra-credit questions are denoted by a circular screen.

1. Why do you suppose the "bronze age" occurred before the "iron age"? What age do you think we are in now?
2. What kind of ores and separation procedures would you expect for the lanthanides? Why?
3. What compounds function as collectors in the flotation process? Give reasons for your suggestions.

4. Why is the degree of spontaneity of the reaction of one metal with the oxide of another largely independent of temperature?
5. Why is carbon such an effective reducing agent for metal ores? What are its limitations as a reducing agent for the refining of metals?
6. What are the problems involved in using standard-state thermodynamic tables for calculating the potentials needed for producing metals by electrolysis?
7. Show that the refining processes for gold, iron, copper, titanium, and tungsten are consistent with the positions of these metals on the periodic table.
8. The most common ore of lead is galena (PbS) What chemical reactions are necessary for conversion of this ore to Pb metal?
9. Propose a step-by-step process for converting junk cars into scrap that could be recycled. List the kinds of problems you would encounter and estimate if the process would be economical.
10. Why do you think that aluminum is cheaper to produce in Norway than in the Netherlands?
11. What are the likely fates of the CaS and CO in the process for removing FeS in the production of steel?

PROBLEMS

Problems marked with a bullet (•) are answered in Appendix A, in the back of the text.

Chemical Reductions [1–18]

•1. Calculate $\Delta G°$ for the reaction of $Hg(\ell)$ with $O_2(g)$ to form $HgO(s)$ at 298K. $\Delta H_f°$ is -90.83 kJ/mol for HgO(s), and $S°(HgO)$ is 70.29 J/mol·K at 298K.

2. What is $\Delta G°$ for the decomposition of 1 mol of ZnO(s) to the elements?

•3. Calculate the minimum temperature necessary for the decomposition of CuO to metallic Cu. Assume the values of $\Delta S°$ and $\Delta H°$ for the reaction are independent of temperature.

4. At what temperature is the value of $\Delta G°$ equal to zero for the decomposition of Fe_2O_3 to its elements, assuming that $\Delta S°$ and $\Delta H°$ for the reaction are constant over the range of temperature involved?

•5. Calculate the equilibrium constant for the reduction of Fe_2O_3 by Al at 1000.°C. Use data from Tables 12.4 and 13.1 together with the following standard entropies in J/mol·K: $Al_2O_3(s)$, 50.92; Al(s), 28.33. Make the assumption that the values of $\Delta S°$ and $\Delta H°$ hold constant over the range of temperatures involved.

6. Calculate the value of $\Delta G°$ for the reduction of CaO with Ba at 400.°C The standard enthalpies of formation for CaO(s) and BaO(s) are -635.09 kJ/mol and -558.1 kJ/mol, respectively. Make the same assumptions as were made in the previous problem.

7. What is the minimum temperature at which PbO can be reduced to the metal, using carbon as the reducing agent? Assume constant $\Delta S°$ and $\Delta H°$ values. Standard entropies in J/mol K for PbO(s) and Pb(s) are, respectively, 68 and 65.

8. Calculate the minimum temperature necessary for reducing CaO with carbon, making the standard assumptions used in the preceding problem.

9. Calculate the value of K for the reduction of PbO with carbon (Problem 7) at 1000.K.

10. Determine the equilibrium constant for the reduction of calcium oxide (Problem 8) with carbon at 1000.K. Compare this value with the result of Problem 9 and comment on the difference.

•11. Calculate the value of $\Delta G°$ for the reduction of CuO with hydrogen at 500.°C and

1000.°C, making the assumptions made in the problems above.

12. Calculate the value of the equilibrium constant for the reduction of ZnO with hydrogen at 1250°C. Use the value of −348.3 kJ/mol for the ΔH_f° of ZnO, and make the usual assumptions.

13. Plot the equilibrium constant for reducing CaO at five different temperatures with hydrogen and with carbon. Plot these values and comment on the differences between the two graphs.

14. Compare the reduction of PbO with hydrogen and with carbon by plotting the equilibrium constants for each of the reactions at several temperatures. Standard entropies for PbO(s) and Pb(s) are 68 J/mol·K and 65 J/mol·K, respectively.

•15. Write balanced chemical equations using two different reducing agents that will produce the following results in each case:
(a) $Fe_2O_3 \longrightarrow Fe$ (b) $CuO \longrightarrow Cu$
(c) $Ag_2S \longrightarrow Ag$

16. Write balanced chemical equations using two different reducing agents that will produce the following results:
(a) $TiO_2 \longrightarrow Ti$ (b) $WO_3 \longrightarrow W$
(c) $As_2S_3 \longrightarrow As$

•17. Antimony (Sb) is freed from the sulfide ore known as stibnite (Sb_2S_3) by pulverizing it and heating a mixture with scrap iron turnings. The molten antimony is drawn off the bottom of the reactor in this *batch* process. Write a balanced chemical equation for the reduction process.

18. Suppose the reactor is charged with 6 kg of stibnite and 2.5 kg of scrap iron turnings, and that 2 kg of antimony are recovered. Determine the limiting reagent and the percent conversion in this batch.

Electrolytic Reductions [19–20]

•19. How much sodium will be deposited in 1.00 hour by electrolysis of molten NaCl using a current of 25.0 A?

20. What mass of Cu metal will be deposited in the electrolytic refining of the metal using a current of 155 A for 15 hours?

Additional Problems [21–23]

21. Derive a general equation to calculate the minimum temperature at which a metal oxide MO can be reduced to the metal M, using carbon as the reducing agent. Assume that the carbon is oxidized to CO, and specify any other assumptions that you make.

22. Calculate the equilibrium constant for the following reaction at 650°C.

$$2ZnS(s) + 3O_2(g) \rightleftharpoons 2ZnO(s) + 2SO_2(g)$$

The value of ΔH_f° in kJ/mol for ZnS(s) is −201. Make the usual assumptions that the values of ΔH° and ΔS° are independent of temperature.

•23. Suggest how to complete and balance the following equations:
(a) $TiO_2(s) + Cl_2(g) + C(s) \longrightarrow$
(b) $Al_2O_3(s) + C(s) \longrightarrow$

Cumulative Problems [24–25]

24. How much limestone and carbon would be necessary to convert 1000. kg of an iron ore containing 50.% Fe_3O_4 and 50.% SiO_2 by mass to pure Fe?

25. What volume of $H_2(g)$ at 800°C and 1.00 atm is necessary to produce 1.00 kg of tungsten?

Applied Problems [26]

26. Tantalum metal has a standard entropy of 41 J/mol·K, and $Ta_2O_5(s)$ has values of 2092 kJ/mol and 143 J/mol·K for ΔH_f° and S°, respectively. Analyze the possible methods of chemically reducing the oxide to the metal and chose the one that you consider the most reasonable. Give clear reasons for your choice.

ESTIMATES AND APPROXIMATIONS [27]

27. Estimate the electric potential necessary to produce sodium metal using the electrolysis of molten NaCl.

FOR COOPERATIVE STUDY [28]

28. Manganese oxides can be reduced to manganese metal using aluminium as the reducing agent. The enthalpies of formation of the manganese oxides and Al_2O_3 are as follows:

	ΔH_f°(kJ)
MnO	−385
MnO_2	−520
Mn_2O_3	−947
Mn_3O_4	−1387
Al_2O_3	−1674

It is desirable for the enthalpy of the reaction to be sufficient to raise the temperature of the product mixture to about 2300K, where the products are all liquids and separation is facilitated. The average heat capacities of the products for the range 298K to 2300K are

	J/mol·K
Mn	50.
Al_2O_3	178.

The reaction is considered to be self-sustaining if its enthalpy is sufficient to maintain the products in their molten states. Analyze this reaction for the different manganese oxides and decide what kind of a mixture of oxides could just produce a self-sustaining reaction.

WRITING ABOUT CHEMISTRY [29–30]

29. Most of the elements in the periodic table are metals. Describe how much the properties of metals vary compared with the variation of the properties of nonmetals. Pick a metal to use as an example of a "typical metal" and look at a wide range of its properties. You should go to a standard reference source such as *CRC Handbook of Chemistry and Physics* to help you with this. Is your metal really typical? Explain why or why not. Is there such a thing as a typical metal or nonmetal? Give carefully considered reasons for your answer.

30. In about 500 words or less, summarize the fundamental principles of separating metals from their ores. This summary should be sufficient for a person to get a reasonable idea of the separation method of any metal in which he or she is interested.

21

chapter twenty-one

ORGANIC CHEMISTRY

21.1 BEGINNING ORGANIC CHEMISTRY

Organic chemistry, the study of most of the compounds of carbon, is arguably one of the most vigorous and exciting subjects in all of natural science and one of science's most remarkable achievements. The element carbon forms literally millions of different compounds, and these compounds fill an incredible range of different uses. In this chapter we shall try to explain just what organic chemistry is and what organic chemists do, what questions they ask of nature, and what kinds of answers they seek. We will explore the practical aspects of organic chemistry, because the social and economic implications of the work of the chemist are especially significant in organic chemistry. We also need to seek answers to the question, "Why is there a separate branch of chemistry devoted exclusively to the organic compounds of carbon?"

Until the early nineteenth century, chemists commonly distinguished between plant and animal matter, in contrast to mineral matter. **Organic substances** were those produced by living organisms—plant or animal—or obtainable from living organisms by chemical reactions. **Inorganic substances** occurred in minerals and could be isolated from them. Rock salt and marble were classically understood to be inorganic compounds, as they are today; vinegar, grape sugar, and alcohol were known as organic chemicals, then as now.

It was not thought possible 200 years ago to prepare organic compounds from inorganic substances, leading to the widely held view that there was a vital force unique to living matter. Scientists believed organic compounds could not be prepared in the laboratory from lifeless inorganic materials. However, in 1828 Friedrich Wöhler (German, 1800–1882) laid that erroneous notion to rest when he obtained urea from ammonium sulfate and potassium cyanate and succeeded in showing that his product was in all ways the same as that isolated from animal urine. The first step in the reaction chemistry is the formation of ammonium cyanate:

$$\underset{\text{ammonium sulfate}}{(NH_4)_2SO_4(aq)} + \underset{\text{potassium cyanate}}{2KOCN(aq)} \longrightarrow \underset{\text{ammonium cyanate}}{2NH_4OCN(aq)} + \underset{\text{potassium sulfate}}{K_2SO_4(aq)}$$

followed by its thermal rearrangement to urea:

$$\underset{\text{ammonium cyanate}}{NH_4OCN(aq)} \longrightarrow \underset{\text{urea}}{NH_2CONH_2(aq)}$$

Wöhler's synthesis is now generally accepted as the starting point for organic chemistry. As he wrote in a letter to his Swedish contemporary, Jöns Berzelius, who was perhaps the leading authority in chemistry in the world at that time, "I must tell you that I can prepare urea without requiring a kidney of an animal, either Man or dog." Although the distinction between organic and inorganic no longer retains its original significance—and in fact has become still more vague in recent years—we continue to distinguish compounds of carbon from all others.

Organic compounds of carbon that are both interesting and useful to humans have been obtained from nature throughout human history. Today, our main resources of organic starting materials are still (1) coal and petroleum, the fossilized remains of once-living matter; (2) agricultural products from plants such as potato and cereal crops, cane and beet sugars, corn, rice, and wheat, vegetable fats and related tree and plant materials; and (3) animal fats and proteins, as well as traditional by-products such as milk and hides. From these natural resources, organic chemists extract or construct thousands of products—from pharmaceuticals to plastics—either by direct-isolation techniques or by synthetic methods. As the twentieth century began, local anesthetics such as aspirin and novocaine had been synthesized—not just extracted or isolated from plants—along with colorful and important dyes, the first synthetic plastics, and chemotherapeutic agents. Today there are millions of known organic compounds.

Carbon has a special chemistry, and this is based principally on three related factors: (1) Carbon has four electrons and four valence orbitals, and so its compounds have neither too few electrons, requiring electron-deficient structures, nor too many electrons, resulting in excessive lone pair–lone pair repulsions; (2) carbon can form compounds containing stable carbon–carbon bonds; and (3) carbon can form stable bonds with a number of important elements, particularly hydrogen, oxygen, nitrogen, sulfur, and the halogens (Group VIIA). The net result of all three factors is the existence of an almost endless range of compounds with these elements bonded to, or included in, straight chains, branched chains, and rings of carbon atoms. These are the compounds of organic chemistry.

21.2 ALKANES

The Simplest Hydrocarbons

The simplest organic compounds are the **hydrocarbons**, which are molecules containing only carbon and hydrogen atoms. These important compounds are classified according to the types of covalent bonds—single, double, triple—that connect their carbon atoms. **Saturated** hydrocarbons are those with no carbon–carbon multiple bonds. **Unsaturated** hydrocarbons contain at least one multiple bond. They are also classified by whether the carbon atoms are arranged in chains or rings.

Alkanes are hydrocarbons containing only single bonds, that is, they contain only sigma bonds—no pi bonds. Alkanes that consist of chains with no rings follow the general formula C_nH_{2n+2}, where n is the number of carbon atoms in the compound. If an alkane contains a ring, it is called a **cycloalkane**, for which the general formula is C_nH_{2n}.

EXAMPLE 21.1

Write the formulas for the alkanes without rings containing 6, 12, and 15 carbon atoms.

COMMENT Use the general formula C_nH_{2n+2}.

SOLUTION C_6H_{14}, $C_{12}H_{26}$, and $C_{15}H_{32}$

EXERCISE 21.1

Write the formulas for cycloalkanes containing five, six, and eight carbon atoms.

ANSWER C_5H_{10}, C_6H_{12}, C_8H_{16}

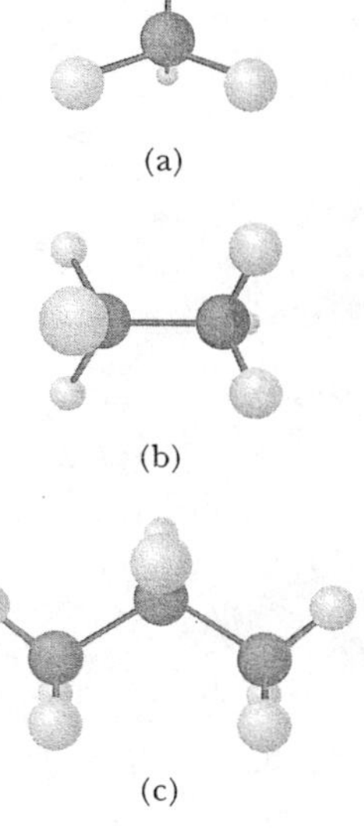

FIGURE 21.1 Methane, ethane, and propane structures. Shown as ball-and-stick models: (a) methane (CH_4) is the simplest, with one carbon atom (black sphere) at the center of a regular tetrahedron; (b) the carbon atoms in ethane (CH_3CH_3) and (c) propane ($CH_3CH_2CH_3$) maintain tetrahedral geometry. Note that the size that an atom is drawn indicates the closeness of that atom to the viewer.

The simplest alkane is **methane** (CH_4), the principal component of natural gas and the ignitable gas known as fire damp that coal miners the world over fear. **Ethane** (C_2H_6) is the second; then **propane** (C_3H_8) and **butane** (C_4H_{10}; Table 21.1). In these compounds, each carbon atom is sp^3 hybridized and is located at the center of a tetrahedron, with bond angles very close to 109.5°. In the vast majority of organic compounds, each carbon has four covalent bonds and no nonbonding electrons. The structures of the first three alkanes are shown in Figure 21.1. Ethane may be considered to be methane in which one of the hydrogens is replaced by CH_3. Propane may be considered to be related to ethane in a similar manner. As a consequence, formulas of adjacent alkanes differ by "CH_2," and the repeating tetrahedral structures result in a zig-zag molecular arrangement (Fig. 21.2).

There are several ways of writing formulas with which you should be familiar. Even though the structures are three-dimensional, we will typically represent

TABLE 21.1 Normal Alkanes

Common Name	Formula	Condensed Structure	m.p. (°C)	b.p. (°C)	Petroleum Fraction
Gases at Room Temperature					
methane	CH_4	CH_4	−183	−162	natural gas
ethane	C_2H_6	CH_3CH_3	−172	−89	
propane	C_3H_8	$CH_3CH_2CH_3$	−187	−42	
n-butane*	C_4H_{10}	$CH_3CH_2CH_2CH_3$	−135	0	
Liquids at Room Temperature					
n-pentane	C_5H_{12}	$CH_3(CH_2)_3CH_3$	−130	36	gasoline components
n-hexane	C_6H_{14}	$CH_3(CH_2)_4CH_3$	−94	69	
n-heptane	C_7H_{16}	$CH_3(CH_2)_5CH_3$	−91	98	
n-octane	C_8H_{18}	$CH_3(CH_2)_6CH_3$	−54	126	
n-nonane	C_9H_{20}	$CH_3(CH_2)_7CH_3$	−30	151	
n-decane	$C_{10}H_{22}$	$CH_3(CH_2)_8CH_3$	−26	174	
Solids at Room Temperature					
n-octadecane	$C_{18}H_{38}$	$CH_3(CH_2)_{16}CH_3$	28	308	lube oils

*The prefix *n*- stands for "normal" and means that the carbon atoms are arranged in a straight chain with no branches and no rings.

them by flat structural diagrams or by condensed formulas, as shown for propane:

```
    H   H   H
    |   |   |
H — C — C — C — H
    |   |   |
    H   H   H
```

structural formula

$CH_3CH_2CH_3$

$CH_3{-}CH_2{-}CH_3$

condensed formulas

As carbon atoms are added to lengthen the chain, the molar mass increases, and melting and boiling points increase, resulting in a steady progression from gases to increasingly viscous liquids, then oils, and finally the familiar solid hydrocarbons called waxes. At that point the chain lengths are about 25 carbon atoms or more.

Starting with butane and beyond, there are families of alkane **isomers**. Isomers are compounds of the same atoms arranged differently. That is, isomers have the same formula but a different structure or geometry. Structural isomers in alkanes are the result of chain-branching, which first becomes possible when there are four carbon atoms in a chain. Each member of a family of isomers has the same composition but a different structure and therefore different properties. Two different C_4H_{10} structures, isomeric butanes, exist and they differ in their physical and chemical properties (Table 21.2). The branched isomer is commonly called isobutane.

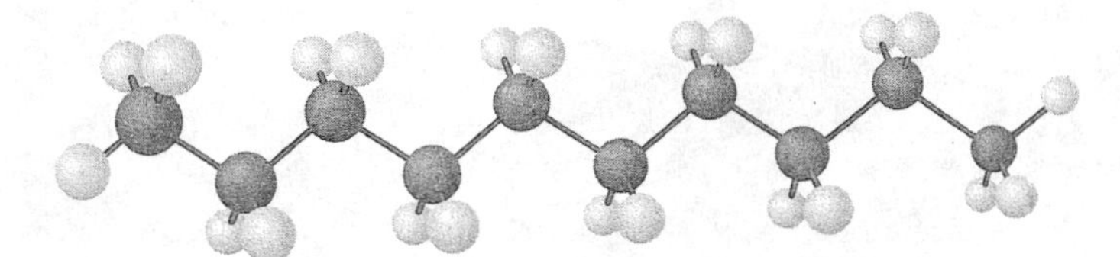

FIGURE 21.2 ***n*-decane is a linear hydrocarbon.** Except for the two terminal carbon atoms, each of which has three hydrogen atoms (CH_3—), *n*-alkanes beyond propane differ from each other by the number of —CH_2— groups along the chain. *n*-decane ($C_{10}H_{22}$) has eight —CH_2— (methylene) groups in between terminal CH_3 (methyl) groups. This is a ball-and-stick representation of *n*-decane. Eight connected methylene (CH_2) groups constitute the central architecture, which can be represented spatially: The main chain of carbon atoms is depicted in the plane of the page, the smaller spheres represent H atoms below the plane of the page and the larger spheres extend above it. Note as well that "straight chain" hydrocarbons are really zig-zag chains of connected carbon atoms.

$$\underset{n\text{-butane}}{CH_3{-}CH_2{-}CH_2{-}CH_3} \qquad \underset{\text{isobutane}}{CH_3{-}\overset{\overset{\displaystyle CH_3}{|}}{C}H{-}CH_3}$$

The *n*- prefixed to most of the alkanes found in Table 21.1 is an abbreviation for "normal." Normal means that the carbon atoms are all linked in a head-to-tail chain arrangement, with no branches or side groups. The effect of branching in the carbon chains on physical properties can be seen in Table 21.2.

Pentane (C_5H_{12}) is the next member of the alkane family and the first liquid hydrocarbon at room temperature. There are three structural isomers of the C_5H_{12} hydrocarbon: *n*-pentane has the linear chain, whereas isopentane and neopentane are branched structures:

$$\underset{n\text{-pentane}}{CH_3{-}CH_2{-}CH_2{-}CH_2{-}CH_3}$$

$$\underset{\text{isopentane}}{CH_3{-}CH_2{-}\overset{\overset{\displaystyle CH_3}{|}}{C}H{-}CH_3} \qquad \underset{\text{neopentane}}{CH_3{-}\underset{\underset{\displaystyle CH_3}{|}}{\overset{\overset{\displaystyle CH_3}{|}}{C}}{-}CH_3}$$

TABLE 21.2 Melting and Boiling Points of the Isomeric Butanes and Pentanes

	Butanes		Pentanes		
	n-	*iso-*	*n*-	*iso-*	*neo-*
m.p. (°C)	−138	−145	−130	−160	−20
b.p. (°C)	0	−10	36	28	10

APPLICATIONS OF CHEMISTRY

Hydrocarbons, Octane Rating, and Engine Knock Natural gas and petroleum are the chief sources of hydrocarbons and will continue to be so well into the twenty-first century. Natural gas is mostly methane, and it is mainly used as a fuel, although, lately it has become increasingly interesting to scientists and engineers as a resource for the synthesis of large numbers of other organic compounds. *Petroleum* is a brownish-black, generally viscous mixture of organic compounds in varying proportions, along with small percentages of sulfur, nitrogen, and oxygen.

Methane and ethane together make up about 95% of what we call natural gas. Propane is bottled and sold for a variety of heating needs, ranging from outdoor cooking and home heating to industrial uses. Butane is the gas used in cigarette lighters. The liquid hydrocarbons from the C_5 pentanes to the C_{12} dodecanes are gasoline-range hydrocarbons. Higher boiling fractions are kerosenes, turbine fuels, and jet fuels (boiling range 175 to 275°C); gas oil, fuel oil, and diesel oil (boiling range 250 to 400°C); and lubricating oils, greases, and the paraffins or waxes. In the United States, the most important feedstocks available to the chemical industry are the gasoline-range hydrocarbons:

Hydrocarbon gases	10%
Gasoline-range	50%
Kerosenes/diesel fuel	25%
Heavy end oils, bitumens	15%

The first distillation products from petroleum, in the boiling range from 0 to 200°C, are referred to as "straight run" gasoline. It has been estimated that this fraction contains more than 500 identifiable compounds, and it is impossible to separate them completely. Large amounts of gasoline-range hydrocarbons are prepared by cracking higher boiling, longer chain fractions down to size. At first, cracking of hydrocarbons was done strictly by heating. But as new "cracker reactors" were built, thermal cracking was replaced by catalytic cracking in which petroleum vapors are passed over heated beds of aluminum silicate. Not only are the yields of gasoline increased, but the octane rating is also improved because the process tends to produce branches along the main hydrocarbon chain.

Engine knock (discussed in Chapter 15) is the audible "ping" caused by premature ignition of the fuel in an internal combustion engine. Knocking varies with the fuel and is a function of the hydrocarbon structures present. As a general rule, the more side chains or branches on the parent hydrocarbon, the less pinging. To measure the knocking characteristics of gasolines, an octane rating system was established in 1938. It is based on a zero-octane rating for *n*-heptane, which has the greatest tendency to cause knocking because it has no branches, and a 100-octane rating for highly branched 2,2,4-trimethylpentane, better known as isooctane, which was one of the high octane fuels used for aircraft fuel in World War II. At the time, there was no better motor fuel for a high-compression, internal-combustion engine. By blending these two hydrocarbons, mixtures can be made that match the knocking characteristics of ordinary gasolines in a standard test engine. For example, a fuel with a typical 87-octane rating has knocking characteristics equivalent to a mixture of 87% isooctane and 13% *n*-heptane. The enthalpies of combustion of *n*-heptane and isooctane are 4817 kJ/mol (48.1 kJ/g) and 5461 kJ/mol (47.8 kJ/g). The higher octane fuel does not provide more energy per unit mass and provides no benefit if it is not necessary to prevent knocking.

For many years, the octane rating of gasolines was improved by addition of tetraethyllead, $Pb(C_2H_5)_4$. Ethylene dibromide, $BrCH_2CH_2Br$, was added to react with and remove the lead by-products, mostly lead oxides and metallic lead, and prevent scarring of the cylinder walls in the engine. As you might expect, because lead is a severe environmental contaminant, especially in densely populated, urban areas, tetraethyllead has been completely phased out of the current motor-fuel mix.

Questions

Which of the following two isomeric C_7H_{16} hydrocarbons would you expect to have the higher "octane rating" and why: 2,4-dimethylpentane or 3-ethylpentane?

How would you prepare a "90-octane" motor fuel?

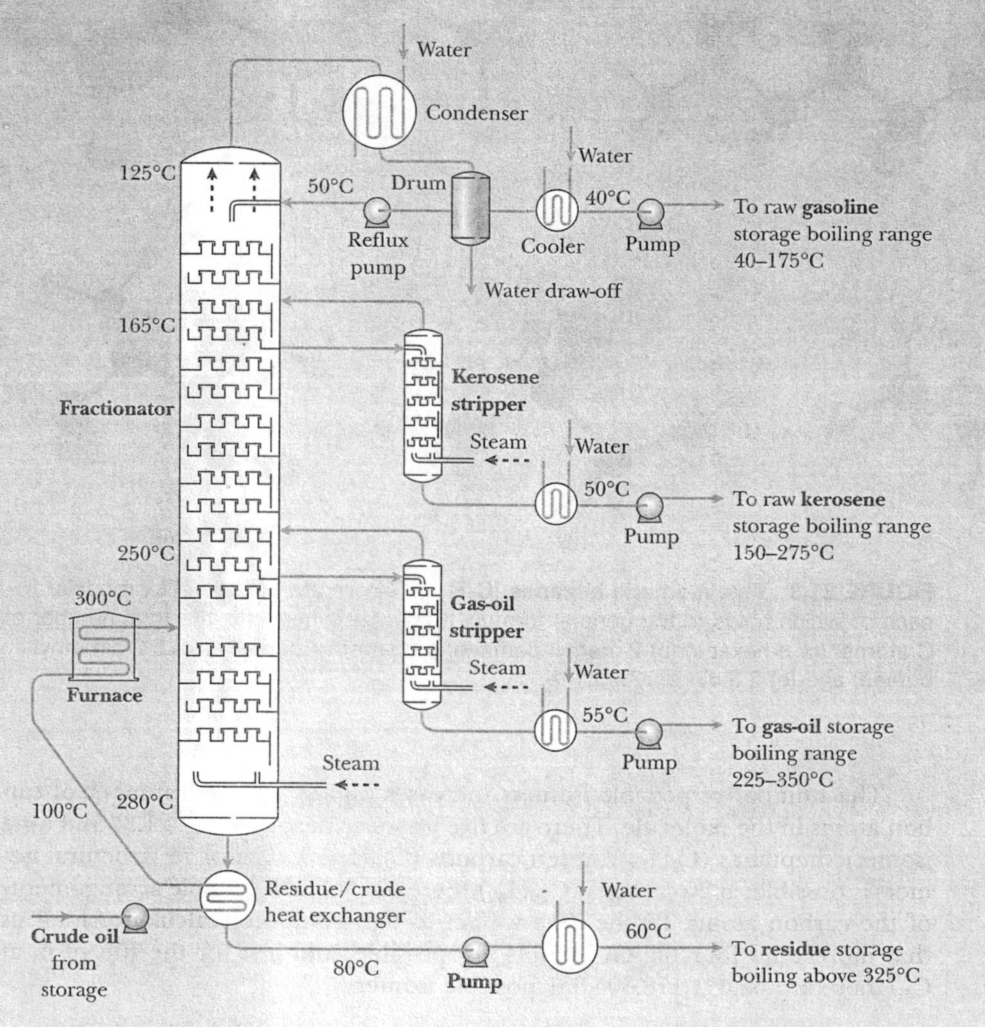

Fractional distillation of crude oil. This is one of the most commonly used separation processes. The heart of the distillation, is the central column (tower), which contains a series of trays (stages) that are usually perforated metal plates. Vapor passes up the column countercurrent to condensed liquid, which flows across each tray and down. As vapor bubbles up through descending liquid, mass is transferred between phases and the ascending vapor gets richer and richer in the more volatile components.

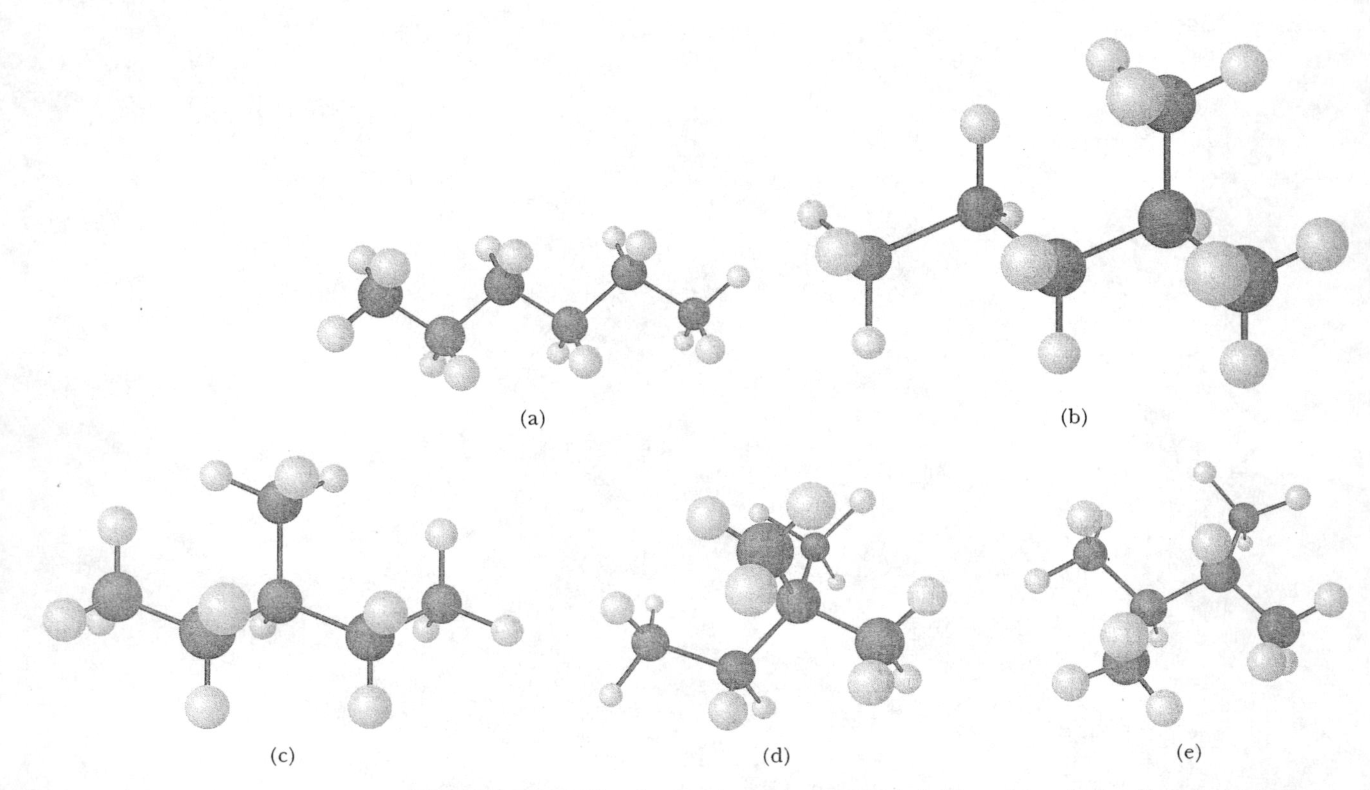

FIGURE 21.3 Five isomeric hexanes (C_6H_{14}). Shown are ball-and-stick models, each of which holds to the general formula (C_nH_{2n+2}), where *n* is the total number of C atoms: (a) *n*-hexane; (b) 2-methyl pentane; (c) 3-methyl pentane; (d) 2,2-dimethyl butane; and (e) 2,3-dimethyl butane.

The number of possible isomers increases rapidly with the number of carbon atoms in the molecule. There are five isomeric hexanes (Fig. 21.3) and nine isomeric heptanes (C_7H_{14}). At ten carbons ($C_{10}H_{22}$) a total of 75 structural isomers is possible; at 20 carbons ($C_{20}H_{42}$) there are 366,319 possible arrangements of the carbon atoms. By the time we get to C_{30}, computer calculations tell us that more than four billion isomers are possible, and just for the fun of it, at C_{40} there are 62,491,178,805,831 possible isomers.

Naming Organic Molecules

A common or trivial nomenclature system for naming hydrocarbons has grown up with organic chemistry. In **trivial nomenclature**, each compound is named independently with no general system. But there are 2 butanes, 3 pentanes, 5 hexanes, 9 heptanes, and 18 octanes. You can see what is happening! Thirty-seven names are needed to uniquely distinguish the hydrocarbons from four to eight carbon atoms. Clearly, naming is going to get quite complicated and rapidly out of hand if we rely on a trivial system of nomenclature. To help clear things up, a **systematic nomenclature** has been established. This is a regular system that gives a unique name to every compound and is based on selection of a main chain and identification of branch locations along it. This system is approved by the International Union of Pure and Applied Chemistry (IUPAC).

1. Identify the longest continuous chain of carbon atoms that can be followed nonstop throughout the entire molecule; the chain need not be linear. This parent chain is given the name corresponding to its number of carbon atoms.

The chain has five carbons, so the parent hydrocarbon is pentane:

$$\underset{5}{CH_3}-\underset{4}{\overset{\displaystyle CH_3}{\overset{|}{C}H}}-\underset{3}{CH_2}-\underset{\displaystyle \overset{|\,2}{} \atop CH_3}{\overset{\displaystyle CH_3}{\overset{|}{C}}}-\underset{1}{CH_3}$$

2. Identify the hydrocarbon branches on this longest chain and name them as alkyl groups. An **alkyl group** is an alkane that has lost a hydrogen, and the ending *-ane* is replaced by *-yl*. Thus, C_2H_6 is ethane, and C_2H_5— is ethyl. Others are given in Table 21.3.

There are three "methyl" branches.

3. Beginning at the terminal carbon closest to the branching, number the main chain of carbon atoms. The name of the side chain or group precedes the name of the parent chain and is itself preceded by the location of the branch. If the alkyl group is attached more than once, use the prefixes *di-*, *tri-*, *tetra-*, and so forth, to indicate the number of places.

Number from the right as shown. The systematic (IUPAC) name is 2,2,4-trimethylpentane. The common name is iso-octane.

4. If there are two or more different branches, number each one and place them all before the parent name, in alphabetical order. Number so that the sum of all numbered branches is as small as possible.

Numbering the molecule from the other direction would result in 2,4,4-trimethylpentane, which adds up to a larger number and is therefore wrong.

TABLE 21.3 Common Radicals Used in Systematic Naming of Hydrocarbons

Parent		Radical	
CH_4	methane	CH_3—	methyl
CH_3CH_3	ethane	CH_3CH_2—	ethyl
$CH_3CH_2CH_3$	propane	$CH_3CH_2CH_2$—	propyl
		$CH_3\underset{\vert}{C}HCH_3$	isopropyl
C_6H_{12}	cyclohexane	C_6H_{11}—	cyclohexyl
C_6H_6	benzene	C_6H_5—	phenyl

Very large, straight-chained hydrocarbon molecules such as *n*-triacontane ($C_{30}H_{62}$) and hexacontane ($C_{60}H_{122}$) are nature's $(—CH_2—)_x$ hydrocarbon polymers. The polymer chain can be made much longer, and because of the way this polymer is produced commercially from ethylene (ethene), it is called polyethylene. Synthesis of polyethylene can be carried out to produce an almost linear product, tens of thousands of methylene ($—CH_2—$) units long. Polyethylene can also be prepared with a branched structure. Branching alters the properties of the large, polymeric hydrocarbons as it does for the smaller alkanes. Linear polyethylene is rigid and tough because the long strands can lie very close to each other in a regular, almost crystalline pattern. Branching produces a less rigid, less dense, less crystalline structure because the molecules can no longer arrange themselves in as orderly a fashion (Fig. 21.4).

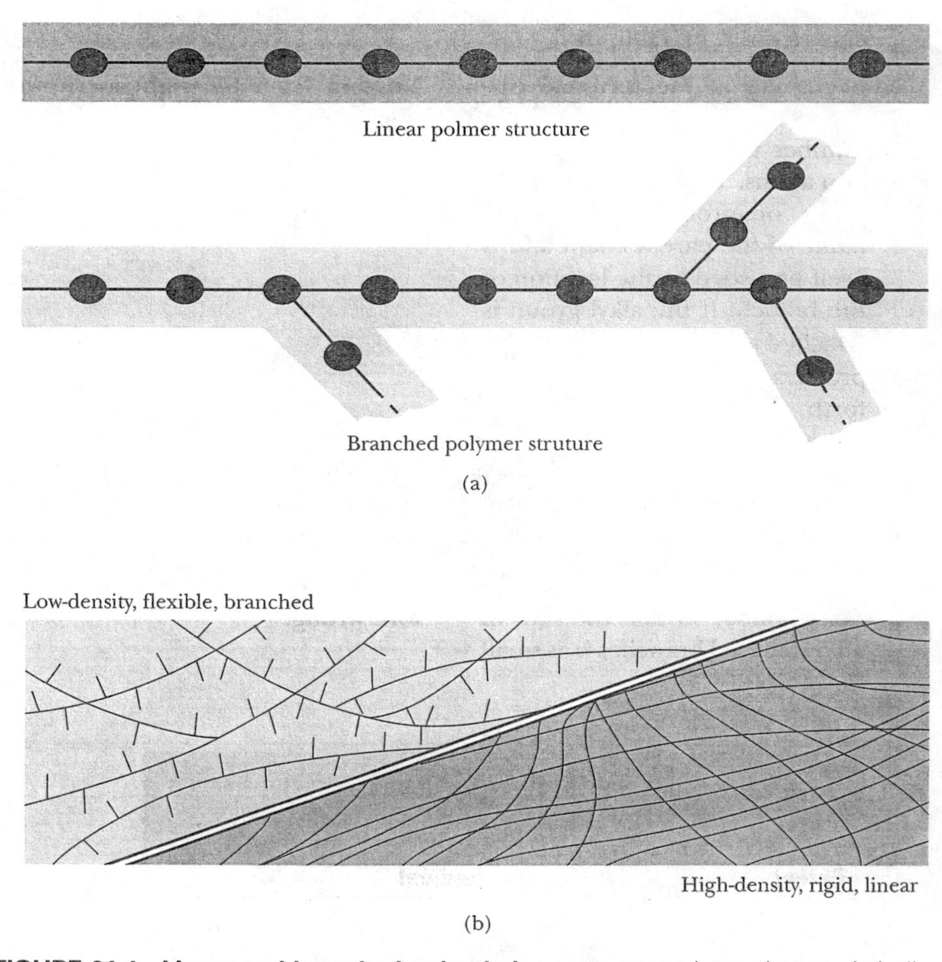

FIGURE 21.4 Linear and branched polyethylene structures have characteristically different properties, such as density and toughness. (a) A regular arrangement of the methylene units characterizes linear polyethylene; small segments along the backbone characterize the branched polyethylene structure. (b) Without branching, the chains can pack closely together in a more dense arrangement, leading to a more rigid material; branching results in a more flexible structure of lower density.

Whether we are talking about hydrocarbon chains of candle-wax length in the range of C_{25} to C_{30} or polymer chains on the order of $C_{25,000}$, consider what we mean by **linear** and **branched**. First, because of the tetrahedral geometry about carbon, linear chains of C atoms are not 180° straight-line or linear chains. The approximately 109° bond angle between carbon atoms results in a zig-zag arrangement (Fig. 21.2). Nevertheless "linear" still provides a satisfactory general description for the overall interlocking arrangement of many connected carbon atoms. "Branched," on the other hand, refers to carbon atoms that stick out from the main chain. Branching generally produces impediments to molecular associations, disrupting the already weak van der Waals forces. That results in lower boiling points for the liquid hydrocarbons and pliability in the polymeric hydrocarbons.

EXAMPLE 21.2

Beyond the five-carbon alkane compounds, it is necessary to use a systematic method of naming. Sketch the structural formulas for the five isomeric hexanes and give their systematic names in approximate order of their boiling points.

COMMENT Draw all possible structures, being careful that no structures are duplicated. The same structure can look quite different if it is written two or more different ways. Name the structures by the given rules. Check Table 21.2 to see the effect of branching on the boiling points of isomeric hydrocarbons.

SOLUTION The formulas are shown in Table 21.4. Branching lowers boiling points of isomeric alkanes.

TABLE 21.4 The Five Isomeric Hexanes

Formula	Name	Boiling Point (°C)
$CH_3CH_2CH_2CH_2CH_2CH_3$	*n*-hexane	69
$CH_3CH_2-CH(CH_3)-CH_2CH_3$	3-methylpentane	63
$CH_3CH_2CH_2-CH(CH_3)-CH_3$	2-methylpentane	60
$CH_3-CH(CH_3)-CH(CH_3)-CH_3$	2,3-dimethylbutane	58
$CH_3CH_2-C(CH_3)_2-CH_3$	2,2-dimethylbutane (*neo*-hexane)	50

TABLE 21.5 The Isomeric Heptanes

Formula	Name	Boiling Point (°C)
$CH_3CH_2CH_2CH_2CH_2CH_2CH_3$	*n*-heptane	93.6
$CH_3CH_2-CH-CH_2CH_3$ $\vert$ CH_2CH_3	3-ethylpentane (triethylmethane)	93.5
$CH_3CH_2CH_2-CH-CH_2CH_3$ $\vert$ CH_3	3-methylhexane	92
$CH_3CH_2CH_2CH_2-CH-CH_3$ $\vert$ CH_3	2-methylhexane (*iso*-heptane)	90
$CH_3CH_2-CH-CH-CH_3$ $\vert \quad \vert$ $CH_3 \; CH_3$	2,3-dimethylpentane	89.8
CH_3 $\vert$ $CH_3CH_2-C-CH_2-CH_3$ $\vert$ CH_3	3,3-dimethylpentane (*neo*-hexane)	86
CH_3 $\vert$ $CH_3CH-C-CH_3$ $\vert \quad \vert$ $CH_3 \; CH_3$	2,2,3-trimethylbutane (triptane)	80.9
$CH_3-CH-CH_2-CH-CH_3$ $\vert \qquad \vert$ $CH_3 \qquad CH_3$	2,4-dimethylpentane	80.5
CH_3 $\vert$ $CH_3CH_2CH_2-C-CH_3$ $\vert$ CH_3	2,2-dimethylpentane	79.2

EXERCISE 21.2

Draw the structural formulas for the nine isomeric heptanes, give their systematic names, and order them approximately according to their boiling points.

ANSWER See Table 21.5.

As we noted earlier, cycloalkanes are hydrocarbons in which the carbon chains are arranged into ring structures. (Table 21.6). They have the general formula C_nH_{2n} and use the alkane name preceded by *cyclo.* The fact that carbon normally has a tetrahedral bond angle of 109.5° suggests at least one basic difference between alkanes and cycloalkanes—the presence of ring strain in the latter. Cyclopropane is the smallest cycloalkane, and the internal C-to-C bond

TABLE 21.6 Cycloalkanes

Compound	Structural Formula	Skeleton Structure
cyclopropane	CH_2 / \\ CH_2—CH_2	
cyclobutane	CH_2—CH_2 | | CH_2—CH_2	
cyclopentane	CH_2 CH_2 CH_2 CH_2—CH_2	
cyclohexane	CH_2—CH_2 CH_2 CH_2 CH_2—CH_2	and

angle is 60°. The molecule exists but is highly strained. With an internal bond angle of 90°, cyclobutane is also highly strained relative to *n*-butane or isobutane.

In contrast, the five- and six-membered rings are almost free of strain. Cyclohexane has a "puckered" ring that is capable of existing in two configurations, one boat-like and the other chair-like (Fig. 21.5). The chair is the preferred geometry because it keeps the atoms as far apart as possible. Note that if the ring were not puckered, there would be strain because the internal bond angle would be 120° rather than the 109.5° tetrahedral bond angle. Cyclohexane is puckered because all the ring bonds are sigma bonds, with sp^3 hybridization.

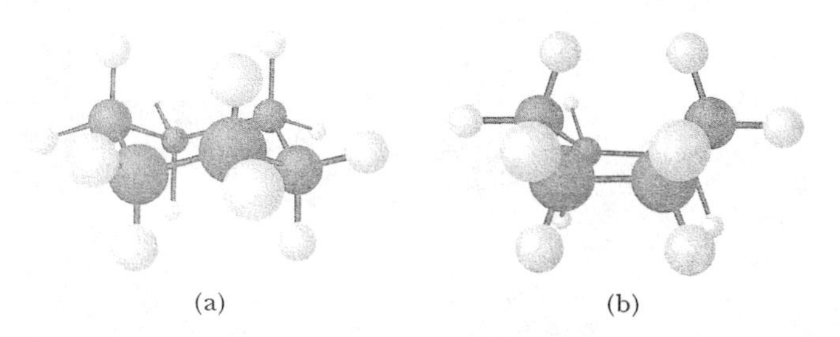

FIGURE 21.5 Boat (a) and chair (b) forms of cyclohexane (C_6H_{12}). The more stable of the two conformations (configurations) is the chair form. They are interchangeable by simply "flexing" the rings, with not much energy being required. Industrial demand for cyclohexane is driven by nylon synthesis, of which it is one of the principal raw materials. Originally obtained from crude oil by distillation, it is now widely prepared by direct (catalytic) hydrogenation of benzene: $C_6H_6(g) + 3H_2(g) \longrightarrow C_6H_{12}(g)$.

SEEING THINGS

Robert Wilhelm Bunsen and His Famous Burner Bunsen's ingenious physical insights that led to the founding of the modern study of spectroscopy were mentioned in Chapter 5. He was a chemist of considerable reputation, and his famous "Bunsen" burner is known to all students in chemistry laboratories. After completing graduate studies at Göttingen, Bunsen (German, 1811–1899) began research on the organic compounds of arsenic, nearly lost an eye in a laboratory explosion, and collapsed twice in the laboratory, probably because of inhalation and slow absorption of arsenic. Safety in the laboratory was not easily or often practiced and certainly not well-understood in those early days. His student Edward Frankland (British, 1825–1899) was inspired to continue Bunsen's work, which in turn provided the foundations needed by Paul Ehrlich (German, 1854–1915) in the first decade of the twentieth century to produce Salvarsan-606 (arsphenamine), an organoarsenic compound that proved effective against syphilis and opened the field of chemotherapy:

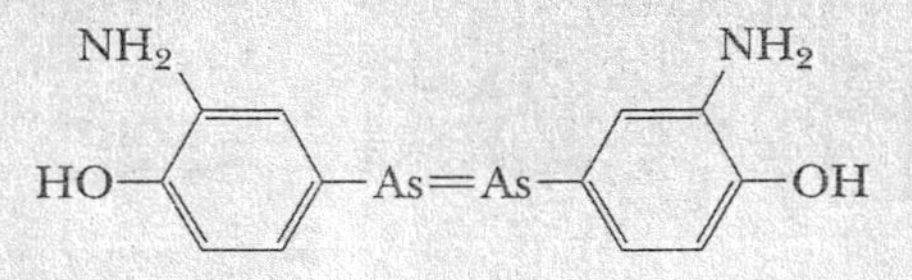

Bunsen's burner, first used in 1855, remains an icon of modern chemistry and is likely the single most-remembered feature of the introductory course. When a gas is to be used for heating, the complete combustion of the gas, without any intermediate production of elemental carbon in the form of "soot," is necessary. And that is just what Bunsen's burner accomplished. Air is drawn into perforations at the bottom of the vertical tube by the flowing gas used as the combustion fuel *before* reaching the flame at the top. The air cools the middle zone of the flame so that the temperature required for carbon production is not reached and there is sufficient oxygen for CO_2 production. A flame of this type is *not* luminous, and this characteristic was needed for research in obtaining emission, line spectra of atoms (Chapter 5).

The "carburated" fuel–air mixture burns steadily, providing uniform heat, little light, not much smoke, and hardly any flickering. It revolutionized laboratory work in chemistry. Placed in a horizontal position, the ordinary bunsen burner serves as the cooking source of the domestic gas cooking range (stove).

Although it was not the first of its kind—Faraday had designed something similar in 1830—it was Bunsen, one of the great early teachers of chemistry, who popularized it and justly gets proper credit for the "Bunsen burner."

21.3 CHLORINATED HYDROCARBONS AND STRUCTURAL ISOMERS

Reactions of Alkanes

Alkanes are not particularly reactive except at high temperatures. Of the few reactions they do undergo, combustion and halogenation are the most familiar. Complete combustion of any hydrocarbon eventually produces carbon dioxide and water. The Bunsen burner is a good example of a device for the burning of hydrocarbon fuels in oxygen:

Questions

A candle can serve as a light source as well as a heat source. Likewise, a Bunsen burner, which can be adjusted to burn the gaseous fuel with little air mixed in. But a properly adjusted flame with an optimal air–fuel mixture is only a heat source. Almost no light! Why?

On a gas stove burner where there has been a food spillage that has not been completely cleaned up, the normally blue and nearly invisible flame becomes yellow and luminous. Why?

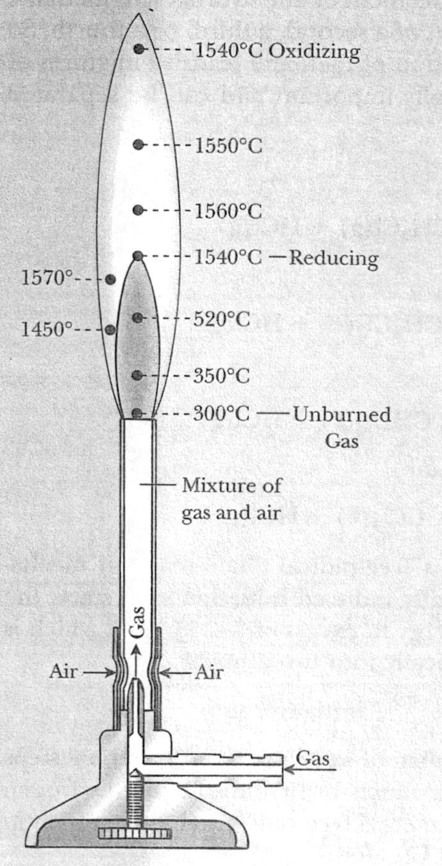

Bunsen's burner and how it works. Mixtures of methane and oxygen are explosive. The maximum effect is obtained when two volumes of oxygen (ten volumes of air) are mixed with one volume of methane. Something close to that condition gives the blue flame of the bunsen burner and delivers maximum heat. Air enters the burner and mixes with the fuel supply, both of which can be regulated—air by swiveling the main burner tube that opens or closes the orifice at its stem, and fuel by adjusting a pin valve typically under the base.

$$CH_4(g) + 2O_2(g) \longrightarrow CO_2(g) + 2H_2O(\ell) \qquad \Delta H^\circ = -890 \text{ kJ}$$

All hydrocarbons are potential fuels.

Much more important to organic chemistry are chlorination and bromination reactions, in which hydrogen atoms on alkanes are replaced with chlorine or bromine atoms. Unlike combustion, these reactions maintain the structure of the hydrocarbon chain and are among the most important industrial processes involving alkanes.

The Chlorination of Methane

Under appropriate conditions, a molecule of methane can react with a molecule of chlorine, yielding methyl chloride (CH_3Cl) and hydrogen chloride:

$$CH_4(g) + Cl_2(g) \longrightarrow CH_3Cl(g) + HCl(g) \qquad \text{monochlorination of methane}$$

Unfortunately, chlorine is so reactive that all types of C-to-H bonds are attacked with nearly equal vigor, and the replacement of one hydrogen in methane does not appreciably affect the replacement of a second, a third, or a fourth. So the thermal or radiation-induced chlorination of methane leads to mixtures of four products, all of which are commercially important and can be separated from each other by distillation:

Methyl chloride (chloromethane)

$$CH_4(g) + Cl_2(g) \longrightarrow CH_3Cl(g) + HCl(g)$$

Methylene chloride (dichloromethane)

$$CH_3Cl(g) + Cl_2(g) \longrightarrow CH_2Cl_2(\ell) + HCl(g)$$

Chloroform (trichloromethane)

$$CH_2Cl_2(\ell) + Cl_2(g) \longrightarrow CHCl_3(\ell) + HCl(g)$$

Carbon tetrachloride (tetrachloromethane)

$$CHCl_3(\ell) + Cl_2(g) \longrightarrow CCl_4(\ell) + HCl(g)$$

These reactions are known to follow a free-radical chain-reaction mechanism (Chapter 15). A radiation or thermally induced **initiation step** starts the reaction as radiation or heat-providing energy in excess of 242 kJ/mol, which is absorbed, breaking the Cl_2 bond homolytically into two atoms:

$$Cl{-}Cl \longrightarrow 2\ Cl\cdot \qquad \text{initiation step}$$

Once chlorine atoms are present, a number of very rapid **propagation steps** push the reactants along the pathway to products by first abstracting hydrogen atoms from methane, producing HCl and methyl free radicals ($CH_3\cdot$), which in turn are capable of producing $CHCl_3$ and $Cl\cdot$ atoms:

$$Cl\cdot + CH_4 \longrightarrow CH_3\cdot + HCl \qquad \text{propagation steps}$$

$$CH_3\cdot + Cl_2 \longrightarrow CH_3Cl + Cl\cdot$$

A huge number of CH_3Cl molecules are formed by only a very few $Cl\cdot$ atoms, but **termination steps** eventually stop the process through coupling reactions that leave no free radicals to continue the process:

$$CH_3\cdot + CH_3\cdot \longrightarrow CH_3CH_3 \qquad \text{termination steps}$$

$$Cl\cdot + Cl\cdot \longrightarrow Cl_2$$

$$CH_3\cdot + Cl\cdot \longrightarrow CH_3Cl$$

In industry, the reaction is carried out thermally by passing the mixture of methane and chlorine through hot tubes or photochemically in special reactors fitted with mercury lamps, which provide the necessary light quanta.

EXAMPLE 21.3

How would you go about maximizing the yield of methyl chloride, the monochlorination product in the photochlorination of methane, minimizing the yields of the di-, tri-, and tetrachloromethanes?

COMMENT Think about how you would fix the conditions to maximize the collisions you want and minimize the collisions you do not want.

SOLUTION The answer lies in having a very large excess of methane, so that chlorine molecules almost invariably encounter CH_4 and not CH_3Cl molecules.

EXERCISE 21.3

Would the photobromination of methane require light of longer or shorter wavelength to initiate the reaction? Your answer should be based on the bond dissociation energy for Br_2. Briefly explain.

More than ten million tons of chlorine are produced in the United States yearly, three-fourths of which is used in the synthesis of organic compounds. The electrochemical manufacturing process for preparing chlorine on a commercial scale was described in Chapter 14. Chloromethane (methyl chloride, b.p. −24°C) is mostly used for the synthesis of silicone polymers and plastic materials such as butyl rubber. It is also used to a small extent as a pesticide. For many years it was used in the synthesis of tetramethyllead, a combustion-improving additive for gasoline. It has also been used in the manufacture of methyl cellulose for the food industry, which, among other things, is used as a bulk producer in dietary foods.

Dichloromethane (methylene chloride, b.p. 40°C), the least toxic of the chlorinated methanes, is used extensively as a reaction medium and solvent, especially in the synthesis of triacetate films and fibers, and as a paint thinner and varnish remover. Trichloromethane (chloroform, b.p. 61°C) is the principal intermediate in the synthesis of Freon aerosol propellents and refrigerants. Tetrachloromethane (carbon tetrachloride, b.p. 77°C) is the most important chlorinated methane. It is converted to solvents for the dry-cleaning industry and into Freon aerosols. In common with some other chlorinated hydrocarbons, carbon tetrachloride is quite toxic, causing liver damage on long-term exposure and posing a hazard in the workplace.

Structural Isomers and Chlorinated Alkanes

We encountered structural isomerism in our earlier discussion of butanes, pentanes, and larger hydrocarbons. Chlorinating alkanes provides more structural isomers. For ethane, only one monochlorination product is possible because of the symmetry. However, if you replace an H atom on ethyl chloride by another Cl atom, two structural isomers are possible. Both chlorine atoms may be attached to the same carbon atom, or one chlorine may be attached to each carbon. As you might suspect, important properties such as melting and boiling points are clearly different. The number prefixes indicate the carbon atom to which each chlorine atom is bonded:

$$\begin{array}{c} \text{H}\quad\text{H} \\ | \quad\; | \\ \text{H}-\text{C}-\text{C}-\text{Cl} \\ | \quad\; | \\ \text{Cl}\quad\text{H} \end{array} \qquad\qquad \begin{array}{c} \text{H}\quad\text{H} \\ | \quad\; | \\ \text{H}-\text{C}-\text{C}-\text{Cl} \\ | \quad\; | \\ \text{H}\quad\text{Cl} \end{array}$$

	1,2-dichloroethane	1,1-dichloroethane
m.p. (°C)	−35	−96
b.p. (°C)	84	57

Two structural isomers are also possible for trichloroethane:

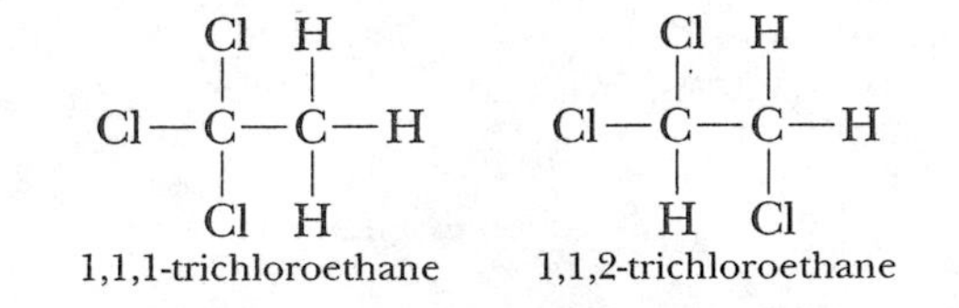

With four chlorine atoms in the molecule, two isomeric structures are again possible:

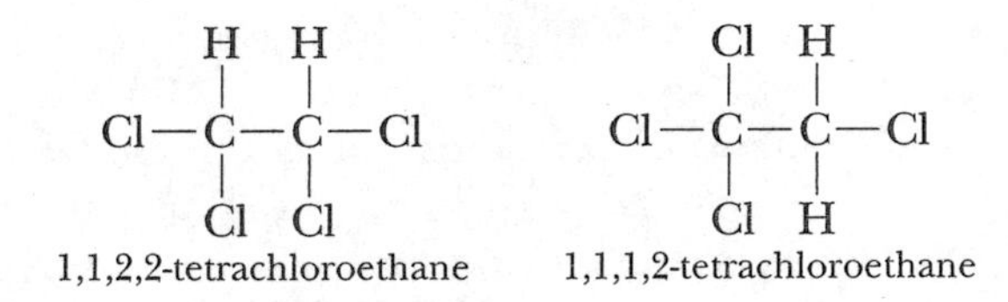

With five and six chlorine atoms, only one bonding arrangement is possible in each case, as you can easily verify. Summing up the results for ethane, nine distinctly different chloroethanes are possible.

Chlorinated hydrocarbons are especially important because they are considerably more reactive than the parent hydrocarbons. As a consequence, they are the starting point for a wide variety of new organic compounds, which in turn lead to literally thousands of products used in today's world. Let us look at some reactions of just one chlorinated hydrocarbon, chloroethane. Note that it is common practice in organic chemistry to just write the reagents directly above the arrow.

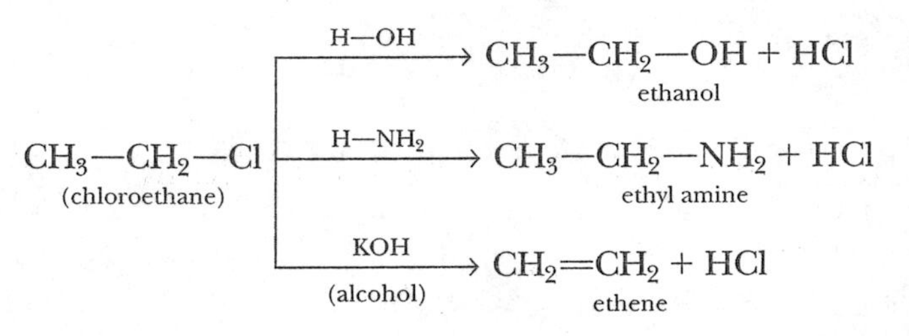

In the first two examples, the —Cl group attached to the hydrocarbon is replaced by —OH from water and by $-NH_2$ from ammonia, yielding new derivatives of ethane. In the third example, H and Cl atoms are removed from adjacent carbons to yield a carbon-to-carbon double bond.

Let us note some examples of what are known as functional groups: A **functional group** is an atom or group of atoms on a chain that will behave more or less the same in different compounds. Shown below are some functional groups, where R represents a hydrocarbon chain.

R—**X** are **halides**, where X = F, Cl, Br, I

R—**OH** are **alcohols**

R—$\mathbf{NH_2}$ are **amines**

R—**C=C**—*R* are **alkenes**

Thus, any hydrocarbon chain containing the specific group —OH would have the general properties and reactivity of alcohols. We will introduce several other functional groups in Section 21.7.

21.4 ALKENES

Alkenes are hydrocarbons containing a double bond between two carbon atoms (Fig. 7.13). Also called olefins, alkenes contain only carbon and hydrogen, but the double bond allows for reactions not possible for alkanes. The general formula for straight chain alkenes containing one double bond is C_nH_{2n}. Cycloalkenes containing one double bond have the general formula C_nH_{2n-2}. The first members of the alkene series are ethene and propene, commonly referred to as ethylene and propylene:

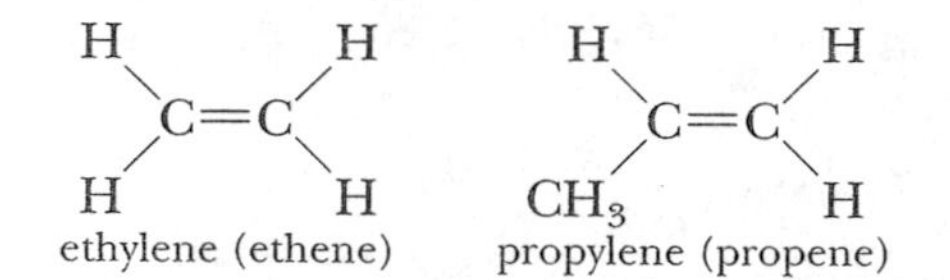
ethylene (ethene) propylene (propene)

Beginning with the butenes, alkenes are named systematically to locate the double bonds unambiguously. Systematic nomenclature for the alkenes begins with the parent hydrocarbon chain of longest continuous length that contains the double bond. Drop the *-ane* ending of the corresponding alkane and replace it with the *-ene* ending for alkene. Then number the carbon atoms in the chain so that the carbon atom at which the double bond begins has the lowest possible number, and include this number before the name:

$$\underset{4}{CH_3}-\underset{3}{CH_2}-\underset{2}{CH}=\underset{1}{CH_2} \qquad \underset{4}{CH_3}-\underset{3}{CH}=\underset{2}{CH}-\underset{1}{CH_3}$$

1-butene 2-butene

The double bond in an alkene imposes a barrier to rotation about the carbon–carbon bond, which can also lead to isomers. Thus, two isomeric 2-butenes are possible.

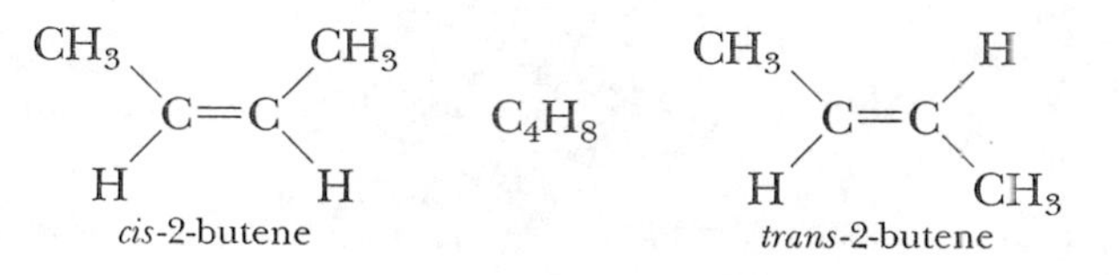
cis-2-butene C_4H_8 *trans*-2-butene

TABLE 21.7 Melting and Boiling Points of the C_2—C_4 Alkenes

	m.p. (°C)	b.p. (°C)
ethylene (ethene)	−169	−102
propylene (propene)	−185	−48
1-butene	−195	−6
trans-2-butene	−106	1
cis-2-butene	−139	4
2-methylpropene	−141	−7

To distinguish between the isomeric 2-butenes, we name them in a special way. Consider the substituents on the two carbon atoms of the centrally located double bond. When the hydrogen atoms (or the methyl groups) are on the same side of the double bond, we refer to the geometry as *cis-*. We refer to the geometry as *trans-* when these same groups lie on opposite sides of the plane. Thus, we have *cis*-2-butene and *trans*-2-butene. There is also another butene isomer, in which both methyl groups are attached to the carbon atom at one end of the double bond. It is named isobutene or 2-methylpropene. Note the difference in properties between ethylene, propylene, and the four possible butene isomers (Table 21.7). You may recall *cis-trans* isomers, discussed in Chapter 19.

Preparation of Alkenes

Alkenes can be synthesized in the laboratory and on an industrial scale in several ways. As examples, consider the following:

1. **Dehydrogenation** of an alkane creates a double bond by removing hydrogen from adjacent carbons. The process, carried out at elevated temperatures, is illustrated for ethane and propane:

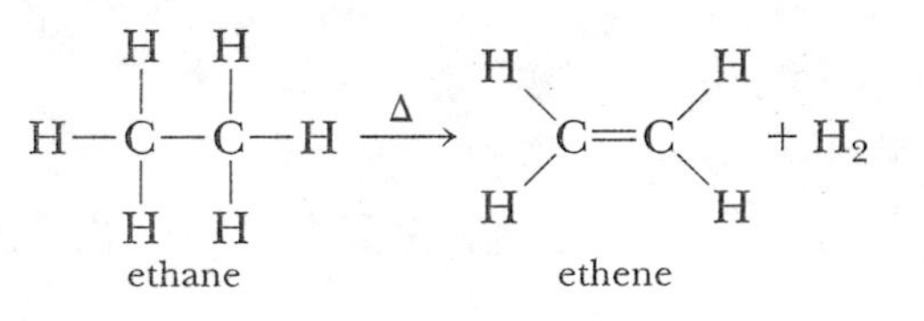

A delta Δ placed over an arrow is commonly used to indicate adding heat to a chemical reaction:

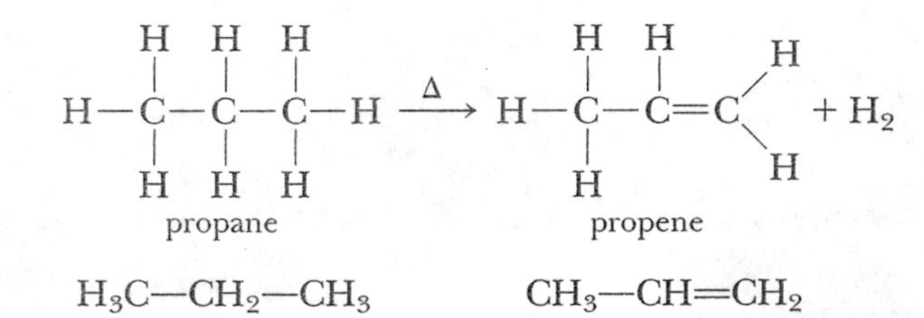

2. **Elimination of hydrogen halides**, typically HCl and HBr, from adjacent carbon atoms of halogenated alkanes using an alcohol (ethanol) solution of KOH introduces double bonds. This is called dehydrohalogenation, because a hydrogen atom and a halogen atom are both eliminated:

$$CH_3CH_2Cl + KOH \longrightarrow CH_2{=}CH_2 + KCl + H_2O$$

3. **Dehydration of alcohols** using strongly dehydrating agents such as sulfuric acid results in formation of double bonds. An alcohol is a hydrocarbon in which an H atom has been replaced by an —OH group:

$$CH_3CH_2OH \xrightarrow{H_2SO_4} CH_2{=}CH_2 + H_2O$$

EXAMPLE 21.4

Synthesize propene from propane by two different chemical reaction schemes.

COMMENT We know how to chlorinate a hydrocarbon and then dehydrohalogenate the monochlorination product; and we know that dehydrogenation of hydrocarbons can be used to introduce double bonds into a molecule.

SOLUTION

1. Photochlorination of propane, followed by dehydrohalogenation:

$$CH_3CH_2CH_3 + Cl_2 \xrightarrow{h\nu} CH_3CHClCH_3 + HCl \qquad \text{photochlorination}$$

$$CH_3CHClCH_3 \xrightarrow{\text{alcoholic KOH}} CH_3CH{=}CH_2 + HCl \qquad \text{dehydrohalogenation}$$

2. Dehydrogenation of propane:

$$CH_3CH_2CH_3 \xrightarrow[750°C]{\text{catalyst}} CH_3CH{=}CH_2 + H_2$$

EXERCISE 21.4

Write equations for the synthesis of 2-methylpropene from 2-methylpropane.

ANSWER Photochlorination of 2-methylpropane, followed by dehydrohalogenation

Addition Reactions

Double bonds are much more reactive than single bonds. As a consequence, alkenes are characterized by the ease with which reagents add to the carbon atoms on either side of the double bond at the expense of the higher energy π-bond. The σ-bond is left intact. For example, a simple test for distinguishing between an alkane and an alkene is the fading of the characteristic red–brown color of bromine dissolved in carbon tetrachloride after the bromine solution is added:

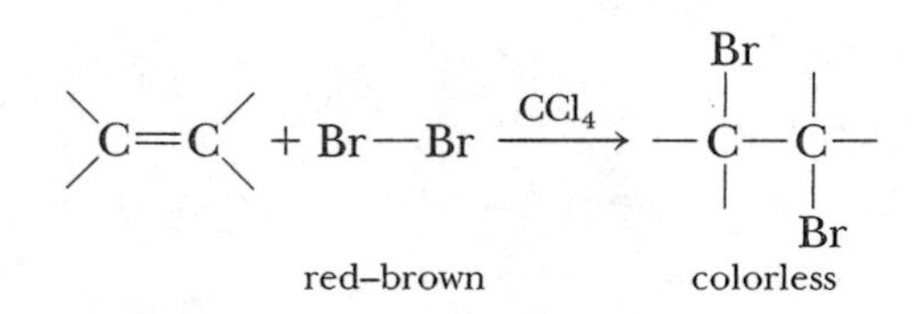

For example, the addition of bromine to 1-pentene:

$$CH_3CH_2CH_2CH{=}CH_2 + Br_2 \xrightarrow{CCl_4} CH_3CH_2CH_2CH(Br)CH_2Br$$

1-pentene 1,2-dibromopentane

No reaction occurs on addition of bromine to *n*-pentane in the absence of light or heat because there is no double bond present:

$$\underset{n\text{-pentane}}{CH_3CH_2CH_2CH_2CH_3} + Br_2 \xrightarrow{CCl_4} \text{no reaction}$$

Hydrogen, under high pressure and in the presence of catalysts such as nickel, palladium, or platinum metals, also adds to double bonds:

$$\rangle C{=}C\langle + H_2 \xrightarrow{\text{catalyst}} -\overset{|}{\underset{H}{C}}-\overset{|}{\underset{H}{C}}-$$

For example, the catalytic hydrogenation of 1-pentene produces *n*-pentane:

$$CH_3CH_2CH_2CH{=}CH_2(g) + H_2(g) \longrightarrow CH_3CH_2CH_2CH_2CH_3(\ell)$$

In fact, this is one of the best ways of synthesizing alkanes, providing the corresponding alkenes are readily available. It is because hydrogen can be added to alkenes that they are known as unsaturated hydrocarbons, in contrast to alkanes, which are known as saturated hydrocarbons.

The hydrogenation reaction can be used to measure the stabilities of alkenes by comparing respective ΔH values. Let us look at the enthapies of hydrogenation for *cis*- and *trans*-2-butene:

$$cis\text{-2-butene} + H_2 \longrightarrow n\text{-butane} \qquad \Delta H = -115 \text{ kJ/mol}$$

$$trans\text{-2-butene} + H_2 \longrightarrow n\text{-butane} \qquad \Delta H = -112 \text{ kJ/mol}$$

There is a difference of 3 kJ/mole, even though both isomers yield *n*-butane, the same product, on adding one mole of hydrogen. Therefore, we conclude that the *trans*-isomer must be stabilized, relative to the *cis*-isomer, to the extent of 3 kJ/mol.

Hydrogen chloride and hydrogen bromide can be added directly to the double bond, yielding the corresponding alkyl halides:

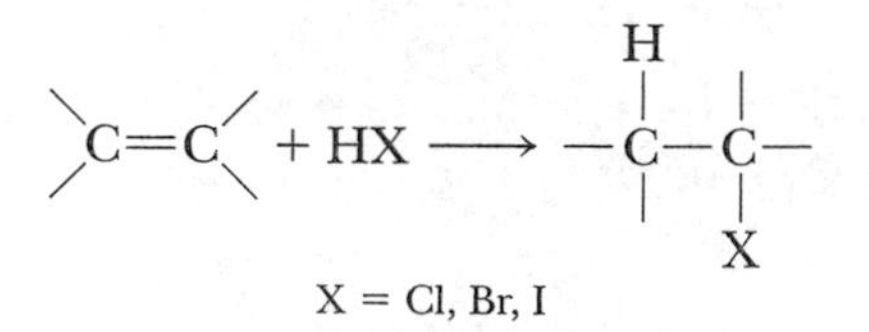

However, with the exception of symmetrical alkenes such as ethylene and 2-butene, a special problem arises. Which way do the elements H and X add? For example, H—Cl can add to propylene in two different ways:

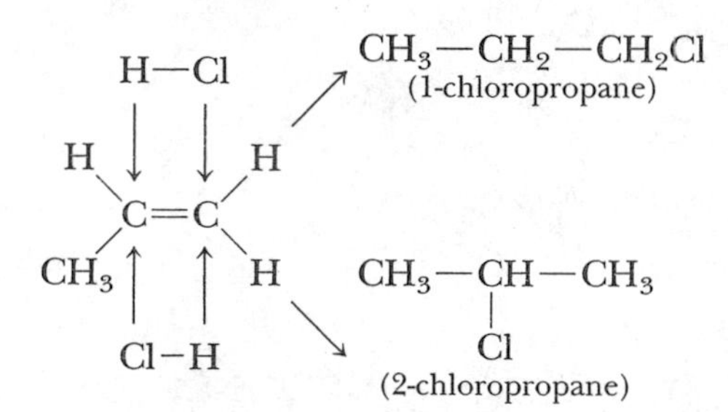

The answer is provided by an empirical rule established by a Russian chemist, Vladimir Markovnikov, after studying many such additions to alkenes: *The H atom always goes to the carbon atom of the double bond that already has the most H atoms.* Therefore, 2-chloropropane would be produced in the previous reaction.

The addition of water to double bonds yields alcohols in which an alkyl group now has an OH group instead of a halogen attached to it:

$$\rangle C{=}C\langle \; + \; HOH \xrightarrow[H_2SO_4]{\text{dilute}} -\overset{H}{\overset{|}{C}}-\underset{OH}{\underset{|}{C}}-$$

The reaction is catalyzed by acids, particularly sulfuric acid. On adding water, ethylene yields ethyl alcohol (ethanol) and 2-butene gives rise to a molecule called secondary-butyl alcohol or 2-butanol. The direction of addition to ethylene is not important because of the symmetry of the starting material. But with propylene, Markovnikov's rule needs to be consulted to predict the product. It could be either 1-propanol (*n*-propyl alcohol) or 2-propanol (isopropyl alcohol):

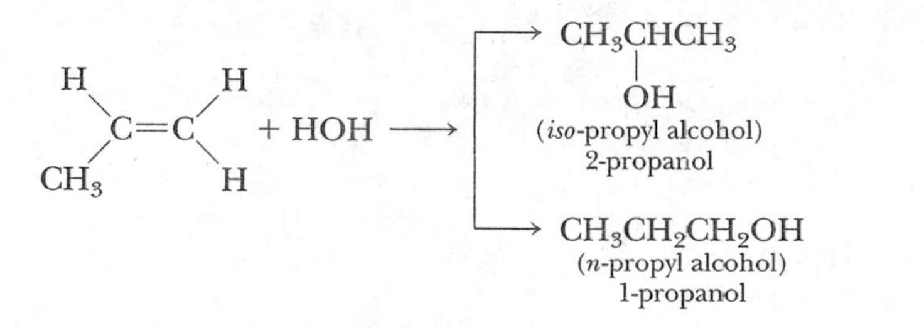

The dominant product is 2-propanol (isopropyl alcohol), just as Markovnikov's rule predicts.

EXAMPLE 21.5

Using Markovnikov's rule, predict the principal products of the reactions of HBr and H_2O with 1-butene and 3-hexene.

COMMENT Remember, the H atom goes to the carbon atom of the double bond having the most H atoms to begin with.

SOLUTION For 1-butene:

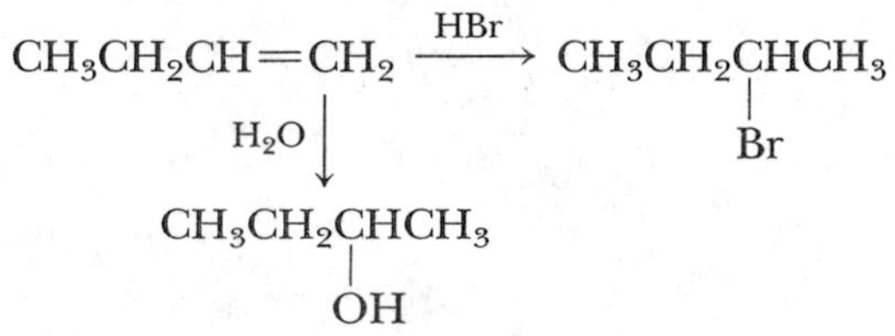

$$CH_3CH_2CH{=}CH_2 \xrightarrow{HBr} CH_3CH_2\underset{\displaystyle Br}{\underset{|}{C}}HCH_3$$

$$CH_3CH_2CH{=}CH_2 \xrightarrow{H_2O} CH_3CH_2\underset{\displaystyle OH}{\underset{|}{C}}HCH_3$$

For 3-hexene:

$$CH_3CH_2CH_2\underset{\displaystyle Br}{\underset{|}{C}}HCH_2CH_3$$

$$CH_3CH_2CH_2\underset{\displaystyle OH}{\underset{|}{C}}HCH_2CH_3$$

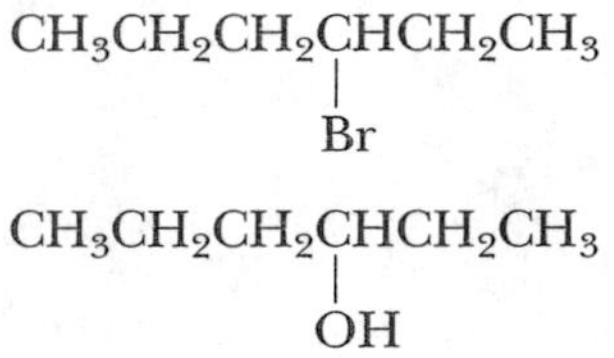

EXERCISE 21.5

Name each of the alkyl halides in Example 21.5 unambiguously as a bromoalkane.

ANSWER 2-bromobutane, 3-bromohexane

Addition Polymerization Reactions

One of the more important reactions alkenes undergo is the self-addition known as **chain-growth (addition) polymerization**, the combination of many small molecules called monomers into much larger molecules called polymers. When a monomer such as ethylene is heated under pressure, the high-molecular-mass, alkane-like polymer called polyethylene is obtained. The process by which the monomer units are snapped together is an addition polymerization reaction in which the π-bond electron pair is redistributed to the two carbon atoms. That alters the hybridization about carbon from sp^2 to sp^3, allowing formation of two new σ bonds between adjacent molecules. In that way, the monomer becomes a dimer ($n = 2$), then a trimer ($n = 3$), a tetramer ($n = 4$), and eventually a polymer (n = very many):

$$\underset{\text{monomer}}{nCH_2{=}CH_2} \xrightarrow{\text{catalyst}} \underset{\text{polymer}}{\left[CH_2{-}CH_2\right]_n}$$

Natural rubber is composed of many hundreds of isoprene (2-methyl-1,3-butadiene) monomer units, added to each other end-to-end, forming a high-molecular-mass polymeric structure with a *cis*-geometry about the double bond. This is the soft, elastic material isolated from the sticky sap of the rubber tree:

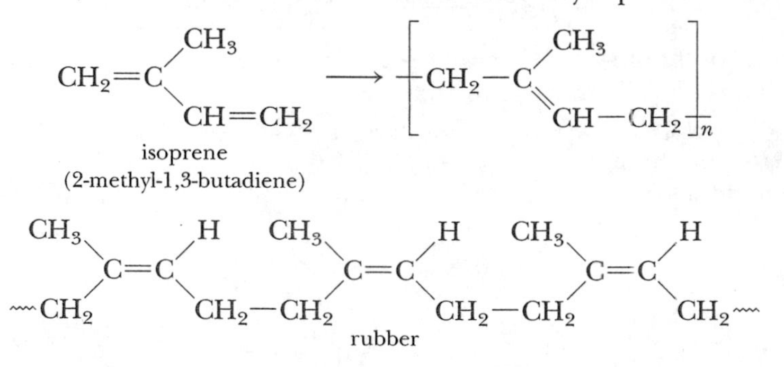

If the orientation about the double bond is *trans*, there is quite a change in the properties of the polymeric product. Gutta-percha is an all *trans*-polyisoprene, and it has none of the elastic properties of natural rubber with the *cis* structure:

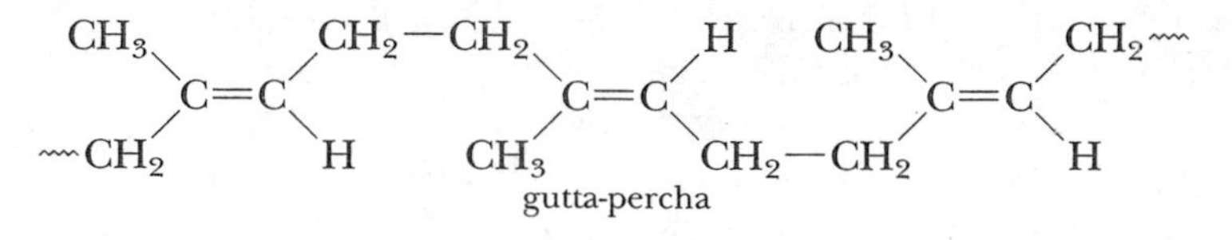

21.5 ALKYNES: SYNTHESIS AND REACTIONS

Alkynes are hydrocarbons that contain a triple bond between adjacent carbon atoms. Like alkanes and alkenes, alkynes contain only carbon and hydrogen atoms, and the reactions they undergo are the result of the *sp* hybridization on carbon, which gives rise to the triple bond (Fig. 7.14). The general formula is C_nH_{2n-2}, and the simplest and best-known member of the family is acetylene, $H{-}C{\equiv}C{-}H$, for which the IUPAC name is ethyne. Acetylene melts at $-82°C$ and boils at $-75°C$ and can be produced from calcium carbide, which is made from coke and calcium oxide (CaO) in an electric furnace at 2000°C:

$$\underset{\text{coke}}{3C(s)} + \underset{\text{lime}}{CaO(s)} \xrightarrow{2000°C} \underset{\text{calcium carbide}}{CaC_2(s)} + CO(g)$$

$$CaC_2(s) \xrightarrow{H_2O} \underset{\text{acetylene}}{H{-}C{\equiv}C{-}H(g)} + Ca(OH)_2(aq)$$

Because of the high percentage of carbon in acetylene compared with other hydrocarbons, acetylene burns with a very luminous flame. The luminosity of the flame is actually the result of tiny carbon particles glowing to incandescence in the heat. Acetylene was at one time widely used for illumination, particularly for movable lamps on cars, bicycles, and miners helmets. Portable acetylene generators consisted of a canister of calcium carbide and a water reservoir that could be controlled to drip at a measured rate. The rate of the water drip into the calcium carbide controls the rate of evolution of the acetylene and regulates the brightness of the lamp.

Newer methods of preparing acetylene involve natural gas or liquid hydrocarbon feedstocks at high temperatures:

$$2CH_4(g) \xrightarrow{1800^\circ} C_2H_2(g) + 3H_2(g)$$

Reaction conditions are critical. At sufficiently high temperatures, C_2H_2 is more stable than other hydrocarbons that can form.

A general method for the laboratory preparation of alkynes is based on the same kind of elimination reaction used to prepare alkenes—namely dehydrohalogenation—but to generate the second π bond, a second molecule of HX must be removed:

$$-\underset{\underset{X}{|}}{\overset{\overset{H}{|}}{C}}-\underset{\underset{X}{|}}{\overset{\overset{H}{|}}{C}}- \xrightarrow[\text{KOH}]{\text{alcohol}} \left[-\overset{\overset{H}{|}}{C}=\underset{\underset{X}{|}}{C}-\right] \longrightarrow -C\equiv C- + 2HCl$$

Systematic names of alkynes are formed by changing the parent hydrocarbon name to *-yne* and locating the triple bond by the smallest possible number. Thus, $CH_3—C\equiv C—CH_3$ is 2-butyne.

Just as alkene chemistry revolves around addition to the double bond, alkyne reactions are largely additions to the triple bond. Stepwise addition to the triple bond in alkynes is possible, but the intermediate double-bonded compound is hard to isolate, because the double bond is actually more reactive than the triple bond. Thus, reacting excess HI or Cl_2 with acetylene leads to the addition of two moles of reactant:

$$HC\equiv CH \xrightarrow{2HI} H_3C—CHI_2$$

$$HC\equiv CH \xrightarrow{2Cl_2} Cl_2HC—CHCl_2$$

21.6 AROMATIC HYDROCARBONS

Resonance Stabilization in Benzene

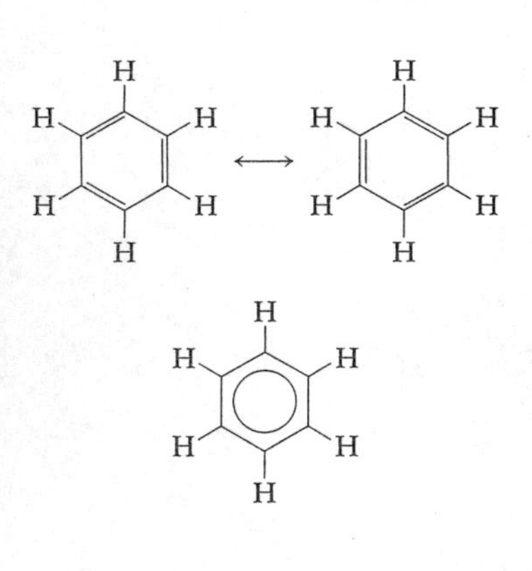

Aromatic hydrocarbons are based on the cyclic hydrocarbon known as benzene (C_6H_6), one of the most important compounds in all of organic chemistry. As we saw in Chapter 7, it is a six-membered ring of carbon atoms, each of which is also bonded to a hydrogen atom in a trigonal planar environment (Fig. 7.15). All six carbon atoms constituting the hexagon lie in the same plane. The six carbon-to-carbon bonds are all 1.39 Å long, considerably shorter than the C-to-C σ bond in ethane (1.54 Å) but longer than the C-to-C π bond in ethylene (1.33 Å). All bond angles are 120°. The two resonance structures, each having alternating double and single bonds, are best represented as a resonance hybrid in which the π electrons are delocalized throughout the ring (Fig. 7.15). The delocalization of the electrons is continuous because of the ring structure, making for an extraordinarily symmetric molecule. By convention, the C_6H_6 structure is represented by a hexagon; each corner is understood to consist of a carbon atom attached to a single hydrogen atom. The π electrons can be represented by a ring drawn in the center of the hexagon, which emphasizes the fact that the π-bonding electrons are not the property of any individual carbon–carbon bonds but are delocalized.

Benzene rings appear again and again as parts of other organic molecules and contribute significantly to their properties. They impart rigidity and structure to many of the polymeric materials discussed in Chapter 18. Because some of the compounds that contain benzene rings have fragrant odors, compounds containing these rings have come to be known as aromatic compounds. An **aromatic** compound is any compound that contains one or more benzene rings. A benzene molecule with one hydrogen atom removed appears in many compounds and is called a phenyl group. Thus, a **phenyl group** has the formula C_6H_5—.

To illustrate how aromaticity alters chemical behavior, let us compare the reactivity of the bonds in benzene to the double bond in cyclohexene and the single bonds in cyclohexane. Because it is an alkane, cyclohexane is characteristically unreactive. Cyclohexene, on the other hand, is an alkene and typically undergoes addition to the double bond. The bonds in benzene, however, are not susceptible to the same addition reactions (Tab. 21.8). Resonance imparts a pronounced stability to benzene with respect to hydrogenation and oxidation. A measure of this added stability is the diminished enthalpy change for hydrogenation and combustion reactions.

Substitution Reactions

The reactivity that benzene and benzene-like hydrocarbon molecules exhibit can be understood in terms of preserving the integrity of the stable aromatic ring. Thus, substitution reactions are typical; addition reactions are not. Benzene, for example, reacts with chlorine and bromine by undergoing ring substitution rather than double-bond addition. But, a catalyst such as iron or $FeCl_3$ is nec-

TABLE 21.8 Comparing Cyclohexane, Cyclohexene, and Benzene*

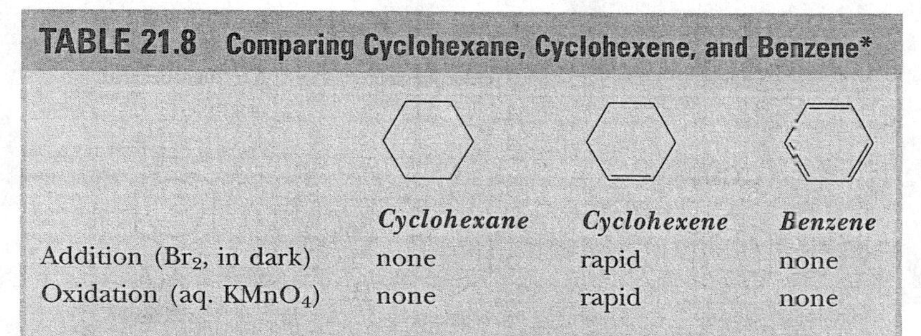

	Cyclohexane	*Cyclohexene*	*Benzene*
Addition (Br_2, in dark)	none	rapid	none
Oxidation (aq. $KMnO_4$)	none	rapid	none

**Note:* The convention for writing structural formulas of ring compounds is as follows. Each corner in the ring represents a carbon atom. The hydrogen atoms bonded to these carbons are not shown. It is understood that there are enough hydrogens attached to each carbon to bring the number of bonds on the carbon to four. Thus, the formula shown for cyclohexane represents six carbon atoms in a ring, each bonded to two hydrogen atoms. For benzene rings, a circle drawn within the hexagon can be used to represent the delocalized π electrons.

essary for these halogenations. In contrast, such reactants add to the isolated double bonds found in alkenes. In this respect, benzene is more like an alkane.

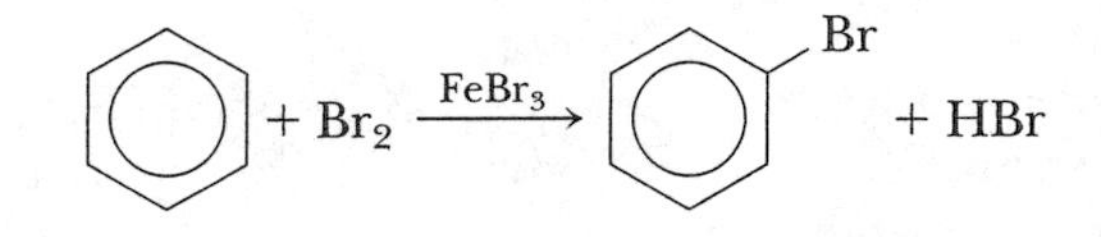

(In reactions of benzene, it is customary not to show the six hydrogen atoms attached to benzene's ring of carbon atoms, but remember that they are there.)

Benzene also undergoes a number of other important ring-substitution reactions, leading to many useful compounds (Table 21.9). Whole industries—especially drugs and dye manufacturing—have been built on these reactions, which were largely developed in Germany, Great Britain, and Switzerland in the late nineteenth century.

1. **Nitration**, with nitric acid yields nitrobenzene:

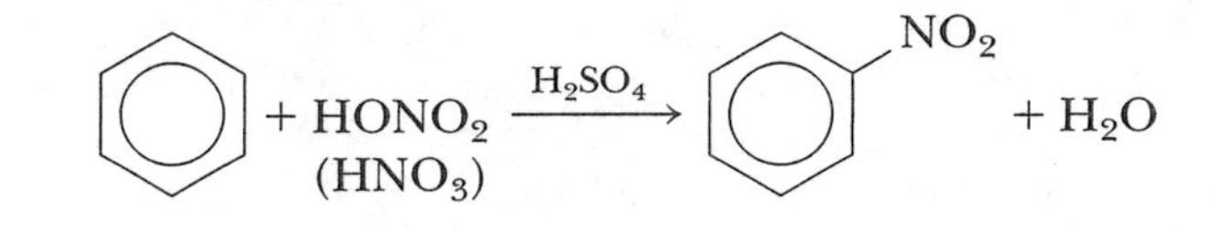

TABLE 21.9 Benzene and Its Derivatives

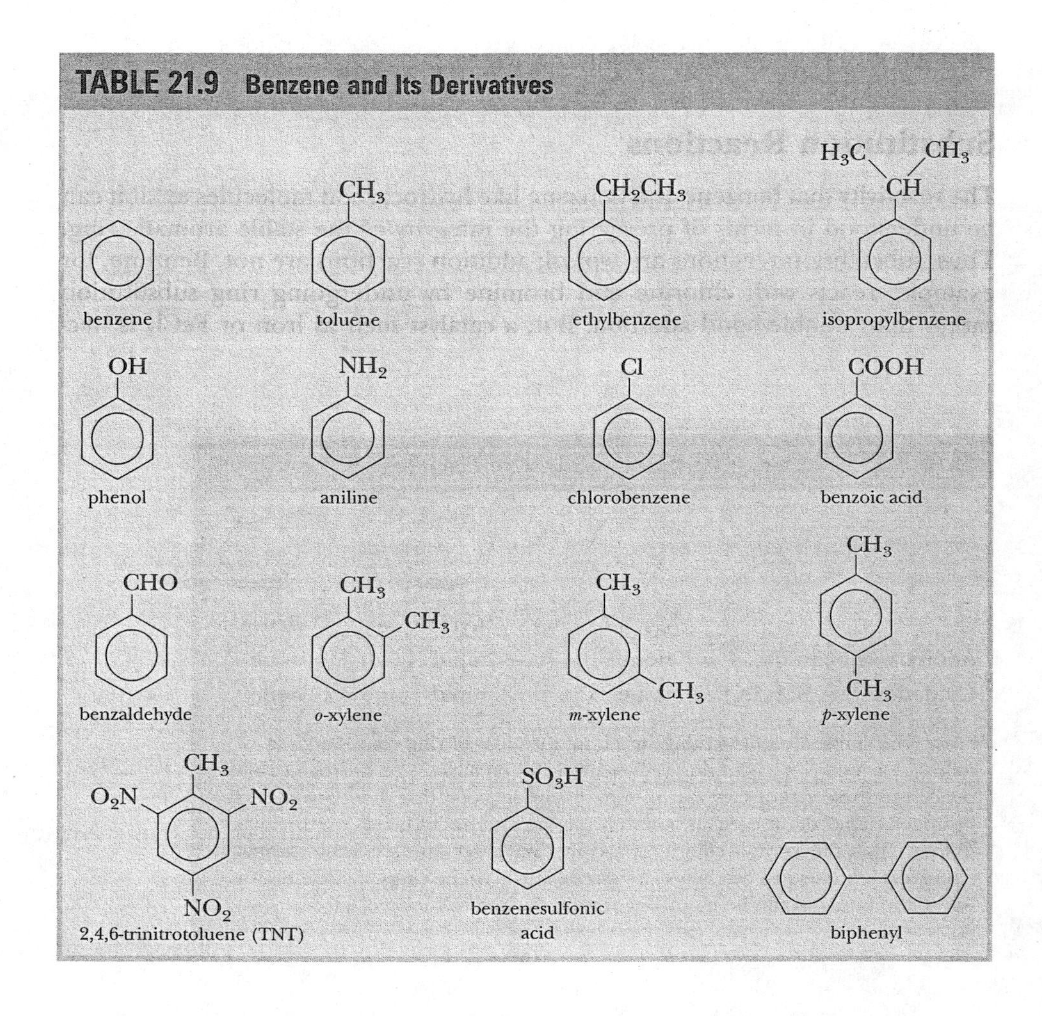

APPLICATIONS OF CHEMISTRY

Polycyclic Aromatic Hydrocarbons and Carcinogenesis Considerable research is currently being devoted to the properties of condensed or polycyclic aromatic hydrocarbons. They are among a small but growing number of compounds known to induce cancer in humans. It was well-known in the eighteenth century that soot was a factor leading to a high incidence of testicular cancer among chimney sweeps. Furthermore, workers in the coal-tar industry frequently developed skin cancer. Recent studies conducted with laboratory animals under carefully controlled conditions have shown that cancers can be caused by a polycyclic aromatic compound, 3,4-benzpyrene, which is isolated from coal tar, tars from cigarette smoke, and the soots found in heavily polluted urban environments. These observations have led to the synthesis and study of a large number of related compounds that have also turned out to be potent carcinogens.

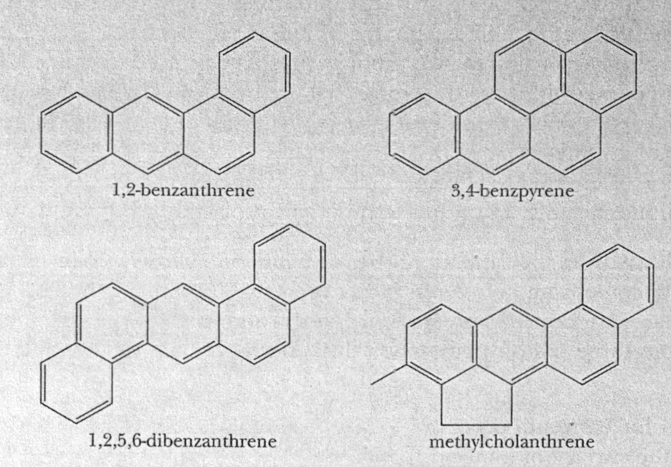

Question

Considering the structures of the polycyclic aromatic hydrocarbons known to be carcinogenic, a prudent person would be cautious in handling benzene, naphthalene, and anthracene in the laboratory. Why?

2. **Sulfonation**, with fuming sulfuric acid yields benzene sulfonic acid:

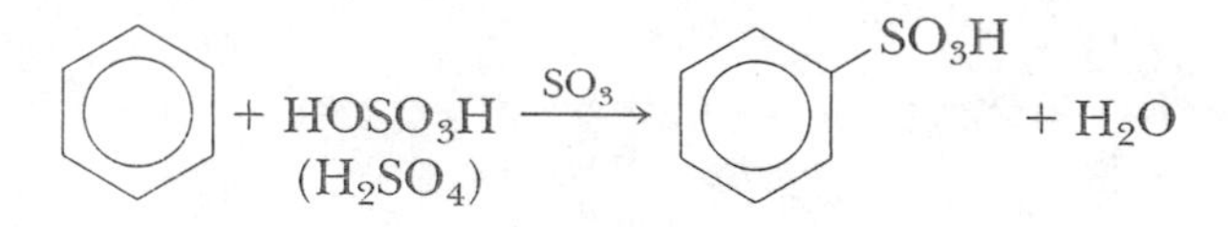

3. **Alkylation**, with alkyl halides, such as methyl chloride, yields the corresponding alkyl benzene. The reaction product is methyl benzene, better known as toluene:

$$C_6H_6 + CH_3Br \xrightarrow{AlBr_3} C_6H_5CH_3 + HBr$$

PROCESSES IN CHEMISTRY

Benzene: Its Principal Sources and Derivatives, Coal, Coke, and Coal Tar Coal is a complex mass of organic compounds derived from plants that have partially decayed over millennia due to the action of heat and pressure. The substances called peat, lignite, bituminous or soft coal, and anthracite or hard coal are successively more advanced stages in a metamorphosis leading to an ever-higher ratio of carbon to all other elements present. When bituminous coal is heated to temperatures near 1000°C in the absence of air in a process called destructive distillation, volatile products are given off and a residue of impure carbon, called coke, remains. On cooling, a fraction of the volatile products condenses to a black, viscous liquid called coal tar, leaving the uncondensed gases.

- **Coal gas** is a 1:1 mixture of hydrogen and methane along with trace compounds such as CO, H_2S, HCN, oxides of nitrogen, ammonia, and water vapor. With the impurities removed, it can be piped directly for use as illuminating gas, in domestic heating and as a source of industrial energy.
- **Coal tar** varies in composition according to the carbonization process used to create it in the first place. Table 21.10 lists the chief compounds and those of commercial importance. The percentages are approximate. Although the numbers may seem small, consider the huge quantities of coal that are processed annually, leading to nearly a billion gallons of coal tar in the United States.
- **Coke** is used for reduction of ores in blast furnaces and as a smokeless industrial fuel. As recently as the 1940s, it was a major fuel for home heating, but it has been largely replaced by oil and natural gas.

Benzene and Its Principal Derivatives Of the more than 800 million gallons of benzene produced annually in the United States, more than 85% come from petroleum rather than coal. Of the 500 million gallons of toluene (methyl benzene) produced, more than 95% came from petroleum, as did almost all of the 400 million gallons of xylenes (dimethyl benzenes). Most of the three billion pounds of ethylbenzene for use in the synthesis of styrene was synthesized from benzene.

- Because of the demand for benzene, processes for its production from toluene have been developed. One process takes place on the surface of a mixed Cr_2O_3—Al_2O_3—NaOH catalyst at 600°C and 55 atm:

$$\underset{\text{toluene}}{C_6H_5CH_3(g)} + H_2(g) \longrightarrow \underset{\text{benzene}}{C_6H_6(g)} + CH_4(g)$$

- Because the benzene ring has six equivalent hydrogen atoms positioned around the ring, positional isomers arise when two substituents are present. Xylenes (dimethylbenzenes) are important industrial solvents and intermediates. They can be conveniently distinguished by numbering the positions around the ring. Thus, we would have 1,2-, 1,3-, and 1,4-dimethylbenzene. However, the trivial names *ortho-* for 1,2-disubstituted benzenes, *meta-* for 1,3-disubstituted benzenes, and *para-* for 1,4-disubstituted benzenes are commonly used (Table 21.9).
- A typical example is *para*-dichlorobenzene, which is the substance in moth balls.

Question

Burning "coked" coal in an oxygen atmosphere produces carbon dioxide:

$$C(s) + O_2(g) \longrightarrow CO_2(g)$$

A second combustion product forms on further burning of the coke in the carbon dioxide–rich atmosphere. Write the balanced equation for this chemical reaction. Why is this product important industrially?

21.7 FUNCTIONAL GROUPS

The prospect of studying the chemical and physical properties of the 300,000 or so isomers of eicosane, $C_{20}H_{42}$, is beyond comprehension. However, most hydrocarbons and their derivatives—eicosane and its isomers included—follow a

TABLE 21.10 Principal Components of Coal Tar

Compound	Percentage
benzene	0.1
toluene	0.2
mixed xylenes	1.0
naphthalene	10.9
α- and β-methylnaphthalenes	2.5
dimethylnaphthalenes	3.4
acenaphthylene	1.4
fluorene	1.6
phenanthrene	4.0
anthracene	1.1
carbazole	1.1
coal tar bases/pyridine (0.1)	2.0
coal tar acids/phenol (0.7)	2.5

general pattern of physical and chemical behavior. One need only study the family characteristics to know a fair amount about any member of the family. When other atoms or groups of atoms such as Cl, OH, or NH_2 are introduced, the resulting compound tends to exhibit patterns of reactivity that are characteristic of that atom or group. Because these groups react as functional units, they have come to be called functional groups. A **functional group** is a group of atoms that imparts predictable characteristics to an organic molecule. We already encountered a few functional groups in Section 21.3. To know the chemistry of the variety of hydrocarbon derivatives containing a particular functional group, one need only generalize from the chemistry of a few examples of that functional group. A number of those most frequently encountered are given with specific examples showing the actual functional group in bold type:

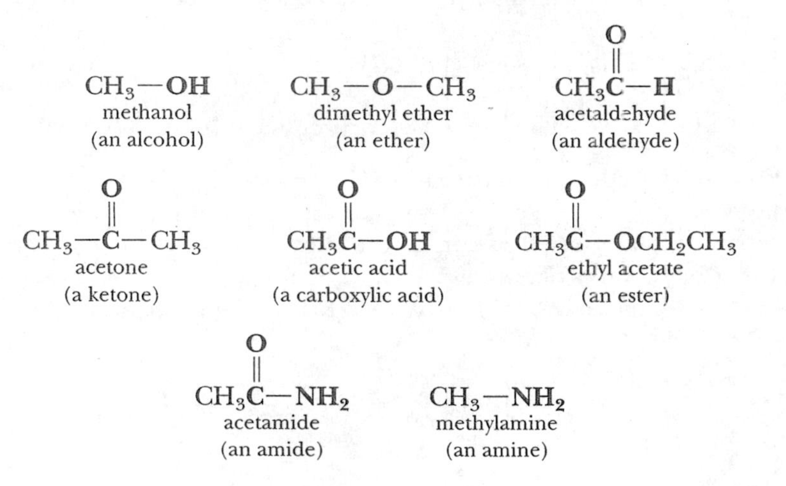

Alcohols and Ethers

Alcohols are compounds of the general formula R—OH, where R is any alkyl or substituted alkyl group, and OH is the **hydroxyl** functional group. Alcohols are named by adding the suffix -ol to the name of the parent hydrocarbon. The position of the —OH on a chain is given by the number of the carbon atom in that chain. Counting along the chain is done to minimize the number of the —OH position. The properties of the alcohols depend on how the hydrocarbon chain is arranged. They are classified as primary (1°), secondary (2°), or tertiary (3°) alcohols, depending on the number of carbon atoms bonded directly to the carbon atom to which the OH group is attached. Ethanol (ethyl alcohol) is a primary alcohol; 2-propanol (isopropyl alcohol) is a secondary alcohol; and 2-methyl-2-propanol (tertiary butyl alcohol) is a tertiary alcohol:

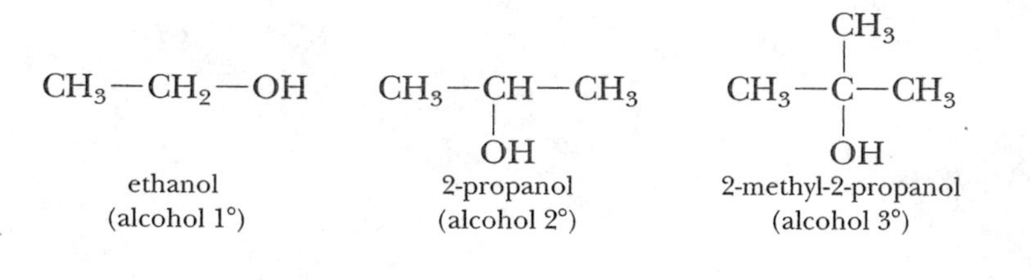

The physical properties of the lower molar-mass alcohols differ markedly from those of the corresponding hydrocarbons. They may be looked on as derivatives of both water and a hydrocarbon. Because of the presence of the polar OH group, there is considerable intermolecular hydrogen bonding. However, as the carbon chain becomes longer, the OH group becomes less important, and the character of the alcohol becomes increasingly hydrocarbon-like rather than water-like. Alcohols are not good proton donors and are only very weakly acidic.

One general method already noted for preparing alcohols is the sulfuric acid-catalyzed hydration (addition of water) to the double bonds of alkenes. Thus, adding water to propene produces 2-propanol. Hydrolysis of an alkyl halide with strong base will also result in alcohol formation. For example, methyl bromide or ethyl bromide reacts with aqueous potassium hydroxide to produce the corresponding alcohol as the hydroxide ion displaces bromide ion from carbon:

$$OH^- + CH_3Br \longrightarrow CH_3{-}OH + Br^- \qquad \text{(methanol)}$$

$$OH^- + CH_3CH_2Br \longrightarrow CH_3CH_2{-}OH + Br^- \qquad \text{(ethanol)}$$

Ethanol is produced industrially by the acid-catalyzed hydration of ethylene, but it gets its common name of grain alcohol from a rather different method of preparation, the fermentation of the sugar glucose with yeast. Starch from grain that has been broken down with enzymes is the most common source of glucose, and other enzymes in the yeast promote the production of ethanol and carbon dioxide:

$$\text{starch} \xrightarrow[\text{catalyst}]{\text{enzyme}} \underset{\text{glucose}}{C_6H_{12}O_6} \xrightarrow[\text{catalyst}]{\text{enzyme}} 2CH_3CH_2OH + 2CO_2$$

The chemistry behind the manufacture of beer and wine depends directly on this reaction. As you might anticipate, the carbon dioxide by-product is important in natural carbonation of these beverages. Beer contains about 4% alcohol; wines near 12%; gins, brandies, and whiskies are distilled to raise their alcohol content to 40 to 50% (80 to 100 proof). Ethanol is the only alcohol that can be consumed by the human body with a modicum of safety, though it, too, should be recognized for the toxic substance that it is. Fortunately, in the case of ethanol, the human body has the capacity to metabolize the substance, although at a limited rate. Contrary to popular belief, it is a depressant, not a stimulant, to the central nervous system.

Ethanol is an important industrial solvent for a variety of industries, ranging from paints, inks, and dyes to perfumes, cosmetics, and pharmaceuticals. It is an important intermediate in the manufacture of many other chemicals. As a fuel for internal combustion engines, ethanol can also be used either neat (pure) or as a gasoline extender in mixtures of various compositions.

Methanol, or wood alcohol, is prepared by reacting carbon monoxide with hydrogen in the presence of a mixed-metal oxide catalyst:

$$CO(g) + 2H_2(g) \xrightarrow{\text{catalyst}} CH_3OH(g)$$

The common name "wood alcohol" derives from the destructive distillation of wood, an earlier method of preparation. At about 250°C, in the absence of air, wood decomposes to charcoal and a volatile fraction that is 3% methanol. Widely used as a paint solvent, it is a particularly insidious poison, producing first blindness and eventually death on prolonged exposure to the liquid or vapor. These symptoms were reported for World War II submarine personnel who drank the methanol used as torpedo propellent. There has been considerable interest in methanol in recent years as an extender for gasolines, and even pure as a motor fuel. But perhaps most important, it is an intermediate in the synthesis of many other organic compounds that find their way into pharmaceuticals, agricultural chemicals, and the polymers of plastics for uses ranging from carpet fibers and paint brush bristles to office machines and television sets.

EXAMPLE 21.6

Until about 1965, methanol was produced on an industrial scale almost exclusively by the reaction of synthesis gas (CO and H_2) over a 3:1 mixture of ZnO and Cr_2O_3 at 350°C and 30 atm, under which conditions the conversion is 19.5%:

$$CO(g) + 2H_2(g) \rightleftharpoons CH_3OH(g) \qquad \Delta H = -94.5 \text{ kJ/mol}$$

(a) If the pressure drops to 10 atm, the conversion drops to 3.5%. Briefly explain.
(b) Will the conversion be favored (or inhibited) by a temperature increase? Briefly explain.

COMMENT Consider LeChatelier's principle. You will probably need to look up the conversion factor for pascals and atmospheres.

SOLUTION (a) High pressure favors the products and low pressure favors the reactants, according to LeChatelier's principle.
(b) The exothermic reaction is inhibited by high temperatures, again according to LeChatelier.

EXERCISE 21.6

The industrial synthesis of ethanol from ethylene and $H_2O(g)$, carried out in the gas phase over phosphoric acid on celite, a form of clay, takes place at 300°C and 7 atm and gives a 5% conversion. If the enthalpy of reaction is −44.1 kJ/mol, how would you go about varying temperature and pressure to increase the conversion?

Aromatic alcohols have the same general properties, except for the special case in which the hydroxyl group is substituted directly on one of the ring-carbon atoms. Such aromatic alcohols are called **phenols** (Table 21.11). The delocalization of the electrons in the ring of a phenol exerts a marked effect on the —OH group. One characteristic difference between alcoholic OH groups and phenolic OH groups is the marked acidity of the latter. The K_a of phenol, traditionally known as carbolic acid, is about 10^{-10}, whereas that of ethanol is about 10^{-16}. Thus, phenol is 10^6 times as strong an acid.

TABLE 21.11 Phenol and Phenolic Compounds

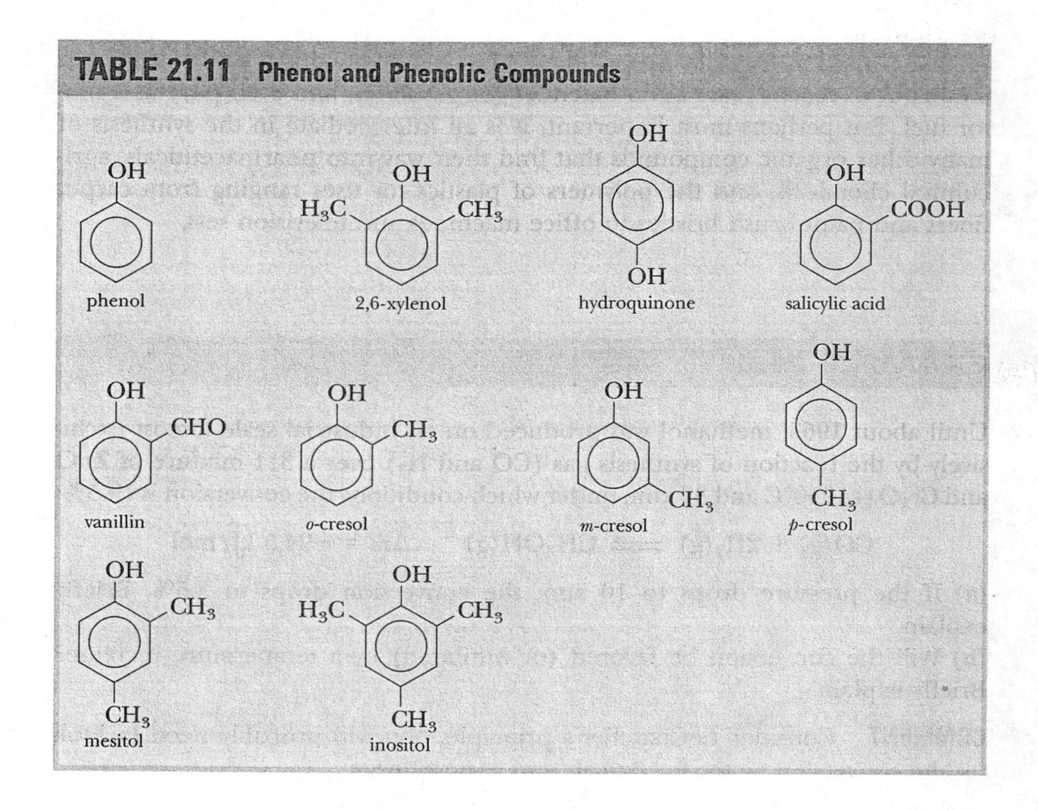

PROCESSES IN CHEMISTRY

Phenol, Acetone, and the Battle of Britain The availability of phenol from coal is limited by the demand for other coal-tar products and coal gas, and the demand outstripped availability long ago. Phenol is one of the highest volume organic chemicals manufactured worldwide. More than four billion pounds were manufactured in 1993. Fortunately, phenol is easily—and therefore economically—available by means of some chemical technology left over from World War II via the British Petroleum–Hercules cumene process. Cumene is a 92-octane aviation fuel developed by the British in the darkest days of World War II. In the battle for air supremacy over Britain, cumene gave their fighter aircraft a considerable advantage over the Axis powers. When the war ended, Britain had an enormous capacity for producing cumene and nothing to do with the product. But it was quickly found that cumene could be air-oxidized to cumene hydroperoxide (CHP), an unstable intermediate easily decomposed in the presence of trace quantities of acid to phenol and an important solvent called acetone:

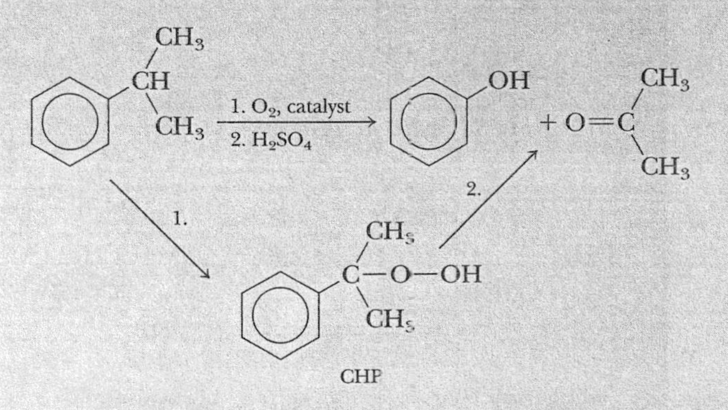

Phenols are important intermediates in the synthesis of resins for use in the manufacture of plastics, particularly Bakelite. They show up in the supermarket in alkylated form as antioxidants in foods and as surface-active agents in detergent compositions. Chlorinated phenols have been used as antiseptics, plant-growth regulators, and timber preservatives. A significant fraction of the worldwide synthesis of nylon comes from compounds obtained from phenol. Aspirin and photographic chemicals also depend on phenol (Table 21.12).

Question

Alkylated phenols are used as antioxidants, inhibiting spoilage of foods and expanding their shelf life, or preventing reactions of polymers and plastics at molding temperatures. 2,6-di-tertiary-butyl phenol is widely used for that application. Draw the structure.

Ethers are compounds of the general form R—O—R′, where R stands for an alkyl group or an aromatic group such as the phenyl C_6H_5- group and R′ is the same or another group. They are named by calling —O— "ether" and identifying the groups attached to it in the usual way.

dimethyl ether	CH_3—O—CH_3
diethyl ether	C_2H_5—O—C_2H_5
methylethyl ether	CH_3—O—C_2H_5
methylphenyl ether	C_6H_5—O—CH_3

TABLE 21.12 Industrial Uses of Phenolic Compounds

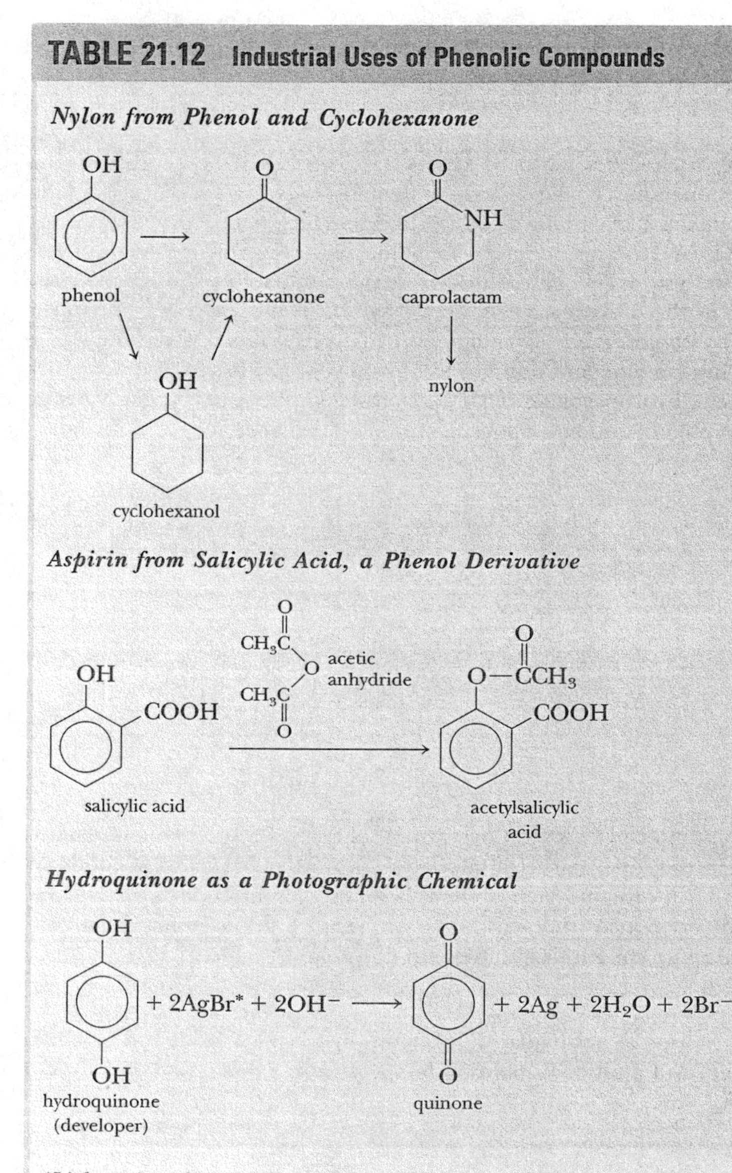

*Light-activated silver bromide crystals in the AgBr emulsion of the photographic film. The developer is oxidized to a quinone and the silver is reduced to the metal, which precipitates on the emulsion. The unactivated AgBr is removed with a "fixer" solution, often aqueous thiosulfate ($S_2O_3^{2-}$), leaving the precipitated silver as the black part of the "black-and-white" negative.

Low molar-mass ethers are volatile liquids, prepared by the loss of a molecule of water from two molecules of alcohol. This dehydration happens during distillation in the presence of sulfuric acid under carefully controlled conditions:

$$2R\text{—}OH \xrightarrow{H_2SO_4} R\text{—}O\text{—}R + H_2O$$

The temperature is critical. For example, the preparation of diethyl ether takes place smoothly at 140°C. But at 180°C dehydration of the alcohol to ethylene is the preferred reaction:

$$\begin{array}{c} CH_3{-}CH_2{-}OH \\ + \\ CH_3{-}CH_2{-}OH \end{array} \xrightarrow[140°C]{H_2SO_4} \begin{array}{c} CH_3{-}CH_2 \\ \\ CH_3{-}CH_2 \end{array}\! \phantom{CH_3{-}CH_2}\;\;\; O + H_2O$$

$$\downarrow H_2SO_4,\ 180°C$$

$$2CH_2{=}CH_2 + 2H_2O$$

In an earlier discussion (Chapter 15), we noted these competing reactions as an example of the competition between kinetic and thermodynamic control of reaction pathways.

Diethyl ether, commonly called just plain "ether," is the best-known member of the family because of its early use as an anesthetic and for its wide laboratory use as a solvent. Ether is a notorious fire hazard because of its very high vapor pressure (b.p. 35°C), vapor density, and flammability. Having a density greater than air, ether fumes can roll along bench tops and over floors to be ignited by burners, hot surfaces, and electrical equipment long distances from the source. Its use as a general anesthesia after the Civil War changed the practice of surgery.

Aldehydes and Ketones

The intermediate products in the controlled oxidation of primary alcohols are **aldehydes**. When secondary alcohols are oxidized, **ketones** result. The general formulas of aldehydes and ketones are shown in Table 21.13. Aldehyde and ketone chemistry is predominantly the chemistry of the **carbonyl group**, a carbon atom double bonded to an oxygen atom:

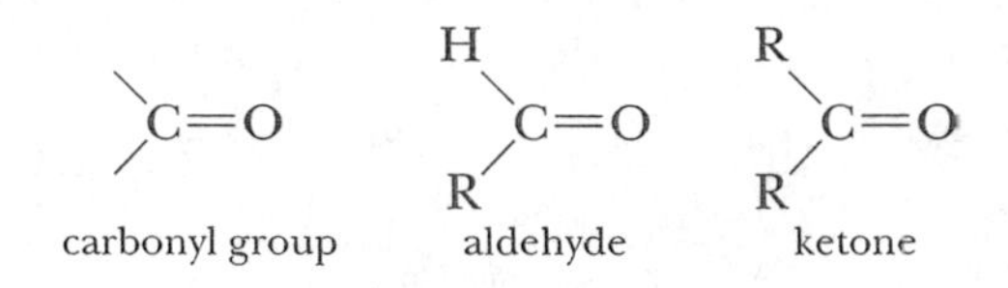

carbonyl group aldehyde ketone

The properties of these two classes of compounds are similar, but differences do arise because of the hydrogen atom on the aldehyde that is not present in the ketone. The aldehyde hydrogen atom leads to a marked susceptibility toward further oxidation to carboxylic acids (next section), whereas under similar conditions the further oxidation of the ketone does not occur:

$$\underset{\text{(ethanol)}}{\underset{1°\text{ alcohol}}{CH_3CH_2OH}} \xrightarrow{\text{oxidized}} \underset{\text{(acetaldehyde)}}{\underset{\text{aldehyde}}{CH_3CHO}} \xrightarrow{\text{oxidized}} \underset{\text{(acetic acid)}}{\underset{\text{carboxylic acid}}{CH_3COOH}}$$

$$\underset{\text{(2-propanol)}}{\underset{2°\text{ alcohol}}{(CH_3)_2CHOH}} \xrightarrow{\text{oxidized}} \underset{\text{(dimethyl ketone)}}{\underset{\text{ketone}}{(CH_3)_2CO}} \xrightarrow{\text{oxidized}} \text{no reaction}$$

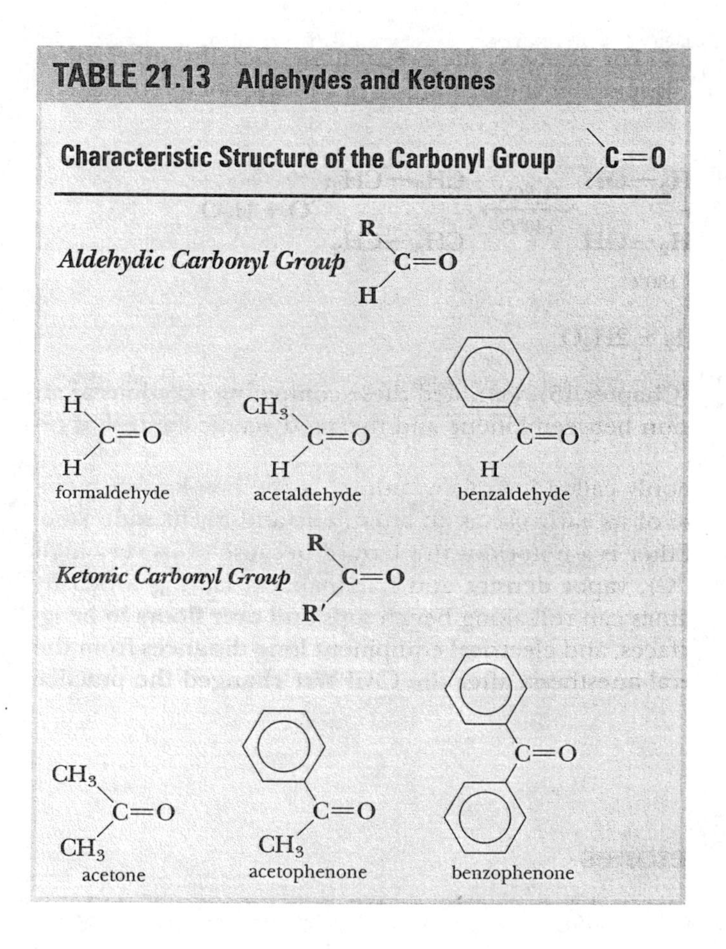

Formaldehyde (HCHO), the simplest aldehyde and the only one with two hydrogen atoms on the carbonyl group, is prepared by the controlled, air oxidation of methanol over a mixed oxide catalyst at about 600K.

$$2CH_3OH(g) + O_2 \longrightarrow 2H{-}\underset{\displaystyle H}{\underset{|}{C}}{=}O(g) + H_2O(g) \qquad \Delta H = -154\text{ kJ}$$

At room temperature, formaldehyde is a gas with a particularly irritating and offensive odor. Because of its high solubility in water, it is often handled as a 40% aqueous solution known as formalin. It is widely used in the synthesis of low-cost, durable plastic materials, the best known of which is Formica. It is also used to make a polyether polymer called Delrin that has largely replaced a number of metals in light engineering applications, including outdoor, in-the-ground plumbing. In its clinical applications, formaldehyde is a general antiseptic, a widely used preservative agent, and the major active ingredient in most embalming fluids because it stiffens protein.

Most of the other lower molar-mass aldehydes and ketones are liquids with pleasant odors. Acetaldehyde can be synthesized by the controlled oxidation of ethanol:

$$2CH_3CH_2OH(g) + O_2(g) \longrightarrow 2CH_3CHO(g) + 2H_2O(g)$$

The trick is to have the more volatile aldehyde reaction product distill as it forms so as not to be subjected to further oxidation to acetic acid.

One of the characteristic reactions of aldehydes is their oxidation by a silver-ammonia complex known as Tollen's reagent, which is reduced to silver. The silver deposits as a telltale mirror on the walls of the reaction container:

$$RCHO + 2Ag(NH_3)_2^+ + 2H_2O \longrightarrow RCOO^- + 2Ag(s) + 4NH_4^+ + OH^-$$

Ketones can be prepared from alcohols by direct oxidation. For example, oxidation of 2-propanol leads to dimethyl ketone, better known as acetone:

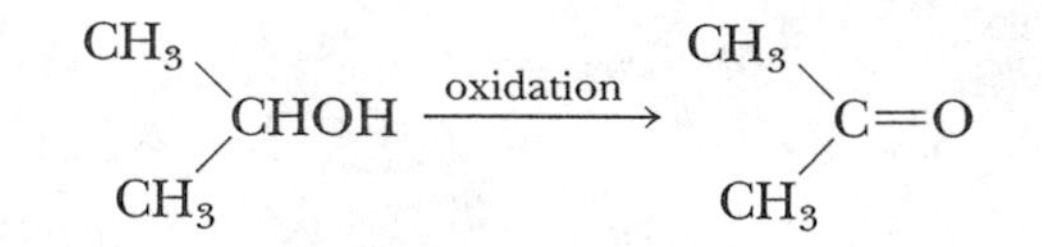

However, the commercial synthesis of acetone, at a rate of 2.5 billion pounds per year, is as a coproduct with phenol in the cumene process. Acetone is an industrial solvent, chemical intermediate, and common laboratory reagent.

In the human body, acetone shows up in the blood in very small concentrations as a by-product of certain metabolic pathways. Diabetics produce excessive amounts of acetone, which can be detected in the urine, and in severe cases the characteristic odor of acetone is obvious on the patient's breath.

Carboxylic Acids, Esters, and Amines

Carboxylic acids are weak acids characterized by the presence of the —COOH functional group. Only slightly dissociated in aqueous solution, the K_a for acetic acid is about 10^{-5}. The acidity of carboxylic acids lies between that of phenol and inorganic acids such as HF and H_3PO_4. Formic acid (HCOOH) and acetic acid (CH_3COOH), the first members of the family of organic acids, are soluble in water. But as was the case with alcohols, extension of the carbon chain results in a diminishing contribution by the functional group and the solubility in water begins to drop off rapidly. The low molar-mass carboxylic acids are liquids with sharp, unpleasant odors (Table 21.14):

- acetic acid [CH_3COOH] — vinegar
- butyric acid [$CH_3CH_2CH_2COOH$] — rancid butter
- caprylic [$CH_3(CH_2)_6COOH$] and caproic [$CH_3(CH_2)_4COOH$] acids — essence of goat (on a humid day)

Beyond ten carbons in the chain, the carboxylic acids are all waxy solids with low vapor pressures. One of the most fascinating uses to which nature puts the carboxylic acids, especially formic acid, is as the active component in the chemical defense mechanisms of a variety of insects such as ants, cockroaches, beetles, and bees. Industrial applications include soaps and detergents.

Carboxylic acids are mainly synthesized commercially by the oxidation of intermediates such as aldehydes. For example, acetic acid can be prepared from acetaldehyde by a catalyzed oxidation:

$$\underset{\text{acetaldehyde}}{CH_3CHO} \xrightarrow[Mn^{2+}]{O_2} \underset{\text{acetic acid}}{CH_3COOH}$$

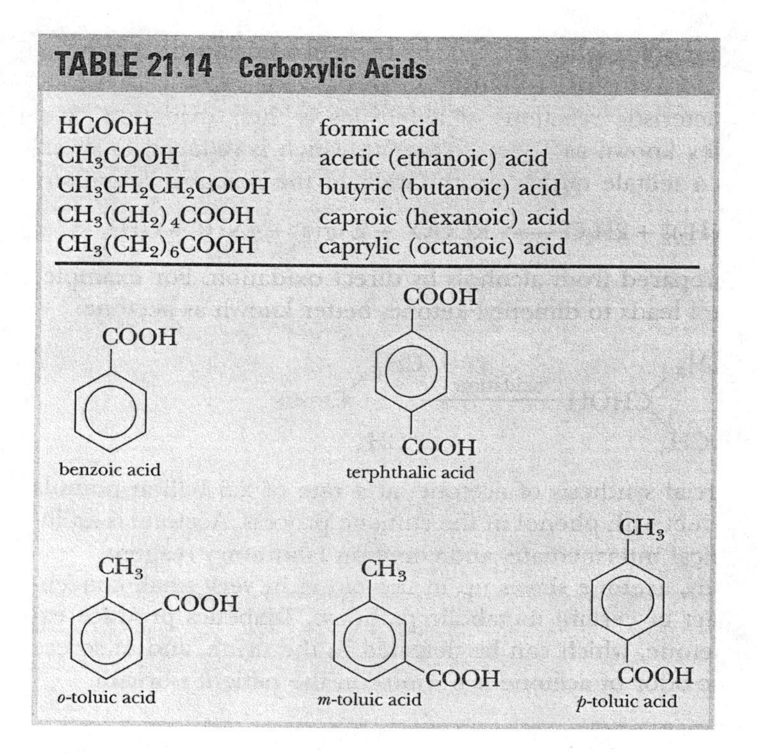

TABLE 21.14 Carboxylic Acids

Formula	Name
$HCOOH$	formic acid
CH_3COOH	acetic (ethanoic) acid
$CH_3CH_2CH_2COOH$	butyric (butanoic) acid
$CH_3(CH_2)_4COOH$	caproic (hexanoic) acid
$CH_3(CH_2)_6COOH$	caprylic (octanoic) acid

Benzoic acid can be prepared from toluene:

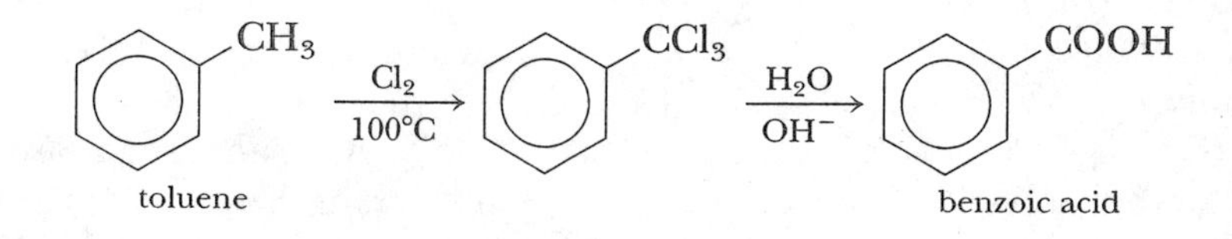

The most important reactions of carboxylic acids involve their ready conversion into one of a number of other functional group derivatives (Table 21.15). One can generalize the structures in Table 21.15 in terms of the characteristic *acyl* group that is common to all:

$$R{-}\underset{|}{C}{=}O \qquad \text{(acyl group)}$$

The fourth bond to the carbonyl carbon in carboxylic acids and their derivatives is to an electronegative atom—halogen, N, or O—rather than to C or H, as in aldehydes and ketones. The acid halides are used primarily as precursors to amides and esters. Acid chlorides are prepared by substituting Cl for the OH in the carboxyl group. $SOCl_2$ or PCl_5 are the reagents commonly employed:

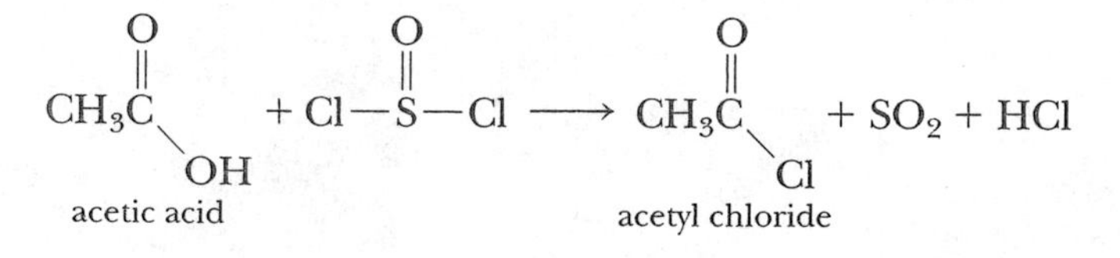

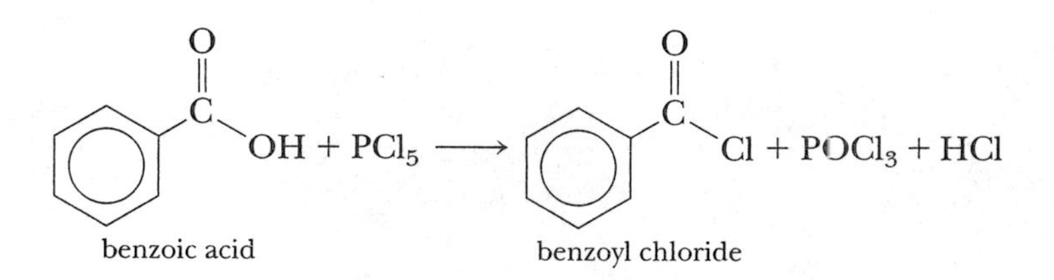

Esters can be prepared by reaction of an acid chloride with an appropriate alcohol:

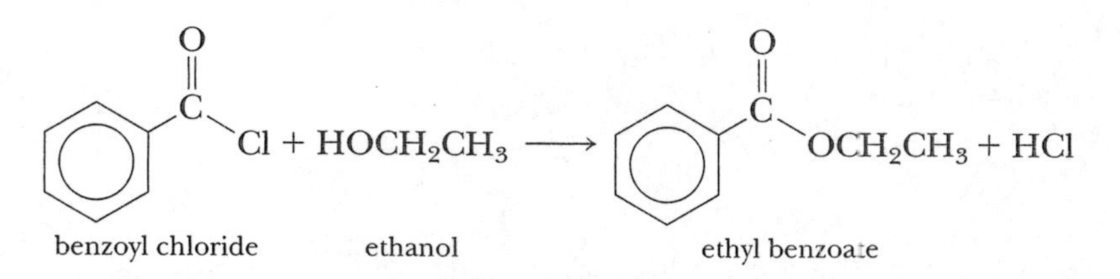

Alternatively, the ester can be prepared by direct reaction of alcohol and acid, but the equilibrium usually is not as favorable for preparing esters as in the acid chloride reaction. The direct reaction is catalyzed by acids:

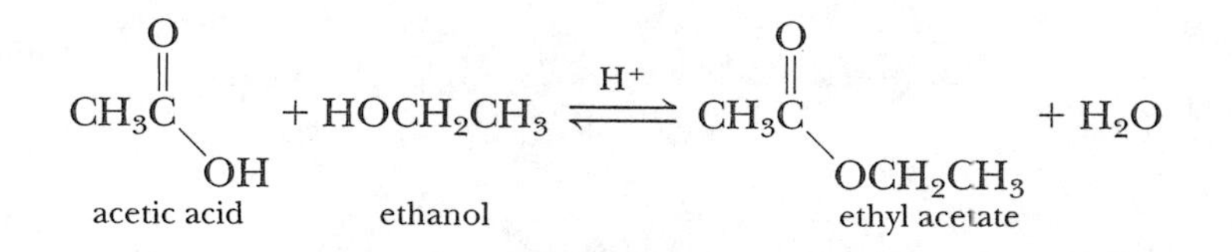

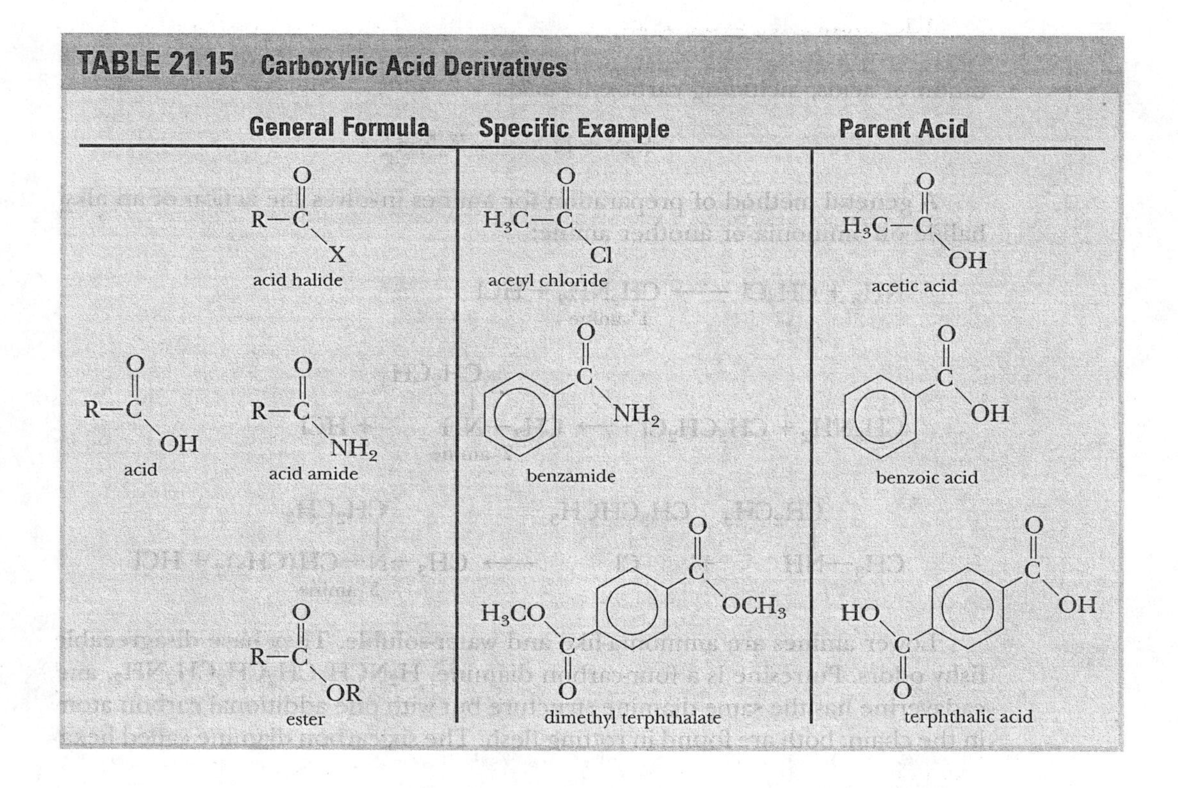

TABLE 21.15 Carboxylic Acid Derivatives

	General Formula	Specific Example	Parent Acid
	$R-C(=O)-X$ acid halide	$H_3C-C(=O)-Cl$ acetyl chloride	$H_3C-C(=O)-OH$ acetic acid
$R-C(=O)-OH$ acid	$R-C(=O)-NH_2$ acid amide	$C_6H_5-C(=O)-NH_2$ benzamide	$C_6H_5-C(=O)-OH$ benzoic acid
	$R-C(=O)-OR$ ester	$H_3CO-C(=O)-C_6H_4-C(=O)-OCH_3$ dimethyl terphthalate	$HO-C(=O)-C_6H_4-C(=O)-OH$ terphthalic acid

Many esters are pleasant-smelling substances used as perfumes and flavoring agents. Among the more important applications of compounds characterized by the presence of the ester group are industrial solvents and the universal analgesic known as aspirin (sodium acetylsalicylate), the sodium salt of acetylsalicylic acid:

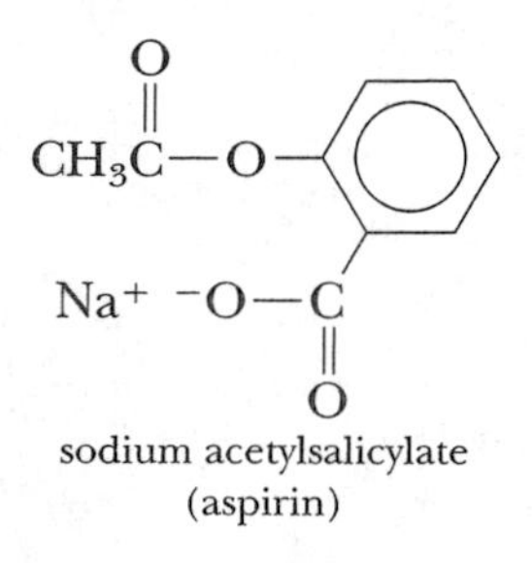

sodium acetylsalicylate
(aspirin)

It is useful to think of **amines** as being derived from ammonia by replacing one or more of the ammonia N—H bonds by an N—C bond. As was true for alcohols, amines may be classified as primary (1°), secondary (2°), and tertiary (3°), according to the number of carbon atoms attached to nitrogen.

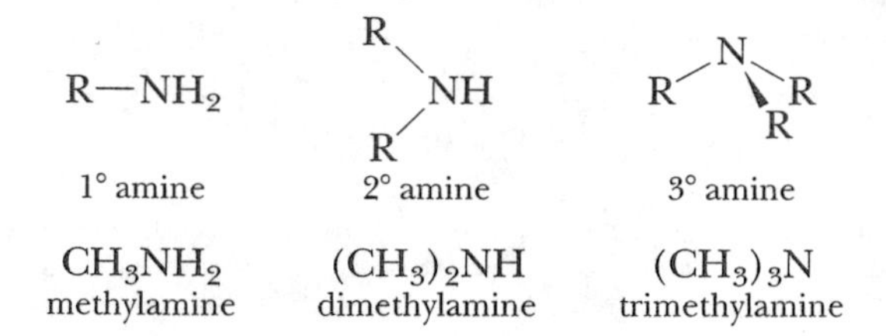

The amines are weak bases with strengths generally on the order of that of NH_3, for which $K_b = 1.8 \times 10^{-5}$. They form amine and ammonium salts on addition of acids, including carboxylic acids:

$$R_3N + H^+ \longrightarrow R_3NH^+$$

A general method of preparation for amines involves the action of an alkyl halide on ammonia or another amine:

$$NH_3 + CH_3Cl \longrightarrow \underset{\text{1° amine}}{CH_3NH_2} + HCl$$

$$CH_3NH_2 + CH_3CH_2Cl \longrightarrow \underset{\text{2° amine}}{CH_3\text{—}\overset{\displaystyle CH_2CH_3}{\overset{|}{N}}H} + HCl$$

$$CH_3\text{—}\overset{\displaystyle CH_2CH_3}{\overset{|}{N}}H + CH_3\overset{}{\underset{\displaystyle Cl}{\underset{|}{C}}}HCH_3 \longrightarrow \underset{\text{3° amine}}{CH_3\text{—}\overset{\displaystyle CH_2CH_3}{\overset{|}{N}}\text{—}CH(CH_3)_2} + HCl$$

Lower amines are ammonia-like and water-soluble. They have disagreeable fishy odors. Putresine is a four-carbon diamine, $H_2NCH_2CH_2CH_2CH_2NH_2$, and cadaverine has the same diamine structure but with one additional carbon atom in the chain; both are found in rotting flesh. The six-carbon diamine called hexamethylenediamine is used on a large scale for the synthesis of nylon.

Amides form when ammonia or an amine is caused to react with an acid chloride:

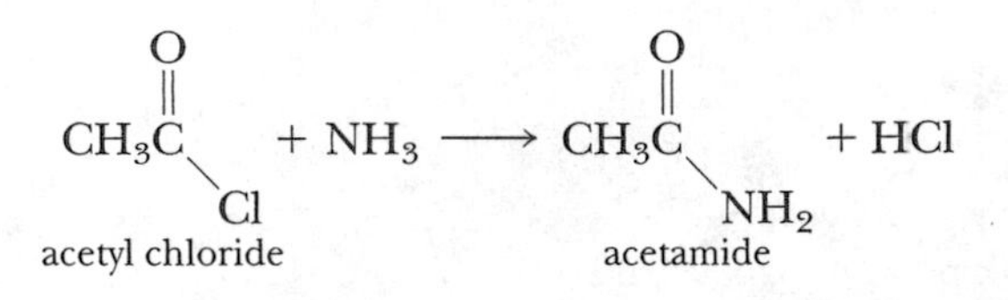

They can also be prepared by making the ammonium salt of the carboxylic acid followed by heating to drive off a molecule of water.

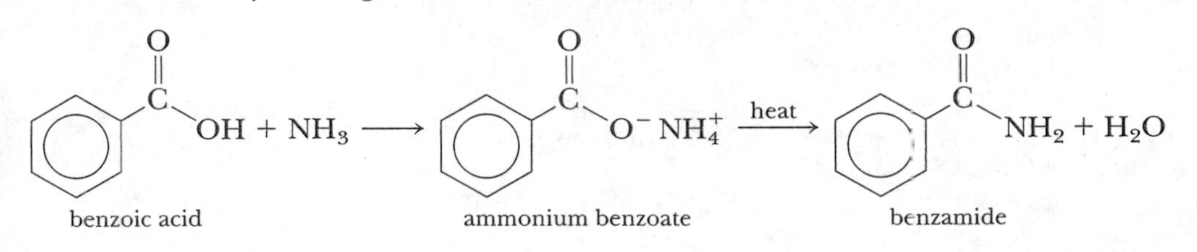

EXAMPLE 21.7

Why do you think the direct reaction of ammonia or amines with alkyl halides might be of limited preparative value?

COMMENT Consider the possibility of competing reactions.

SOLUTION The reaction is not specific to a single reaction product, because primary, secondary, and tertiary products can all form.

EXERCISE 21.7

If you are using the direct reaction of ammonia with an alkyl halide to produce an amine, how would you go about minimizing the amounts of secondary and tertiary amine products? How would you produce largely the tertiary amine product?

Amide and Ester Polymerization Reactions

If one of the hydrogen atoms on the methyl group of an acetic acid molecule is replaced by a $-NH_2$ functional group, the new compound is α-aminoacetic acid, a difunctional molecule better known as glycine:

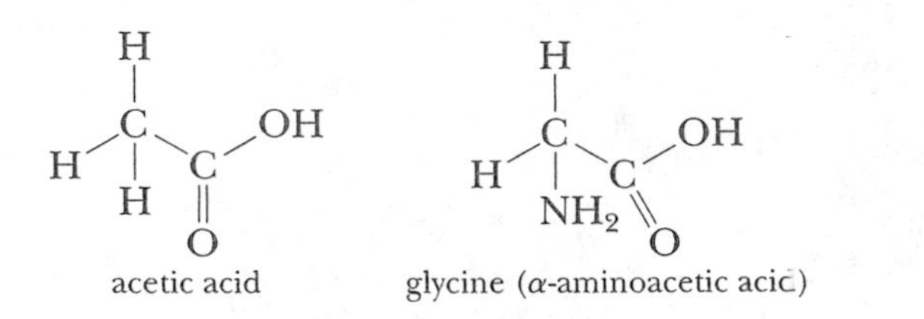

(text continued on p. 978)

PROCESSES IN CHEMISTRY

Cellulose and the Kraft Process for Paper Wood is a major material of construction and protection, and an important resource for the production of paper and paper products. The main components of wood are cellulose and lignins. The cellulosic materials are polysaccharides—polymerized sugars—that hydrolyze in aqueous acid to simple sugars such as glucose and fructose. Lignins are the compounds that endow wood with its unique structural properties by binding the cellulosic fibers together, but they must be separated from the cellulosic materials in manufacturing paper. In the Kraft process this is usually done in a digester (cooker) containing mostly Na_2S, NaHS, and NaOH after chipping logs into wood chips. The cooking converts the chips into a pulp and degrades the lignins, which are then separated. After washing and neutralizing—a critical step, because residual acid affects the lifetime of the paper product—the pulp is bleached to enhance its brightness. Before recycling the spent cooking liquors, they are neutralized, concentrated, and treated to remove the tall oils and rosin acids, which have commercial values.

Another by-product of considerable commercial value derives from mercaptans (RSH), the sulfur analogs of alcohols, and disulfides (R_2S), the sulfur analogs of ethers. As is the case for so many organo-sulfur compounds, these substances are exceedingly unpleasant smelling and account for the characteristic odors often associated with pulp and paper plant sites. They are also the same classes of compounds associated with the essence of agitated skunks.

An organic sulfide of special importance is the simplest, dimethyl sulfide. Approximately 2 kg of dimethyl sulfide/ton cellulose can be recovered from the exhaust gases and then oxidized to produce a versatile solvent known as dimethyl sulfoxide (DMSO), a substance that has been strangely effective at treating arthritis. But it must also be handled carefully because of its ability to dissolve many potentially toxic substances and transport them through otherwise protecting layers of skin into the human body. The oxidation process is catalyzed by nitrogen dioxide, which is regenerated on reaction of nitrogen monoxide with oxygen:

$$(CH_3)_2S(g) + NO_2(g) \longrightarrow (CH_3)_2S{=}O(\ell) + NO(g) \qquad \text{synthesis of DMSO}$$

$$NO(g) + O_2(g) \longrightarrow NO_2(g) \qquad \text{regenerating the catalyst}$$

Treating cellulose from wood pulp with concentrated solutions of nitric and sulfuric acids produces a variety of nitrated cellulose products used as lacquers for coatings and finishes, for printing inks, in bookbinding applications, and as rocket propellants and explosives. In 1869, John Wesley Hyatt (American, 1837–1920) discovered that mixtures of cellulose nitrate and camphor could be molded and hardened. Under the trademark "Celluloid," this first commercial plastic material was fabricated into everything from billiard balls to shirt collars. However, its extreme flammability made it a hazardous material. For example, silk-like fabrics woven from these cellulosic fibers became known as "mother-in-law's silk." Early motion picture film used celluloid as a base material, and many cinema epics were totally destroyed by fire in studio warehouses.

Questions

Suggest an oxidizing agent other than the nitrogen dioxide to carry out the conversion of dimethyl sulfide to dimethyl sulfoxide, based on chemistry learned in this chapter. Write the reaction.

Can you suggest one major drawback to the use of nitric acid in the Kraft process for making pulp for paper?

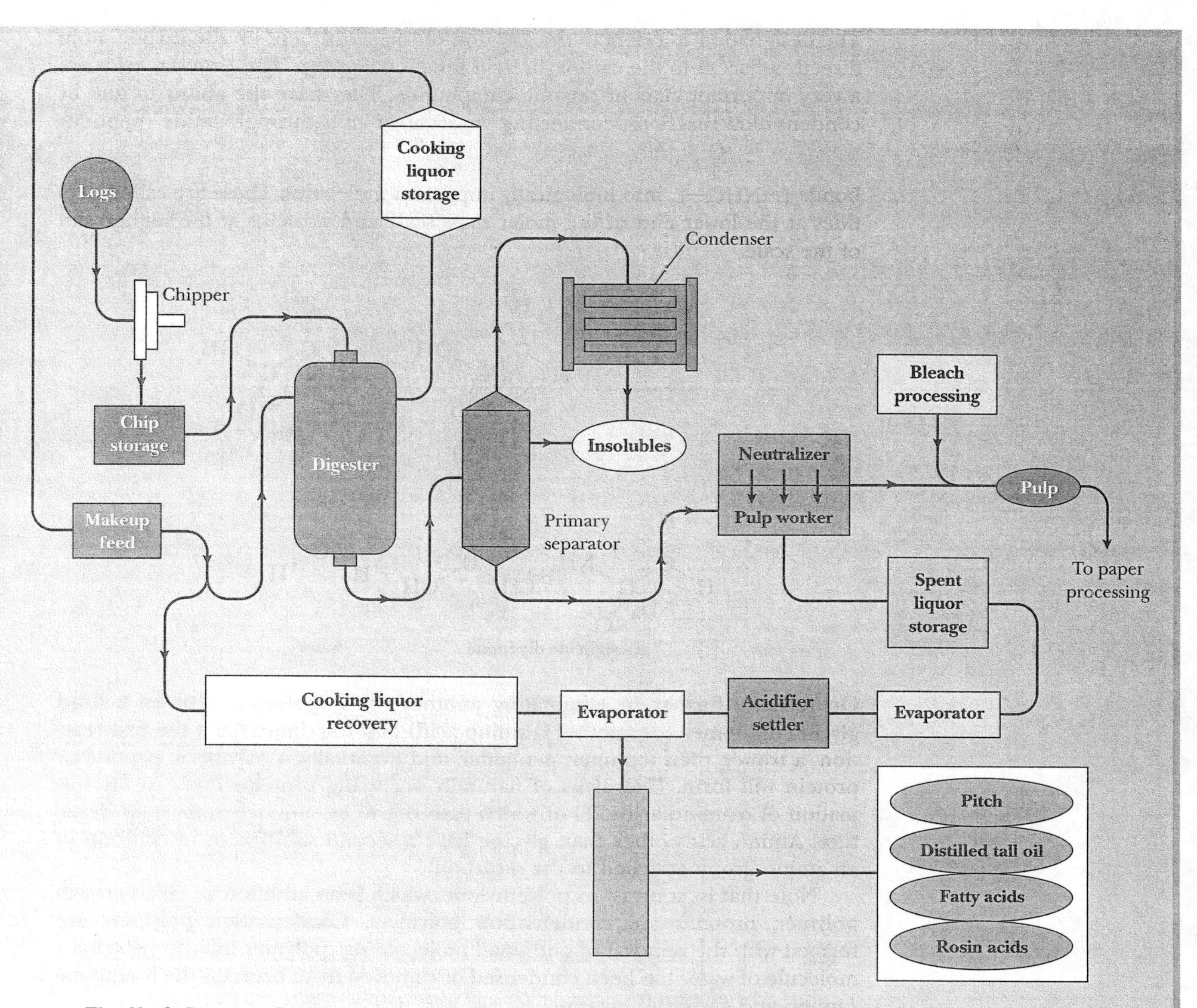

The Kraft Process. Begins with logs and a chipper. The digester, or cooker, takes the complex cellulosic material and separates it from the lignins that bind these carbohydrate polymers together, giving wood its unusual strength. Pulp for paper is the principal product, along with secondary (by-product) chemicals, and a lot of materials in need of recycling. Historically, the pulp and paper industry has been responsible for many severe environmental problems associated with rivers and lakes downstream because of their ravenous appetite for processing water and their release of toxic wastes into the environment.

The designation α- refers to the location of the substituent on the carbon atom directly adjacent to the carboxylic acid functional group. The α-amino acids are a very important class of organic compounds. They have the ability to link by condensation reactions, connecting consecutive units through amide (peptide) bonds ($-\mathrm{NH}\overset{\overset{\displaystyle O}{\|}}{C}-$), into biologically important molecules. These are called peptides at the lower end of the molar mass scale and proteins at the higher end of the scale:

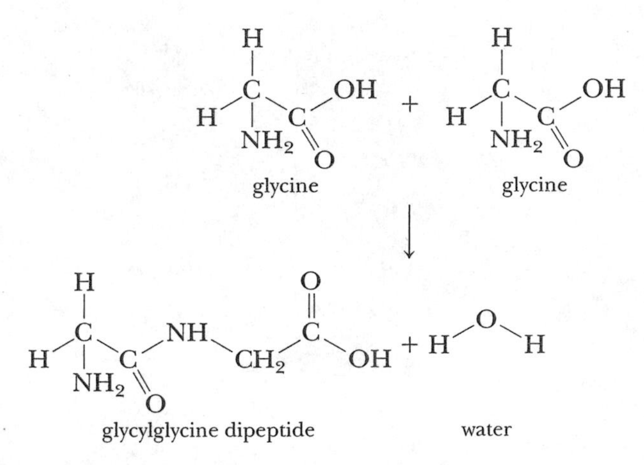

On reacting further by eliminating another water molecule between a third glycine monomer (or another α-amino acid) and the dimer from the first reaction, a trimer, then tetramer, pentamer, and eventually, a polymeric peptide or protein will form. Hydrolysis of naturally occurring proteins leads to the formation of α-amino acids, 20 of which turn out to be uniquely important in nature. Amino acids other than glycine have a second substituent in addition to an amino group attached to the α-carbon.

Note that in contrast to polyethylene, which is an addition or chain-growth polymer, proteins are condensation polymers. **Condensation polymers** are formed with the removal of one small molecule per polymer link. In proteins a molecule of water has been condensed or removed from between the functional (amino and carboxyl) groups.

Polypeptides are polyamides, the amide functional group being the characteristic connecting unit between the monomers. The synthetic polyamide known as nylon-6,6 is prepared by eliminating water molecules between hexamethylenediamine and adipic acid monomer units. Both raw materials are themselves obtained synthetically from cheap sources such as oat hulls, corncobs, and petroleum chemicals. Blending the difunctional amine with the difunctional acid produces a salt. When heated to about 250°C, the salt yields a condensed amide molecule containing all the original atoms of the two monomers, less one molecule of water per amide bond. These condensed amide molecules themselves condense; about 45 to 50 units linking head to tail, releasing water molecules at each condensation site, and forming a polymer strand of about 10,000 molecular mass:

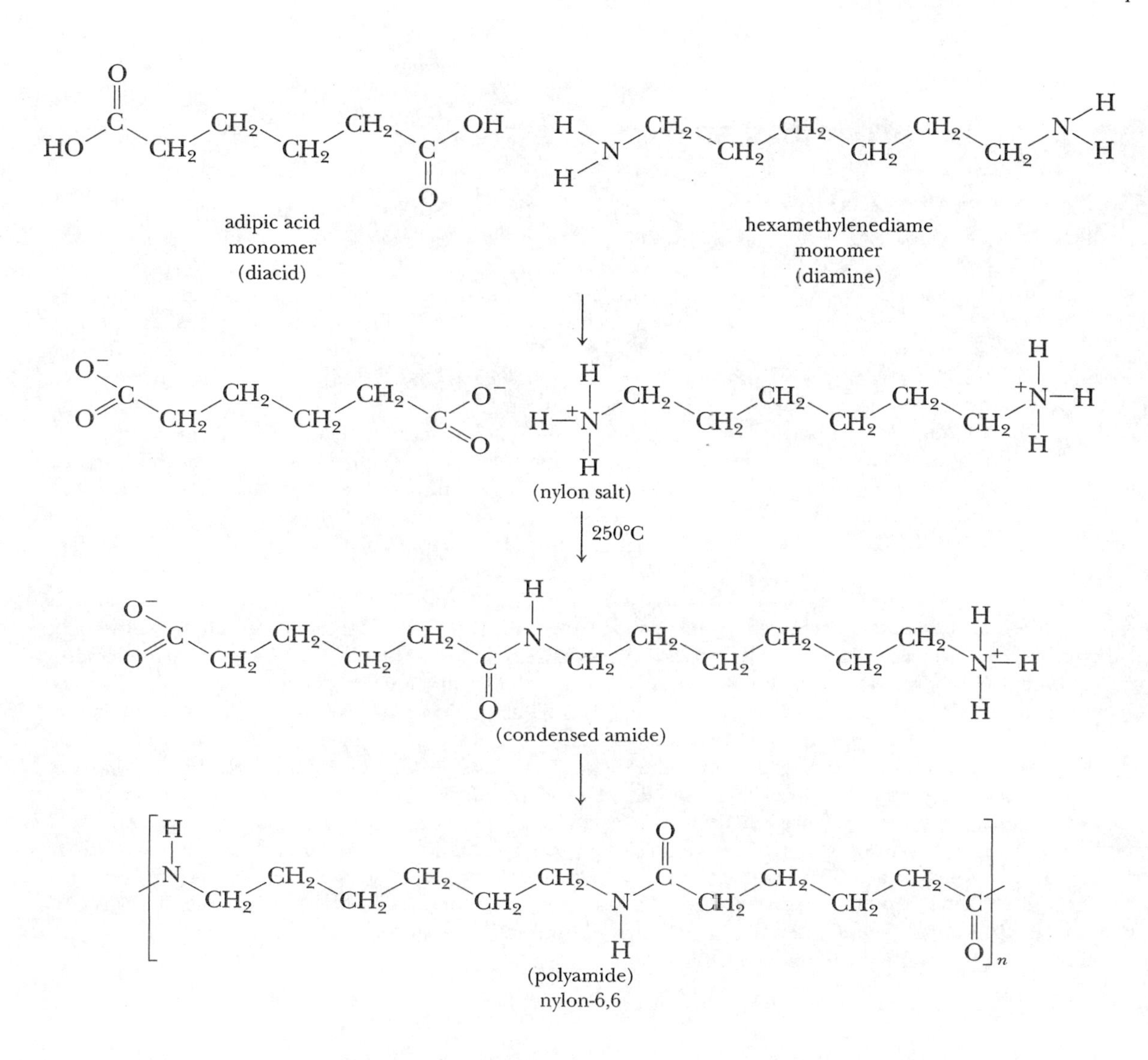

One of the best-known polyamide fibers is Kevlar, the DuPont trade name for its aromatic polyamide. Because of its extremely high tensile strength, greater resistance than steel to elongation, and high-energy absorption, it is an excellent material for belting radial tires, for which it was originally developed, in applications requiring high-heat deflection and in bulletproof vests.

Polyesters

Dacron fibers and Mylar film are made of the same condensation polymer, poly(ethylene terphthalate). It is prepared from the ester, dimethyl terphthalate, and ethylene glycol by a process called transesterification:

PROFILES IN CHEMISTRY

Macromolecules and Chemical Bonds In the 1920s, two chemists—one in Europe, the other in America—brought to light one of the most significant advances in our understanding of chemical bonding. In its own way, their work was as significant as Dalton's introduction of the atomic theory and Planck's quantum of action. Working in Zurich and Freiburg, Hermann Staudinger (German, 1881–1965) championed the idea that molecules composed of normal covalent bonds could indeed be grown to high molecular masses. At the time, it was thought that molecules such as peptides, the largest molecules then known, could not have molecular masses beyond a limit of about 5000 amu. Staudinger showed that not only was it possible to produce molecular masses well beyond that arbitrary limit but that their physical properties were often a function of molecular mass.

After the work of Staudinger, the polymers of plastics were no longer to be understood as collections or aggregations of smaller units. Rather, the regularly repeating monomer units were best understood as chemically bonded into chains by normal covalent bonds, the same kind that held hydrogen, nitrogen, oxygen, and carbon atoms together in simple molecules. These were molecules with molecular masses commonly in excess of 10,000 amu and chain lengths on the order of 10,000 Å, rather than 10 Å. These were true macromolecules, or polymers. Staudinger literally forced the scientific community to change its view of what was meant by the word "molecule."

Wallace Carothers (1896–1937) began his industrial career in 1927 at the Du Pont Company, the year after it launched its program of fundamental research. In the last decade of his short life, Carothers was able to synthesize a large number of important macromolecules with interesting properties. He found that these polymeric (macromolecular) structures could be fashioned into materials that in many cases had marvelous engineering properties, suitable for replacing metal, wood, and glass, our traditional materials of construction and protection. He studied reaction mechanisms involved in their synthesis and the thermodynamics of these macromolecular processes, laying the foundation for the plastics revolution that began after World War II and has yet to cease. His work showed that low molar-mass substances could react to form high molecular-mass polymers. By careful selection of difunctional monomers as reactants, molecules with great length compared to their cross-section (fiber-forming polymers) could be obtained. Furthermore, Carothers's studies were entirely consistent with and complementary to Staudinger's work in Europe at the same time. In a brilliant professional career that lasted fewer than ten years, he and his research group brought to the world neoprene, the first commercially successful synthetic rubber, and nylon, the first synthetic textile fiber and the granddaddy of all engineering plastics. He suffered deepening and lengthening periods of depression, however, and he eventually took his own life.

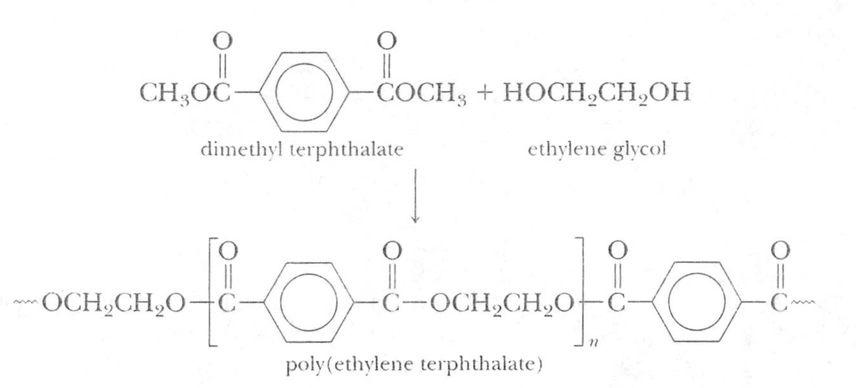

One methanol (CH_3OH) molecule is given off per chain link.

Questions

What is your best estimate of the number of monomers needed to prepare a polypeptide of about 5000 molecular mass from valine $(CH_3)_2CHCH(NH_2)COOH$, one of the "essential" α-amino acids? Or, to put the question another way, how would you determine the degree of polymerization in the formula for the polypeptide?

What is your best estimate of the fraction of monomer mass lost in the condensation (step-growth) process of the preceding question?

Wallace Hume Carothers brought polymer chemistry to the attention of American chemists and "nylons" to the world. This caricature was drawn by Philip Burke, an undergraduate chemistry student.

SUMMARY

The chemistry of carbon is very different from that of its nearest neighbors in the periodic table, especially so because of its tendency to form stable covalent bonds with itself and with hydrogen, oxygen and sulfur, nitrogen and phosphorus, and the halogens. From a few natural sources, such as petroleum and coal, and plant and animal matter, literally millions of well-defined organic compounds have been isolated or synthesized, characterized, and identified.

Hydrocarbons can be saturated or unsaturated, straight chained or cyclic. The alkanes or saturated hydrocarbons, having no multiple C-to-C bonds, are characteristically unreactive except to halogenation and combustion in air. If the alkane has no ring it falls into a series with general formula C_nH_{2n+2}, where n

is the number of C atoms in the molecule. Cycloalkanes have the general formula C_nH_{2n}. The hydrocarbon chains found in petroleum range from gases to lube oils and candle waxes with 20 or more carbon atoms. Branching alters physical properties; thus, isobutane has a lower boiling point than *n*-butane. The IUPAC system of nomenclature is based on the number location of the alkyl branch along the main chain of the carbon skeleton; hence, isobutane would be named 2-methylpropane.

Alkenes are unsaturated hydrocarbons. Referred to as olefins, their enhanced reactivity over the alkanes is due to the presence of the double bond. Bromine and hydrogen bromide typically add to the double bonds in ethylene and propylene. An especially important reaction of the π electrons in alkenes is the self-addition or polymerization reaction, leading to high molecular mass, polymeric products such as polyethylene and polypropylene. Alkynes contain triple bonds; the best-known example is acetylene, the first member of the family. Benzene, C_6H_6, has the structure contained in aromatic compounds. Resonance in the benzene molecule confers stability to the ring structure, and this is reflected in the reactions that benzene undergoes.

When functional groups are present in the molecule, replacing hydrogen atoms along the carbon chains in the backbone or in the branches, the number of possible isomeric forms increases dramatically, along with general reactivity. For example, there are two isomeric C_4H_{10} hydrocarbons, but there are four isomeric C_4H_9OH alcohols. From alcohols, it is possible to synthesize alkenes or ethers, depending on the reaction conditions by acid-catalyzed dehydration. Among the most widely encountered functional groups are the alcoholic (R—OH) and phenolic (C_6H_5OH), aldehydes (RCHO) and ketones (RCOR), carboxylic acids (RCOOH) and esters (RCOOR′), ethers (ROR′), amines (RNH_2), and amides (RCONHR′). Industrial and biological applications of these compounds abound.

Difunctional monomers produce macromolecules by condensation (step-growth) polymerization, whereby a small molecule is lost for every polymer link formed. Addition (chain-growth) polymerization leads to high molecular-mass polymers from compounds containing available π electrons as in alkenes.

TERMS

Organic chemistry (21.1)
Organic substances (21.1)
Inorganic substances (21.1)
Hydrocarbon (21.2)
Saturated (21.2)
Unsaturated (21.2)
Alkane (21.2)
Cycloalkane (21.2)
Methane (21.2)
Ethane (21.2)
Propane (21.2)
Butane (21.2)
Isomer (21.2)
Pentane (21.2)
Trivial nomenclature (21.2)
Systematic nomenclature (21.2)
Alkyl group (21.2)
Linear (21.2)
Branched (21.2)
Initiation step (21.3)
Propagation step (21.3)
Termination step (21.3)
Alcohol (21.3)
Amine (21.3)
Alkene (21.3)
Functional group (21.3)
Halide (21.3)
Cis- (21.4)
Trans- (21.4)
Dehydrogenation (21.4)
Chain-growth (addition) polymerization (21.4)
Alkynes (21.5)

Aromatic (21.6)
Phenyl group (21.6)
Hydroxyl (21.7)
Phenol (21.7)
Ether (21.7)
Aldehyde (21.7)
Ketone (21.7)
Carbonyl group (21.7)
Carboxylic acid (21.7)
Ester (21.7)
Amide (21.7)
Condensation polymer (21.7)

QUESTIONS

Conceptual questions are denoted by a square screen.
Extra-credit questions are denoted by a circular screen.

1. What is your understanding of each of the following?
   (a) C_nH_{2n+2}
   (b) C_nH_{2n}
   (c) Coal gas and coal tar
   (d) Dehydrogenation
   (e) Alkanes, alkenes, and alkynes
   (f) Cycloalkanes and aromatic hydrocarbons
   (g) Isomeric pentanes
   (h) *Cis-* and *trans-* isomers
   (i) Octane rating
2. Distinguish between each of the following:
   (a) Proteins and peptides
   (b) An alkane and an alkyl group
   (c) Primary, secondary, and tertiary carbon atoms
   (d) Addition and substitution reactions
   (e) Addition and condensation reactions
3. Which of the following sets of structures represent identical compounds?
   (a)

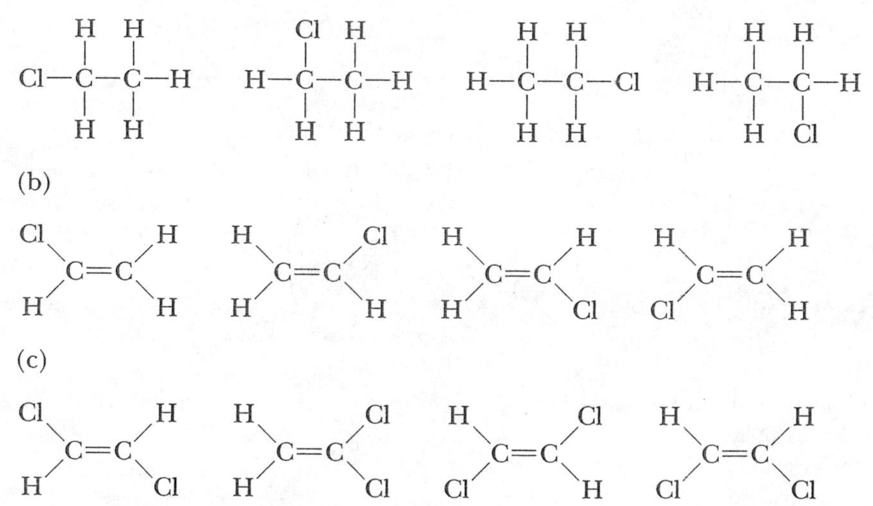

(d)

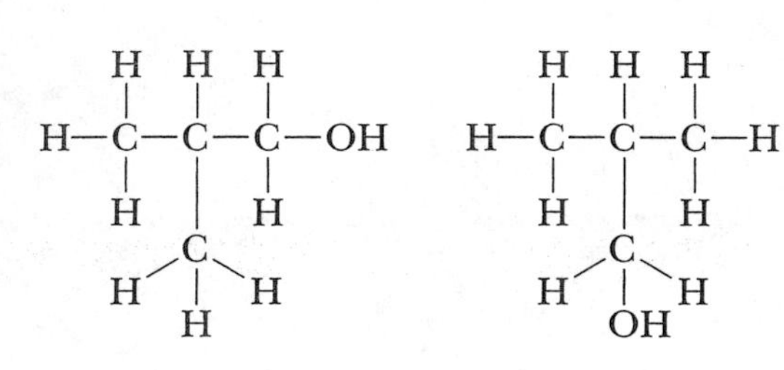
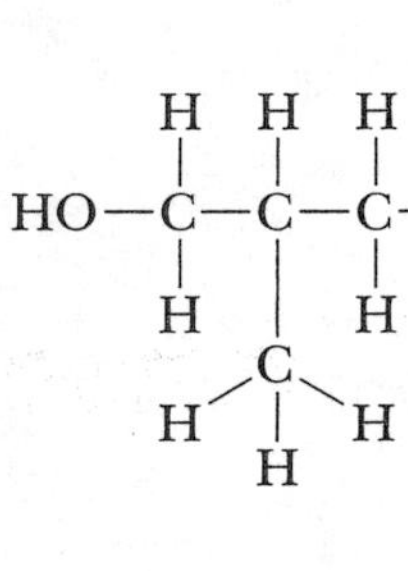
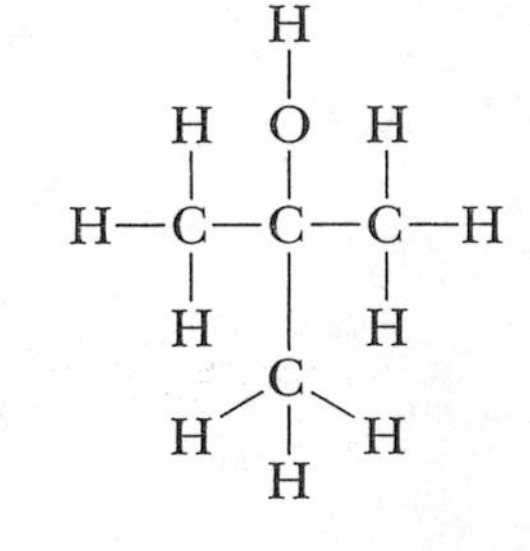

(e)

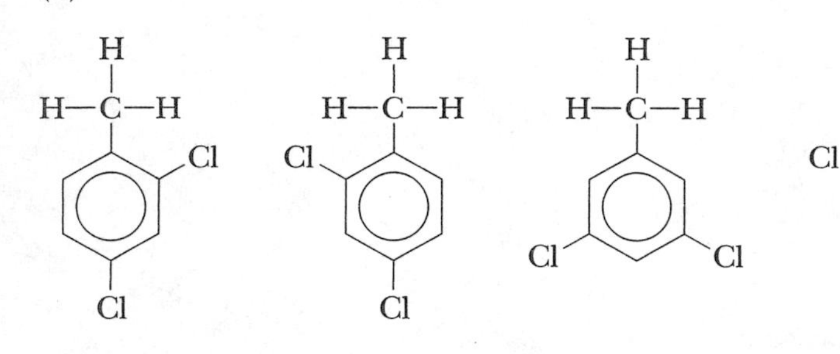

4. What simple chemical tests can be used to distinguish between *n*-hexane and 1-hexene?
5. How many different brominated benzenes are possible?
6. Why is the direct chlorination of ethane in sunlight not a useful way of preparing 1,1,1-trichloroethane, an important industrial solvent for paints and finishes?
7. Why is the addition reaction to double bonds synthetically important?
8. How do aldehydes differ from ketones? How are amides different from amines?
9. How would you distinguish between each of the following?
   (a) An alcoholic OH and a phenolic OH
   (b) An alcoholic OH and a carboxylic OH
   (c) An alcohol and an ether
   (d) An acid and an ester
   (e) An aldehyde and a ketone
   (f) An amine and an alcohol
10. What are the formulas for the following compounds?
   (a) Hydrazine (b) Isobutane
   (c) Acetylene (d) Isopropyl alcohol
   (e) Acetic acid (f) Formaldehyde
   (g) Acetone (h) Phenol
   (i) Chloroform (j) Glycine
11. What are the molecular formulas for the following?
   (a) Methane
   (b) Ethylene
   (c) Trichloromethane
   (d) Ethyl alcohol
   (e) Urea and ammonium cyanate
   (f) Toluene
   (g) The natural rubber monomer
   (h) Sodium acetate
12. An organic compound reacts readily with bromine to form the compound $CH_3CHBrCH_2Br$. What was the original compound?
13. An organic compound reacts readily with hydrogen bromide to form the compound $CH_3CHBrCH_3$.
   (a) What was the original compound?
   (b) What principal product(s) would be formed when the original compound reacts with water in the presence of acid?
14. By what general process—oxidation, reduction, cracking, heating, dehydration, halogenation, hydrohalogenation, dehydrohalogenation—would you accomplish each of the following conversions?
   (a) Isopropyl alcohol to acetone
   (b) 2-butene to 2-bromobutane

(c) Ethyl alcohol to ethylene
(d) Kerosenes to gasolines
(e) Coke to carbon monoxide
(f) Coal to coke
(g) Methane to chloroform
(h) Glycine to glycylglycine

PROBLEMS

Problems marked with a bullet (•) are answered in Appendix A, in the back of the text.

Structural Formulas [1–4]

•**1.** Draw structural formulas for the isomeric hexanes C_6H_{14}, and then name the compounds by the IUPAC system.

2. Draw structural formulas for the isomeric pentenes C_5H_{10}, and then name the compounds by the IUPAC system.

•**3.** There are nine possible isomeric heptanes:
(a) Write out structural formulas for each.
(b) Name each by its systematic name.
(c) Which might be commonly called isoheptane?
(d) Which might be commonly called neoheptane?
(e) Are names such as "isoheptane" or "neoheptane" unambiguous? Explain.

4. There are 17 possible isomeric hexenes:
(a) Write out the structural formulas for each.
(b) Which is 3-hexene?
(c) Explain why there are two isomeric 3-hexenes.

Reactions and Preparation of Alkanes, Alkenes, and Alkynes [5–8]

•**5.** Using whatever inorganic reagents you might need, write a sequence of chemical reactions that would produce
(a) Isopropyl alcohol from propane
(b) Propane from isopropyl alcohol

6. Write chemical reactions, using necessary inorganic reagents, to produce the following transformations:
(a) Acetylene from ethylene
(b) Polyvinyl chloride $+CH_2-\underset{Cl}{\underset{|}{CH}}+_n$ from acetylene
(c) Phenol from cumene

•**7.** Write Lewis structures for acetylene, the acetylide anion (acetylene with one proton removed), and the acetylene dianion (acetylene with two protons removed).

8. Explain why the carbon-to-carbon bond length changes from 1.54 Å in ethane to 1.33 Å in ethylene and 1.20 Å in acetylene.

Aromatic Derivatives [9–10]

9. Write structures for all the possible methylbenzenes.

10. For the various substituted chlorobenzenes, fill in the chart.

Substitution	Number of Isomers	Examples
mono-		
di-		
tri-		
tetra-		
penta-		
hexa-		

Alcohols and Ethers [11–12]

•**11.** Place the following compounds in order of their increasing solubility in water:

CH_3CH_2OH

$CH_3(CH_2)_5OH$

$HO(CH_2)_6OH$

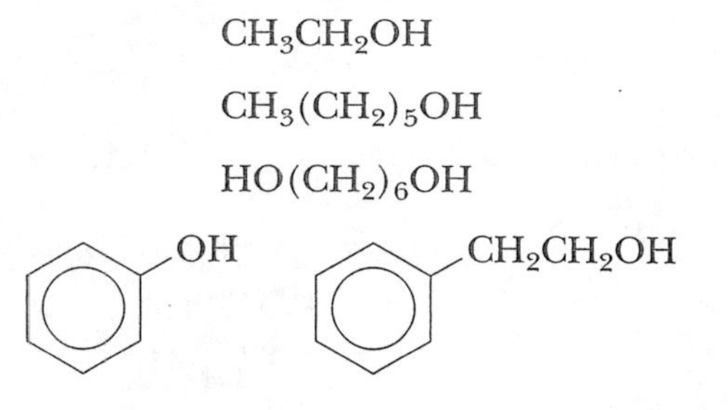

12. Considering the data in the following table, explain why ethanethiol (CH_3CH_2SH) boils at an appreciably lower temperature than ethanol (CH_3CH_2OH) although its molar mass is nearly 50% greater.

	Molar Mass	b.p. (°C)
CH_3CH_2OH	46	78.5
CH_3CH_2SH	62	37.0

Aldehydes, Ketones, Carboxylic Acids, Esters, and Amines [13–20]

•**13.** There are a number of isomeric monobromobutanes.
(a) Draw structural formulas for all that are possible.
(b) Indicate by equations which one(s) will form primary alcohols on hydrolysis with aqueous alkali. Which one(s) will form secondary alcohols? Which ones will form tertiary alcohols?
(c) Indicate which of these alcohols
(i) can be oxidized to carboxylic acids.
(ii) can be oxidized to an aldehyde.
(iii) can be oxidized to a ketone.
(iv) cannot be further oxidized without disrupting the carbon skeleton.
(d) Draw structural formulas for the alkenes that might form on dehydrohalogenation of the bromobutanes.

14. What is the common feature or functional group of aldehydes and ketones? How are they both structurally related to carboxylic acids? Explain why aldehydes are susceptible to oxidation but ketones are not.

•**15.** The pK_a values for acetic acid and α-bromoacetic acid are, respectively, 4.7 and 2.5.
(a) Write out the structural formulas for both acids.
(b) What is the pH of a 0.10-M solution of each?
(c) How might you account for the considerable acidity of the brominated acetic acid as compared to acetic acid itself?

16. Carboxylic acids are weak acids, such as we dealt with in Chapter 11. Consider a solution that is 0.10 M in acetic acid and 0.10 M in sodium acetate.
(a) Show by equations the effect of adding acid and base to the solutions.
(b) In what pH range might a solution of α-bromoacetic acid and its sodium salt be an effective buffer?

17. Complete the following reactions as indicated for (a) acetic acid, (b) methyl acetate, (c) acetyl chloride, and (d) acetamide:

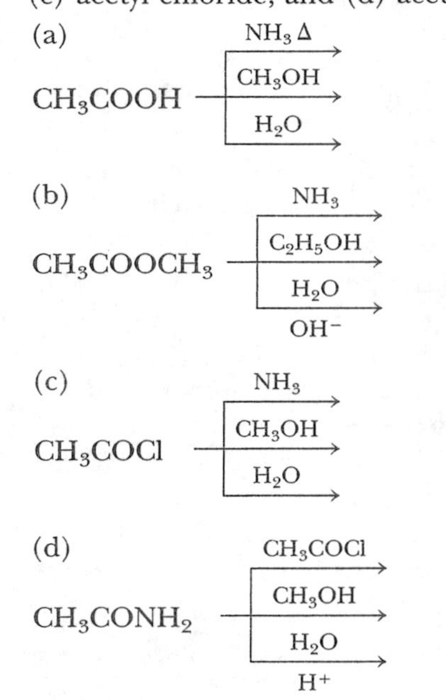

18. Write the products of NH_3, CH_3OH, and H_2O with each of (a) benzoic acid, (b) methyl benzoate, (c) benzoyl chloride (acid chloride of benzoic acid), and (d) benzamide (amide of benzoic acid and ammonia).

•**19.** Write the structural formulas for all the possible isomeric butylamines.

20. Write the formulas for all the amines having the formula C_3H_9N. Label each amine as primary, secondary, or tertiary.

Organic Reactions [21–24]

•**21.** Using the organic material indicated and whatever inorganic reactants you might require, show how the following conversions can be accomplished:

(a) $CH_3CH_2CH_2OH \longrightarrow CH_3CH(OH)CH_3$

(b) $CH_3CH(Cl)CH_3 \longrightarrow CH_3CH_2CH_2Br$

(c) $CH_3CH(OH)CH_3 \longrightarrow CH_3CH(NH_2)CH_3$

(d) $CH_3CH_2OH \longrightarrow CH_3C(=O)OCH_2CH_3$

(e) $CH_3CH_2OH \longrightarrow CH_3COOH$

22. Considering the previous reactions, write equations for the following chemical conversions using only inorganic substances where necessary:
(a) 1-butanol into 2-butanol
(b) 1-chlorobutane into 2-bromobutane
(c) Formic acid into methyl formate
(d) 2-propanol into acetone

•**23.** Consider each of the following and determine the lettered compounds in question. Write equations for the reactions taking place:
(a) *A* contains only C, H, and O, has a molar mass of 30, and reduces Tollen's reagent.
(b) *B* has the molecular formula C_3H_8O. An alkaline permanganate solution oxidizes it to a new compound *C* whose molecular formula is C_3H_6O. Compound *C* is not further oxidized readily.

24. Determine the compounds identified by letters. Write the reactions that take place.
(a) *D* is an aromatic compound that can be chlorinated at elevated temperatures, yielding *E* in a process in which a mole of HCl is produced for every mole of chlorine absorbed. *D* has a molecular formula of C_7H_8. *E* is $C_7H_5Cl_3$.
(b) Compound *F*, whose molecular formula is C_2H_6O, is treated with aqueous sulfuric acid to yield a new compound *G*, whose formula is $C_4H_{10}O$. If the reaction were carried out under somewhat harsher conditions, *H* with the formula C_2H_4 is formed instead of *G*.

Additional Problems [25–29]

25. Draw structural formulas for all of the isomers of $C_5H_{11}Cl$.

26. Consider the following thermochemical data:

cis-2-pentene $\longrightarrow$ *n*-pentane $\quad \Delta H = -118$ kJ/mol

trans-2-pentene $\longrightarrow$ *n*-pentane $\quad \Delta H = -115$ kJ/mol

(a) Write the structural formulas for *cis*- and the *trans*-2-pentene.
(b) By examining the structural models would you conclude that the *cis*- form is more stable (thermodynamically) than the *trans*- form? Why (or why not)?
(c) Is your answer to (b) borne out by the thermochemical data given previously? Explain.
(d) Based on the fact that ΔH is -126 kJ/mol for the reduction (hydrogenation) of 1-pentene to *n*-pentane, compare the relative stabilities of the three isomeric butenes.

•**27.** Because both propylene and toluene can be characterized as having a methyl (CH_3) group adjacent to a C-to-C double bond, how do you account for the obvious differences in their reactivity with respect to bromination and oxidation? Where reactions occur, write the equations.

28. Explain why the hydrolysis of an ester to yield alcohol and acid is best carried out in basic rather than acidic solution. Use equations to illustrate your answer.

29. By proper use of boiling-point elevation data, the molar mass of benzoic acid can be determined. When measurements are made in acetone solution, the molar mass is close to 122; in carbon tetrachloride, it is nearly 244. How do you rationalize the data?

Cumulative Problems [30–33]

30. Which of the compounds, thiophenol, C_6H_5SH, or phenol, C_6H_5OH, should boil at the lower temperature? Explain why.

•31. A sample contains a mixture of propane and propene. A 10.0-g sample of this mixture is found to react immediately with 24.7 g of bromine. What is the percentage of propane in this mixture?

32. You are running a reaction of ammonia with methyl chloride, CH_3Cl, to produce a mixture of methyl amine, dimethyl amine, and trimethyl amine. Suppose you know the mass of the ammonia and the mass of the methyl chloride that react and the total mass of the mixture of amines produced. Write three equations in three unknowns that could be used to determine the amount of each of the three amines in the mixture. It is not necessary to solve the equations.

•33. 0.100 L of carbon monoxide at 4.50 atm and 25°C is reacted with excess sodium hydroxide to produce sodium formate, which is acidified to produce formic acid. The formic acid produced is titrated 0.100M HCl, and it is found that the inflection point occurs at 123 mL. What was the percentage yield of the formic acid synthesis?

Applied Problems [34–35]

34. What volume of water vapor is produced for every kg of nylon-6,6 at the reaction temperature of 250°C and a pressure of 1.00 atm?

•35. How many liters of methanol, or CH_3OH, are produced as a by-product for every 100. kg of DuPont Dacron or Mylar poly(ethylene terphthalate)? The density of methanol at 25°C is 0.79 g/cm^3.

ESTIMATES AND APPROXIMATIONS [36–37]

36. During one intermediate step in the photochlorination of methane, chlorine atom extraction of an H atom leaves a methyl radical that experiment shows to be planar:

$$CH_4 + Cl\cdot \longrightarrow CH_3\cdot + HCl$$

(a) What hybridization is most consistent for the orbitals on carbon in the methyl radical in light of the observed planar geometry?
(b) Based on the hybridization on carbon, construct a simple energy-level diagram for the carbon atom and place the valence electrons so as to correctly represent the electronic structure for the methyl-free radical.

37. List boiling point data (in °C) for the following saturated hydrocarbons from the appropriate table in the chapter: ethane, propane, hexane, octane, and nonane. Now consider the cracking of a saturated hydrocarbon that has a boiling point of +174°C and gives among the products compounds with boiling points of −250°C, −162°C, and +98°C, respectively. None of these compounds is capable of being polymerized, and none decolorizes bromine water.
(a) Suggest what the original saturated hydrocarbon might have been.
(b) Suggest what the three products might have been, being sure to show how you arrived at your conclusions.
(c) Some saturated hydrocarbons are gases, some are liquids, and some are solids (at 25°C). Which of the liquids has the lowest mass for one mole of molecules?

FOR COOPERATIVE STUDY [38–40]

38. Draw structures that clearly depict all the C_4H_8 isomers capable of causing the red-brown color of bromine in carbon tetrachloride to fade.

39. Draw structures that clearly indicate all the C_4H_8 isomers that do not alter the purple color of dilute aqueous permanganate solutions. (*Note:* Permanganate solutions can oxidize alkenes to compounds with —OH groups on two adajacent carbon atoms.)

40. Although the periodic table suggests that silicon should bear a close family resemblance to carbon because both are Group-IVA family members with four valence electrons in four *sp*^{3} hybrid-bonding atomic orbitals, this is not the case. Develop an argument in support of that fact based on the shielding effect due to silicon's extra inner shell of electrons and repulsive forces that do not exist in carbon atom chemistry but do exist between silicon atoms.

WRITING ABOUT CHEMISTRY [41–42]

41. Chemistry is traditionally divided into four broad areas—analytical, physical, inorganic, and organic chemistry. Biochemistry is often added as a fifth area. Analytical chemistry is concerned with determining the presence and amounts of substances in a sample. Physical chemistry studies the theories by which chemistry is understood and provides new tools and techniques for further developing and understanding those theories. Inorganic and organic chemistry are defined by the compounds under study (as illustrated, for example, by this chapter). How good a classification system is this? In what kinds of cases does it help make good classifications and where (if anywhere) does it fail? Use examples from throughout this book to make this analysis.

42. The system of nomenclature for organic compounds is considerably more complex than nomenclature that we have discussed previously. Explain why this is the case. Then come up with an alternative system for naming alkanes and describe it. Explain carefully why you believe that it is better than, equivalent to, or worse than, the existing nomenclature system.

43. Imagine for a moment you have stepped back in time to 1828. You are a young upstart professor of organic chemistry named Friedrich Wöhler writing a letter to the leading authority of the day, the Swedish chemist Jöns Berzelius. You must describe your experiments and convince him that the urea you have obtained from purely inorganic chemicals is identical to that isolated from urine. Write such a letter.

chapter twenty-two

22

NUCLEAR CHEMISTRY

22.1 SUBATOMIC PARTICLES AND THE NUCLEAR ATOM

The nature of ordinary chemical reactions can be understood in terms of the electron configuration of the atoms of the elements involved. However, ordinary chemical reactions involving the loss, gain, and sharing of electrons are not the only reactions that atoms can undergo. The nucleus of the atom has a definite structure and is capable of undergoing reactions in which the energies involved are several orders of magnitude greater than ordinary chemical reactions. Reactions releasing energy on that scale lead to applications ranging from weapons of extraordinary destructive power to modern electric power generation. Beyond energy generation, nuclear reactions have led to applications in fields as diverse as medicine, agriculture, engineering, and archeology.

One of the first experiments designed to probe the structure of the atom was conceived by Ernest Rutherford and his students (Chapter 2). Bombarding thin metal foils with a stream of alpha particles, doubly positively charged helium nuclei (He^{2+}) emanating from a radioactive source, they made two highly significant observations:

- Most of the alpha particles streamed right through as if the metal foil were not even there.
- A tiny fraction of the particles were deflected from their original path, including a very few that were deflected in the direction of their original source.

From these results, Rutherford concluded that matter must be mostly empty space, with almost all the mass concentrated in very dense regions, and that these very dense regions must be positively charged to deflect the like-charged alpha particles. Deflection toward the source would not occur if the dense regions were either negatively charged or neutral.

The basic features of the Rutherford atom derived from these conclusions survive to the present. An atom has a very dense positively charged nucleus surrounded by a much larger region of empty space, which contains the negatively charged electrons. These give the atom its overall neutral charge. The particles that make up the atom are called **subatomic particles** because they are smaller than atoms themselves. The first of the subatomic particles that we will consider is the electron.

Electrons

Electrons were the first subatomic particles known, and they were discovered by J. J. Thomson and his students (Chapter 2). Their discovery emerged from practical experiments designed to study the nature of electricity. When highly evacuated glass tubes fitted with metal electrodes were connected to a source of high potential, a glow between the electrodes was observed. As the internal pressure was further reduced, this gradually converted to a green glow on the walls of the tube farthest from the cathode, or negative electrode. The glow appeared to be caused by some type of radiation coming from the cathode. Called cathode rays by the scientists who first observed them, they were shown to be streams of negatively charged particles because they migrated to the positive electrode. **Cathode rays** were eventually found to be streams of electrons.

A stream of charged particles in a cathode ray tube can be deflected from its line of flight by an electric field or a magnetic field. Thomson observed this deflection of the cathode rays using a phosphor-coated screen that glowed when struck by the electrons. Using data from these experiments, he obtained a value for the charge (e) to mass (m) ratio of the electron, that is, the e/m ratio (Fig. 22.1). Thomson's experiments were simple yet brilliantly conceived, yielding the proper magnitude for e/m and showing these particles to be fundamental, universal particles of matter. They were found to be independent of the composition of the cathode material or the chemical nature of the residual gas in the tube. The accepted value of e/m for the electron is 1.76×10^8 coulombs/gram (C/g).

The tiny charge on a single electron was determined by Robert Millikan (American, 1868–1953) in an equally simple yet ingenious experiment. His apparatus consisted of a pair of metal plates within a chamber containing air (Fig. 22.2). These plates could be charged to produce an electrostatic field between them. Droplets of oil produced by an atomizer were electrically charged by collisions with X-rays and then allowed to fall through a small hole in the top plate. With the aid of a microscope eyepiece, an observer could watch an individual oil droplet passing between the plates. Downward motion was due to the gravitational field of the Earth acting on the mass of the droplet. Upward motion resulted from the action of the electric field on the charged droplet. With the electrostatic field turned on and properly adjusted, the droplet could be held motionless. Under these conditions, the downward force F_{down} and the upward

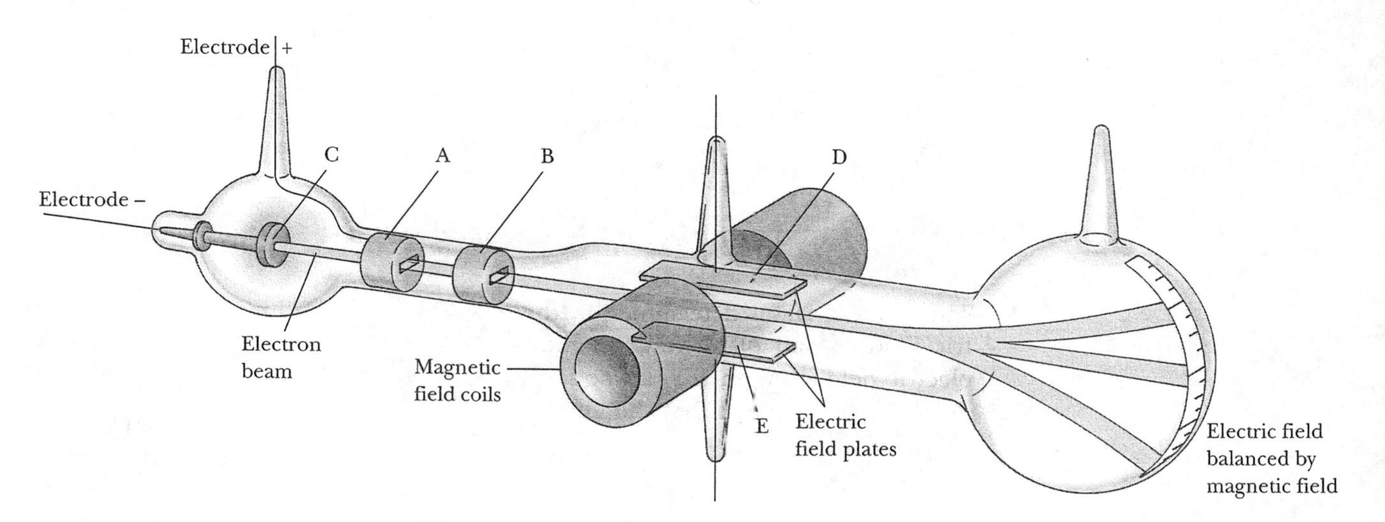

FIGURE 22.1 Measuring the charge-to-mass (e/m) ratio for the electron. This depended on being able to deflect a stream of charged particles by an electric or magnetic field (J. J. Thomson). A beam of rays from the cathode (C) narrowed on passing through a slit in the anode (A). The beam passed through a second slit (B) and fell on a screen at the end of the tube producing sharply defined, fluorescent lines. Upward deflections of the beam were brought about by connecting the upper plate (D) to the positive pole of the battery; downward deflections were produced by connecting the lower plate (E) to the positive pole. When placed between the poles of an electromagnet, the effect of a magnetic field on the rays could be tested.

force F_{up} could be assumed to be the same. The downward force is the mass m of the droplet times the acceleration of gravity g:

$F_{down} = mg$ downward force on droplet produced by gravitational field

The upward force depends on the charge q on the droplet and the strength of the electrostatic field E:

$F_{up} = qE$ upward force on droplet equals charge times field strength

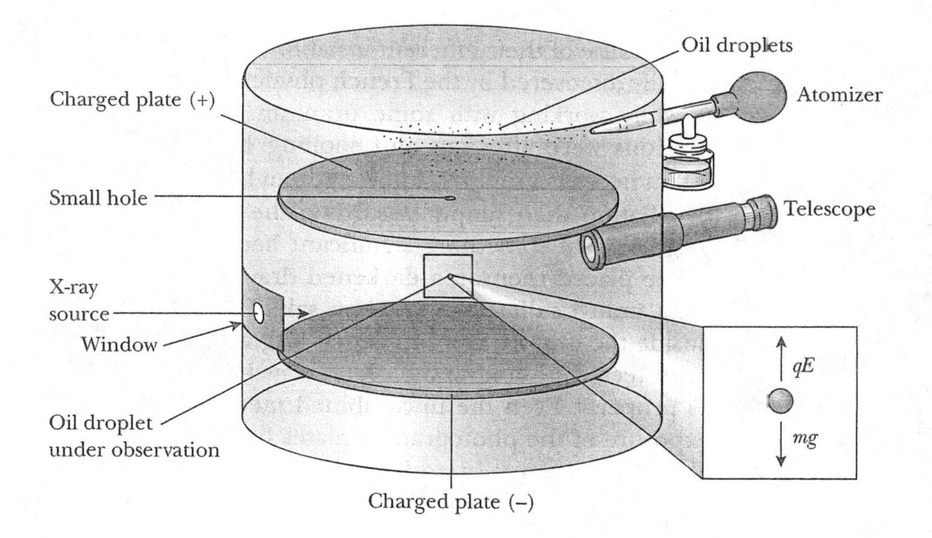

FIGURE 22.2 Measuring the charge on the electron (Millikan). Tiny oil drops produced by an atomizer fall between the plates where they are charged by X-rays. By controlling the voltage between the plates, it is possible to balance the electric (qE) and gravitational (mg) forces and determine the charge on the droplet. When the droplet stands suspended in space, the downward force acting on the droplet of mass m equals the upward force due to the strength of the electric field E, and $mg = qE$.

Setting $F_{down} = F_{up}$ gives

$$qE = mg \qquad \text{upward and downward forces are equal}$$

After rearranging to solve for the value of q, the charge on the droplet is

$$q = \frac{m \cdot g}{E} \qquad \text{charge on droplet}$$

The value of m was determined by the droplet's rate of fall in air with the electric field turned off. The acceleration due to gravity (g) is known, and the electric field strength (E) can be set to the desired value.

Millikan assumed that the charge on any droplet was an integral multiple of e; that is, $q = ne$, where n is an integer and e is the unit charge on the electron. On the basis of a very large number of measurements, Millikan concluded that q was an integral multiple of a number very close to 1.602×10^{-19} C, the accepted value for the charge on the electron. Knowing the charge-to-mass ratio and the electron charge, the mass of the electron (m_e) could be determined. The current value of m_e is 0.0005486 amu, or 9.109×10^{-28} g.

Radioactivity

X-Rays, and their amazing penetrating properties were reported by the German physicist Wilhelm Röntgen (1845–1923) a year before Thomson's observations. Experiments with cathode rays accidentally led him to the discovery. Röntgen found that unknown radiations coming from his cathode ray tube could stimulate certain compounds, causing them to fluoresce, that is, emit visible light. This occurred even when the compound he was studying was placed some distance away from the cathode ray tube with heavy black paper in between. These cathode ray tube emanations (or radiations) were called **X-rays**. When these rays were allowed to pass through matter onto a protected photographic plate, they produced strange and important results—the first X-ray photographs. But, when the electron beam in the cathode ray tube was shut down, so too was the stimulating radiation—no fluorescence and no exposed photographic plates.

Radioactivity is the production of radiation through the spontaneous disintegration of certain nuclei because of their inherent instability (Chapter 2). This phenomenon was accidentally discovered by the French physicist Henri Antoine Becquerel (1852–1908) while working with some uranium compounds. He found that invisible radiations were affecting and spoiling his photographic plates in spite of the fact that he had carefully protected them by a double thickness of heavy black paper. Rather than simply discarding the films, Becquerel was alert enough to recognize that something significant had happened. His films were exposed when he placed them in a darkened drawer beside a sample of crystals of potassium uranyl sulfate—a uranium salt. The uranium salt, which had been sealed inside the drawer, had succeeded in giving off sufficient radiation to expose the protected photographic plates. Furthermore, all uranium salts exhibited this property. Even the uncombined metal produced the effect, with the rate of exposure of the photographic plates being solely a function of how much uranium was present. In addition, these strange radiations persisted, undiminished by time or temperature.

Did nuclei other than uranium exhibit this phenomenon? How widespread was radioactivity? What was the nuclear driving force propelling these curious uranium emanations? Marie Sklodowska Curie (Polish, 1867–1934) and Pierre Curie (French, 1859–1906) undertook a systematic study of the new phenomenon in Becquerel's laboratory and began to examine these questions. Two more radioactive elements—polonium and radium—were discovered and isolated by the Curies from two tons of pitchblende, a uranium ore. Later, thorium was isolated from monazite sands, natural phosphate salts of metals, particularly lanthanides or rare earths. Radon, a radioactive gas, also was discovered. Polonium proved to be the most intensely radioactive of these natural elements. Eventually, an international army of scientists contributed to this robust new science, but it was the Curies (especially Marie) who were justifiably recognized for doing the pioneering work.

Alpha, Beta, and Gamma Radiation

Radioactivity involves the spontaneous disintegration of nuclei, accompanied by the emission of one or more of three principal radiations. These radiations can be distinguished from each other experimentally (1) by observing their range and penetrating power in passing through metal foils of various thicknesses and (2) by observing the behavior caused by magnetic and electric fields to which the rays are subjected. The three radiations were arbitrarily labeled alpha (α) and beta (β) particles and gamma (γ) rays (Table 22.1). The **gamma rays** turned out to be electromagnetic radiation similar to X-rays—highly penetrating and not deflected by electric or magnetic fields. **Beta particles** were identified as high-speed electrons. **Alpha particles** were shown to be the nuclei of helium atoms and, like all nuclei, positively charged. Note that the name "beta particle" is reserved for electrons ejected from nuclei as distinguished from the electrons surrounding the nuclei of atoms.

The identity of alpha particles was established in another ingenious experiment conducted by Ernest Rutherford (British, 1871–1937) and his students. A sample of the newly discovered radioactive gaseous element radon, which was known to emit α-particles, was sealed in a thin-walled glass capillary tube contained within a larger tube. The capillary glass wall was approximately 0.01 mm thick, thin enough to allow fast-moving α-particles, but not radon atoms, to pass through. Over a week's time a gas accumulated in the space between the two tubes (Fig. 22.3). The trapped gas was then identified as helium. Thus, emitted α-particles had picked up electrons to form neutral helium atoms, proving that they were simply helium nuclei.

TABLE 22.1 Relative Penetrating Power, Charge, and Mass Data for Radium Radiation

Type	Foil Thickness	Relative Penetrating Power	Charge	Rest Mass
α	0.005 mm	1	+2	4 amu
β	0.50 mm	100	−1	1/1823 amu
γ	5.0 mm	1000	0	0

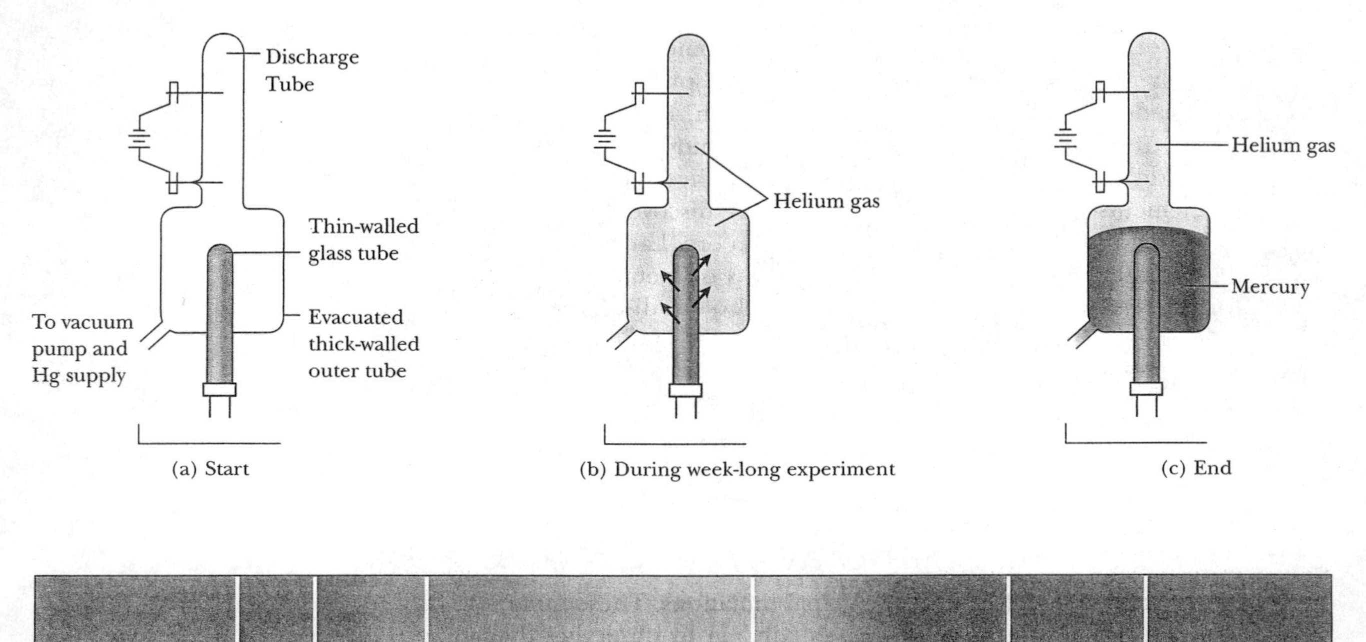

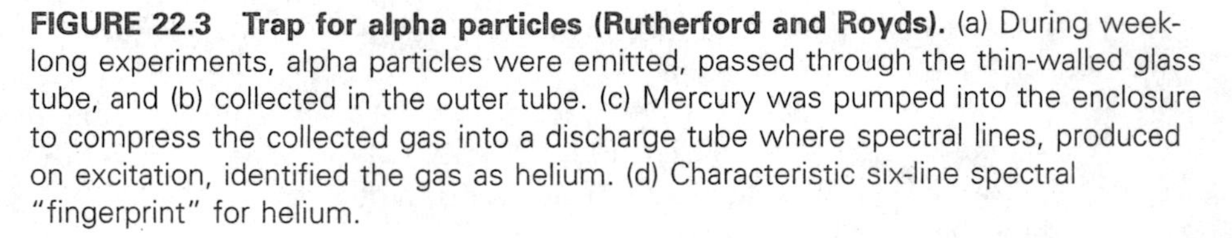

FIGURE 22.3 Trap for alpha particles (Rutherford and Royds). (a) During week-long experiments, alpha particles were emitted, passed through the thin-walled glass tube, and (b) collected in the outer tube. (c) Mercury was pumped into the enclosure to compress the collected gas into a discharge tube where spectral lines, produced on excitation, identified the gas as helium. (d) Characteristic six-line spectral "fingerprint" for helium.

The fact that α-particles and β-particles are charged particles emerged largely as a result of experiments on magnetic and electrostatic deflection. Beams of these particles were directed through a magnetic field to see whether they deviated from the incident path. The technique is based on fundamental principles of physics and is still widely used in studying the behavior of nuclear events. When a charged particle moves across a magnetic field, the force of the field acts at right angles to the direction of motion of the charged particle. The particle experiences a continual deflection and, if directed into a uniform field, will move along the arc of a circle.

We begin an experiment by placing a radium sample at the base of a narrow passage through a lead block. A pencil-thin beam of α-particles, β-particles, and γ-rays is then directed along the only escape path. As the beam passes into a strong, uniform magnetic field, it separates into the three types of radiation (Fig. 22.4). The uncharged γ-rays continue along the original line of flight, whereas the β-particles are deflected to one side, and the α-particles to the other, in circular arcs of different radii. Now the charge-to-mass ratio (e/m) for each type of particle can be determined. For β-particles it coincides with known results for electrons from J. J. Thomson's previous work. Very strong magnetic fields must be used for the α-particles because the ratio of charge to mass is only 3×10^{-4} times that of a β-particle.

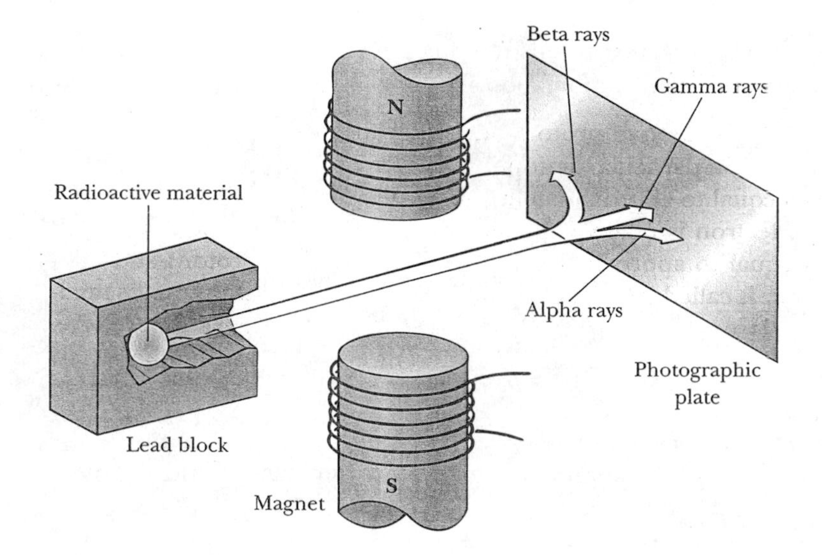

FIGURE 22.4 Separating α-particles, β-particles, and γ-rays: The three kinds of natural radiation, or "radioactivity." The central beam of gamma rays is undeflected; the beta particles are deflected by a large amount, being negatively charged and having little mass, the alpha particles, being more massive and positively charged, are deflected less and in the opposite direction.

These different radiations have different powers of penetration: α-particles can be stopped by little more than a sheet of tissue paper or a few centimeters of air; β-particles, although more penetrating, will be stopped by several meters' travel through an air space or by passing through a thin sheet of aluminum foil; γ-rays pass unhindered through several inches of lead or several feet of concrete. When γ-rays are involved in experiments, heavy shielding is necessary for protection of the experimenters. In the early days of radiation research, many workers were exposed to dangerous levels of radiation, which led to serious health problems later in life.

Nuclear Protons and Neutrons

When an electric discharge is passed through a gas at low pressure, cathode rays are not the only particles produced. Almost from the outset, investigators were aware of the presence of positively charged particles as well, particularly the proton. The **proton** is equal in magnitude of charge to the electron and opposite in sign, but with a mass 1836 times larger.

Rutherford was the first to predict the existence of a neutral *nuclear* particle. One of his students, James Chadwick (British, 1891–1974), first identified the **neutron** in 1932 as a product of bombarding beryllium atoms with α-particles. The mass of the neutron was found to be very slightly less than the sum of the masses of a proton and an electron.

The protons and the neutrons are the dominating particles in the nucleus of an atom and are referred to as **nucleons**. A nucleon is a particle that is part of a nucleus. Because the proton has an equal but opposite charge to the electron, the number of each is equal in any electrically neutral atom, and this number is known as the **atomic number** (Z). Because electrons determine ordinary chemistry, all atoms with the same Z undergo the same chemical reactions and are considered to be atoms of the same element. If these atoms have different numbers of neutrons, they are called isotopes. **Isotopes** are atoms with different numbers of neutrons, but the same number of protons; they are still atoms of the same element. Because neutrons are uncharged, their number in a

TABLE 22.2 Masses of Subatomic Particles and Selected Isotopes (amu)

Electron	0.00054858
Proton	1.00727647
Neutron	1.0086649
Positron	0.00054858
^{1}H	1.00782505
^{2}H	2.0141079
^{3}H	3.016049
^{4}He	4.0026033
^{7}Li	7.016005
^{9}Be	9.0121825
^{10}Be	10.013535
^{10}B	10.012938
^{11}B	11.009305
^{12}C	12.000000
^{13}C	13.003354
^{14}C	14.003242
^{14}N	14.003074
^{16}O	15.9949146
^{17}O	16.999131
^{18}O	17.9991594
^{19}F	18.9984033
^{20}Ne	19.992435
^{21}Ne	20.993843
^{22}Ne	21.991383
^{23}Na	22.989770
^{30}P	29.97832
^{32}S	31.972072
^{56}Fe	55.9349
^{60}Ni	59.9308
^{235}U	235.0439
^{238}U	238.050784

particular nucleus is unrelated to maintaining neutrality of charge. However, as we will discuss, only certain combinations of protons and neutrons actually lead to stable nuclei.

A particular combination of protons and neutrons is called a nuclide. A **nuclide** is the nucleus of a particular isotope of an element. Protons and neutrons each have a mass equal to approximately one atomic mass unit (amu). Because the mass of the electron is relatively small, about 1/1836 amu, the mass of an atom in amu is equal to approximately the total number of protons and neutrons. This number is called the atomic mass number (A). The **atomic mass number** is an integer close to the actual atomic mass of the nuclide. Isotopes therefore have different numbers of neutrons, different atomic mass numbers, and different atomic masses (Table 22.2).

Nuclides are given by a symbol based on the chemical symbol of the element with the atomic mass number given as a left superscript and the atomic number given as a left subscript; that is, $^{A}_{Z}\text{Symbol}$. For example $^{18}_{8}O$ is the symbol for the nuclide with atomic number $Z = 8$, which makes it an isotope of oxygen, and atomic mass number $A = 18$. Because the number of protons equals eight, this nuclide must contain ten neutrons. Notice that, because all nuclides of equal Z are isotopes of the same element, the value of Z can be omitted in the symbol. Isotopes are sometimes identified by the name of the element followed by its atomic mass number. Thus, $^{18}_{8}O$ can be written oxygen-18.

Positrons and Neutrinos

Nature's collection of subatomic particles includes **antiparticles**: particles that are exact opposites of each other. When antiparticles meet their particle counterparts, they annihilate each other, and their total mass is converted to energy. For example, **positrons** (β^+) are antiparticles of electrons (β^-), identical in every way except that they are opposite in charge. When an electron and a positron come together, they annihilate each other, and the energy liberated is equivalent to their combined masses and two γ-ray photons are produced:

$$\beta^- + \beta^+ \longrightarrow \gamma + \gamma$$

Electron/positron annihilation produces two 8.1×10^{14} J γ-ray photons, representing complete conversion of mass to energy, in agreement with the predictions of the Einstein equation $E = mc^2$. Positrons were discovered in 1932 as a fourth form of radiation from radioactive substances by Carl Anderson, a student of Robert Millikan at Caltech. Why did the discovery of positron emission have to wait 35 years beyond the discovery of radioactivity? Because positron emissions were masked by their very rapid annihilation with electrons, which are very abundant. The isotope $^{18}_{9}F$ emits positrons, and organic molecules tagged with ^{18}F have been used to explore the recesses of the brain by positron emission tomography (PET). As the ^{18}F isotope decays, positrons are emitted that almost immediately encounter electrons, resulting in annihilation and production of pairs of gamma-ray photons going in opposite directions. These distinctive photon pairs are detected and computer analyzed to determine the location in the brain where the $^{18}_{9}F$ has migrated. The technique has been particularly successful in the study and treatment of Parkinson's disease.

Neutrinos (ν) are unusual particles, to say the least. **Neutrinos** are described as having zero charge and near-zero mass. Their existence was predicted from theory, and they were demonstrated experimentally only after antineutrinos

($\bar{\nu}$) had been found. **Antineutrinos** are antiparticles to neutrinos. The neutrino is associated with the emission of positrons (β^+) and the antineutrino with the emission of electrons (β^-).

Recently, an underground facility in Japan recorded the most convincing evidence yet that this ephemeral subatomic particle has mass. This could alter our view of the universe. The actual mass of the neutrino has yet to be determined and is likely to be a minute fraction of the mass of the electron. Small as it is, scientists believe this new result may affect calculations of the total mass of the universe, with implications for understanding its origin and eventual fate. Early estimates put the lower limit of the neutrino mass at approximately one ten-millionth the electron mass.

The evidence for mass is the latest twist in the strange saga of the neutrino, a particle originally suggested by the Austrian physicist Wolfgang Pauli in 1930. Created in staggering numbers by the "big bang" and by the nuclear processes driving the Sun and the other stars, the chargeless neutrinos flow through matter like sunlight through glass. Detecting these particles is difficult, and more data will be necessary to stitch the results into a consistent picture of neutrino masses. Meanwhile, whatever the outcome, this exciting corner of nuclear science is likely to have a *weighty* impact on the theories of the structure and formation of the universe.

In our discussion of radioactivity and atomic and nuclear structure, we have defined a number of important subatomic particles, electrons and positrons, protons and neutrons, neutrinos, and antineutrinos. We have also identified the four forms of emitted radiation: α-particles (He nuclei), β-particles (electrons), γ-rays, and positrons. It is now time to consider which isotopes are radioactive and why they emit the radiation that they do.

22.2 SPONTANEOUS NUCLEAR DECAY AND NUCLEAR REACTIONS

Radioactivity is the outward evidence of nuclear decay, and the radiation that is observed is simply the by-product of nuclear reactions, resulting in the formation of different nuclides. One step toward understanding nuclear decay is to understand nuclear stability—which nuclei decay and which do not. The chart of stable nuclei, which we will examine in the next section, is a most important tool for advancing this understanding.

Charting Nuclei

Because nuclides are characterized by their proton and neutron numbers, we can draw a chart (Fig. 22.5) consisting of vertical columns of protons and horizontal rows of neutrons for all the known stable nuclides. The result is a grid in which each square represents a single nuclide, one stable combination of protons and neutrons. Such a chart turns out to be very useful as an organizing scheme and for predicting pathways for radioactive decay and nuclear transformations. To the nuclear chemist, this systematic display of the stable nuclides is as important as the periodic table of the elements itself. The stable isotopes form a **band of stability** on the chart. This band of stability has a slope corresponding to about a 1:1 proton–neutron ratio for the lightest nuclei. Typical

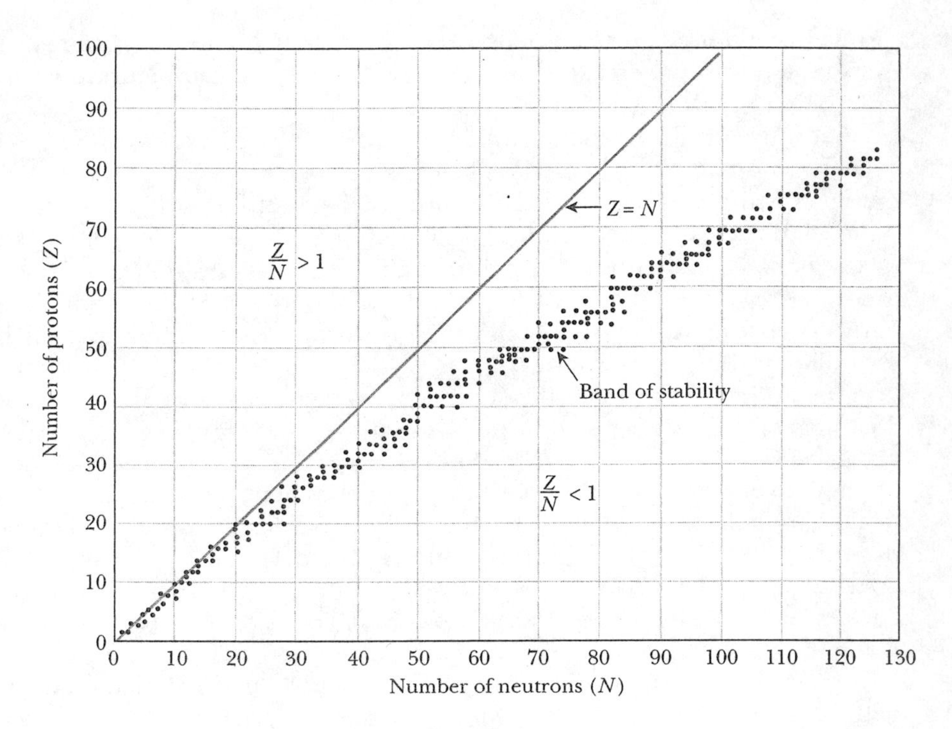

FIGURE 22.5 Plot of proton (*Z*) versus neutron (*N*) number. Each dot on this graph represents a stable combination of protons and neutrons. These dots trace out a "band of stability." Other combinations of protons and neutrons are unstable and can be expected to decay.

examples are $^{4}_{2}He$, $^{14}_{7}N$, and $^{40}_{20}Ca$. For heavier nuclei, the slope of the band of stability gradually decreases toward a limiting value of about a 1:1.5 proton–neutron ratio; for example, $^{208}_{82}Pb$. Note that the band of stability stops at 83 protons and 126 neutrons. No known nuclides with larger numbers of protons or neutrons are stable. All nuclides with proton–neutron combinations off the band of stability are unstable and will decay. There are also some nuclides along the band of stability that are not stable. In the next section we will discuss how the mode of decay is dependent on the relationship of the position of the nuclide on the chart to the band of stability.

Unstable Nuclides

Unstable nuclear arrangements arise from three general sets of circumstances, and the instability tends to be self-correcting through a number of decay processes.

1. **Excessively large number of protons and neutrons (beyond the end of the band of stability).** In this case there is just too much mass. Nature's solution to the problem is to have the unstable nuclide lose an α-particle. The driving force for this process is the high stability of the α-particle and the fact that the expelled α-particle carries away considerable mass compared to other radioactive emissions. With only a few exceptions, α-decay occurs among nuclides with mass 200 or greater, such as ^{235}U:

$$\underset{\text{parent}}{^{235}_{92}U} \longrightarrow \underset{\text{daughter}}{^{231}_{90}Th} + \underset{\alpha\text{-particle}}{^{4}_{2}He} \qquad \alpha\text{-decay process}$$

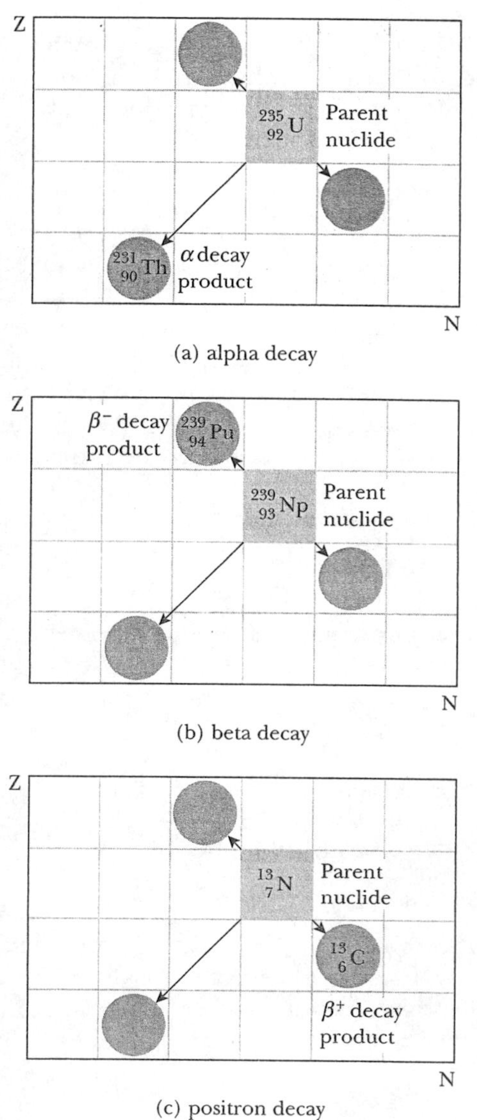

Parent nuclides and their decay products. (a) Alpha decay, (b) beta decay, and (c) positron decay.

The mass number is reduced by four units and the atomic number is reduced by two units as the heavier parent nuclide produces a lighter daughter nuclide. Note that "parent" and "daughter" are the nuclear chemistry equivalents of reactant and product.

This is the first of a number of nuclear reactions that we will write. In writing these reactions it is convenient to give the mass number and the atomic (proton) number for every nuclide. Nuclear equations must balance, just as ordinary chemical equations must. The sum of the mass numbers must have the same total on each side of the equation to ensure that the total number of nucleons has not changed. In addition, the atomic numbers must have the same total on either side to show that there has been an overall conservation of charge.

2. **An excess of neutrons over protons (to the right of the band of stability).** Unstable nuclides tend to rectify this situation by converting a neutron to a proton and emitting an electron and an antineutrino:

$$\underset{\text{parent}}{^{130}_{53}\text{I}} \longrightarrow \underset{\text{daughter}}{^{130}_{54}\text{Xe}} + \underset{\text{electron}}{^{0}_{-1}\beta} + \underset{\text{antineutrino}}{\bar{\nu}} \qquad \beta^{-}\text{-decay process}$$

Notice that the beta particle is written with a mass number of zero because it contains no protons or neutrons and an atomic number of -1 because that is its charge. Writing the beta particle this way simplifies balancing the nuclear equation. By the process of beta decay the atomic number increases to the next element and at the same time an electron or β-particle is ejected, conserving electrical charge of the nucleus. The nuclear mass number is unchanged. Many artificial isotopes are produced in nuclear reactions by bombarding target nuclides with neutrons giving isotopic products that have too many neutrons and generally decay by beta emission.

3. **An excess of protons over neutrons (to the left of the line of stability).** Unstable nuclides correct this situation by two different mechanisms as follows.
   (a) **Conversion of a proton into a neutron and emission of a positron and a neutrino:**

$$\underset{\text{parent}}{^{13}_{7}\text{N}} \longrightarrow \underset{\text{daughter}}{^{13}_{6}\text{C}} + \underset{\text{positron}}{^{0}_{+1}\beta} + \underset{\text{neutrino}}{\nu} \qquad \beta^{+}\text{-decay process}$$

   (b) **Capture of an electron by the nucleus and conversion of a proton to a neutron:**

$$\underset{\text{parent}}{^{79}_{36}\text{Kr}} + \underset{\text{electron}}{e^{-}} \longrightarrow \underset{\text{daughter}}{^{79}_{35}\text{Br}} + \underset{\text{neutrino}}{\nu} \qquad \text{electron capture process}$$

Either route leads to daughters with the atomic number reduced by one but with the same mass number as the parent. Note that a neutron is produced from the excess proton and the captured electron. **Electron capture** is especially interesting to chemistry because this decay process is known to be influenced by chemical composition.

Half-Life and Rate of Nuclear Decay

Radioactive decay is an event governed by the laws of probability. There is a statistical probability that a given atom will decay within a given time without regard to the origin or present environment of the nuclide in question. The decay rate for a sample of a particular atomic nucleus is directly proportional to the number of nuclei present. If twice as many are present, there will be twice the number of disintegrations per unit time. The process follows first-order kinetics (Chapter 15). The rate for a first-order chemical reaction is

$$\text{rate} = k[\text{A}] \qquad \text{first-order rate law}$$

where k is the specific rate constant, and [A] is the concentration of species A. As A is used up in the reaction, the concentration [A] decreases, eventually approaching zero. The equation that gives the concentration of component A as a function of time for a first-order reaction is

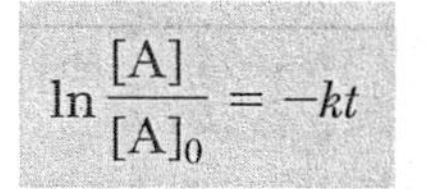

$$\ln \frac{[A]}{[A]_0} = -kt$$

first-order dependence of concentration on time

In this equation, k is the specific rate constant, $[A]_0$ is the initial concentration of A, and [A] is the concentration of component A after the reaction has proceeded for a time t. The **half-life** $t_{1/2}$ is the value of t when [A] is one-half of $[A]_0$ and is a particularly useful concept for a first-order reaction. At $t = t_{1/2}$, the preceding equation becomes

$$\ln \frac{\frac{1}{2}[A]_0}{[A]_0} = \ln \tfrac{1}{2} = -0.693 = -kt_{1/2}$$

Solving for $t_{1/2}$ gives

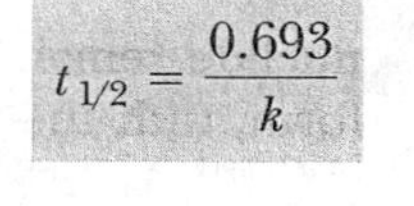

$$t_{1/2} = \frac{0.693}{k}$$

Relationship of half-life to *k*: The equation for half-life for any process with first-order kinetics, including first-order chemical reactions and nuclear decay. Notice that the half-life is independent of the amount of material.

The concepts of first-order kinetics and half-life can be directly applied to nuclear decay. However, the number of nuclei (N) available to decay is ordinarily used rather than [A], the concentration of A. Therefore, N replaces [A] and N_0, the number of nuclei present at $t = 0$, replaces $[A]_0$. The equation for the dependence of concentration on time becomes,

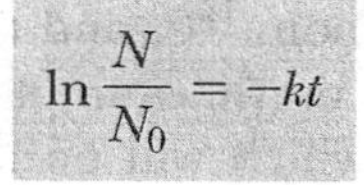

$$\ln \frac{N}{N_0} = -kt$$

Dependence of number of nuclei on time: The equation written for nuclear decay. N_0 is the number of nuclei at the start of the process and N is the number of nuclei remaining at time t. Thus, N/N_0 is the fraction of nuclei that still have not decayed at time t.

The half-life expression for nuclear decay is the same as for first-order chemical reactions,

$$t_{1/2} = \frac{0.693}{k}$$

Half-lives for radioactive nuclides range from small fractions of a second to well over 10^{14} years. As the nuclide decays, half of its starting mass disappears in $t_{1/2}$; half of what is left disappears in a second $t_{1/2}$; half of what is still left disappears in a third $t_{1/2}$, and so on. Thus, the fraction of the original mass that remains is $\frac{1}{2}$, $\frac{1}{4}$, $\frac{1}{8}$ and so forth after each successive $t_{1/2}$ period, and the fraction remaining after n half-lives is $(\frac{1}{2})^n$. The radioactive emissions are proportional to the amount of radioactive nuclide, and so these also drop off at the

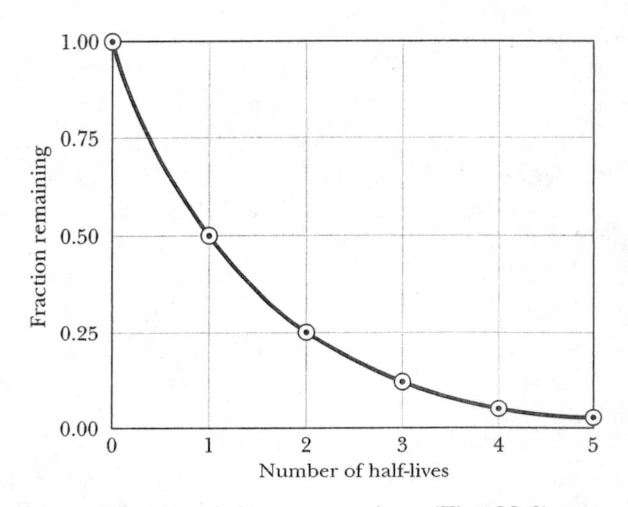

FIGURE 22.6 Radioactivity follows first-order decay. The number of nuclei of a radioactive isotope that remain is described by an exponential decay plot.

same rate. A plot of fraction of nuclide remaining versus time (Fig. 22.6) can be seen to have an exponential form, with the fraction dropping most rapidly at the beginning and asymptotically approaching zero.

The rate constant for radioactive decay is inversely proportional to $t_{1/2}$; that is, if $t_{1/2}$ is short, the fraction of the mass of the radioactive nuclide remaining will drop quickly and the initial production of radiation—γ-rays, α-particles, β^--particles, or positrons (β^+)—will be great. Thus, nuclides with short half-lives will produce high outputs of radiation for relatively short times, and nuclides with long half-lives will produce low outputs of radiation for long times.

Half-life is especially applicable to methods of radioactive dating, of which the most familiar is ^{14}C dating. The technique of ^{14}C dating is made possible by high-energy nuclear reactions of cosmic rays with the gases at the highest region of the atmosphere. These produce a number of products, including neutrons, protons, α-particles, ^3H, and, particularly, ^{14}C. The ^{14}C isotope mixes thoroughly with the common isotope of carbon, ^{12}C, and these are taken up together into the tissues of plants. Particularly useful examples are cotton, which will be converted into fabrics, and wood, which will be used for fashioning objects and structures.

The radioactive ^{14}C nucleus is a β-particle emitter with a half-life of 5720 years. The specific rate constant k for the decay of ^{14}C is calculated from half-life:

$$k = \frac{0.693}{t_{1/2}} = \frac{0.693}{5720 \text{ yr}} = 1.211 \times 10^{-4} \text{ yr}^{-1}$$

It has been shown that the fraction of ^{14}C in living plant tissue generally has been constant at 1.3×10^{-12} ^{14}C atoms/^{12}C atom since the very distant past. This shows that the rate of decay of the ^{14}C nuclei has been equal to the rate at which new ^{14}C atoms are formed in the upper atmosphere. The ^{14}C atoms in living plant tissue disintegrate at the rate of 15 nuclei per gram of carbon per minute, giving 15 β-particles per gram every minute. After a plant has been harvested and stops growing, no new ^{14}C is assimilated and the rate of emission of β-particles will slow as the ^{14}C decays. In 5720 years, the ^{14}C half-life, the number of emissions per gram, will have dropped to half the original value, and at twice 5720 years they drop to one-quarter. Thus, $(\frac{1}{2})^n$ is the fraction of the emission of 15 particles per gram per minute, which is obtained after n half-lives. We can now look at examples of the use of the ^{14}C dating technique.

EXAMPLE 22.1

Consider a bone known to contain 200. grams of carbon for which the decay rate is found to be 400. disintegrations/minute. How old is the bone?

COMMENT Determine how much the decay rate has decreased from the initial value. Then use the basic relationship for a first-order process and the known rate constant for ^{14}C decay. Alternatively, recall that the decay rate will be $(\frac{1}{2})^n$ of the original rate, where n is the number of half-lives that have elapsed since the bone was formed. Determine n, the number of half-lives elapsed, and then multiply this times $t_{1/2}$, the time of the half-life.

SOLUTION If the bone were still part of a living system, the decay rate would be

$$(15 \text{ decays/g min})(200.\text{ g}) = 3.0 \times 10^3 \text{ decays/min}$$

The decay rate has decreased by a factor of 400/3000.

The dating of the bone can now be handled in a direct fashion using the basic relationship

$$\ln \frac{N}{N_o} = -kt$$

where N/N_o, the fraction of ^{14}C remaining, is

$$\frac{N}{N_o} = \frac{400.}{3.0 \times 10^3} = 0.133$$

because the decay rate is proportional to the number of nuclei. Then

$$\ln \frac{N}{N_o} = \ln(0.133) = -2.0 = -(1.211 \times 10^{-4} \text{ yr}^{-1})\,t$$

$$t = \frac{2.0}{1.21 \times 10^{-4} \text{ yr}^{-1}} = 1.7 \times 10^4 \text{ yr}$$

Alternatively, the problem can also be solved using the half-life concept: After n half-lives the decay rate decreases by $(\frac{1}{2})^n$. Therefore,

$$\left(\frac{1}{2}\right)^n = \frac{1}{2^n} = \frac{400.}{3.0 \times 10^3}$$

$$2^n = \frac{3.0 \times 10^3}{400.} = 7.5$$

Take the logarithm of each side of the equation (either natural or common [base 10] logarithms can be used):

$$\log 2^n = \log 7.5$$

$$n \log 2 = \log 7.5$$

$$n = \frac{\log 7.5}{\log 2} = \frac{0.875}{0.301} = 2.91 \text{ half lives}$$

Finally, solve for the age of the bone:

$$t = nt_{1/2} = (2.91 \text{ half lives})(5720 \text{ yr/half life}) = 1.7 \times 10^4 \text{ yr}$$

EXERCISE 22.1(A)

A bone found in an archaeological dig is believed to be about 12,000 years old. What is the approximate decay rate for the ^{14}C radionuclide if that is true?

ANSWER Approximately 3–4 decays/min

EXERCISE 22.1(B)

A wood sample contains 10. g carbon and exhibits a decay rate of 10. disintegrations/min. How old is the sample?

ANSWER 2.2×10^4 yr

Anthropologists have successfully used radioisotope content to study and date fossils. The skeleton of an eleventh-century inhabitant of a South African village, for example, posed a significant problem. Physically, the man's skeleton was different from those of the other villagers, suggesting he was not a native of the area. However, when the skeleton was analyzed for isotopes, its $^{14}C/^{12}C$ ratio was found to be like that of other skeletons recovered from the same village. Therefore, this man lived with the other individuals at about the same time. The anthropologists concluded that the man in question probably migrated to the village from a distant region and then spent most of his life after that in the village.

Now that we have considered which nuclei are unstable and the rates of nuclear decay, we will consider the energy of nuclear reactions and the origin of this tremendous energy.

Mass Defect, Binding Energy, and Nuclear Stability

Knowing the exact mass of atomic nuclei provides a key to understanding their stability and the energy released in nuclear reactions, like radioactive decay. Although atomic masses are close to integral numbers of atomic mass units, they are not quite integers. All except ^1_1H are slightly less than the sum of the masses of the nucleons present, and this difference is called the mass defect. Thus, the **mass defect** is the difference between the actual mass of a nucleus and the sum of the masses of its nucleons (subatomic components). According to Einstein and the special theory of relativity, this mass deficiency or mass defect has an energy equivalent known as the nuclear binding energy. The **nuclear binding energy** is the energy calculated by $E = mc^2$ where m is the mass defect. As an illustration, consider the following balance sheet for the α-particle:

The masses for the isolated (or free) nucleons involved are

$$\begin{aligned} 2 \text{ neutrons} &= 2(1.008665 \text{ amu}) = 2.017330 \text{ amu} \\ 2 \text{ protons} &= \underline{2(1.007276 \text{ amu}) = 2.014552 \text{ amu}} \\ \text{Total mass} &= 4.031882 \text{ amu} \end{aligned}$$

This is what one would expect for the mass of one α-particle. However, the actual measured mass is 4.00151 amu.

Total mass of separated nucleons = 4.031882 amu

Combined mass of the α particle = 4.00151 amu

Mass defect = 0.03037 amu

The mass defect, 0.03037 amu, is the difference between what the α-particle mass "should" be and what it is experimentally known to be. The binding energy is the energy equivalent of the mass difference using the Einstein relationship between mass and energy, $E = mc^2$. First determine the defect for a single α-particle,

$$m = \frac{0.03037 \text{ g/mol}}{6.022 \times 10^{23} \text{ particles/mol}} \times \frac{1 \text{ kg}}{1000 \text{ g}} = 5.043 \times 10^{-29} \text{ kg}$$

Now calculate the binding energy,

$$E = mc^2 = (5.043 \times 10^{-29} \text{ kg})(3.00 \times 10^8 \text{ m/s})^2 = 4.54 \times 10^{-12} \text{ J}$$

Thus, 4.54×10^{-12} J is the total binding energy for one α-particle. Note that in nuclear chemistry energies are often given per particle rather than per mole of particles as is common in ordinary reaction chemistry.

The stability of a nucleus is generally expressed in binding energy per nucleon because the more nucleons there are to be held together in the nucleus, the more binding energy that is needed. There are four nucleons in the α-particle—two protons and two neutrons—leading to an atomic mass number of 4. Thus, the binding energy per nucleon for an α-particle is

$$\frac{4.54 \times 10^{-12} \text{ J}}{4 \text{ nucleons}} = 1.13 \times 10^{-12} \text{ J/nucleon}$$

The mass defect provides the energy that holds together the α-particle or any other nuclide. To help understand this, consider bringing two protons and two neutrons together to form an α-particle. To do that, energy equivalent to the mass defect, 4.54×10^{12} J, would have to be released. The nucleus stays together as an α-particle until such time as the energy necessary to break it up into its nucleons is suddenly available. This would be the same as the energy given off when the α-particle is formed from its nucleons. So the binding energy is the energy equivalent of the missing mass, the energy with which the nucleus is held together. In the next example, we will look at the binding energy in the nucleus of an atom.

EXAMPLE 22.2

Use the masses of particles and isotopes in Table 22.2 to calculate the mass defect and binding energy per nucleon for the ^{16}O atom.

COMMENT From the table, the mass of the ^{16}O atom is 15.994915 amu, including its eight electrons. The total mass of the nucleons plus the electrons can be determined by summing the masses of eight neutrons and eight hydrogen atoms. The eight hydrogen atoms have eight protons and eight electrons. Then, find the mass defect for the oxygen atom, the mass of the nucleons and the electrons minus the mass of the ^{16}O atom.

SOLUTION

$$\begin{aligned} &8 \text{ neutrons} &&= 8(1.008665) = 8.069320 \text{ amu} \\ &8 \text{ hydrogen atoms} &&= \underline{8(1.007825) = 8.062600 \text{ amu}} \end{aligned}$$

$$\text{Total mass} = 16.131920 \text{ amu (8 neutrons, 8 protons, 8 electrons)}$$

Therefore, the atom of oxygen-16 has less mass than the sum of its nucleons and electrons. The difference is

$$\begin{aligned} &\text{sum of the parts} &&= 16.131920 \text{ amu} \\ &\text{known mass} &&= \underline{15.994915 \text{ amu}} \\ &\text{difference} &&= 0.137005 \text{ amu} \end{aligned}$$

Converting amu to kg and applying Einstein's equation,

$$\frac{0.137005 \text{ g/mol}}{6.022 \times 10^{23} \text{ particles/mol}} \times \frac{1 \text{ kg}}{1000 \text{ g}} = 2.28 \times 10^{-28} \text{ kg}$$

$$E = mc^2 = (2.28 \times 10^{-28} \text{ kg})(3.00 \times 10^8 \text{ m/s})^2 = 2.05 \times 10^{-11} \text{ J}$$

There are 16 nucleons in an ^{16}O nucleus, 8 protons and 8 neutrons. The binding energy per nucleon is

$$\frac{2.05 \times 10^{-11} \text{ J}}{16 \text{ nucleons}} = 1.28 \times 10^{-12} \text{ J/nucleon}$$

EXERCISE 22.2(A)

Calculate the mass defect and the binding energy per nucleon for ^{19}F.

ANSWER Mass defect = 0.157832 amu; binding energy per nucleon = 1.24×10^{-12} J/nucleon

EXERCISE 22.2(B)

Calculate the mass defect and binding energy for a tritium atom, ^3_1H.

ANSWER Mass defect = 0.00910 amu; binding energy = 1.36×10^{-12} J

The binding energy per nucleon is generally 1.3×10^{-12} J or less. A plot of binding energy per nucleon versus mass number illustrates the point (Fig. 22.7). The maximum in the curve of binding energy per nucleon occurs at a mass number in the vicinity of the elements iron and nickel. According to this curve of binding energy per nucleon, the isotopes of iron and nickel include the most stable nuclei in the periodic table of the elements.

The existence of a maximum in the middle of the binding energy curve is the basis for understanding how we can generate energy from nuclear fission, the splitting of the heaviest elements, and nuclear fusion, the combining of the lightest elements.

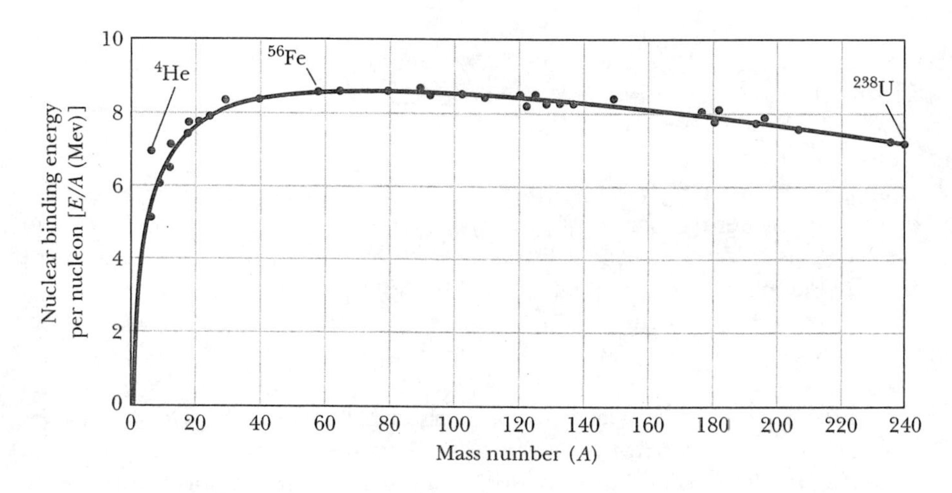

FIGURE 22.7 The curve of nuclear stability. Measured as a plot of binding energy per nucleon versus mass number (*A*). The maximum in the curve corresponds to the most stable nuclei, those of the elements close to iron in the periodic table.

Synthetic Nuclear Reactions

Most of our information about nuclei has been obtained by hurling particles at them and watching what happens. Early experiments were limited. One nuclide could be converted to another by bombarding the parent with an α-particle or a γ-ray photon emitted from a naturally radioactive source. Rutherford and his students in 1911 carried out the first such transformations, successfully injecting α-particles from a radium or polonium source into nitrogen nuclei:

$$^{14}_{7}\text{N} + {}^{4}_{2}\text{He} \longrightarrow {}^{18}_{9}\text{F}^* \longrightarrow {}^{17}_{8}\text{O} + {}^{1}_{1}\text{H}$$

The written reaction includes an intermediate fluorine nucleus in an excited or unstable state, as indicated by the asterisk. This is typical of nuclear bombardment reactions. The target nucleus, or parent, absorbs the α-particle and forms a compound nucleus in an excited state, which subsequently decays by one or more routes to more stable daughters.

The first transformations using accelerated particles with high kinetic energies were carried out with protons in the early 1930s. Because protons carry an electric charge, they can be accelerated through an evacuated tube by the application of a very high electric potential. The kinetic energy per particle can be expressed in electron volts, eV, the product of the particle's charge in units of electronic charge times the accelerating voltage in volts.

$$KE = eV$$

Kinetic energy of an accelerated electron: The product of the charge of the particle (*e*) and the accelerating potential (*V*).

Kinetic energy is the energy of motion,

$$KE = \tfrac{1}{2}\, mv^2 \qquad \text{kinetic energy of any object in motion.}$$

The units often used to express the kinetic energy (KE) of the particle are electron volts (eV) and kiloelectron volts (keV) but mostly megaelectron volts (MeV). These correspond respectively to the energy acquired by a charged particle carrying a unit electric charge when it is accelerated through potential differences of 1 volt, 10^3 volts, 10^6 volts. The conversion factor to joules is

$$1 \text{ MeV} = 1.60 \times 10^{-13} \text{ J}$$

A kinetic energy of 1 MeV is approximately equivalent to the energy needed to overcome the potential barrier a proton experiences when it collides with light nuclei such as lithium and beryllium. This barrier is simply the repulsion between the like charges of the proton and the nucleus. For heavier nuclei, which often carry much larger nuclear charges, the proton kinetic energy required to cause a nuclear reaction is correspondingly higher.

The development of van de Graaff generators, linear accelerators, and the first cyclotrons greatly improved the experimental scope of nuclear reactions. The van de Graaff generator is an electrostatic generator capable of producing carefully controlled potentials of several million volts. Among other things, the device served as an early source of neutrons by making it possible to bombard beryllium nuclei with deuterons, the nuclei of deuterium ${}^2_1\text{H}$ atoms. Neutrons have the symbol ${}^1_0\text{n}$ since their atomic mass number is 1 and their charge is 0.

$${}^9_4\text{Be} + {}^2_1\text{H} \longrightarrow {}^1_0n + {}^{10}_5\text{B}$$

In a **linear accelerator** (Fig. 22.8), an ion is accelerated through a tube within a magnetic field, which serves to hold it in a straight line. As the ion particle emerges from the tube, the polarity is reversed, accelerating the particle into a second tube. By repeating the process, the charged particle is subjected to a series of electric kicks. Each tube must be longer than the one before it, so that the alternating polarity of all the tubes can be synchronized.

Cyclotrons (Fig. 22.9) also operate on the principle of multiple acceleration of ions, and in many ways they can be thought of as linear accelerators wrapped into a spiral configuration. The tubes are replaced by a pair of flat, hollow, semicircular electrodes called "dees" because they resemble the letter D. Every time the particle crosses from one dee to the other, it gets an energy boost. During the past 50 years, largely because of the development of this kind of technology for accelerating and detecting particles, synthesis of new nuclei has become a distinct subfield of chemistry.

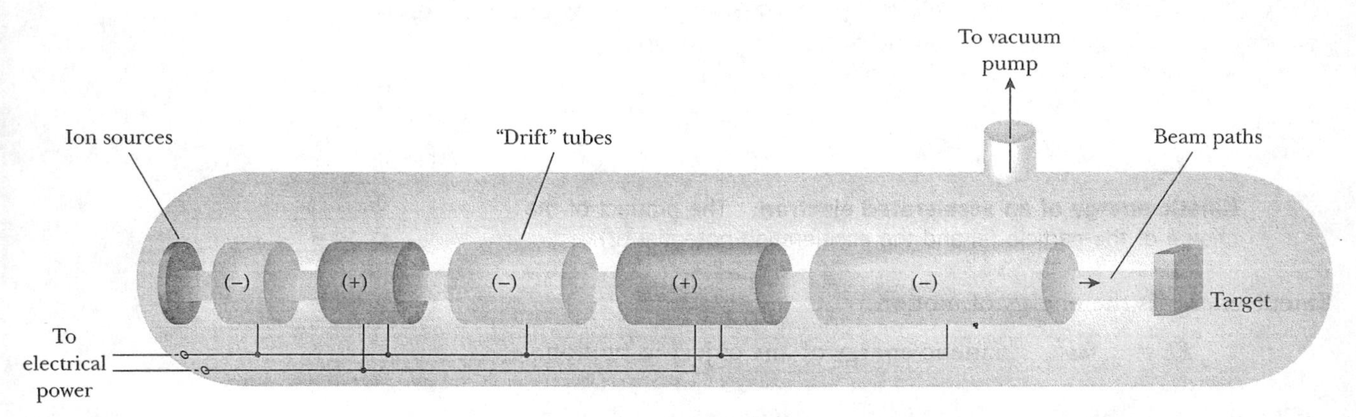

FIGURE 22.8 Linear accelerator. An accelerated ion is given a boost in energy each time the polarity is reversed.

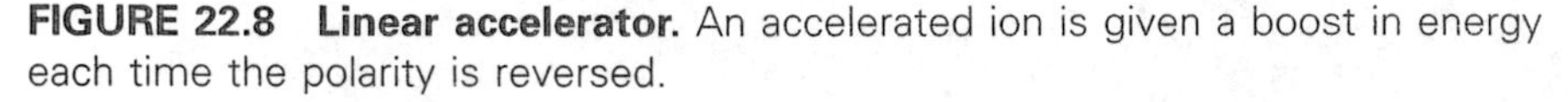

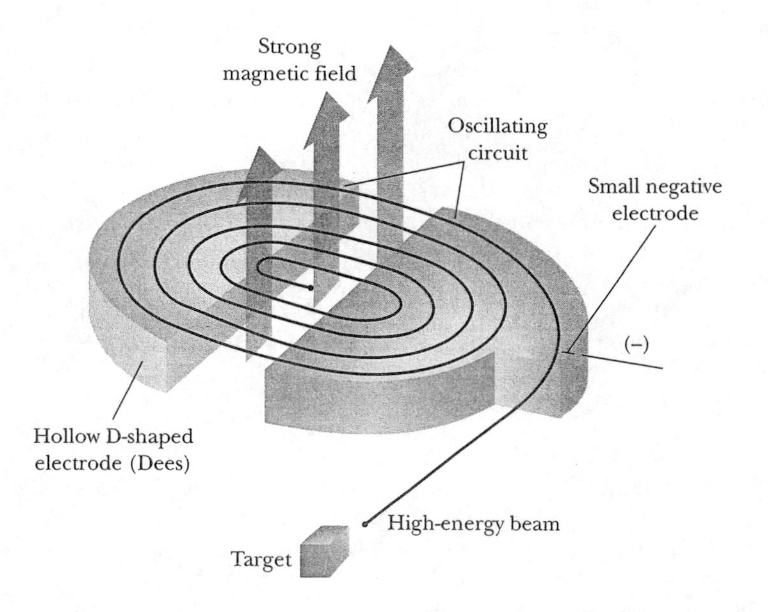

FIGURE 22.9 Cyclotron. Multiple acceleration of ions wrapped into a spiral configuration.

As with all chemical reactions, nuclear reactions may either release energy or absorb it. Consider, for example, the Cockcroft–Walton experiment. Using a high-voltage generator, 6.4×10^{-14} J or 0.40 MeV protons (that is, protons accelerated to that kinetic energy) were hurled into lithium nuclei, producing two 1.4×10^{-12} J or 8.8 MeV α-particles. The energy accounting is interesting:

$$^{7}_{3}\text{Li} + {}^{1}_{1}\text{H} + \text{energy} \longrightarrow {}^{8}_{4}\text{Be}^* \longrightarrow 2\,{}^{4}_{2}\text{He} + \text{energy}$$

$$\begin{aligned} \text{energy in} &= 6.4 \times 10^{-14}\text{ J} \\ -\text{energy out} &= \underline{-2.8 \times 10^{-12}\text{ J}} \\ \text{energy overall} &= -2.7 \times 10^{-12}\text{ J} \end{aligned}$$

Hence, the process is net energy-releasing overall, and there should be an equivalent amount of mass lost in going from reactant nuclides to product nuclides:

$$\begin{aligned} &\text{reactant nuclides } 7.016005 + 1.007825 = 8.023828\text{ amu} \\ &\text{product nuclides } 2(4.00260)\text{ amu} = \underline{8.00520\text{ amu}} \\ &\text{mass difference} = 0.01863\text{ amu} \end{aligned}$$

The binding energy is calculated from the mass difference and $E = mc^2$:

$$E = \left(\frac{0.01863\text{ g/mol}}{6.022 \times 10^{23}\text{ particles/mol}}\right)\left(\frac{1\text{ kg}}{1000\text{ g}}\right)(3.00 \times 10^{8}\text{ m/s})^2 = 2.8 \times 10^{-12}\text{ J}$$

Notice that atomic masses were used in this calculation without regard to the mass of the electrons in the atoms. This is allowable because the Li + H reactants and the He + He products both have four electrons. Consequently, the masses of the electrons cancel each other.

The worldwide synthetic radioisotope preparation has expanded rapidly due, in large part, to the booming radio-pharmaceuticals industry, which produces a steadily increasing number of radio-labeled drugs for diagnostic and therapeutic use in medicine. All synthetic isotopes, with the exception of those derived from uranium decay, must be produced by neutron irradiation or ion bom-

(text continues on pg. 1014)

PROCESSES IN CHEMISTRY

Modern Alchemy—Synthesis of New Elements By the early years of the twentieth century, it was clear to scientists of the day that a limit had been reached in the search for new or missing elements by conventional means. Through studies of optical spectra, rubidium, cesium, indium, helium, and gallium had been discovered. X-ray spectral studies led to hafnium and rhenium.

Investigations of natural radioactivity revealed polonium (84), radon (86), francium (87), radium (88), actinium (89), and protactinium (91). Nuclear reactions and artificially induced radioactivities produced technetium (43), promethium (61), and astatine (85). The transuranium elements to 106 were largely discovered between 1934 and 1974 by research groups in California at the Lawrence Berkeley Laboratory and Lawrence Livermore Laboratory by Glenn Seaborg and coworkers. At the same time, scientists were making similar and sometimes conflicting discoveries of transuranium elements at the Joint Institute for Nuclear Research in Russia. These new elements were produced by fusing the heaviest elements known at the time, such as curium, californium, or einsteinium, with ions of neon, carbon, oxygen, or helium.

Because heavier projectiles are needed to produce still heavier elements, powerful particle accelerators needed to be used. Cyclotrons, for example, have been used to accelerate increasingly heavy ions to fusion energies. The target elements into which these accelerated ions are slammed have been microgram-to-milligram quantities of lead and bismuth and, more recently, the actinide elements. The actinides are a series of chemically similar elements named for actinium, the beginning member of the series (element 89), and ending at lawrencium, element 103.

Using lead and bismuth with beams of chromium and iron produced elements 107, 108, and 109 in the 1980s. The first atoms of element 110 were documented on November 9, 1994, at GSI, the heavy ion research center at Darmstadt, Germany, by a team headed by Peter Armbruster and Sigurd Hofmann. During the course of a two-week experiment in which a lead-208 target was irradiated with nickel-62 ions, four atoms of the 271 isotope were observed.

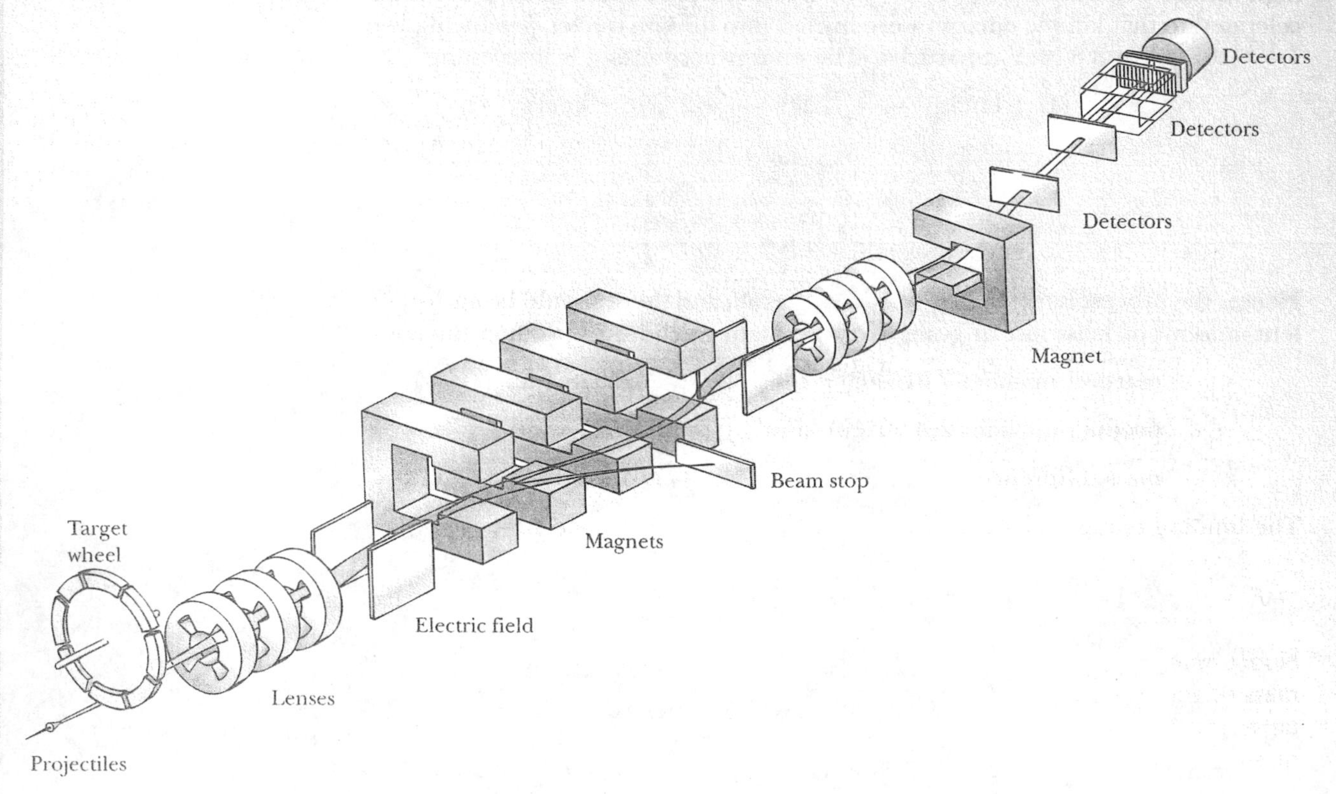

Heavy-ion separator. Schematic diagram of the 11-meter-long GSI Separator for Heavy Ion Products (SHIP). Note the projectile beam and the rotating target wheel.

Four atoms! Not 4 grams, or even micrograms, but four atoms. That accomplishment was followed by the preparation of the first atom—not atoms, but atom—of element 111 on December 8, 1994, using a nickel-64 projectile and a bismuth-209 target. It was an atom of the 272 isotope.

Since then, Armbruster and Hofmann have reported the synthesis of element 112 by fusing zinc ions into a lead target. The newly formed isotope has 165 neutrons and decays after a brief 280-microsecond existence by emitting an alpha particle to form a new isotope of element 110, which in turn decays to an isotope of hassium, element 108, in about 110 microseconds, and finally into an isotope of seaborgium, element 106. After emission of more alpha particles, fermium, element 100, is reached. In 1999 element 114 was reported by Russian and American workers. Even more recently, elements 116 and 118 have been announced.

It is also interesting to note that just as the atoms of the noble gases—helium, neon, argon, krypton, and xenon—owe their special stability to closed shells of electrons, certain atomic nuclei exhibit special stability, which may be attributable to closed shells of neutrons and protons. For example, unusual stability due to closed shells is observed for proton number Z or neutron number N, equal to 2, 8, 20, 28, 50, and 82 and for neutron number 126 and, possibly, 162. These numbers have been designated as "magic numbers." Lead is a doubly magic nucleus with 82 protons and 126 neutrons and is the stable end-product of several decay chains of unstable heavier nuclei. It now appears that nuclei centered around $Z = 108$ and $N = 162$ are stable enough to be observed.

Questions

Why did the alchemists fail to transmute elements, whereas it is now done routinely?

What is the isotopic end-product of the element 112 decay series if it proceeds by successive alpha particle emission to an isotope of fermium?

It has been proposed that element 116 can be prepared using a selenium-82 projectile and a lead-208 target. Which isotope of 116 will be produced?

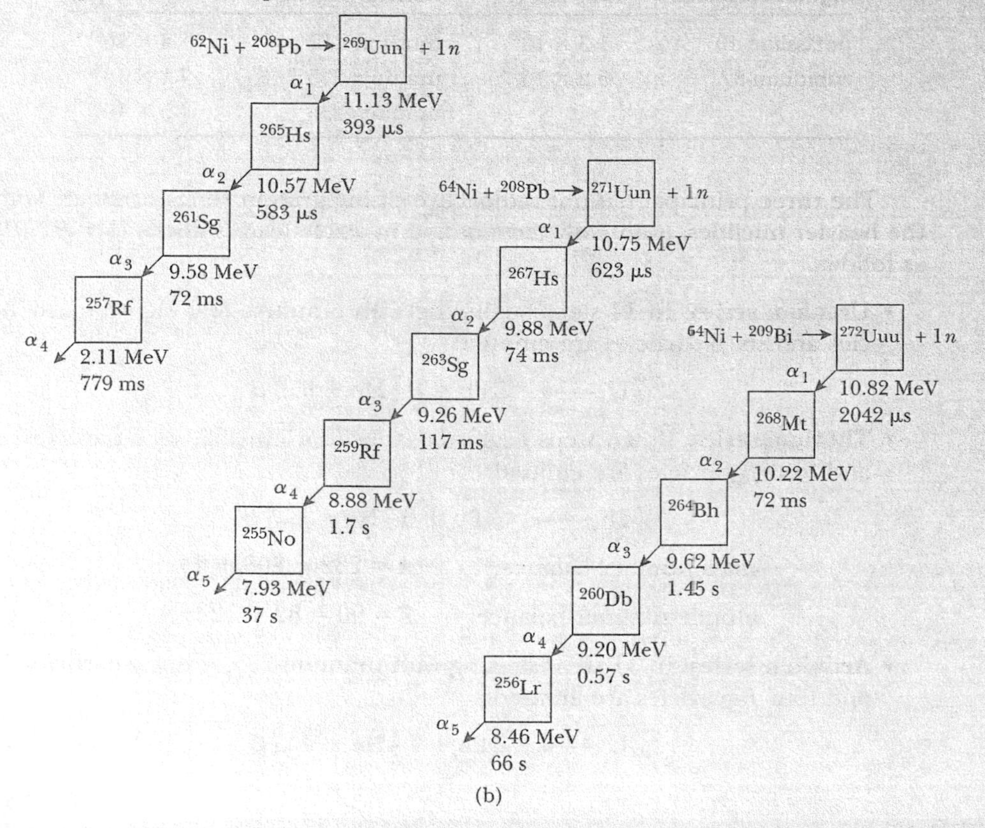

Decay series for the new elements 110 and 111. These series end at lawrencium, nobelium, and rutherfordium. The 112 series (not shown here) ends at seaborgium. Energies and half-lives are given for each process.

bardment. Both procedures require cyclotrons and other costly equipment. For 50 of the elements in their 240 isotopes there is only one reliable source in the world, a bank of 30 electromagnetic separators owned by the U.S. Department of Energy (DOE) at the Oak Ridge National Laboratories in Tennessee.

22.3 NUCLEAR DECAY IN NATURE

Nuclides of *naturally* radioactive elements disintegrate or decay into other elements. With the notable exception of $^{14}_{6}C$, which is formed in the upper atmosphere, and perhaps tritium, $^{3}_{1}H$, these nuclides fall into two main categories. There are those with half-lives about as long as the age of the Earth, estimated to be about 4.5×10^9 years, and those that are nuclear daughters, being continually produced from one of three disintegration or decay series, each beginning with particularly long-lived parent nuclides. Representatives of the first group, with half-lives on the geologic time scale, can be found both early in the periodic table and among the heavy elements:

Lighter Nuclides	Half-life (yr)	Heavier Nuclides	Half-life (yr)
potassium-40	1.3×10^9	thorium-232	1.4×10^{10}
rubidium-87	5.2×10^{10}	uranium-235	7.1×10^8
		uranium-238	4.5×10^9

The three principal natural radioactive disintegration series originate with the heavier nuclides, eventually terminating in stable lead isotopes (Fig. 22.10) as follows.

- **Uranium series:** In 14 steps beginning with uranium-238, eight α-particles and six β-particles are emitted:

$$^{238}_{92}U \longrightarrow {}^{206}_{82}Pb + 8\,{}^{4}_{2}He + 6\,{}^{\;0}_{-1}\beta$$

- **Thorium series:** In ten steps beginning with thorium-232, six α-particles and four β-particles are emitted:

$$^{232}_{90}Th \longrightarrow {}^{208}_{82}Pb + 6\,{}^{4}_{2}He + 4\,{}^{\;0}_{-1}\beta$$

mass number balance: $A = 232 = 208 + 24$

atomic number balance: $Z = 90 = 82 + 12 - 4$

- **Actinium series:** In 11 steps starting with uranium-235, seven α-particles and four β-particles are emitted:

$$^{235}_{92}U \longrightarrow {}^{207}_{82}Pb + 7\,{}^{4}_{2}He + 4\,{}^{\;0}_{-1}\beta$$

EXAMPLE 22.3

The actinium series, beginning with ^{235}U, ends up eleven steps later at the stable ^{207}Pb isotope. The first three steps are α-decay, then β-decay, and then α-decay:

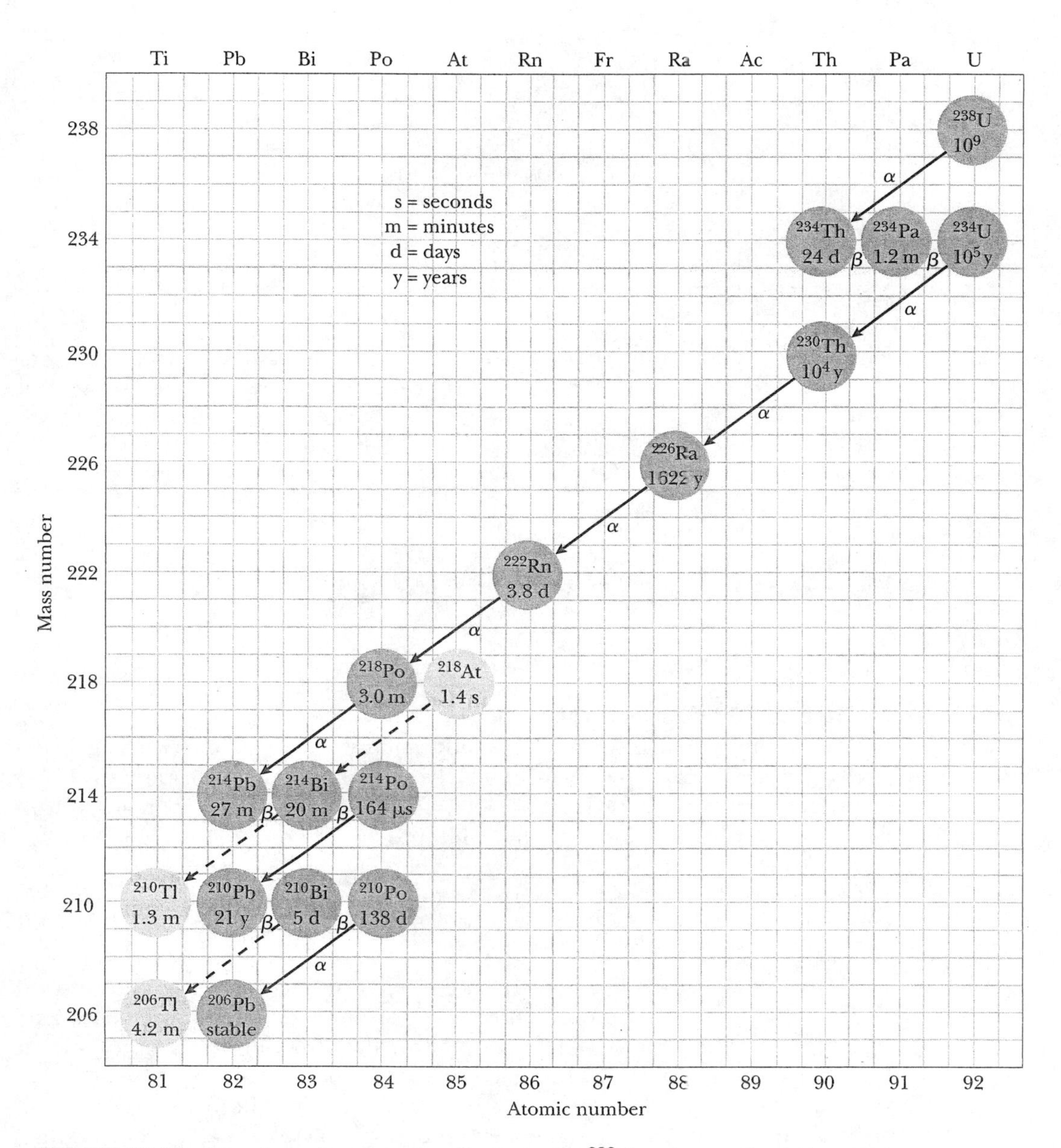

FIGURE 22.10 Uranium decay series. Begins with ^{238}U and extends to the stable lead isotope ^{206}Pb.

$$^{235}_{92}\text{U} \longrightarrow {}^{231}_{90}\text{Th} + {}^{4}_{2}\text{He}$$

$$^{231}_{90}\text{Th} \longrightarrow {}^{231}_{91}\text{Pa} + {}^{0}_{-1}\beta$$

$$^{231}_{91}\text{Pa} \longrightarrow {}^{227}_{89}\text{Ac} + {}^{4}_{2}\text{He}$$

If the next three steps of the decay pattern are α-decay, β-decay, α-decay, write the nuclear reactions.

COMMENT Make sure that the equations are balanced in terms of mass number and proton number and remember that loss of an electron converts a neutron to a proton.

SOLUTION

$$^{227}_{89}\text{Ac} \longrightarrow\ ^{223}_{87}\text{Fr} + \,^{4}_{2}\text{He}$$

$$^{223}_{87}\text{Fr} \longrightarrow\ ^{223}_{88}\text{Ra} + \,^{0}_{-1}\beta$$

$$^{223}_{88}\text{Ra} \longrightarrow\ ^{219}_{86}\text{Rn} + \,^{4}_{2}\text{He}$$

EXERCISE 22.3

The ^{233}U isotope, an important nuclide that could be used in nuclear reactors, does not exist in nature. However, it has been successfully produced by nuclear reactions beginning with ^{232}Th. The process begins with neutron capture, followed by successive β-emissions. Write the nuclear reactions.

22.4 ENERGY FROM NUCLEAR REACTIONS

Fission

In the fall of 1938, scientists working in Germany performed one of the most profoundly important experiments in modern times. Otto Hahn (German, 1879–1968), Fritz Strassmann (German, 1902–1985), and Lise Meitner (Austrian, 1879–1969) accidentally discovered that when uranium nuclei were bombarded with neutrons, they became unstable and split into two fragments, forming more stable nuclei with intermediate atomic numbers. This process is called **fission**. The fission process that followed absorption of the neutron is analogous to the behavior of a liquid drop as it contracts, elongates, and eventually comes apart (Fig. 22.11). This is the liquid drop model of nuclear fusion.

In addition to the enormous release of energy, the most significant feature of nuclear fission is that more neutrons are produced than consumed. As a result, a **chain reaction** is feasible, with neutrons released in one fission reaction bringing about a second fission reaction, and so on. A chain reaction is one in which a product of one reaction event or step starts another reaction event. In general, in a chain reaction, one reaction event starts several new events so that, once started, the extent of the reaction grows very rapidly. To guarantee a self-sustaining chain reaction, such as that in a nuclear reactor, each fission must produce excess neutrons that can initiate successive fission reactions. It is necessary to achieve maximum neutron production, along with minimum loss due to nonfission absorption or escape from the surface of the fuel. The number of neutrons produced is proportional to the volume of reactor fuel, which is a function of r^3 where r is a characteristic dimension of the fuel, such as its radius if it is in the shape of a sphere. However, neutron losses from the surface are proportional to r^2, which is a function of the surface area. For example, if we double the size of r for the fuel, we increase neutron production eightfold, whereas neutron loss only increases fourfold. As a consequence, there must be a certain minimum or critical size of fuel mass at which the chain reaction produces enough neutrons to become self-sustaining. Applications of nuclear fission for

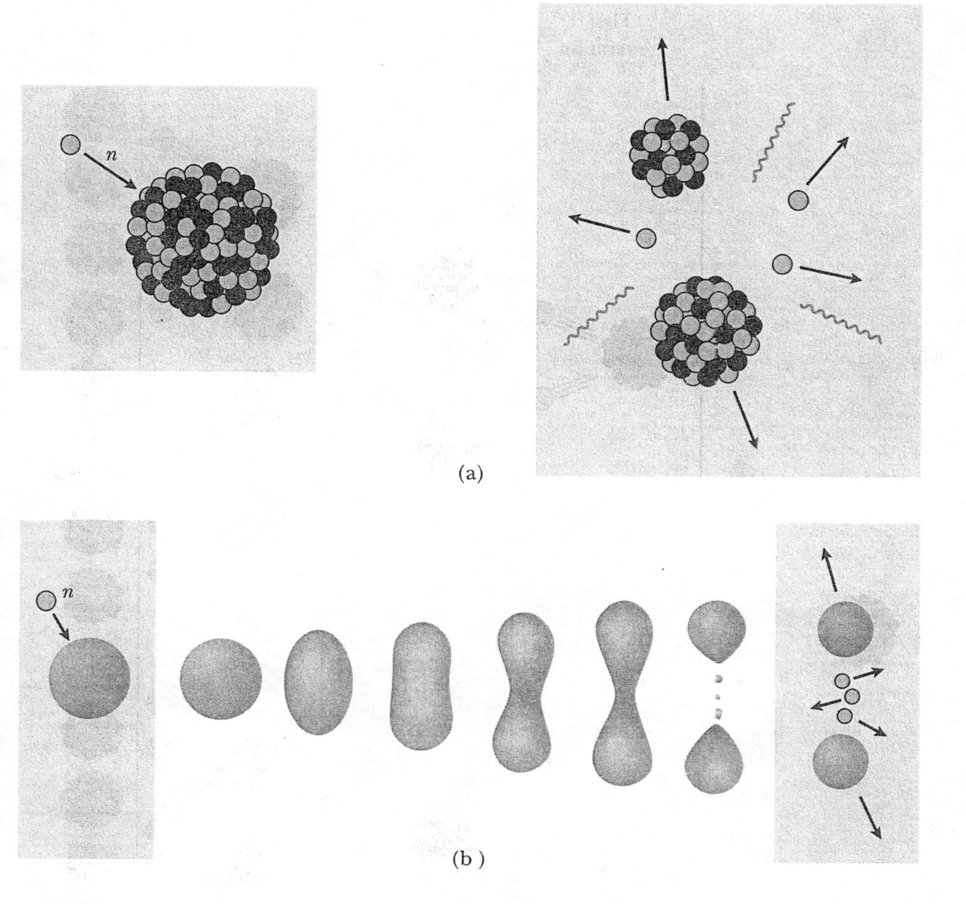

FIGURE 22.11 Nuclear fission. (a) Schematic diagram representing the uranium fission process. (b) Bohr–Wheeler liquid drop model for heavy nuclei.

weapons and electric power were immediately evident. Table 22.3 lists three nuclides used as fission reactor fuels.

Uranium as found in nature is composed of two isotopes: ^{235}U (0.711%), is fissionable, or supports a self-sustaining chain reaction, whereas the principal component, ^{238}U (99.3%), is not fissionable. Therefore, methods for separating, converting, and enriching the isotopes are important. They cannot be separated by chemical means because as isotopes their chemistry is almost identical, but they can be separated physically, and fuels that are even slightly enriched have proved satisfactory for power reactors. Enrichment of ^{235}U to only 2 to 4% is sufficient for what is called a light-water reactor (LWR) nuclear power plant. Weapons-grade fuels need to be enriched considerably more.

The nuclear reactor that has become the common commercial source of nuclear electric power and isotope production is known as the thermal reactor (Fig. 22.12). It consists of assemblies of fuel rods, all surrounded by a medium containing atoms of low atomic number. This medium can be a solid, a liquid, or a gas and is called a **moderator**. The role of the moderator is to make propagation of the chain reaction possible by slowing the neutrons to the point where they can be absorbed by the heavy nuclei. Neutrons, being neutral, are not repelled by nuclei and can pass right through without being absorbed if they are going too fast. The probability that a neutron will be absorbed in a collision with

FIGURE 22.11 (continued) (c) The beginning of a chain reaction; the center nucleus has undergone fission, coming apart into two pieces and releasing gamma rays and more neutrons. Some of the neutrons are captured by other nuclei, propagating the growing chain reaction.

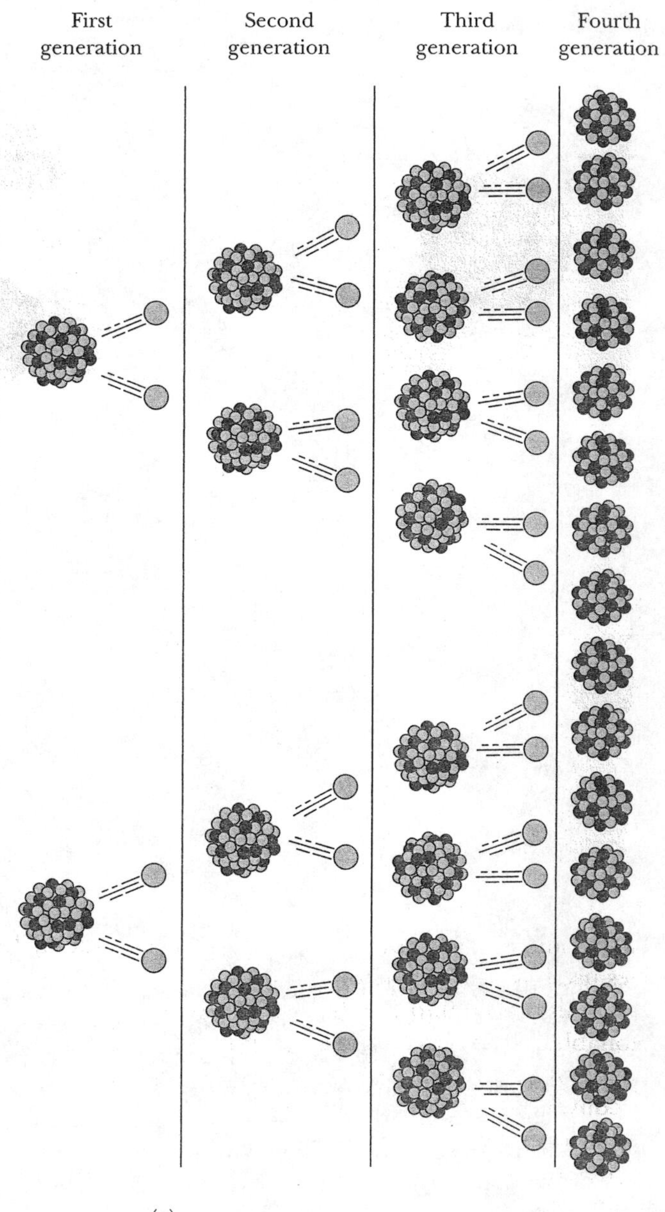

(c)

TABLE 22.3 Fission Fuels

Uranium-235	Occurs in nature and constitutes 0.7% of natural uranium
Uranium-233	Formed by neutron capture in thorium
Plutonium-239	Produced by neutron capture in ^{238}U

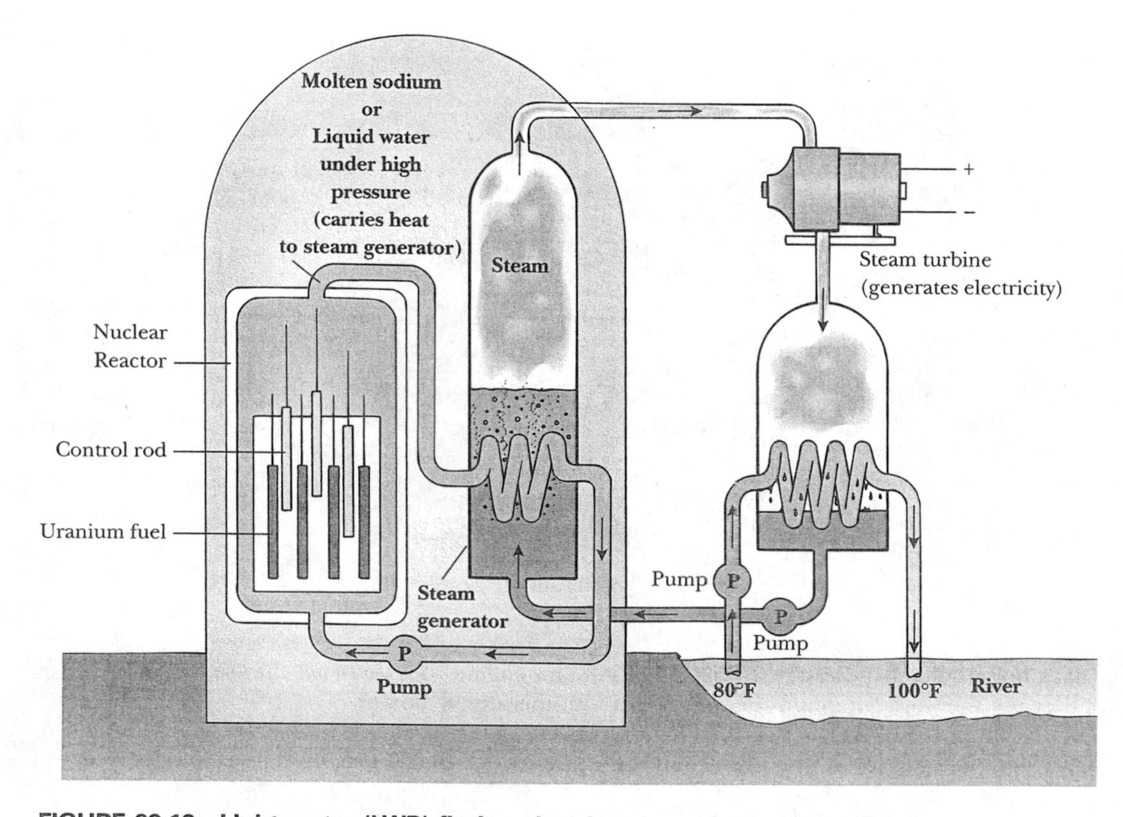

FIGURE 22.12 Light water (LWR) fission electric power plant reactor. This is the reactor that is typical of the American nuclear industry. Heat produced in the fission reaction does not directly turn water into steam. As this simplified diagram indicates, the water is heated in an "exchanger" by a fluid that circulates through the reactor core.

a nucleus increases as the neutron's energy decreases. When the high-energy neutrons collide with light nuclei, some of their energy is transferred, and the same neutrons are sufficiently slowed to be captured by the fuel rods. Graphite and heavy water are commonly used as the moderators in uranium reactors. Ordinary water can be used with enriched uranium reactors. Beryllium can also be used, but the cost is prohibitive.

The first central-station nuclear electric power plant operated by a public utility in the United States was built in Shippingport, Pennsylvania and commissioned in 1957. Its capacity was about 100 megawatts. Some 40 years later, there are about 100 nuclear electric power stations in the United States with capacities of about 1000 megawatts each, which is approximately the size of the largest fossil fuel–fired electric power stations. Together they provide about 15% of the electrical needs of the United States and place the United States 19th on the world list of nuclear power users. Most are located east of the Mississippi and concentrated along the Northeast Corridor and near the Great Lakes. Smaller concentrations exist on the West Coast. It normally takes ten or more years to build a 1000-MW reactor station at costs approaching $5 billion. Nuclear plants are more expensive to construct than coal- or oil-fired stations, and

(text continues on pg. 1022)

PROFILES IN CHEMISTRY

Bomb Physics at Farm Hall, August 6, 1945 During the waning days of World War II in April 1945, many of the German nuclear scientists, their equipment, and their papers were captured by allied troops. A dozen of the principal scientists, including Otto Hahn and Werner Heisenberg, were singled out for continued internment and were isolated for 6 months at Farm Hall, an English manor near Cambridge. The manor house had been wired by U.S. and British scientific intelligence officers to record the conversations. Ever since that time, a debate has raged among scientists and historians as to what the German scientists really knew and understood. Could they have built a bomb? Were they in the process of doing so and only the Allied victory prevented their completing the task? And on hearing for the first time of the dropping of an atomic bomb on Hiroshima on August 6, 1945, what was their response? Before dinner on that day, Hahn was informed that the British Broadcasting Corporation (BBC) was about to announce that an atomic bomb had been dropped. Hahn was reported to be "completely shattered by the announcement and said he felt personally responsible for the deaths of hundreds of thousands of people . . . as it was his original discovery that had made the bomb possible." He was calmed down from his original agitated state and went down to dinner and to tell the others. Here are the recorded comments of Hahn, Heisenberg, and the other German scientists.

Hahn: They can only have done that [referring to the news of an atomic bomb being dropped] if they have uranium isotope separation.

Max von Laue: [Uranium] 235?

Paul Harteck: That's not absolutely necessary. If they let a uranium engine [reactor] run, they separate 93 . . .

Heisenberg: Did they use the word "uranium" in connection with this atomic bomb?

All: No.

Heisenberg: Then it's got nothing to do with atoms, but the equivalent of 20,000 tons of high explosives is terrific.

Karl Friedrich von Weizsäcker: It corresponds exactly to the factor 10^4 [that one kilogram of uranium would correspond to about 10^4 tons of high explosives if fission were complete].

Walther Gerlach: Would it be possible that they have got an engine running pretty well, that they have had it long enough to separate 93?

Hahn: I don't believe it.

Heisenberg: All I can suggest is that some dilettante in America who knows very little about it has bluffed them by saying, "If you drop this it has the equivalent of 20,000 tons of high explosives," and in reality doesn't work at all.

Hahn: At any rate, Heisenberg, you're just second-raters and you might as well pack up.

Heisenberg: I quite agree.

Hahn: They are fifty years further advanced than we are . . .

Von Weizsäcker: I think it is dreadful of the Americans to have done it. I think it is madness on their part.

Heisenberg: One can't say that. One could equally well say, "That's the quickest way of ending the war."

Hahn: That's what consoles me.

Heisenberg: I still don't believe a word about the bomb, but I may be wrong. I consider it perfectly possible that they have about ten tons of enriched uranium, but not that they can have ten tons of pure uranium-235.

At 9:00 PM after dinner, all present assembled to hear the official BBC announcement of the Hiroshima bombing and something of the Manhattan Project and its scale and scope. British intelligence monitoring the recording devices at Farm Hall remarked that the listeners were completely stunned when they realized that the news was genuine.

Hahn and von Laue were both senior German scientists, neither of whom were involved in weapons-related nuclear research during the war. Hahn had made the original discovery of nuclear fission. Max von Laue, whose name is uniquely connected with X-ray crystallography, was at the time one of the preeminent German scientists and leaders to whom the allies would look in rebuilding the German scientific establishment after the war. Of all the German scientists, especially among the physicists, Heisenberg was the most important, influential, and charismatic for his work in quantum mechanics and his long association with Neils Bohr. Others present included Paul Harteck, a university professor and a very effective member of the German nuclear physics community, especially in the development of the heavy-water project and their first nuclear reactor, which they called a nuclear *engine.* Karl Friedrich von Weizsäcker was an outstanding young physicist, a protegé of Heisenberg and the son of the German foreign minister in the Hitler regime. Walther Gerlach was a distinguished physicist and director of the German nuclear research effort. Karl Wirtz was an expert on heavy-water and isotope separation.

The conversation prior to the news of the bombing continued as Hahn responded to Heisenberg's statement that the Americans could not have ten tons of pure uranium-235.

Hahn: I thought that one needed only very little 235.

Heisenberg: If they only enrich it slightly, they can build an engine that will go, but with that they can't make an explosive that will . . .

Hahn: But if they have, let us say, 30 kilograms of pure 235, couldn't they make a bomb with it?

Heisenberg: But it still wouldn't go off, as the mean free path is still too big.

Hahn: But tell me why you used to tell me that one needed fifty kilograms of 235 to do anything. Now you say one needs two tons.

Heisenberg: I wouldn't like to commit myself for the moment, but it certainly is a fact that the mean free paths are pretty big. . . . If it has been done with uranium-235 then we should be able to work it out properly. It just depends upon whether it is done with 50, 500, or 5000 kilograms, and we don't know the order of magnitude. We can assume that they have some method of separating isotopes of which we have no idea. [Perhaps surprising, if not astonishing, is the fact that a scientist like Heisenberg had presumably never bothered to "work it out properly"].

Wirtz: I would bet that it is a separation by diffusion with recycling [which was a correct assumption].

Heisenberg: Yes, but it is certain that no apparatus of that sort has ever separated isotopes before.

Von Weizsäcker: Do you think it is impossible that they were able to get element 93 [neptunium] or 94 [plutonium] out of one or more engines?

Wirtz: I don't think that is very likely.

Von Weizsäcker: I think the separation of isotopes is more likely, because of the interest [the allies on questioning us] showed in it to us and the little interest they showed in the other things.

Hahn: Well, I think we'll bet on Heisenberg's suggestion that it is a bluff . . .

The emphasis of the conversation shifted after the news of the bombing had reached this gathering of scientists.

Hahn: Of course we were unable to work on that scale.

Heisenberg: One can say that the first time large funds were made available in Germany was in the spring of 1942 when we convinced [the government] that we had absolutely definite proof that [a bomb] could be done. On the other hand, the whole heavy-water business, which I did everything I could to further, cannot produce an explosive.

Harteck: Not until the engine is running [to produce plutonium].

Hahn: They seem to have made an explosive before making an engine, and now they say, "In the future we will build engines."

Harteck: If it is a fact that an explosive can be produced by means of the mass spectrograph . . . we would never have done it, as we could never have employed 56,000 workmen. . . .

Von Weizsäcker: How many people were working on V-1 and V-2 [rockets]? Thousands worked on that.

Heisenberg: We wouldn't have had the moral courage to recommend to the government in the spring of 1942 that they should employ 120,000 men just for building the thing up.

Von Weizsäcker: I believe the reason we didn't do it was because all the physicists didn't want to do it, on principle. If we had all wanted Germany to win the war, we would have succeeded.

Hahn: I don't believe that, but I am thankful we didn't succeed.

Later that evening, Hahn and Heisenberg spoke alone [but again recorded by British intelligence officers]. The substance of their conversation suggested clearly a high level of understanding of bomb physics and the distinction between nuclear reactors [engines] and bombs. At the same time, it seems quite unlikely that Nazi Germany could have even come close to building a bomb. Yet there is clear evidence as well of the beginnings of conversations have led some to suggest the German nuclear physicists really worked to prevent Hitler from obtaining the bomb and perhaps winning the war, on moral grounds. However, the weight of the facts of the recorded conversations suggest otherwise, that in fact German failure was partly based on economic unfeasibility, and, more significantly, on overestimation of certain technical difficulties.

(text continues on pg. 1022)

PROFILES IN CHEMISTRY (continued)

Questions

Why did some of the German scientists doubt that the United States had constructed a nuclear bomb?

Reference is made to the separation of element 93 from a uranium reactor. How might neptunium (element 93) come about from uranium in the first place?

The 10^4 factor is correct and indeed 1 kg uranium corresponds to 10,000 kg of high explosives. Why this huge difference in energies between nuclear and chemical reactions involving bond-breaking and formation?

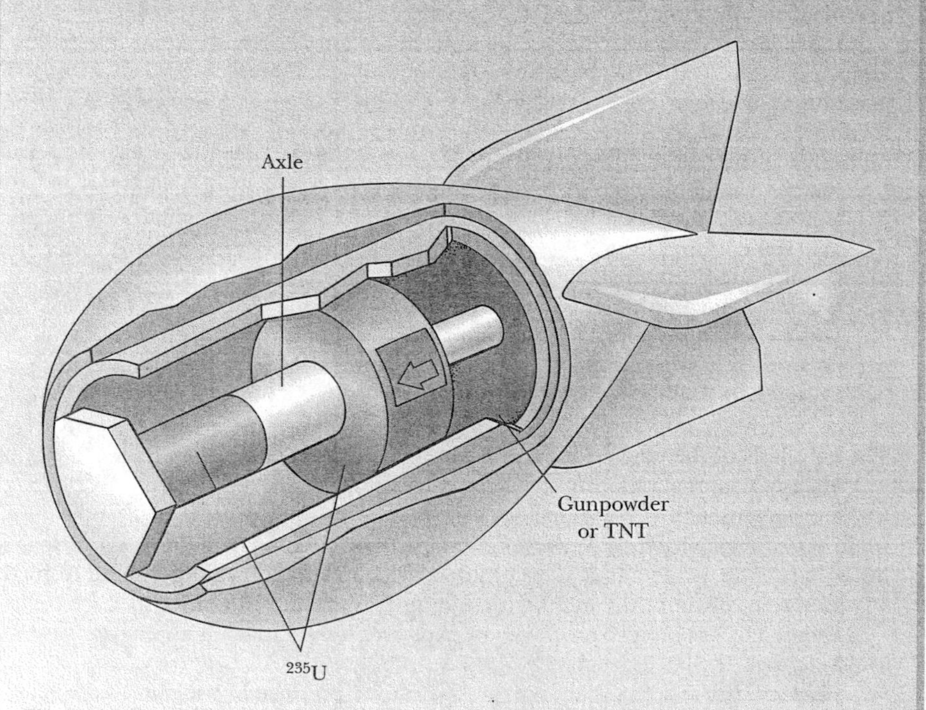

Fission weapon. Sketch of a nuclear fission weapon. Note the simplicity in the overall design.

nuclear power has risen from having the lowest cost among all power sources to become not much different from those of other sources, despite earlier predictions of producing electricity almost free.

A type of nuclear reactor called a breeder reactor can convert nonfissionable ^{238}U to fissionable ^{239}Pu while it produces energy. Various types of breeder reactors have been proposed in anticipation of the time when the world's limited supplies of ^{235}U have been depleted. Natural uranium contains 99.3% ^{238}U, but this isotope can undergo several kinds of nuclear reactions with neutrons, one of which is neutron capture. Thus, a blanket of $^{238}_{92}U$ that surrounds the breeder reactor can capture excess neutrons:

$$^{238}_{92}U + ^{1}_{0}n \longrightarrow ^{239}_{92}U$$

PROCESSES IN CHEMISTRY

Enriching Uranium Uranium ore is mined by both open pit and underground operations, mostly in the western United States and especially in New Mexico. Major worldwide resources are found in Australia, Canada, South Africa, and Russia. Typical U.S. deposits assay about 0.25% uranium metal. Despite the low content of metal, uranium ore is 30 to 50 times more efficient than coal on the basis of available energy per ton. Because part of environmental impact is proportional to the amount of ore mined, this is one of the advantages of nuclear electric power over coal-fired power plants.

Uranium with increased abundance of ^{235}U is needed for use in weapons or nuclear reactors. The enrichment process begins with purification of U_3O_8 or yellow cake—so named because of its color. Reaction with fluorine produces uranium hexafluoride (UF_6). This compound is a gas at temperatures above 56°C at atmospheric pressure, and it has been widely used by the United States in a number of nuclear enrichment schemes. The most famous is the gaseous effusion method in which UF_6 is forced against a porous barrier. The lighter isotope ($^{235}UF_6$) molecules penetrate the barrier in greater numbers than do the heavier ($^{238}UF_6$) molecules, as predicted by Graham's law of effusion (Chapter 4). By passing the gas successively through many barrier stages, highly enriched uranium can be obtained. Because the amount of enrichment per stage is very small, thousands of individual stages are required, and the power consumption of the interstage pumps is tremendous. Yet, the power produced in return by such a plant is on average 20 times greater than the expended energy. Newer technologies of separation and enrichment are under development, including laser separation techniques that can be expected to compete favorably with the energy-intensive gaseous effusion process.

Questions

Explain the principle of the gaseous diffusion method of separating uranium.

Calculate the enrichment factor for gaseous diffusion separation of the uranium-235 and -238 isotopes, based on Graham's law.

The excess of neutrons in $^{239}_{92}U$ results in β-emission,

$$^{239}_{92}U \longrightarrow \,^{239}_{93}Np + \,^{0}_{-1}\beta$$

The $^{239}_{93}Np$ still has an excess of neutrons and emits another β-particle,

$$^{239}_{93}Np \longrightarrow \,^{239}_{94}Pu + \,^{0}_{-1}\beta$$

The resulting $^{239}_{94}Pu$ is fissionable and can be used as a reactor fuel.

The troubled Russian reactor at Chernobyl used graphite as its moderator. Graphite is chemically quite resistant to reaction, but once it starts to react with oxygen and burn, it is very difficult to extinguish. It is speculated that an interruption in the flow of gaseous coolant caused overheating in the nuclear reactor and started the fire, which burned uncontrolled, causing great damage to the reactor and releasing tremendous quantities of radiation and radioactive nuclei.

One of the most attractive developments in nuclear power technology today is the modular high-temperature gas-cooled reactor. Because it uses helium instead of water as the coolant, the reactor can operate at higher temperatures, hence the description "high-temperature" reactor. Because it can operate at higher temperatures, it can achieve efficiencies closer to 50% rather than 30%. But the key safety feature of the new design is that the fuel is more dilute than usual. It is made by forming uranium into billions of tiny sand-sized grains that are each covered with a tough ceramic shell that can withstand temperatures higher than any reactor temperatures. Therefore, the reactor cannot have a "meltdown," in which the heat of the reaction melts the fuel. In conventional reactor designs, the dense, reactive fuel is concentrated in rods that begin to fail at temperatures of 1500 to 1800°C. Temperatures that high can be reached in seconds if the complex cooling system should fail, as it did partially at Three Mile Island (Harrisburg, Pennsylvania, 1979). In the high-temperature reactor, these little glassy spheres are packed into billiard ball–sized containers that transfer heat and trap fission waste products but remain intact up to temperatures of 3300°C (Fig. 22.13). Because the fuel is more dilute, the reactor would have to be much larger to achieve the power output of more conventional reactors.

Fusion

Nuclear **fusion** combines smaller nuclei into larger nuclei and is the opposite of nuclear fission. Production of electric power from the energy produced by the fusion of light nuclei, in effect duplicating the furnace of the Sun here on Earth, holds great promise for inexpensive, safe, clean, nuclear electric power. The hydrogen isotopes that will be used as fuels are freely available. The reactions produce little in the way of hazardous wastes compared with fission reac-

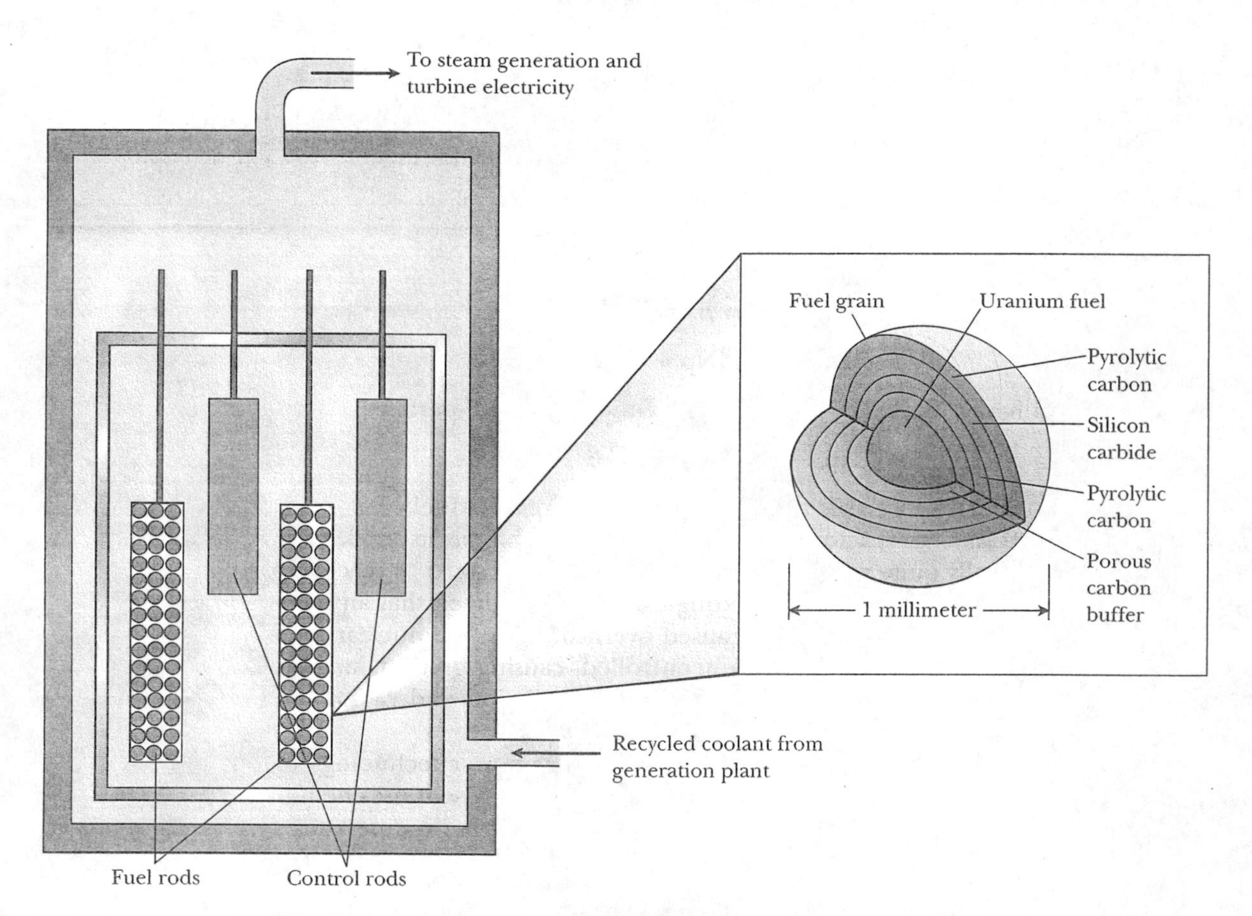

FIGURE 22.13 High-temperature gas-cooled reactor. The prototype of such a reactor is being built in Idaho. Note the modular design and the novel fuel capsules.

PROCESSES IN CHEMISTRY

Nucleogenesis Fusion reactions are the postulated and reasonable reactions by which the elements in the periodic table were created. According to the currently well-accepted "big bang theory," the initial processes in the creation of the universe resulted in the formation of elements 1 and 2, hydrogen and helium. How the remaining elements were formed from these two elements is not totally clear. One explanation, the theory of nucleogenesis, is receiving a lot of attention. Nucleogenesis is the creation of the nuclei of the elements. According to this theory, the gravitational forces within huge clouds of hydrogen nuclei in space caused these clouds to condense on themselves. As gravity caused the mass of gas to contract, it heated until the nuclei fused, forming more helium and releasing a great deal of energy:

$$^{1}_{1}\text{H} + ^{1}_{1}\text{H} \longrightarrow ^{2}_{1}\text{H} + ^{0}_{+1}\beta$$

$$^{1}_{1}\text{H} + ^{2}_{1}\text{H} \longrightarrow ^{3}_{2}\text{He}$$

$$^{3}_{2}\text{He} + ^{3}_{2}\text{He} \longrightarrow ^{4}_{2}\text{He} + ^{1}_{1}\text{H} + ^{1}_{1}\text{H}$$

This process converts hydrogen to helium. It is referred to as the hydrogen-burning reaction and is similar to the processes that fuel the stars, including our own furnace, the Sun.

When a star's hydrogen supply is nearly exhausted, the diminished energy release from fusion can no longer keep the nuclei separated against the force of gravity. The resulting gravitational collapse raises the temperature to the point at which helium fuses, which happens much faster than hydrogen burning, and carbon and oxygen are formed in abundance,

$$^{4}_{2}\text{He} + ^{4}_{2}\text{He} + ^{4}_{2}\text{He} \longrightarrow ^{12}_{6}\text{C}$$

$$^{12}_{6}\text{C} + ^{4}_{2}\text{He} \longrightarrow ^{16}_{8}\text{O}$$

As most of the helium is consumed, the temperature rises again and in a still faster phase, carbon nuclei fuse:

$$^{12}_{6}\text{C} + ^{12}_{6}\text{C} \longrightarrow ^{24}_{12}\text{Mg}$$

As gravitational collapse continues, the pace of events quickens, and heavier nuclei are formed:

$$^{16}_{8}\text{O} + ^{4}_{2}\text{He} \longrightarrow ^{20}_{10}\text{Ne}$$

$$^{24}_{12}\text{Mg} + ^{4}_{2}\text{He} \longrightarrow ^{28}_{14}\text{Si}$$

These processes continue toward production of the most stable nuclides in the highest yields.

The ultimate fate of a star depends on how much mass it started with. Small stars, up to about four times the mass of our Sun, pass through a "red giant" phase and eventually become stable "white dwarf" stars. Much more massive stars, however, become "red supergiants" that eventually form substantial amounts of iron in their cores. When the iron nuclei begin to fuse, they absorb rather than release energy because iron has the highest binding energy per nucleon and is the most stable nucleus (Fig. 22.7). This cools the core of the star, leading to gravitational collapse and sudden reheating. The outer layers of the star are blown off in a tremendous supernova explosion. Most of the star's matter is flung into space, where the atoms may eventually become part of another star or its planets.

Although our theory of stellar synthesis is incomplete, it does successfully account for the natural abundance of the elements throughout the universe as we know it today.

Questions

How can the curve of binding energy per nucleon help to explain the theory of nucleogenesis?

The Earth's core is known to be largely composed of molten iron, the element with the greatest nuclear stability. Along with cobalt and nickel, iron stands at the maximum in the curve of binding energy. Conceive of a scheme of nuclear reactions that would lead to the nucleogenesis of iron.

tors, and there is no possibility for producing nuclear fuels for weapons in any direct way. Attempts have been underway for more than 40 years to develop such power stations, but unfortunately only the fundamentals are reasonably known at present. In all likelihood we will be well into the twenty-first century before the technology is fully developed and we can look forward to tapping fusion energy for electric power.

There are several possible fusion reactions. The experimental prototype is the so-called deuterium–tritium or DT reaction,

$$^{2}_{1}H + ^{3}_{1}H \longrightarrow ^{4}_{2}He + ^{1}_{0}n + 2.8 \times 10^{-12}\ J$$

Because of its short half-life, the tritium must be generated as needed by the bombardment of lithium deuteride (LiH) with neutrons,

$$^{6}_{3}Li\,^{2}_{1}H + ^{1}_{0}n \longrightarrow ^{3}_{1}H + ^{4}_{2}He + ^{2}_{1}H$$

The DT reaction is the best bet for use in the first generation of fusion power plants. Deuterium exists in naturally occurring hydrogen, but a deuterium–deuterium DD reaction requires a much higher temperature to start.

Because of the electrical repulsion between nuclei of like charge, the hydrogen nuclei must have kinetic energies on the order of 1.6×10^{-12} J, corresponding to a temperature of about 100,000,000K. Because of the high temperatures involved, these nuclear reactions are called **thermonuclear** processes. Thermonuclear fusion requires an addition of energy to produce this high temperature. At such temperatures, which are indeed found in the interior of stars, atoms separate into nuclei and electrons, a state of matter called a **plasma**. The main problem that scientists face in achieving controlled thermonuclear fusion is confining the plasma long enough and at sufficient density to get the nuclei to fuse. At present, magnetic fields are used to confine the hot plasma in a doughnut-shaped device called a tokamak. Scientists and engineers have reached the break-even condition at which they are able to extract at least as much energy as is required to generate the fusion conditions in the first place.

In a second scheme, which may well seem as though it was taken from "Buck Rogers in the 24th Century," solid DT fuel pellets are struck by many intense pulsed laser beams, which compress the pellet to a density four orders-of-magnitude greater than normal (Fig. 22.14). The sudden compression heats the fuel pellet to fusion temperatures. This approach, called inertial confinement fusion, is in a much earlier stage of development than tokamak fusion. Nevertheless, it has been demonstrated experimentally.

22.5 RADIATION CHEMISTRY AND THE ENVIRONMENT

Ionizing radiations have been present in the environment since the Earth's formation approximately 4.5 billion years ago. Since that ancient time, atoms of primordial, long-lived radioisotopes and their short-lived daughters have turned the Earth into a complex, ionizing radiation environment. In addition, the Earth is continuously assaulted by high-energy cosmic radiations. Humans have now added significantly to the radiation in the environment, especially in the period since the discovery of nuclear fission in 1938.

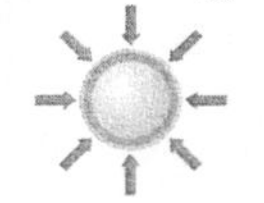

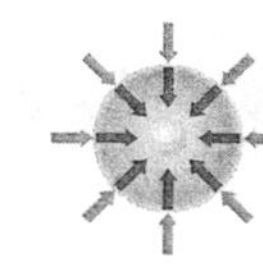

Laser energy
Inward-transported thermal energy
Atmosphere Formation
Laser or particle beams rapidly heat the surface of the fusion target, forming a surrounding plasma envelope.

Compression
Fuel is compressed by rocket-like blowoff of the surface material.

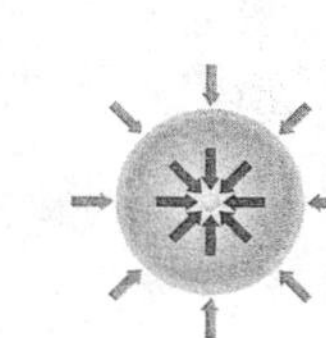

Ignition
With the final driver pulse, the full core reaches 1000–10,000 times liquid density and ignites at 100,000,000°C.

Thermonuclear burn spreads rapidly through the compressed fuel, yielding many times the driver input energy.

LiH ablator
Polymer shield
Frozen DT-main fuel
Low-density DT gas
Au pusher
DT-igniter fuel
Diameter 0.5 cm

FIGURE 22.14 Fusion energy produced in an implosion. The fuel pellet is no larger than a tiny grain of sand, yet the design is very intricate. Essentially a source of the DT fuel, the pellets are struck by intense laser beams that produce a uniform burn of the surface material, causing an implosion that heats the fuel to fusion temperatures.

Among the important concepts we need to keep in mind regarding the effect of radioactive nuclides on the environment is half-life $t_{1/2}$. If $t_{1/2}$ for a nuclide is short, the effect of emitted radiation, although possibly severe at first, will diminish quickly. The effect of long-lived radioactive isotopes such as the fission fuels, ^{235}U and ^{239}Pu, is particularly serious because both continue emitting radiation for extended periods of time. Furthermore, many of the large array of fission products also have long half-lives and safe disposal is a long-term problem. Thus, an accident at a fission fuel-based electric power plant such as Chernobyl creates a serious radiation hazard that can be a problem for many years.

As nuclear radiations pass through matter, atoms and molecules can be ionized. That is, the high energy of the radiation causes electrons to be shifted in a way that creates pairs of oppositely charged ions. A single α-particle—being highly ionizing—is capable of producing as many as 10^5 to 10^6 of these ion pairs. Radiation damage is due primarily to this displacement of electrons and the release of a great deal of energy in a small space. It can affect biological systems, as well as material systems. At a chemical level, radiation damage is largely due to the ability of ionizing radiation to disrupt chemical bonds. Unfortunately, the chemical bonds most sensitive to radiation often turn out to be the very bonds responsible for fine-tuning the functioning of biological systems. At the cellular level, damage is found to be greatest in cells that multiply most rapidly, especially in the lymphatic system (Table 22.4). A summary of hazardous isotopes produced in nuclear reactors is given in Table 22.5.

Two distinct types of radiation hazards are generally encountered in handling radioactive materials: contamination hazards and radiation hazards. **Contamination hazards** involve ingestion of radioactive materials, usually through food cycles. The α-emitters and β-emitters are the worst risks. Because of their chemical properties, isotopes may be deposited in critical tissues in the body.

TABLE 22.4 Cellular Effects of Ionizing Radiation

Tissue	
Lymph	↑
Blood	
Bone	increasing susceptibility
Nerve	
Brain	
Muscle	

Isotopes of elements in the same group as calcium, for example, accumulate in the bone where they can irradiate blood-forming cells. For that very reason, ^{90}Sr and ^{226}Ra are highly radiotoxic. **Radiation hazards** result from the exposure to radiation from sources outside the body. Except for workers in the nuclear industry, most people are not subject to radiation hazards other than background radiation, the topical effects of a sunny day at the beach, television viewing, and exposure to occasional medical or dental X-rays. However, contamination of water supplies by radon from minerals and exposure of the pilots of high-altitude flights to radiation are exceptions. Radon tends to collect in well-insulated homes built over rocks that contain its parent nuclides.

The effects of ionizing radiation on the human body can be divided into two categories: somatic effects and genetic effects. **Somatic effects** are produced in the body after exposure to radiation. The somatic effects of long-term exposure to radiation can include skin cancer, sterility, cataracts, and blood disorders. Exposure is expressed in terms of *rems,* the unit of radiation dose equivalent (roentgen **e**quivalent, **m**an). At 25 rems or less, effects are not generally detectable, although the number of leukocytes (white blood cells) may decrease. At 200 rems, nausea, fatigue, and increased susceptibility to infectious disease occur. The LD-50, a radiation dose lethal to 50% of those receiving it, is 400 rems; the others recover over a period from weeks to months. At 600 rems, only

TABLE 22.5 Hazardous Isotopes Produced in Nuclear Reactors

Isotope	Half-life	Problems
Long-lived fission products	>1000 yr	Long-term storage necessary to reduce radioactivity and heat output
^{85}Kr and ^{133}Xe	10.7 yr	Noble gases not easily removed from off-gases
^{3}H	12.5 yr	Not easily removed chemically; incorporates into biosphere
^{131}I	8 days	Damages thyroid
^{137}Cs	30 days	Replaces Na in tissue
^{90}Sr	28 days	Replaces Ca in bones
^{239}Pu	24,400 yr	Deposits in bones and lungs

APPLICATIONS OF CHEMISTRY

The Geiger Counter and the Smoke Detector From the very beginning of the study of radiation chemistry, the ionizing effects of radiations have been used as detectors. Early researchers used low-power optical magnifiers, actually counting individual hits, as radiations fell onto a fluorescent screen. Geiger and Marsden did just that, making the tedious measurements that provided Rutherford with the raw data for the scattering experiments that led to the nuclear atom model. A Geiger counter senses the ionizations produced by emissions. It consists of a wire electrode within a tube filled with neon or argon. The central wire electrode is positive at a potential of 1000 to 5000 volts with respect to the walls of the cylinder, which serve as the negative electrode. When a high energy particle or γ-ray photon enters the tube, the gas becomes ionized at that point, producing positive ions and negative electrons. They accelerate toward the opposite electrodes, ionizing gas particles along the way. The result is a cascade of particles that forms an electrical pulse, which is capable of activating a counting device. An audible click records each event as the pulse is amplified and fed through a speaker.

Smoke detectors sense the presence of small particles of smoke that rise from a smoldering material, tripping off an alarm and providing early warning. There are two kinds: (1) optical detectors that use a light beam and a light sensor that responds to anything that obscures the beam; and (2) ionization detectors, which are electrical sensors capable of detecting smaller particles than their optical counterparts by taking advantage of emissions from weak radioactive sources, specifically the 241 isotope of americium.

Alpha particles emitted by the americium-241 radioactive isotope ionize the air within the detector chamber, setting up a low-level current between an anode and a cathode. Smoke particles entering the chamber compete for the ions produced by the radioactive source, thereby reducing the current and setting off the alarm. The isotope itself has a half-life of 432.2 years and is readily available in kilogram quantities from plutonium processing in nuclear reactor fuels.

Questions

The Geiger counter tube is filled with neon or argon. Give reasons to explain how the counter would work if there were a vacuum in the tube.

What is the plutonium isotope that gives rise to americium-241?

What possible explanation(s) can you offer for false alarms that occasionally occur with ionizing radiation detectors.

about 20% of exposed populations can be expected to survive, recovering from the effects only after many months.

Genetic effects are the result of radiational damage or alterations in the reproductive cells. Genetic effects are observed in later generations. It should be noted that ionizing radiations are capable of causing gene mutations and chromosome abnormalities at any exposure.

22.6 ISOTOPES IN INDUSTRY, AGRICULTURE, AND MEDICINE

In addition to the best-known applications of isotopes, such as the use of ^{235}U as a fission fuel for power reactors and nuclear weapons, separated isotopes serve many practical purposes in basic and applied research, industry, and medicine. A few of the many interesting and important uses are mentioned here. We have already discussed dating techniques. The remaining applications can be roughly divided into two categories: **tracer applications**, in which the nuclear properties of the isotope provide a label for studying the properties or behavior of another substance or material, and **radiation applications**, in which radioactivity is used to affect a substance or provide the basis for certain measurements on the substance.

Tracer Applications

The Army Corps of Engineers has used water-insoluble $^{140}BaSO_4$ to study the movement of silt in waterways, at the mouth of the Mississippi, for example. After a "tagged" compound is released to the riverbed, the movement of the material under various tidal conditions can be followed with a suitable radiation detector hung over the side of a boat. Hydrologists sometimes add a water-soluble radioisotope to underground water supplies through a bore hole, and then sample the water supply from adjacent bore holes. In this way they can study the direction and velocity of large water tables. Iodine-131 has been used for this purpose.

Hemoglobin labeled with ^{59}Fe has been used in studies of red blood cells and bone marrow. Casein, a milk product, labeled with ^{32}P has been used to study metabolism and nutrition in laboratory animals, as well as in human beings; and important drug absorption and metabolite studies of penicillin have been carried out with ^{35}S-labeled penicillin-G. Painstaking synthetic procedures are often required in applications such as these to prepare the properly labeled compounds. Sometimes biosynthetic techniques are used, as in the preparation of the labeled penicillin, which is made by adding ^{35}S-labeled sulfate ion to the medium in which the penicillium mold is growing.

Iodine-131 has been used in studies of thyroid activity and brain tumors, and ^{58}Co isotopes are used in the detection of pernicious anemia and the inability of the body to absorb vitamin B_{12}. Phosphorus-32 and sulfur-35 were used in experiments that demonstrated that DNA is indeed the carrier of genetic information. Nitrogen-15 was used in studies that showed DNA was passed from generation to generation in a semiconservative fashion. Radioisotopes have been used to induce mutations and for the purpose of carrying out genetic studies.

Suitably labeled ^{32}P phosphates have been used to study the uptake rates and mechanisms whereby plants absorb nutrients, leading to the development of improved fertilizers and crops and a better understanding of root systems and plant foliage. Plant respiration and photosynthetic studies have been carried out using $^{14}CO_2$.

Radiation Applications

Devices called "go-devils" have been used for many years for detecting leaks in long pipelines for water and oil. The go-devil is a packaged radiation detector that is inserted into the pipeline and is pushed along by the fluid. Cobalt-60 sources are attached along the pipeline at fixed intervals, and as the go-devil passes each marker a signal is recorded.

Thickness gauging, based on the reduction of the intensity of radiation as it passes through a medium, is widely used in industry to control the uniformity of films, sheets, and laminates. Transmission thickness gauges use β-emitters for

PROFILES IN CHEMISTRY

The Strange Legacy of the Radium Dial Painters—A Tragedy of Innocence The Environmental Protection Agency is trying to remove thousands of tons of contaminated earth from a toxic dump site in New Jersey without destroying the affected neighborhoods. The cleanup, at a cost of $200 million, is one of the most expensive ever undertaken by the EPA. Because it is in a suburban area, this New Jersey site is sometimes compared to New York State's notorious Love Canal, where buried chemical wastes forced mass evacuations two decades ago. However, in this situation, the EPA is dealing with a more complicated response to a more subtle threat, radium. Radium is radioactive, an element left over when uranium, the material that fuels nuclear reactors, decays. As radium itself decays, it emits hazardous gamma rays and leaves behind a radioactive calling card, radon gas, which has received a lot of media attention for being linked to cancer.

Radium, however, has another interesting property: It glows! And that led to the development early in the twentieth century of a cheap radium paint, which could be used to light up watch dials. Thus, the U.S. Radium Company was born in 1917 and began to produce glow-in-the-dark watches at what has now become a toxic, radioactive waste dump site in Orange, New Jersey.

Over the years, U.S. Radium hired scores of teenage girls to paint the watch faces. None had any idea about the dangers of radiation. They worked assiduously and they painted hundreds of dials a day. They pointed their paint brushes between their lips in order to do fine work, and every time they pointed their brushes, they literally ate some radium. First these women developed what was called "rot jaw," in which the jaw became diseased. Then their teeth would start to fall out. And eventually, more severe manifestations of radiation poisoning, such as anemia, various forms of cancer, and eventually death would occur. Initially, the company denied radium had anything to do with their medical problems, but by the late 1920s, public health records for these women were filled with comments about mouths that actually glowed in the dark, exhaled air containing radioactive material, and, at autopsy, bones that glowed and exposed X-ray film when simply placed on them for a period of time.

Twenty-nine women died! Their story has become a landmark in the study of workplace hazards. Although U.S. Radium is long gone, its legacy is not. The company extracted each ounce of radium from up to a ton of radioactive ore, hazardous material that became waste. Much of that waste may have been dumped in the towns in the vicinity of West Orange, New Jersey. Houses were subsequently built on the dump sites, and only in the 1980s did it become clear that these sites and the homes built on them were in some cases dangerously radioactive. Some 350 homes are currently affected. The cost to the EPA . . . and the taxpayers? Estimated at $220 million. A tragedy of innocence. Sue to recover the costs? But at this late date, who is there to sue? Meanwhile, although the radiation is not nearly the level that caused the deaths of 29 women, people cannot live on these sites without removal of the radioactive wastes.

Questions

A radioactive waste dump will contain a considerable number of different radioactive nuclides. What is the difference in the dangers resulting from nuclides with short half-lives compared with nuclides with long half-lives?

Look up the half-life for radium-226 and estimate how long the Orange, New Jersey, site of U.S. Radium Corporation might have to sit idle before it could be safely inhabited again if the radioactivity needs to be reduced to 0.1% of current levels.

paper sheets, plastic films, linoleum, and thin metal foils, or γ-emitters for heavy-gauge metal sheet. The source is placed on one side of the material to be measured and the detector is placed on the other (Fig. 22.15). Backscatter gauges, having both the source and the detector on the same side of the material to be measured, allow one to control the thickness of backing materials and laminates.

Cobalt-60 has been used for sterilization of foods, pharmaceuticals, packaging materials and containers, and increasing the shelf-life of fruits, such as strawberries. It is widely used as a chemotherapeutic agent, particularly for cancer.

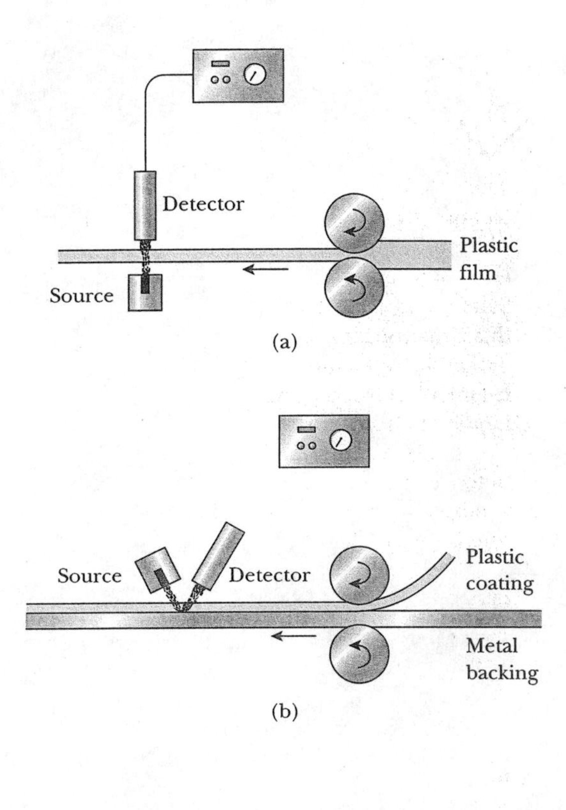

FIGURE 22.15 Transmission thickness gauge. Industrial applications include tracking the uniformity of plastic films and paper stock. (a) By placing a β-emitter source on one side of the material and a detector on the other, and measuring the reduction in the intensity of the radiation, one can gauge the thickness. (b) Backscatter gauges have both the detector and the source on the same side of the film or stock whose thickness is to be measured.

SUMMARY

In a period spanning a professional lifetime, our knowledge of nuclear properties expanded from the basics of structure and properties of the atom to enormously important applications. These range from the beneficial applications of electric power and medicine to weapons of imponderable destruction.

Nucleons are neutrons and protons in an atomic nucleus. Nuclear binding energy is related by the Einstein equation to the mass deficit, the difference between the mass of the nucleons that make up a nucleus and the mass of the nucleus itself. The measure of the stability of a nucleus is the binding energy per nucleon. The maximum in the curve of binding energy per nucleon versus atomic mass number is reached at the atomic numbers of iron and nickel. This means that both fission of heavier elements and fusion of lighter elements can be energy releasing.

Some atomic nuclei are unstable and spontaneously decay to other nuclei. This decay is called radioactivity and is accompanied by the emission of radiation. Three types of radiation are detected in natural radioactivity, alpha particles or helium nuclei, beta particles or electrons, and gamma rays, which, like X-rays, are short-wavelength, high-energy photons. A fourth kind of particle produced by a decaying nucleus is the positron, the antiparticle of the electron, which is annihilated on the first electron contact. Three situations cause nuclei to be unstable, and each situation results in a specific emission: (1) too high a total mass results in alpha emission, (2) too many neutrons results in beta emission, and (3) too many protons results in positron emission. The rate constant for radioactive decay is determined by the half-life, with short half-lives corresponding to rapid decay.

Nuclear fission begins with capture of a neutron by one of a few nuclides, such as ^{235}U, leading to an unstable nuclide that comes apart. Two more stable nuclides of intermediate elements nearer the middle of the periodic table are formed and two or three neutrons are emitted. A chain reaction can result if the emitted neutrons are captured by other fissionable nuclei causing them to come apart in a similar fashion. Practical methods for nuclear-fusion energy production have not yet been found, though it has been demonstrated in tokamak and inertial confinement experiments with deuterium and tritium.

TERMS

Subatomic particle (22.1)
Cathode rays (22.1)
X-rays (22.1)
Radioactivity (22.1)
Gamma rays (22.1)
Beta particles (22.1)
Alpha particles (22.1)
Proton (22.1)
Neutron (22.1)
Nucleon (22.1)
Atomic number, Z (22.1)
Isotope (22.1)
Nuclide (22.1)
Atomic mass number, A (22.1)
Antiparticle (22.1)
Positron (22.1)
Neutrino (22.1)
Antineutrino (22.1)
Band of stability (22.2)
Electron capture (22.2)
Half-life (22.2)
Mass defect (22.2)
Nuclear binding energy (22.2)
Linear accelerator (22.2)
Cyclotron (22.2)
Fission (22.4)
Chain reaction (22.4)
Moderator (22.4)
Fusion (22.4)
Thermonuclear (22.4)
Plasma (22.4)
Contamination hazards (22.5)
Radiation hazards (22.5)
Somatic effects (22.5)
Genetic effects (22.5)
Tracer applications (22.6)
Radiation applications (22.6)

IMPORTANT EQUATIONS

$t_{1/2} = \frac{0.693}{k}$ First-order reaction: relationship of half-life to k

$\ln \frac{N}{N_0} = -kt$ Dependence of number of nuclei on time for nuclear decay

$KE = eV$ Kinetic energy of an accelerated electron

$E = mc^2$ Mass energy conversion/conservation

QUESTIONS

Conceptual questions are denoted by a square screen. Extra-credit questions are denoted by a circular screen

1. What is the proper order of penetrating power for α-, β-, and γ-rays? Why is penetrating power an inverse function of ability to cause ionization?
2. What was the evidence in favor of β-particles being electrons?
3. What was Rutherford's experimental basis for establishing the identity of the α-particle as a helium nucleus?
4. Why is it necessary to propose the existence of special nuclear forces to hold the nucleus together?
5. How are radioactive emissions from an element affected by chemical combination?
6. Why is it so difficult to separate radioisotopes and decay products?
7. By how many units of mass does a radioisotope change after undergoing α-decay? β-decay?
8. If nuclei do not contain electrons, how is β-emission possible?
9. Why do protons generally make better bullets for nuclear reactions than α-particles? What makes neutrons particularly useful in that role?
10. Why is it inaccurate to speak about the lifetime of a sample of a certain radioisotope or of a certain radioactive nucleus?
11. What happens to the total mass if a stable nucleus is broken up into its subatomic particles? Is this process endothermic or exothermic? Why?
12. Chain reactions are made possible by what fission product?
13. Briefly discuss the role of the moderator in a nuclear reactor.
14. How is the rate of reaction controlled in a nuclear reactor?
15. Why are especially high temperatures required to bring about fusion reactions?
16. What are plasmas and under what conditions do they exist?
17. What is the relationship between the half-life of an isotope and the kind of dating applications for which it can be used? Briefly explain.
18. In studying the movement of silt in river estuaries and harbors using isotopic tracers, why must the tracer be a γ-emitter with a moderate half-life?
19. In studying the movement of underground waters with radioisotopes, why must the tracer be water-soluble, with a relatively short half-life?
20. Why does it make little sense to try to date the following by ^{14}C methods: coal deposits, Texas crude oil, and the fossilized remains of dinosaurs?
21. What problems do you envision in trying to carbon date a sample of recent vintage, say, 1988 Bordeaux?
22. What is the overall reaction in the hydrogen burning process that occurs in the Sun?

PROBLEMS

Problems marked with a bullet (•) are answered in Appendix A, in the back of the text.

Atomic Structure and the Subatomic Particles [1–2]

•1. Supply the following information for an unknown nuclide $^{10}_{5}M$.
   (a) the element's mass number
   (b) the element's atomic number
   (c) the proton number
   (d) the number and identity of the extranuclear particles
   (e) the number and identity of the nucleons (nuclear particles)
   (f) the identity of the element M

2. Provide the information requested in Problem 1 for the nuclide $^{37}_{17}M$.

Unstable Nuclides, Nuclear Reactions [3–14]

•3. Given that the stable isotope of sodium is ^{23}Na, what kind of radioactivity would you expect for ^{22}Na and ^{24}Na respectively, and why?

4. Predict the modes of decay for $^{26}_{13}Al$ and for $^{28}_{13}Al$, and write the nuclear equations for the processes. Naturally occurring aluminum is 100% $^{27}_{13}Al$.

•5. Rutherford's source of α-particles was the decay of radium-226. Write the nuclear equation.

6. The nuclide $^{210}_{84}Po$ decays by alpha emission. Write the nuclear equation for the process.

•7. Carbon-14 decays by beta emission. Write the nuclear equation for that process.

8. Arsenic-81 is an unstable nuclide, which decays by β-emission. Write down the nuclear equation for this decay process.

•9. Complete the following nuclear equations:
   (a) $^{10}_{5}B + ^{4}_{2}He \longrightarrow \_\_\_ + ^{1}_{1}H$
   (b) $^{9}_{4}Be + ^{1}_{1}H \longrightarrow \_\_\_ + ^{2}_{1}H$
   (c) $^{26}_{13}Al + ^{1}_{0}n \longrightarrow \_\_\_ + ^{28}_{13}Al$

10. Complete the following nuclear transformations by listing the missing nuclide:
   (a) $^{2}_{1}H + ^{4}_{2}He \longrightarrow \_\_\_ + ^{1}_{0}n$
   (b) $^{9}_{4}Be + ^{4}_{2}He \longrightarrow \_\_\_ + ^{1}_{0}n$
   (c) $^{197}Au \longrightarrow \_\_\_ + \beta^-$

11. Predict the decay products for ^{33}Mg.

12. Vanadium-43 is a radioactive isotope. Predict the products.

•13. The first artificial transmutation of an element occurred when Rutherford bombarded a sample of ^{14}N with α-particles producing a proton and an isotope of oxygen. Write the nuclear equation for this artificial transmutation.

14. Write nuclear equations for each of the following processes:
   (a) A deuterium atom absorbs gamma radiation and then emits a neutron.
   (b) The $^{40}_{20}Ca$ isotope of calcium absorbs a neutron and emits an α-particle.
   (c) The $^{238}_{92}U$ isotope is bombarded with the ^{12}C isotope, yielding four neutrons and an isotope of californium.

Rate of Nuclear Decay [15–22]

15. What fraction of a certain radioisotope will remain unchanged after a time equal to $4t_{1/2}$ has passed?

16. Suppose that after many half-lives have passed, only two atoms of a certain radioisotope remain. Is it possible to predict what happens after one more half-life elapses? Why or why not?

•17. A sample of radioactive silver foil was placed adjacent to a Geiger counter and an initial decay rate of 1000. counts per second was observed. The half-life for the silver isotope is known to be 2.4 minutes.
   (a) What is the decay rate 4.8 minutes later?
   (b) When will the decay rate be 30. counts per second?

18. The initial decay rate for a certain radioisotope is 8000. counts per second. After ten

minutes have elapsed, the decay rate is observed to have fallen to 1000. counts per second.

(a) What is the half-life of the radioisotope?

(b) What is the rate constant for the decay of the isotope?

•**19.** Calculate the number of disintegrations per second for a gram of ^{226}Ra. The half-life for the radioisotope is 1620 years.

20. One radioisotope in the uranium series has a decay rate such that after 53.6 minutes only 25% of the original sample was still present.

(a) Calculate the rate constant for decay.

(b) Calculate the half-life of the radioisotope.

•**21.** A sample taken from a wooden artifact was found to contain 10. g of carbon. The decay rate was found to be 100. counts per minute. How old was the specimen?

22. An animal bone, supposedly more than 10,000 years old, contained 15 g of carbon. Calculate the maximum value for the decay rate of the radiocarbon in this old bone.

Mass Defect, Binding Energy [23–30]

•**23.** How much energy would be released in a proton/antiproton annihilation event?

24. What is the energy released during an electron/positron annihilation event?

•**25.** Compute the total binding energy and the binding energy per nucleon for each of the following nuclides:

(a) $^{2}_{1}H$ (b) $^{12}_{6}C$ (c) $^{56}_{26}Fe$

26. Calculate the total binding energy and binding energy per nucleon for each of the following:

(a) $^{7}_{3}Li$ (b) $^{60}_{28}Ni$ (c) $^{235}_{92}U$

•**27.** Calculate the mass defect and binding energy per nucleon for ^{18}O.

28. Compute the mass defect, binding energy, and binding energy per nucleon for the sulfur isotope with 16 neutrons.

•**29.** Calculate the energy released during the deuterium/tritium reaction:

$$^{2}_{1}H + ^{3}_{1}H \longrightarrow ^{4}_{2}He + ^{1}_{0}n + \text{energy}$$

30. Calculate the energy released in the following reaction:

$$^{1}_{1}H + ^{1}_{0}n \longrightarrow ^{2}_{1}H + \text{energy}$$

Decay Series and Fission and Fusion [31–38]

31. The nucleus $^{235}_{92}U$ is unstable and decays by sequentially emitting α- and β-particles in the following order: $\alpha\ \beta\ \alpha\ \beta\ \alpha\ \alpha\ \alpha\ \alpha\ \beta\ \beta\ \alpha$. Write down the series of complete nuclear symbols of the nuclei that are produced by the disintegration process.

32. One important fissionable nuclide, the $^{239}_{94}Pu$ isotope, does not exist in nature, but it has been successfully produced by nuclear reactions beginning with $^{238}_{92}U$. The process begins with neutron capture, followed by successive β-emissions. Write the nuclear reactions.

•**33.** In the following set of nuclear reactor neutron irradiation chains, take the horizontal arrows to indicate neutron capture and the vertical arrows to indicate β-emission. Fit the appropriate nuclides in the boxes:

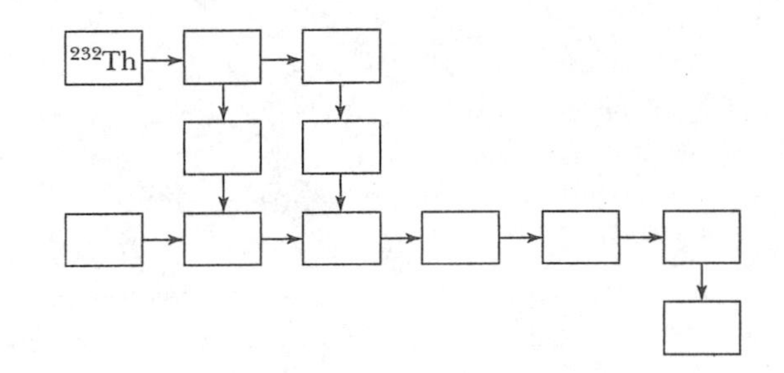

34. Here is a set of nuclear reactor neutron irradiation chains. If the horizontal arrows indicate neutron capture, and the vertical arrows indicate β-emission, fit the appropriate nuclides in the boxes:

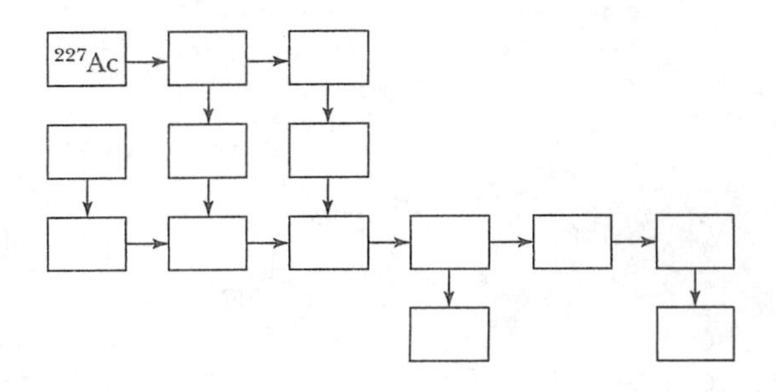

35. A hypothetical nucleus whose mass is 152 amu undergoes a fission reaction and two fission fragments of mass numbers 80 and 70 result. Using the curve of binding energy (Fig. 22.7), estimate the total binding energy of the original nuclide and the two fission fragments. Is energy released or must it be supplied? In either case, estimate how much.

36. Use the curve of binding energy versus mass number to estimate the energy evolved or absorbed when a nuclide of mass 82 amu divides into two equal fission fragments of mass 40 amu.

•37. Fission of $^{235}_{92}U$ is induced by neutron capture. The fission fragments are elements in the middle of the periodic table, along with neutrons that provide the means for generating a self-sustaining chain reaction. One mode of fission produces ^{90}Sr and ^{143}Xe. Write the nuclear equation.

38. Write a possible nuclear reaction for a fission of $^{239}_{94}Pu$ that produces two nuclides of smaller mass and two neutrons.

Additional Problems [39–46]

39. A medically important isotope of iodine, ^{131}I, has a radioactive half-life of 193.2 hours, decaying to a stable xenon isotope.
(a) Write the nuclear reaction of the indicated decay process.
(b) If the iodine isotope has an initial decay rate of 1500. counts per minute, what would the count or rate be after 24.00 hours have elapsed?

40. The nuclide ^{40}K is known to decay to ^{40}Ca by beta decay (89%) and to ^{40}Ar by electron capture (11%). Write nuclear equations for the two processes.

•41. From the curve of binding energy, the mass defect for tin (Sn) is known to be about 0.00915 amu per nucleon. Using values for the neutron mass and the mass of the hydrogen atom (or the proton and the electron), estimate the mass of an atom of the ^{120}Sn isotope of tin.

42. Find the binding energy for the ^{7}Li isotope.

•43. (a) Yttrium-90 and copper-66 are both beta emitters. Identify both daughter isotopes.
(b) Plutonium-236 and proactinium-226 are alpha emitters. Identify both daughter isotopes.

44. Plutonium-246 successively emits two β-particles, and then two α-particles along with several γ-rays. What is the isotope that results?

•45. Actinium-231 emits two β-particles, then four α-particles, then one β-particle, and then one α-particle as it decays to what nuclide?

46. Chlorine-36 is an unstable artificial isotope. It decays by either β-decay or electron-capture. Write nuclear equations for both possible processes.

Multiple Principles [47–50]

•47. (a) There are three known neon isotopes that occur naturally: ^{20}Ne (90.9%), ^{21}Ne (0.30%), and ^{22}Ne (8.8%). (See Table 22.2.) Find the average mass for the natural abundance of the isotopes of neon. Compare your value to that given in the periodic table.
(b) There are two principal isotopes of uranium that occur naturally, ^{235}U and ^{238}U. Use the average mass given in the periodic table or the table of atomic masses to find the approximate percent composition of the natural abundance of these two uranium isotopes.

48. The radius of the Earth is 6×10^6 m and its density is 6 g/cm^3. If the Earth shrunk to the point where its density was of the same order-of-magnitude as nuclear densities, what would its radius have to be?

•49. In a particle accelerator experiment, the common isotope of magnesium, $^{24}_{12}Mg$, is bombarded by two different nuclear particles, producing the uncommon isotope of sodium $^{22}_{11}Na$. Write possible nuclear equations for the two-step synthesis of sodium-22.

50. Recall that fusion energy is being considered as an alternative for the twenty-first century. Determine whether it is a viable alternative to nuclear fission by comparing the energy release per kilogram for deuterium-tritium fusion with that for nuclear fission.

Applied Principles [51–54]

51. Calculate the energy released in J/kg for the fission of uranium-235, assuming the energy release to be 3.2×10^{-11} J per fission.

52. Calculate the masses of ^{235}U and ^{238}U per metric ton (1000 kg) of uranium ore, assuming uranium is 0.25% by mass of the total content.

•**53.** It is useful to make some comparisons between coal and uranium as fuels for power, considering coal to have a typical energy content of 19 to 28 GJ/metric ton of mined material. Uranium, as used in the most widely used commercial reactor design, the LWR reactor, has an energy content of 460 GJ/kg of uranium metal. On that basis, calculate the energy content of one metric ton of a typical uranium ore and its ratio to that of the value for coal. One metric ton is 1000 kg. See previous problem.

54. It is also useful to make a comparison between the power from coal and nuclear sources. Considering that electrical conversion efficiencies ($J_{electrical\ power}/J_{heat}$) are about 32% for an LWR nuclear electric power plant and 38% for a modern coal-fired electric power plant, calculate the electric energy ratios in J/metric ton for coal and nuclear using the results of the previous problem.

ESTIMATES AND APPROXIMATIONS [55–57]

55. Consider a fusion reaction involving two ^{11}B nuclei. Write the reaction(s) leading to the likely final product. Approximately how much energy would be released?

56. The decay of one gram of ^{14}C to ^{14}N gives how much energy compared with reacting the one gram of carbon with oxygen to produce CO_2?

57. What is about the shortest time that can be approximated by carbon dating? What is about the longest?

COOPERATIVE STUDY [58–59]

58. In this chapter, we mentioned several times that the energies of nuclear reactions are much greater than the energies of ordinary chemical reactions. Give carefully considered reasons *why* you think this is the case and use numbers to back up your argument.

59. The isotope $^{10}_{4}Be$ decays with the emission of only a beta particle. Calculate the effective wave length of this ejected particle.

WRITING ABOUT CHEMISTRY [60–61]

60. Throughout this chapter, particularly in Section 22.1, there are descriptions of a number of important discoveries in nuclear chemistry. Make a list of these discoveries describing each briefly. Then discuss these discoveries in terms of the flow of the advancement of nuclear science. What did each experimenter or team of experimenters start with? How did the knowledge progress? Was there a key discovery more significant than the others and absolutely indispensable to the scientific advance? Were there any of the discoveries described that were not really necessary? Give well-reasoned explanations for all of your conclusions.

61. The nuclear binding energy resulting from the mass defect makes all of nuclear chemistry, in fact all of chemistry, possible. It is not, however, an easy concept to understand or explain. Explain this concept to a bright person who happened to know essentially no science and no mathematics beyond arithmetic. Be sure to convince your student how important the nuclear binding energy is.

62. Return in time to August 1945. You are a British intelligence officer listening in on the conversations of German nuclear scientists held as prisoners at Farm Hall, England, as they learn about the Allies' dropping an atomic bomb on Hiroshima, Japan. Write to your commanding officer, reporting on the issue of whether there was a substantial German atomic bomb threat or not.

Appendices

A. Answers to Selected Odd-Numbered Problems

B. Nobel Prize Winners

C. Scientific Notation

D. Logarithms

E. Quadratic Equations

F. Graphing Equations

G. Derivations Using Calculus

H. Physical Constants

I. Conversion Factors

J. Vapor Pressure of Water as a Function of Temperature

Appendix

Answers to Selected Odd-Numbered Problems

Chapter 1

1.1 20. 2.9×10^2 0.34 2.0

1.3 9.47×10^{12} km

1.5 (a) 2.0500×10^3 (b) 2.050×10^{-5} (c) 1.0×10^0 (d) 7.27×10^{-1} (e) 6.35×10^1

1.7 0.000000000300 625 0.0625 10,000 0.0001

1.9 (a) four (b) three (c) three

1.11 0.10 mm/sheet

1.13 6×10^{-4}

1.15 357°C, 675°F

1.17 (a) 160.9 km/hr (b) 91.4 m (c) 481 yd (d) 1.00000×10^7 m (e) 2.99×10^{10} cm/s

1.19 (a) 100.0 yd is shorter than 100.0 m (b) It will take longer to swim 400.0 m

1.21 5×10^4 g sulfur

1.23 7.9×10^2 g

1.25 (a) 56.2 g (water + flask) (b) 51.0 g (alcohol + flask) (c) 371 g (mercury + flask)

1.27 $d = 19.3$ g/mL; gold

1.29 $1.20/kg

1.31 $15.52/g

1.37 (a) 1.625×10^3 kg/m^3 (b) 70 lb

1.41 $d = 19$ g/cm^3, which is consistent with the density of gold

1.43 2.59×10^4 kg/hr

1.47 $R = 4Q + 842$

Chapter 2

2.1 (a) 9 electrons, 9 protons, and 10 neutrons. (b) 20 electrons, 20 protons, and 25 neutrons (c) 19 electrons, 19 protons, and 20 neutrons (d) 13 electrons, 13 protons, and 14 neutrons

2.3 (a) $Z = 238 - 146 = 92$; the element is ^{238}U (b) $Z = 22 - 13 = 9$; the element is ^{22}F (c) $Z = 26$; element 26 is Fe (d) $Z = 43$; element 43 is Tc

2.5 (a) $^{51}_{28}Ni$ (b) $^{29}_{12}Mg$ (c) $^{71}_{35}Br$ (d) $^{87}_{37}Rb$

2.7 (a) sulfur dioxide (b) silicon tetrachloride (c) iodine monochloride

2.9 (a) boron tribromide (b) carbon disulfide (c) phosphorus pentafluoride

2.11 (a) CO_2 (b) BrF_3 (c) CCl_4

2.13 (a) ICl (b) XeF_4 (c) CH_4

2.15 (a) Na^+ (b) P^{3-} (c) Br^- (d) Sr^{2+}

2.17 (a) Na_2S (b) BaO (c) MgO

2.19 (a) magnesium iodide (b) potassium fluoride (c) barium chloride

2.21 (a) magnesium oxide (b) lithium bromide (c) radium bromide (d) cesium oxide

2.23 (a) potassium sulfate (b) magnesium hydroxide (c) barium carbonate (d) lithium nitrate

2.25 (a) $Mg(NO_2)_2$ (b) $CaCO_3$ (c) K_2SO_4 (d) $NaClO_3$

2.27 (a) 123.90 (b) 60.05 (c) 267.7 (d) 811.1

2.29 207 amu

2.31 55.85 amu

2.35 (a) 31 electrons, 39 neutrons, 31 protons (b) 46 electrons, 58 neutrons, 46 protons (c) 94 electrons, 148 neutrons, 94 protons (d) 18 electrons, 22 neutrons, 18 protons

2.37 (a) dinitrogen pentoxide (b) disilicon hexachloride

2.39 (a) K^+ (b) Te^{2-} (c) Br^- (d) Mg^{2+}

Chapter 3

3.1 (a) 6.02×10^{24} molecules

3.3 (a) 1.055×10^{-22} g (b) 1.048×10^{-22} g (c) 2.821×10^{-16} g

3.5 (a) 1.00 mol compound (b) 1.00 mol Na, 2.00 mol H, 1.00 mol P, 4.00 mol O (c) 23.0 g Na, 2.02 g H, 31.0 g P, 64.0 g O (d) 6.02×10^{23} atoms Na, 1.20×10^{24} atoms H, 6.02×10^{23} atoms P, 2.41×10^{24} atoms O

3.7 1.8×10^{24} molecules

3.9 8.4×10^{25} atoms

3.11 C_3O_2

3.13 $CaCO_3$

3.15 (a) CrO_3 (b) Cr_2O_3 (c) CrO

3.17 (a) $FeC_{10}H_{10}$ (b) $FeC_{10}H_{10}$

3.19 (a) $C_3H_8 + 5O_2 \longrightarrow 3CO_2 + 4H_2O$
(b) $Fe_2O_3 + 3CO \longrightarrow 2Fe + 3CO_2$
(c) $Mg_3N_2 + 6H_2O \longrightarrow 2NH_3 + 3Mg(OH)_2$
(d) $Ca_3(PO_4)_2 + 2H_2SO_4 \longrightarrow 2CaHPO_4 + Ca(HSO_4)_2$
(e) $3K_2CO_3 + Al_2Cl_6 \longrightarrow Al_2(CO_3)_3 + 6KCl$
(f) $8KClO_3 + C_{12}H_{22}O_{11} \longrightarrow 8KCl + 12CO_2 + 11H_2O$
(g) $KOH + H_3PO_4 \longrightarrow KH_2PO_4 + H_2O$

3.21 (a) $3Fe + 4H_2O \longrightarrow Fe_3O_4 + 4H_2$
(b) 2.40×10^{-2} mol H_2 (c) 14.1 g Fe_3O_4

3.23 (a) $2C_8H_{18} + 25O_2 \longrightarrow 16CO_2 + 18H_2O$
(b) 3.08 g CO_2
(c) O_2 is the limiting reagent; you are left with 115 g C_8H_{18}, 28 g CO_2, and 13 g H_2O.

3.25 2120 g SO_2

3.29 (a) 0.75 g of metal (b) 5.75 g MX_2 will be formed

3.31 24.0 g MnO_2 required

3.33 (a) 1.00 mol HCl can yield 0.500 mol H_2 and 0.500 mol Cl_2; 1.00 mol H_2O can yield 1.00 mol H_2 and 0.500 mol O_2; 1.00 mol NH_3 can yield 1.50 mol H_2 and 0.500 mol N_2
(b) 2.02 g H_2; 4.03 g H_2; 6.05 g H_2

3.35 (a) $3LiAlH_4 + 4BF_3 \longrightarrow 3LiF + 3AlF_3 + 2B_2H_6$
(b) BF_3 is the limiting reagent; 45.9 g B_2H_6

3.37 (a) 20.0 mol HCl (b) 5.00 mol Cl_2 (c) 5.00 mol Cl_2

3.39 C_3H_8

3.41 CH_2O

3.43 57.45% Ag in the alloy

3.45 50.0% K_2SO_4 and 50.0% Na_2SO_4

3.47 A mixture of 0.43 g NaCl and 0.57 g KCl

3.49 45% Cu_2O and 55% CuO

3.51 Sulfur is in excess; 21.3% unreacted material

3.53 Empirical formula is HSO_4; molecular formula is $H_2S_2O_8$

3.55 (a) $C(s) + CO_2(g) \longrightarrow 2CO(g)$
(b) H_2O is the limiting reagent (c) 64.5%
(d) 2.16 g H_2

3.57 (a) 25% $NaClO_3$
(b) 7.0 mol ClO_2 (4.8×10^2 g) and 7.0 mol Cl_2 (5.0×10^2 g)

3.59 Empirical formula: $AlCl_3$
Balanced reaction: $AlCl_3 + 3Na \longrightarrow Al + 3NaCl$

3.61 (a) $KClO_4 + 4C \longrightarrow KCl + 4CO$

$KClO_4 + 2C \longrightarrow KCl + 2CO_2$

(b) 1.7×10^2 g C

3.63 4.54×10^3 mol $KHCO_3$

3.65 The sums of molar masses of reactants and products are 147.03 g and 288.83 g, respectively, for the unbalanced reaction. The balanced reaction is

$$CS_2 + 3Cl_2 \longrightarrow CCl_4 + S_2Cl_2$$

3.67 (a) 225.7 g C required (b) 300.6 g O_2 required

3.69 61% $CaWO_4$

Chapter 4

4.1 52 m

4.3 13.4 m

4.5 715.3 torr

4.7 1.2 atm; 20. kPa

4.9 (a) 10.0 atm (b) 1.01×10^4 kPa (c) 1.00×10^3 L (d) 2.5×10^2 atm

4.11 1.43 g/L

4.13 6.33 g O_2

4.15 1665 mL

4.17 439K

4.19 3.55×10^{13} particles

4.21 212 L CO_2

4.23 5.02 g/L

4.25 131 g/mol

4.27 1.09 atm

4.29 $C_2H_4Cl_2$

4.31 1.96 g/L

4.33 2140K

4.35 $p_{N_2} = 602$ torr, $p_{Ar} = 8$ torr

4.37 76% N_2, 23% O_2, 1% Ar; ave. molar mass = 29 g/mol

4.39 (a) 0.21 atm (b) 0.78 atm (c) 0.79 atm

4.41 $CH_4 = 8.51\%$, $C_2H_6 = 4.0\%$, $CO_2 = 87.5\%$

4.43 C_3H_8

4.45 3.72 kJ

4.47 1.37×10^3 m/s

4.49 (a) Molar KE is equal for the two gases.
(b) Twice as many molecules of B.
(c) Average speeds are nearly the same.

4.51 $\text{rate}_{Ne}/\text{rate}_{Ar} = 1.41$

4.53 5.68 cm^3/s

4.57 Small; strong; low; moderately low

4.61 0.56 L

4.63 47.27 g/mol

4.65 $P_x/P_y = 4$

4.67 (a) 97 g/mol (b) 4.0 g/L

4.71 18.8K, or −254.4°C

4.73 (a) 4×10^2 g/mol (b) $M_x = 184$ g/mol (c) WCl_6

4.75 1.29 g/L

4.77 4.22×10^6 L air required

4.81 Price at 70°F will be 13% higher.

Chapter 5

5.1 3.6×10^{14} J

5.3 3×10^{10} hr

5.5 (a) 4×10^{-15} J (b) 2×10^{-19} J (c) 7×10^{-14} J

5.7 2.74×10^{-18} J

5.9 8.02×10^{-19} J or 5.00 eV/molecule

5.11 260 nm

5.15 (a) 1.10×10^{-18} J (b) 7.69×10^{-19} J

5.19 0.2493 Å

5.21 (a) 5.3×10^{-24} Å (b) 1.36×10^{-27} Å

5.23 2.4×10^{-24} cm

5.25 10^7 m/s

5.27

n	ℓ	m_ℓ	m_s
1	0	0	$+\frac{1}{2}$
1	0	0	$-\frac{1}{2}$
2	0	0	$+\frac{1}{2}$
2	0	0	$-\frac{1}{2}$

5.29 (a) $\ell = 0, 1, 2$ (b) $m_\ell = -3, -2, -1, 0, 1, 2, 3$ (c) $m_s = +1/2, -1/2$

5.33 (a) $[Ar]4s^23d^{10}4p^3$ (b) $[Ar]4s^23d^2$ (c) $[Kr]5s^24d^{10}5p^5$

5.35 $n = 4$, $\ell = 2$, and

$m_\ell =$	−2	−1	0	1	2
$m_s =$	$\pm\frac{1}{2}$	$\pm\frac{1}{2}$	$\pm\frac{1}{2}$	$\pm\frac{1}{2}$	$\pm\frac{1}{2}$

5.37 (a) rhodium (b) boron

5.39 potassium

5.41 (a) 3*d* is higher than 4*s* (b) 4*p* is higher than 4*s* (c) 5*s* is higher than 4*p*

5.45 Na is $[Ne]3s^1$ K is $[Ar]4s^1$ Ag is $[Kr]5s^14d^{10}$ Au is $[Xe]6s^15d^{10}$

5.47 (a) Neutral, ground state (b) Anion, ground state (c) Neutral, ground state (d) Impossible (e) Anion, ground state

5.49 Five parallel spins and one opposing spin

5.53 5.3×10^5 m/s

5.57 (a) $n = 2$, $\ell = 0$ (b) $n = 2$, $\ell = 1$
(c) Ground state: $n = 3$, $\ell = 0$; first excited state: $n = 3$, $\ell = 1$

5.59 Cr: $[Ar]4s^13d^5$ Mn: $[Ar]4s^23d^5$ Cu: $[Ar]4s^13d^{10}$ Zn: $[Ar]4s^23d^{10}$

5.61 (a) 1 (b) 3 (c) 5
(d) Se^{2-} is [Kr]; Cs is $[Xe]6s^1$

5.63 (a) $\nu = 9.31 \times 10^{12}$/s (b) $\lambda = 3.22 \times 10^{-5}$ m
(c) radiowave (d) no

5.65 $\Delta E = 109.0$ kJ/mol

5.67 (a) 15 W is more energy than a water molecule going over Niagara falls.
(b) Similar to one carbon atom being burned
(c) Much less than ^{235}U fission

5.75 5.53 mm; radiowave

Chapter 6

6.1 (a) $^{\ominus}\!:\ddot{\text{O}}\!-\!\text{H}$ (b) $\text{H}\!-\!\ddot{\text{O}}^{\oplus}\!-\!\text{H}$ (with H below O) (c) $\text{H}\!-\!\text{N}^{\oplus}\!-\!\text{H}$ (with H above and below N)

(d) $\text{H}\!-\!\ddot{\text{N}}^{\ominus}\!-\!\text{H}$ (e) $^{\ominus}\!:\ddot{\text{S}}\!-\!\text{H}$ (f) $\text{H}\!-\!\ddot{\text{S}}^{\oplus}\!-\!\text{H}$ (with H below S)

6.3 (a) $:\ddot{\text{F}}\!-\!\text{B}\!-\!\ddot{\text{F}}:$ (with $:\ddot{\text{F}}:^{\ominus}$ above B and $:\ddot{\text{F}}:$ below B) (b) $\text{H}\!-\!\ddot{\text{O}}\!-\!\ddot{\text{O}}:^{\ominus}$

(c) $\text{H}\!-\!\ddot{\text{O}}\!-\!\text{S}\!-\!\ddot{\text{O}}:$ (with $:\ddot{\text{O}}:^{\ominus}$ above S and $:\ddot{\text{O}}:$ below S)

6.5 (a) $\text{H}\!-\!\ddot{\text{O}}\!-\!\text{H}$ (b) $\text{H}\!-\!\ddot{\text{O}}\!-\!\ddot{\text{O}}\!-\!\text{H}$ (c) $\ddot{\text{O}}\!=\!\text{C}\!=\!\ddot{\text{O}}$

(d) $:\text{O}:$ double-bonded to C; C bonded to two $\ddot{\text{O}}$, each bonded to H (e) $\text{H}\!-\!\text{C}\!-\!\ddot{\text{O}}\!-\!\text{H}$ (with H above and below C)

(f) $\text{H}\!-\!\text{C}\!-\!\text{C}\!-\!\ddot{\text{O}}\!-\!\text{H}$ (each C with H above and below) or $\text{H}\!-\!\text{C}\!-\!\ddot{\text{O}}\!-\!\text{C}\!-\!\text{H}$ (each C with H above and below)

6.7 (a) $\text{H}\!-\!\ddot{\text{N}}\!=\!\text{N}\!=\!\ddot{\text{N}}$ (b) $\text{H}\!-\!\ddot{\text{N}}\!-\!\text{C}\!\equiv\!\text{N}:$ (with H below N)

(c) $\text{H}\!-\!\text{C}\!-\!\ddot{\text{N}}\!-\!\text{H}$ (with H above and below C, H below N)

6.9 $\left[:\ddot{\text{O}}\!-\!\ddot{\text{S}}\!-\!\ddot{\text{O}}:\right]^{2-}$ (with $:\ddot{\text{O}}:$ below S)

6.17 (a) Ionic (b) Ionic (c) Covalent (d) Covalent

6.21 HF

6.23 (a) O is positive; F is negative.
(b) N is positive; O is negative.
(c) S is positive; O is negative.
P—H, C—S and N—Cl are three practically nonpolar bonds.

6.33 (a) $2SO_2 + O_2 \longrightarrow 2SO_3$
$SO_3 + H_2O \longrightarrow H_2SO_4$
(b) 50 tons SO_2 per hour
(c) Too low

Chapter 7

7.1

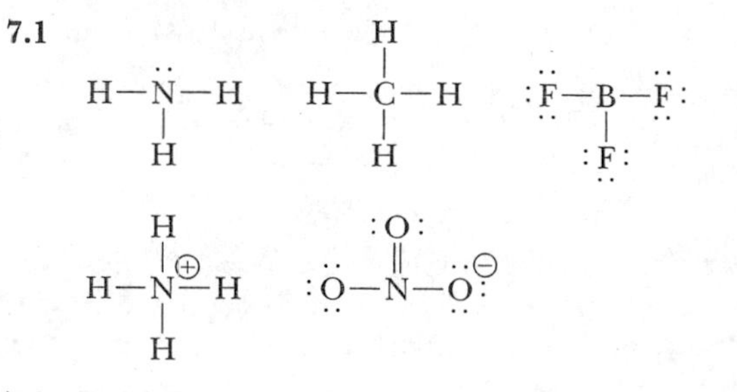

7.3 Yes; both S and O have two bonds to the hydrogens and both have two nonbonding electron pairs.

7.5

	(a)	(b)
SCl_2	bent	sp^3
SF_4	seesaw shaped	sp^3d
Cl_2PF_3	trigonal bipyramidal	sp^3d
SbF_3	trigonal pyramidal	sp^3
$AsCl_5$	trigonal bipyramidal	sp^3d
BrF_2^-	linear	sp^3d

7.7

	(a)	(b)
KrF_2	linear	sp^3d
XeF_4	square planar	sp^3d^2
BrF_3	T-shaped	sp^3d
IF_5	square pyramidal	sp^3d^2
BeI_2	linear	sp

7.9 $:\ddot{\text{Cl}}\!-\!\text{C}\!-\!\ddot{\text{Cl}}:$ (with $:\text{O}:$ double-bonded above C) $\text{H}\!-\!\text{C}\!-\!\ddot{\text{O}}:^{\ominus}$ (with $:\text{O}:$ double-bonded above C)

7.11 (a) 7 σ-bonds and 3 π-bonds
(b) 8 σ-bonds and 2 π-bonds

7.13 The bond order is zero, so Be_2 doesn't exist.

$\uparrow\downarrow$ σ^*; $\uparrow\downarrow$ 2s; $\uparrow\downarrow$ 2s; $\uparrow\downarrow$ σ

7.15 The valence bond view: Four electrons are shared by the two oxgen atoms, resulting in a double bond. The molecular orbital view: There are eight bonding electrons and four antibonding electrons, for a net bond order of 2. A double bond results.

7.17 O_2^+ has 11 valence electrons:
$(\sigma_{2s})^2(\sigma^*_{2s})^2(\pi_{2p})^4(\sigma_{2p})^2(\pi^*_{2s})^1$

7.19 He_2^+ and O_2

7.23 (a) The species does exist.
(b) The apparent bond order is 1.
(c) It is diamagnetic.
(d) The bond length is shorter than in Li_2^+.
(e) The bond length is longer than in H_2.

(f) The energy released is greater than in Na_2 formation.
(g) The energy released is less than in B_2 formation.

7.29 Most polar: NH_3 and ClF_3 (unsymmetrical molecules)
Least polar: BCl_3 and PCl_5 (symmetrical molecules)

7.31 BCl_3, because it is electron deficient;
$BCl_3 + :NH_3 \longrightarrow Cl_3B - NH_3$

7.33 Both species have 10 valence electrons, 1 triple bond, and 2 lone pairs. Both have one σ bond and two π bonds, and both are *sp* hybridized.

7.35 SF_4 is more reactive than SF_6 because of its lone pair of electrons and molecular polarity.

7.37 $:N{\equiv}C{-}\ddot{\underset{..}{O}}:^{\ominus}$

8.33 (a) Ionic solid (b) Covalent gas (c) Covalent gas
(d) Covalent gas (e) Covalent gas
(f) Covalent gas (g) Covalent gas (h) Ionic solid
(i) Covalent gas

8.35 $Pb + O_2(\text{excess}) \longrightarrow PbO_2$
$PbO_2 + 4HCl \longrightarrow PbCl_4 + 2H_2O$

8.37 496 kJ/mole

8.39 The H—S bond is weaker.

8.41 Be: Group IIA, *s*-block O: Group VIA, *p*-block
Cl: Group VIIA, *p*-block K: Group IA, *s*-block
Ti: Group IVB, *d*-block (all electron configurations are in Table 5.7)

8.43 15.6 L SiF_4

Chapter 8

8.1 (a) Group IVA (b) Group IIA

8.3 (a) $K < Mg < Be$ (b) $Na < Al < Mg$

8.5 $K < Li < I < Cl$

8.7 BaO, KF, and SrO

8.9 (a) H is +1 and N is −3.
(b) K is +1, O is −2, and Mn is +7.
(c) O is −2 and Br is +5.
(d) Ca is +2, O is −2, and C is +4.
(e) Na is +1, F is −1, and Cl is +3.

8.11 (a) $2Cs(s) + Br_2(\ell) \longrightarrow 2CsBr(s)$
(b) $2Na(s) + H_2(g) \longrightarrow 2NaH(s)$

8.13 (a) $C_3H_8 + 5O_2 \longrightarrow 3CO_2 + 4H_2O$
(b) $N_2H_4 + O_2 \longrightarrow N_2 + 2H_2O$

8.15 (a) BaH_2 (b) $Ba(OH)_2$ and H_2 (c) BaI_2

8.17 (a) $Al^{3+} < Mg^{2+} < Na^+$ (b) $F^- < Cl^- < Br^-$
(c) $Mg^{2+} < Na^+ < Cl^-$

8.19 (a) $Li < Na < Rb$ (b) $F < N < B$
(c) $Cl < P < Ga$

8.21 1.45 Å

8.23 $Al < Ca < Sr < Rb < Cs$

8.25 $Al^{3+} < Mg^{2+} < Ca^{2+} < O^{2-} < Cl^- < S^{2-}$

8.27 $NO + O_3 \longrightarrow NO_2 + O_2$;
$NO_2 + O \longrightarrow NO + O_2$; (cycle repeats)

8.29 $4Li + O_2 \longrightarrow 2Li_2O$; $2Na + O_2 \longrightarrow Na_2O_2$;
$K + O_2 \longrightarrow KO_2$; $Rb + O_2 \longrightarrow RbO_2$;
$Cs + O_2 \longrightarrow CsO_2$

8.31 $CaCO_3 \longrightarrow CaO + CO_2$
$CaO + H_2O \longrightarrow Ca(OH)_2$

Chapter 9

9.1 1 atm

9.3 77°C

9.5 100°C

9.7 *ca.* 97°C and 91°C

9.9 0.01°C and 4.6 torr

9.11 (a) Increasing pressure will increase boiling point.
(b) Decreasing *T* will cause freezing.
(c) 345K, 390K, 0.25 atm at 300K

9.13 (a) 75 atm (b) solid R will sink
(c) 8°C and 25 atm
(d) Substance does not have a normal boiling point.

9.15 0.638 moles

9.17 $X(\text{ethanol}) = 0.232$; $X(\text{water}) = 0.595$;
$X(\text{ethylene glycol}) = 0.172$

9.19 (a) 2.5 m NaOH
(b) 0.10 m CH_3OH
(c) 0.10 m naphthalene

9.25 137 torr

9.27 (a) 0.0574 (b) 0.780 m naphthalene
(c) 94.3 torr

9.29 −3.1°C

9.31 −26.4°C

9.35 1.0×10^2 g/mol

9.37 −33.5°C

9.39 (a) 4.20 torr (b) 697 torr (c) 102.6°C
(d) −9.30°C

9.41 5.0×10^2 g/mol

9.43 6.7 atm

9.45 2.7 atm

9.47 (a) 2.0×10^4 g/mol (b) $\Delta T_f = 9.30 \times 10^{-4}$C° (c) Osmotic pressure is the most accurate technique.

9.49 i = 3, 2, and 3 for the three salts

9.53 3.5

9.55 P(benzene) = 37.5 torr; P(toluene) = 11 torr

9.57 $P_A = P_B$ = 16 torr; 1:1 mole ratio A:B

9.59 (a) X(ethyl alcohol) = 0.51; X(methyl alcohol) = 0.49 (b) X(ethyl alcohol) = 0.34; X(methyl alcohol) = 0.66

9.61 P_B° = 0.70 atm, P_A° = 1.90 atm

9.65 X(acetic acid) = 0.0292; $X(H_2O)$ = 0.972; 1.67 m

9.67 $X(H_2) = 1.46 \times 10^{-5}$; $K = 6.85 \times 10^4$ atm

9.69 1.3×10^4 g/mol

9.71 (a) i = 4.0 (b) 100.021°C (c) Ions are Ti^{3+} and $3Cl^-$.

9.73 P_A° = 450 torr; P_B° = 150 torr

9.77 188 g/mol

9.79 16 m ethanol; X(ethanol) = 0.22; 8.5 M ethanol; no

9.81 1.4×10^4 atm

9.83 (a) 450K (b) 375K (c) 340K (d) 0.4 atm (all values are approximate)

9.85 500 m H_2SO_4

9.89 0.41 L

9.91 More than 25 atm; −1.9°C

Chapter 10

10.1 raising the temperature will produce more reactants

10.3 (a) Right (b) Right

10.5 (a) Left (b) Left

10.7 (a) $K = \dfrac{[Zn^{2+}]p_{H_2}}{[H^+]^2}$ (b) $K = [F^-]^2[Ca^{2+}]$

(c) $K = \dfrac{1}{[Cl^-]^2[Cu^{2+}]}$ (d) $K = \dfrac{[V^{2+}]^3}{[Cr^{3+}]^2}$

10.9 (a) Left (b) Right (c) Left (d) Right (e) Right (f) Unchanged

10.11 (a) $K_2 = (K_1)^{1/2}$ (b) $K_2 = 1/K_1$

10.13 0.111

10.15 K_c = 0.078

10.17 To the right (more products)

10.19 Not at equilibrium and will proceed spontaneously to the left

10.21 K = 1.6, to the right

10.25 $p_{PCl_3} = p_{Cl_2}$ = 0.012428 atm; $p_{PCl_5} = 7.2 \times 10^{-5}$ atm

10.27 p_{SO_3} = 0.019 atm; p_{O_2} = 0.091 atm, and p_{SO_2} = 0.081 atm

10.29 $p_{CO_2} = p_{H_2}$ = 0.133 atm; and $p_{CO} = p_{H_2O}$ = 0.067 atm

10.31 (a) $K = \dfrac{p_{H_2}p_{CO_2}}{p_{H_2O}p_{CO}}$ (b) 0.24 mol each of CO_2 and H_2; 0.76 mol each of CO and H_2O (c) The reaction will favor CO formation.

10.33 (a) $K_c = 9.38 \times 10^{-3}$; K_p = 0.326; K_c' = 107; K_p' = 3.07 (b) Left (c) Left

10.35 Solubility = 9.2×10^{-9} mol/L

10.37 4.0×10^{-7}

10.39 molar solubility of NiS = 1.4×10^{-22} mol/L $[Ni^{2+}] = 1.4 \times 10^{-22}$ M, $[S^{2-}]$ = 0.010 M

10.43 $[CO_3^{2-}] = 4.0 \times 10^{-4}$ M, $[Pb^{2+}] = 3.8 \times 10^{-10}$ M

10.45 0.99917

10.47 (a) $Mg^{2+}(aq) + 2OH^-(aq) \longrightarrow Mg(OH)_2(s)$ (b) $3Ca^{2+}(aq) + 2PO_4^{3-}(aq) \longrightarrow Ca_3(PO_4)_2(s)$

10.49 (a) $K_p = \dfrac{p_{UF_6}}{p_{F_2}}$ (b) $K_p = \dfrac{p^2_{BrCl}}{p_{Cl_2}}$

(c) $K_p = p_{SiF_4}$

10.51 $p_{CO_2} = 2.44 \times 10^{-2}$ atm; $p_{NH_3} = 4.87 \times 10^{-2}$ atm

10.53 Not at equilibrium

10.55 p_{NO_2} = 0.079 atm; $p_{N_2O_4}$ = 0.056 atm

10.57 6.3×10^{-8}%

10.59 V = 300 L

10.61 (a) Solubility = 0.457 g/L; 2.07×10^{-3} M (b) adding lanthanum nitrate would decrease the solubility

10.67 K_p = 67

10.69 0.520

Chapter 11

11.1 1.8×10^{-7} M

11.3 (a) 1.89 (b) 1.636 (c) 7.10

11.5 (a) $[H_3O^+]$ = 0.06 M; $[OH^-] = 2 \times 10^{-13}$ M (b) $[H_3O^+] = 2 \times 10^{-7}$ M; $[OH^-] = 5 \times 10^{-8}$ M (c) $[H_3O^+] = 4 \times 10^{-14}$ M; $[OH^-]$ = 0.25 M

11.7 (a) Arrhenius acid (b) Neither
(c) Arrhenius acid

11.9 (a) Acid (b) Base (a) Acid

11.11 (a) CN^- (b) SO_4^{2-} (c) NH_3

11.13 (a) NO_2^- is a stronger base.
(b) CO_3^{2-} is a stronger base.
(c) HSO_3^- is a stronger base.

11.15 (a) Forward direction (b) Reverse direction

11.17 0.0010 M $[H_30^+]$; 1.0×10^{-11} M $[OH^-]$

11.19 (a) $[H_3O^+] = 6.1 \times 10^{-5}$ M;
$[OH^-] = 1.6 \times 10^{-10}$ M; pH = 4.21
(b) $[H_3O^+] = 6.7 \times 10^{-3}$ M;
$[OH^-] = 1.5 \times 10^{-12}$ M; pH = 2.17

11.21 (a) $[OH^-] = 0.15$ M; $[H_3O^+] = 6.7 \times 10^{-14}$ M,
pH = 13.17
(b) $[OH^-] = 8.1 \times 10^{-3}$ M;
$[H_3O^+] = 1.2 \times 10^{-12}$ M, pH = 11.92

11.23 $[H_3O^+] = 6.5 \times 10^{-9}$ M, $[OH^-] = 1.5 \times 10^{-6}$ M

11.27 $[H_3O^+]$ is decreased by 8.45×10^{-6} M

11.29 $[H_2SeO_3] = 0.46$ M; $[HSeO_3^-] = 0.037$ M;
$[SeO_3^{2-}] = 5.0 \times 10^{-8}$ M; $[H_3O^+] = 0.037$ M

11.31 (a) 1.8×10^9 (b) 1.4×10^{-5} (c) 1.5×10^{-11}

11.33 $[H_3O^+] = 4.3 \times 10^{-3}$ M; 0.43%

11.35 $[H_3O^+] = 3.9 \times 10^{-5}$ M, $[OH^-] = 2.5 \times 10^{-10}$ M,
pH = 4.4

11.39 pH = 7.35

11.41 pH = 0.726; pH = 0.602

11.43 pH = 4.74

11.45 $[OHAc]/[OAc^-] = 0.002$; not effective

11.47 (a) pH = 9.22 (b) pH = 9.14 (c) pH = 9.4

11.49 $[B] = 7.32 \times 10^{-2}$ M

11.51 80.0 g/mol

11.55 $K_a = 3.7 \times 10^{-8}$

11.57 (a) 1.1×10^{-3} M (b) 4.9×10^{-9} M (c) 0.91 M

11.65 pH = 3.47

11.67 (a) pH = 7.00 (b) pH = 7.00 (c) pH = 11.11
(d) pH = 2.89 (e) pH = 5.12 (f) pH = 8.89
(g) pH = 9.25

11.69 $[H_3O^+] = 4.3 \times 10^{-5}$ M; $[OH^-] = 2.3 \times 10^{-10}$ M;
pH = 4.37

11.75 (a) 0.0339 M (b) pH = 3.92

11.77 1.8%, 6.2×10^{-5}

11.81 0.874%

Chapter 12

12.1 (a) 1.9×10^5 J (b) 3 J

12.3 $q = -650.$ J

12.5 −3.1 kJ

12.7 (a) $\Delta E = 0$, $\Delta H = 0$, $q = 9.0 \times 10^2$ J,
$w = -9.0 \times 10^2$ J
(b) $-w = q = \Delta E = \Delta H = 0$

12.9 $w = -1.52$ kJ; $\Delta E = -1.02$ kJ

12.11 (a) $w = 0$ (b) $T_f = 33°C$

12.13 $q_p = -110.$ kJ $q_v = -111$ kJ

12.19 (a) $C_6H_5CO_2H(s) + (7.5)O_2(g) \longrightarrow$
$7CO_2(g) + 3H_2O(\ell)$
(b) $w = 0$, $q = -32.27$ kJ, $\Delta E = -32.27$ kJ
(c) $\Delta H = -3228$ kJ/mol

12.21 −44.3 kJ/mol

12.23 (a) −283 kJ/mol (b) −286 kJ/mol
(c) 44 kJ/mol (d) $\Delta H_{rxn} = -41$ kJ/mol

12.25 7.64 kJ

12.27 −235 kJ

12.29 −985 kJ

12.31 3480 kJ of heat are released

12.33 −298 kJ/mol

12.37 $\Delta H = 40.66$ kJ $\Delta E = 37.56$ kJ

12.41 −359 kJ

12.43 $\Delta H = -175$ kJ; $C_2H_2 + H_2 \longrightarrow C_2H_4$

12.45 C = 10.8 kJ/°C; ΔE = −2310 kJ/mol

12.49 0.878, −15,600 kJ/kg

12.51 360 kJ of heat is given off

12.53 −136.8 kJ/mol, −137 kJ/mol

12.55 $\Delta H_f^\circ(WC(s)) = -872$ kJ/mol

12.57 42.0 kJ/g

12.59 (a) −2204 kJ/mol (b) $\Delta H_c^\circ = -2220$ kJ/ft^3
(c) 40 ft^3

Chapter 13

13.1 -1.5×10^2 J; this is less than the w for a reversible process

13.3 One step: −67.6 J; two steps: −74.3 J;
three steps: −76.8 J

13.5 $\Delta S_{fus} = 22.1$ J/mol·K $\Delta S_{univ} = 0$

13.7 $\Delta S_{univ} = 0$ $\Delta S_{fus} = 172$ J/K

13.9 $\Delta G° = -1181$ kJ; reaction is spontaneous

13.11

	$\Delta S°$	$\Delta G°$	$\Delta H°$
(a)	−208.2 J/mol·K	−1208 kJ/mol	−1270 kJ/mol
(b)	−269.6 J/mol·K	−23.5 kJ/mol	−103.8 kJ/mol
(c)	92.7 J/mol·K	403.9 kJ/mol	431.5 kJ/mol
(d)	−161.5 J/mol·K	−115.4 kJ/mol	−163.5 kJ/mol
(e)	42.4 J/mol·K	28.5 kJ/mol	41.1 kJ/mol
(f)	−184.3 J/mol·K	−90.8 kJ/mol	−145.7 kJ/mol

13.13 (a) −1208 kJ/mol (b) −23.5 kJ/mol (c) 403.9 kJ/mol (d) −115.4 kJ/mol (e) 28.5 kJ/mol (f) −90.8 kJ/mol

13.15 (a) −216 kJ/mol (b) 14.3 kJ/mol (c) −86.3 kJ/mol

13.17 (a) 7.29×10^{37} (b) 3.11×10^{-3} (c) 1.34×10^{15}

13.19 $\Delta G°_{683} = -264$ kJ/mol, $K = 1.55 \times 10^{20}$

13.21 1103 kJ; reverse direction is spontaneous

13.23 (a) 969K (b) 79K

13.25 81.6 kJ/mol

13.27 High temperatures and low pressures; low temperatures and not significantly affected by pressure variations

13.29 High pressure and low temperature

13.31 33.0 kJ/mol

13.35 $\Delta H = q = 44$ kJ, $\Delta S = 120$ J/mol·K, $\Delta G = 0$, $w = -3.1$ kJ, $\Delta E = 41$ kJ

13.39 230 J/mol·K

13.41 $K = 1.5$, $p_{O_2} = 0.55$ atm

13.43 (a) $K_p = p_{SO_3}/p_{SO_2}p_{O_2}^{1/2}$
(b) $\Delta H° = -95.3$ kJ/mol, $\Delta S° = -90.5$ J/mol·K
(c)

T (K)	1/*T* (K^{-1})	ln *K*
800	1.25×10^{-3}	3.444
850	1.18×10^{-3}	2.625
900	1.11×10^{-3}	1.879
950	1.05×10^{-3}	1.176
1000	1.00×10^{-3}	0.615
1100	9.09×10^{-3}	−0.465

13.45 at 0°C: 1.64×10^7; at 200°C: 0.60; at 400°C: 5.70×10^{-4}

Chapter 14

14.1 (a) +1 in Cl_2O; +4 in ClO_2; +7 in Cl_2O_7
(b) −2 in HS^-; +4 in HSO_3^-; +7 in $HS_2O_8^-$; +2.5 in $HS_4O_6^-$; +6 in HSO_4^-
(c) 0 in P_4; −3 in PH_3; −2 in P_2H_4; +1 in H_3PO_2; +3 in H_3PO_3; +5 in H_3PO_4; +5 in $(NH_4)_4P_2O_7$

14.3 (a) Na is the reducing agent. Cl_2 is the oxidizing agent.
(b) Zn is the reducing agent. H_2SO_4 is the oxidizing agent.
(c) Fe is the reducing agent. Cl_2 is the oxidizing agent
(b) S is the reducing agent. F_2 is the oxidizing agent.

14.5 (a) oxidized (b) oxidized (c) oxidized (d) reduced

14.7 (a) $14H^+ + Cr_2O_7^{2-} + 6Cl^- \longrightarrow 2Cr^{3+} + 3Cl_2 + 7H_2O$
(b) $4H^+ + MnO_2 + 2Hg + 2Cl^- \longrightarrow Mn^{2+} + Hg_2Cl_2 + 2H_2O$
(c) $4H^+ + 3Ag + NO_3^- \longrightarrow 3Ag^+ + NO + 2H_2O$
(d) $8H^+ + H_3AsO_4 + 4Zn \longrightarrow AsH_3 + 4Zn^{2+} + 4H_2O$
(e) $18H_2O + 10Au^{3+} + 3I_2 \longrightarrow 10Au + 6IO_3^- + 36H^+$
(f) $6H^+ + IO_3^- + 8I^- \longrightarrow 3I_3^- + 3H_2O$
(g) $H^+ + 3HS_2O_3^- \longrightarrow 4S + 2HSO_4^- + H_2O$
(h) $4H^+ + 2O_2^{2-} \longrightarrow O_2 + 2H_2O$

14.9 (a) $OH^- + 2Co(OH)_3 + Sn \longrightarrow 2Co(OH)_2 + HSnO_2^- + H_2O$
(b) $3ClO_4^- + I^- \longrightarrow 3ClO_3^- + IO_3^-$
(c) $OH^- + H_2O + PbO_2 + Cl^- \longrightarrow ClO^- + Pb(OH)_3^-$
(d) $OH^- + H_2O + NO_2^- + 2Al \longrightarrow NH_3 + 2AlO_2^-$
(e) $2ClO^- \longrightarrow 2Cl^- + O_2$
(f) $2OH^- + HXeO_4^- + 3Pb \longrightarrow Xe + 3HPbO_2^-$
(g) $2H_2O + 2Ag_2S + 8CN^- + O_2 \longrightarrow 2S + 4Ag(CN)_2^- + 4OH^-$
(h) $8H_2O + 2MnO_4^- + 7S^{2-} \longrightarrow 2MnS + 5S + 16OH^-$
(i) $2OH^- + 2ClO_2 \longrightarrow ClO_2 + ClO_3^- + H_2O$

14.11 Na > Al > Co > Ni > H_2 > Ag

14.13 (a) I_2 (b) I_2 (c) Br_2 (d) H_4XeO_6 (Note: there are many correct answers to this question. Each answer given is just one possibility.)

14.15 (a) 0.74 V; spontaneous (b) 1.01 V; spontaneous (c) 0.406 V; spontaneous (d) 0.002 V; spontaneous

14.17 $Mn + Ni^{2+} \longrightarrow Mn^{2+} + Ni$ $E° = 0.93$ V

14.19 $2Cr + 3Cu^{2+} \longrightarrow 2Cr^{3+} + 3Cu$ $E° = 0.407$ V

14.21 Anode: $Zn + 2KOH + 2OH^- \longrightarrow K_2ZnO_2 + 2e^- + 2H_2O$
Cathode: $H_2O + 2e^- + HgO \longrightarrow Hg + 2OH^-$
Overall: $Zn + 2KOH + HgO \longrightarrow K_2ZnO_2 + Hg + H_2O$

14.25 $\Delta G° = -2.32$ kJ/mol; $K = 2.55$

14.27 27.8 kJ

14.29 (a) $2Fe^{3+} + Zn \longrightarrow 2Fe^{2+} + Zn^{2+}$ (b) Zn is negative, Pt is positive (c) $E° = 1.53$ V; $K = 10^{51.9}$ (d) If $[Fe^{3+}]$ were increased, E would become more positive

14.39 When first connected, $E = 0.102$ V; as the reaction proceeds, E will drop; at equilibrium, $[Pb^{2+}] = 0.216$ M

14.41 pH = 0.85

14.45 27 g chromium

14.47 0.380 A

14.49 3700 kJ

14.51 (a) 1.32 g (b) 0.475 g (c) 12.7 g

14.53 (a) 0.522 L O_2 (b) 3.1 min

14.55 27,600 C

14.57 0.2945 g Cu; 0.1606 g Cr; 0.104 L H_2; 0.0519 L O_2

14.59 (a) $4OH^- + VO^{2+} \longrightarrow VO_3^- + 2H_2O + e^-$
(b) $7H_2O + 2Cr^{3+} \longrightarrow Cr_2O_7^{2-} + 14H^+ + 6e^-$
(c) $2H_2O + Mn^{2+} \longrightarrow MnO_2 + 4H^+ + 2e^-$
(d) $4OH^- + NO \longrightarrow NO_3^- + 2H_2O + 3e^-$
(e) $Fe^{3+} + e^- \longrightarrow Fe^{2+}$

14.63 (a) −0.79 V (b) 0.338 V (c) 1.43 V

14.65 2.36; to the right

14.67 (a) $E° = 0.16$ V, $E = 0.15$ V. When $E = 0$, $[Fe^{2+}]/[Co^{2+}] = 2.6 \times 10^5$. (b) 1.628 V

Chapter 15

15.1 (b) and (c)

15.3 (a) $-\Delta[A_2]/\Delta t = -\Delta[B_2]/\Delta t = \frac{1}{2}\Delta[AB]/\Delta t$
(b) $-\Delta[A_2]/\Delta t = -\Delta[B_2]/\Delta t = \Delta[A_2B_2]/\Delta t$

15.7 Rate = 3.7×10^{13} $L^3/mol^3 \cdot s$ $[H_3AsO_2][H_3O^+]^2[I^-]$

15.9 (a) Third-order overall
(b) First-order in [A] and second-order in [B]
(c) 0.4 M^{-2} min^{-1}

15.15 $k = 0.010$ min^{-1}; $t_{1/2} = 69.3$ min

15.17 (a) $-\Delta[A]/\Delta t = k[A]$ (b) $t_{90\%} = 2.303/k$

15.21 $E_a = 95$ kJ/mol; $t_{1/2} = 403$ s

15.23 $E_a = 97$ kJ/mol; $k = 1.67 \times 10^{-3}$ min^{-1}

15.25 (a) $A + B \longrightarrow C$; $C \longrightarrow P$ (b) rate = $k_1[A][B]$

15.27 $NO_2 + F_2 \longrightarrow NO_2F + F$ slow
$F + NO_2 \longrightarrow NO_2F$ fast
rate = $k_1[NO_2][F_2]$
$F + F \rightarrow F_2$ $NO_2 + NO_2 \longrightarrow N_2O_4$

15.31 (a) first-order in [C] (b) 0.039 min^{-1}

15.33 (a) 52.2% (b) 113 min

15.35 (a) elementary processes higher than third-order are unknown
(b) first-order in A and fourth-order overall. $\Delta[X]/\Delta t = k[A][B][C][D]$

Chapter 16

16.3 6.08×10^{23}

16.5 107 g/mol

16.7 8.97 g/cm^3

16.9 1.96 Å

16.11 1.28 Å

16.13 25.9%

16.17 1.62 Å

16.19 5.95×10^{23}

16.21 8 Si atoms/unit cell

16.25 2.07 g/cm^3

16.27 0.613 g/L

16.29 1.43 Å; there is little bonding between planes of carbon in graphite.

16.31 2.144×10^{13} atoms

16.39 (a) 4 lattice points (b) 1 lattice point
(c) 2 lattice points

Chapter 17

17.1 (a) A simple metal or a polymer
(b) A composite with minimum tendency to creep
(c) A polymer or composite material
(d) A variety of materials (metals, polymers, composites)
(e) A structural ceramic, a heat-resistant polymer, or a composite

17.3 Van der Waals forces

17.5 (a) SiC forms a covalent lattice similar to that of diamond, so SiC should melt at a temperature somewhat less than 3500°C. Its actual melting point is around 2700°C.

(b) MgO should have an ionic structure similar to CaO, so MgO should melt at a somewhat higher temperature due to the smaller size of the Mg^{2+} ion as compared with the Ca^{2+} ion. Its actual melting point is 2800°C.
(c) Cu is held together by metallic bonding and is close to Fe on the periodic table, so Cu should melt at a temperature somewhat close to that of Fe. The actual melting point of copper is 1083°C.

17.9 Cobalt

17.11 (a) Liquid solution
(b) Heterogeneous mixture of solid solution and liquid solution

17.17 (a) $Si_3AlO_8^-$ (b) $Si_2Al_2O_8^{2-}$

17.19 0.74

17.21 $—(CH_2—CH_2)_n— + 3nO_2 \rightarrow 2nCO_2 + 2nH_2O$

17.23 850°C; two phases: a silver-rich solid solution (containing about 93% Ag, 7% Cu), and a copper-rich solid solution (containing about 7% Ag, 93% Cu)

17.25 $\varepsilon = 0.0435$; $E = 1.0 \times 10^7$ psi

17.27 The number of flaws will decrease.

Chapter 18

18.1 Synthetic: nylon; partial: cotton blend; natural: wool

18.3 Rubber mat, plastic wrap, nylon serving spoon, plastic forks

18.5 Vinyl siding, garden hose, roofing tar, concrete drive, plastic hooks

18.7 Steering wheel, seat belt, air bag, floor mat, upholstery

18.9 150

18.11 $M_n = 154{,}000$, $M_w = 158{,}000$

18.13 $M_w/M_n = 1.03$

18.15 $M_n = M_w$

18.17 1.9×10^4

18.19 890.

18.23 Propylene is a thermoplastic (straight-chain polymer).
Polyisoprene is a thermoplastic elastomer (polymer with some cross-linking).

18.25 (a) Ethylene glycol (b) Glycerol

18.29 Polymeric materials have more individual bonds; side chains can rotate, vibrate, stretch, all of which are means to absorb energy.

Chapter 19

19.3 Ni^{2+} is $[Ar]3d^8$; Cu^+ is $[Ar]3d^{10}$
Zn^{2+} is $[Ar]3d^{10}$; Ti^{2+} is $[Ar]3d^2$

19.7 $Pd(NH_3)_2Cl_2$ has no free chloride ions and no total ions.
$[Pd(NH_3)_3Cl]Cl$ has one free chloride ion and two total ions.
$[Pd(NH_3)_4]Cl_2$ has two free chloride ions and three total ions.

19.9 Two isomers are available to this octahedral structure

19.11 (a) Potassium tetracyanonickelate(II)
(b) Sodium trioxalatochromate(III)
(c) Pentaammine chloroplatinum(IV) chloride
(d) Tetraammine dinitroiron(III) sulfate
(e) Hexaaquacobalt(III) iodide
(f) Pentaamminechlororuthenium(III) bromide
(g) Potassium tetrachlorocobaltate(II)

19.13 (a) 4, +2 (b) 2, +1 (c) 4, +2

19.17 Either five or one; either zero or four; two

19.19 The complex is high-spin.

19.21 3.3×10^{-19} J; blue

19.23 (a) Paramagnetic (b) Paramagnetic
(c) Paramagnetic
(d) Paramagnetic (high spin) or diamagnetic (low spin)
(e) Paramagnetic (f) Paramagnetic

19.27 pH = 2.559

19.29 6.7×10^{-28} M

19.31 (a) Hexaquacopper(II) chloride
(b) Tetramminedichloronickel(II)
(c) Sodium tetrachloroplatinate(II)
(d) Potassium hexacyanoferrate(II)

19.33 8 formula units per unit cell, 16 Fe^{3+}, and 8 Fe^{2+} per unit cell

19.35 −30. kJ/mol

Chapter 20

20.1 −58.6 kJ

20.3 1710 K

20.5 7.6×10^{32}

20.11 $\Delta G_{500} = -124$ kJ; $\Delta G_{1000} = -148$ kJ

20.15 (a) $2Fe_2O_3 + 3C \longrightarrow 4Fe + 3CO_2$
$Fe_2O_3 + 2Al \longrightarrow 2Fe + Al_2O_3$
(b) $CuO + H_2 \longrightarrow Cu + H_2O$
$3CuO + 2Al \longrightarrow 3Cu + Al_2O_3$

20.17 $Sb_2S_3 + 2Fe \longrightarrow Fe_2S_3 + 2Sb$

20.19 21.4 g

20.23 (a) $TiO_2(s) + 2Cl_2(g) + C(s) \longrightarrow TiCl_4(\ell) + CO_2(g)$
(b) $2Al_2O_3(s) + 3C(s) \longrightarrow 4Al(s) + 3CO_2(g)$

Chapter 21

21.1 There are five isomeric hexanes. Their IUPAC names are *n*-hexane, 2-methyl pentane, 3-methyl pentane, 2,2-dimethyl butane, and 2,3-dimethyl butane.

21.3 (a) There are nine isomeric heptanes.
(b) Their IUPAC names are *n*-heptane; 2-methylhexane; 3-methylhexane; 2,2-dimethylpentane; 2,3-dimethylpentane, 2,4-dimethylpentane, 3,3-dimethylpentane; 3-ethylpentane; and 2,2,3-trimethylbutane.
(c) 2-methylhexane might be called isoheptane because it has the simplest branching pattern.
(d) 2,2,3-trimethylbutane might be called neoheptane because it is the isomer with the greatest amount of branching.
(e) The names used in (c) and (d) are *not* unambiguous.

21.5 (a) $CH_3CH_2CH_3$ + heat (750°C) $\longrightarrow CH_2{=}CHCH_3 + H_2$
$CH_2{=}CHCH_3 + H_2O$ (in dilute acid) $\longrightarrow CH_3{-}CH(OH){-}CH_3$
(b) $CH_3{-}CH(OH){-}CH_3 + H_2SO_4$ (180°C) $\longrightarrow CH_3CH{=}CH_2 + H_2O$
$CH_3CH{=}CH_2 + H_2$ (and catalyst) $\rightarrow CH_3CH_2CH_3$

21.7 $H{-}C{\equiv}C{-}H$ $[H{-}C{\equiv}C{:}]^-$ $[{:}C{\equiv}C{:}]^{2-}$

21.11 $CH_3CH_2OH > C_6H_5OH > HO(CH_2)_6OH > CH_3(CH_2)_5OH > C_6H_5(CH_2)_2OH$

21.13 (a) There are four monobromobutanes. Their IUPAC names are 1-bromobutane, 2-bromobutane, 1-bromo-2-methylpropane, and 2-bromo-2-methylpropane.
(b) $CH_3(CH_2)_3Br + OH^- \longrightarrow CH_3(CH_2)_3OH + Br^-$
$(CH_3)_2CHCH_2Br + OH^- \longrightarrow (CH_3)_2CHCH_2OH + Br^-$
$(CH_3CH_2{-}CH(Br){-}CH_3 + OH^- \longrightarrow CH_3CH_2{-}CH(OH){-}CH_3 + Br^-$
$(CH_3)_3CBr + OH^- \longrightarrow (CH_3)_3COH + Br^-$
(c) $CH_3(CH_2)_3Br$ + alcoholic KOH $\longrightarrow CH_3CH_2CH{=}CH_2$
$CH_3CH_2{-}CH(Br){-}CH_3$ + alc. KOH $\longrightarrow CH_3CH{=}CHCH_3$ or $CH_3CH_2CH{=}CH_2$
$(CH_3)_2CHCH_2Br$ + alc. KOH $\longrightarrow (CH_3)_2C{=}CH_2$
$(CH_3)_3C{-}Br$ + alc. KOH $\longrightarrow (CH_3)_2C{=}CH_2$

21.15 (a) $CH_3{-}C({=}O){-}OH$ (acetic acid) $H{-}C(H)(Br){-}C({=}O){-}OH$ (α-bromoacetic acid)
(b) Acetic acid: pH = 2.85, α-bromoacetic acid: pH = 1.78
(c) The electronegative bromine atom in α-bromoacetic acid stabilizes the α-bromoacetate anion that forms as protons dissociate. In general, electronegative substituents tend to increase the acidity of carboxylic acids. This is called the inductive effect.

21.19 There are four isomeric butylamines. Their IUPAC names are 1-aminobutane, 2-aminobutane, 1-amino-2-methylpropane, and 2-amino-2-methylpropane.

21.21 (a) $CH_3CH_2CH_2OH + H_2SO_4$(catalyst) $\longrightarrow CH_3CH{=}CH_2 + H_2O$
$CH_3CH{=}CH_2 + H_2O \rightarrow (CH_3)_2CHOH$
(b) $(CH_3)_2CHCl$ + alcoholic KOH $\longrightarrow CH_3CH{=}CH_2 + KCl + H_2O$
$CH_3CH{=}CH_2 + HBr \longrightarrow (CH_3)_2CHBr$
(c) $(CH_3)_2CHOH + H_2SO_4$(catalyst) $\longrightarrow CH_3CH{=}CH_2 + H_2O$
$CH_3CH{=}CH_2 + HCl \longrightarrow (CH_3)_2CHCl$
$(CH_3)_2CHCl + NH_3 \longrightarrow (CH_3)_2CH{-}NH_2 + HCl$
(d) $(CH_3CO)_2O + H_2SO_4$(catalyst) $+ CH_3CH_2OH \longrightarrow CH_3CO_2CH_2CH_3$
(e) $CH_3CH_2OH + O_2 \longrightarrow CH_3COOH$

21.23 (a) Aldehydes reduce Tollen's reagent so "A" must be formaldehyde.
$H_2CO(\ell) + 2Ag(NH_3)_2^+(aq) + 2H_2O \longrightarrow HCO_2^-(aq) + Ag(s) + 4NH_4^+(aq) + OH^-(aq)$
(b) "B" and "C" must be isopropanol and acetone, respectively.
$5(CH_3)_2CHOH(\ell) + 2MnO_4^-(aq) \longrightarrow 5(CH_3)_2CO + 2Mn^{2+}(aq) + 6OH^-(aq) + 2H_2O$

21.27 Halogenation (e.g., bromination) and oxidation of propylene result in addition across the double bond.
$CH_3CH{=}CH_2 + Br_2 \longrightarrow CH_3CH(Br){-}CH_2Br$
$CH_3CH{=}CH_2 + KMnO_4 \longrightarrow MnO_2 + CH_3CH(OH){-}CH_2OH$
Halogenation (e.g., bromination and oxidation of toluene do *not* involve chemistry of the double bond. Benzene rings are stabilized by the aromatic nature of their cyclic, conjugated pi-systems. It would take a lot of energy to disrupt such a stable system. Instead, aromatic chemistry is predominantly substitution chemistry or the chemistry of appended functional groups.
$C_6H_5CH_3 + Br_2 + Fe \longrightarrow$ brominated toluenes (e.g., $C_6H_4BrCH_3$)
$C_6H_5CH_3 + KMnO_4 \longrightarrow C_6H_5CO_2H$

21.31 35.2%

21.33 66.8%

21.35 18 L

Chapter 22

22.1 (a) A = 10 (b) Z = 5 (c) Z = 5
(d) 5 electrons (e) 5 neutrons (f) Boron-10

22.3 ${}^{22}_{11}Na \longrightarrow {}^{0}_{1}\beta + {}^{22}_{10}Ne$ β^+ emission to correct for having excess protons
${}^{24}_{11}Na \longrightarrow {}^{0}_{-1}\beta + {}^{22}_{12}Mg$ β^- emission to correct for having excess neutrons

22.5 ${}^{226}_{88}Ra \longrightarrow {}^{4}_{2}He + {}^{222}_{86}Rn$

22.7 ${}^{14}_{6}C \longrightarrow {}^{0}_{-1}\beta + {}^{14}_{7}N$

22.9 (a) ${}^{10}_{5}B + {}^{4}_{2}He \longrightarrow {}^{13}_{6}C + {}^{1}_{1}H$
(b) ${}^{9}_{4}Be + {}^{1}_{1}H \longrightarrow {}^{8}_{4}Be + {}^{2}_{1}H$ or
${}^{9}_{4}Be + {}^{1}_{1}H \longrightarrow {}^{8}_{4}Be + {}^{2}_{1}D$
(c) ${}^{27}_{13}Al + {}^{1}_{0}n \longrightarrow \gamma + {}^{28}_{13}Al$

22.13 ${}^{14}_{7}N + {}^{3}_{2}H \longrightarrow {}^{1}_{1}p + {}^{17}_{8}O$

22.17 (a) Decay rate = 250 counts per second
(b) 12 minutes

22.19 3.61×10^{10} counts/s

22.21 3400 yr

22.23 1.88×10^{3} MeV

22.25 (a) 2.3 MeV, 1.2 MeV/nucleon;
(b) 92.2 MeV, 7.68 MeV/nucleon;
(c) 492.4 MeV, 8.793 MeV/nucleon

22.27 0.1325 amu, 7.714 MeV/nucleon

22.29 17.5 MeV of energy are released

22.33 ${}^{232}_{90}Th$ ${}^{233}_{90}Th$ ${}^{234}_{90}Th$
${}^{233}_{91}Pa$ ${}^{234}_{91}Pa$
${}^{232}_{92}U$ ${}^{233}_{92}U$ ${}^{234}_{92}U$ ${}^{235}_{92}U$ ${}^{236}_{92}U$ ${}^{237}_{92}U$
${}^{237}_{93}Np$

22.37 ${}^{235}_{92}U + {}^{1}_{0}n \rightarrow {}^{143}_{54}Xe + {}^{90}_{38}Sr + 3{}^{1}_{0}n$

22.41 Mass ${}^{120}Sn$ = 119.90 amu

22.43 (a) ${}^{90}_{40}Zr$ and ${}^{66}_{30}Zn$
(b) ${}^{232}_{92}U$ and ${}^{222}_{89}Ac$

22.45 ${}^{211}_{82}Pb$

22.47 (a) 20.18 amu (b) 0.73% ${}^{235}U$ and 99.27% ${}^{238}U$

22.49 ${}^{24}_{12}Mg + {}^{4}_{2}He \rightarrow {}^{28}_{14}Si$ ${}^{28}_{14}Si + {}^{1}_{0}n \rightarrow {}^{22}_{11}Na + {}^{7}_{3}Li$

22.53 1.6×10^{4} to 2.4×10^{4} times the energy content of coal.

Appendix

Nobel Prize Winners

Year	Chemistry	Physics	Medicine/Physiology
1901	JACOBUS HENRICUS van't HOFF (*Dutch*)—Laws of chemical dynamics and osmotic pressure	WILHELM K. RÖNTGEN (*German*)—Discovery of X-rays	EMIL von BEHRING (*German*)—Diphtheria antitoxin
1902	EMIL FISCHER (*German*)—Studies on sugars, purine derivatives, and peptides	HENDRIK ANTOON LORENZ and PIETER ZEEMAN (*Dutch*)—Effect of magnetism on radiation-the Zeeman effect	Sir RONALD ROSS (*British*)—Malaria, the malarial parasite, and the mosquito
1903	SVANTE A. ARRHENUS (*Swedish*)—Electrolytic dissociation theory	ANTOINE HENRI BECQUEREL and PIERRE and MARIE CURIE (*French*)—Natural (spontaneous) radioactivity	NIELS R. FINSEN (*Danish*)—The light treatment of disease, especially Lupus vulgaris
1904	Sir WILLIAM RAMSAY (*British*)—Discovery of inert gaseous elements in the atmosphere	LORD RAYLEIGH, JOHN WILLIAM STRUTT (*British*)—Discovery of argon and studies of gaseous densities	IVAN P. PAVLOV (*Russian*)—The physiology of digestion
1905	ADOLPH von BAEYER (*German*)—Organic dyes, aromatic compounds, and the synthesis of indigo	PHILIPP LENARD (*German*)—The properties of cathode rays	ROBERT KOCH (*German*)—Tuberculosis, discovery of tubercule bacillus and tuberculin
1906	HENRI MOISSAN (*French*)— Fluorine; the development of the electric furnace	JOSEPH JOHN THOMSON (*British*)—Electrical discharge through gases	SANTIAGO RAMON y CAJAL (*Spanish*) and CAMILLIO GOLGI (*Italian*)—Studies of the nervous system and the structure of nerve tissue
1907	EDUARD BUCHNER (*German*)—Cell-free fermentation	ALBERT A. MICHELSON (*American*)—Spectroscopic studies, optical instrumentation, and the speed of light	CHARLES L. A. LAVERAN (*French*)—Studies on protozoa in the generation of disease

1908	ERNEST RUTHERFORD (*British*)—Behavior of alpha rays; physics and chemistry of radioactive substances	GABRIEL LIPPMANN (*French*)—Color photography; the phenomenon of interference	PAUL EHRLICH (*German*) and ELIE METCHNIKOFF (*French*)—Immunity
1909	WILHELM OSTWALD (*German*)—Catalysis, chemical equilibrium, and rates of chemical reactions	GUGLIELMO MARCONI (*Italian*) and KARL FERDINAND BRAUN (*German*)—Wireless telegraphy	THEODOR KOCHER (*Swiss*)—The thyroid gland; physiology, pathology, and surgery
1910	OTTO WALLACH (*German*)—Alicyclic substances	JOHANNES D. van der WAALS (*Dutch*)—Studies of relationships between gases and liquids	ALBRECHT KOSSEL (*German*)—Cell chemistry, especially proteins and nucleic substances
1911	MARIE CURIE (*French*)—Radium and polonium and the compounds of radium	WILHELM WIEN (*German*)—Heat radiation by black bodies	ALLVAR GULLSTRAND (*Swedish*)—Studies in the refraction of light through the eye (the dioptrics of the eye)
1912	VICTOR GRIGNARD and PAUL SABATIER (*French*)—The Grignard reaction for synthesizing organic compounds	NILS G. DALÉN (*Swedish*)—Automatic gas regulators for coastal lighting	ALEXIS CARREL (*French*)—Studies on vascular seams and the grafting of organs and blood vessels
1913	ALFRED WERNER (*Swiss*)—Coordination theory for arrangements of atoms in molecules	HEIKE KAMERLINGH-ONNES (*Dutch*)—Properties of matter at low temperatures; liquefaction of helium	CHARLES RICHET (*French*)—Studies on allergies caused by foreign substances; anaphylactic test
1914	THEODORE W. RICHARDS (*American*)—Atomic weight determination of the elements	MAX T.F. von LAUE (*German*)—X-ray diffraction in crystals	ROBERT BARANY (*Austrian*)—Studies on the physiology and pathology of the human vestibular system
1915	RICHARD WILLSTATTER (*German*)—The coloring in plants, especially chlorophyll	Sir WILLIAM H. BRAGG and Sir WILLIAM L. BRAGG (*British*)—Crystal structure study by X-ray methods	*No award*
1916	*No award*	*No award*	*No award*
1917	*No award*	CHARLES G. BARKLA (*British*)—Studies on the diffusion of light and X-radiations from elements	*No award*
1918	FRITZ HABER (*German*)—*The Haber* process for synthesizing ammonia	MAX PLANCK (*German*)—The quantum theory of light (the element of action)	*No award*
1919	*No award*	JOHANNES STARK (*German*)—The Stark effect of spectral lines in electric fields; the Doppler effect	JULES BORDET (*Belgian*)—Discoveries on immunity
1920	WALTHER NERNST (*German*)—Studies on heat changes in chemical reactions	CHARLES-EDOUARD GUILLAUME (*Swiss*)—The special properties of nickel alloys and their importance in precision physics	AUGUST KROGH (*Danish*)—Studies on the regulating action and behavior of the blood capillaries
1921	FREDERICK SODDY (*British*)—Radioactive substances and the origin and nature of isotopes	ALBERT EINSTEIN (*German*)—Contributions to mathematical physics and the photoelectric effect	*No award*
1922	FRANCIS W. ASTON (*British*)—The behavior of isotope mixtures, whole-number rule on atomic weights, and the mass spectrograph	NEILS BOHR (*Danish*)—Atomic structure and atomic radiations	ARCHIBALD V. HILL (*British*) and OTTO MEYERHOF (*German*)—Discoveries on heat production in muscles and lactic acid production in muscles
1923	FRITZ PREGL (*Austrian*)—Methods of microanalysis for organic substances	ROBERT A. MILLIKAN (*American*)—The elementary electric charge and photoelectric effect	Sir FREDERICK G. BANTING (*Canadian*) and JOHN J. R. MacLEOD (*Scottish*)—Discovery of insulin

1924	*No award*	KARL M. G. SIEGBAHN (*Swedish*)—X-ray spectra	WILLEM EINTHOVEN (*Dutch*)—Discovery of the electrocardiogram
1925	RICHARD ZSIGMONDY (*German*)—Studies on the nature of colloids	JAMES FRANCK and GUSTAV HERTZ (*German*)—Laws governing the collision of an electron and an atom	*No award*
1926	THEODOR SVEDBERG (*Swedish*)—Dispersions and the chemistry of colloids	JEAN B. PERRIN (*French*)—The discontinuous structure of matter; measurements on sizes of atoms	JOHANNES FIBIGER (*Danish*)—Discovery of Spiroptera carcinoma, a cancer-producing parasite
1927	HEINRICH WIELAND (*German*)—Bile acids and related substances	ARTHUR H. COMPTON (*American*) and CHARLES T.R. WILSON (*British*)—Discovery of dispersion of X-rays reflected from atoms (the Compton effect)	JULIUS WAGNER-JAUREGG (*Austrian*)—Fever treatment (malaria vaccination) in treating paralysis
1928	ADOLPH WINDAUS (*German*)—Studies on sterols and their connection with vitamins	OWEN W. RICHARDSON (*British*)—Studies on the thermionic effect and electrons emitted by hot metals	CHARLES NICOLLE (*French*)—Typhus exanthernaticus
1929	Sir ARTHUR HARDEN (*British*) and HANS von EULER-CHELPIN (*German*)—Fermentation and enzyme action	Prince LOUIS-VICTOR de BROGLIE (*French*)— The wave nature of the electron	CHRISTIAAN EIJKMAN (*Dutch*) and Sir FREDERICK G. HOPKINS (*British*)—Growth-promoting and antineuritic vitamins
1930	HANS FISCHER (*German*)—Chemistry of pyrrole and the synthesis of hemin	Sir CHANDRASEKHARA V. RAMAN (*Indian*)—The diffusion of light and the Raman effect	KARL LANDSTEINER (*American*)—The four main human blood types
1931	CARL BOSCH and FRIEDRICH BERGIUS (*German*)— High-pressure methods for chemical manufacture	*No award*	OTTO H. WARBURG (*German*)—Enzyme role in tissue respiration
1932	IRVING LANGMUIR (*American*)—Surface chemistry	WERNER K. HEISENBERG (*German*)—Quantum mechanics	EDGAR D. ADRIAN and Sir CHARLES S. SHERRINGTON (*British*)—Function of neurons
1933	*No award*	PAUL A. M. DIRAC (*British*) and ERWIN SCHRODINGER (*Austrian*)—New forms of atomic theory	THOMAS H. MORGAN (*American*)—The hereditary function of the chromosomes
1934	HAROLD C. UREY (*American*)—Discovery of deuterium	*No award*	GEORGE R. MINOT, WILLIAM P. MURPHY, and GEORGE H. WHIPPLE (*American*)—Liver treatment for anemia
1935	FREDERIC and IRENE JOLIOT-CURIE (*French*)—Synthesis of new radioactive elements	JAMES CHADWICK (*British*)—Discovery of the neutron	HANS SPEMANN (*German*)—Studies in embryonic growth and development
1936	PETER J. W. DEBYE (*Dutch*)—Studies on dipole moments; diffraction of electrons and X-rays in gases	CARL D. ANDERSON (*American*) and VICTOR F. HESS (*Austrian*)—The positron and cosmic rays	Sir HENRY H. DALE (*British*) and OTTO LOWEI (*Austrian*)—Chemical transmission of nerve impulses
1937	Sir WALTER N. HAWORTH (*British*) and PAUL KARRER (*Swiss*)—Research on carbohydrates, vitamins C, A, and B_2 carotinoids, and flavins	CLINTON J. DAVISSON (*American*) and GEORGE P.T HOMSON (*British*)—Diffraction of electrons by crystals	ALBERT SVENT-GYÖRGYI (*Hungarian*)—Research on biological oxidation, especially vitamin C and fumaric acid
1938	RICHARD KUHN (*German*)—Carotinoids and vitamins	ENRICO FERMI (*Italian*)—New elements beyond uranium; nuclear reactions by slow electrons	CORNEILLE HEYMANS (*Belgian*)—Regulation of respiration

1939	ADOLPH BUTENANDT (*German*) and LEOPALD RUZICKA (*Swiss*)—Sex hormones and polymethylenes	ENREST O. LAWRENCE (*American*)—The cyclotron and artificially radioactive elements	GERHARD DOMAGK (*German*)—Prontosil, the first sulfa drug
1940	*No award*	*No award*	*No award*
1941	*No award*	*No award*	*No award*
1942	*No award*	*No award*	*No award*
1943	GEORG VON HEVESY (*Hungarian*)—Isotopes as tracers in chemical studies	OTTO STERN (*American*)—Molecular beam method for the study of the atom; the magnetic moment of the proton	HENRIK DAM (*Danish*) and EDWARD DOISY (*American*)—Discovery and synthesis of vitamin K
1944	OTTO HAHN (*German*)—Atomic fission of heavy nuclei	ISADOR I. RABI (*American*)—Magnetic properties of atomic nuclei	JOSEPH ERLANGER and HERBERT S. GASSER (*American*)—Studies on single nerve fibers
1945	ARTTURI I. VIRTANEN (*Finnish*)—Discovery of new methods for agricultural biochemistry	WOLFGANG PAULI (*American*)—The exclusion principle (Pauli principle) of electrons	ERNST B. CHAIN (*German*), Sir ALEXANDER FLEMING, and Sir HOWARD W. FLOREY (*British*)—Penicillin
1946	JOHN H. NORTHROP, WENDELL M. STANLEY, and JAMES B. SUMNER (*American*)—Crystallizing enzymes; preparation of pure enzymes and virus proteins	PERCY W. BRIDGMAN (*American*)—High-pressure apparatus; studies at very high pressures	HERMANN J. MULLER (*American*)—X-ray-induced mutations
1947	Sir ROBERT ROBINSON (*British*)—Alkaloids and other plant substances	Sir EDWARD V. APPLETON (*British*)—Physical properties of the ionosphere	CARL F. and GERTY T. CORI (*American*) and BERNARDO A. HOUSSAY (*Argentinian*)—Animal metabolism and the study of the pituitary gland and pancreas
1948	ARNE TISELIUS (*Swedish*)—Nature of the serum proteins	PATRICK M. S. BLACKETT (*British*)—Cosmic radiation and nuclear physics	PAUL MULLER (*Swiss*)—Insect-killing properties of DDT
1949	WILLIAM F. GIAUQUE (*American*)—Chemical thermodynamics; effects due to extreme cold	HIDEKI YUKAWA (*Japanese*)—The meson	WALTER R. HESS (*Swiss*) and ANTONIO E. MONIZ (*Portuguese*)—Studies in brain control and brain surgery techniques
1950	OTTO DIELS and KURT ALDER (*German*)—Diene synthesis	CECIL F. POWELL (*British*)—Photographic techniques for atomic nuclei; discoveries concerning mesons	PHILIP S. HENCH, EDWARD C. KENDALL (*American*), and TADEUS REICHSTEIN (*Swiss*)—Cortisone and ACTH
1951	EDWIN M. McMILLAN and GLENN T. SEABORG (*American*)—Plutonium and other transuranium elements	Sir JOHN D. COCKROFT (*British*) and ERNEST WALTON (*Irish*)—Transmutation of atomic nuclei through artificially accelerated atomic particles	MAX THEILER (*American*)—Discovery of yellow fever vaccine
1952	ARCHER J. P. MARTIN and RICHARD L. M. SYNGE (*British*)—Partition chromatography	FELIX BLOCH and EDWARD M. PURCELL (*American*)—Magnetic moment method for atomic nuclei measurements	SELMAN A. WAKSMAN (*American*)—Streptomycin
1953	HERMANN STAUDINGER (*German*)—Studies in the synthesis of giant molecules	FRITZ ZERNIKE (*Dutch*)—Phase contrast microscope for cancer research	HANS A. KREBS (*British*) and FRITZ A. LIPMANN (*American*)—Metabolic studies and biosynthesis

1954	LINUS PAULING (*American*)—Chemical bonds in protein molecules and forces in matter	MAX BORN and WALTHER BOTHE (*German*)—Quantum mechanics and cosmic radiation	JOHN F. ENDERS, FREDERICK C. ROBBINS and THOMAS H. WELLER (*American*)—Method for growing polio viruses in test tubes
1955	VINCENT du VIGNEAUD (*American*)—Synthetic hormones	POLYKARP KUSCH and WILLIS E. LAMB (*American*)—The structure of the hydrogen spectrum; the magnetic moment of the electron	HUGO THEORELL (*Swedish*)—The nature of oxidation enzymes
1956	Sir CYRIL N. HINSHELWOOD (*British*) and NIKOLAJ N. SEMENOV (*Russian*)—Reaction kinetics and chemical reaction mechanics	JOHN BARDEEN, WALTER H. BRATTAIN and WILLIAM SHOCKLEY (*American*)—The transistor	ANDRE F. COURNAND, DICKINSON W. RICHARDS, Jr., (*American*), and WERNER FORSSMANN (*German*)—The use of the catheter in heart research
1957	ALEXANDER R., LORD TODD (*British*)—Work on nucleotides and nucleotide coenzymes	TSUNG DAO LEE and CHEN NING YANG (*American*)—Disproving the laws of conservation of parity	DANIEL BOVET (*Italian*)—Antihistamines
1958	FREDERICK SANGER (*British*)—Structure of proteins, especially the insulin molecule	PAVEL A. CHERENKOV, ILYA M. FRANK, and IGOR TAMM (*Russian*)—Study of high-energy particles and the Cherenkov effect	GEORGE W. BEADLE, JOSHUA LEDERBERG, and EDWARD L. TATUM (*American*)—Genetic mechanisms and heredity
1959	JAROSLAV HEYROVSKY (*Czechoslavakian*)—Polarographic analysis and techniques	OWEN CHAMBERLAIN and EMILIO SEGRÈ (*American*)—The antiproton	ARTHUR KORNBERG and SEVERO OCHOA (*American*)—Artificial nucleic acids
1960	WILLARD F. LIBBY (*American*)—Radiocarbon dating	DONALD A. GLASER (*American*)—The bubble chamber for subatomic particles	Sir FRANK MACFARLANE BURNET (*Australian*) and PETER B. MEDAWAR (*British*)—Acquired immunological tolerance
1961	MELVIN CALVIN (*American*)—Photosynthesis	ROBERT HOFSTADTER and RUDOLPH MÖSSBAUER (*American*)—The nucleons and gamma ray research; the Mossbauer effect	GEORGE von BÉKÉSY (*American*)—Physical mechanisms of stimulation in the cochlea
1962	JOHN C. KENDREW and MAX F. PERUTZ (*British*)—Structure of complex globular proteins	LEV D. LANDAU (*Russian*)—Theories for condensed matter, especially liquid helium	FRANCIS H. C. CRICK, MAURICE H. F. WILKINS (*British*), and JAMES D. WATSON (*American*)—Study of structure of deoxyribonucleic acid (DNA)
1963	GIULIO NATTA (*Italian*) and KARL ZIEGLER (*German*)— Polymers of simple hydrocarbons; improved plastics	J. HANS JENSEN (*German*) and MARIA GOEPPERT-MAYER, and EUGENE P. WIGNER (*American*)—Structure of atomic nuclei and elementary particles	Sir JOHN C. ECCLES (*Australian*), ALAN L. HODGKIN, and ANDREW F. HUXLEY (*British*)— Ionic mechanisms and behavior of nerve impulses
1964	DOROTHY CROWFOOT HODGKIN (*British*)—Structures of vitamin B_{12} and penicillin by X-ray methods	NIKOLAY BASOV, ALEXANDER M. PROKHOROV (*Russian*), and CHARLES H. TOWNES (*American*)—Masers and lasers	KONRAD BLOCH (*American*) and FEODOR LYNEN (*German*)—Cholesterol and fatty acid metabolism
1965	ROBERT B. WOODWARD (*American*)—Contributions to synthetic organic chemistry	RICHARD P. FEYNMAN, JULIAN S. SCHWINGER (*American*), and SCHINICHIRO TOMONAGA (*Japanese*)—Basic studies in quantum electrodynamics	FRANCOIS JACOB, ANDRE LWOFF, and JACQUES MONOD (*French*)—Genetic control in the synthesis of enzymes and viruses
1966	ROBERT S. MULLIKEN (*American*)—Molecular orbital theory for chemical structure	ALFRED KASTLER (*French*)—Studies on the energy levels of atoms	CHARLES B. HUGGINS and FRANCIS P. ROUS (*American*)—Use of hormones in treating cancer and discovery of a cancer-producing virus

1967	MANFRED EIGEN (*German*), RONALD G. W. NORRISH, and GEORGE PORTER (*British*)—Very fast chemical reactions	HANS ALBRECHT BETHE (*American*)—Theory of nuclear reactions, especially on the source of energy in stars	RAGNER GRANIT (*Swedish*), H. KEFFER HARTLINE, and GEORGE WALD (*American*)—Chemical and physiological visual processes in the eye
1968	LARS ONSAGER (*American*)—Various types of relationships for thermodynamic activity	LUIS W. ALVAREZ (*American*)—Studies on subatomic particles and techniques for detecting them	ROBERT W. HOLLY, H. GOBIND KHORANA, and MARSHALL W. NIRENBERG (*American*)—Role of the genes in cell function
1969	DEREK H. BARTON (*British*) and ODD HASSEL (*Norwegian*)—Conformations and the relationship between chemical shape and chemical reactivity	MURRAY GELL-MANN (*American*)—Contributions toward the understanding and classification of elementary particles	MAX DELBRÜCK, ALFRED D. HERSHEY, and SALVADOR E. LURIA (*American*)—Viruses, viral diseases, and the foundations of molecular biology
1970	LUIS F. LELOIR (*Argentinian*)—Sugar nucleotides	HANNES ALFVEN (*Swedish*) and LOUIS NEEL (*French*)—Theoretical basis for magnetohydrodynamics; antiferromagnetic materials and behavior	JULIUS AXELROD (*American*), ULF S. von ELLER (*Swedish*), and BERNARD KATZ (*British*)—Discoveries related to the search for remedies for nervous and mental disturbances
1971	GERHARD HERZBERG (*Canadian*)—Contributions to electronic structure and geometry of molecules	DENNIS GABOR (*British*)—Invention and development of the holographic method of three-dimensional imagery	EARL W. SUTHERLAND, Jr. (*American*)—Discoveries concerning the mechanisms of the actions of hormones
1972	CHRISTIAN B. ANFINSEN, STANFORD MOORE, and WILLIAM H. STEIN (*American*)—Ribonuclease	JOHN BARDEEN, LEON N. COOPER, and JOHN R. SCHRIEFFER (*American*)—jointly developed the theory of superconductivity	GERALD M. EDELMAN (*American*) and RODNEY R. PORTER (*British*)—Research on the chemical structure and nature of antibodies
1973	ERNST OTTO FISCHER (*West German*) and GEOFFREY WILKINSON (*British*)—Chemistry of the organometallic, so-called sandwich compounds	IVAR GIAEVER (*American*), LEO ESAKI (*Japanese*), and BRIAN D. JOSEPHSON (*British*)—Experimental discoveries regarding tunneling phenomena in semiconductors and superconducters	KARL von FRISCH, KONRAD LORENZ (*Austrian*), and NIKOLAAS TINBERGEN (*Dutch*)—Studies of individual and social behavior patterns
1974	PAUL J. FLORY (*American*)— Fundamental achievements in the physical chemistry of the macromolecules	MARTIN RYLE and ANTHONY HEWISH (*British*)—Radio astrophysics and the discovery of pulsars	GEORGE E. PALADE, CHRISTIAN de DUVE (*American*), and ALBERT CLAUDE (*Belgian*)—Contributions to understanding the inner workings of living cells
1975	JOHN CORNFORTH (*Australian-British*) and VLADIMIR PRELOG (*Yugoslavian-Swiss*)—Stereochemistry of enzyme-catalyzed reactions	JAMES RAINWATER (*American*), BEN MOTTELSON (*American-Danish*), and AAGE BOHR (*Danish*)—Discovery of the connection between motion in atomic nuclei and the development of the theory of the structure of the atomic nucleus	DAVID BALTIMORE, HOWARD M. TEMIN, and RENATO DULBECCO (*American*)—Interaction between tumor viruses and the cell's genetic material
1976	WILLIAM N. LIPSCOMB (*American*)—Studies on the structure of boranes illuminating problems of chemical bonding	BURTON RICHTER and SAMUEL C. C. TING (*American*)—Discovery of a heavy elementary particle of a new kind	BARUCH S. BLUMBERG and D. CARLETON GAJDUSEK (*American*)—Discoveries concerning new mechanisms for the origin and dissemination of infectious diseases
1977	ILYA PRIGOGINE (*Belgian*)—Contributions to nonequilibrium thermodynamics	JOHN H. van VLECK, PHILIP W. ANDERSON (*American*), and NEVILL F. MOTT (*British*)— Fundamental theoretical investigations of the electronic structure of magnetic and disordered systems	ROSALYN S. YALOW, ROGER C. L. GUILLEMIN, and ANDREW V. SCHALLY (*American*)—Research in the role of hormones in the chemistry of the body

1978	PETER MITCHELL (*British*)—Contribution to the understanding of biological energy transfer through formulation of the chemiosmotic theory	PYOTR KAPITSA (*Russian*), ARNO PENZIAS, and ROBERT WILSON (*American*)—Basic inventions and discoveries in the area of low-temperature physics	DANIEL NATHANS, HAMILTON SMITH (*American*), and WERNER ARBER (*Swiss*)—Discovery of restriction enzymes and their application to problems of molecular genetics
1979	HERBERT C. BROWN (*American*) and GEORGE WITTIG (*German*)—Development of boron- and phosphorus-containing compounds	STEVEN WEINBERG, SHELDON L. GLASHOW (*American*), and ABDUS SALAM (*Pakistani*)—Contributions to the theory of the unified weak and electromagnetic interaction between elementary particles	ALLAN McLEOD CORMACK (*American*) and GODFREY NEWBOLD HOUNSFIELD (*British*)—Development of computed axial tomography (CAT scan) X-ray technique
1980	PAUL BERG, WALTER GILBER (*American*), and FREDERICK SANGER (*British*)—Fundamental studies of the biochemistry of nucleic acids, with particular regard to recombinant DNA	JAMES W. CRONIN and VAL L. FITCH (*American*)—Discovery of violations of fundamental symmetry principles in the decay of neutral K-mesons	BARUJ BENECERRAF, GEORGE D. SNELL (*American*), and JEAN DAUSSEY (*French*)—Discoveries that explain how the structure of cells relates to organ transplants and diseases
1981	KENICHI FUKUI (*Japanese*) and ROALD HOFFMANN (*American*)—Theories concerning the course of chemical reactions	NICOLASS BOEMBERGEN, ARTHUR SCHLAWLOW (*American*) and KAI M. SIEGBAHN (*Swedish*)—Contribution to the development of laser spectroscopy	ROGER W. SPERRY, DAVID H. HUBEL (*American*), and TORSTEN N. WIESEL (*Swedish*)—Studies vital to understanding the organization and functioning of the brain
1982	AARON KLUG (*South African*)—Development of crystallographic electron microscopy and structural elucidation of biologically important nucleic acid–protein complexes	KENNETH G. WILSON (*American*)—Theory for critical phenomena in connection with phase transitions	SUNE BERGSTROMM, BENGT SAMUELSSON (*Swedish*), and JOHN R. VANE (*British*)—Research in prostaglandins, a hormone-like substance involved in a wide range of illnesses
1983	HENRY TAUBE (*Canadian*)—Studies of the mechanisms of electron transfer reactions, particularly of metal complexes	SUBRAHMANYAN CHANDRASEKHAR and WILLIAM FOWLER (*American*)—Theoretical studies of the physical processes of importance to the structure and evolution of stars	BARBARA McCLINTOCK (*American*)—Discovery of mobile genes in the chromosomes of a plant that change the future generations of plants they produce
1984	BRUCE MERRIFIELD (*American*)—Methodology for chemical synthesis on a solid matrix	SIMON van der MEER (*Dutch*) and CARLO RUBBIA (*Italian*)—Discovery of field particles of W and Z, communicators of the weak interaction	CESAR MILSTEIN (*British-Argentinian*), NIELS K. JERNE (*British-Danish*), and GEORGES J. F. KOHLER (*West German*)—Work in immunology
1985	HERBERT A. HAUPTMANN and JEROME KARLE (*American*)—Development of direct methods for the determination of crystal structures	KLAUS von KLITZING (*West German*)—Discovery of the quantum Hall effect	MICHAEL S. BROWN and JOSEPH L. GOLDSTEIN (*American*)—Contributions to our understanding of cholesterol metabolism, prevention and treatment of atherosclerosis and heart attacks
1986	DUDLEY HERSCHBACH, YUAN T. LEE (*American*), and JOHN C. POLANYI (*Canadian*)—Fundamental contributions to the development of the dynamics of chemical reactions	ERNEST RUSKA (*German*), GERD BINNIG (*West German*), and HEINRICH ROHRER (*Swiss*)—Design of the scanning tunneling microscope	RITA LEVI-MOTALCINI (*American-Italian*) and STANLEY COHEN (*American*)—Contributions to understanding the substance that influences cell growth

1987	DONALD J. CRAM, CHARLES J. PEDERSON (*American*), and JEAN-MARIE LEHN (*French*)—Discovery of how to make relatively simple molecules that mimic the functions of much more complex molecules produced by living cells	K. ALEX MUELLER (*Swiss*) and J. GEORGE BEDNORZ (*West German*)—Discovery of superconductivity in a new class of ceramics at temperatures higher than had previously been thought possible	SUSUMU TONEGAWA (*Japanese*)—Discoveries of how the body can marshal its immunological defenses against millions of new and different disease agents
1988	HARTMUT MICHEL, ROBERT HUBER, and JOHANN DEISENHOFER (*West German*)—Mapping the structure of protein molecules essential in photosynthesis	LEON LEDERMAN, MELVIN SCHWARTZ, and JACK STEINBERGER (*American*)—Experiments with the subatomic neutrino particle	GEORGE H. HITCHINGS, GERTRUDE B. ELION (*American*), and Sir JAMES W. BLACK (*British*)—Work that had led to the introduction of drugs widely used to treat heart disease, ulcers, and leukemia
1989	SIDNEY ALTMAN and THOMAS CECH (*American*)—Discovered the active role played by RNA in catalyzing chemical reactions in cells	NORMAN RAMSAY (*American*)—Techniques for studying the structure of atoms, the hydrogen atom and hydrogen maser, and development of the cesium atomic clock; HANS DEHMELT (*American*) and WOLFGANG PAUL (*German*)—The ion-trap method for separating charged particles, especially the electron and ions	MICHAEL BISHOP and HAROLD VARMUS (*American*)—Studies leading to understanding of how normal genes that control cell growth can cause cancer
1990	ELIAS JAMES COREY (*American*)—The theory and methodology of organic synthesis	JEROME I. FRIEDMAN (*American*), HENRY W KENDALL (*American*), and RICHARD E. TAYLOR (*Canadian*)—Investigations concerning deep inelastic scattering of electrons on protons and bound neutrons, which have been of essential importance for the development of the quark model in particle physics	JOSEPH E. MURRAY (*American*) and E. DONNALL THOMAS (*American*)—Discoveries concerning organ and cell transplantation in the treatment of human disease
1991	RICHARD R. ERNST (*Swiss*)—The development of the methodology of high resolution nuclear magnetic resonance (NMR) spectroscopy	PIERRE-GILLES de GENNES (*French*)—Discovered that methods developed for studying order phenomena in simple systems can be generalized to more complex forms of matter, in particular to liquid crystals and polymers	ERWIN NEHER (*German*) and BERT SAKMANN (*German*)—Discoveries concerning the function of single ion channels in cells
1992	RUDOLPH A. MARCUS (*American*)—Contributions to the theory of electron transfer reactions in chemical systems	GEORGES CHARPAK (*French*)—The invention and development of particle detectors, in particular the multiwire proportional chamber	EDMOND H. FISCHER (*American*) and EDWIN G. KREBS (*American*)—Discoveries concerning reversible protein phosphorylation as a biological regulatory mechanism
1993	KARY B. MULLIS (*American*)—The invention of the polymerase chain reaction (PCR) method. MICHAEL SMITH (*Canadian*)—Fundamental contributions to the establishment of oligonucleiotide-based, site-directed mutagenesis and its development for protein studies	RUSSELL A. HULSE (*American*) and JOSEPH H. TAYLOR Jr (*American*)—The discovery of a new type of pulsar, a discovery that has opened up new possibilities for the study of gravitation	RICHARD J. ROBERTS (*British*) and PHILLIP A. SHARP (*American*)—The discovery of split genes

1994	GEORGE A. OLAH (*American*)—Contributions to carbocation chemistry	BERTRAM N. BROCKHOUSE (*Canadian*)—The development of neutron spectroscopy. CLIFFORD G. SHULL (*American*)—The development of the neutron diffraction technique	ALFRED G. GILMAN (*American*) and MARTIN RODBELL—The discovery of G-proteins and the role of these proteins in signal transduction in cells
1995	PAUL J. CRUTZEN (*Dutch*), MARIO J. MOLINA (*American*), and F. SHERWOOD ROWLAND (*American*)—Work in atmospheric chemistry, particularly concerning the formation and decomposition of ozone	MARTIN L. PERL (*American*)—The discovery of the tau lepton. FREDERICK REINES (*American*)—The detection of the neutrino	EDWARD B. LEWIS (*American*), CHRISTIANE NÜSSLEIN-VOLHARD (*German*) and ERIC F. WIESCHAUS (*American*)—Discoveries concerning the genetic control of early embryonic development
1996	ROBERT F. CURL Jr. (*American*), Sir HAROLD W. KROTO (*British*) and RICHARD E. SMALLEY—The discovery of fullerenes	DAVID M. LEE (*American*), DOUGLAS D. OSHEROFF (*American*), and ROBERT C. RICHARDSON (*American*)—The discovery of superfluidity in helium-3	PETER C. DOHERTY (*Australian*) and ROLF M. ZINKERNAGEL (*Swiss*)—Discoveries concerning the specificity of the cell mediated immune defense
1997	PAUL D. BOYER (*American*) and JOHN E. WALKER (*British*)—The elucidation of the enzymatic mechanism underlying the synthesis of adenosine triphosphate (ATP) JENS C. SKOU (*Danish*),—The first discovery of an ion-transporting enzyme, Na^+, K^+-ATPase	STEVEN CHU (*American*), CLAUDE COHEN-TANNOUDJI (*French*) and WILLIAM D. PHILLIPS (*American*)—Development of methods to cool and trap atoms with laser light	STANLEY B. PRUSINER (*American*)—Discovery of Prions—a new biological principle of infection
1998	WALTER KOHN (*American*)—Development of the density-functional theory. JOHN A. POPLE (*American*)—Development of computational methods in quantum chemistry	ROBERT B. LAUGHLIN (*American*), HORST L. STÖRMER (*German*) and DANIEL C. TSUI (*American*)—Discovery of a new form of quantum fluid with fractionally charged excitations	ROBERT F. FURCHGOTT (*American*), LOUIS J. IGNARRO (*American*), and FERID MURAD (*American*)—Discoveries concerning nitric oxide as a signaling molecule in the cardiovascular system
1999	AHMED ZEWAIL (*American*)—Studies of the transition states of chemical reactions using femtosecond spectroscopy	GERARDUS 'T HOOFT (*Dutch*), MARTINUS J. G. VELTMAN (*Dutch*)—Elucidating the quantum structure of electroweak interactions in physics	GÜNTER BLOBEL (*American*)—Discovery that proteins have intrinsic signals that govern their transport and localization in the cell
2000	ALAN J. HEEGER (*American*), ALAN G. MacDIARMID (*American*), and HIDEKI SHIRAKAWA (*Japanese*)—Discovery and development of conductive polymers	ZHORES I. ALFEROV (*Russian*) and HERBERT KROEMER (*American*)—Developing semiconductor heterostructures used in high-speed- and opto-electronics JACK ST. CLAIR KILBY (*American*)—For his part in the invention of the integrated circuit	AVRID CARLSSON (*Swedish*), PAUL GREENGARD (*American*) and ERIC KANDEL (*American*)—Discoveries concerning signal transduction in the nervous system

Appendix

Scientific Notation

Scientific notation of numbers is often also referred to as **exponential notation** because numbers are expressed in terms of exponents or powers of ten. This notation serves well in science because it is convenient for expressing the very large and very small numbers often encountered. In addition it gives a way of stating precision or number of significant figures without ambiguity.

It is worthwhile to review some powers of ten.

10000000000.	10^{10}
100000.	10^{5}
100.	10^{2}
10.	10^{1}
1.	10^{0}
0.01	10^{-2}
0.0000001	10^{-7}
0.000000000001	10^{-12}

For these numbers, the exponent on the ten equals the number of places that the decimal point must be moved to reach the number 1. Movement of the decimal point to the left gives a positive exponent, whereas movement to the right gives a negative exponent.

To express numbers other than even powers of ten it is necessary to multiply ten raised to some power by a coefficient. This coefficient is normally chosen to be between one and ten. Thus

$$269000 = 2.69 \times 100000 = 2.69 \times 10^{5}$$

$$0.000000318 = 3.18 \times 0.0000001 = 3.18 \times 10^{-7}$$

The convenience of scientific notation for very large and very small numbers is clear. Avogadro's number, the number of atoms or molecules in a mole of a substance, is always given in scientific notation, 6.022×10^{23}, to avoid having to write 602200000000000000000000. The mass of an electron in kilograms is always given as 9.1094×10^{-31} so that we do not have to write 0.00000000000000000000000000000091094.

Significant figures are very easily expressed in scientific notation. The rule is that all figures given in the coefficient are significant. As examples,

$3. \times 10^5$	1 significant figure
3.00×10^5	3 significant figures
3.00000×10^5	6 significant figures

When the number is expressed as 300000, there is no standard way of stating the number of significant figures it contains.

Appendix

Logarithms

The logarithm of a number is the power to which some base number must be raised to give the original number. The most familiar base is 10, giving common logarithms that have the symbol **log** or $\mathbf{log_{10}}$. The other common base is **e** (2.71828), giving natural logarithms with the symbol **ln** or $\mathbf{log_e}$, Common logarithms have traditionally been the more usually tabulated. However, natural logarithms show up in scientific derivations and are now easily available on electronic calculators.

As two examples of common logarithms,

$$\log 1000 = 3, \text{ because } 10^3 = 1000$$

$$\log 513 = 2.71, \text{ because } 10^{2.71} = 513$$

An example of a natural logarithm is

$$\ln 10 = 2.303, \text{ because } e^{2.303} = 10$$

As shown in these examples, the common or natural logarithm of a number greater than one has a positive value. The common or natural logarithm of exactly one is zero, because the zero power of any base equals one:

$$\log 1 = 0, \text{ because } 10^0 = 1$$

$$\ln 1 = 0, \text{ because } e^0 = 1$$

Numbers less than one have negative common or natural logarithms.

$$\log 0.1 = -1, \text{ because } 10^{-1} = 1$$

$$\log 0.0209 = -1.68, \text{ because } 10^{-1.68} = 0.0209$$

$$\ln 0.0030 = -5.81, \text{ because } e^{-5.81} = 0.0030$$

Conversion between common and natural logarithms uses the equation

$$\ln a = 2.303 \log a$$

Formerly logarithms were determined from extensive tables, but either common or natural logarithms are now directly obtained using an electronic calculator.

The antilogarithm is the inverse of a logarithm. Thus

$$\log n = x$$

$$\text{antilog } x = n$$

$$10^x = n$$

Logarithms of numbers are convenient for multiplying or dividing numbers and determining powers and roots of numbers. Either common or natural logarithms can be used for these purposes. Because logarithms are exponents, multiplication of two numbers can be accomplished by adding their logarithms and then taking the antilogarithm of the sum:

$$m \cdot n = \text{antilog}(\log m + \log n)$$

$$m/n = \text{antilog}(\log m - \log n)$$

$$m^y = \text{antilog}(y \cdot \log m)$$

Similarly, taking the logarithm of each side of these three equations,

$$\log (m \cdot n) = \log m + \log n$$

$$\log (m/n) = \log m - \log n$$

$$\log (m^y) = y \cdot \log m$$

In the logarithm of a number, the digits to the left of the decimal point merely determine the place of the decimal point in the original number. Therefore, the number of significant figures in the original number should equal the number of digits to the *right* of the decimal point in its logarithm.

Appendix

Quadratic Equations

A quadratic equation is one in which the unknown value to be solved for appears as the second power or square. These equations are easy to solve in cases such as

$$x^2 = 9.86$$
$$x = \sqrt{9.86}$$

In obtaining the solution to this equation it is only necessary to recall that the square root can be either a positive or a negative number. Thus there are two solutions to this equation,

$$x = +3.14$$

$$x = -3.14$$

If the equation is simply a pure, mathematical expression with no physical significance, it is impossible to make a choice between the two solutions. Either solves the equation. However, in actual problems it is often possible to discard one solution. For example, one might be solving for a pressure, a volume, or an absolute temperature for which a negative value is impossible.

Several methods of solution are available if the unknown appears in an equation to both the first and second power. The most dependable method uses the quadratic formula,

$$x = \frac{-b \pm \sqrt{b^2 - 4ac}}{2a}$$

To use this formula it is necessary to arrange the quadratic equation to be solved in the form

$$ax^2 + bx + c = 0$$

The values of a, b, and c can then be substituted into the quadratic formula. Using formula will give two solutions for x, one of which may possibly be discarded because of physical restrictions imposed by the nature of the problem. Examples of the use of this formula appear in Chapter 10.

Appendix

Graphing Equations

Graphing experimental data is a useful technique for determining the relationships between measured variables and the values of parameters. The most useful form in which to graph an equation is that which produces a linear result, a straight line. The general form of a straight line is

$$y = mx + b$$

The variables are x and y, with y plotted on the vertical axis, or ordinate, and x plotted on the horizontal axis, or abscissa. The slope of the line, y/x, is m and the intercept on the y-axis is b.

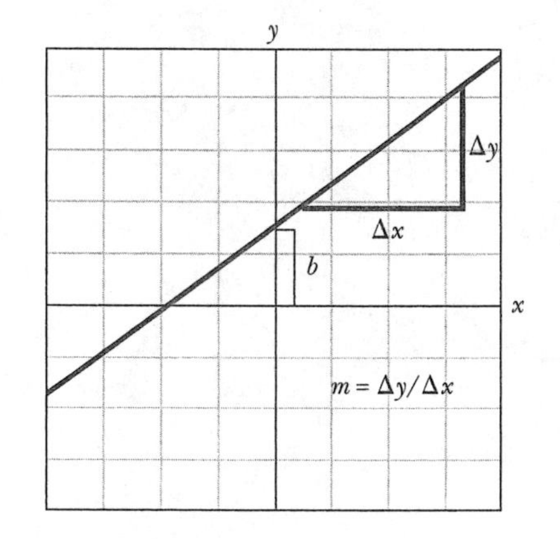

As an example, consider the following equation as a straight line.

$$\ln K = -\frac{\Delta H^\circ}{RT} + \frac{\Delta S^\circ}{R}$$

The variables in this equation are K and T. The values of R, ΔH°, and ΔS° are considered to be constants under the conditions involved. This equation will give a linear plot if

$$y = \ln K$$

$$x = \frac{1}{T}$$

$$m = -\frac{\Delta H^\circ}{R}$$

$$b = \frac{\Delta S^\circ}{R}$$

Thus experimental values of K and T can be plotted in a graph of $\ln K$ as a function of $1/T$. R is a standard constant that can be looked up. The values of ΔH° and ΔS° can be determined from the measured values of the slope and the intercept.

Appendix

Derivations Using Calculus

A number of equations derived or given without proof in the chapters on thermodynamics and kinetics can be arrived at easily using fundamental calculus.

Thermodynamics

If a gas is contained in a container of variable volume and its volume is changed by the infinitesimal increment dV against an external pressure P_{ex}, the infinitesimal increment of work done is

$$dw = -P_{ex}dV \qquad (1)$$

If the volume is changed from V_1 to V_2 against a constant at external pressure, the work can be determined by integrating both sides of equation 1, giving

$$\int dw = w = -P_{ex}\int_{V_1}^{V_2} dV = -P_{ex}\,\Delta V \qquad \text{(for constant } P_{ex}\text{)}$$

This is the same result given in the text.

For the reversible change of volume of an ideal gas the external pressure will be changed by infinitesimal increments and nRT/V can be substituted for P_{ex} in equation 1. In this case integration between V_1 and V_2 gives

$$\int dw = w = -\int nRTdV/V = -nRT\int_{V_1}^{V_2} dV/V$$
$$= -nRT(\ln V_2 - \ln V_1) = -nRT\ln(V_2/V_1)$$

This equation was given in the text without proof.

The first law of thermodynamics is

$$\Delta E = q + w \quad \text{or} \quad E = q + w + \text{const}$$

Differentiation of this gives

$$dE = dq + dw = dq - p_{ex}dV \tag{2}$$

If V is a constant, then $P_{ex}dV = 0$ and

$$dE = dq_v$$

Integration of both sides of this equation gives

$$\Delta E = q_v$$

Enthalpy was defined as

$$H = E + PV$$

Differentiation of this expression gives

$$dH = dE + PdV + VdP \tag{3}$$

Substitution of equation 2 into equation 3 gives

$$dH = dq - P_{ex}dV + PdV + VdP$$

If $P = P_{ex}$ and is a constant, then

$$dH = dq_p$$

Integration of this expression gives

$$\Delta H = q_p$$

For the infinitesimal change of temperature of a substance, dT, the heat transferred is

$$dq = CdT \tag{4}$$

where C is the heat capacity of the substance. Integration of the equation between T_1 and T_2 gives

$$\int dq = q = C\int_{T_2}^{T_1} dT = C\,\Delta T$$

For the infinitesimal transfer of heat at temperature T the entropy change is given as

$$dS = dq_{\text{rev}}/T \tag{5}$$

If the temperature is held constant for a finite heat transfer,

$$\int dS = \Delta S = \int dq_{\text{rev}}/T = q_{\text{rev}}/T$$

For the reversible change of temperature of a substance, substitution of equation 4 into equation 5 gives

$$dS = CdT/T$$

Integration of this between T_1 and T_2 gives

$$\int dS = \Delta S = \int_{T_1}^{T_2} CdT/T = C\int_{T_1}^{T_2} dT/T = C \ln(T_2/T_1)$$

The Gibbs free energy is defined by the equation

$$G = H - TS = E + PV - TS$$

Differentiation of this equation gives

$$dG = dE + PdV + VdP - TdS - SdT \qquad (6)$$

If all of the work is pressure–volume work,

$$dG = dq - P_{ex}dV$$

If the process is reversible,

$$dS = dq_{\mathrm{rev}}/T \qquad \text{and} \qquad dE = TdS - PdV$$

Therefore, under these conditions

$$dG = VdP - SdT$$

If T is constant, $SdT = 0$ and

$$dG = VdP$$

Substitution of $(RT)/P$ for one mole of an ideal gas gives

$$dG = (RT)dP/P$$

Integration of this expression between the standard pressure, $P_0 = 1$ atm and P gives

$$\int_{G^0}^{G} dG = RT\int_{P^0}^{P} dP/P$$
$$= G - G^0 = RT \ln P/P^0 = RT \ln P$$

For electrochemical cells

$$w = w_{\mathrm{pv}} + w_{\mathrm{elec}} \qquad \text{and} \qquad dw = dw_{\mathrm{pv}} + dw_{\mathrm{elec}}$$

Substitution of $dE = dq + dw$ into equation 6 gives

$$dG = dq + dw + PdV + VdP - TdS - SdT$$

For a reversible process at constant pressure and temperature, VdP and SdT are zero and $TdS = q$. Thus

$$dG = -dw + PdV$$

and for an electrochemical cell

$$dG = dw_{\mathrm{pv}} + dw_{\mathrm{elec}} + PdV$$

$$dG = dw_{\mathrm{elec}}$$

Kinetics

Because the rate of a reaction is normally constantly changing, it is more accurately expressed in differential notation with infinitesimal increments of time and concentration rather than finite increments as expressed by the use of Δt. Thus the rate for the disappearance of reactant A is given by

$$\text{rate} = -d[\text{A}]/dt$$

For a first-order reaction this gives a rate law of

$$-d[\text{A}]/dt = k[\text{A}] \text{ or } -d[\text{A}]/[\text{A}] = kdt$$

This can be rearranged to:

$$\int_{[\text{A}^0]}^{[\text{A}]} (d[\text{A}]/[\text{A}]) = k\int_0^t dt$$

Integration of this expression between the initial time and the initial concentration, $[\text{A}_0]$, and the final concentration [A] gives

$$= -\ln\,[\text{A}] + \ln\,[\text{A}_0] = kt$$

And

$$-\ln\,([\text{A}]/[\text{A}_0]) = kt \qquad \text{or} \qquad \ln\,([\text{A}_0]/[\text{A}]) = kt$$

For a second-order reaction the rate law is

$$-d[\text{A}]/dt = k[\text{A}]^2 \qquad \text{or} \qquad -d[\text{A}]/[\text{A}]^2 = kdt$$

A similar integration of this gives

$$-\int_{[\text{A}_0]}^{[\text{A}]} (d[\text{A}]/[\text{A}]^2) = k\int_0^t dt$$
$$= (1/[\text{A}]) - (1/[\text{A}_0]) = kt$$

For a zero-order reaction the rate law is

$$-d[\text{A}]/dt = k \qquad \text{or} \qquad -d[\text{A}] = kdt$$

Integration of this expression gives

$$-\int_{[\text{A}_0]}^{[\text{A}]} d[\text{A}] = k\int_0^t dt = [\text{A}_0] - [\text{A}] = kt$$

Appendix

Physical Constants

Acceleration Due to Gravity	$9.807\ m/s^2$
Atomic Mass Unit (amu)	1.66×10^{-27} kg
Avogadro's Number, N_A	6.022×10^{23} particles/mol
Boltzmann's Constant, k	1.38×10^{-16} erg/K
	1.38×10^{-23} J/K
Electron Charge	4.8033×10^{-10} esu
Electron Mass	9.1094×10^{-31} kg
Faraday's Constant	9.6485×10^{4} C/mol
Gas Constant, R	0.0821 L·atm/mol·K
	1.987 cal/mol·K
	8.31451×10^{7} erg/mol·K
	8.31451 J/mol·K
Pi (π)	3.1416
Planck's Constant, h	6.626×10^{-34} J·s
Speed of Light	2.9979×10^{8} m/s

Appendix

Conversion Factors

Density

$1\ g/cm^3 = 10^3\ kg/m^3$
$1\ lb/ft^3 = 16.0185\ kg/m^3$

Electrical Charge

1 esu = 3.33560×10^{-10} C (coulombs)
1 electron = 4.8033×10^{-10} esu
1 Faraday = 96485 C = 6.022×10^{23} electrons

Electrical Potential

1 V = 1 J/C

Energy

1 J = 1 N·m (newton·meter) = 10^7 erg
1 erg = 1 dyne·cm = 6.2415×10^{11} eV
1 cal = 4.184 J
1 L·atm = 24.2 cal
1 kWh (kilowatt hour) = 3.6×10^6 J
1 BTU = 1.055×10^3 J

Force

1 newton = 1×10^5 dynes = 1 kg·m/s^2
1 dyne = 1 g·cm/s^2

Frequency

$1\ \text{Hz} = 1\ \text{cycle/s}$

Length

$1\ \text{mile} = 1609\ \text{m}$
$1\ \text{in} = 2.54\ \text{cm}$
$1\text{Å} = 10^{-8}\ \text{cm}$
$1\ \mu\ (\text{micron}) = 10^{-6}\ \text{m}$

Mass

$1\ \text{metric ton} = 1000\ \text{kg}$
$1\ \text{ton (short ton)} = 2000\ \text{lb} = 907.2\ \text{kg}$
$1\ \text{amu} = 1.66 \times 10^{-24}\ \text{g}$
$1\ \text{lb} = 454\ \text{g}$

Power

$1\ \text{W (watt)} = 1\ \text{J/s}$

Pressure

$1\ \text{atm} = 101{,}325\ \text{Pa (pascal)}$
$1\ \text{atm} = 760\ \text{torr} = 760\ \text{mm of Hg}$
$1\ \text{lb/in}^2 = 6894\ \text{Pa}$
$1\ \text{bar} = 10^5\ \text{Pa}$

Volume

$1\ \text{m}^3 = 1000\ \text{L}$
$1\ \text{cm}^3 = 1\ \text{mL}$
$1\ \text{ft}^3 = 28.317\ \text{L}$
$1\ \text{gal (U.S.)} = 3.785\ \text{L}$

Appendix

Vapor Pressure of Water as a Function of Temperature

T(°C)	P_{vap}(torr)	T(°C)	P_{vap}(torr)
0	4.58	35	42.2
5	6.54	40	55.3
10	9.21	45	71.9
12	10.52	50	92.5
14	11.99	55	118.0
16	13.63	60	149.4
17	14.53	65	187.5
18	15.48	70	233.7
19	16.48	80	355.1
20	17.54	90	525.8
21	18.65	92	567.0
22	19.83	94	610.9
23	21.07	96	657.6
24	22.38	98	707.3
25	23.76	100	760.0
26	25.21	102	815.9
27	26.74	104	875.1
28	28.34	106	937.9
29	30.04	108	1004.4
30	31.81	110	1073.6

Glossary

absolute error The difference between a measured value and its true value.

absolute temperature A temperature scale that has its zero value at absolute zero.

absolute zero of temperature The lowest possible temperature. Molecular motion is slowed to the minimum.

absorption spectrum A **spectrum** of radiation absorbed by a sample.

acceleration The rate of change of velocity per unit time.

accelerator *See* **catalyst.**

accuracy The degree to which measured and true values correspond; how correct a measurement is. *See* **precision.**

acid–base titration A **titration** in which the known solution is a base and the unknown solution is an acid or vice versa.

acid ionization constant, K_a The equilibrium constant for the ionization of an acid dissolved in water.

acid rain Precipitation that has been made more than naturally acidic as a result of sulfur and nitrogen oxides from polluted air.

acids Substances noted for their sour taste and ability to dissolve some metals and metal oxides, change the colors of vegetable dyes, and neutralize bases (*see* **bases**). Acids are defined chemically as being proton donors or electron acceptors.

actinides The 14 elements, thorium through lawrencium, that follow actinium in the periodic table.

activation energy The energy barrier to reaction; the excess over the ground-state level that has to be overcome for a chemical reaction to take place.

addition polymerization = chain growth polymerization.

adhesion Molecular forces of attraction between unlike particles acting to hold them together.

adsorption Adhesion at the surface of a solid; "sticking" of a few monolayers of particles on the solid surface.

air pollution Fouling of the atmospheric environment by the introduction of natural and artificial contaminants.

alcohol A class of organic molecules generally consisting of a hydrocarbon portion and a hydroxy (OH) group; for example, ethanol (ethyl alcohol), $CH_3CH_2—OH$.

aldehyde A class of organic compounds resulting from oxidation of alcohols and having the characteristic —CHO group; for example, formaldehyde (H—CHO); acetaldehyde (CH_3—CHO).

aliquot The measured volume of the solution being **titrated**.

alkali metals The IA family of metallic elements, from lithium through cesium and francium; readily give up an electron, forming stable cations; are strongly reducing.

alkaline earth metals The Group IIA elements beryllium through radium.

allotropes Element substances that occur in more than one crystalline form: diamond and graphite, for example.

alloy A solid solution of two or more metals.

alpha particle (α), alpha ray Positively charged particle emitted by certain radioactive substances and present in certain emanations and in cathode ray tubes; composed of two neutrons and two protons, a doubly charged helium cation. An alpha ray is a stream of alpha particles.

alternating copolymer A polymer consisting of two kinds of monomer units alternating with each other.

amalgam A metal alloy of mercury.

amides A general class of organic nitrogen compounds characterized by the functional group—CONH—or peptide linkage.

amines Organic ammonia derivatives in which successive hydrogen atoms have been replaced by hydrocarbon groups.

amino acids Organic compounds containing an amine and a carboxylic acid group. Specifically the 20 α-amino acids that are the constituents of proteins.

amorphous Noncrystalline; having neither definite shape nor structure.

amorphous polymer A polymer with no ordered regions.

ampere An electric current flow of one **coulomb** per second.

amphoteric Having the properties of both an acid and a base.

amplitude The magnitude of oscillation of a sine wave above or below zero. The square of the amplitude is the **intensity** of the wave.

angular momentum The momentum of an object in a circular trajectory equaling mvr.

angular momentum quantum number The quantum number 1 with values ranging from 0 to $n - 1$.

anion *See* **ion**.

anisotropic Having different properties in different directions; opposite of **isotropic**.

anneal point The temperature at which atoms in glass can move so as to relieve any strains that are present.

anode The electrode where **oxidation** (removal of electrons) takes place in an electrochemical cell.

antibiotics Chemical substances that inhibit the action of bacteria. Most antibiotics are produced from microorganisms.

antibonding molecular orbital A molecular orbital higher in energy than the two atomic orbitals that were combined to form it. Electrons in an antibonding molecular orbital are not generally between the atoms involved.

antineutrino Antimatter to a **neutrino**.

antiparticles Pairs of particles that can annihilate each other, such as electrons and positrons.

aromaticity The general chemical properties and behavior characteristic of benzene and organic molecules containing the benzene ring or related benzene ring systems.

Arrhenius equation A relationship between the specific reaction rate constant and the activation energy for a given reaction: $k = Ae^{-Ea/RT}$.

Arrhenius plot A graph of the logarithm of the rate constant for a reaction as a function of the reciprocal of the absolute temperature.

atmophiles Elements appearing as gases in the atmosphere or dissolved in water.

atmospheric pressure The **pressure** exerted as a result of the mass of the atmosphere.

atom Smallest particle of an element substance that still retains the characteristic properties of the element.

atomic mass (atomic weight) The mass of an atom of an element relative to the mass of the ^{12}C isotope of carbon taken as 12.

atomic mass number (*A*) An integer that gives the total number of nucleons (protons and neutrons) within the nucleus of an atom.

atomic mass units (amu) A unit of mass for atoms, molecules, and ions such that a ^{12}C atom has a mass of exactly 12 amu.

atomic number (*Z*) The number of protons in the atomic nucleus (unit positive charges on the nucleus); equal to the number of electrons in a neutral atom.

atomic packing factor The fraction of a solid substance actually occupied by atoms.

atomic pile An old-style nuclear reactor built by piling up bricks of graphite and natural uranium. Provides a self-sustaining fission process that releases a great deal of energy.

atomic radius The effective radius of an atom considered as a hard sphere.

aufbau principle A system whereby the electronic structure of an atom is built up by adding electrons to hydrogen-like atomic orbitals.

autoionization, autoprotolysis An acid–base interaction in a liquid involving proton transfer. In the case of water, the resulting ions are hydronium and hydroxide.

Avogadro's law (hypothesis) Under the same conditions of temperature and pressure equal volumes of gases contain identical numbers of particles.

Avogadro's number (N_A) The number of carbon atoms in exactly 12 grams of the ^{12}C isotope of carbon, 6.022×10^{23}. The number of molecules in a mole of molecules.

axial The two corners of a bipyramid at each apex; contrasted with **equatorial**.

balanced equation A chemical equation written so that the correct number of moles of each substance is indicated by coefficients.

band gap The energy gap in a semiconductor between the occupied molecular orbitals and the unoccupied molecular orbitals.

band of stability The band of stable nuclei in a plot of number of neutrons versus number of protons.

band spectrum A **spectrum** consisting of bands separated by blank regions.

barbiturates A general class of organic nitrogen compounds related to barbituric acid; they act as sedatives, depressing the central nervous system.

barometer A device for measuring atmospheric pressure consisting of a liquid in a vertical tube sealed at the top and with its open end under the surface of the same liquid.

base ionization constant, K_b The equilibrium constant for the ionization of a base dissolved in water.

bases Classically, ionic substances that dissociate into hydroxide ions in aqueous solutions; they are defined chemically as proton acceptors or electron donors. *See also* **acids**.

beta particle (β) Elementary particle emitted from a radioactively decaying nucleus, identical to an electron; also generated in certain vacuum tube discharges.

bidentate Said of a ligand with two coordinating sites.

binary compound A compound containing only two different elements.

biodegradation The decomposition of a discarded object by natural microorganisms.

black-body radiation Spectral distribution of radiant energies described by Planck's modified version of the distribution law for a perfect black body; a perfect radiator, for which the distribution of energies is strictly a function of the temperature of the radiation.

block copolymer A polymer consisting of more than one kind of monomer unit with the different monomers grouped together in blocks.

body-centered A **unit cell** having an extra lattice point in the center of the unit cell.

Bohr theory An atomic structure theory based on the existence of discrete electron orbits (energy levels) in which emission and absorption of electromagnetic energy occurs when electrons move between orbits or levels.

boil The phenomenon observed when the vapor pressure of a liquid is greater than the prevailing pressure; characterized by rapid bubbling and evaporation.

boiling point The temperature at which the vapor pressure of a liquid equals the pressure of the atmosphere. It is called the **normal boiling point** at a pressure of one standard atmosphere.

boiling-point elevation The increase in boiling point of a liquid resulting from dissolving a nonvolatile solute in it.

Boltzmann constant, *k* A constant equal to the universal gas constant divided by Avogadro's number.

bond angle The angle inscribed in a chemical structure by two atoms, each bonded to a third atom.

bond energy The energy required to break a bond; the measure of bond strength.

bond length The mean distance between the centers of two atoms in a bond.

bond order The extent of bonding between two atoms, for example, bond order is 1 for a single bond, 2 for a double bond, etc.

bonding molecular orbital For the hydrogen molecule, when two atomic orbitals are mixed in phase, the electron density is concentrated between the nuclei and the energy of nuclei and electrons is lowered. In a manner of speaking, increased electron density is the nuclear adhesive holding things together. *See* **orbital**.

bonding pair A pair of electrons in a **covalent bond**.

boundary That which separates a system from its surroundings.

Boyle's law The relationship between pressure and volume for ideal gases; at constant temperature the product of pressure times volume is a constant.

branched Having a chain of atoms with one or more side chains.

branches Relatively short polymer chains branching from the main chain of a polymer.

buffers Solutions prepared in such a way that they are protected from drastic changes in acidity or basicity. A solution of a weak acid and its salt is the most common form of buffer.

buret A device for dispensing a precisely measured volume of solution in a titration.

calorie A common unit of heat, classically defined as the quantity of heat needed to raise the temperature of one gram of water by one degree Celsius or Kelvin. The calorie can be expected to be gradually replaced by the SI unit, the joule (1 cal = 4.184 joules).

calorimeter Any device for measuring change in enthalpy or internal energy.

carbohydrates Sugars and the more complex polysaccharides: cellulose, starch, glycogen.

catalyst A substance that increases the rate of a chemical reaction without being permanently consumed in the process; substances that slow a reaction are commonly called **inhibitors**.

catenation The formation of chains or rings of an element by means of covalent bonds between atoms. Carbon has the highest tendency to catenation.

cathode The electrode where **reduction** (loss of electrons) takes place in an electrochemical cell.

cathode rays Beams of electrons emitted from the cathode of a discharge tube.

cation *See* **ion**.

ceramic A nonmetallic material composed of crystalline substances. Ceramics are characteristically brittle but capable of resisting high temperatures.

chain-growth polymerization A polymerization reaction in which all atoms of each monomer unit become part of the polymer.

chain reaction A self-perpetuating series of reactions involving an initiation step, a series of chain-propagation reactions, and one or more chain-termination processes; for example, the photochemical chlorination of methane. Also, nuclear chain reactions involving neutron absorption and production.

chalcophiles Elements appearing as sulfides in the Earth's crust.

change in entropy The entropy of a system after a process minus the entropy of the system before the process.

Charles's (Gay-Lussac's) law The relationship between volume and temperature of an ideal gas at constant pressure; the volume occupied by an ideal gas is proportional to the absolute temperature.

chelate A complex formed by a chelate capable of bonding two coordinating sites to the same metal ion.

chelate effect The observation that chelating ligands form more stable complexes than nonchelating ligands.

chemical A substance or mixture of substances produced or used in a chemical process.

chemical bonds The interactions holding atoms together in molecules.

chemical equation A shorthand description of a chemical reaction showing the reactants and products of the process.

chemical equilibrium A situation in which opposite chemical processes are proceeding by the same rate; no change in the composition of the reaction mixture occurs.

chemical formula A shorthand for describing substances that includes the symbols of the substituent element with the number of each given as a subscript.

chemical process *See* **chemical reaction**.

chemical property A property of a substance that results in its being changed into a different substance.

chemical reaction A process that changes the identity of one or more substances.

chemistry The science of substances, dealing with the investigation of composition of matter and interaction between matter and energy in the universe.

chirality The characteristic quality of an asymmetric molecule that it cannot be superimposed on its reflection in a mirror.

chlorofluorocarbons Compounds containing chlorine, fluorine, and carbon, which are used as propellants, refrigerants, and foaming agents. They are generally stable and nontoxic but are a serious threat to the ozone layer. Also called CFCs and Freons.

chromatography A technique for physically separating and purifying substances by selective adsorption, usually on a solid support, as components pass over.

cis A geometric arrangement in which two chemical groups are located next to each other.

clathrates Included or caged compounds, trapped within a cavity formed by a crystal lattice or present in a large molecule. Generally the properties of the clathrate are those of the enclosing species.

Clausius–Clapeyron equation An equation for calculating the vapor pressure of a liquid.

cleavage plane A favored direction in which a crystal can be split.

closed-ended manometer A **manometer** having the end closed to the atmosphere.

closed system A system limited to exchange of energy with its surroundings; no matter can be transferred.

closest packed The arrangement that places the most atoms or ions in a unit volume; the **coordination number** is 12.

cloud chamber A device for detecting nuclear particles based on the fact that their passage through a saturated water vapor atmosphere leaves a trail.

coherent Radiation with all waves in phase.

cohesion Intermolecular forces between particles of a substance that act to bind them together.

coke The residue that remains after coal is heated in the absence of air, principally carbon and ash.

colligative properties Properties of a solution that depend on the nature of the solvent and the number of solute particles present without concern for the nature of the solute particles themselves.

colloid A chemical system in which one phase is composed of particles whose dimensions range from 1 to 1000 nanometers.

combining capacity The relative amounts of different elements that will combine with one common element such as hydrogen or oxygen.

combining volumes, law of Gay-Lussac For reactions involving ideal gases as reactants and/or products, the ratios of the respective gas volumes can be expressed as ratios of small whole numbers.

commodity plastics Polymers used for producing ordinary consumer products.

common ion effect The effect whereby the presence of an additional concentration of an ion that is a product in a reaction shifts the equilibrium to the reactant side of the chemical equation.

complementarity Where two different and conflicting descriptions of nature can both be made to partially fit the facts or data; for example, the classical and quantum mechanical descriptions of the atom.

complex ion An ion composed of a central metal ion complexed to **ligands**.

composites Materials of two distinctly different and reinforcing substances.

compound Homogeneous chemical combination of two or more different chemical elements in definite proportions.

compressibility The extent to which volume can be reduced by the application of pressure.

compressibility factor A measure of the extent to which a gas is ideal, PV/nRT. It equals 1.00 for an ideal gas.

Compton effect The scattering observed on collision of photons and electrons. Because collisions are elastic, the electron gains energy and recoils, and the photon loses energy, resulting in a change in wavelength that depends on the scattering angle.

concentrated Said of a solution with a high concentration.

concentration The measure of amount of **solute** in a solution.

concentration cell An **electrochemical cell** driven by a difference of concentration of solute in the two half-cells.

condensation Transformation from vapor (gaseous) to liquid state; the reverse of vaporization. Sometimes also used for the reverse of **sublimation**.

condensation polymerization = step growth polymerization.

conductivity A measure of the effectiveness of a substance in transferring heat (**thermal conductivity**) or electricity (**electrical conductivity**).

conductor A substance that allows the passage of heat or electrical charge.

conjugate base, conjugate acid A **conjugate acid–base pair** is made up of two species that differ from each other only in that one contains a proton that the other does not.

conservation of mass The principle that mass can neither be created nor destroyed in a chemical reaction.

constant composition, law of *See* **definite proportions, law of.**

contamination hazard A hazard caused by the ingestion of radioactive substances.

continuous spectrum A **spectrum** consisting of a continuous range of wavelengths.

conversion factor The ratio of one set of units to another, which can be used for converting between the two sets of units. The conversion factor has a numerical value of one and so multiplication by it does not change the numerical value of another quantity.

coordination complex (compound) A compound or an ion consisting of a central atom or ion surrounded by **ligands**.

coordination number The number of nearest neighbors of an atom or ion in a structure.

copolymer A polymer consisting of more than one kind of monomer unit.

core electrons The unreactive electrons in an atom contained within a noble gas electron configuration.

coulomb A unit of electrical charge; a steady flow of one coulomb per second is called one **ampere**.

coulombic forces The electrical forces of attraction or repulsion between charged objects.

covalent Bonded by shared pairs of electrons in **covalent bonds**. Contrasted with **ionic**.

covalent radius The effective radius of an atom in a covalent bond.

creep Slow deformation of an object under a constant stress.

critical point The temperature and pressure on the phase diagram of a substance that represents the highest temperature at which the liquid and gaseous states can be distinguished. Above this temperature the substance is known as a **supercritical fluid**.

critical pressure The pressure at the **critical point** of a substance.

critical temperature The temperature at which a substance can no longer exist as a liquid.

cross-links Relatively short polymer chains linking two main chains of the polymer.

crystal, crystalline matter Matter consisting of atoms in systematically repeating, three-dimensional units periodically extended through space.

crystal field theory A bonding theory for **coordination complexes** that assumes bonding by purely electrostatic forces in which the **ligands** can be considered to be point charges.

crystal lattice A collection of all equivalent points in a crystal structure.

crystal structure The arrangement of atoms in a **crystal**.

crystalline polymer A polymer containing **crystallites**.

crystallite A very small crystalline or ordered region within a polymer.

cubic closest packing A three-dimensional arrangement of closest packed spheres with every third layer in an equivalent position.

current A flow of charged particles such as electrons or ions; current is measured in amperes—that is, coulombs per second.

current rectification Conversion of alternating electric current to direct current.

cycle A portion of a wave starting and ending at identical points of the wave.

cyclotron An accelerator consisting of two **dees** for charged particles that take a circular path.

***d* block** The elements in periodic table Groups IIIB to IIB where *d* orbitals are being filled; known as the transition elements.

Dalton's law of partial pressures The observation that in a mixture of ideal gases the total pressure equals the sum of the partial pressures of all gases in the mixture, and the partial pressure of each gas is proportional to its mole fraction.

de Broglie wavelength The effective wavelength of an object in motion.

dee *See* **cyclotron**.

defect An irregularity in a crystal structure. Can lead to a **nonstoichiometric** compound.

definite proportions, law of All pure chemical compounds have a fixed, definite percent-mass composition, no matter the source of the substance. Also called law of constant composition.

degenerate Having the same energy.

degree of polymerization The average number of monomer units in a polymer chain.

delocalized Said of electrons; spread out among more than two atoms.

density Ratio of mass to volume for a homogeneous substance.

deposition The phase change for a gas to a solid.

derived units Units that are combinations of other units, such as distance divided by time.

diamagnetic Not attracted to a magnetic field; not having unpaired electrons.

diatomic Consisting of two atoms.

diffusion The mixing process that takes place among particles in the fluid state because of their random thermal motions.

dilute Said of a solution with a low concentration.

dipole A separation of opposite charges in a bond or in a molecule.

dipole moment A measure of the extent of a dipole calculated as the product of the magnitude of the separated charges and the distance separating them.

diprotic acid An acid that produces two hydronium ions when dissolved in water.

discreteness The concept that matter and energy consist of definite units (atoms, molecules, photons).

dispersion (of light) Separation of a light wave into its spectral components.

disproportionation A redox reaction in which the same species is both oxidized and reduced.

dissociation A usually reversible decomposition (often by heat or solvent interaction) resulting in separation of a given chemical composition into atoms, ions, molecules or radicals.

dissolution To form a solution; normally said of the **solute**.

distillation A method of separating components of a mixture by taking advantage of relative differences in volatility among the various components.

doped Said of a **semiconductor** that has had very small amounts of impurities added to it.

double bond Four pairs of shared electrons between two atoms. One of the two bonds is a **sigma bond** and the other is a **pi bond**.

ductile The ability to be drawn into a wire. A characteristic of metals.

effective nuclear charge The effective charge exerted on an electron by the nucleus of an atom corrected for **shielding** by inner electrons.

effusion The passage of a gas through a small opening.

elastic deformation (elasticity) Tendency for an object to return to its original dimensions after a **stress** is removed.

elastic limit The amount of stress needed to make an elastic material fail.

elasticity The ability of a material to be deformed and then return to its original dimensions.

elastomer An **elastic** polymer.

electrical conductivity Ability to accommodate a flow of electrical charges.

electrical potential A separation of charge that can result in a flow of electric current.

electrochemical cell A reaction system that separates the oxidation and reduction half reactions so that current can flow through a conductor connecting the cells.

electrochemical series A series of elements in order of their ability to displace each other in chemical reactions under **standard conditions.**

electrochemistry The study of reactions that involve flow of electrons along a conductor. The reactions studied in electrochemistry are oxidation-reduction reactions.

electrode The conducting material that draws electrons into and out of an electrolyte solution, a molten medium, a gas, or through a vacuum. Electrodes can be reactive (participate in the electrochemical process) or inert.

electrolysis Decomposition process brought about by passing an electric current through a compound substance in solution or in the molten state.

electrolyte Substance that dissolves in a solvent to produce a solution that conducts electricity, usually by movement of ions; acids, bases, and soluble salts are electrolyte substances in aqueous media.

electrolytic cell An **electrochemical cell** that is driven by an outside source of electrical potential.

electromagnetic radiation Radiation ranging from radio waves to cosmic rays and including visible light all propagated at the **speed of light**.

electron An elementary, subatomic particle with unit-negative electrical charge (4.8×10^{-10} esu) and mass 1/1837 of a proton (9.1×10^{-28} g); electrons constitute the nonnuclear particles within the atom and determine chemical properties.

electron affinity The energy when an electron is added to an atom or an anion.

electron capture A means by which nuclides with an excess of protons can achieve stability.

electron cloud A region around an atom where the probability of finding an electron is high.

electron configuration The arrangement of electrons around an atom expressed in terms of the orbitals occupied.

electron shells The spheres or orbits, in which electrons are said to be located in fixed numbers; the principal energy levels in the electronic structure model for the atom in which each shell has a fixed value of the major quantum number *n*.

electron spin The intrinsic angular momentum or spin orientation of an electron; the spin-quantization number, *s*, is $\pm\frac{1}{2}$.

electronegativity The relative ability of each atom in a molecule to attract bonding electrons to itself.

electroplating Using an **electrochemical reaction** to plate a layer of metal.

electrovalence The valence or combining capacity of an atom involved in ionic bonding; a theory of chemical combination based on electron transfer.

element One of the 115 presently known, pure chemical substances that cannot be decomposed (divided) into any other pure substance by ordinary chemical processes. A substance that has atoms of all the same atomic number.

elementary particles The growing list of particles of matter and radiation, often referred to as "fundamental" particles, including electrons, protons, neutrons, and neutrinos and particles that do not exist independently under normal conditions, such as mesons, muons, baryons, and antiparticles.

elementary process A reaction mechanism with only one step.

emission spectrum A **spectrum** of radiation emitted from an excited sample.

empirical formula A symbolic statement of the relative ratios of the elements in a molecular arrangement of atoms expressed as integers; the simplest possible molecular formula in keeping with the percent-by-mass relationships among the combined elements.

emulsion A more or less uniform suspension (dispersion) of two immiscible liquids.

end point The point at which a **titration** is terminated as a result of some signal such as **pH**.

endothermic Absorbing heat from the surroundings. Said of a system, reaction, or process.

energy The ability to do work or give off heat.

energy conservation The principle that energy is conserved in any chemical process or that the total amount of energy in the universe is constant. See **mass conservation**.

energy state A state of a system at a particular energy depending on the electron structure.

engineering plastics Polymers intended for high performance uses.

enthalpy (H) A thermodynamic state function of a system ($H = E + PV$) that can be used to measure processes taking place at constant pressure; the change in enthalpy of a system, ΔH, is the heat of a process at constant pressure.

enthalpy of combustion The enthalpy change that occurs when one mole of a substance is totally combusted in oxygen or air.

enthalpy of fusion The enthalpy change involved when melting one mole of a substance.

enthalpy of sublimation The enthalpy change involved when subliming one mole of a substance.

enthalpy of vaporization The enthalpy change involved when vaporizing one mole of a substance.

entropy (S) A measure of the disorder of a system; the change in entropy of a system is ΔS.

enzymes Biochemical catalysts; generally complex protein structures, operating in living systems at the cellular level.

equation of state for an ideal gas An equation, $PV = nRT$, expressing the relationships among pressure, volume, temperature and amount of an **ideal gas**. Also known as ideal-gas law or ideal-gas equation.

equatorial The three corners around the belt of a bipyramid; contrasted with **axial**.

equilibrium A system after it has reached a time-independent state with no changes in the composition of the system occurring; characterized by the fact that no spontaneous reaction is taking place.

equilibrium constant The constant value for an expression involving the reactants and products of a chemical process in terms of their concentrations or partial pressures. This expression will have a constant value for a particular chemical process at a particular constant temperature. The equilibrium constant has the symbol K. Symbols of equilibrium constants for specific processes are: K_w ion product of water; K_a acid dissociation constant; K_b base dissociation constant; K_{sp}, solubility product constant; K_f formation constant.

equilibrium state A state, particularly, of a system that is not undergoing change.

equivalence point The point in a **titration** at which the number of moles of solute in the added solution equals the number of moles of the solute in the solution being titrated.

error The difference between the actual and measured value of a quantity. Usually of two types: systematic and random.

ester The class of organic compounds formed from organic acids and alcohols by elimination of a molecule of water; they are often characteristically pleasant, sweet-smelling substances.

ether Organic molecules in which an oxygen atom is bound to two carbon atoms, as in diethylether ($CH_3CH_2—O—CH_2CH_3$), the classic anesthetic.

eutectic The composition of a mixture with the lowest melting point or boiling point.

eutrophication Decay of organic matter that has accumulated in a body of water due to the presence of excessive amounts of dissolved nutrients, resulting in oxygen depletion of the water.

evaporation The liquid-to-vapor transformation at temperatures generally below the boiling point of the liquid.

excited state An **energy state** higher than the **ground state**.

exclusion principle, Pauli exclusion principle The rule that no two electrons in a given atom can have the same set of quantum numbers.

exothermic Evolving heat to the surroundings. Said of a system, reaction, or process.

extensive property A property such as mass that depends on the amount of a substance.

extraction A physical method of separation and purification that involves dissolving a substance in a solvent. Liquid–liquid extraction takes advantage of differences in solubilities of one substance between a pair of immiscible solvents.

extrusion Shaping a material by forcing it through a die.

***f* block** The rare earths; the block in which *f* orbitals are being filled.

face-centered A **unit cell** having an extra lattice point in the center of every face.

family *See* **group**.

faraday (*F*) The electric charge on a mole of electrons, or the charge associated with one gram-equivalent of electrochemical reaction; 96,500 coulombs.

fatty acid An organic acid with a long carbon chain; components of fats.

fermentation The enzyme-catalyzed reactions of certain organic substrates; for example, the hydrolysis of starches and the fermentation of the resulting sugars forming alcohol; in a sense, the biological equivalent of oxidation taking place in the absence of air.

fiber optics A method of communication in which signals are sent by light sources along fine glass fibers.

filtration A physical method for separating solids from liquids or gases by allowing the liquid or gas to pass through a semipermeable membrane (for example, filter paper), leaving the solid phase behind.

first law of thermodynamics A statement of the conservation of heat, work, and internal energy, $\Delta E = q + w$.

first-order reaction A reaction for which the sum of the exponents in the rate law is 1.

fission Splitting of a heavy nucleus into nuclei of lighter elements, releasing considerable quantities of energy and usually emitting one or more neutrons; can occur spontaneously, but usually does not.

flotation A separation procedure depending on whether or not a desired ore is wet by water.

fluid Aggregated state of matter in which particles (usually molecules) flow past each other as in gases and liquids.

foam A colloidal system in which a gas phase is uniformly dispersed through a liquid phase.

formal charge The charge on an atom in a compound calculated on the assumption that the electrons in each bond are shared equally between the two atoms involved.

formula mass A relative mass calculated from the numbers of atoms given in a formula that is not necessarily the **molecular formula**.

fractional distillation A multiple-step **distillation** in which the distilled product of each step is distilled again.

free energy The enthalpy change of a process minus the entropy change times the absolute temperature; determines whether a process is spontaneous or not.

free expansion An expansion of a gas in which no work is done.

free radical A chemical species having an unpaired electron.

freely soluble ionic compound An **ionic compound** generally with **solubility** in water of more than 1 g solute per 100 mL of water.

freezing point The temperature at which the liquid and solid forms of a substance are at equilibrium. At a pressure of one atmosphere this is called the **normal freezing point**.

freezing-point depression The decrease in freezing point of a liquid resulting from dissolving a solute in it.

Frenkel defect A crystal **defect** resulting from an atom or ion out of its usual position.

freons *See* **chlorofluorocarbons**.

frequency The number of cycles or repetitions of a periodic process that are completed per unit of time; usually repetitions per second, when dealing with electromagnetic radiations.

friction The resistance to the flow or motion of one object or substance over or through another.

fuel cell An **electrochemical cell** that oxidizes a gaseous or liquid fuel such as hydrogen.

functional group An atom or cluster of atoms that bestows particular properties on a compound.

fundamental mode The longest wavelength possible for a vibrating system.

fundamental units Units such as meters and kilograms from which all other units in the **International System of Units** can be derived.

fusion (1) The phase change from a solid to a liquid. (2) Synthesis of heavier nuclei by fusing lighter nuclei together; process occurs with release of energy.

galvanic cell An **electrochemical cell** that makes use of a spontaneous chemical reaction to generate a flow of electricity, as in a battery.

gamma rays (γ) High-energy and short-wavelength electromagnetic radiation; frequently accompanies alpha and beta emission; always accompanies fission.

gangue Unwanted rock material in an ore.

gas A state of matter with particles moving independently at distances sufficiently great that interactions between them are minimal. Gases are characterized by great compressibility, high rates of diffusion, and the ability to fill their container completely.

Gay-Lussac, law of *See* **combining volumes**.

Geiger counter A counter for high-energy particles that depends on the ability of these particles to ionize the gas in the counter.

gel An essentially solid or semisolid two-phase colloidal system composed of the solid phase and a liquid phase; the liquid component is generally the principal component of the system.

gene (DNA) The hereditary unit of the chromosome that controls the development of inherited traits and characteristics.

genetic effects Damage caused to reproductive cells by radiation that show up in succeeding generations.

geochemistry The study of the chemistry of the earth and the adjacent portion of the atmosphere.

glass A liquid with viscosity so high that it appears not to flow; a rigid amorphous substance; a supercooled liquid.

glass transition temperature The temperature at which the behavior of a polymer becomes more like a liquid and less like a solid.

graft copolymer A polymer having short chains of one polymer grafted on to the long chains of another.

Graham's law The observation that **effusion** and **diffusion** of particles is inversely proportional to the square root of their molar masses.

grain A single crystal in a metal structure.

greenhouse effect An atmospheric effect caused by infrared absorbers such as CO_2. The projected result is global warming.

ground state The state of an atom, atomic nucleus, molecule, or ion associated with its lowest energy—that is, with the electrons occupying their lowest energy levels.

group A family of elements existing in one column of the periodic table.

half-life Time required for half of a substance to decompose or disintegrate. Measured half-lives of radioactive elements vary from billionths of a second to billions of years.

half-reaction A reaction that uses or produces a species that cannot exist in a free state under the reaction conditions such as an electron or a proton; it must be combined with another half-reaction.

halogens The Group VIIA family of nonmetallic elements, including fluorine, chlorine, bromine, iodine, and astatine; atoms easily gain one electron, forming halide anions; elements are strongly oxidizing.

harmonic motion Periodic motion that can be described as a sinusoidal function of time; a plucked guitar string; a swinging pendulum; the regular motion described by a sine curve.

heat Transfer of energy between a system and its surroundings because of an existing temperature differential between the two.

heat capacity The amount of heat needed to raise the temperature of a system one degree, usually measured either at constant volume or constant temperature.

heat of combustion, fusion, sublimation, vaporization *See* **enthalpy**.

hemoglobin The iron–protein complex found in red blood cells that is responsible for oxygen transport.

Henry's law The observation that under ideal conditions the concentration of a gas dissolved in a liquid is proportional to the partial pressure of the gas.

Hess's laws The statement that **enthalpy** changes are additive as a result of enthalpy being a **state function**.

heterogeneous Not homogeneous; having a composition that is not constant over the entire sample.

heterogeneous catalyst A catalyst in a different phase than the reactants.

heterolytic bond dissociation The mode of bond breaking that leaves the bonding electrons on one fragment resulting in two charged species.

heteronuclear Composed of different atoms.

hexagonal closest packing A three-dimensional arrangement of closest packed spheres with every other layer in an equivalent position.

high spin An arrangement of electrons in orbitals so as to leave some unpaired rather than paired.

hole A space in a crystal structure that can accommodate an atom or ion.

homogeneous Having continuous properties; having the same composition throughout the entire sample.

homogeneous catalyst A catalyst in the same phase as the reactants.

homolytic bond dissociation The mode of bond breaking that splits the bonding electrons evenly between the two fragments.

homonuclear Composed of identical atoms.

homopolymer A polymer consisting of only one kind of monomer unit.

Hume–Rothery rules A set of rules for predicting what combinations of metals will produce solid solutions.

humidity Moisture content of the air expressed as mass/volume is the **absolute humidity**. The ratio of the partial pressure of water in air to the vapor pressure of water at the given temperature is the **relative humidity**.

Hund's rule The rule that electrons in degenerate orbitals have **parallel** spins if possible.

hybridization The equalization of energies to explain equivalency of chemical bonds formed by combination of orbitals of apparently different energies; for example, hybridization of *s*- and *p*-type orbitals on carbon in methane.

hydrate A compound containing water of hydration. *See* **hydration**.

hydration The binding of water molecules to a substance, usually through hydrogen bonding and in crystalline hydrates, at definite lattice sites. Binding or association of water molecules with ions in solution.

hydrocarbons A general class of organic compounds consisting of only atoms of carbon and hydrogen. Subclasses are alkanes, alkenes, alkynes, aromatics, and alicyclics.

hydrogen bond A weak chemical bond formed by dipole–dipole attraction between a hydrogen atom in one molecule with an electronegative atom in the same or another molecule; hydrogen bonds can form when H is covalently bonded to N, O or F.

hydrolysis Decomposition or other chemical transformations brought about by the action of water; for example, hydrolysis of certain salts to form acidic and basic substances.

hydronium ions A hydrated proton of general formula H_3O^+ found in pure water and in all aqueous solutions.

hydrophilic Water-loving; attracted to water.

hydrophobic Water-hating; repelled by water; said of compounds or parts of compounds.

hypothesis A generalization made from observations but based on as yet incomplete evidence; not as certain as a theory.

ideal gas law (equation) *See* **equation of state for an ideal gas.**

ideal gas One that behaves according to the ideal gas law, $PV = nRT$; the particles in an ideal gas have no volume and neither attract nor repel each other.

ideal solution A solution in which all particles act independently of each other.

immiscible Liquids not capable of forming a homogeneous solution; insoluble.

induced dipole A **dipole** induced on an otherwise nonpolar atom or molecule resulting from proximity to another dipole.

inertia The property of all matter in motion or at rest representing the resistance to any alteration in the existing state of affairs.

inflection point The point in a titration curve at which the sign of the slope changes—that is, is most nearly vertical.

infrared spectrum A **spectrum** encompassing the infrared region of electromagnetic radiation.

inhibitor A substance that slows a chemical reaction without itself being changed.

initiation step The step that starts a **chain reaction**.

inorganic substances Compounds of all the elements except carbon.

insulator A substance that allows very little passage of heat or electrical charge.

intensity The energy flux of a wave in photons per unit time.

intensive property A property such as color or density that is independent of the size of a sample.

interference phenomena The result of the combination of two or more waves. If the waves are exactly in phase, the interference is said to be **constructive** and the amplitude of the resulting wave will be larger than that of the component waves. If the waves are out of phase, the interference is said to be **destructive** and the amplitude of the resulting wave will be smaller than that of the component waves.

intermolecular forces The forces between molecules that hold them in the liquid or solid state.

International System of Units (SI) A universally accepted system of units having seven fundamental units.

interstitial solid solution A solid solution in which the solute atoms are located in holes in the crystal structure.

ion An atom or molecule that has lost or gained one or more electrons, becoming electrically charged in the process: the loss of electrons produces a positive ion, or **cation**; the gain of electrons produces a negative ion, or **anion**.

ionic Consisting of ions. Contrasted to **covalent**.

ionic radius The effective radius of an ion in an ionic solid.

ionization constant of water The product of the concentrations of hydronium and hydroxide ions in water, 1.0×10^{-14} at 25°C.

ionization energy, ionization potential The energy per-unit charge required to remove an electron an infinite distance from an atom or molecule.

ionization The process of removing or adding electrons or atoms and molecules, creating electrically charged particles called ions in the process. High temperatures, electrical discharges, or nuclear radiation can cause ionization. So can dissolving in water.

irreversible process *See* **spontaneous process**.

isoelectronic Having the same number of valence electrons.

isolated system (thermodynamic) A system that cannot exchange matter or energy with its surroundings.

isomers One of two or more substances whose chemical compositions are identical but whose structures, and therefore properties, differ. *See also* **chirality**.

isomorphism Different compositions with the same crystalline form.

isotherm A curve representing some function at a constant temperature—for example, the relationship between pressure and volume at a constant temperature.

isotope One of two or more forms of an element with the same atomic number (Z) but different mass number (A); ${}^{12}_{6}C$ ${}^{13}_{6}C$, and ${}^{14}_{6}C$ are all isotopes of carbon and have very nearly identical chemical properties.

isotropic Having identical properties in all directions; no distinguishable direction through the material. *See also* **anisotropic**.

kelvin The unit of temperature in the **International System of Units**.

ketones Organic molecules characterized by the presence of the carbonyl group: acetone, for example

kilogram The fundamental unit of mass in the **International System of Units**.

kinetic (-molecular) theory A statistical explanation of the behavior of gases. Assumption: They are composed of very large collections of particles (atoms or molecules) in a state of random, ceaseless motion, involving collisions in which energy and momentum are conserved.

kinetic control Relating to a process for which the products formed depend on the rate of possible reaction paths.

kinetic energy The energy a body possesses as a result of its being in motion; a particle of mass m moves with speed v and has a kinetic energy equal to $\frac{1}{2}mv^2$.

knock inhibitors Compounds added to motor fuels to retard the rate of burning and alleviate knocking in the engine.

lanthanides A group of 14 metallic elements, cerium through lutitium, that follow lanthanum in the periodic table. Their properties are very similar to those of lanthanum.

laser (light amplification by stimulated emission of radiation) A device producing visible radiation by stimulated emission of already excited atoms.

lattice A system of identical points called **lattice points** located in a crystal structure. *See* **unit cell.**

law of definite proportions The observation that the composition of a pure substance is always the same.

Le Chatelier's principle Changes in a system at equilibrium will result in internal compensation (changes in the position of the equilibrium) to partially restore the system to its original, stable, equilibrium state.

Lewis (dot) structure A depiction of the structure of an atom using lines for bonds and dots for nonbonded electrons.

Lewis symbol The symbol of an element with its valence electrons depicted by dots.

ligand An atom, molecule, ion, or other complex group of atoms chemically bound to a central atom, often a metal atom; groups coordinated to central atoms.

limiting reagent The reactant in a chemical product that limits the amounts of products that can be formed.

line spectrum A **spectrum** consisting of separated lines.

linear Having a chain of atoms with no branches.

linear accelerator An accelerator for charged particles in which the particle path is a straight line.

lipids Biomolecules that are soluble in organic solvents. The most familiar lipids are fats, triglyceride esters of glycerol, and fatty acids.

liquid A condensed form of matter without the regular long-range structure of a crystal. Liquids flow but have low compressibility and do not fill their containers completely.

liquid crystal A liquid having some of the order of a solid crystal.

lithophiles Elements appearing as oxides in the Earth's crust.

lone pair A pair of valence electrons on an atom in a molecule not participating in bonding and confined to a single nonbonding orbital.

low spin An arrangement of electrons in orbitals so as to pair them.

lustrous Capable of reflecting light.

Lyman series A series of lines in the near ultraviolet region of the spectrum of atomic hydrogen with emissions terminating at principal quantum number 1.

macroscopic Observable directly.

macroscopic property Directly observable property of a substance or system.

magnetic quantum number number The quantum number m_ℓ with values ranging from $-\ell$ to $+\ell$.

main groups The eight groups of the periodic table in the *s*- and ***p*-blocks.**

malleable The ability to be pounded into a thin sheet; a characteristic property of metals.

manometer A device for measuring the pressure of a gas within a container consisting of a liquid in a U-shaped tube.

maser (microwave amplification by stimulated emission of radiation) A device producing microwave radiation by stimulated emission of already excited atoms.

mass The quantity of matter constituting any substantive body; the physical measure of the inertial property of a substance.

mass average molecular mass An average molecular mass of a polymer that emphasizes the chains with higher molecular masses.

mass conservation The principle that the mass of a system is constant in the universe through any chemical transformation and that the total mass of the universe is constant. Mass–energy conservation broadens the interpretation of this law to include energy conservation as well.

mass defect The difference between the mass of a nucleus and the sum of the masses of its component nucleons. This difference is the mass equivalent of the **nuclear binding energy**.

material A combination of **matter**; called a **raw material** if used as a starting point for the production of substances or other materials.

matter Any collection of particles having mass and occupying space.

Maxwell–Boltzmann distribution A calculated graph of probability versus particle speed.

melting point The temperature at which a solid and a liquid of the same substance are at equilibrium.

meniscus The curvature of a liquid surface in a container of narrow diameter due to surface tension; concave if the liquid wets the walls of the container; convex otherwise.

metal One of the majority of elements on all but the upper right side of the periodic table, characterized in general by tendency to form cations, conductivity of electricity and heat, luster, ductility, malleability, and solubility in mineral acids.

metalloid An element with properties intermediate between the metals and nonmetals. Also called **semimetal** and **semiconductor**.

metallurgy The science and technology of extracting metals from their ores.

metathesis A substitution reaction that does not involve the transfer of electrons with accompanying change in oxidation states.

meter (m) The fundamental unit of length in the **International System of Units**.

micelle A highly charged, hydrated colloidal aggregate in which a polar group is oriented toward the aqueous medium and a nonpolar group toward the nonaqueous medium in solutions of electrolytes such as soaps, or in detergent systems.

microscopic property A property at the atomic level of matter, which can only be inferred indirectly.

miscible Said of two or more liquids that are soluble in each other.

mixture An aggregate collection of different substances and materials, separable by simple mechanical or physical means.

model A simplified construction used to understand a more complex concept.

modulus of elasticity The ratio of **stress** to **strain**.

molality The concentration of a solution in moles of solute per kg of solvent.

molar mass The mass in grams of one mole—that is, 6.022×10^{23} particles of a substance.

molarity The concentration of a solution in moles of solute per L of solution.

molding Shaping a material by forming it in a mold.

mole (mol) Avogadro's number of atoms or molecules of a substance. An amount of a substance equal to its **atomic** or **molecular mass**.

mole fraction A unit of concentration for mixtures equal to the number of moles of a particular substance divided by the total number of moles in the mixture.

molecular compound A compound existing as discrete molecules.

molecular formula The formula of a compound that includes the exact number of every atom present in a molecule.

molecular mass The sum of the relative or atomic mass of the component atoms in a molecule; can be calculated directly from the **molecular formula**.

molecular orbital model A **model** of chemical bonding in which orbitals are constructed for an entire molecule and then populated with electrons analogously to the **aufbau principle**.

molecule The smallest particle of any chemical combination of atoms that still retains all the properties of the original collections of particles.

momentum For any moving body, the product of its mass and speed.

monatomic Said of ions consisting of a single atom such as Na^+ or S^{2-}.

monodentate Said of a ligand with one coordinating site.

monodisperse A polymer having all molecules with the same chain length.

monomers The molecules used to produce a **polymer**.

monoprotic acid An acid that produces one hydronium ion when dissolved in water.

most probable speed The speed at the maximum of a **Maxwell–Boltzmann distribution**.

multiple bond A **double** or **triple bond**.

multiple proportions, law of The small whole-number relationship that exists in the mass ratios of elements between two elements when combined in different ways (i.e., H_2O and H_2O_2).

negative charge A charge with the same sign as the charge on the electron. An ion has a negative charge if it contains more electrons than protons.

nematic Said of a liquid crystal with all molecules lined in the same direction.

Nernst equation An equation relating **electrical potential** at given conditions with electrical potential at **standard conditions**.

net ionic equation A chemical equation showing all reactants and products in their ionized form and not including spectator ions.

neutralization Classically, the reaction of equivalent amounts of an acid with a base, producing a salt and water.

neutrino Particle of zero charge and near-zero mass.

neutron Neutral elementary particle of mass slightly greater than the proton; found in the nucleus of all atoms heavier than hydrogen. A highly penetrating, interactive form of matter.

neutron number The number of neutrons in the nucleus of an atom.

nitrogen fixation The formation of nitrogen compounds from free, molecular nitrogen.

noble gas One of the gases in Group VIIIA.

node A point, line, or plane on a wave or wave function at which the amplitude is zero.

nonelectrolyte A substance that does not conduct electricity when dissolved in water—that is, does not dissociate into ions.

nonmetals The elements at the extreme upper right of the periodic table with properties contrasting to those of the **metals**.

nonprimitive cell Any **unit cell** in a crystal larger than the **primitive cell**.

nonstoichiometric Not having atoms in a strict ratio of small whole numbers. Normally encountered in **defect** structures.

nonvolatile Having a low vapor pressure.

normal boiling point The **boiling point** of a liquid at 1 atm of pressure.

***n*-type semiconductor** A **semiconductor** in which the charge is carried by electrons.

nuclear binding energy *See* **mass defect**.

nuclear magnetic resonance spectroscopy (nmr) A spectroscopic technique based on a property of many nuclei that, when placed in a magnetic field, absorb characteristic energies from a radio frequency field superimposed on them.

nucleic acids Polymeric strands composed of nucleotide units in which phosphoric acid is combined with a sugar (ribose, for example) and a nitrogen base (purine or pyrimidine derivatives); prominent cellular components, functioning in information storage and transfer. Ribonucleic acid (RNA) arid deoxyribonucleic acid (DNA).

nucleon One of the components of the nucleus, particularly **protons** and **neutrons**.

nucleus The massive center of an atom.

nuclide A specific atomic nucleus.

number average molecular mass The ordinary arithmetic average molecular mass of a polymer.

octahedral hole A **hole** in a crystal structure for which the nearest neighbors are at the vertices of an octahedron.

octet rule The principle that elements in the *s* and *p* blocks tend to achieve noble gas electron configurations when forming compounds either by donating and receiving electrons or sharing them.

open-ended manometer A **manometer** having one end open to the atmosphere.

open system (thermodynamic) A system that can freely transfer matter and energy with its surroundings.

orbital A one-electron wave function having particular values of n, ℓ, and m_ℓ expressing the probability of finding the electron in a given region in space.

order of a reaction The exponent of the concentration of a particular reactant in the rate law for a chemical reaction or the sum of the exponents of the overall reaction.

ore A natural solid source of a desired element.

organic substances Compounds containing carbon atoms bonded to hydrogen.

osmosis Dilution of a solute by passage of solvent molecules through a semipermeable membrane due to osmotic pressure; a colligative property.

osmotic pressure Pressure resulting from having two solutions of unequal concentration on either side of a semipermeable membrane.

overlap The common space shared by two orbitals in a bonding configuration.

overtone A wavelength that is an integral fraction of the **fundamental mode**.

overvoltage Potential needed to perform an electrolysis in excess of that calculated using standard potentials and the **Nernst equation**.

oxidation Chemical reactions involving removal of electrons resulting in a higher **oxidation state**; **reduction**, or gain in electrons by another species must accompany the oxidation process.

oxidation state The charge on an atom in a molecule calculated on the assumption that bonding electrons between atoms are assigned to the more electronegative atom. Calculation is normally by a set of rules that assumes certain oxidation states for certain atoms.

oxidation-reduction (redox) reaction A reaction that involves a transfer of one or more electrons from one reactant to another.

oxidizing agent, oxidant A reactant that causes another reactant to be oxidized in a redox reaction; the oxidant is reduced.

oxo anions Anions containing one or more oxygen atoms around a central atom—for example, SO_4^{2-} and ClO^-.

oxyacids The acids of **oxo anions**—for example, H_2SO_4 and HClO.

ozone layer Layer in upper atmosphere containing a high ozone concentration that serves as a filter for ultraviolet radiation. Ozone is currently threatened by atmospheric pollutants, particularly chlorofluorocarbons.

parallel spins Electrons in singly occupied degenerate orbitals having the same values of m_s.

paramagnetic Being attracted to a magnetic field as a result of unpaired electrons.

partial charge A charge of less than one electronic unit resulting from uneven sharing of bonding electrons.

partial pressure The effective pressure of a gas in a mixture. The partial pressure p_a of component a in a mixture of gaseous components can be stated as the product of the total pressure (P_T) and the mole fraction of a.

Paschen series A series of lines in the infrared region of the spectrum of atomic hydrogen with emissions terminating at principal quantum number 3.

passivated surface A metal surface protected from corrosion by a tightly held layer of metal oxide.

path The route taken by a **process** described by a series of very small steps.

path function A quantity such as work or heat for which the value depends on the path taken between the initial and final states. *See* **state function.**

Pauli exclusion principle *See* **exclusion principle**.

***p*-block** The six columns of the periodic table farthest to the right; characterized by filling of p orbitals.

peptization Transformation of a colloidal gel into a **sol**.

percent yield The percentage of the stoichiometrically possible yield in a reaction that is actually obtained—% yield = 100%·actual yield/theoretical yield.

performance characteristics The useful properties of material or the properties demanded for a specific application.

period One horizontal row of the **periodic table** of elements.

periodic table A systematic arrangement for the elements according to increasing atomic number and periodic recurrence of chemical properties that suggests family classification of the elements in vertical columns.

periodicity Having the property of repeating over and over.

permeability The ability to allow passage of a substance through a membrane under given conditions.

pH The negative logarithm (base 10) of the hydrogen ion concentration; a measure of relative acidity in aqueous solutions: above 7, basic; below 7, acidic.

phase Visibly distinct portion of a heterogeneous system; one that is separated by a clearly defined surface.

phase diagram A diagram on which the phase of a substance or a mixture can be determined as a function of temperature, composition, and pressure.

photoelectric effect Liberation of electrons by a substance (usually a metal) on irradiation by an incident beam of electromagnetic radiation, usually light of short wavelength (UV, for example); the speed of ejected electrons is proportional to the frequency of radiation. Einstein photoelectric law: $E = h\nu - W$.

photon The carrier of the quantum of electromagnetic energy; photons have momentum but not mass or electrical properties.

photosynthesis The process whereby green plants effect conversion of carbon dioxide and water to carbohydrates and oxygen in the presence of sunlight.

physical methods Methods of separating, purifying, and otherwise processing a substance that do not change the identity of the substance.

physical process A process that does not change the identity of a substance. For example, melting.

physical property A property of a substance that is exhibited without any change in the identity of the substance—for example, melting point or viscosity.

pi-bond A bond formed by lateral (rather than linear) combination of p or d atomic orbitals; the second bond in a double bond, for example, in ethylene; the second and third bonds in a triple bond, for example, in acetylene and nitrogen.

Planck's constant The proportionality constant *h* in the Planck relationship between energy and the frequency (or wavelength) of radiation. *See also* **quantum hypothesis**.

plasma A collection of atoms that have been freed from their electrons by a very high temperature.

plastic deformation (plasticity) Permanent deformation of a sample by application of a **stress**.

plastic materials Polymeric materials used for making commercial items.

plasticity The ability of a material to be deformed permanently without returning to its original dimensions.

plasticizer A compound added to a polymer to increase its flexibility.

point defects Crystal **defects** restricted to a small point in the structure.

polar molecule A molecule that has a permanent electrical dipole moment due to unequal distribution of electric charge between bonded atoms within the molecule.

polarizability The ease with which a dipole is induced on a nonpolar atom or molecule; this increases with the size of the atom or molecule.

polyatomic Having more than one atom present; said of ions.

polydisperse A polymer having molecules with different chain lengths.

polydispersity A measure of the extent to which a polymer has molecules with different chain lengths.

polymer A compound formed of giant molecules generally constructed from one or two monomer units that have been chemically combined in polymerization processes.

polymeric materials Materials composed mainly of polymers.

polymorphism The same composition of a substance exhibiting two or more different crystalline forms.

polyphase system An alloy containing a mixture of crystals having different compositions and structures.

polyprotic acid An acid that produces more than one hydronium ion when dissolved in water.

Portland cement A mixture made using clay, limestone ($CaCO_3$), and gypsum ($CaSO_4 \cdot 2H_2O$) that will harden and mix with water; a principal component of concrete.

positive charge A charge with the same sign as the charge on the proton. An ion has a positive charge if it contains more protons than electrons.

positron The antiparticle of the electron having the same mass as the electron but a positive charge.

potential energy Energy that can be converted to kinetic energy. It depends on position and on attraction or repulsion.

precipitate A solid that separates from a solution because the existing properties at that moment make it insoluble: often due to temperature change, loss of solvent through evaporation, or addition of a reactive reagent.

precipitation To form a solid solute from a solution.

precision The degree to which measured observation can be judged reliable; the reproducibility of results.

pressure Uniform stress in all directions produced by a gas, measured as a force per unit area; the pressure exerted by a gas in a container results from collisions of the gas molecules with the container walls; **atmospheric pressure** is due to the mass of the atmosphere on the surface of the Earth.

primary cell An **electrochemical cell** that cannot be recharged.

primitive cell The **unit cell** for a crystal having the smallest volume.

principal quantum number The quantum number on which the energy of a state most depends, symbol n.

process The changing of a system from one **equilibrium state** to another.

product A substance resulting from a chemical reaction.

propagation step A step in a **chain reaction** that keeps the process going.

property An aspect of a substance or mixture by which it can be identified. *See* **chemical properties** and **physical properties**.

proteins A biologically important class of complex, long-chained organic nitrogen compounds of high molecular mass made up of many alpha-amino acids linked through peptide (—CONH—) bonds.

proton Positively charged elementary particle in the nucleus of all atoms; the nucleus of the ^1H atom; carries a charge equal to the charge on the electron but of opposite sign. Atomic number (Z) of an atom equals the number of protons in its nucleus.

p-type semiconductor A **semiconductor** in which the charge is carried by holes that can be viewed as being positively charged.

quantized Consisting of quanta.

quantum (*plural* quanta) The smallest possible unit of mass or energy.

quantum hypothesis The principle that emission or absorption of radiant energy in atomic events can assume only discrete values; the units of energy of quanta are equal to the product of the frequency of the radiation and h, the Planck constant (6.626×10^{-34} J·s).

quantum jump A transition from a lower **energy** state to higher one.

quantum number A number characteristic of a particular **energy state**. Quantum numbers arise in determinations of the electron structures of atoms.

quantum relaxation A transition from a higher **energy** state to lower one.

radiation applications Direct use of the radiation from a radioactive isotope.

radiation hazard A hazard caused the presence of radioactive radiation.

radical A classical name for a group of atoms (charged or uncharged) that enter into chemical combination as a unit: NH_4^+; CH_3; SO_4^{2-} (ammonium, methyl, and sulfate, respectively).

radioactive decay (disintegration) Spontaneous transformation of one nuclide into another or into a different energy state of the same nuclide by emission of alpha- or beta-particles or gamma rays. The decay process follows a first-order rate law.

radioactivity Spontaneous disintegration or decay of an unstable atomic nucleus, usually accompanied by ionizing radiation.

random copolymer A polymer consisting of a random arrangement of more than one kind of monomer unit.

Raoult's law The principle that in an ideal solution the vapor pressure of a component is proportional to its concentration in mole fraction.

rare earths *See* **lanthanides**.

rate of a chemical reaction The speed at which reactant or product concentrations change.

rate constant *See* **specific rate constant**.

rate-determining step A step in a reaction mechanism that is sufficiently slower than the other steps that it governs the rate of the reaction.

rate law A mathematical expression of the rate of a reaction shown as the rate of change of the concentration of a reactant or product as a function of the concentration of one or more reactants and products.

reactant One of the starting materials in a chemical reaction.

reaction coordinate The extent to which a reaction is complete; used when plotting the energy of a particular reaction path.

reaction mechanism The steps in the process of going from reactants to products.

reaction quotient A quantity calculated with the same expression as that for the **equilibrium constant** but at any conditions, not necessarily equilibrium.

recrystallization A physical method for separating and purifying substances that involves dissolving the substance and then allowing it to form crystals again slowly.

redox *See* **oxidation-reduction**.

reducing agent, reductant A reactant that causes another reactant to be reduced in a redox reaction; the reductant is oxidized.

reduction Chemical reaction accompanied by a gain in electrons by one or more participating reagents. *See also* **oxidation**.

refractories Materials useful because of their hardness and tolerance of high temperatures.

relative error The difference between a measured value and the true value divided by the measured value.

relative humidity The ratio of the partial pressure of water vapor in the air to the vapor pressure of water at the prevailing temperature.

resistivity An **intensive property** that measures the tendency of a substance not to conduct electricity.

resonance A stabilizing quality of certain molecules that can be represented by considering the electron distribution in an ion or molecule as a composite of two or more forms, in those cases where a single form is an inadequate representation; for example, benzene and the carbonate ion.

resonance hybrid The composite structure resulting from resonance.

resonance structure One of the forms that make up a **resonance hybrid**.

reversible Said of a process for which the direction can be changed by an infinitesimal change of an outside variable. In general, equilibrium constitutes a reversible situation.

root-mean-square The square root of an average of squared values, applied particularly to particle speeds.

Rydberg equation The empirical spectroscopist's equation that gave the frequencies for the lines in the spectrum for hydrogen and from which Bohr was able to derive considerable support for his theory.

sacrificial anode A reactive metal anode that protects a less active metal from corrosion.

salt A neutral combination of an **anion** and a **cation.** The ionic product of the reaction of an acid–base neutralization reaction, forming a salt and water; also produced by the displacement of acidic hydrogen atoms by metallic ions.

salt bridge A bridge that allows ions to flow from one half cell to another in an **electrochemical cell**; needed to maintain electrical neutrality in each half cell.

saponification A reaction that yields a soap. The reaction of an ester with a strong base in water yielding an alcohol and the salt of an organic acid.

saturated (1) A solution of maximum solute concentration. (2) A hydrocarbon chain containing no double bonds.

saturation (1) The dissolution of the maximum quantity of solute in a particular solvent at a given temperature; results in a saturated solution under equilibrium conditions. (2) Having the full number of hydrogen atoms possible; said of hydrocarbons and hydrocarbon radicals. Having no carbon–carbon double bonds.

***s*-block** The first two groups of the periodic table, IA and IIA.

Schottke defect A crystal **defect** resulting from the absence of two or more ions of opposite charge.

second The unit of time in the **International System of Units**.

second law of thermodynamics The law that the entropy of the universe increases for every spontaneous process, is unchanged for a reversible process, and can never decrease.

second order reaction A reaction for which the sum of the exponents in the rate law is 2.

secondary cell A **galvanic cell** that can be recharged.

selective precipitation A method of separating salts by precipitating most of the more soluble salt before any appreciable amount of the other salt precipitates.

selective settling A method of separating ores according to density.

semiconductor Certain crystalline substances that have electrical conductivities between metals (conductors) and insulators (nonconductors). *See* **metalloid.**

semimetal *See* **metalloid.**

semipermeable Allowing passage of some molecules but not others.

shell The energy states of an atom represented by a particular value of the **principal quantum number**, n.

shielding The situation in which outer orbitals are shielded from the positive charge of the nucleus by inner orbitals.

siderophiles Metal alloys in the Earth's core.

sigma bond Chemical bond formation involving linear combination of s- and p-type atomic orbitals between adjacent atoms. Such bonds have their bonding electron density concentrated directly between the bonding atoms and have molecular orbitals that do not change sign when crossing the internuclear axis.

significant figures The useful digits in measurement, typically including the first doubtful digit as a reasonable estimate.

simple cell = primitive cell.

simple metal A pure metal or a **single-phase alloy**.

single bond A single pair of shared electrons forming a covalent bond between two atoms. This is normally a **sigma bond**.

single-phase alloy An alloy that is homogeneous throughout.

slippage Movement of layers of a crystal against each other as a result of stress.

smectic Said of a liquid crystal with all molecules lined in the same direction and with the centers of the molecules also lined up.

soap The salt of a fatty acid. It has **hydrophilic** and **hydrophobic** ends and forms **micelles** in water.

softening point The temperature at which a glass will bend under its own weight.

sol A colloidal solution of a gas, liquid, or solid in a suitable dispersion medium; usually applies to liquid systems; if the continuous phase is water, the term *hydrosol* is used.

solid An aggregated state of matter in which the substance exhibits definite volume and shape and is resistant to distorting forces that might alter its volume and shape. Solids have regular crystalline structures.

solubility The property of substances that allows them to form homogeneous mixtures called **solutions**.

solubility product An equilibrium constant that is the product of the ion concentration in a saturated solution; it defines the degree of solubility of a given substance in a specific solvent at a specific temperature.

solute That which is dissolved in a solvent to make a **solution**.

solution Mixture of substances forming a homogeneous phase.

solvent Usually the principle component of a **solution**.

somatic effects Damage caused to the body by radiation.

sparingly soluble ionic compound An **ionic compound** generally with **solubility** in water of considerably less than 1 g solute per 100 mL of water.

specific energy The energy available from an **electrochemical cell** per unit mass.

specific gravity The ratio of the **density** of a substance or material to that of some standard substance. Water is the common standard for liquids and solids.

specific heat The quantity of heat required per unit mass per degree to change the temperature of a substance without causing a phase change; heat capacity per gram.

specific rate constant A multiplier used to calculate the rate of a reaction; it is a constant for a particular reaction at a particular temperature.

spectator ion An ion in a chemical reaction present in the same quantities before and after reaction.

spectral lines Single wavelengths of electromagnetic radiations arising from electronic transitions (emissions and absorptions) in atoms.

spectrochemical series A series of **ligands** that occurs in **coordination complexes** in order of the extent to which they split the energy of otherwise **degenerate** valence orbitals on the central atom.

spectrophotometer An instrument for recording a **spectrum**.

spectrum A record of absorption or emission of electromagnetic radiation as a function of frequency or wavelength.

speed of light The speed of electromagnetic radiation in a vacuum, $3.00 \times 10^8 m/s$.

spin quantum number The quantum number for the electron m_s having values $-\frac{1}{2}$ and $+\frac{1}{2}$.

spontaneous process A process occurring in a nonreversible manner. *See* **reversible**.

stainless steel One of a number of alloys of iron, chromium and other elements that is highly resistant to corrosion.

standard conditions A set of prescribed conditions for chemical processes, particularly all concentrations of reactants and products at 1 M and all partial pressures of reactants and products at 1 atm.

standard enthalpy (heat) of formation The enthalpy change involved in producing one mole of a compound in its standard state from its constituent elements in their standard states.

standard free energy of formation The free energy change involved in producing one mole of a compound in its standard state from its constituent elements in their standard states.

standard potential The potential in volts for a cell or a half cell under **standard conditions. Standard half-cell potentials** are based on the assignment of 0.0000 V for the hydrogen half cell.

standard state The most stable form of a substance at 1 atm pressure and a specified temperature, often 298K.

standard temperature and pressure (STP) Conditions defined as 273.15K and 1 atm.

standing wave A wave that is not moving forward, such as the vibration of a string.

state The form of a substance—that is, solid liquid or gas. For a system the state is a sufficient list of its properties to describe it completely.

state function A quantity such as entropy or enthalpy for which the value depends only on the initial and final states of the system and is independent of the path taken. *See* **path function**.

step-growth polymerization A polymerization reaction in which one small molecule is released for each monomer unit that joins the polymer chain.

steroid A lipid with a characteristic four-ring structure. Includes important hormones such as progesterone and estradiol.

stiffness Resistance to deformation under stress; having a high **elastic modulus**.

stimulated emission Emission of radiation from an atom in an excited state stimulated by radiation of the same wavelength as that to be emitted.

stoichiometric quantities Quantities of reactants and products in a chemical reaction that are in exact proportion to the balanced equation.

stoichiometry That aspect of chemical science dealing with proportions of reactants and products involved in chemical transformations.

strain The change in a dimension divided by the original dimension for a sample put under **stress**.

strain point The temperature below which a glass exhibits **elastic** behavior.

stress The load put on a sample-per unit cross-section of the sample.

strong acid An acid that produces a high concentration of hydronium ions when dissolved in water—that is, ionizes almost completely.

strong base A base that produces a high concentration of hydroxide ions when dissolved in water.

strong electrolyte A substance that produces a high concentration of ions when dissolved in water—that is, dissociates almost completely.

structural formula A molecular formula indicating at least some aspects of the geometry of a molecule.

subatomic particles Particles that compose atoms such as protons, neutrons, electrons, and so forth.

sublimation Direct transformation from solid to vapor without passing through the liquid state.

subshell The energy states of an atom represented by a particular value of **angular momentum quantum number**, ℓ.

substance A homogeneous form of matter of definite composition; a pure chemical compound or element.

substitutional solid solution A solid solution in which the positions of some of the atoms have been taken up by other atoms.

supercritical fluid *See* **critical point**.

superheat To heat above the boiling point; an unstable situation.

supersaturated A solution of greater than maximum solute concentration; an unstable situation that can sometimes be maintained for a long time.

surface tension A tension at the surface of a liquid resulting from forces of attraction acting normally to the surface on molecules close by; this causes the liquid surface to behave as a tense membrane.

surroundings All of the universe not included in a system, particularly that part of the universe affected by the system.

symmetry A property of an object whereby it can be operated on in some fashion—for example, rotated—and appear unchanged.

system (thermodynamic) That sample or limited segment of the universe under investigation.

systematic nomenclature Names of chemical substances based on the chemical formula of each substance.

temperature A property of systems that establishes thermodynamic equilibrium: Two systems are in equilibrium with each other when they are at the same temperature. Temperature is proportional to the average kinetic energy for an ideal gas.

temperature inversion An atmospheric condition characterized by increasing temperature with altitude; little interchange and mixing of materials results in buildup of atmospheric pollutants.

termination step A step in a **chain reaction** that terminates a chain.

tetrahedral hole A **hole** in a crystal structure for which the nearest neighbors are at the vertices of an tetrahedron.

theoretical yield The highest yield allowed by the stoichiometry of a reaction.

theory An underlying principle governing natural phenomena; experimentally defined doctrine that has been established by confirmation of predictions based on the theory.

thermal conductivity Ability to accommodate the transfer of heat.

thermochemical equation A chemical equation giving information about the change in enthalpy or internal energy involved in the reaction.

thermochemistry The study of the heat exchanges that accompany chemical reactions and phase transformations.

thermodynamic control Relating to a process for which the products formed can be determined by thermodynamic calculations.

thermodynamics The branch of physical science dealing with the transfer of heat and energy interconversions. The laws of thermodynamics are based solely on experience.

thermonuclear Referring to nuclear **fusion** reactions.

thermoplastic polymer A polymer that softens on heating; can be reheated and reshaped.

thermoset polymer A polymer that sets into a rigid form on heating and that cannot be shaped after setting.

third law of thermodynamics The law that the entropy of a perfect crystal of a pure substance at a temperature of absolute zero is zero.

time A system for identifying the order and duration of events.

titration A technique for analyzing the composition of a solution by careful addition of a reagent solution of accurately known concentration (standard solution) until the indicator present confirms the end-point.

titration curve A graph of some measurable signal such as pH versus volume of solution added in a **titration.**

torr A unit of pressure equal to 1 mm of mercury or 1/760 of 1 atm.

total energy All of the energy of a system including **kinetic energy** and **potential energy.**

tracer application Use of radioactive isotopes to follow the course of a process.

trans A geometric arrangement in which two chemical groups are located opposite each other.

transition elements (transition metals) The block of metallic elements in the middle of the periodic table characterized by the progressive buildup of *d* electrons. Can also include **lanthanides** and **actinides.**

transition state An unstable state (high energy) that exists in a **reaction mechanism.**

translation Simple movement in a straight line.

translational kinetic energy The **kinetic energy** resulting from **translation.**

triple bond A covalent bond between two atoms consisting of three pairs of shared electrons. Of the three bonds, one is a **sigma bond** and two are **pi bonds.**

triple point The temperature and pressure at which all three phases of a pure substance can exist at equilibrium.

triprotic acid An acid that produces three hydronium ions when dissolved in water.

trival nomenclature The common or historic names of chemical substances usually based on chemical properties.

uncertainty principle The principle that the position and the velocity of a particle can be identified but not simultaneously. The same problem arises in the simultaneous determination of energy and time.

unit cell Basic repeating volume unit in a sample of crystalline matter that can be used to generate the entire structure by translation along the coordinate axes. The unit cell can be constructed by connecting specific **lattice points.**

units The actual quantity being measured, as kilograms when determining mass.

universal gas constant The proportionality constant, R, in the **ideal gas law,** $PV = nRT$. The value of R depends on its units.

unsaturated (1) A solution of less than maximum solute concentration. (2) A hydrocarbon chain containing one or more double bonds.

vacuum distillation A **distillation** performed under reduced pressure.

valence A number representing the general combining capacity of an element in its chemical combination or the proportions in which atoms combine relative to hydrogen taken as 1.

valence bond model A **model** of chemical bonding in which bonds are created by atoms sharing their electrons.

valence shell electron pair repulsion model (VSEPR) A scheme for predicting molecular structure based on the principle that electron pairs around an atom will separate as much as possible.

valence (shell) electrons The electrons in an atom that are involved in chemical reactions. These are normally those outside the next lowest noble gas configuration.

valence shell expansion The placing of more than four pairs of electrons in the valence shell of an atom.

van der Waals equation An equation of state for a real gas including corrections for particle volume and attractions or repulsions among particles.

van der Waals forces Attractive forces between atoms or among molecules that arise because of the presence of a dipole moment; in nonpolar molecules, weak intermolecular forces other than the usual valence bonding.

vapor A gas; usually a gas in contact with the same substance in the liquid state.

vapor pressure The equilibrium pressure due to the vapor of a solid or liquid substance at a given temperature.

vaporization Change of state from a liquid to a gas.

viscosity The internal resistance of fluids to flow.

volatile Having a high vapor pressure.

volt An electrical potential of one joule per coulomb.

volume The space occupied by a body of matter at any given temperature and pressure.

vulcanization A process for cross-linking the polymer chains in natural and synthetic rubber with sulfur to harden the rubber and make it more resistant to higher temperatures.

wave function A solution to the wave equation for the electronic structure of an atom having a specific energy.

wavelength The distance between two identical points in consecutive cycles of a periodic wave, along the direction of propagation of the wave; usually measured in nanometers or angstroms when dealing with electromagnetic radiations in chemistry.

wave number A measure of frequency equaling the reciprocal of the **wavelength**.

wax An ester of a fatty acid and an alcohol.

weak acid An acid that produces a low concentration of hydronium ions when dissolved in water.

weak base A base that produces a low concentration of hydroxide ions when dissolved in water.

weak electrolyte A substance that produces a low concentration of ions when dissolved in water.

weight The force exerted on an object as a result of its mass and the prevailing acceleration of gravity. If two objects are subject to the same gravitational acceleration, their weights are proportional to their masses. Therefore, weight is commonly, if loosely, used as a synonym for **mass**.

work An activity proceeding against some opposing force; the classical definition is the product of a force and the distance through which the force operates.

work function In the photoelectric effect, the minimum energy that will remove an electron from the surface of a metal.

work hardened Having been subjected to stresses to minimize the effect of crystal defects; said of metals.

working point The temperature at which a glass can be shaped into useful or decorative objects.

X-rays Penetrating form of electromagnetic radiation with high energy and short wavelength emitted either when inner-orbital electrons of an excited atom fall back to their normal state or when a metal target is bombarded with high-speed electrons. X-rays are nonnuclear in origin.

yield *See* **percent yield** and **theoretical yield**.

zeroth-order reaction A reaction for which the sum of the exponents in the rate law is 0.

zone refining A method of purifying a solid by repeatedly heating section by section.

INDEX

(Pages in *italics* indicate illustrations; pages followed by t indicate tables; pages in **bold** indicate profiles of scientists/discoveries.)